Biology of Women

BIOLOGY OF WOMEN

Second Edition

ETHEL SLOANE
Chair, Department of Biological Sciences
University of Wisconsin—Milwaukee
Milwaukee, Wisconsin

DELMAR PUBLISHERS INC.®

NOTICE TO THE READER

Publisher and author do not warrant or guarantee any of the products described herein or perform any independent analysis in connection with any of the product information contained herein. Publisher and author do not assume, and expressly disclaim, any obligation to obtain and include information other than that provided to them by the manufacturer.

The reader is expressly warned to consider and adopt all safety precautions that might be indicated by the activities described herein and to avoid all potential hazards. By following the instructions contained herein, the reader willingly assumes all risks in connection with such instructions.

The publisher and author make no representations or warranties of any kind, including but not limited to, the warranties of fitness for particular purpose or merchantability, nor are any such representations implied with respect to the material set forth herein, and the publisher and author take no responsibility with respect to such material. The publisher and author shall not be liable for any special, consequential or exemplary damages resulting, in whole or in part, from the readers' use of, or reliance upon, this material.

For information, address Delmar Publishers Inc.
2 Computer Drive West, Box 15-015
Albany, New York 12212

Printed in the United States of America
Published simultaneously in Canada
By Nelson Canada
A Division of The Thomson Corporation

10 9 8 7 6 5 4 3 2 1

ISBN 0-8273-4366-3

Preface to the Second Edition

Since the publication of the first edition of *Biology of Women*, there have been many changes that have bearing on women's biology and health. Some have been startling advances in science and medicine; some have been modest breakthroughs in the treatment and diagnosis of disease; and some have again proved the old adage that the more things change, the more they stay the same.

Life expectancy for women has increased slightly, the birth rate has again decreased, and infant mortality has reached an all-time low. "New" diseases and health hazards have appeared in a burst of publicity; in some instances, the headlines have decreased but the hazards remain and the cases still occur. The hysterectomy rate has declined modestly; the cesarean rate has risen significantly. The list of sexually transmitted infections has lengthened. Their increased incidence and related problems, such as adverse effects on newborn infants or infertility in women, are further unhappy by-products of the sexual revolution. The incidence of breast cancer has increased, but the rate of lung cancer linked to smoking is rapidly overtaking that of breast cancer in women, additional confirmation that many health problems are caused by, or at least strongly influenced by, life-style—the way we eat, whether we drink or smoke, how much exercise we get.

Women are still the greater consumers of health care. They continue to see more doctors, have more surgery performed on them, and receive more prescription drugs than do men. Health costs have also continued their annual double-digit inflationary climb, threatening to become so insupportable that now even the health care system is trying to change the system. Still unchanged and unresolved, for the most part, are questions and controversies about which treatment, procedure, contraceptive, or medicine provides the greatest benefits. Also unchanged is the need for women and men to have the necessary information to assume the responsibility for, and make decisions about, their bodies in health and disease.

To further this knowledge and provide a current view of all aspects of the biology of women, the earlier book has been fully revised and updated. Two new chapters have been added to the second edition. Chapter 4 discusses the causes and treatment of eating disorders and amenorrhea, premenstrual syndrome, dysmenorrhea, toxic shock syndrome, and endometriosis. Chapter 14 contains information on the effects of toxic agents in the workplace, work during pregnancy, and women, work, and stress. Many new or expanded areas have also been added to the book. These include current theories of hormone action and biological mechanisms at the cellular and molecular levels; female sexuality; breast cancer diagnosis and treatment; sexually transmitted diseases; urinary tract infections; the effects of drugs, alcohol, and smoking on pregnancy; diagnostic ultrasound and prenatal therapy; in vitro fertilization, surrogate mothers, freezing embryos, and embryo transfer; new contraceptives; and additional information on skin and hair care.

I am grateful to the reviewers who provided me with suggestions for this second edition and to my encouraging and patient editors at John Wiley & Sons, Andrea Stingelin and Janet Walsh Foltin. I again express my gratitude to John Alley for his excellent, exciting photographs that have done much to enhance the quality of both editions. I was fortunate to have the additional assistance of David Mackin and Allan Magayne-Roshak of the photographic services and artists Lana Lewandowski and Carol Davis of Visual Design Services at the University of Wisconsin—Milwaukee. I thank Marilyn Schaller for her scanning electron micrographs of hair and my geneticist colleague Ruth Phillips for the human karyotypes, and I appreciate the help and cooperation of Mount Sinai Medical Center, St. Francis Hospital, St. Michael Hospital, and Dr. Richard O'Malley. I owe a special thank-you for the assistance of Edith Bjorklund, Assistant Director of the Golda Meir Library at the University of Wisconsin—Milwaukee. The women and men who are my students deserve my appreciation for their interest, suggestions, sharing of their experiences as health care professionals and patients, and their willingness to let me bounce my ideas off them. Finally, I thank my husband and daughters, who recognized that the second edition was no less difficult for me than the first, and gave me the help and encouragement I so often needed.

Ethel Sloane

Preface to the First Edition

This book is about the human female throughout her entire lifespan. It explains biological sex differentiation and fetal development. It is concerned with all the events of a woman's reproductive life from menarche to menopause. It deals with sexuality, birth control, infertility, and pregnancy. It discusses controversies in the treatment of breast cancer and gynecological difficulties. It examines sociological and cultural factors that influence a woman's nutrition, physical activity, and use of cosmetics.

This book is also about health care. Medical information is essential to women—and men—who want to interact intelligently with health care professionals. Without it, many find themselves in an uncomfortable "you doctor, me patient" relationship—a relationship that, as consumers, they resent. Bolstered with information, however, they are able to understand the functioning of their bodies and take responsibility for maintaining and enhancing their health. They can take an active role in a partnership with their doctors. They can make decisions that affect their

health based on their own knowledge and personal preferences as opposed to those of their physicians.

Making informed choices is not simple in our society. Health care today is as three-ringed as a circus—consumers are confronted with traditional, alternative, and controversial methods, all purporting to help them become and stay healthy. How are decisions to be made? What *are* the facts, when newspapers almost daily report new health information different from information published the week before, when magazine articles refute one another on what is right and wrong for our health? We all must know enough to choose the methods of being and staying healthy appropriate for us. This book, therefore, examines traditional attitudes toward women's health—common assumptions concerning anatomy, physiology, female reproduction, sexuality, and behavior. It presents the most important research studies on which these assumptions are founded, describes their methodology and results, and encourages readers to form conclusions about their validity. It discusses relevant current research, including clinical studies now in progress, and suggests other areas in which new studies are necessary. It thus enables readers to weigh reports of medical research that seem to show *both* sides of an issue to be true.

There is nothing mysterious about medicine and human structure and function; it is all knowable, if a woman understands medical jargon and the technical terms that describe human beings. Learning the language of medicine and health enables women to talk to health professionals on a professional level. Many students have told me that as a result of their new knowledge and demands their doctors have stopped using such words as "womb," "plumbing," and "pipes" during medical examinations. This is gratifying to them; it is another small but symbolic indication that control of their own bodies truly can be theirs.

That medicine can be demystified—that informed men and women are fully capable of participating in medical decisions about their own health and that it is their *right* to do so—is a recent recognition. Only 10 or 15 years ago the notion of patients' rights, of autonomy in health care, was unheard of. Certainly on college compuses there were no undergraduate academic courses that taught women to be well-informed health care consumers. A course focusing fully on female anatomy and physiology in states of health and disease would have been viewed as appropriate only in a nursing or medical school. During the late 60s and early 70s, however, with the rise of the women's movement, came a breakthrough. Today colleges have changed, and women's studies courses, including courses in the biology of women, have proliferated. The biology of women is no longer a radical or barrier-breaking subject. Each year enrollment and interest increase; each year I find my classes filled with more students who

have better questions and ideas. Clearly, there is no need now to justify the existence of a course—or a book—on the biology of women. Interest in women's health, with all its political, moral, and emotional ramifications, has extended beyond college classrooms into society as a whole: it has become a concern of everyone, female or male.

It is in response to the expressed needs of women and men, in and out of college, eager to learn about women's bodies in health and disease and to be responsible for their own health care, eager to examine the ways in which scientists produce "knowledge," eager to question orthodox assumptions about women, evaluate the bases of research, and stimulate new thinking and research, that I structure my course in the biology of women. But because no book available parallels my concerns, until now I have had to jerry-rig perspectives for my students; I have had to assign readings piecemeal, from a dozen different sources. Obstetrical-gynecological medical school texts, even in "core" or synopsis versions, require an academic background that many of my students do not have. Moreover, these texts (written by doctors for other doctors or for future doctors) often draw subjective inferences about women as patients; most contain questionable assumptions about women's behavior and needs. Books written by doctors for the layperson—the ask-your-doctor variety—are, to me, condescending and patronizing. Books and pamphlets that have arisen out of the women's movement, worthwhile as they are in demonstrating that women can educate themselves about topics traditionally considered too complicated for their pretty little heads, lack necessary detailed anatomical and physiological information. Many tend, also, to be too "alternative," too polemic for a subject of study that must not be regarded as an alternative. Recently books on one or more aspects of women in health and disease have become available, but the need for a comprehensive text has not yet been filled. It was with this goal in mind that I have written this book.

It has also been my goal that this book be of value both to people with little or no background in biology and to students in health professions who have taken many courses in anatomy, physiology, and chemistry. I have tried to achieve both broadness of scope and inherent flexibility, in order to provide a book adaptable to various needs: for class use in such courses as biology of women and human reproduction; as a resource in such courses as maternity nursing and gynecological nursing; for health professionals in continuing education programs, members of community-based women's health groups, and individual readers who want to know more about a woman's body and how to care for it. My intention has been to write a book that can be meaningful to any woman at different times of her life, one that any woman can share with her mother, her friend, her husband. It is exactly this sharing of knowledge which is so very important to the women's movement today—women with peer women, women and the generation of women who gave birth to them, women and men, and most certainly, women and their daughters. It is in this spirit, finally, that I have written this text.

Ethel Sloane

Contents

Biology of Women

1. *Women and Their Health*

"Taking Our Bodies Back," a movie produced by Cambridge Documentary Films, Inc., makes an eloquent and powerful statement concerning the dissatisfaction of women about the health care they receive, and expresses their growing assertiveness in trying to regain control of their bodies. To people unacquainted with the activities of the women's health movement, perhaps the most startling and controversial portion of the film occurs at the beginning, when a young woman representing the Boston Women's Community Health Center demonstrates vaginal self-examination. In the opening scene, she is standing on a platform in front of a large audience showing and describing a plastic speculum, an instrument that is inserted into the vagina to spread apart the vaginal walls. She jokes about its "duck-bill" bivalves, and says that she had to buy one for her little boy as well as herself because he wanted it as a "quack-quack" toy. The audience laughs, and she goes on, talking rapidly. Moving very quickly now, she climbs up on a table, puts herself into the familiar gynecological examination position, inserts the speculum, and the movie camera and lights focus in to frame . . .

absolutely incredible! Both the film audience and live audience are being invited to look at her cervix! The women solemnly file past her, their excitement and interest apparent on their faces. The movie audience too, in the darkened room, is fascinated. Surprisingly, there is no embarrassed laughter, but complete absorption. All the women are seeing a part of a woman's body they have never seen before, a portion of their reproductive tracts that until that moment has been visible only to their doctors—and they are evidently captivated by the sight.

To doctors and many others, this demonstration is a prime example of the lunatic fringe of the women's health movement. Peering into body orifices has always been the prerogative of the physician, and most physicians would like to keep it that way. Why, say the doctors, would a woman want to look at her own, or another woman's cervix? For a woman to buy a plastic speculum and twist herself into a pretzel to do a vaginal examination—what a strange idea! Why would she do it?

The answer to a woman's desire or need for self-examination perhaps has less to do with

1

Figure 1.1. At every stage of the life cycle, women have special health needs and interests. They must have the factual information to aid them in maintaining and promoting their own health as well as easy access to high quality health care.

seeing the cervix, but more with the demystification of one's own body. Most women have an appalling ignorance and hence an uncomfortable feeling about their own reproductive anatomy. Most men do not. The genital organs of a man are exposed, easily visible; they can be seen and touched, and are, many times a day. If any changes occur, the man can note and describe them himself. In a woman, however, the reproductive organs are not easily seen. They are internal and not easily subject to examination. What cannot be seen—a woman's cervix, uterus, ovaries—can almost be forgotten, and yet a woman knows that such aspects of her body are at the very core of her sexuality, of her womanhood. Thus, such organs are there-but-not-there, like secrets that women, the owners of these things, cannot understand. What women are, reproductively, remains hidden to them, enigmatic and strange.

If all women took a mirror and a diagram of their anatomy and then viewed and examined their own external genitalia, they would progress tremendously in self-awareness and in the reassurance that everything is "normal." And, since the cervix is the most accessible part of the internal reproductive tract of the female, looking at it, for many women, could be a way of feeling more comfortable about their reproductive organs, a way of dispelling much of the mystique that surrounds them. Some may find it even more helpful if self- and mutual examination is performed within a warm and supportive group atmosphere; others prefer to try it alone. The advocates of vaginal self-examination claim that when it is performed frequently, it is possible to quickly recognize changes indicating pregnancy or a developing pathological condition. Whether *regular* vaginal self-examination is necessary has not been established. The benefits are certainly not as definitive as they are for regular self-examination of the breasts, which is mandatory for women and can mean the difference between life and death in the early detection of breast cancer. For a woman's emotional well-being, however, examining herself even once can be a way of increasing her self-confidence about her body.

The choice of whether or not to look at the cervix is up to the woman. For many of us, knowing *about* our reproductive organs is enough, and we really do not choose to examine them. There are strong cultural taboos stemming from childhood that discourage touching or tampering with one's self "down there." Only *men* are legitimately allowed to touch, and when they are doctors, to examine and describe the reproductive anatomy of females. Not only are the almost completely (88%) male members of the medical profession permitted to do so—continued health and good preventive medicine make it a requirement. And so, once or twice a year, a woman is encouraged to see her doctor for her pelvic examination, and the position, size, shape, and general health of her reproductive organs are checked.

The specialist in obstetrics and gynecology has become the acknowledged expert, the authority within the medical profession on the aspect of a woman's life so highly valued in our society, that pertaining to the sexual organs. Nonpregnant healthy women consult an Ob/Gyn for their routine gynecological examination, and an increasing number of women have no family doctor and rely on their obstetrician-gynecologist for an evaluation of their general health as well. A woman may ask her physician for advice about any medical problems, sexual matters, becoming pregnant, or about avoiding pregnancy. She exposes her most intimate self and her most intimate problems to her obstetrician-gynecologist, the doctor a healthy woman sees most frequently.

When a woman visits her doctor, however, the experience is likely to be an ordeal. A female attendant conducts her to one of the small rooms in the doctor's office and prepares her for examination. Stripped of her clothing and much of her dignity, she lies on her back draped in white sheets on the examining table, her feet up in stirrups.

"The doctor will be with you shortly," the attendant says, and leaves. The woman stares at the ceiling until the doctor enters the room. Almost immediately, he (nearly nine out of ten are men) sets the tone for their relationship.

"Hello, Julie (or Nancy, or Carol, or Susan)."

"Hello, Dr. Blank."

He raises the draped sheet and inserts a cold metal speculum into her vagina. (Would it be so difficult to warm it first?) He does not tell her what he is doing, or offer to let her see what he is seeing, which would not be impossible with a mirror and a little extra time. Instead, he takes a smear from the cells of the cervix, and withdraws the speculum. Then he puts two fingers of one hand in her vagina, places the other hand on her abdomen, and hurriedly pokes and prods, either saying nothing at all, or making such chatty remarks as "Hmmm, you're kind of small," or ". . . you have a tipped womb," or "an infantile uterus," or "an eroded cervix," without offering any further explanation of his statements.*

When the doctor completes the examination, he tells the woman to get dressed, and that he will see her in his office. She will sit there while he takes or makes a few phone calls. She senses that his attention to her is perfunctory, that she is taking up too much of his valuable time. She forgets most of the questions she had prepared. When he does respond to her, he may make judgmental recommendations concerning her life-style, her decisions about becoming pregnant, or what kind of contraceptive she should use. She

---

* One young woman indicated that after hearing "eroded," she thereafter visualized her cervix, which she thought was located somewhere up around her navel, looking like the side of a mountain after a rainstorm. Another young woman was told "My, your vagina is long and narrow," and wanted to respond, "Well, doctor, maybe your hand is short and fat," but was too intimidated. She now regrets not having said it.

feels depersonalized and thinks she is being treated as a set of reproductive organs. She leaves, dissatisfied, hostile, and further alienated from her body.

Is the foregoing unfair to doctors? Certainly it is, to some of them. Many, especially the younger ones, are less arrogant and authoritarian and more understanding. Not all doctors have sexist attitudes about women, or are brusque, hurried, and unable to communicate with their patients. Not all women mind if they are. Some are so convinced of their physician's competence that, willingly passive, they are able to ignore the apparent lack of concern. Many more women, however, who care about and want to care for their bodies, have come to realize that a doctor who is unable or unwilling to communicate with his patients or to treat them as intelligent adults is not a competent doctor, no matter what his capabilities or credentials.

BASIC ISSUES IN WOMEN'S HEALTH

The gynecological examination is only one aspect of health care about which women have become highly critical. Women are angry with the medical profession, and most of their frustration and dissatisfaction—right or wrong—has focused on the obstetrician-gynecologist, the arbiter of women's health care. They had trusted their doctors, and they believed that their physicians were giving them the quality of health care they wanted and needed. Ten years ago, a small group of women's health activists broke new ground by calling attention to the failure of traditional medicine to deliver quality health services. With the success of books like *Our Bodies, Ourselves*, by the Boston Women's Health Book Collective, and the publication of other books, articles, pamphlets, newsletters, and even television specials on women's health, women gradually became aware of the evident lack of concern and insensitivity that medical professionals, in general, had for their female patients. Women began to realize the health hazards inherent in the treatments they received. Throughout the world, women perceived their mutual plight as they began to recognize that their basic health rights had been virtually ignored. They saw a woman-exploiting tradition of health care within which:

• Women have been given hormones to prevent miscarriage, to prevent pregnancy, to keep them "feminine forever," and were not fully, or perhaps not at all, informed about the potential risks of such treatment while their doctors ignored or minimized scientific evidence of such risks.

• Women have been used, often without their full knowledge and consent, as subjects for new medical devices, surgical procedures, and drugs. They recognized the full extent of "woman-as-guinea-pig" as research done on minority women, Third World women, and institutionalized women became known.

• Surgery on women' bodies has been frequently unnecessary and excessive, and women were not given the opportunity to consider alternatives.

• Women have not been given the information or the opportunity to make informed choices about birth control. Women who chose sterilization as a contraceptive option were discouraged and found it difficult or even impossible to obtain this procedure if they were young and childless; if they were minors, black, or members of other minority groups or on welfare, they may have been pressured or even forced without consent into the operation.

• Pregnancy and the childbirth experience, a natural and normal function, has been turned into a medical problem to be technologically "managed," frequently for the convenience of the hospital and medical staff and to the possible detriment of the child—certainly to the psychological detriment of the mother.

• Women's mental health has often been defined in terms of the social and cultural expectations of the stereotyped feminine role. It remains obvious that male bias exists in classical psychiatry, and women's illnesses are frequently perceived as psychosomatic, their reproductive disorders as manifestations of psychiatric disorders, and their demands for treatment as neurotic. Pills have been the medical panacea for women's emotional problems, and there is evidence of the overprescription of tranquilizing and mood-elevating drugs.

A decade of consciousness-raising about women's health issues and increased attention to the shortcomings of traditional medical practice has not gone unnoticed. There is now federal legislation providing guidelines for the protection of human subjects from risk during clinical investigations of drugs or medical devices. Since 1978, there have been federal regulations that protect women from forced sterilization; they require women to give informed consent to sterilization and a 30-day waiting period from the time of consent to the actual procedure. Public funds can no longer be used to sterilize institutionalized women or the mentally retarded. In response to the demands of pregnant women in some parts of the country, there are alternatives to the usual hospital-based delivery of infants. There are enlightened doctors and other medical professionals who now treat women with greater respect and sensitivity, and there are women, now with greater knowledge, who are able to talk back to their doctors, reserving the right to reject advice or seek a second opinion. For the most part, however, traditional medicine has not been transformed. For the vast majority of women, underlying health concerns are unchanged, and the problems that they have had in the areas of reproductive rights, unnecessary surgery, prescription drug abuse, pregnancy and childbirth interventions, hormonal therapy, and mental health treatment have increased rather than been solved. Additional issues have emerged—the health of women in the labor force, especially when they are single parents coping with the stresses of work and child care; environmental effects on women's reproductive health; and health in the middle and later years of life.

The control of women's reproductive processes and capacities, the definitions of their physical and mental sickness and health, and the kinds of treatment they are given are still determined by a relatively small group of almost exclusively male physicians. Those determinations may reflect more of the personal biases and prejudices of male doctors than the particular physiological and psychological needs of women. Even some women who would shrink from the label of "feminist" cannot help but recognize that some of their doctors make decisions about a woman's mind and body not only on the basis of her actual health status, but also for what the doctor presumes that she is or thinks that she should become. Minority women, lesbian women, teenage and older women, and poor women often feel the lack of dignity and respect in medical treatment even more keenly.

There is evidence that misconceptions and unwarranted assumptions about female anatomy, physiology, and psychology are included in medical education. In a widely quoted study called "A Funny Thing Happened on the Way to the Orifice: Women in Gynecology Textbooks," Diane Scully and Pauline Bart surveyed 27 gynecology books published between 1943 and 1972 (1973). They discovered that the books were consistently biased toward a greater concern for the husband of the patient than for the patient herself and that women were described as anatomi-

cally designed to reproduce, to raise and nurture children, and to keep their husbands happy. At least half the books stated that the female sex drive was weaker than the male's, that a woman was more interested in sex for procreation than for enjoyment, that most women were frigid, and that the vaginal orgasm was the only true response. The majority of physicians practicing today received that kind of textbook information during their training.

Even if subsequent editions of those texts edited out the offensive opinions that would necessarily hamper a physician in dealing objectively and effectively with women patients, current medical training in obstetrics and gynecology, according to Dr. Michelle Harrison, instills inhumane attitudes and promotes inadequate treatment—undertreatment of poor women and overtreatment of the middle class. In her book describing her medical residency program at Boston's Beth Israel Hospital (1982), she writes: "The future of women's health care, however, does not lie in the domain of current obstetrics and gynecology, which is founded on certain assumptions about women's bodies and women's lives . . . women's health care will not improve until women reject the present system and begin instead to develop less destructive means of creating and maintaining a state of wellness."

WOMEN AS HEALTH CARE CONSUMERS

Women are not alone in their disenchantment with the medical profession. A Louis Harris survey in 1966 reported that 73% of Americans had confidence in their doctors; by 1983 the figure had dropped to 35%, and it is likely to be less today. The golden age of American medicine, during which physicians enjoyed unprecedented social esteem and prestige, may be over.

Almost daily, the media report that doctors are involved in health insurance fraud, receive kickbacks from pacemaker companies to use a particular company's device, or pay enormous malpractice settlements because of incompetence. But unfavorable publicity is not the only reason doctors are off their pedestals. There is an almost universal dissatisfaction with the cost, quality, and kind of health care that people receive, and, as the predominant consumers of health care, women are more frequently at the receiving end. By all indices of measurement of illness, women evidently get sick more often than men. They have more days of restricted activity associated with acute conditions, more days of bed rest, more physician visits, and more discharges from short-stay hospitals than men. They take more prescription drugs in all categories and receive two thirds of all the prescriptions for psychoactive (mood-elevating or tranquilizing) drugs. It is not, however, to be inferred that women are less healthy than men. Women live longer—a female baby born in 1984 has a life expectancy at birth of 78.3 years, exceeding that of a male baby by more than 8 years—and women experience lower death rates than men for all causes except diabetes mellitus. However, women do report symptoms of both physical and mental illness more frequently than men. Of course, it may be that they report more illness than men because it is culturally more acceptable for them to do so. Women are thought of as the weaker sex, and illness is perceived as weakness, whereas strength, vigor, and good health are typically macho qualities, and men are held to a more rigid standard. As explanation for the greater number of psychogenic disorders in females, Jean and John Lennane speculated that doctors may perceive some complaints of women, such as menstrual cramps, morning sickness, labor pains, and "colic" in their babies, for which scientific evidence clearly indicates an organic cause, as arising from women's frustrations, anxieties, or depression (1973). Such dismissal of disorders as "neurotic," said the Lennanes in the *New England Journal of Medicine*, may be a form of sexual prejudice on the part of their physicians.

A number of studies have attempted to document such prejudice (physician bias), but as pointed out by Verbrugge and Steiner (1981), it is very difficult to scientifically *prove* sex bias in health care. It is evident, however, that for whatever reasons, males and females utilize health services differently. Women, moreover, have a unique problem within the health care system because they are, as indicated, the predominant *consumers* of health care, while men (the doctors) are the predominant *providers*.

If health insurance, pharmaceutical, and government employees are included, there are more than 10 million Americans who make their living in the health care industry. Although this labor force is more than 75% female, most of the women in the field are poorly paid, poorly organized, and have virtually no decision-making power. Policy is set by a relatively small group of male doctors, hospital administrators, medical school deans, and pharmaceutical and insurance industry executives. These workers in the medical–industrial complex are extremely well organized through their professional organizations, are able to mount extensive lobbying efforts, and are very well paid, their incomes having increased inordinately over the salaries of other health care workers in the past decades.

WOMEN AS HEALTH CARE PROVIDERS

Women are practically nonexistent in the power positions of the health care system. From 1980 census figures, 88% of the physicians are male , as are 96% of the dentists, 96% of the optometrists, and 75% of the pharmacists. Women form the large group of nurses (more than 1.4 million), dieticians, occupational and physical therapists, social workers, and medical technologists, and the even larger group of clerical and service workers. The participation of women in the health care labor force has been highly segregated and has been chiefly limited to supportive or auxiliary positions. The health team is a hierarchy, and at the top are the unquestioned leaders: The upper-middle-class, predominantly white, male physicians who control the health care delivery system.

It is this elite structure that defines the illness, decides whether or not hospitalization should take place, prescribes the drugs, and determines the nature and extent of the treatment. If the status of women as both consumers and producers of health care is to improve, a major change will be necessary —the admittance of women to the prestigious health professions and the participation of women from all economic levels in decision-making jobs in the health care system.

There is evidence that substantial change has occurred in the familiar "my son-the-doctor, my-daughter-the-nurse-therapist-dietician" tradition. In 1968–1969, women made up only 6% of the students entering medical schools. According to the Association of American Medical Colleges, however, by 1983–1984 32% of the 15,978 freshmen were women, and of the total population of medical students, 30% were women. If the number of women in medical schools continues to increase, or even if it levels off at around 30%, one could expect that the processes of medical education that socialize physicians toward sexist attitudes will necessarily be modified. Moreover, by the end of the twentieth century, these women will be in their forties and should be having considerable influence

on medical practice with a resulting improvement of medical services for women and all of society.

At present, however, women medical students still face discrimination, although they may be finding that their increased numbers have made their medical training easier for them than it was for their counterparts a decade ago. There is less overt discrimination in recruitment, admissions, financial aid, health services, and lodging. Such bias is unquestionably illegal and specifically prohibited by Title IX of the 1972 Education amendments to the Civil Rights Act. Women students still encounter problems, but perhaps now in a more subtle way. They may be teased, baited, called on in class too much or not at all, or they may be asked how many hours they will work, or how many years they will take off for childbearing and rearing. They have gained equal opportunity to become physicians, but once they graduate, they may find that certain subspecialty residencies are virtually closed to them. The traditional expectation that women will want to link family responsibilities with their professional duties has resulted in the specialty orientation of women in areas of pediatrics, allergy, psychiatry, and anesthesiology, which are viewed as being more compatible with the traditional female image. Women graduates themselves, aware of the potential conflicts produced by marriage, motherhood, and medicine, self-select into those residencies. According to a 1983 report by the Task Force on Opportunities for Women in Pediatrics, anesthesiology, psychiatry, and pediatrics are the specialties with the greatest number of women residents. Although there are no programs that are exclusively male (women are training in every specialty) they continue to be aggregated in the less prestigious, lower-paying areas. In 1980, the women residents in pediatrics numbered 2,432, or around 18%; but there was only one woman resident in the country in colon and rectal surgery, five in pediatric surgery, and eight in thoracic surgery—less than 0.01%.

One of the most insidious forms of discrimination against women medical students is the lack of role models during their medical education because of the few, if any, senior faculty women as instructors or administrators. The expectation that the increasing numbers of women doctors will eventually infiltrate medical faculty ranks is belied by data accumulated in the last decade. Although the number of women in the student body has increased threefold, a corresponding increase in the number of women physicians on medical school faculties has not occurred (Wallis et al., 1981). Moreover, those female physicians who want to do research and teach in medical academia are clustered in the nontenured, lower academic ranks. A 1982 American Association of Medical Colleges report states that only 28% of women were full professors compared with 61% of the men. Only 2% of medical school department chairs are women, 10% of associate deans, and 17% of assistant deans. The latter administrative positions were most often in student or minority affairs. In 1983, there was no woman dean of a medical school, and, given such bleak statistics, the prospects of getting one are dim.

In 1974, when few women were accepted to medical school, Dr. Sonia Bauer, a California physician writing in the *New England Journal of Medicine,* stated: "The sexism is so deeply ingrained in the (medical) profession that it takes an effort to notice it, and having noticed it, it will take an extraordinary effort to eliminate its effects." In 1983, when women were being accepted in greater numbers, the problem is a more subtle form of sexism. Thus, Dr. Lila Wallis, associated clinical professor of medicine at Cornell University, was quoted in *Medical News* (1982) ". . . The most damaging myth around is that the inclusion of numbers of women medical students is going to be incisive in the way that medicine is taught and practiced." If women do not gain greater access to research and academic medicine, she warns, "all these gains in the last decade are likely to get lost in the next backlash."

On the basis of numbers alone, however, women doctors and women administrators may eventually receive a greater share of the power and wealth at the top. Moreover, formerly submissive groups like nurses, technicians, nurse-midwives, and nurse-practitioners, with a growing assertiveness of their professional status and rights, will insist on their significant impact on the composition of the policy-making bodies. The system of control and governance has to change with the more vocal presence of women in the health hierarchy, and health care is bound to be more responsive to those women who receive it and those who work in it.

THE HEALTH CARE INDUSTRY

All segments of society have a great deal to gain by changing the health care system. The emphasis through the years seems to have shifted from the "caring" to the "system," and the medical care industry may become the biggest business in the United States before the end of the 1980s. In 1950, the cost of health care was $12 billion, but it was $322 billion by the early 1980s, having risen faster than any other item in the cost of living. According to the Department of Health and Human Services, 42% of the health care dollars went to hospitals, 19% to doctors, 9% to nursing homes, and 6% to dentists. The entire complicated system—the 500,000 doctors, the 10 million health workers, the 700 hospitals, the clinics, the nursing homes, the drug industry, the supporting insurance plans, the sophisticated equipment, the new and costly techniques of care—is apparently exempt from the usual economic laws that govern supply and demand. There is little competition in hospital service charges, physician and dentist fees, prescriptions, and health insurance rates, and people do not ordinarily shop around for those services—they take what is available to them. A 1980 report issued by the Graduate Medical Education Advisory Committee to the Department of Health and Human Services predicted a 15% doctor surplus by the end of the decade, and, if nothing were done to stop the oversupply, there could be a 29% surplus by the end of the century. But more doctors would not necessarily create cost-cutting competition; several studies have shown that additional doctors mean additional medical care. Although patients are referred to as consumers, they are not the ones to decide, as they may with other services and products, what and how much medical care to buy. It is the physicians who determine the nature and extent of medical services, and, under the traditional doctor–patient relationship, the patient does not question the decisions. If the patient is hospitalized for tests, treatment, or surgery, a third party pays: either the government, if the individual is older than 65 or poor or disabled, or the private insurance plans if the patient is wealthy enough to pay for them or has an employer who picks up the cost of premiums. An inherent potential for abuse exists in this kind of system. Not only do the physicians, hospitals, and other providers of health care have an apparently insatiable capability to absorb reimbursement for services, there is also ample opportunity for the greedy, unscrupulous, or downright incompetent health professional to function under few restraints.

MEDICAL INCOMPETENCY

Poor-quality medical care, that which falls far below acceptable standards and is delivered by careless or inept physicians, can cause serious damage and actually be life threatening.

All people, men and women, children and adults, can suffer from bad medical practice. There are indications that billions of dollars are wasted on unneeded hospitalizations and that thousands of lives may be needlessly lost as a result of unnecessary surgery, useless and ineffective treatment, and adverse drug reactions from excessive and irrational prescribing.

At the Third National Conference on the Impaired Physician, held in 1978 by the American Medical Association, it was estimated that between 10% and 15% of the nation's doctors are chemically dependent, that is, in some state of alcoholism or other drug addiction. This figure, based on the known rates of drug addiction and alcoholism in professional groups and on cases reported by patients, pharmacists, narcotics agents, or doctors themselves, means that, potentially, more than 50,000 physicians may be impaired because of some degree of dependency. This number does not reflect or consider the doctors who are incompetent and unfit to practice medicine because they are suffering from mental illness, physical disability, or senility. In recognition of the problem of incompetency, the American Medical Association, in 1974, sponsored a model bill aimed at "providing for the restriction, suspension, or revocation of the license of any physician to practice medicine because of his inability to practice medicine with reasonable skill and safety to patients, due to physical or mental illness, including deterioration through the aging process or loss of motor skill, or abuse of drugs, including alcohol." By 1977, 35 states had passed such "disabled doctor" legislation, but only an average of 150 licenses per year are currently revoked nationwide by state licensing boards.

The thoroughly unfit doctors are presumably guilty of great damage to their unsuspecting patients, but there may be inestimable harm done by the much greater number of doctors who, for whatever reasons, do not practice the very best medicine they can. Doctors who are arrogant or careless, who ignore recent medical advances, whose prescribing practices reflect medical advertising more than scientific evaluation, who give the wrong drug or the right drug in the wrong dosage, who operate too much or not soon enough, or who bungle the surgery when it is performed—these are examples of substandard medical practices frequently unrecognized because consumers are ignorant of what actually constitutes quality and competence in health care. Even if people who are healthy get sick, or if people who are sick get worse or die because of the medical treatment they have received, there is still not much formal recourse available to the survivors or relatives.

The number of medical malpractice suits has increased in the past few years, as people attempt to attain a favorable verdict and compensation for damages; however, cases of malpractice by physicians and hospitals are difficult to evaluate and prove. Unfortunately, physicians now view each patient who walks in the door as a potential litigant, and they practice more defensive medicine, and by ordering more diagnostic tests and more hospitalizations, pass these costs along to the patients, who then pay even more for their health care.

MONITORING OF QUALITY BY THE MEDICAL PROFESSION

Doctors affirm that they have always had a major interest in policing their own profession. In hospitals, for example, there have always been committees that surveyed the treatment delivered and the surgeries performed, and it has been maintained that the medical societies have adequate mechanisms to control incompetence. Physicians conventionally claim that the percentage of incompetent professionals and the magnitude of abuses are highly exaggerated by the mass media and government agencies, and these physicians probably make such claims in good faith. Doctors, however, inherently lack objectivity. They are evaluating their own colleagues and, in the medical societies, their own dues-paying members. They have no particularly effective disciplinary sanctions to impose should they find abuses, and they generally do not actively search them out.

Responding to public complaint, legislators in many states have demanded that doctors take a more active role in protecting the public against medical incompetency. New statutes have been passed, or old ones have been revised to put more clout in physician-policing laws. Forty-four states now provide legal immunity for doctors who report colleagues to licensing boards, and some states make such reporting by doctors and/or hospitals, medical societies, and malpractice insurers mandatory. There are now new grounds for disciplining doctors, a greater flexibility in penalties so that various degrees of punishment may be used, and, in general, stricter sanctions. Some state boards use an informal hearing process and plea bargaining—action on the license will be stayed if the doctor voluntarily seeks therapy, for example. Disabled doctors' programs with goals of prevention, intervention, and rehabilitation of impaired physicians have been initiated by a number of state medical societies. The effectiveness of such activities is controversial.

"Physician helping physician" is one solution to the problem of medical incompetency, but identification of the impaired physician remains the difficulty. A hospital administrator, writing in a state medical journal, described some reasons why the medical staff may procrastinate or refuse to take action against an incompetent physician. An impaired doctor's popularity and the nurses' and aides' unwillingness to cause him personal or professional harm may result in silence. The medical staff members could be reluctant to set a precedent; someone may blow a whistle on them next. Moreover, fellow professionals are generally not directly affected by incompetency; it is the patients who are prey to the actions of the impaired physician. Finally, physicians may not want to get involved because it is they who might be "vulnerable to economic reprisals from the impaired physician and his friends—referral patterns may be altered" (Kane, 1982).

UNNECESSARY SURGERY

In January 1976, the congressional Subcommittee on Oversight and Investigations reported that of the 14 million nonemergency operations performed annually, 2.4 million (17%) were unnecessary. As a result of this unnecessary surgery, charged the Subcommittee, 11,900 people die annually.

The American Medical Association emphatically disagreed and accused the Subcommittee of factual errors and of "undermining the confidence of the medical profession in the eyes of the public." Undeniably, statistics like these are frightening, and publishing them means that some people will then view all surgery with consternation, even if it is necessary to improve their health or save their lives. Since even the most conscientious and competent doctors might disagree on the need for a particular operation at a particular time, the committee's samplings may have been inaccurate and their statistics exaggerated. That rates of surgery have increased, however, is incontestable. Open-heart surgeries have increased almost 100% in the last few years, and almost 200,000 coronary bypass surgeries annually were being performed by 1984, an increase of almost 600% since 1970. Whether

such expansion of surgical rates means that a certain proportion is unnecessary surgery is unclear. The following section presents data concerning the most prevalent major surgical procedure of all types and the most common operation performed on women—the hysterectomy, or removal of the uterus.

It has been claimed that the United States is the hysterectomy capital of the world. More than 800,000 such surgeries are performed here annually, more than in any other country (Barber, 1982). Accompanying the increasing number of hysterectomies is the controversy surrounding the necessity for this kind of surgery. Currently, there is not any completely acceptable method available that has been developed to assess what proportion of surgery is necessary or that would effectively decrease the number of unnecessary operations.

During the early 1970s, Eugene McCarthy and Geraldine Widmer (1974) at Cornell University Medical College made an extensive study of the need for elective, nonemergency surgery in a group of union members in New York. They set up a screening program in which a person who had been recommended for an operation would go to a board-certified physician as a consultant in order to get a second opinion on the need for the operation. It was discovered that when a second gynecologist examined the patients recommended for hysterectomies, nearly one out of three, 32%, were not confirmed as being necessary. A subsequent investigation of the New York Hospital–Cornell Medical Center patients studied through 1980 produced similar results—41.3% of the women who voluntarily sought a second opinion and 30.7% of those who were required to seek a second opinion were not confirmed for the necessity of the hysterectomy by the board-certified consultant (Finkel et al., 1982). These studies, done preoperatively or prospectively, confirm data found by other workers who had conducted retrospective audits (i.e., after the hysterectomies had been performed). Truesell et al., (1962) found the same proportion of hysterectomies (32%) to be unnecessary to unsatisfactory in their study done on Teamster's Union families in the New York area. Perhaps not unexpectedly, opposite and contrasting results were reported by the Medical Society of the County of New York, which used the same procedure of retrospective review and maintained that only 3.2% of hysterectomies were unnecessary and that 7.1% were questionable.

There have been estimates that, on the average, gynecologists perform 37 hysterectomies a year. At a fee of $1,000–$1,500 for each, it is possible for the physician to earn between $37,000 and $55,000 annually on hysterectomies alone. Some feminist groups have claimed that women are frightened into having hysterectomies and are exploited to make money for the doctors, even though there are alternative methods of treatment. Some doctors, however, maintain that there has been an increasing demand on the part of women for hysterectomies. These doctors claim that as a reflection of the women's movement, more women request the operation as a means of sterilization or because they are unwilling to put up with the "nonsense" of menstruation each month, and that if one doctor will not remove their uteri, these women will find one who will. If this is actually a trend, it is another indication of how ill-informed women are about their own bodies and major medicial procedures. Using hysterectomy to achieve sterilization is excessive for the purpose and somewhat equivalent to using decapitation to cure a migraine.

It seems clear that second-opinion programs for all surgeries might decrease the number of operations. After all, some people get several estimates on their cars before undergoing a major repair job; their own bodies deserve no less. Many health insurance companies will now pay for consultants to verify a diagnosis recommending surgery, and, if the consultant differs from the first opinion, a patient may get a third opinion. Both the patient and the doctor can then be more certain about the decision to operate, and the final choice is up to the patient.

CONSUMERISM IN HEALTH CARE

The second-opinion-in-surgery plan, a process by which a patient can make an informed choice concerning a proposed treatment, is one step toward a different kind of doctor–patient relationship, different from the traditional one in which the doctor makes all the decisions and the patient rarely questions them. To many people, the old way is no longer tenable. They view doctors not as deities with superhuman powers, but as highly skilled, highly educated human beings who provide services that are purchased by consumer-patients. They believe that all individuals have the right to decide what shall be done with or to their bodies and their personalities and that the essence of a just and adult physician–patient relationship in a democratic society is the sharing of information and opinion to reach an intelligent judgment together. Patients do not want the entire burden of decision-making placed on them, but they do want to be invited to participate in the process. They assume that a mature and competent physician has no desire for total control but feels a responsibility to proceed in the best interest of the patient.

Some members of the medical profession have a great deal of difficulty accepting the idea of sharing decision-making with nonprofessionals. Hippocrates himself advised doctors to conceal most things from their patients and to reveal nothing of the patient's present and future condition. This attitude on the part of physicians has frequently resulted in having mature, tax-paying citizens relegated to the status of passive, childlike individuals with less than normal intelligence. Women in particular have felt themselves in a double-bind, since they have been treated not only within such an authoritarian paternalistic relationship, but also within the social framework of the stereotyped feminine role—dependent, unquestioning, accepting.

WOMEN'S HEALTH MOVEMENT

With the goal of increased self-determination, groups of women in the 1970s, under the impetus of the larger women's liberation movement, began to organize into the women's health movement. Banding together to provide themselves and other women with health information and advocacy services, they formed grass-roots women's health collectives, some publishing newsletters and pamphlets. By the middle of the 1970s, there were more than 1,200 such local groups in the United States. A handful of women's health activists combined to form the National Women's Health Network (NWHN), in order to influence federal policy on women's health issues. Currently, the Washington-based public interest group with 14,000 members and 420 organizational members representing about 500,000 women is one of the most significant national voices for women's health needs. The NWHN is not only the watchdog on federal health policy affecting women, but is distributes health information at the local level through a speaker's bureau, appearances on radio and TV programs, and publication of a newsletter, as well as resource guides, on various aspects of women's health.

A number of the women's health advocates hired their own physicians to start their own women-controlled health centers. Some, like the New Hampshire Feminist Health Center, Elizabeth Blackwell Women's Health Center

in Minneapolis, Bread and Roses Women's Health Center in Milwaukee, Feminist Women's Health Center in Los Angeles, Vermont Women's Health Center in Burlington, and the Women's Health Service of Colorado Springs offer a wide range of gynecological, contraceptive, childbearing, and feminist therapy services. There are an estimated 100 or more such women's clinics in major urban areas in Canada and the United States. They all have similar aims: the increase of knowledge concerning female anatomy and physiology, a strong emphasis on preventive health care, an improved patient–practitioner communication, a call for more professional accountability and competence, and an understanding and evaluation of what constitutes quality health care so that it can be expected and demanded. All of the clinics encourage maximum community participation and community control. In many of them, the women served by the clinic not only sit on its Board of Directors, but frequently are trained as staff members or in occupations needed by the clinic and by the community.

In some of the women's health centers, the goals have changed from the development of better-informed patients and improved doctor–patient relationships to more radical self-help activity. Self-help clinics emphasize self-examination, self-medication, and the management and control of reproduction and common gynecological problems with little physician interaction. Strong justification can be made for the medical consciousness-raising and the sharing of, rather than restricting of, medical knowledge in this alternative approach to health care; however, even advocates of this plan express concern that self-diagnosis and self-treatment can carry medical risk, should a serious disease fail to be detected, for example. In a modification of the more radical route, but with the same conviction that increased knowledge is the way to power, some clinics have chosen the term "self-health" to denote that the patient retains control over decisions concerning her own health.

The successful existence and expansion of women's health clinics demonstrates that women's health needs have not been met in most communities. In many areas of the country, local government and private agencies have adopted the methods of the women's clinics, using them as models for an innovative and different delivery of health care—a clinical situation in which women help, support, and share knowledge with other women. This adaptation of the principles of woman-oriented health care delivery is only one example of how the goals of the women's health movement, with emphasis on self-care, medical self-awareness, patient control, and improvement and change in the health care delivery system have become a part of a general social trend for consumers of health care. Through forums and conferences, community "know-your-body" courses and "health fairs," through university and college courses on biology of women and women's health, through magazine and newspaper articles, and television programs, men and women are being given the information they need to interact with the health professionals, and not be acted upon.

There have been irrevocable changes that have already intruded upon the customary and age-old personal relationship between doctor and patient. Modern medical care has become fragmented, highly specialized, institutionalized, and enormously expensive, a condition considered counterproductive and "sickening" by many social critics. But as improved public education further demystifies medical practices and individuals realize that health is also their own responsibility and not the sole responsibility of the medical profession, patients can become their own consumer advocates and demand and understand:

- What the doctor is doing, and why the doctor is doing it
- A complete explanation of their medical condition, and the right to privacy and confidentiality
- A thorough description of all medications prescribed, their side effects, and their potential risks
- Participation as a partner in decisions concerning their treatment, all the alternatives to treatment, and their right to refuse treatment

Figure 1.2. Lack of knowledge can make a woman feel vulnerable and afraid. Knowing how the body functions in health and disease can give women the self-confidence to participate in the decisions concerning their own health care.

People will no longer tolerate being patronized or treated with condescension; they will insist on being treated with dignity and respect. In this way a further *positive* modification of the traditional doctor-dominant, patient-subordinate role can be accomplished to the mutual benefit of both (Fig. 1.2).

REFERENCES

Bauer, S. Letter. *N Engl J Med* 291(21):1141–1142, November 1974.

Barber, H. R. K. The training of the gynecologist. *Obstet Gynecol* 60(6):708–710, 1982.

Committee on Interstate and Foreign Commerce. Cost and Quality of Health Care: Unnecessary Surgery. Report by a subcommittee of the House of Representatives. Government Printing Office, Washington, D.C., 1976.

Finkel, M. L., McCarthy, E. G., and Ruchlin, H. S. The current status of surgical second opinion programs. *Surg Clin North Am* 62(4):705–719, August, 1982.

Harrison, M. *A Woman in Residence.* New York, Random House, 1982.

Kane, D. Role of hospital administrator. *Wis Med J* 81:35–36, April, 1982.

Lennane, K. J. and Lennane, R. J. Alleged psychogenic disorders in women—A possible manifestation of sexual prejudice. *N Engl J Med* 288(6):288–292, February, 1973.

McCarthy, E. G. and Widmer, G. Effects of screening by consultants on recommended elective surgical procedures. *N Engl J Med* 291:1331–1335, 1974.

Participation of women and minorities on medical school faculties. Association of American Medical Colleges, Washington, D.C., 1982.

Report of the Task Force on Opportunities for Women in Pediatrics. *Pediatrics* 71(4)(suppl): 679–714, April, 1983.

Scully, D. and Bart, P. A funny thing happened on the way to the orifice: Women in gynecology textbooks, in *Changing Women in a Changing Society*, Huber, J. (ed). Chicago, University of Chicago Press, 1973.

Truesell, R. E., Morehead, M. A., and Ehrlich, J. The quantity, quality, and costs of medical and hospital care secured by a sample of teamsters families in the New York area. New York, Columbia University School of Public Health and Administrative Medicine, 1962.

Verbrugge, L. M. and Steiner, R. P. Physician treatment of men and women patients. Sex bias or appropriate care? *Med Care* 19:609–632, 1981.

Wallis, L. *Medical News and International Report* May 10, 1982.

Wallis L. A., Gilder, H., and Thaier, H. Advancement of men and women in medical academia. *JAMA* 246(20):2350–2353, Nov. 20, 1981.

Wolfe, S. M. Statement of Sidney M. Wolfe, M.D. Health Research Group, Washington, D.C. Presented at the hearings before the Subcommittee on Health. Committee on Labor and Public Welfare, U.S. Senate, 93rd Congress, Second Session, Examination of the Pharmaceutical Industry, 1975.

Wolfe, S. M. and Johnson, A. Statement of Sidney M. Wolfe and Anita Johnson, Health Research Group, Washington, D.C. Presented at the hearings before the Subcommittee on Public Health and Environment of the Committee on Interstate and Foreign Commerce, House of Representatives, 93rd Congress, First session, Medical Devices, 1975.

2. Reproductive Anatomy

Historically, the female body and, more specifically, the female reproductive tract, because it is internal and therefore hidden, have always been subject to much romanticism, fantasizing, and descriptive error. Until the publication of the writings and illustrations of the most outstanding anatomist of the Renaissance, Andreas Vesalius, there was virtually no anatomically correct knowledge of female structure. The reason for the lack of information until the sixteenth century has been attributed to the lack of material for dissection. Even when courses in human anatomy were recognized as part of the curriculum in medical schools all over Europe, corpses for dissection were difficult to obtain because only the bodies of executed criminals who came from an area at least 30 miles away could be used. One or two dissections a year were performed, and female cadavers were rarely available. Vesalius' unprecedented graphic visualization of anatomy, *De Humani Corporis Fabrica*, was not only an accurate representation of the structure of males but was based on the dissections of at least nine female cadavers as well and, therefore, formed the foundation for modern anatomical knowledge

of both men and women. In the 400 years since Vesalius, if there were any women physicians and anatomists who contributed to the advancement of knowledge concerning the female reproductive tract, it would not be obvious from the nomenclature. The discoverers of the female anatomical parts, the recognizers of clinical syndromes, signs, tests, and phenomena, the developers of instruments, techniques, operations, and therapies were evidently all men or, at any rate, only men have received acknowledgment (Table 2.1).

Even after anatomical knowledge of the human female was available, there was relatively scant knowledge about her physiology until the last 50 years, and the dissemination of such information to men and women has been minimal. Some of the myths that have arisen concerning the physical and mental abilities of women have had remarkable persistence among the public and among educators and physicians who presumably should know better. Fallacies and generalizations about female anatomy, physiology, and sexuality have arisen based more on cultural assumptions than on accurate observations. Even when it is pointed out that insufficient

Table 2.1 SOME CONTRIBUTORS TO NOMENCLATURE IN GYNECOLOGY AND OBSTETRICS

Individual	Eponym	Description
Caspar Bartholin 1655–1738	Bartholin's glands	Greater vestibular glands
James Read Chadwick 1884–1905	Chadwick's sign	Color changes in the pregnant vulvovaginal mucosa
Albert Döderlein 1860–1941	Döderlein's bacilli	Lactobacilli of vagina
Gabriele Fallopius 1523?–1562	Fallopian tubes	Oviducts
Regnier de Graaf 1641–1673	Graafian follicle	Preovulatory follicle
Alfred Hegar 1830–1914	Hegar's sign	Softening of the lower uterine segment during pregnancy
John Braxton Hicks 1823–1897	Braxton Hicks' contractions	Contraction of the pregnant uterus
Hugh Lenox Hodge 1796–1873	Hodge pessary	Vaginal support for uterine displacement
Max Huhner 1873–1947	Huhner test	Postcoital semen examination
William Fetherstone Montgomery 1797–1859	Montgomery's tubercles	Breast areolar changes during pregnancy
Johannes Müller 1801–1858	Müllerian ducts	Embryonic paired ducts; give rise to uterus and vagina
Martin Naboth 1675–1721	Nabothian cysts	Cervical mucous cysts
Franz Carl Nägele 1777–1851	Nägele's rule	Formula for estimation of date of delivery
Anton Nuck 1650–1692	Canal of Nuck	Inguinal canal
George Papanicolaou 1883–1962	"Pap" smear	Cervical cancer detection
Isidor Rubin 1883–1958	Rubin test	Tubal insufflation
Alexander Skene 1838–1900	Skene's ducts	Paraurethral ducts
Friedrich Trendelenburg 1844–1924	Trendelenburg position	Elevated pelvic position
Henry Turner 1892–	Turner's syndrome	Ovarian dysgenesis
Thomas Wharton 1614–1673	Wharton's jelly	Umbilical cord mucous matrix
Caspar Wolff 1733–1794	Wolffian duct	Embryonic mesonephric duct

evidence exists for a belief, or that scientific data invalidate a previously held opinion, there is a tendency to cling to outmoded misinformation. The ignorance of men and women concerning the female body has helped to perpetuate the continuation of fallacy and superstition; this chapter provides the anatomical basis to distinguish the myths from the realities.

THE PELVIC GIRDLE

One widely held notion, for example, is that there is greater risk in participation in contact sports for females because they are more vulnerable to injury than males and because their internal reproductive organs are more vulnerable to damage. Actually, there are dif-

ferences in the susceptibility of females and males to the impact of direct contact in sports, but the resultant injuries are to ligaments and muscles, for reasons to be described later, and not to the internal organs. A protective cage for the sexual organs of women is formed by the strong bony pelvic girdle, and the pelvic viscera are seldom damaged, even in the crushing injuries of accidents. The exposed genitalia of the male are far more likely to be injured in contact sports (see Fig. 2.1).

The pelvic girdle is the general name given to the two broad, heavy hip bones that provide an attachment for the leg and support the lower spine in order to transmit the weight of the body from the vertebral column to the limbs. Each hip bone, also called the *innominate bone*, is composed of three fused bones, the *ilium*, the *ischium*, and the *pubis*. A

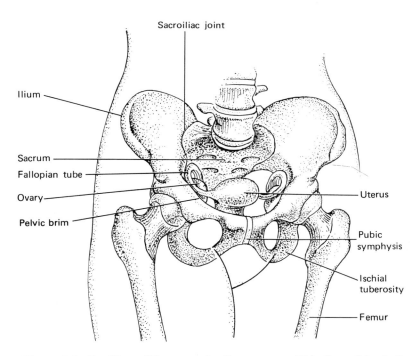

Figure 2.1. Position of the reproductive organs within the pelvic girdle.

cup-shaped socket, the *acetabulum* (Latin for little saucer that holds vinegar), receives the head of the thigh bone or femur to form the hip joint (see Figs. 2.2, 2.3).

Ilium

The large ilium flares upward and outward from the acetabulum. When hands are placed on hips, one is feeling the broad and thick crest of the ilium. Follow it forward along the forefinger. The tip of the forefinger is on the *anterior superior iliac spine.* The tip of the thumb is approximately in the region of the *posterior superior iliac spine,* easily palpable through the skin of a thin person, and always marked by a dimple. The inner surface of the iliac bones is concave, and forms the origin of

a powerful muscle of thigh and trunk movement, the *iliacus.* On the right side, the concavity in the iliac bone, the iliac *fossa,* accomodates the *cecum,* which is the pouchlike, blind end of the large intestine formed at its junction with the small intestine. From the cecum extends the vermiform (wormlike) appendix, and because of its anatomical position, the pain of acute appendicitis is felt in the right iliac fossa. The fallopian tubes or oviducts are also in anatomical proximity to the concavities of both iliac bones, and pain originating from the tubes when they are inflamed or infected is often referred to (i.e., felt in) the fossae.

The rear, or posterior part of the inner surface of each ilium has an *auricular* (ear-shaped) *surface* that forms an articulation or

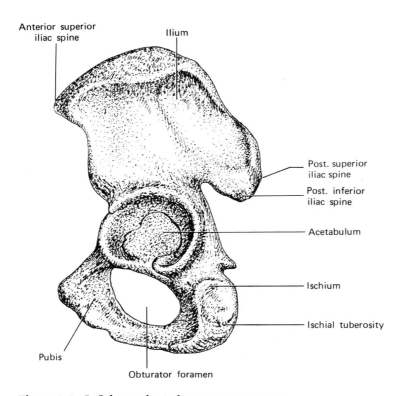

Figure 2.2. Left innominate bone, outer aspect.

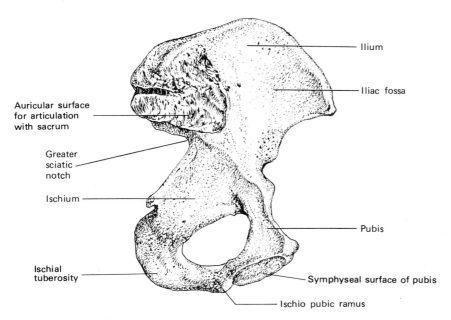

Ilium

Iliac fossa

Auricular surface
for articulation
with sacrum

Greater
sciatic
notch

Ischium

Pubis

Ischial
tuberosity

Symphyseal surface of pubis

Ischio pubic ramus

Figure 2.3. Left innominate bone, inner aspect.

junction with the *sacrum*, the fused vertebrae at the lower end of the spinal column. When one bone articulates with another, a joint is formed, whether or not it is movable. The joint here is the *sacroiliac joint*, one of the most important joints of the body because the body's full weight is transmitted through it to the legs when a person is standing upright. The great load placed on the sacrum by the entire vertebral column would tend to cause it to rock back and forth between the hip-bones if it were not for enormously strong ligaments that surround the sacroiliac joint to form an interlocking mechanism to brace and reinforce it.

Ischium

Below the sacroiliac joint, the rear border of each ilium indents to form a huge greater *sciatic notch* through which pass blood vessels and nerves. The lower part of the greater sci-

atic notch is formed by the *ischium*, which ends in a large knob, the *ischial tuberosity*. When one is sitting up straight, one sits on the ischial tuberosities, and they, instead of the legs, receive the weight of the body through the sacroiliac joints. The hamstrings, that group of large muscles on the back of the thigh, are attached to the ischial tuberosities.

Pubis

From the tuberosity of the ischium extends a flattened *ischial ramus*, or bar, which meets the flattened ramus of the *pubis*. The ramus of the pubis flares out to form the body of the pubis, which meets the other pubic bone from the opposite side. The union of the two pubic bones is called the *pubic symphysis*, a joint that is united by cartilage and held together by strong ligaments. During pregnancy, both the pubic symphysis and the sacroiliac joints are softened and stretch as a result of the tremendous amounts of hormones that are produced. The joints become mobile and make delivery easier.

Bony Pelvis

The term bony pelvis refers to the bowl–like structure (pelvis means "basin" in Latin) that is formed by the hip bones at the sides and the front and the sacrum and coccyx at the back. It is divided anatomically into the:

1. false or greater pelvis, made up of the upper flared parts of the two iliac bones with their concavities, and by the two wings of the base of the sacrum.
2. true or lesser pelvis, formed by the rest of the ilium, pubis, and ischium on both sides, and the sacrum and the coccyx.

The boundaries of the opening to the true pelvis, or *pelvic inlet*, are called the *pelvic brim*. The diameters of the pelvic brim have particular obstetric significance. The dimensions of the *pelvic outlet*, bounded by the ischial tuberosities, the lower rim of the pubic symphysis, and the tip of the coccyx, are also of great importance obstetrically (see Fig. 2.4).

At birth the three parts of the hip bone, the ilium, ischium, and pubis, are composed mainly of cartilage and are separate bones. The ischium and the pubis become bony and fuse at approximately 7 or 8 years of age, but total ossification of all the cartilaginous portions is not completed until sometime between 17 and 25. Although pregnancy is possible after puberty, the pelvic ring may not be as capable of withstanding the stresses and strains of childbearing until all the weaker cartilaginous links among the three bones have fused and become bony.

SEX DIFFERENCES IN THE PELVIS

The body measurements (stature, sitting height, head circumference, and so on) of an adult female average approximately 92% of the body measurements of the adult male. This same proportionality may be applied to skeletal measurements. Generally, the male pelvis as a whole is larger than the female pelvis, except for the dimensions of the *true pelvis*, which has to accommodate the dimensions of the full-term fetal head, since 95% of babies are born head first. Pelvic measurements, however, show considerable variation (as do all other measurements in humans), and there is actually as much variation in the size and shape of the pelvis among women as there is between women and men. No two pelves are alike; it is the individual pelvis and the particular fetal head involved that become important obstetrically.

Sexual differences in the adult pelvis have been studied extensively, and there are different classifications that have been used to describe the normal range of variation in the morphology of the male and female pelvis. The most commonly quoted are those described by W. E. Caldwell and H. C. Moloy, based on X-ray determinations of the dimensions of the pelvic inlet, or superior opening of the true pelvis. These authorities said that female pelves are divided into four main groups:

1. The *anthropoid* pelvis is common in men, and occurs in 20%–30% of white women and nearly 50% of black women. The pelvic inlet is oval and the sacrum is long, producing a deep pelvis.
2. The *android* pelvis is also common in men, but one-third of white women and 10%–15% of black women also have this type, in which the inlet is heart shaped and the side walls are narrow. This classification, also called the "funnel" pelvis, produces difficulty in delivery of the baby.
3. The *gynecoid*, or true female, pelvis is less common in males. About 50% of all women have this type. The inlet is round, the outlet is roomy, and the subpubic angle or pubic

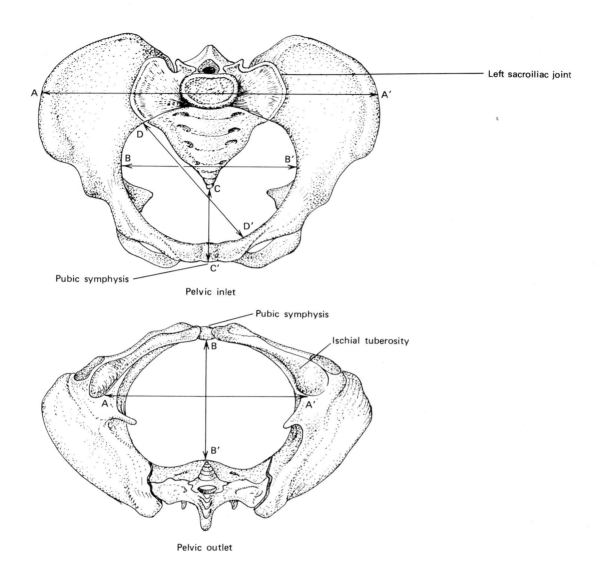

Left sacroiliac joint

A ————————————— A'

D

B ————————————— B'

C

D'

C'

Pubic symphysis

Pelvic inlet

Pubic symphysis

Ischial tuberosity

B

A ————————————— A'

B'

Pelvic outlet

Figure 2.4. Diameters of the pelvic inlet and the pelvic outlet. Inlet: A–A' = false pelvis; B–B' = true pelvis, transverse diameter; C–C' = anterior-posterior diameter, from tip of the coccyx to the pubic symphysis; D–D' = oblique diameter from sacroiliac joint to the iliopubic eminence.
Outlet: A–A' = transverse diameter, between inner edges of ischial tuberosities; B–B' = anterior-posterior diameter, from pubic symphysis to tip of coccyx.

arch is almost a 90° angle. The gynecoid is the best pelvic type for an easy normal delivery and is the one selected for contrast with the android pelvis in anatomy books to depict typical male and female pelvic differences.

4. The *platypelloid*, or flat, pelvis is the least common type of pelvic structure among most men and women. In this rare type, the pelvic cavity is shallow but widens at the pelvic outlet, permitting a delivery that is not difficult, as long as the fetal head can pass through the pelvic inlet (see Fig. 2.5).

Many women have a combination of these four basic types, and the anterior part of the pelvis may be one classification, whereas the posterior segment is another. These classifications are based on average values obtained

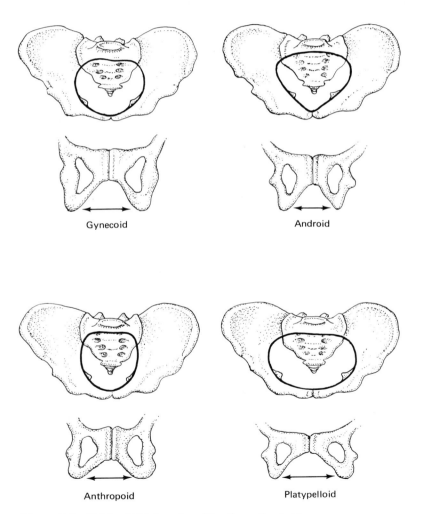

Gynecoid

Android

Anthropoid

Platypelloid

Figure 2.5. Caldwell-Moloy classification of pelves.

from skeletal material and, as indicated, are obviously not as important as the individual woman's measurements compared to the measurements of the head of the child she is bearing. Measuring the dimensions of the true pelvis is called pelvimetry, and it is usually performed as part of the physical examination of a pregnant woman to determine if she will have any difficulty in delivery.

Pelvimetry can be done by X-ray, by external measurements made with a *pelvimeter*, and by internal examination through the vagina. Because of the danger of radiation to the ovaries of the mother and the fetus, radiography is used rarely, delayed until the time of delivery, and then done only if there is some apparent difficulty in labor. Most X-ray pelvimetry has currently been replaced by ultrasonography.

If a woman has very narrow pelvic dimensions or if some distortion of the normal pelvis has occurred as a result of poor nutrition, injury, or disease, the decrease in size may be enough to interfere with normal labor. Such a pelvis is said to be contracted and could cause *dystocia*, or long, difficult labor. It may be necessary for the fetus to be removed by an incision into the uterus, or *cesarean section*.

Pelvic Tilt

When we stand in an upright position, the whole pelvis is tipped forward so that the plane of the pelvic brim forms an angle of approximately 50°–60° with the horizontal. The plane of the pelvic outlet, an imaginary line drawn from the tip of the coccyx to the inferior part of the pubic symphysis, forms an angle of about 15° with the horizontal. This means that the pelvic surface of the pubic symphysis faces upwards as much as it does backwards, and the concavity of the sacrum is directed both downwards and forwards. If one were to stand upright, facing flat against a wall, the anterior superior iliac spines and the upper border of the pubic symphysis would both almost touch the wall; that is, they would be in the same vertical plane. This tilt of the pelvis is called the angle of pelvic inclination, and although subject to great individual variation, it is frequently exaggerated in women due to the difference in dimensions of the true pelvis. As a result of the greater tilt of the pelvis in females, the spinal curvature in the lumbar region of the spine is increased in an anterior (forward) direction to compensate and maintain the center of gravity. Otherwise, a woman might fall forward and be unable to maintain an erect posture. Since this forward lumbar curve is greater in some women, their buttocks are usually more prominent than those of men, depending, of course, on the shape of the pelvis and amount of the pelvic tilt (see Fig. 2.6).

Another general statement, again subject to great individual variation, is that women are more frequently seen with knock-knees than are men. This is true because the head of the femur fits into the cup-shaped acetabulum to form an angle of approximately 125° at the neck of the femur, so that the shaft of the thighbone can swing clear of the pelvis when the leg moves. That angle determines the position of the knees. The more oblique the angle, the more the shafts of the femur will slope inward, and the closer the knees will meet. In women, the pelvis is generally wider and the femurs shorter, and hence the greater tendency to have knock-knees (see Fig. 2.7). Some orthopedic physicians have maintained that because of the abovementioned body mechanics, women's knees are much more vulnerable to injury and that they should, therefore, avoid participation in contact sports. This may be true for some women who have a gynecoid pelvis and a smaller, lighter bone structure. It may be equally valid for some men.

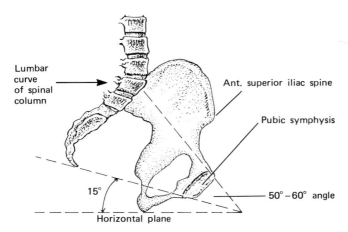

Lumbar curve of spinal column

Ant. superior iliac spine

Pubic symphysis

15°

50° – 60° angle

Horizontal plane

Figure 2.6. Angle of pelvic inclination. When standing erect, the whole pelvis is tilted forward, and the pelvic canal is directed backwards relative to the abdominal cavity and the torso. The greater the pelvic tilt of the pelvis, the greater the curve in the lower back.

Backache and Its Relation to the Pelvis

If a woman has a greater angle of pelvic inclination, she will also have a greater increase in the curvature of the lumbar spine. Then anything that places further strain on the lumbar area, such as the protruding abdomen of pregnancy or obesity, is almost certain to result in lower back pain. There are estimates that 50%–60% of the population has had back trouble at one time or another, but more females than males suffer with chronic backache.

Sometimes the very process of delivering a baby, particularly if the labor is long and difficult, puts unusual strain on the muscles and ligaments surrounding the sacroiliac joints, which have "unlocked" and stretched under hormonal action to have much more mobility at full-term than in a nonpregnant woman. Although these sacroiliac joint changes regress in the period after the birth of the baby, it can take a long time for them to get back to

normal, and the woman will have chronic low back pain in the interim.

With the exception of the period of *puerperium* (after delivery of a baby), most backaches in women are muscular aches resulting from poor posture and are not gynecological in origin. If lower back pain is a result of pelvic pathology, it is felt in the area over the sacrum; postural backache is higher—between the top of the sacrum and the bottom of the rib cage.

When one is standing erect with good posture, the weight of the body is passed down the lumbar vertebrae through the sacrum, divided through each sacroiliac joint to the acetabula of the hip bones, and then to the legs. The most important supporting spinal muscles, those that keep the line of gravity in appropriate balance, arise from the front and back of the pelvis. When good posture is disturbed, those muscles are placed under stress, and they become sore and ache. Numerous factors can act to destroy good posture. One, mentioned above, is the strain on

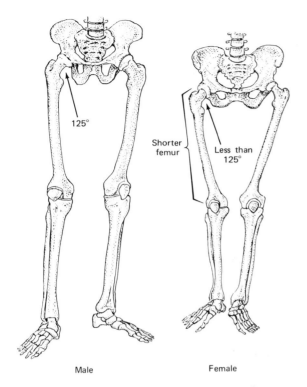

Figure 2.7. The femoral angle is less than 125° in some women, giving women a greater tendency to knock-knees.

the muscles of the spinal column caused by pregnancy, labor, and the puerperium. Or there may be congenital (present at birth) defects in the spine, pelvis, or feet. Sometimes sleeping on the wrong kind of mattress, wearing shoes or clogs with rigid soles, or sitting all day in a particular type of chair can result in backache. These factors, however, once recognized, are not difficult to eliminate. When minor defects in the spine, hips, knees, or feet result in poor posture, there are exercises than strengthen the tone of the abdominal and back muscles so that the spine can be supported in a corrected position.

When the ligaments and tendons that support the spinal column are traumatically stretched, the sudden strain can produce muscle spasm and backache. This happens when heavy objects are lifted incorrectly by bending forward with the knees straight. Bed rest and moist heat will relax the muscles and relieve the pain. Another cause of back pain is osteoarthritis or degenerative joint disease, which usually occurs after the fourth decade of life. Should the intervertebral disc between two adjacent vertebrae begin to degenerate, great stress is placed on the surrounding ligaments. If the disc herniates or protrudes, it may impinge on the nerves that exit in-between the vertebrae. A "slipped disc" may also be associated with numbness or weakness in one or both legs because of the pressure on the spinal nerves. These kinds of symptoms may require hospitalization, physical therapy, or even surgery for treatment.

The point is, most backaches arise in the back, that is, they are musculoskeletal in origin. They are much more rarely the result of ulcers, kidney disease, or pelvic pathology. Some of the gynecological difficulties that do result in back pain will be covered later.

EXTERNAL GENITALIA

The Vulva

The vulva, sometimes called the pudendum, is the term for the visible external genitalia. The name vulva means "covering" in Latin and refers to the area bounded by the *mons pubis* anteriorly, the *perineum* posteriorly, and the *labia minora* and *majora* laterally. The vulva is erotic and highly sensitive to touch; it also serves to protect the urethral and vaginal openings (see Fig. 2.8).

Mons Pubis

Another name for the mons pubis is the *mons veneris;* literally, the "the mountain of Venus."

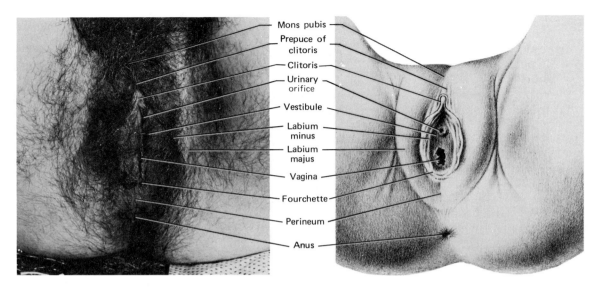

Mons pubis
Prepuce of clitoris
Clitoris
Urinary orifice
Vestibule
Labium minus
Labium majus
Vagina
Fourchette
Perineum
Anus

Figure 2.8. External genitalia.

Suitably named, the ancients thought, because Venus was the goddess of love (venereal disease has the same linguistic root). It is the cushion of fatty tissue and skin that lies over the pubic symphysis and, after puberty, is covered with pubic hair. Pubic hair varies in texture, as does the hair on the head, depending on the race of the individual. In many woman, the upper border of pubic hair is straight across, forming a triangle, and this is the so-called "female escutcheon." In males, the pubic hair supposedly grows upward toward the umbilicus as a result of androgenic activity, thus defining maleness and femaleness in the configuration of pubic hair growth. In actuality, some 25% of women and most men do have an apparent upward continuation of pubic hair to the umbilicus; it is not pubic hair proper, but ordinary body hair. Some people are just genetically hairier than other people. When a woman has this abdominal growth of hair, it is very rarely a sign of some virilizing hormonal influence.

Labia Majora

Extending down from the mons pubis are two longitudinal folds of skin, narrowing to enclose the vulvar cleft, and meeting posteriorly in the *perineum*, that area of skin between the junction of the labia major and the anus.

These *labia majora* (major lips) protect the inner parts of the vulva. The outer surface of the labia is covered with pubic hair. The inner surface is not, but has many sebaceous and sweat glands. The tissue inside the labia majora is loose connective tissue with pads of subcutaneous fat. The fat, like the fat on the hips and in the breasts, is particularly sensitive to estrogen. This is why the labia and the rest of the vulva become enlarged (hypertrophy) from puberty on, and shrink (atrophy) after the menopause. Underneath the subcutaneous fat and deep within the substance of the labial tissue are masses of erectile tissue, tissue filled with large blood spaces that engorge with blood during sexual excitement. These masses, which encircle the vaginal opening, are called the *bulbs of the vestibule;* they are equivalent to the corpora spongiosa of the male penis.

Under the skin of the labia majora, there are fibers of smooth muscle that are similar to the dartos muscle of the male scrotum. This subcutaneous muscle is temperature sensitive, and causes the labia to wrinkle up when exposed to cold and to appear larger and softer in warm weather.

Labia Minora

The labia minora are the delicate inner folds of skin that enclose the urethral opening and the vagina. They are also called *nymphae,* from the Greek word for "maiden," referring to the goddesses of the fountain.* The labia minora grow down from the anterior inner part of the labia majora on each side. Each fold joins above and below the *clitoris.* The joining of the folds above the clitoris forms the *prepuce;* the junction below the clitoris forms the *frenulum.* Each labium minus then extends downwards to surround the vagina and join again at the posterior end of the vagina, where it blends into the skin of the labia majora. At this junction, there is a slightly raised ridge of skin, the *fourchette.* After the birth of a baby, the fourchette flattens out. There are no pubic hairs on the labia minora, but there are many sebaceous glands that feel like tiny grains of sand when pressed between the thumb and forefinger.

The large numbers of sebaceous glands on the vulvar skin produces sebum, a mixture of oils, waxes, triglycerides, cholesterol, and cellular debris. Sebum lubricates the skin, and in combination with the secretions from the sweat glands and the vagina forms a water-proofing protective layer that enables the vulvar skin to repel urine, menstrual blood, and bacterial infections. Because of the many sebaceous glands, however, the labia minora, particularly in the area of the clitoris, are frequently the site of sebaceous cysts: painful nodules about the size of a pea in the skin. A vulvar sebaceous cyst usually spontaneously drains and disappears within a few days; however, it may become secondarily infected and require treatment or removal.

There are wide variations in the size and shape of the labia minora, and one is generally larger than the other. Sometimes they are completely hidden by the labia majora, or they may be enlarged so that they project forward. Enclosed within the skin of the labia minora are venous sinuses or blood spaces that become engorged with blood during sexual excitement, causing a color change and an increase in the thickness of the labia, sometimes as much as two to three times their diameter.

Vestibule

The vestibule is the area enclosed by the labia minora. Opening into the vestibule are the urethra from the urinary bladder, the vagina, and the two ducts of *Bartholin's glands,* also called the greater vestibular glands. Bartholin's glands produce a few drops of mucus during sexual excitement in the female in order to moisten the vestibule in preparation for intercourse. This amount of secretion is not significant in the lubrication of the vagina. The duct of a Bartholin's gland may become obstructed for no apparent reason, and the gland continues to secrete behind the duct. The result is a large Bartholin's cyst, which usually produces no symptoms, but which occasionally may form an abscess and have to be removed. A gonorrhea infection may sometimes cause a Bartholin's cyst.

* Everyone knows that a nymphomaniac is a woman with an excessive sex drive. Why is it that hardly anyone knows the same condition in males is satyriasis? Think about it.

Clitoris

The word "clitoris" is from the Greek word for "key," indicating that the ancient anatomists considered it the key to a woman's sexuality, a perception that has been largely ignored until recently. It is always referred to as the homologue of the penis; that is, similar in embryological origin and structure, but not necessarily in function. Some texts describe the clitoris as a *vestigial* homologue of the penis; a vestige being a small, degenerate, or incompletely developed structure. That is a truly erroneous statement except for one thing—the clitoris is smaller than the penis and is usually more easily felt than seen. It has, however, for its size, a generous blood and nerve supply relatively greater than that of the penis. There are more free nerve endings of reception located on the clitoris than on any other part of the body, and it is, unsurprisingly, the most erotically sensitive part of the genitalia for most females.

The clitoris consists of two *crura*, or roots, a *shaft*, or body, and *glans*. The two crura arise from the lower borders of the ischiopubic rami and join at the pubic symphysis to form the shaft of the clitoris. Within the shaft are the two *corpora cavernosa*, the "cavernous bodies," consisting of erectile tissue that, when engorged with blood, causes the clitoris to become erect and double in size. A the end of the shaft is the rounded glans, extremely sensitive to the touch.

There are two muscles on each side important in clitoral erection. The *ischiocavernosus* muscles arise from the ischium and insert into the corpora cavernosa, and the *bulbocavernosus* muscles arise from the area around the vestibular bulbs of the labia majora and also insert into the corpora cavernosa of the clitoris. During sexual excitement, these muscles contract and compress the dorsal vein of the clitoris, the only vein that drains the blood from the spaces in the corpora cavernosa. The arterial blood continues to pour in and, having no way to drain out, fills the venous spaces until they become turgid and engorged with blood. This mechanism causes the stiffening and erection of the clitoris (see Fig. 2.9).

The bulbocavernosus muscles compress the vestibular bulbs during sexual excitement, and they become congested and erect as well, contributing to what Masters and Johnson call the "orgasmic platform" (see Chapter 6). Like the penis, the clitoris is suspended from the lower border of the pubic arch by a ligament called the *suspensory ligament*. The prepuce, or foreskin, is a little hood over the glans formed by the anterior junction of the labia minora. As in the male, a sebaceous gland secretion is produced under the prepuce and can cause irritation and itching if not washed away.

Urethra

Approximately 2.5 cm below the clitoris, there is a small elevation like a dimple. In its center is the opening of the urethra, called the *external urethral orifice*. The urethra is the passageway for urine from the urinary bladder to the outside. When a catheterization is performed to empty the bladder or to obtain an uncontaminated urine specimen, a flexible tube or catheter is inserted into the external orifice and passed upward into the bladder.

On either side of the midline, just posterior to the external urethral orifice, are the openings from the paraurethral or Skene's glands, the female homologue to the prostate glands in the male. They secrete a small amount of mucus and, along with the secretion of other small mucus-secreting glands in the wall of the urethra, function to keep the opening moist and lubricated for the passage of urine.

When Skene announced his discovery of the glands that bear his name in the nine-

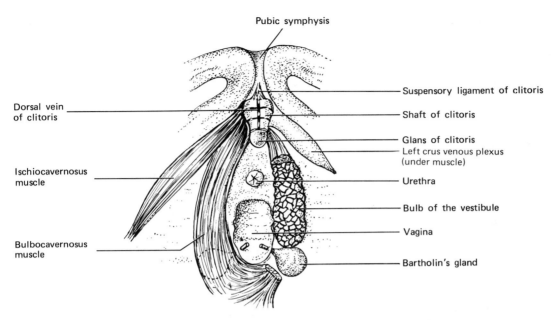

Pubic symphysis

Dorsal vein
of clitoris

Ischiocavernosus
muscle

Bulbocavernosus
muscle

Suspensory ligament of clitoris

Shaft of clitoris

Glans of clitoris
Left crus venous plexus
(under muscle)

Urethra

Bulb of the vestibule

Vagina

Bartholin's gland

Figure 2.9. Mechanism of clitoral erection. Tactile stimulation of the clitoris results in an increased blood flow to the erectile tissue (corpora cavernosa) in the shaft of the clitoris. The contraction of the two clitoral muscles (bulbocavernosus and the ischiocavernosus) compresses the only vein that drains the corpora cavernosa. The blood, trapped in the erectile spaces of the cavernous bodies, causes their engorgement and thus the enlargement and erection of the clitoris.

teenth century, they had already been described by de Graaf 200 years earlier as the producers of "female semen," the lubricating fluid discharged during sexual stimulation. That in some women these glands may produce a secretion emitted from the urethra during coitus was momentously "rediscovered" in 1982 by Alice Ladas, Beverly Whipple, and John Perry. The three reported in their instant best-selling book, *The G Spot: & Other Recent Discoveries about Human Sexuality,* that the secretion, dubbed the female ejaculate, occurred in response to stimulation of the Grafenberg spot, located on the anterior wall of the vagina. Ernst Grafenberg, a German gynecologist, had been, until then, better known for the Grafenberg ring, the first widely used intrauterine device. His 1950 description of a "zone of erogenous feeling located along the anterior vaginal wall" was largely ignored, but the publication of Ladas', Whipple's, and Perry's book revived controversial inquiry about the phenomenon. The G spot aroused a flurry (some would say a frenzy) of interest on TV talk shows and in newspaper and magazine articles, but whether it has profound and wide-ranging implications for female sexual response remains to be demonstrated. Some women may be concerned about expelling what seems to be a small gush of urine during intercourse. They are probably experiencing a greater discharge from the paraurethral (Skene's) and vulvovaginal (Bartholin's) glands.

The female urethra, in contrast to that of a male, is short: only about 4 cm in length when a woman is standing erect. This anatomical difference predisposes a woman to bladder infection much more frequently than a man, whose urethra is 18–20 cm in length. Not only is the external meatus exposed to vaginal discharges containing bacteria, but organisms like *Escherichia coli* from the rectum can easily ascend up the short urethra and multiply tremendously within hours, leading to a condition called *cystitis*, or urinary tract infection (UTI). For this reason, women should always wipe from front to back after moving their bowels to avoid transferring the bacteria from the anus to the urethra. For similar reasons, internal tampons inserted into the vagina are preferable to external pads. Although a tampon can be associated with other health problems, a pad can provide a direct link from the anus and vagina to the urethra, as well as an environment in which microorganisms can thrive.

"Honeymoon cystitis," a term that has lost considerable relevance to present life-styles, is the name given to the bladder infections that occur within a short time after the initiation of regular intercourse in a woman, particularly after a period of little sexual activity. The proximity of the urethra and the base of the bladder to the anterior vaginal wall can cause those structures to become swollen and irritated as a result of repeated coitus, causing an urge to urinate more frequently. Vigorous thrusting movements of the penis can also be responsible for forcing microorganisms up into the urethral orifice and into the bladder to cause cystitis, which must be treated with an antibiotic. One way of preventing or at least decreasing susceptibility to this cause of bladder infection is to maintain a high fluid intake and to urinate before and after sexual intercourse.

Symptoms of cystitis are burning pain on urination, and an urge to urinate frequently, even immediately after voiding. The urine is cloudy and dark, and contains pus cells, red blood cells, and many bacteria. There may be fever, backache, and lower abdominal pain. If the bladder infection is not treated, it can spread to the kidneys and cause a condition called *pyelonephritis*, which is a potentially very serious complication. Additional causes, prevention, and treatment of cystitis and UTI are discussed in Chapter 9.

Hymen

Below the external urethral orifice in the vestibule is the opening to the vagina. Around the vaginal opening there is a small insignificant membrane with no known function called the *hymen*, after Hymen, the god of marriage in Greek mythology. There are probably few other parts of the female reproductive tract as subject to folklore and misconception as this little membrane. Most people do not even know where it is—they think it is somewhere up in the vagina near the cervix. It is commonly believed that it tears at the first coitus, copious and visible bleeding occurs, and that the virgin has been "deflowered," defloration being a curious and romantic term for rupture of the hymen. If no bleeding occurs, this is taken as evidence of nonvirginity. Many also believe that the hymen makes it impossible for a virgin to wear tampons during the menstrual period, or that if the attempt is made, this sign of virginity will disappear.

All of this is nonsense. An intact hymen is not proof of virginity; a ruptured hymen is not indicative that sexual intercourse has occurred, and no one, including a doctor, can tell whether or not there has been initial coitus by just looking at the vaginal opening. It is usually possible to determine with accuracy whether or not a woman has had a child, but generally impossible to say whether or not she is a virgin. This is because the hymen may be:

- Thin as a spiderweb, or thick and fleshy
- Quite vascular, with a good blood supply, or relatively avascular
- Extremely variable in how much of the vaginal opening is covered
- Sometimes so pliable and flexible that it never ruptures but only stretches, even after childbirth

The hymen very rarely completely occludes the opening to the vagina. This is called an *imperforate* hymen and is usually discovered during adolescence. A girl with an imperforate hymen will menstruate into the vagina month after month, and the discharge will accumulate in the vagina, a condition called hematocolpos. If it is not recognized, menstrual blood my fill the uterus and the fallopian tubes as well. Cutting of the hymen cures the problem, and there is no further difficulty.

Another variation is the septate hymen. If the septum is not too thick and is stretchy enough, intercourse is not hampered. There is usually no problem with insertion of tampons, although there may be some difficulty encountered in removal.

After childbirth, the hymen usually no longer has a continuous rim, but remains as isolated remnants with gaps in-between. These are referred to as *carunculae hymenales*. The hymen is present throughout life, has no function, and is merely an embryological vestige. Its only signficance, if one believes the mythology, is psychological, sociological, or cultural.

Vagina

The vagina is a tube that passes upwards to the uterus at an approximate 45° angle from the vulva.

The size of the vagina is so variable and so capable of distension that it is difficult to measure its dimensions. This great distensibility enables the vagina to withstand vigorous stresses during intercourse or childbirth. Despite all the stories, a normal vagina can accommmodate any size penis with ease. There is no relationship between the size of a woman's vagina and her general body size or shape; the same is true of the size of the penis of a man.

Actually, the tube of the vagina is only a potential space, because its anterior and posterior walls are thrown up into transverse folds that are in close apposition. In a woman who has never had a child, there are many folds and the vaginal walls are firm. After parturition (giving birth), the walls are more or less smoothed out, but they retain their firmness, especially near the opening of the vagina. This is not always true for women who have had many children.

Projecting into the upper part of the vagina is the lower part of the uterus, the conical *cervix*. The circular gutter formed all around the cervix is anatomically divided into the *anterior, posterior,* and *lateral fornices* (see Fig. 2.10). The walls of the fornices are thin, since they consist only of the vaginal wall with the pelvic cavity on the other side. During an internal pelvic examination, the position and relations of the various pelvic viscera can be palpated and outlined through the fornices. Normally, the posterior fornix is empty, and the body of the uterus can be felt through the anterior fornix; the fallopian tubes and the ovaries, through the lateral fornices.

The posterior fornix extends deeper into the pelvis and is, therefore, larger and longer than the anterior fornix. This anatomical arrangement favors the passage of sperm into the cervix during intercourse because when a women lies on her back, the opening in the cervix is not only directly exposed to the male ejaculation, but the pool of ejaculated semen collects in the posterior fornix and bathes the cervix, which is resting in it (see Fig. 2.11). The posterior fornix also takes the brunt of penile

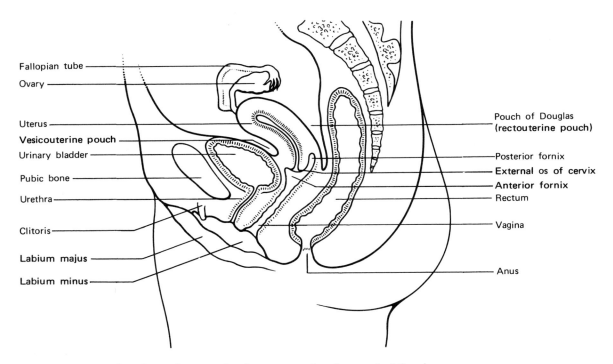

Fallopian tube

Ovary

Uterus

Vesicouterine pouch

Urinary bladder

Pubic bone

Urethra

Clitoris

Labium majus

Labium minus

Pouch of Douglas
(rectouterine pouch)

Posterior fornix

External os of cervix

Anterior fornix

Rectum

Vagina

Anus

Figure 2.10. Section through reproductive organs showing vaginal fornices.

thrusting during coitus and thus prevents injury or jarring of the cervix.

Only a thin partition of vaginal wall separates the posterior fornix from the lining *(peritoneum)* of the pelvic cavity, which dips down to form the *pouch of Douglas*, or the rectouterine pouch. The posterior fornix therefore forms the route for several kinds of diagnostic and surgical procedures. In *culdocentesis*, a needle is inserted into the pouch of Douglas to determine the nature of any fluid that might be present—blood, pus, excess tissue fluid. *Colpotomy* is surgical incision through the posterior fornix into the peritoneum of the pelvic cavity. Through this opening, the pelvic viscera can actually be visually explored by means of a *culdoscope*, which is a tube equipped with optical devices and light—a sort of internal microscope. Colpotomy can also provide access for some sur-

gical procedures, such as tubal ligation (cutting the oviducts) or, sometimes, even removal of a tubal pregnancy.

The Vaginal Lining

The lining of the vagina is called the vaginal epithelium and consists of layers (as many as 40) of cells resting on connective tissue, containing blood vessels and nerves. This epithelium is like skin but has no hair follicles, sweat glands, or sebaceous glands to weaken it or to provide a passageway for entrance of microorganisms. It is, therefore, a very tough, resistant protective lining (see Table 2.2). The deeper basal layers of the cells are closer to the blood vessels in the connective tissue underneath and proliferate faster, pushing up to replace the superficial layers that are sloughed off. Estrogen, the female sex hormone, stimulates the growth of the cell. Before puberty and af-

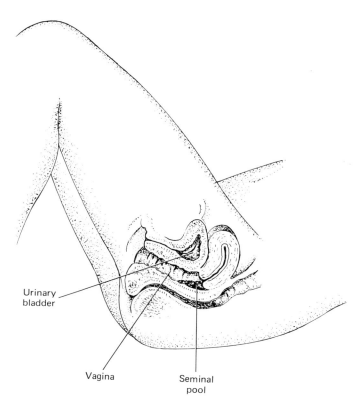

Urinary
bladder

Vagina

Seminal
pool

Figure 2.11. When a woman is on her back during intercourse, the semen collects in the posterior fornix, providing sperm with easy access to the cervical opening.

ter menopause when there is little estrogen present, the vaginal epithelium is thin and made up almost entirely of basal cells. When the ovaries are actively producing estrogen during the reproductive years, a smear of cellular material taken from the thick vaginal epithelium will show large numbers of sloughed-off cells, and it is possible to recognize phases of the menstrual cycle by the shape and staining qualities of these cells. Án index of estrogenic activity can, therefore, be determined by the relative numbers of basal, intermediate, or superficial cells present in a stained smear of the vaginal lining. The appearance of cells of the vaginal lining is also utilized in tests for

cancer detection (Pap test, Schiller test), to be described later.

Vaginal Discharge

The contents of the vagina after puberty and before menopause are normally quite acid, with a pH of approximately 3.5–5.5 (pH is a measure of the acidity or alkalinity of a solutioin on a scale of 1–14. The lower numbers indicate acidity, the higher numbers, alkalinity, and 7 is neutral). The mechanism for the production and maintenance of this acidity is that the estrogen produced by functioning ovaries during sexual maturity causes rapid

Table 2.2 CHANGES IN THE VAGINAL EPITHELIUM WITH AGE

	Newborn Child	One Month Old	Puberty	Sexually Mature Child-bearing Years	Post-Menopause
Amount of estrogen present (causes changes in) →	Much estrogen from placenta	None	Begins to appear	High levels	Relatively very low levels
Appearance of Epithelium	Similar to mature	Very thin	Develops thickness	Thick, 20–40 layers	Thinner, may be atrophied
Amount of Glycogen in cells (causes) →	+	−	From − to +	+	−
pH of Vaginal Secretions (when acted upon by)	Acid, pH 4 to 5	Alkaline, pH 7	Becomes acid	Acid pH 4 to 5	Neutral, may be alkaline
Type of bacteria present	Sterile but Doderlein's bacilli present after 12 hours, from mother	Very sparse, some cocci	Goes from sparse varied to Doderlein's bacilli	Doderlein's bacilli and other microorganisms	Various bacteria

proliferation of the basal cells of the vaginal epithelium. These actively growing cells accumulate glycogen granules, the stored form of glucose, in their cytoplasm. Certain bacteria called Döderlein's bacilli (after Albert Döderlein, a German obstetrician) that are normal residents of the vagina are lactobacilli; that is, they have the ability to break down the glycogen to form lactic acid. The presence of the lactic acid is responsible for the lowered pH of the vaginal contents.

The normal vaginal discharge is a clear acid material consisting of fluid arising from the capillaries in the vaginal walls with lesser amounts contributed from the cervical glands, the uterine cavity, and the fallopian tubes. In the fluid are mucus, superficial sloughed-off cells, Döderlein's bacilli, other microorganisms, and various fat, protein, and carbohydrate compounds. The acid discharge combined with the toughness of the thick epithelium protects the vagina from infection by harmful bacteria and makes the vagina much more vulnerable to infection before puberty and after menopause. Döderlein's bacilli are absolutely essential to normal vaginal physiology. If they are disturbed or destroyed by chemical contraceptives, antibiotics, or excess douching, vaginal infections will occur more easily.

Douching

Douching is a procedure of vaginal irrigation in which fluid in a bag is permitted to run through a tube, entering the vagina under slight pressure and ballooning it out slightly. As the fluid runs out or is expelled through muscular action, the vaginal contents are washed out. Douching to prevent pregnancy after intercourse is totally ineffective, since it has been determined that sperm can be recovered from the uterus within seconds after being deposited in the vagina—and no woman can get the douche bag apparatus set up and going that fast.

Some women like to douche after their menstrual period because it makes them feel clean. As long as the bag is not suspended too high so that the fluid enters the vagina at too great a pressure, there is no harm in occasional douching for general hygiene. Habitual and vigorous douching may force fluid, and a possible infection, upward into the uterus and into the fallopian tubes. Douching is seldom essential for normal health because the vagina is self-cleansing through the process of normal discharge. But if sexual intercourse is frequent, and gels, foams, or creams are used for contraception, there may be vaginal leakage, which can feel unpleasantly wet. The spermicidal gel used with a diaphragm nightly may have to be douched out weekly, for example. To douche more than every 4 or 5 days is excessive, however, and it will destroy the normal physiology of the vagina.

For a douching medium, plain warm water or water with one tablespoon of vinegar (or bicarbonate of soda) to a quart is certainly as effective as the commercial solutions, and certainly less expensive. Perfumed, colored, and flavored solutions are a waste of money. The manufacturers of such products prey on a woman's concern, reinforced by massive advertising campaigns, that natural genital odor is offensive, and that to feel "feminine fresh all over," one must use perfumed and deodorizing douches, suppositories, or feminine hygiene sprays.

There is a specific odor produced by the healthy female genitalia that comes from the secretions of the sebaceous and apocrine sweat glands of the vulva, the secretions of Bartholin's and Skene's (urethral) glands, and from the vaginal discharge. It is not an unpleasant odor and is in no way unclean, unless it has been permitted to remain on the skin too long or if there is an infection in the vagina causing an abnormal discharge.

Feminine deodorant sprays are particularly unnecessary and should not be used. If vaginal infection is the source of a perceived malodorous condition, the spray cannot possibly get to the source of the odor and may only cover it up and delay treatment. In some women, allergic sensitivity to the sprays has resulted in a very painful reaction to their use. Some of these products were taken off the market, and others were required to include labels with printed warnings about their use.

Europeans, particularly in France and Italy, have always been more advanced than Americans in sensible genital hygiene for both men and women. There is, in most home and hotel bathrooms, a bidet. The bidet, which most Americans think is a funny-looking toilet (or perhaps a foot bath), is a porcelain basin that one straddles, so that a stream of warm water flows over genital and perineal areas after urination or bowel movement. This provides an excellent and convenient method of cleansing the small folds and crevices of the genitalia.

Vaginal Odor

Chemical compounds secreted by an organism into the environment in order to evoke various developmental, behavioral, or reproductive responses in another member of the same species are called *pheromones*. Sex attractant pheromones have been studied extensively in insects and synthetic pheromones or their analogues are increasingly being used in biological (rather than pesticide) control of insect pests. Nonhuman mammals have glands that produce a variety of secretions, some odor-free and some distinctly malodorous, and the importance of such pheromones in establishing territoriality or in affecting reproductive behavior is well known. Ask anyone with a male dog near an unspayed female in heat. But while the evidence for the existence of sex attractant pheromones in nonprimates is clear-cut, the case for such a phenomenon operating in primates like rhesus monkeys is controversial, and it is even more speculative in humans. To date, it has been determined that both human women and female monkeys have the same kind of organic acids (short-chain carbon aliphatic acids, such as lactic, butyric, pentanoic, and hexanoic) present in their vaginal secretions, and that levels of these acids change during the menstrual cycle. Although the fluctuation of these vaginal acids is reportedly dependent upon fluctuating ovarian hormone levels in the monkeys, any attempts to correlate changing acid levels with hormonal changes in women have, thus far, been unsuccessful (Bauman et al., 1982). If such a biological marker could be found, however, it might be possible to predict and determine ovulation, with obvious implications for contraception or infertility problems. Another possibility, that vaginal acids may act as sex attractant pheromones in humans, has been pursued by several investigators, but the evidence is slim. There has been some effort to relate possible changes in the intensity and pleasantness of vaginal odor to the day of the menstrual cycle with the suggestion that the odorous components of the vagina not only fluctuate, but that they may be perceived as mild and pleasant around the time of ovulation (Doty et al., 1975; Keith et al., 1975). Another intriguing bit of information on "human pheromones" has been contributed by Tonzetich and colleagues (1978), who studied the volatile sulfur compounds in the mouth ("bad breath") throughout the menstrual cycle in five subjects. They determined that bad breath increased fourfold near ovulation and also during menstruation. The paradox of "good" smells in one place and "bad" smells in another at the time most conducive to conception further illustrates how little is known of the role, or even the very existence, of human response to such body odors.

Laypersons have probably always been aware of the individuality of what has been termed the *olfactory signature* of a person. The erotic potential of odor is certainly well known to the manufacturers and advertisers of perfumes and incense. Musk oil-based scents have become very popular with both men and women and to many people musk smells very much like healthy body odor. One enterprising British firm has allegedly put sex appeal in a bottle with its production of a scent that contains a synthetic pheromone. Called Muskone pH5, the perfume is available for men and women at a cost of $20–$200 in a gold-plated bottle. Another manufacturer, Jovan, has put a compound made from sweat and tears on the market. It costs $10–$12 and is sold in department and discount stores as Andron. Evidently, the deodorant manufacturers would have us repress or eliminate all of our body odors so that we can be persuaded to buy them back from the perfume manufacturers.

Further investigation of the role of human olfactory stimuli on human behavior is needed. The suggestion has been made that personal odors may be translated into hormonal effects that control menstrual cycles and reproductive behavior (see "Menstrual Synchrony," Chapter 3).

Vaginal Lubrication

There is always a certain amount of lubrication of the vagina from vaginal discharge, but sexual stimulation produces considerable wetness, which is usually noticeable. Since the vaginal epithelium has neither sweat glands nor sebaceous glands, where does this moisture come from? Before the observations of Masters and Johnson, it was always incorrectly assumed that Bartholin's glands and cervical mucus from the cervical glands produced the vaginal lubrication. Now it is believed that the blood vessels of the wall of the vagina actually play the essential role. The wall of the vagina is supplied with arterial blood by four branches of the internal iliac artery: the vaginal, uterine, internal pudendal, and middle rectal arteries. They branch into capillary networks, and the blood then drains from the vagina through veins that are large, thin walled, and form an interweaving plexus at the sides of the vagina. This venous plexus ultimately drains through the vaginal veins into the internal iliac vein. Under the conditions of sexual stimulation, the veins around the vagina become engorged with blood. The resulting congestion of blood in the venous plexus results in pressure that forces a mucoid kind of liquid, or *transudate*, to pass from the veins through the epithelium of the vagina. This liquid at first forms individual droplets and then, as the droplets coalesce, forms a coating for the entire vagina. Because this appeared to Masters and Johnson similar to drops of perspiration beading on a forehead, they called this the "sweating phenomenon" of the vagina. This occurs very early in the sexual response of the female and provides sufficient lubrication for intercourse. The vaginal response tends to be prevented when certain kinds of vaginal infections are present, and some women who are on the pill may also find that their ability to lubricate is diminished. Without lubrication, of course, penetration during coitus may be very uncomfortable.

Nerves of the Vagina

The upper two-thirds of the vagina is supplied almost entirely by nerve fibers from the autonomic nervous system, which means that they control the constriction and dilation of blood vessels in the walls of the vagina. There are very few sensory receptors for touch or pain located in the vagina. Sensations of awareness or pressure within the vagina are mainly received by nerve receptors in the rectum or urinary bladder and not in the vagina itself. The upper part of the vagina is hence relatively insensitive. The lower third of the vagina does contain some touch and pain receptors from the pudendal nerve but such innervation is scanty. In contrast, the vulva has a rich supply of fibers from the pudendal nerve and is very sensitive. If the internal wall of the vagina is inflamed or infected (vaginitis), although the site of the irritation is inside, the itching or pain is referred to the outside on the vulva, because of the way that the nerves are distributed. If the area around the vaginal opening itches and burns and an abnormal discharge is present, it very likely means that undesirable organisms are inhabiting the vagina.

UTERUS

From the medical records that have remained intact, it is evident that the uterus, more than all the female reproductive organs, was the most subject to fanciful and erroneous description by ancient physicians. Evidently unwilling or unable to credit women for owning an organ in which the fetus developed, early accounts described the uterus as an independent animal capable of independent activity. A Greek physician in the second century A.D. wrote:

In the middle of the flanks of women lies the womb, a female viscus, closely resembling an animal, for it is moved of itself hither and thither in the flanks . . . and in a word, is altogether erratic. It delights also in fragrant smells and advances toward them, and it has an aversion to fetid smells and flees from them; and on the whole the womb is like an animal within an animal.*

Six centuries earlier, Hippocrates had written that the uterus went wild unless it was often fed with male semen. Even by the fourteenth century, the descriptions of the uterus were equally inaccurate, but perhaps more ingenious. The prevailing opinion of female anatomy was still that it was a poor second to the obviously superior architecture of the male,

* Quoted in Speert, H. *A Pictorial History of Gynecology and Obstetrics.* Philadelphia, F. A. Davis Co., 1973

and a French surgeon compared the uterus to a male organ turned inside out!—"It has in its upper part two arms with the testicles . . . like the scrotum," he wrote, and went on to compare the body of the uterus, with a canal in it, to the shaft of the penis with the urethra running through it.

It is not surprising that extravagant errors were made in explaining the "mysteries of the womb"—the uterus is a very unique organ. Every month it prepares to receive a fertilized ovum; if it does not, it sheds its lining and starts all over again. If it does receive a fertilized ovum, it shelters it, nourishes and protects it during its development for 9 months, and then expels it at the end of pregnancy. The actual mechanism for doing all this is still unknown. The uterus has the ability to grow from a weight of about 2 ounces when nonpregnant, to 2 pounds, its weight immediately after delivery, and then shrink back to its original size by 6 weeks after delivery.

Hollow and muscular, the uterus has an upper expanded portion called the body or *fundus*, and a lower constricted part called the neck or *cervix*. The cervix projects down into the vagina and has an opening, the *external os*. As with all other parts of the reproductive tract, there is considerable variation in the size of the normal, nonpregnant uterus, but on the average, it is 3 inches long, 2 inches wide at the fundus, 1 inch thick at its thickest part, and the wall of the uterus is $\frac{1}{2}$ inch thick. The exact size of the uterus can only be measured with an instrument called a uterine sound, which is inserted through the external os. Clinical impressions of size derived from pelvic examination are very deceptive, making physicians' remarks such as "you have an infantile uterus" not only unnecessarily disturbing, but usually erroneous.

The walls of the uterus are solid and made of smooth involuntary muscle. They enclose a cavity that is lined with epithelium called the *endometrium*, which undergoes cyclic changes during menstruation and forms the site for implantation of the fertilized egg if pregnancy occurs.

Positions of the Uterus

The normal uterus is a very mobile organ, and its position between the rectum and the bladder varies, depending on posture, how full the bladder or the rectum is, and how many children have been borne. As seen in Figure 2.12, the body of the uterus is typically inclined forward. When the urinary bladder is distended, the backward movement of the uterus is called *retroversion;* when the rectum is distended, the forward movement of the body of the uterus is called *anteversion.* Further and marked anteversion with relation to the cervix is called acute flexion; marked retroversion is called *retroflexion.* The uterus is normally anteverted and anteflexed. It may also be retroverted and retroflexed and still be normal; a "tipped" uterus is of no clinical significance and results in no symptoms. Backache, constipation, menstrual cramps—none of these can be related to retroversion or retroflexion. During the nineteenth century, gynecologists collected a great deal of money for inserting

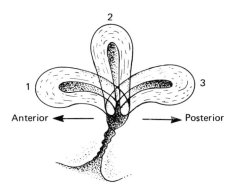

Figure 2.12. The normal uterus can vary in position. (1) Uterus markedly anteverted. (2) Uterus slightly anteverted; the usual position when the urinary bladder is full. (3) Uterus retroverted.

vaginal pessaries, which are devices used to correct retrodeviations of the uterus. During the early part of the twentieth century, there was much unnecessary surgery performed to "correct" the position of the uterus to relieve the symptoms ascribed to its nontypical but perfectly normal position. Of course, it is important for the size, shape, and position of the uterus to be known if any procedure like an abortion or a D and C is to be performed. (D and C: *dilatation,* or enlargement of the cervical opening with dilators, to permit *curettage,* or scraping of the uterine lining with an instrument called a curette.)* Otherwise, an instrument inserted into a markedly anteflexed or retroflexed uterus may perforate the wall.

The lower part of the uterus, the cervix, projects into the vagina and is circumferentially attached to it, thus dividing the uterus into an upper, or supravaginal part, and a lower, or vaginal portion. The vaginal part of the cervix is called the *portio vaginalis.* In an adult woman who has not delivered any children (nullipara), the cervix comprises approximately one-half the length of the uterus; in a woman who has given birth to at least one child (primipara; more than one child, multipara), it may be only one-third the length of the uterus. In its typical position in most women, the cervix is directed downward and backward, with its long axis making an angle of between 80° and 120° in relation to the forward-inclined body of the uterus. This position is supported and maintained by fibrous muscular bands called ligaments, described later.

External Os

The opening of the cervix into the vagina is the *external os.* The opening into the uterine cavity is the *internal os.* The canal between

---

* Dilation and dilatation are synonymous; the dilation of the cervical opening is customarily referred to as dilatation.

the external and the internal os is the *endo-cervical canal*, lined with the endocervical epithelium. It contains many mucus-secreting glands and is approximately 1 inch long.

There is a very visible difference between the nulliparous os and the multiparous one. Before childbirth, the external os is a little round dimple, approximately 3 millimeters in diameter. It is sometimes called the *os tincae*, or "mouth of a small fish." This tiny canal dilates during labor and delivery to accommodate the passage of the head and body of a full-term infant, so that after parturition it never again returns to its previous shape, but becomes a transverse slit with irregular margins. Enlarging the cervical opening with dilators in order to perform an abortion or a D and C has the same effect. Note that although it is not possible to determine whether or not a woman is a virgin, an examiner can tell

whether she has had a baby or an abortion. No matter how small the trauma, very minute lacerations and abrasions occur that change the appearance of the os (see Fig. 2.13).

Cervix

The cervix is predominantly composed of fibrous connective tissue with many smooth muscle fibers. It is firm to the touch, except after approximately 6 weeks of pregnancy, when it softens, owing to the increased blood supply to the uterus and cervix. The feel of the nonpregnant cervix has been compared to the feel of the tip of the nose, or to the glans of the erect penis.

The epithelium on the surface of the vaginal aspect of the cervix is pale pink, whereas the epithelium of the endocervical canal leading into the uterus from the external os is redder

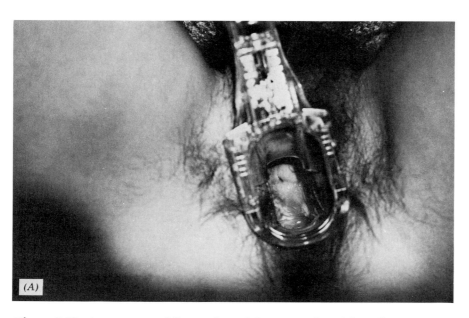

(A)

Figure 2.13. Appearance of the cervix and the external os. (A) Nulliparous os; (B) Parous os; (C) Appearance of the vaginal vault after the cervix has been removed in conjunction with a hysterectomy performed through the vagina. The vertical line represents the location of the suturing of the vaginal mucosa.

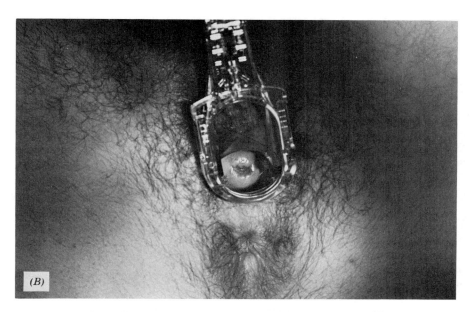

(B)

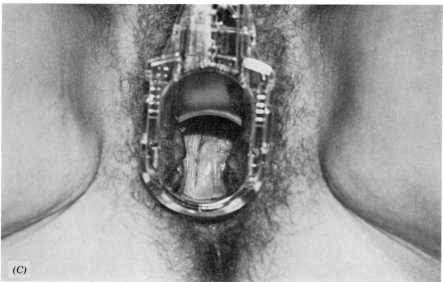

(C)

Figure 2.13. *(continued)*

in color. Because of this color difference, inflammations (cervicitis), extensions of the endocervical epithelium onto the vaginal aspect of the cervix (erosions), benign polyps, or cysts are highly visible on the surface of the portio vaginalis, even though these conditions originate on the mucous membrane of the endocervical canal. The glands on the endocervical canal are highly branched and burrow deeply into the recesses and folds of the cervical mucosa. Once organisms get into this desirable environment, they remain hidden and can stubbornly resist most attempts to get rid of them. This is why infections of the cervix tend to become chronic.

Cervical Mucus

The endocervical glands secrete cervical mucus, which is composed mostly of water, electrolytes like sodium and potassium ions dissolved in the water, blood proteins including immunoglobulins, and mucins, a complex group of glycoproteins. Glycoproteins are compounds made up of carbohydrates and proteins, and in cervical mucus they have been found to contain an extremely high percentage of carbohydrates. During the menstrual cycle, the physical properties of the cervical mucus change as a result of the differing levels of circulating hormones. Before and near the time of ovulation, the mucus is dilute, secreted in greatest quantity, and when it is spread out on a slide it will undergo "ferning" or "arborization," that is, it forms a distinct pattern as it dries that looks like ferns or tree branches. After ovulation, the mucus becomes much thicker, reaches a gel state, and no longer demonstrates ferning. These changes form the basis for tests of both hormonal activity and determination of the time of ovulation.

The thinner, more fluid cervical secretion enhances sperm migration up to the uterine cavity; the gel-like mucin in the latter half of the menstrual cycle acts as a barrier to sperm motility. It is possible that the cervical mucus may function as a bacteriacide to protect the upper reproductive tract from invasion by harmful bacteria. It has also been suggested that it may form a mechanical protection at the junction of the external os and the endocervical canal against the development of cancer.

OVIDUCTS OR FALLOPIAN TUBES

The oviducts or fallopian tubes and the ovaries are frequently referred to as the *adnexa*, because they are adjacent or next to the uterus. The fallopian tubes are named after the sixteenth century anatomist, Gabrielle Fallopius, who thought they resembled tubas or curved trumpets. More ancient anatomists thought the oviduct looked like a straight trumpet, and they called it the "salpinx" (Greek for tube). That prefix appears in the words describing conditions of the oviducts, such as *salpingitis*, or inflammation of the oviducts, or *salpingectomy*, removal of a tube, among other salpingo-type procedures.

The fallopian tube is anatomically divided into four sections:

1. The interstitial or uterine portion: very short, very narrow in diameter, and lying completely within the muscle of the uterus.

2. The isthmus: the straight part with a thick muscular wall and a narrow lumen (passageway); the section that is the usual site of a tubal ligation, a surgical procedure that prevents the sperm from meeting the ovum by cutting or ligating the oviduct.

3. The ampulla: occupying about one-half the entire length of the tube, thin walled, with a highly folded lining.

4. The infundibulum: nearest the ovary, a trumpet-shaped expansion with fingerlike

projections called fimbriae that wave back and forth through muscular action to attract the ovum into the opening, or *ostium*. One fimbria is longer, closer to the ovary, and is called the ovarian fimbria (see Fig. 2.14).

The walls of the tubes contain many blood vessels and much longitudinally and circularly arranged smooth muscle. The lining of the tubes is thrown up into folds that almost fill the lumen, but the number of the folds varies with each segment and is most extensive in the infundibulum. The epithelium of the lining consists of those cells that secrete a nutrient fluid to provide an environment necessary for movement, fertilization, and sustenance of the ovum, and other cells that bear

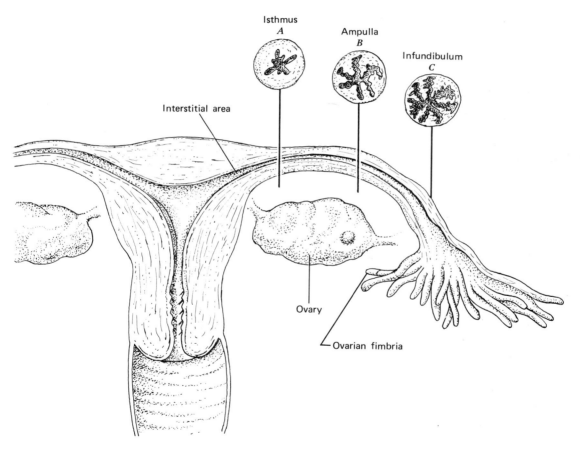

Figure 2.14. The fallopian tube and its relationship to the ovary and the uterus. (A) Isthmus. 25% of ectopic pregnancies occur in this area. The thick muscular wall permits little distension, and early rupture is likely. (B) Ampulla. More than half of ectopic pregnancies implant in this portion of the tube, which is more distensible. The pregnancy is further advanced before rupture. (C) Infundibulum. Fewer than 10% of tubal pregnancies occur here or in the fimbriated end.

cilia, little hairlike processes that beat toward the uterus to create a current to help the egg progress into the uterine cavity. The transport of the egg into the uterus is mainly accomplished by contraction of the circular and longitudinal muscle that creates peristaltic waves, which move the ovum along. This muscular activity is influenced by hormones and is greatest at the time of ovulation.

When an ovum is released from the ovary, it is not in direct contact with the end of the fallopian tube, and the exact mechanisms that keep the egg from falling into the pelvic cavity and getting lost are still not completely known. Culdoscopic observations at the time of ovulation have revealed that the fimbriae are brought closer to the ovary to curve around it by contraction of muscle fibers in the connective tissue, which covers and suspends the fallopian tubes from the abdominal wall. It has also been suggested that muscular contractions in the walls of the fimbriae, coupled with suction action from the ciliary movement of their lining cells, actually actively engulf the ovum, swallowing it up as though the end of the tube were a vacuum cleaner hose.

It is possible that scar tissue resulting from previous pelvic surgery or fallopian tube infection can interfere with the mechanism of picking up the ovum at the time of ovulation, and thereby cause infertility. On the other hand, there are cases on record in which women had a tube removed (salpingectomy) on one side, an ovary removed (ovariectomy) on the other side, and still had several children. Somehow these ova crossed over the uterus to enter the ostium of the opposite side.

Once in the infundibulum of the tube, the egg is rapidly transported through the ampulla until it reaches the ampullary–isthmus junction, where its transport is delayed for about 30 hours. This restraint of movement is referred to as the "tube-locking" mechanism, or the 'isthmic block," and if sperm are present, fertilization occurs here. If conception occurs, the new embryo divides and develops, nurtured and sustained by the secretions of the tubal lining, and passes fairly rapidly through the isthmus and interstitial portion into the uterine cavity. No fewer than 45 hours nor more than 80 hours elapse between the time of ovulation and the ovum's entry into the uterus. Two to 3 more days pass before the timing is optimal for the joining of the embryo and the uterine lining, the process called implantation.

If the egg is not fertilized, it disintegrates, and its remains are cleaned up by those ubiquitous scavenger cells of the body, the macrophages.

Ectopic Pregnancy

After the conceptus (the collective name for the egg and all its derivatives from the time of fertilization until birth of the baby) arrives into the uterine cavity, it normally implants in the upper part of the cavity. Should it, for some reason, fail to get to that destination, it may implant in some other place, like the fallopian tube, the ovary, the cervix, or (rarely) even the abdominal cavity. Such ectopic, or out-of-place, pregnancies are a major public health problem, and their incidence has been increasing dramatically. One in 60 to one in 80 pregnancies are reported to be ectopic, and 98% occur in the fallopian tubes. The tubes are not distensible enough to permit adequate development of the embryo and subsequent increase in size, particularly the closer the tubal implantation is to the uterus. The embryo, therefore, must always be removed surgically before the tube bursts, which can cause severe hemorrhage and infection in the abdominal cavity. Difficulty and delay in diagnosis and treatment have made ectopic pregnancy the major cause of death during the first 3 months of pregnancy.

Symptoms of an Ectopic Pregnancy

Unfortunately, the early diagnosis of tubal pregnancy is difficult because the early symptoms of an ectopic are no different from the symptoms of a normal uterine pregnancy—a missed period, slight uterine enlargement, and softening of the cervix. If the tubal implantation is situated so that fetal tissue is in contact with maternal blood vessels, the hormone tested in pregnancy tests will appear in maternal blood and will yield a positive result. It is possible, however, for the fetal tissue to become separated from the tubal wall, and then a pregnancy test will prove negative, which of course does not exclude the possibility of ectopic pregnancy. A tender and distended tubal mass, palpable on pelvic examination, is a classic symptom, but it is not always detectable. The signs and symptoms of a tubal implantation vary greatly; differentiating it from threatened miscarriage, urinary tract infection, appendicitis, salpingitis, or an ovarian cyst is sometimes very difficult.

Lower abdominal or pelvic pain, with or without vaginal bleeding, is one of the most common symptoms. When there is bleeding or spotting accompanied by the sudden onset of pain after one period has been missed—corresponding to 6 to 8 weeks of gestation—it certainly warrants investigation. Pain and bleeding may be an indication that the tube has ruptured and that there is hemorrhage into the uterus and body cavity. The pain may be increased when a Valsalva maneuver (increasing abdominal and thoracic pressure by a forced expiration against a closed glottis) is performed. If the internal bleeding irritates the peritoneal lining of the body cavity, the pain may be referred to the shoulder.

Special means for diagnosis, such as laparoscopy, ultrasonic imaging, and radioimmunoassay pregnancy tests, have made earlier determination possible, but it could take all three tests to diagnose a questionable ectopic pregnancy. Ultrasound usually cannot determine whether implantation has occurred in the fallopian tube; it can, however, rule out a tubal pregnancy if a clearly defined gestational sac is seen in the uterus. If a highly sensitive pregnancy test, such as the human chorionic gonadotropin beta-subunit radioimmunoassay, is positive for pregnancy, and there is no uterine pregnancy visible with ultrasonography, laparoscopic examination to permit visualization of the internal organs can make it possible to diagnose ectopic pregnancy before rupture occurs.

An ectopic pregnancy must always be terminated because there is no way for the embryo to develop to full term without rupturing the tube. In very rare instances, a fertilized ovum escapes through the ostium of the fimbriated end of the tube, implanting itself and developing in the abdominal cavity. Some of these pregnancies have actually been successful, and a nearly full-term baby can be delivered by an abdominal incision. Even more rarely, a fertilized ovum implants itself directly on a surface within the abdominal cavity, without ever having entered the tube.

If the isthmus of the tube is the site of the implantation, however, an early rupture is more likely to take place because here the thick muscle of the tubal wall permits little distension. In the infundibular part of the fallopian tube or in the ampulla portion, it is possible for the pregnancy to continue longer, but it always ultimately terminates in rupture, usually into the pelvic cavity.

Causes of Ectopic Pregnancy

In many ectopic pregnancies, the etiology is unknown, although endocrine dysfunction has been suggested. In some instances, there are conditions in or around the fallopian tube that prevent the fertilized ovum from getting to the uterine cavity. Thus, a tubal implantation results.

One such condition can occur when a gonorrhea infection that has spread to the fallopian tubes causes a further narrowing, through scar-tissue formation, of the already narrow lumen. The tiny opening may permit the sperm to get up to the egg, but the fertilized developing egg, larger than a sperm, may not be able to descend to the uterus, and implants in the tube. Actually, it is more likely that gonorrheal infection of the upper reproductive tract results in infertility, since the destructive changes in the lining of the oviducts usually do not permit the passage of sperm at all.

Another cause of ectopic pregnancy may be related to the presence of an intrauterine device (IUD) that has been placed in the uterus for purposes of contraception. It has been observed that tubal pregnancy occurs more often in women with IUDs, and it may result from an alteration in tubal motility, causing a delay in the descent of the conceptus, which then goes ahead and implants in the tube. At any rate, since the function of an IUD is not to prevent the sperm from getting to the egg, but somehow to prevent uterine implantation, any woman with a previous history of gonorrhea (and a possibly narrowed tube) should never have an IUD inserted. It places her in double jeopardy for the possibility of an ectopic pregnancy.

Tubal pregnancy has also been associated with sterilization by tubal ligation. According to Mathelier (1983), the younger the woman at the time of sterilization, the greater the risk of ectopic pregnancy. The incidence is rare, however.

It is entirely possible and probable that many tubal pregnancies, like many uterine pregnancies, spontaneously regress or abort at a very early stage, and the woman may never suspect she was pregnant at all.

Treatment of Ectopic Pregnancy

Before 1975, nearly 80% of ectopics ruptured before diagnosis. As a result of the newer diagnostic techniques, currently fewer than 30% rupture before they are recognized, despite the increased incidence of tubal pregnancy. Diagnosed, unruptured ectopic pregnancy allows the surgical procedure of salpingostomy (making an incision in the tube and removing its contents) rather than the former salpingectomy (removal of the entire tube), frequently accompanied by oophorectomy (removal of the ovary on the affected side). Such conservative treatment permits preservation ot the fallopian tube for future fertility. It may even be possible, in some selected cases, for the surgeon to remove the pregnancy from the tube by "milking" it out of the fimbriated end of the tube.

Despite attempts to preserve the tube, women who have had an ectopic pregnancy have a 12% incidence of reoccurrence, and 40% are unable to conceive again. Of the 60% who do become pregnant, approximately 15%–20% will miscarry (DeCherney, 1982; Badawy et al., 1983).

Peritonitis: More Likely in Women

The body cavity is not a cavity at all, but is a potential space that exists between the abdominopelvic viscera, which are covered by a membrane called the *visceral peritoneum*, and the body wall, lined by a continuation of that same membrane, which is reflected backwards and called the *parietal peritoneum*. When bacteria like gonococci, staphylococci, and streptococci gain access to that space and attack the peritoneum, the resulting inflammation and infection is called *peritonitis*, and it can be life threatening. While one may think that all body openings lead into the body cavity, they actually do not. The mouth opening leads into the long tube of the digestive tract and ultimately exits at the other end in the anus; the urethral opening at the tip of the penis in males permits exit of spermatozoa from the testis and urine from the urinary bladder. In females, the urethral opening is for the passage of urine only and leads to the bladder. The only way, in both males and females, that harmful organisms can get into the body cavity is through an opening in the body wall by surgery, by accidental perforation, by rupture of an organ (like a ruptured appendix), or through the bloodstream (blood poisoning). In females, however, there *is* another route, which results in a unique anatomical disadvantage. Since the end of the fallopian tube at the ostium opens directly into the pelvic or body cavity, the outside of the body, by way of the vaginal opening, is in contact with the inside of the body cavity via the fallopian tube.

Women, therefore, are much more subject to pelvic peritonitis than men. Fortunately, the acidity of the vaginal secretions and the cervical mucus act as a bacteriacide for most harmful organisms, with the exception of the gonococci of gonorrhea. But whenever the cervix is dilated for procedures such as an abortion, a D and C, a uterine biopsy (a small bit of tissue is taken for examination), or the insertion of an IUD, the possibility of peritonitis exists unless sterile techniques are very carefully observed.

OVARIES

The ovaries are two glands in the pelvic cavity that produce ova and sex hormones. The ovaries are the size and shape of almonds in the shell; that is, they are approximately 3 centimeters long by $1\frac{1}{2}$ centimeters wide by 1 centimeter thick, although the size varies. They are suspended in the pelvic cavity three ways: attached to the peritoneal covering over the back of the uterus (the broad ligament) by connective tissue called the mesovarium; attached to the uterus by the ovarian ligament; and attached to the lateral body wall by the suspensory ligament of the ovary. The actual position of the ovaries is variable, especially after the birth of a child, since at that time they are displaced from their original position and may never return to it.

The ovaries are the only organs in the pelvic cavity that are not covered with peritoneum, and they are a dull grey color in comparison to the pink, shiny, smooth uterus. The surface of the ovaries is covered with the *germinal epithelium*, a flattened layer of cells that was misnamed because it was thought that it gave rise to ova throughout life. Under the germinal epithelium is a zone or region called the *cortex*, the area in which the ova develop. Inside the cortex is the region called the *medulla*, with many large blood vessels, lymphatic vessels, and nerves. The connective tissue of the ovary is called the *stroma*, a framework of fibrous cells.

When a female baby is born, her ovaries contain a fixed number of primary ovarian follicles, and the ova she is destined to produce throughout her reproductive life will develop from them. These primary follicles arose from special cells called *primordial germ cells*, which were segregated from the rest of the body cells by as early as 10 days after fertilization. After migrating to the site of the primitive ovary, the germ cells, now called *oogonia*, become surrounded with a layer of cells called follicle cells, and the combination is known as the primary follicle. The primary follicles in the fetal ovary divide at a prodigious rate, and by 20 weeks of fetal life, there are more than 7 million. After that, there is no further division either before birth or after it; from that time on and for the next 50 years or so, the majority of ova undergo a process of regression and degeneration called *atresia*.

Various investigators differ in their estimates of the number of follicles remaining at birth. Some say that 400,000 exist, others maintain that roughly 1 million are present to form the stock from which all future eggs to be ovulated will be selected. At birth, the primary follicles contain an ovum arrested in the stage of development known as primary oocyte. By puberty, the number of follicles is further decreased to 50,000 (or fewer), and only one of these is ovulated each month—a total of approximately 400 from puberty to menopause.

A woman is, therefore, born with all the eggs she will ever have. If she is still ovulating at the age of 50, that ovum has been in her ovary for some 50 years and 4 months. All the rest have become atretic. If this seems like an inordinate waste of cells with talent, consider that when a male ejaculates approximately 3 milliliters of seminal fluid, each milliliter contains 40–250 million sperm—yet only one sperm is necessary to fertilize an egg! This apparent overkill is not at all unusual in nature, whose interest is perpetuation of the species.

After puberty, those primary oocytes that will be ovulated go through a monthly scheme of development called the ovarian cycle. The hormones produced by the ovary during its cycle govern the activity of the endometrium of the uterus and result in cyclic menstruation. What is going on in the ovaries and the uterus is determined by hormones released by the anterior pituitary gland, and the anterior pituitary, in turn, is controlled by secretions called releasing factors from a part of the brain, the hypothalamus. This constitutes the hypothalamic–pituitary–gonadal–uterine axis described in detail in the next chapter.

Anatomically, the ovarian cycle is diagrammed in Figure 2.15. At the beginning of each cycle, a group of follicles begins to undergo development, but only one will completely differentiate and mature. The rest become atretic. The follicle cells around the ovum multiply, form many layers, and become known as the granulosa cells. The stroma cells become organized into two layers, the *thecae interna* and *externa*, which produce hormones. The ovum itself enlarges and becomes surrounded by a membrane, the *zona pellucida*, a noncellular clear zone that lies between it and the granulosa cells. The zona pellucida functions during fertilization, permitting the entrance of only one sperm and blocking all others.

The next stage of follicular development is called the secondary or antral follicle. The granulosa cells initially, and later the theca layers, secrete a viscous follicular fluid that accumulates in between the granulosa cells forming cavities that eventually coalesce to form one large fluid-filled space called the antrum. The primary oocyte stays embedded at one side surrounded by a mound of granulosa cells called the *cumulus oophorus* (Latin for heap of egg cells). When the ovum is ovulated, the two or three adhering layers of granulosa cells shed with it form the *corona radiata*, or crown of cells. Perhaps 20–50 primary follicles reach the antral stage, but only one goes on to ripen fully to ovulation. That dominant one is called the mature *graafian follicle*. In the follicles that do not fully mature, the granulosa layers become disorganized, the cells deteriorate, the follicular cavity shrinks, and the ovum itself degenerates.

The mature graafian follicle moves toward the outer part of the ovary and forms a bulge on its surface. It manages to move mostly because of its increased size (2.5 cm or about 1 inch in diameter) and because the theca layers somehow facilitate a pathway. The area on the surface where the bulging follicle has caused a thinning out that looks like a blister is called the *stigma*. At the time of ovulation, the follicle bursts, and the egg, surrounded by its protective and nourishing corona radiata,

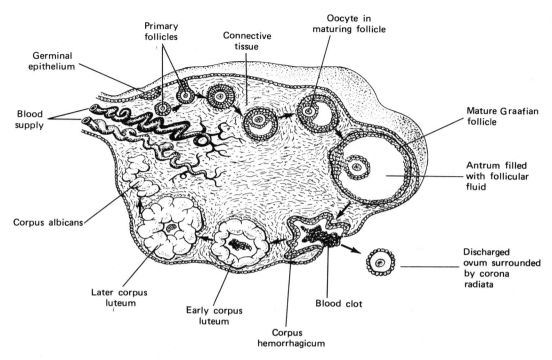

Figure 2.15. The ovarian cycle.

floats out into the peritoneal cavity on a rush of follicular fluid. There is slight bleeding into the center of the follicle, and some of the blood may escape into the pouch of Douglas to produce the ovulatory pain called *mittelschmerz* (German for middle pain), which some women experience.

The ovulated egg (now a secondary oocyte) is swallowed up by the fimbriated end of the fallopian tube and begins to travel down it. It will die if it is not fertilized within 24 hours. Meanwhile, back in the ovary, the wall of the ruptured follicle collapses and the granulosa cells greatly increase in size, accumulating a yellow pigment called lutein in their cytoplasm. The cells are now called luteal cells, and the former follicle is called the *corpus luteum*, which produces hormones. The corpus luteum reaches its peak of activity approximately 5 to 7 days after ovulation, and if

the ovum is not fertilized, begins to regress on the tenth day. Its life is over by the fifteenth day. The entire corpus begins to be invaded by connective tissue and, ultimately, is transformed into the *corpus albicans*, the white scar. After puberty, the formation of white scars from previous corpora lutea each month give the ovary its characteristic pitted, convoluted surface. Before these monthly cycles begin, the ovary in a prepubertal female is smooth; afterwards, it looks something like a peach pit. The complete absorption of the corpora albicantes takes a year or more, so the ovary always has a few in various stages of disappearance if a microscopic section is examined.

How the one follicle destined for ovulation is chosen while all the others die is not completely known. It takes an estimated 25 days for a small antral follicle to grow into a mature graafian preovulatory follicle—considerably longer than the period from the beginning of menstruation to ovulation. There must be, therefore, continuous development of a number of primary follicles progressing along to the antral stage independent of hormonal support. Most of them will become atretic, but it is suspected that the follicle most likely to be saved from atresia by the increasing pituitary hormone levels at the onset of a new ovarian cycle is the one best developed at that time. This advanced follicle, rescued by hormones at a later stage than the others because of its increased development, goes on to become the successful ovulatory follicle. The other antral follicles, less mature and lacking the ability to respond to hormones, become atretic.

There must also be some kind of quantitative relationship functioning between antral follicles doomed to atresia and those ordained to ovulate because even one tiny bit of ovarian tissue left after surgery still manages to produce an ovum each month. There is some evidence from animals that in that situation, fewer follicles become atretic than in a woman with two fully functioning ovaries.

When all the follicles are gone from the ovary by atresia or ovulation, ovarian activity ceases, and that period in a woman's life is called menopause. For the great majority of women, the nonactivity of the ovaries in menopause has been highly overrated; it need not be particularly significant to their well-being, sex lives, or general health.

SUPPORT OF THE PELVIC VISCERA

That we stand erect and walk on two legs instead of four has been of primary importance in the development and progress of human society. There are many marvelously ingenious evolutionary modifications that have taken place in order to produce and maintain this posture. These anatomical modifications have significantly influenced our sex lives and the way in which women bear and deliver children. They also have contributed greatly to the development of some human ills, like backaches, sinus trouble, headaches, hernias, impacted wisdom teeth, fallen and weak arches, varicose veins, and hemorrhoids, to mention only a few. Most authorities agree that these malfunctions are signs of how inadequately the human body has adapted to a vertical biped stance instead of a quadruped posture.

The difficulty with standing erect is that gravity is always trying to pull everything down, and the older one gets, the harder it becomes to maintain all structures in their appropriate balanced arrangement of bone, muscle, and soft tissue. The way in which injury or congenital defects can compromise this balance to produce backache has already been discussed. Similarly, there are potential weaknesses that exist in the support of the internal organs. In a four-footed animal, the abdominal and pelvic viscera hang suspended from a horizontal backbone and are supported by the abdominal wall without much trouble, but in humans the abdominal and pelvic organs are constantly subjected to the downward pull of gravity. Even when a person is standing motionless, the viscera exert pressure on the muscles and connective tissue of the pelvic floor. Any movement increases abdominal pressure, as the viscera are squeezed against each other and against the pelvis. When the abominal pressure is deliberately increased by contraction of the abdominal muscles during such ordinary activities as respiration, urination, and defecation, the stress placed on the pelvic supports that keep the viscera from falling out of the pelvis are enormous.

There are three kinds of supports for the abdominopelvic organs. One is provided by the bony bolstering of the spinal curvatures and the flaring of the ilia of the pelvis. These form shelves to brace portions of the digestive tract. Secondly, there are folds of the peritoneal lining of the body cavity and packings of connective tissue that attach to the viscera and hold them in place. The major support for the viscera, however, is from underneath— the muscles and connective tissues that make up the *pelvic diaphragm*, which stretches like a hammock across the bones of the pelvic outlet.

The inherent weakness in the pelvic diaphragm is that openings in it must exist for the exits of the urethra and the rectum, and in women, for the vagina. In females, childbearing and childbirth can strain the integrity of the pelvic support systems, and sometimes, after many pregnancies, the organs actually may slide out, or prolapse from the pelvis.

Structure of the Pelvic Floor

The pelvic floor consists of all the soft tissues that close the pelvic outlet, from the skin on the outside to the peritoneum on the inside, and all the muscles in between. When skin and superficial connective tissue are removed, the next layer is that of the superficial or *perineal muscles*. They include the ischiocavernosus muscles, important in erection of the clitoris; the bulbocavernosus muscles, which form the external vaginal sphincter and also function in clitoral erection; the superficial and the deep transverse perineal muscles; and the muscles of the external anal sphincter (see Fig. 2.16).

Underneath the perineal muscles are the *levator ani* muscles, major components (and

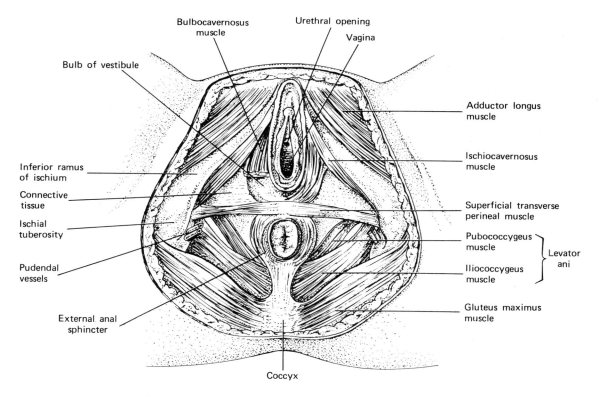

Figure 2.16. Muscles of the pelvic floor, from below.

the largest) of the pelvic diaphragm. In animals, these are the tail-wagging muscles. In humans, since we have no tails to wag, it would be expected that these muscles be diminished and vestigial. Instead, the levator ani muscles are bigger, stronger, and more powerful because their function in us is support, and that makes them the most important muscles in the pelvic floor. They are further strengthened by very strong connective tissue sheaths, the inferior and superior levator fascia. Each half of the levator ani meets its partner from the other side in the midline to surround the urethra, the vagina, and the rectum with fibers that form sphincter muscles, which can powerfully constrict those hollow organs. Of the three component parts of the levator ani, the *pubococcygeus* is the one performing the mainsphincteric action.

Sphincter Actions of the Pubococcygeus on Urinary Control

The pubococcygeus muscle fibers around the base (or trigone) of the urinary bladder form a sphincter mechanism that compresses the urethra as it exits and pulls it upward behind the pubic symphysis so that a sharp angle called the urethrovesical angle is formed between it and the neck of the bladder. Even if the bladder is very full and the pressure within it is much greater than normal, urine can be retained. When the pubococcygeus muscle is voluntarily relaxed, the urethrovesical angle can straighten out and descend. Then the bladder muscle, called the detrusor, involuntarily contracts to let the urine enter the urethra and proceed until the bladder is empty. If the pubococcygeus is voluntarily contracted during urination, the sharp angle between the urethra and the bladder is restored to its previous elevated position, and urination stops. It is, therefore, possible to repeatedly stop and start urinary flow by contracting and relaxing the pubococcygeus muscle.

However, if the fibers of the pubococcygeus muscle are traumatized, stretched, or torn during childbirth, they no longer meet in the midline around the base of the bladder to maintain the integrity of the urethrovesical angle. Then, any increase in intra-abdominal pressure, like sneezing, coughing, or laughing, pushes the viscera against the bladder and results in a dribble or even a gush of urine that cannot be controlled by the now lax pubococcygeus muscle. This annoying and embarrassing situation is experienced by some multiparous women to a slight extent at the time of menstruation, and is called *stress incontinence.* In a more severe form, it is present throughout the month. Some women never experience incontinence until menopause.

Although urinary incontinence may also be a result of a neurological disorder or a mechanical obstruction to the bladder or urethra, the most common cause of difficulty is injury to the soft tissues during delivery. The role of the obstetrician present at a delivery is to minimize the possibility of injury to the mother or to the baby, and the doctor's gentle and presumably expert assistance is certainly required in difficult labors. While it is not reasonable to blame all stress incontinence on incompetent or negligent behavior by the obstetrician, there are some procedures performed by physicians that tend to encourage trauma and tissue injury rather than to decrease it. For example, when labor is induced with hormones for no other reason than the convenience of the physician, the result may be a precipitous delivery that can injure both the baby and the mother. Another example is neglecting to catheterize the bladder during labor if the woman is unable to void on her own. The resultant pressure may then injure the bladder neck area. However, many cases of mild and uncomplicated stress incontinence are unavoidable, even with the best and most conscientious obstetrical help, particularly when large infants are delivered through a small bony pelvis. It is likely that the number of births (parity) of a woman is more significant in causing stress incontinence than any perineal damage associated with an individual delivery.

Kegel Exercises

Most surgical treatment for stress incontinence is aimed at restoring the integrity of the pubococcygeus muscle and the urethrovesical angle. It has been claimed that Kegel exercises can alleviate and even cure many cases of mild to moderate stress incontinence, and they are certainly worth trying before resorting to surgery. These exercises, as described by Kegel, the physician who advocated them, are meant to strengthen the pelvic diaphragm. Fifty to 100 times a day, said Kegel, the pubococcygeus should be contracted and the anus and vagina drawn up into the pelvis. Whenever voiding, the flow of urine should be stopped and started several times. This will strengthen the muscle enough to pull up the neck of the bladder and increase the urethrovesical angle, thereby preventing leakage of urine.

Action of the Pubococcygeus on the Vaginal Wall

Kegel exercises have also been advocated for the improvement of sexual relations between couples because they increase the ability of the vaginal sphincter to contract. The vaginal sphincter is responsible for the clasping action of the vagina around the penis during sexual intercourse, and the sphincter is formed by the fibers of the pubococcygeus muscles blending into those of the superficial perineal muscles around the walls of the vagina.

The involuntary spasm of the pubococcygeus muscle that may occur during sexual intercourse or perhaps during a pelvic examination in a doctor's office is called *vaginismus*. Here, the constrictive action of the pubococcygeus actually shuts off the vaginal opening so that introduction of the penis during coitus or of a speculum during examination produces pain and makes penetration impossible. Vaginismus can be a result of anticipated pain, or it may be caused by fear, guilt, or nervousness. It sometimes develops in a rape victim, particularly if sexual assault has produced severe lacerations to the genital area. Intercourse may then be severely painful to a woman long after the physical damage has healed.

Sometimes a mild and temporary vaginismus, enough to delay penetration or make it uncomfortable, however, may occur as a result of a previously painful intercourse during an episode of vaginitis. A temporary vaginismus can also be associated with a lack of sufficient lubrication.

It is possible to alleviate mild discomfort during intercourse with a surgical, water-based lubricant like K-Y Jelly, but if vaginismus is the primary response to sexual stimulation, some type of therapeutic counseling may be helpful.

No one has ever seen it happen, but almost everyone has heard the story about the couple that became "locked" together during sexual intercourse as a result of vaginismus and had to be separated in the emergency room of the local hospital. There is no medical basis or case on record to verify this persistent myth, called *penis captivus*. It can occur in animals, most frequently in dogs, and this is probably the source of the fantasy.

Action of the Pubococcygeus on the Rectal Wall

The pubococcygeus muscles also have an effect on the walls of the rectum. When strongly contracted, the pubococcygeus pulls the rectum and anus forward toward the pubic symphysis and supplements the action of the anal sphincter.

Other Muscles of the Levatores Ani

In addition to the pubococcygeus muscles, the lateral components of the levator ani are the *iliococcygeus* muscles, which extend (as their name indicates) from the ilium to the coccyx. Posteriorly, the triangular *coccygeus* muscles complete the pelvic diaphragm. The iliococcygeus muscles and the coccygeus muscles are supportive only; they have no sphincteric action.

Internal Support of the Pelvic Viscera

The uterus, the tubes, the ovaries, the urinary bladder, and the rectum are connected to each other and to the walls of the true pelvis by a number of "ligaments," some of which are not ligamentous structures at all, but are merely folds of the peritoneal lining of the pelvic basin. Others do function as ligaments, combinations of smooth muscles and fibrous connective tissue, that tie the pelvic organs together, or function as ropes to suspend them from the body wall. Some additional support for the viscera is actually accomplished by their positions: being all crowded together provides pressure to hold them in place.

The peritoneum is the membrane that lines the body wall and is reflected back to cover all the viscera. The lining part is the *parietal* peritoneum, and the part that covers the organs is the *visceral* peritoneum. In between is the peritoneal cavity, also referred to as the pelvic cavity. It is actually only a potential space, since the parietal and visceral layers are in close contact.

The arrangement of the peritoneum in the abdominopelvic cavity is complicated. Trying to trace it from one organ to another or to the body wall is frustratingly difficult because in their embryonic development the organs pushed into the peritoneum from behind, ac-

quiring for themselves a visceral peritoneal covering, but leaving behind a complex scheme of folds and reflections of the parietal peritoneum. The reflected and folded peritoneal lining forms the route for blood vessels and nerves that supply the organs.

If one imagines the pelvic basin as a huge bowl lined with a membrane, the pelvic peritoneum, it is possible to visualize how the urinary bladder pushes up into it anteriorly, the rectum pushes up into it posteriorly, and the uterus pushes up into it in the middle. The viscera of the pelvis are only partially covered by the peritoneum. They lie mostly underneath it, surrounded by abundant connective tissue, the endopelvic fascia. It is apparent how the two pouches of peritoneum are formed in front and in back of the uterus. The one between the uterus and the bladder is the vesicouterine pouch. The one between the uterus and the rectum is the rectouterine, the pouch of Douglas, or the cul-de-sac, easily reached through the posterior fornix of the vagina.

Each fallopian tube extends laterally from the uterus like an arm, and the peritoneum is draped like a blanket over the uterus and the tubes. Each tube is covered with the blanket except at the ostium, which opens directly into the body cavity. The draped tent of reflected peritoneum is called the *broad ligament*. The ovaries are not covered by the broad ligament, but are suspended from its back (posterior) surface by an extension called the *mesovarium*.

Inside the broad ligament is the connective tissue containing the blood vessels and nerves that supply the pelvic organs. Two cordlike condensations of this connective tissue, the *ovarian ligaments* and the *round ligaments*, are found in the apex of the broad ligament. The ovarian ligaments contain smooth muscle fibers and extend between the ovaries and the lateral angle of the uterus. At the time of ovulation, the fibers contract to change the position of the ovaries and bring them closer to the ends of the fallopian tubes. The round ligaments, almost completely composed of smooth muscle, are flat, narrow bands that extend from the lateral angle of the uterus on either side, cross over many blood vessels and nerves, and travel down through the inguinal canal (a passageway through the anterior abdominal wall) to insert into the tissues of the labia majora. The round ligaments hold the uterus forward in its typical anteverted position over the urinary bladder. As they traverse the inguinal canal, these ligaments become surrounded with connective tissue coverings and are accompanied by blood vessels, nerves, and lymphatic vessels. The lymphatics drain from the upper part of the uterus where the tubes enter and connect with the superficial inguinal lymph nodes, providing easy access for the spread of a possible malignancy from the uterus to the nodes.

Inside the base of the broad ligament, near the cervix, are masses or condensations of connective tissue associated with peritoneum, which extend out to the pelvic wall from the cervix and the wall of the vagina like spokes of a wheel. These are the main supports of the uterus, bladder, and vagina. They are the:

1. *Cardinal* ligaments, extending from the cervix laterally to the pelvic wall
2. *Uterosacral* ligaments, extending from the cervix posteriorly to the sacrum.

All the supports that have been described, the muscular components of the pelvic diaphragm and the broad, round, cardinal, and uterosacral ligaments, do their job of retaining the pelvic viscera inside the pelvis as long as they are strong and firm. If they are repeatedly traumatized in childbirth or if they are lacerated or atrophied, or in any way impaired in their action, they progressively weaken, and

the uterus, bladder, or rectum will be displaced downward and begin to protrude or drop out of position. This dropping down or falling of an organ is called *prolapse*. When the urinary bladder prolapses into the anterior wall of the vagina and causes a bulge, it is called a *cystocele*. When the anterior wall of the rectum bulges into the posterior wall of the vagina, it is called a *rectocele*. Even though the uterus has more supporting structures than the other organs, it is the most likely to prolapse.

All the pelvic organs with their associated connective tissue and muscle are sensitive to estrogen. During pregnancy, when hormone levels are at a maximum, all the tissue greatly increases in size (hypertrophies). After menopause, when estrogen levels are low, muscle and connective tissue are likely to shrink (atrophy). The supports are weakened, and there is an increasing predisposition to prolapse, particularly with a previous history of childbirth damage to the supports. Like a stretched-out rubber band, there may be no tension left. The degree of prolapse may be mild, with the cervix only moderately descended into the vaginal canal (first degree), or the cervix may protrude through the vaginal opening (second degree). In extreme or complete prolapse, the cervix and the entire body of the uterus is pushed outside the vaginal opening.

Treatment depends on the extent of the prolapse but it most usually involves surgical correction and shortening and tightening of the cardinal ligaments and repair of the pelvic diaphragm. If the uterus has actually prolapsed out of the vagina, the cervix extends outside the vaginal opening, and the vaginal canal is inside out. A hysterectomy may then be a necessary part of the surgical procedure. In this situation, in which a woman is virtually walking around with her uterus between her legs, there is justification for the removal of a histologically normal uterus.

For the great majority of women, childbearing and delivery are completely normal physiological processes that result in no permanent injury to the reproductive organs or the surrounding soft tissues. Even after several pregnancies, the uterine supports remain firm and intact, having recovered their full functional capacity even though ligaments were stretched and soft tissues lacerated during delivery. Not all women heal as well as others, however, and some may develop subsequent difficulties. There are women who may have what is called a congenital soft tissue deficiency; that is, they were born with a pelvic support mechanism that is just not as strong as it could be. Despite every effort to eliminate trauma during delivery, these women are predisposed to experience some of the problems that have been described after they have delivered several children. Occasionally, nulliparous women (i.e., those who have never delivered a baby) also experience stress incontinence or uterine prolapse.

REFERENCES

Bauman, J. E., Kolodny, R. C., and Webster, S. K. Vaginal organic acids and hormonal changes in the menstrual cycle. *Fertil Steril* 38(5):572–579, 1982.

Badaway, S. Z., Moses, A., Strecton, D., et al. Ectopic pregnancy—the role of conservative surgical treatment. Abstract, *Int J Fertil* 28(1):29, 1983.

DeCherney, A. H. Ectopic pregnancy: Diagnosis and management. *Del Med J*, 54(6):323–325, 1982.

Doty, R., Ford, M., and Preti, G. Changes in the intensity and pleasantness of human vaginal odors during the menstrual cycle. *Science* 190:1316–1317, 1975.

Keith, L., Stromberg, P., Krotoszynski, B. K., et al. The odors of the human vagina. *Arch Gynak* 220(1):1–10, 1975.

Ladas, A., Whipple, B., and Perry, J. *The G Spot: & Other Recent Discoveries About Human Sexuality.* New York, Holt, Rinehart, & Winston, 1982.

Mathelier, A. Ectopic pregnancy after tubal fulguration. *The Female Patient* 8:32/59–32/61, 1983.

Tonzetich, J., Preti, G., and Huggins, G. R. Changes in concentration of volatile sulphur compounds of mouth air during the menstrual cycle. *J Int Med Res* 6:245–254, 1978.

3. *The Menstrual Cycle and Its Hormonal Interrelationships*

Female primates menstruate. Every month the endometrial lining of the uterus is prepared by hormones to receive and nurture a fertilized egg. Should fertilization not occur, the superficial two-thirds of the lining is shed, and periodic bleeding takes place, emerging through the cervix and out the vagina.

Women, like female monkeys and apes, are primates, but most of them find themselves reluctant to admit the existence and the normality of menstruation. Instead they get "sick," get "the curse," get their "period," the "monthlies," a "visit from a friend," "hoist the red flag," "fall off the roof"—the euphemisms are many.

It is not suprising that many women have difficulty talking about "that time of the month" and feel much more confortable using such euphemisms. Throughout the course of history, society has been less than kind to the menstruating woman. Depending on the era and the culture, she has been regarded with fear and awe, with derision and distaste. Even primitive peoples recognized that blood was essential to life; to bleed and not to die, indeed, to lose blood cyclically and still remain healthy, must be supernatural. Magical powers were ascribed to menstrua-

tion and menstrual blood, and superstitions and taboos surrounding the completely natural physiological function of menstruation endured for centuries.

Some of these myths have persisted in modern society. Few people today, except for the truly uneducated, still believe that a menstruating woman can blight crops, curdle milk, sour wine, wilt flowers, or be responsible for natural disasters like floods or tornados. Many, however, find it completely plausible that a menstruating female should be excused from sports at school, should not take showers after gym, should not get a permanent or have hair tinted because it won't "take," is more vulnerable to illness, and, still believing in the uncleanness of menstrual blood, should not have sexual intercourse. All of these attest to the "sickness" of menstruation, and sickness implies suffering. If a young woman learns that to menstruate is to be sick, that it is a burden she must suffer for 40 years or more, that it is at best an unfortunate, and uncomfortable, nuisance, and that it is physically and psychologically incapacitating, she has then inherited the modern version of the ancient myths, taboos, and superstitions.

Of course, not all women view menstruation in the same way because not all women experience menstruation in the same way. The prevailing attitudes toward the event could be, but are generally not, positive. No one has as yet written a book on the Joy of Menstruation; we get much more information on menstrual disorders and problems. Several books by feminist authors have attempted to change our inherited or learned attitudes toward menstruation—to bring it out of the closet, so to speak. But the reluctance to speak freely about women's physiological processes is founded in entrenched attitudes toward women's "otherness" in a male dominated world, and they are not easily unlearned. When women, who all share the experience, more fully understand the nature, meaning, and function of menstruation, they will talk more openly about it and perhaps develop positive attitudes toward their reproductive cycles, and be less likely to find them difficult and disturbing.

The female reproductive cycle, less accurately called the menstrual cycle, refers to the rhythmic changes that occur in the ovaries and uterus under hormonal influences. While the existence of hormones has been known since the turn of the century, the recognition that there is a complex relationship between the organs of reproduction and the endocrine system is fairly new. Even more recent is the certainty that a part of the brain, the hypothalamus, plays a dominant role in the regulation of reproductive function, and the speculation that the brain should perhaps be viewed as another endocrine organ. Although progress in the field of neuroendocrinology—the study of the integration of the nervous system and endocrine system—has been rapid, there is virtually as much that is theorized or unknown as is known. For this reason, the gynecologist-obstetrician is probably even more of a medical empiricist than other specialists; that is, the gynecologist administers hormones as a treatment because they

work, and not because there is any clearly defined understanding of their action in the body. It is known, for example, that oral contraceptives are highly effective, but there are still only presumptions as to exactly how and why. On the other hand, clomiphene citrate is a drug that was originally administered to prevent ovulation, but it turned out to induce it instead, and it is now given to women who want to become pregnant!

The major efforts in endocrinology today are by neurophysiologists who are studying the hypothalamic control of the pituitary gland, and by biochemists and cell biologists who are beginning to establish how hormones are formed, how they work within the cell, how they can cause synthesis of other substances, and how they are metabolized. In addition to all the unanswered or unanswerable questions, however, there are aspects of human reproductive physiology and the endocrine system that can be described.

ENDOCRINE SYSTEM

The endocrine system was the last organ system to be recognized, probably because there is no anatomical continuity between its parts. The recognized and classic endocrine glands are the pituitary gland, the thyroid gland, the parathyroid glands, the adrenal glands, the islets of Langerhans of the pancreas, and the ovaries and the testes (see Fig. 3.1). Several additions to the list are the hypothalamus, which secretes hormones regulating the pituitary gland, the placenta, which during pregnancy is an endocrine organ, and the pineal gland, which in humans has questionable status as an endocrine gland, and possibly the thymus. The glands of greatest importance to the reproductive cycle and to the physical differences that exist between males and females are the anterior pituitary gland and the

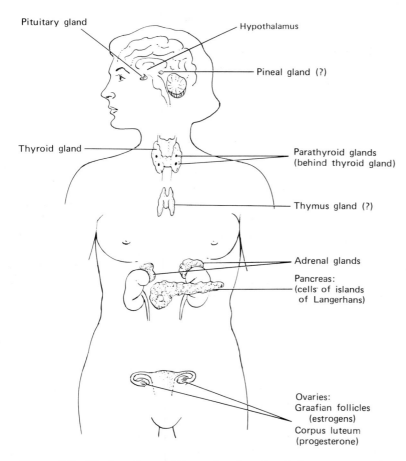

Figure 3.1. The location of the endocrine glands in a woman. The thymus secretes thymic hormone, but little is known about its activity. The pineal gland also has not been proven to be an endocrine gland.

ovaries and testes, but all of the endocrine organs have interrelating physiological effects that are necessary to the proper growth, development, and function of the reproductive system.

Endocrine glands have no ducts; they secrete directly into the tissue fluid that surrounds their cells. The secretion passes into the bloodstream to be transported elsewhere in the body, sometimes a considerable distance from the original gland. Because they secrete into the blood capillaries that surround them, all endocrine glands generally have a very good blood supply. Viewed microscopically, they consist of cords or clumps of cells surrounded by many capillaries and supported by connective tissue.

Hormones

Endocrine secretions are called hormones from the Greek word, *horman,* meaning "to urge on." Hormones secreted by the glands of the endocrine system are all different, and the cells on which they exert their effects are different; however, they are, generally speaking, all of the urging-on type, although exactly how they accomplish this action is still speculative. Some hormones have an effect on the growth and metabolism of all cells of the body; others have an action on just certain tissues, which are then called *target* tissues or cells.

Chemical Composition of Hormones

There are two major types of hormones chemically:

1. Proteins, or derivatives of proteins, such as polypeptides, peptides, amines, and glycoproteins (carbohydrate-protein complexes).
2. Steroids, which are lipid fat-soluble compounds, and part of a large family of substances with the same chemical skeleton. The ovaries, testes, placenta, and the outer part of the adrenal glands produce steroids.

Control of Hormone Secretion

Some endocrine glands are directly connected with the nervous system. Nerve fibers ending on gland cells cause the gland to release its hormones in response to nerve impulses, or the nerve cells themselves produce neurosecretions, which are stored in the gland. Neuroendocrine control may also be indirect, as in the hypothalamic hormonal regulation of the anterior pituitary gland.

The rate and quantity of hormone secretion from an endocrine gland may be controlled by blood levels of another hormone or by levels of organic or inorganic substances other than hormones, like glucose or calcium. If the message of the second hormone or substance is to *inhibit* the further secretion of the endocrine gland, it is said to be a *negative feedback* mechanism; when the message transferred back to the original gland is to *increase* the rate or quantity of secretion, it is called *positive feedback*.

Mechanisms of Hormone Action

Hormones basically control the activities of the cells they affect, and there has been very extensive investigation of the mechanisms by which they provide this control function. Once the mode of action of a hormone at the cellular level is known, it may be possible to interfere with, or manipulate, it in some way—to cure an endocrine defect, to produce a more effective contraceptive, or to increase fertility.

Since hormones circulate in the bloodstream in such very low concentrations compared with some other biologically active substances, such as glucose, there has to be some way for the target cells to recognize, receive, and retain the hormone molecules as they pass by in the capillaries. Accordingly, there are specific receptor sites for specific hormones located on the cell membranes or in the cytoplasm of the target cells. The job of the receptor is to transmit the message of the hormone's arrival to the area of the cell that is involved in the response, either to initiate chemical activities within the cytoplasm or to activate the genes in the nucleus to synthesize intracellular proteins that, in turn, result in specific cellular functions. The cell receptors, almost always composed of glycoprotein, are highly sensitive to the hormones they recognize and bind. It takes only a few molecules of the bound hormone to initiate the full cellular response. Some hormones are even able to regulate the number and activity of their own and other hormone receptors. For example, the pituitary gonadotropins, follicle-stimulating hormone (FSH) and luteinizing hormone (LH) (hormones that act on the gonads), cause the granulosa cells of the primary follicles to develop receptors for FSH and estrogen and stimulate the theca cells to develop LH receptors.

The molecules of a protein or peptide hormones are too large to enter the cell and, therefore, their receptors are located on the cell membrane. Once a protein hormone (the "first messenger") unites with its receptor on the cell surface, the combination of hormone-receptor complex is either *internalized* (engulfed into the cell to a specific location) or activates an enzyme within the cell membrane, adenylate cyclase. Adenylate cyclase passes the message to the inside of the cell by causing the formation of a hormone mediator, called the second messenger. The mediator, 3'5' adenosine monophosphate (cyclic AMP) is formed from the conversion of adenosine triphosphate (ATP). ATP is the high-energy compound stored within all cells to provide a source of energy for cellular activity. Cyclic AMP (sometimes cyclic GMP, guanosine mononucleotide) carries the hormone message throughout the cell and, by initiating a stepwise series of enzyme actions, instructs the cell to respond with any number of cellular activities, depending on its inherent nature. If the cell is in the ovary, testis, or adrenal cortex, cyclic AMP ultimately results in the production of steroid hormones (see Fig. 3.2).

When steroid hormones circulating in the blood reach their target cells, they are small enough molecules to diffuse across the cellular membranes, and they immediately bind with a high-affinity receptor in the cytoplasm specific for that particular steroid. For example, receptors specific for estrogen, a steroid

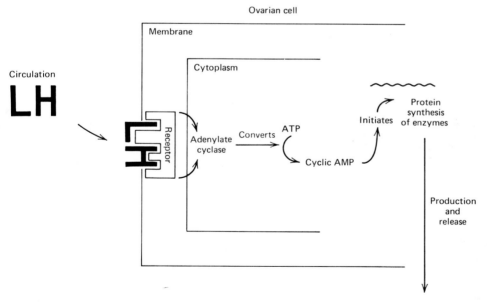

Figure 3.2. The cyclic AMP mechanism by which protein hormones exert their action. A stimulating protein hormone, such as LH, unites with a specific receptor at the cell membrane of an ovarian cell. The combination of hormone and receptor activates adenylate cyclase, which causes the conversion of ATP to cyclic AMP within the cytoplasm. Cyclic AMP, called the second messenger (because the original stimulating hormone is the first messenger), initiates cellular activities that lead to estrogen release.

hormone, exist in the breast, the uterus, the vagina, the pituitary gland, and the hypothalamus of the brain. When estrogen arrives at these sites, it immediately drifts across the membranes of the cells that contain the receptors, and then the receptor-estrogen combination is transported to the nucleus. On the way, the receptor-estrogen molecule undergoes a transformation and becomes a smaller molecule; this changed combination enters the nucleus and activates the transcription of a specific section of DNA. The coded message is transmitted to the cytoplasm and leads to the synthesis of proteins that initiate specific cellular functions (see Fig. 3.3).

RADIOIMMUNOASSAY AND RADIORECEPTOR ASSAY TECHNIQUES

Body cells have receptors not only for protein and steroid hormones, but also possess specific binding sites for a wide variety of natural and introduced substances that may be present in the body fluids. For example, lymphocytes, a type of white blood cell, are known to have surface receptors for viruses, bacterial toxins, histamine, nerve impulse transmitters, various cell-to-cell communication products affecting growth, cell division, or self-recognition, as well as hormone recep-

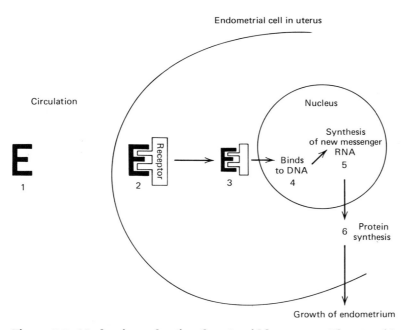

Figure 3.3. Mechanism of action for steroid hormones. The steroid estrogen (1) diffuses across the cell membrane and binds with a cytoplasmic receptor (2). In transit to the nucleus, the estrogen-receptor complex becomes a smaller molecule (3) that enters the nucleus and binds to nuclear DNA (4), activating specific genes to form messenger RNA (5). The mRNA passes into the cytoplasm to result in protein synthesis.

tor sites. It is now possible to employ the binding capacity of such cell surface receptors in order to measure incredibly small concentrations of the substances they bind. The radioimmunoassay, introduced in 1960 by Yalow and Berson (for which Rosalyn S. Yalow received the Nobel Prize in 1977), is based on the technique in which radioactive or "labeled" molecules of a substance and nonlabeled molecules of the same substance compete for a limited number of sites on a receptor protein specific for the substance. The development of this method has made obsolete the hormone analyses that had formerly been made by biological (animal) techniques, in which the hormone was injected into a test animal and the amount of response was noted. Bioassays are relatively reliable, but time consuming and expensive to perform since they require a day or more wait and the destruction and dissection of the animals. Today, it is possible to measure routinely (but not necessarily inexpensively) levels of hormones in billionth of a gram (nanogram), trillionth of a gram (picogram), or even, in the case of estrogen receptor assays, quadrillionths (femtomols) bound per gram.

A radioimmunoassay (RIA) uses the binding sites on immune bodies or antibodies as the binding protein reagent. When the binding protein is not part of the immune system but is a structural component of a cell, the technique is actually a radioreceptor assay (RRA). RIA has virtually become generic, however, for any assay technique that uses radioactive agents that bind to a protein, regardless of whether antibodies or cell receptors are used.

To measure the concentration of a pituitary hormone like LH in the blood, for example, a known quantity of (1) LH antibodies (produced by injecting the antigen LH into animals to cause antibody formation) is mixed with (2) the blood sample containing an unknown amount of LH and (3) a known amount of purified LH that has been labeled with a radioisotope like ^{125}I. The mixture of the three substances is then incubated until the reactions are presumed to have reached equilibrium. Since both the labeled and unlabeled LH in the mixture *compete* for the fixed number of active binding sites on the antibodies, the amount of labeled LH antibody complex formed, as determined by counting the emissions in a scintillation counter, is a function of the LH concentration in the blood sample. That is, the more radioactive LH that was bound, the less natural LH there must have been in the sample. Conversely, if just a small amount of radioactive LH has been bound, there must have been more natural hormone present (see Fig. 3.4). The concentration of LH in the sample can be read directly from a previously constructed standard curve.

Control of normal or abnormal levels of hypothalamic hormones, pituitary gonadotropins, and ovarian steroids can be studied by RIA techniques, and the method has led to a more complete understanding of the endocrine events associated with the menstrual cycle.

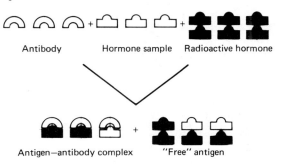

Figure 3.4. The radioimmunoassay technique. A fixed amount of antibody is mixed with an unknown amount of hormone and a known amount of radioactive hormone. The unlabeled nonradioactive hormone competes with the radioactive hormone for attachment to the antibody binding sites. (Sloane, E., Reprinted with permission from The Science Teacher, a publication of the National Science Teachers Association.)

OTHER CHEMICAL MESSENGERS—PROSTAGLANDINS

Prostaglandins are a closely related group of fatty acid derivatives with a variety of effects. They include prostaglandin (PG) D_2, E_2, F_{2a}, thromboxane, prostacyclin, and the leukotrienes. Almost all cells of the body can synthesize prostaglandins by oxygenation of arachidonic acid, a 20-carbon polyunsaturated fatty acid commonly found in food, but the particular type and activity of the prostaglandin formed depends upon the nature and function of the cell producing it. Although they have regulatory effects and are sometimes classified as hormones, prostaglandins are not technically hormones because they are produced by all tissues rather than by special glands. Neither are they trans- ported like hormones in the bloodstream to target cells; most prostaglandins are immediately deactivated by enzymes once they get into the circulation so that they affect cells only a short distance away from their site of production.

These compounds, originally discovered in human semen, were named "prostaglandins" in 1935 by Nobel laureate Von Euler because small amounts were found in the prostate gland. Originally recognized for their potent effects on uterine muscle contraction, subsequent prostaglandin research has now determined their great biological importance as well as their chemical structure and the mechanisms of biosynthesis. Figure 3.5 is a schematic diagram of their formation. There are two main pathways: one, catalyzed by the enzyme cyclooxygenase and leading to pros-

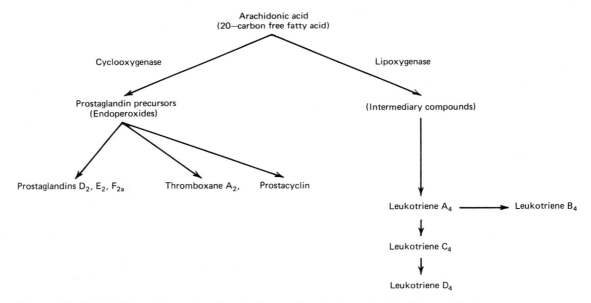

Figure 3.5. Metabolic pathways leading to the synthesis of prostaglandins and leukotrienes. Prostaglandin inhibitors block the enzyme cyclooxgenase and prevent the conversion of arachidonic acid to endoperoxides. The leukotrienes are involved in inflammation and possible in bronchial asthma. The anti-inflammatory effects of corticosteroids are believed to be due to their inhibition of leukotriene synthesis.

taglandin, prostacyclin, and thromboxanes, and the other using the enzyme lipoxygenase to eventually form the leukotrienes. Aspirin and similar anti-inflammatory drugs inhibit the cyclooxygenase enzyme, and a number of other compounds are now known to block synthesis or antagonize the actions of prostaglandins.

The ubiquitous prostaglandins have amazingly diverse physiological effects, ranging from regulatory control in a number of body organs to more general activities associated with pathology, response to trauma, and body defense mechanisms. They may be involved in the maintenance of blood pressure and the rate of blood flow. F-prostaglandins and thromboxane stimulate blood platelet aggregation, necessary for initiation of the clotting mechanism, and cause constriction of tiny blood vessels. Prostacylin and the E-prostaglandins, produced by the cells lining blood vessels, inhibit platelet aggregation and cause vasodilation. Leukotrienes are produced by cells of the immune system and participate in allergic reactions. Prostaglandins of the E-group inhibit gastric secretion and may be protective against ulcer development. High levels of PGs have been shown to be associated with pain and inflammation and possibly with fever. In low concentrations, PGs inhibit transmission of impulses in the nervous system. The earliest work on the nature of PGs, however, was concerned with their effects on the reproductive system. Primarily deriving from their ability to stimulate smooth muscle contraction, prostaglandins are closely associated with reproduction. They may be involved in ejaculation in men, and in facilitation of sperm transport in the uterus and fallopian tubes after ejaculation has occurred. They are now known to be responsible for dysmenorrhea, or menstrual cramps, and play a role in "ripening" of the cervix at the end of pregnancy, as well as in labor and delivery. PGs may be responsible for determin-ing the life span of the corpus luteum and possibly affect hypothalamic and pituitary hormones that trigger ovulation. They are used clinically to induce labor and to cause abortion in the second 3 months of pregnancy. At the cellular level, prostaglandins can cause both stimulation and inhibition of adenylate cyclase, leading to greater or lesser formation of cyclic AMP and suggesting the mechanism for the molecular action of PGs.

The Pituitary Gland

The pituitary gland, oval and about the size of a pea, is attached to the hypothalamus of the brain by a stalk called the infundibulum. The pituitary rests in a depression of the sphenoid bone of the skull and is protected by the bone and by the same tough connective tissue that covers the brain. The anterior part of the pituitary gland (i.e., the section closer to the face) is called the *anterior lobe*, or *adenohypophysis*; the posterior part is the *posterior lobe* or *neurohypophysis*. Each of these parts has a different embryological origin. The adenohypophysis arises from an upgrowth of the roof of the mouth, while the neurohypophysis develops from a downgrowth of the brain. This is why, under the microscope, the neurohypophysis looks like nervous tissue and the adenohypophysis has the typical cell cords and clumps of an endocrine organ (see Fig. 3.6).

Hormones of the Pituitary Gland

At the present time, it is known that there are eight hormones synthesized by the cells of the anterior pituitary gland:

1. Growth hormone (GH), or somatotropin, which controls the growth of all the cells of the body capable of growth, resulting in an increase in the numbers of cells and in enlargement of existing cells.

2. Thyrotropin, or thyroid-stimulating hormone (TSH), which controls thyroid gland activity.

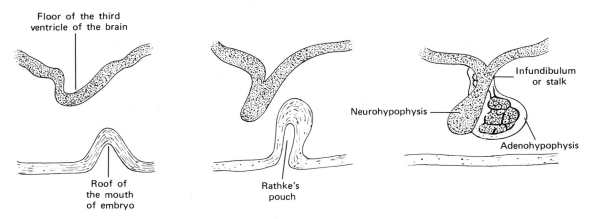

Figure 3.6. Development of the pituitary gland. The neurohypophysis, or part of the gland that arises from brain tissue, never loses its connection and remains attached by the stalk or infundibulum. The portion of the gland that forms from the roof of the primitive mouth of the embryo becomes the adenohypophysis.

3. Adrenocorticotropic hormone, or adrenocorticotropin (ACTH), which is responsible for the activity of the cortex portion of the adrenal glands.

4. Melanocyte-stimulating hormone (MSH), which stimulates pigment formation and dispersal in the pigment-producing cells of the epidermis.

5. β Lipotropic hormone (β-LPH, Lipotropin), which is found in both pituitary and brain (primarily hypothalamic) tissue. Originally believed to be a hormone in search of a function, it is now believed to be a prohormone, since it is known to contain a number of biologically active fragments, the alpha, beta, and gamma endorphins, that have natural pain-killing abilities. The discovery of endorphins, called endogenous opiates because they are innate in the body and mimic the effect of heroin and morphine, created as much excitement among neuroendocrinologists as did the discovery of prostaglandins. Investigations have linked endorphins not only with natural pain-relief, but also with obesity, epilepsy, drug addiction, schizophrenia, the explanation of the effects of hypnosis and acupuncture, and jogger's high or runner's second wind.

6. Follicle-stimulating hormone (FSH), which stimulates the growth and development of the primary follicles and results in hormone production in ovaries and sperm production in testes.

7. Luteinizing hormone (LH), responsible for ovulation, corpus luteum formation, and hormone production in the ovaries, and is the stimulus for hormone production from the interstitial cells of the testes.

8. Prolactin, which is responsible for the initiation and sustaining of milk production from the breasts when they have been previously readied for lactation by the action of other hormones.

The last three hormones, FSH, LH, and prolactin, are called gonadotropins because they regulate the activities of the organs of reproduction. FSH and LH are glycoproteins with a carbohydrate content varying from 13% to

31%. Prolactin has no known carbohydrate content and consists of a single polypeptide chain with a molecular weight of approximately 30,000.

There are two hormones that are synthesized by nerve cells in the hypothalamus and stored in the posterior pituitary gland:

1. Antidiuretic hormone (ADH), also called vasopressin, which affects the excretion of water by the kidneys and increases blood pressure.

2. Oxytocin, a very powerful stimulant of uterine contraction, especially of a pregnant uterus. Oxytocin is believed to participate in uterine contractions during the process of delivery, although the actual mechanism of the initiation of labor is not known. Oxytocin also affects the flow of milk from the breasts in a nursing mother in response to the sucking stimulus from the baby. To a lesser extent than vasopressin, oxytocin affects blood pressure.

Relationship Between the Pituitary Gland and the Hypothalamus

In the basal region of the brain, located underneath the cerebral hemispheres, is the hypothalamus, a part of the brain that consists of neuron cell bodies, fibers, and supporting tissue that make up the floor and part of the side walls of the third ventricle. The hypothalamus is a key portion of a group of brain structures collectively called the *limbic system*, which is believed to be the part of the brain concerned with emotional behavior—fear, anger, feelings of depression and elation, sexual desires, and feelings of reward and pleasure or punishment and pain. The different areas of the limbic system perform different functions, but it is overall thought to be responsible for our emotional and behavioral patterns. The hypothalamus is connected by nerve fiber tracts to all parts of the limbic sys-

tem, which is in turn connected to practically every other section of the brain. The hypothalamus, therefore, can be regarded as the pathway through which not only neural inputs from the limbic system, but also from the higher brain centers in the cerebrum, can influence and control many major functions of the body.

Located in the hypothalamus are groups of nerve cells that control the involuntary activities of the body that are necessary for life, such as regulation of appetite and satiation, body water, body temperature, blood pressure, and heart rate. Moreover, the hypothalamus controls all pituitary gland secretion. Years ago, the pituitary gland was always called the master gland of the body because its hormones were responsible for the activities of many other endocrine glands. The view now is that part of the brain itself—that is, the hypothalamus—is the master gland, providing the integration of the nervous and the endocrine systems. Neurons in the hypothalamus produce *neurosecretions*, which are themselves the hormones, or which cause the pituitary to release its tropic hormones (that is, those that stimulate the growth and function of other endocrine glands). So the pituitary is really the servant of the hypothalamus.

But the question of who is the real boss is complicated by the evidence that regions of the brain outside of the hypothalamus contain peptide hormones. ACTH, MSH, β-LPH, and β-endorphin are apparently all derived from a large precursor protein molecule present not only in the pituitary and hypothalamus but, to a lesser extent, in the rest of the brain, including the limbic areas, the midbrain, cortex, and cerebellum. Evidently, the cells of the pituitary gland and the brain are able to process and split up the same precursor molecule differently in order to produce the different component hormones, each with different activities. Moreover, hormones formerly believed to be secreted only by endocrine cells, like insulin from the pancreas or cholecystokinin and gastrin from the secretory cells of the intestine, have also been discovered in the brain. What pituitary and gut hormones are doing in the brain, and why the brain, in addition to all its other functions, appears to be another endocrine organ, are still not completely known. Some researchers think brain hormones may act as local neuron-to-neuron communicators or as modulators of signals from other transmitters, thus affecting behavioral responses.

Hypothalamus to Posterior Pituitary

The influence of the hypothalamus on posterior pituitary lobe secretions is very direct and evident. Nerve fibers from two groups of neuron cell bodies in the anterior wall of the hypothalamus, the *supraoptic* and the *paraventricular nuclei*, extend down the infundibular stalk and terminate in little bulbous endings in the substance of the neurohypophysis. ADH and oxytocin are secreted by these supraoptic and paraventricular neurons, and these substances pass down their long axon fibers to be stored in the cells of the neurohypophysis. These hormones, therefore, are secretions of the hypothalamic neurons, not of the posterior lobe, and are released into the bloodstream on nerve signal from the hypothalamus.

Hypothalamus to Anterior Pituitary

In contrast to the direct nervous connection of the hypothalamus with the posterior pituitary gland, the message for control of hormone release by the anterior pituitary arrives from the brain in a rather indirect way. It involves a series of blood vessels called the *hypothalamic-hypophyseal portal system*, which consists of two capillary beds, one in the hypothalamus and one in the adenohypophysis, and the veins that connect them. The neurons that secrete the neurosecretory *releasing* and *inhibiting* factors are roughly in the same regions as the supraoptic and paraventricular nuclei; that is, in the medial and lateral parts of the basal hypothalamus. But the fibers of these neurons are not long, and they do not extend down the infundibular stalk. Instead, the axons are short, and they terminate on nearby loops of a capillary bed within the hypothalamus. The releasing and inhibiting factors, actually hormones, are deposited in these capillaries, whose venules drain down the infundibulum and empty into another capillary bed in the anterior lobe of the pituitary gland. The neurosecretions produced by the hypothalamic neurons never get into the general blood circulation, but go first to the anterior pituitary and cause or inhibit the release, and possible the synthesis, of the tropic hormones of the gland. Figure 3.7 shows the relationship of neurosecretory cells in the hypothalamus to the anterior and posterior pituitary glands.

The hypothalamus is known to produce six, and possibly more, releasing or inhibiting factors or hormones regulating pituitary gland secretion. (The secreted substance is generally called a factor rather than a hormone until its chemical structure and its function are completely known.) They include corticotropin releasing factor (CRF), prolactin releasing factor (PRF) (definite in birds, still questioned in humans), prolactin inhibiting factor (PIF) (may actually be a neurohormone related to adrenaline), and growth hormone releasing hormone (GHRH). Thyrotropin releasing hormone (TRH) stimulates the release of thyrotropin and possibly prolactin, and somatostatin has been recognized as a hormone that inhibits growth. Although there has been some evidence for separate factors affecting FSH and LH release from the anterior pituitary gland, it is generally accepted at this time that a single releasing hormone, called luteinizing hormone releasing hormone (LHRH) or gonadotropin releasing hormone (GnRH), causes the release of both FSH and LH.

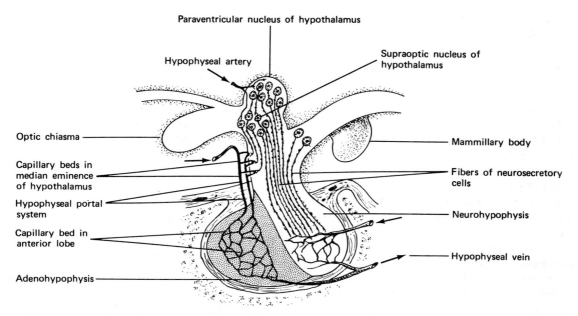

Figure 3.7. The anatomical relationship between the hypothalamic neurons, the hypothalamic-hypophyseal portal system, and the pituitary gland.

What Causes the Release of the Releasing Hormones?

The monthly rhythmic functioning of the female reproductive cycle is totally dependent on the changing concentrations of the steroid hormones estrogen and progesterone, which are secreted in response to the changing concentrations of the pituitary gonadotropins FSH and LH, which are in turn secreted in response to hypothalamic LHRH which, it could reasonably be assumed, also cyclically varies in concentration. This highly complex and intricate relationship is based on both positive and negative feedback mechanisms. For a long time, it has been known that variations in the release of FSH and LH were the result of feedback from gonadal steroids, but just where the site of feedback effects occur— whether at the hypothalamic level, the pituitary level, or both—has been extensively investigated with inconclusive results.

Direct evidence of a hypothalamic site could be obtained by the kinds of experiments that involve destroying the hypothalamic areas where LHRH is produced, cutting the connection between the hypothalamus and the pituitary, implanting steroids directly in the hypothalamus to assess the effects, or sampling the LHRH levels in the portal blood vessels. Obviously, these studies cannot be performed in women, and most research has taken place in female monkeys.

To date, there is no universal acceptance among researchers that a reciprocal feedback relationship does or does not exist between the ovarian steroids and the hypothalamus. It is known that LHRH is secreted in a rhythmic pulsatile fashion, and that FSH and LH are also secreted in rapid, rhythmic pulses that occur every 60–90 minutes throughout most of the cycle, decreasing in frequency closer to menstruation. It is also known that estrogen and progesterone cause variations in the frequency and amplitude of FSH and LH pulses of secretion. The current view on the sites of feedback regulation is that, in primates, the main target of negative and positive feedback effects of estrogen and progesterone is the anterior pituitary, but that these gonadal steroids are able to modulate the pulsatile pattern of LHRH secretion, thus influencing the LHRH signal.

Psychic Effects on Menstruation

Apparently, the releasing hormone can be affected by other internal and external factors. How emotional distress, reaction to a new and strange situation, or concern over a possible unwanted pregnancy can translate itself into a hormonal effect is not clear, but most women are aware that such things exist since the visible effect is delay or cessation of menstrual periods. The clue may be the relationship of the hypothalamus to the entire limbic system of the brain. The amygdala, that part of the limbic system that receives signals from all parts of the cerebral cortex and transmits those signals not only back to the cerebral cortex but also especially to the hypothalamus, is thought to be the monitor of all emotional stimuli that control the overall patterns of behavior. In the incidence of stopped or missed periods that are not a result of pregnancy, it can be theorized that emotional stress may be mediated through the amygdala to suppress LHRH in the hypothalamus. If LHRH is inhibited, LH from the anterior pituitary is not released, ovulation from the ovary does not occur, and menstruation may not take place.

Another explanation for the effect of stress on missed or absent periods may be associated with the influence of endogenous opiates on FSH and LH secretion. There is evidence that beta-endorphins may inhibit gonadotropin release, either by a direct effect on the anterior pituitary or, indirectly, by affecting neurotransmitters in the brain (Quigley and Yen, 1980; Robert et al., 1981; Vrbicky et al., 1982). Since the levels of endogenous opiates have been observed to increase with stress, the possibility of an endorphin effect exists. Endorphins are also known to increase with exercise, and Speroff (1981) has suggested that the rise in endogenous opiate levels may be implicated in the amenorrhea found in women runners.

STEROID HORMONES

Chemical Composition

All steroid hormones in the body are produced by the ovaries, the testes, the cortex of the adrenal glands, and the placenta during pregnancy. Steroid hormones are usually divided into four groups: the estrogens, the androgens, progesterone, and the corticosteroids. The first three are called the sex steroids because they are responsible for the physical and physiological differences that exist between males and females. Estrogens and progesterone are the female sex hormones, and androgens are the male sex hormones.

Steroid is a general term applied to a group of substances that all have a common structural nucleus. Steroids are found in both plants and animals and include a large number of body constituents, vitamins, and drugs, as well as sex hormones. The chemical nucleus they all have in common is called the *cyclopentanoperhydrophenanthrene ring*, which is really not as intimidating as it looks and sounds. Three rings have six carbon atoms each and make up the phenanthrene part; one has five carbon atoms and is the cyclopentane. Conventionally, the rings are designated A through D, and the carbon atoms are numbered C_1 through C_{17}, as shown.

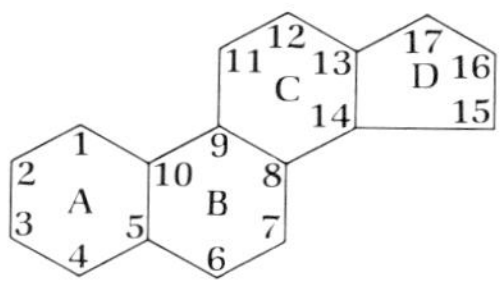

When there is an 18th carbon atom in the form of a methyl group attached at C_{13}, the resulting nucleus is called *estrane* and is the source of all the natural estrogens.

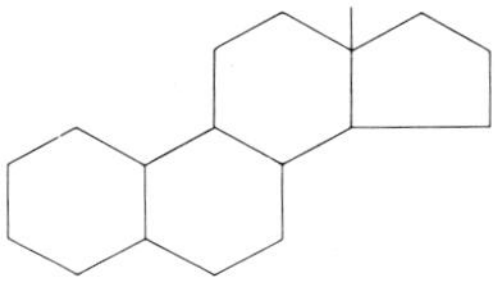

Estrane

All androgens derive from the *androstane* nucleus, which has a 19th carbon atom attached at C_{10}.

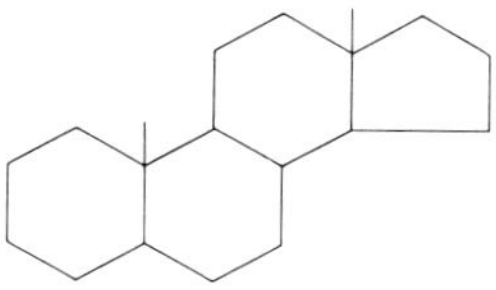

Androstane

Steroids that have 21 carbon atoms are known as the *pregnane* nucleus and give rise to progesterone, its derivatives, and the corticosteroids.

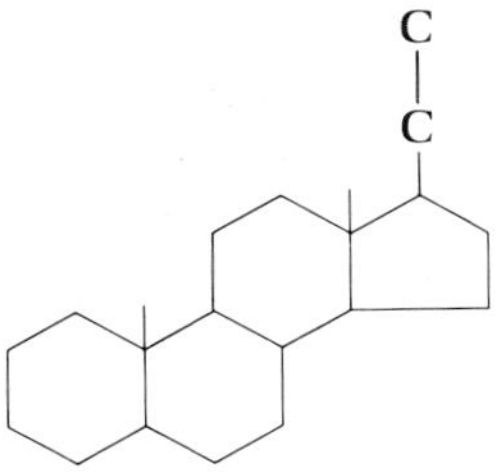

Pregnane

Synthesis of the Steroid Hormones

The basic building block for the synthesis of steroids is *cholesterol*, a 27-carbon compound with the steroid nucleus. Cholesterol has negative associations for most people in relation to its link to hardening of the arteries and cardiovascular disease, and many people are aware of the implications of a high plasma cholesterol level. Although high dietary intake of cholesterol-containing foods can lead to elevated blood cholesterol levels, cholesterol is formed endogenously in all the cells of the body because it is part of their membrane structures, and it is possible that cell synthesis of cholesterol is inversely related to dietary consumption. All the endocrine tissues that produce sex hormones assemble them bit by bit through a biosynthetic pathway, starting with cholesterol.

The ovaries produce all three types of sex steroids: the estrogens, progesterone, and the androgens. The follicular structures (the theca externa and the granulosa cells) tend preferentially to produce estrogens, the corpus luteum cells produce progesterone, and the ovarian stroma or connective tissue, the androgens. Androgens are produced in females by both the ovaries, and to a greater extent, by the adrenal glands. In the ovaries some androgens are precursors in the pathway to the synthesis of estrogens. In the male, the interstitial cells between the seminiferous tubules of the testes produce androgen, but some estrogen is synthesized by the testes and the adrenals.

The pathways for the synthesis of all four types of steroids are shown in Table 3.1. Cholesterol gives rise to pregnenolone, a compound with little known biological activity, that gives rise to progesterone, androgens, and estrogens. The route for the synthesis of estrogens diverges from that of progesterone at pregnenolone. All the steps require energy in the form of ATP and involve one or more

**Table 3.1 THE SYNTHESIS OF ESTROGENS, PROGESTINS, ANDROGENS,
AND CORTICOSTEROIDS FROM CHOLESTEROL**

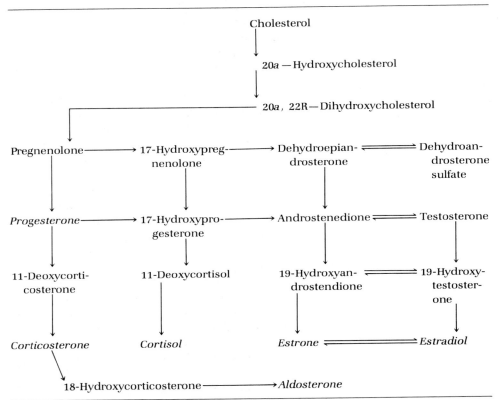

enzymes located in the mitochondria of the cells. The reactions that result in the formation of corticosteroids do not take place in gonadal tissue, but only in the adrenal glands.

Estrogens

Estrogen is a general term. It is used for all those substances that produce the biological effects characteristic of estrogenic hormones. Although there are many naturally occurring estrogens in the body, the three, in order of greatest potency, are *estradiol*, *estrone*, and *estriol*. There are two other types of estrogens that are administered clinically: the conju-

gated estrogens, derived from the urine of pregnant mares, and the synthetic estrogens like diethylstilbestrol (DES), which do not even have a steroid structure but still have an estrogenic effect.

Physiological Effects of Estrogens

Estrogens cause the growth of a thick, folded vaginal epithelium, differentiated into layers, and resistant to trauma and infections. The cells accumulate glycogen, which is acted on by Döderlein's bacilli to produce the acidic vaginal secretions. As a result of estrogen stimulation, cervical glands produce a clear, watery secretion with little mucus and low viscosity, which facilitates sperm passage into the uterus. In addition, the uterine endometrium is stimulated to grow, resulting in both an increase in the numbers of cells (hyperplasia) and an increase in the size of existing cells (hypertrophy). The muscle of the uterus shows an increased tendency to contract. Furthermore, estrogens cause the lining of the fallopian tubes to increase in thickness and there is an enhanced ability of the cilia to beat toward the uterus. In the breasts, estrogen causes growth of the duct tissue and increases fat deposition, thereby enlarging breast size.

Estrogen causes an increase in bone formation. In girls, the growth spurt at puberty is enhanced by it, but estrogen also causes the closing of the epiphyseal cartilages at the ends of the long bones so that cessation of growth occurs. Estrogen influences total body configuration, not only in the skeleton, but also in the increased deposition of fat in all subcutaneous tissue, particularly in the buttocks, thighs, and breasts.

The strength of capillary walls is increased by estrogen, and when estrogen levels are low before and during menstruation, there may be a greater tendency to bruise and have nosebleeds.

Estrogen causes a reduction in blood cholesterol levels and a decrease in the cholesterol-rich fraction of the plasma lipoproteins known as low-density lipoproteins (LDLs).

Liver cells are stimulated by estrogen to produce greater amounts of some blood proteins, particularly those involved in blood clotting, red blood cell formation, and certain of the globulins that bind to some hormones to transport them through the bloodstream.

The steroid-sensitive areas of the hypothalamus are affected by estrogen so that the production of releasing hormone is enhanced or inhibited. The temperature-regulating and vasomotor centers of the hyothalamus controlling the nerves that cause the dilation and constriction of blood vessels are also affected, and the basal body temperature is lowered during the first, or estrogen, half of the menstrual cycle.

Progestins

Progestins is a general term referring to chemical agents, both natural and synthetic, that produce changes in the uterine endometrium after it has previously been primed by estrogen. In vertebrates, the naturally occurring progestin is progesterone, and in women it produces the characteristic secretory and glandular changes in the endometrium after ovulation has occurred. Progesterone is used synonymously with progestin; another term is progestogen.

Physiological Effects of Progesterone
Progesterone prepares the endometrial lining of the uterus for the implantation of a fertilized ovum. To this end, it also inhibits uterine contractions so that an implanted ovum is retained. It also reduces motility in the fallopian tubes. Progesterone causes an increase in the glandular elements of the breasts, but the actual production of milk by the mammary glands is a function of the influence of prolactin after the breasts have been prepared for lactation by estrogen and progesterone. Progesterone causes cervical mucus to become viscous, which tends to prevent the passage of sperm through the cervical os. It also causes a slight rise in basal body temperature.

Progesterone causes an increase in the excretion of water and sodium from the kidneys. This makes it unlikely that progesterone alone is responsible for the water retention and swelling that occurs in many women in the second, or progesterone, half of the menstrual cycle.

Androgens

Androgen is a term generally used synonymously with male sex hormone, but functionally it means any compound that has certain masculinizing effects. The androgen produced by the interstitial cells of the male testes is testosterone, but the adrenal glands produce at least five other androgens.

In the same way that estrogens are responsible in women for those physical and physiological features that distinguish femaleness, androgens are responsible for those features in men that are characteristically male.

Physiological Effects of Androgens

The testosterone secreted by the fetal testes during prenatal development causes male differentiation of the embryonic reproductive tract.

After puberty, testosterone produced by the testes causes an increase in the growth and development of the male genitalia. It is responsible for the distribution of hair in the male pattern—increased on the body and decreased on the top of the head. Testosterone causes enlargement of the larynx and an increase in the length and thickness of the vocal cords, thereby causing a deeper voice. The thickness and texture of the skin is increased, and the skin tone is darkened. Another skin effect is an increase in sebaceous gland secretion, and testosterone is believed to be involved in acne in both males and females. The male sex hormone causes an increase in muscle mass and an increase in the size and strength of the skeleton in general, resulting in the larger body configuration of the male. It causes an increase in the rate of metabolism, heightening activity of all body cells, and results in greater numbers of red blood cells, thus resulting in a greater oxygen capacity in males. Both the ovaries and the adrenals in females produce testosterone, and various precursors to testosterone are produced by those glands and can be metabolized to testosterone by other organs. When plasma levels of testosterone approach the

male range—a situation that can result from diseases of the ovary or adrenals—there are masculinizing effects on females. These can include excessive development and distribution of hair, voice and skin changes, and changes in fat distribution.

PUBERTY

Puberty is the transition period between childhood and adulthood when physical and psychological changes that are associated with the ability to reproduce take place. *Menarche* is the term for the onset of the menstrual periods, but ovulation generally does not take place for a year or more afterwards. Menarche, then, is not the indication of full sexual maturity.

The physical changes associated with puberty will be described later. The endocrine events that control the onset of puberty have been extensively investigated, and there are several hypotheses that attempt to explain it. Of course, attention focuses on the hypothalamus, but the onset of puberty is influenced by genetic factors, social and economic factors relating to nutrition and general health, and possibly (although research is scant) on external environmental factors mediated through the limbic system and the hypothalamus.

The hypothalamus controls more than gonadal function; it also regulates many body activities and a variety of behavioral functions. It may be that all of these activities are integrated with each other to provide a combination of events in addition to the onset of the hypothalamic–hypophyseal–ovarian function that controls the onset of sexual maturity.

The most popular theory of hypothalamic control of puberty suggests that the immature hypothalamus in childhood is extremely sensitive to the relatively small amounts of circulating estrogen produced by the ovaries. LHRH is inhibited through negative feedback control by these low levels of estrogen, and FSH and LH are not produced in appreciable quantities by the anterior pituitary. Then, for some unknown reason, the hypothalamus matures, its sensitivity to steroids is diminished, the releasing hormone causes FSH and LH to be produced by the pituitary, and the subsequent increase in ovarian steroid production results in puberty and menarche.

HORMONES AND THE MONTHLY CYCLES

The menstrual cycle involves the rhythmic fluctuations of the hormones of the hypothalamus, anterior pituitary, and ovaries, and the morphological changes that occur in the ovaries and the endometrium of the uterus. The endometrium is the mirror of the ovaries; whatever is going on in the uterus during the cycle is precisely correlated with whatever is occurring in the ovaries. The purpose of the ovarian cycle is to produce an ovum; the purpose of the endometrial cycle is to prepare a haven to nourish and maintain that ovum should it become fertilized. The ovarian cycle can be divided into three phases: the follicular phase, ovulation, and the luteal phase. The endometrial cycle can be divided into the menstrual and proliferative phases (which corresponds to the follicular phase in the ovaries) and into the secretory or progestational phase, which is synchronized with the luteal phase of the ovaries (see Fig. 3.8).

All that most studies have indicated about the length of the menstrual cycle is its extreme variability. For purposes of description, a 28-day cycle is generally used, with ovulation occurring on the fourteenth day before menstruation.

Ovarian Cycle

Follicular Phase

The primary follicles in the ovary contain the stored oocytes, arrested in the prophase stage of meiosis with all 48 chromosomes still enclosed by a nuclear membrane. As indicated in Chapter 2, throughout the reproductive life of a woman, and without any help from FSH and LH, every day a small number of primary follicles begin to recommence growth. Most of them will undergo atresia, but what happens to a group of developing follicles at any point in time depends on whether adequate levels of FSH and LH in the circulation coincide with the development of receptors for FSH and LH on the granulosa and theca cells of the follicles. At the beginning of the ovarian cycle, the cells of the anterior pituitary respond to the signal of LHRH from the hypothalamus by secreting FSH and LH. The best developed follicles, that is, those with enough granulosa cells, develop receptors for estrogen and FSH on the cells of the granulosa layers and LH receptors on the theca cells. The initial role of FSH is to induce an increase in the development of its own receptors on the granulosa cells so that they can produce estrogen. The initial role of LH is to stimulate theca cell production of androgens, which are then converted to estrogen by the granulosa layers.

The increasing levels of estrogen tend to suppress FSH because of the negative feedback effect. The decline in FSH tends to inhibit the further development of the follicles, but by then a dominant follicle has been selected. This follicle was chosen for ovulation because its rate of granulosa proliferation exceeded that of all the others, and it has the capacity for continued growth despite the decreased levels of FSH because its greater number of granulosa cells necessarily give it greater FSH and estrogen receptor content.

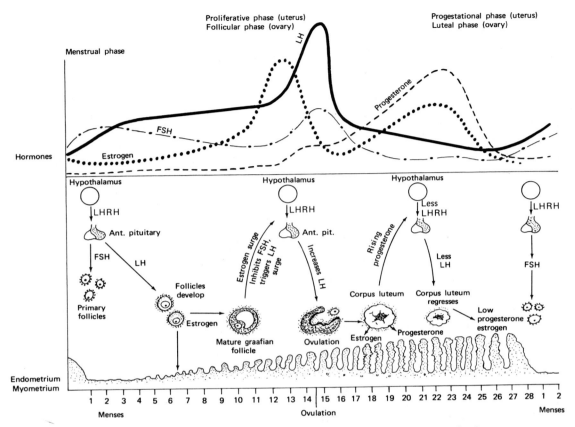

Figure 3.8. The hormonal levels and events of the ovarian and endometrial cycles in the absence of fertilization.

The dominant follicle produces more estrogen than the combined efforts of all the other developing follicles. Its estrogen, in combination with FSH, induces the development of more LH receptors on the outermost granulosa layers. These LH-binding sites are critical for ovulation and for subsequent conversion of the follicle to the corpus luteum.

It has recently been suggested that increased levels of estrogen may not be the only inhibitors of FSH secretion during the follicular phase of the cycle. There is evidence that the granulosa cells secrete a peptide called

inhibin or folliculostatin, which is apparently capable of specifically suppressing FSH from the pituitary gland. One hypothesis for its function is that early in the cycle, the inhibin produced by all of the newly developed follicles limits FSH release from the pituitary and sets the stage for one follicle—the one with the slight edge on development—to emerge as the dominant follicle. The dominant follicle maintains its superiority and continues to flourish despite the initial inhibition of FSH by inhibin and the later negative feedback effect on FSH by the increasing amounts of estrogen.

Ovulation

Despite the waning FSH levels, the accelerated production of estrogen from the dominant follicle continues. Although low levels of estrogen cause a decrease in LH, high levels of estrogen result in a positive stimulatory effect and cause an increase in LH. As estrogen levels increase during the midfollicular phase, they ultimately attain the critical blood level of approximately 200 picograms/ml that is maintained for up to 50 hours. This surge of estrogen affects the pituitary and hypothalamus and is followed by a peak burst of LH, which causes ovulation. The LH surge causes a number of changes in the follicle that has been selected for rupture. First, the nuclear membrane around the oocyte breaks down, the chromosomes progress through the rest of the first meiotic division, and the egg moves on to the secondary oocyte stage. At this point, for some unknown reason, meiosis stops again and will continue only if the ovum is fertilized. LH also results in the luteinization of the granulosa cells, and they begin to produce progesterone. Next, the high levels of LH are believed to cause synthesis of prostaglandins, presumably essential for follicle rupture. It is generally believed that the estrogen surge produces the LH surge and also a smaller peak of FSH by triggering LHRH from the hypothalamus.

Reportedly, ovulation occurs within 24–38 hours after the onset of the LH surge. The variation is believed to be the result of differences in the ways in which luteinizing hormone has been measured. Further research using standardized techniques and time intervals should be able to pinpoint the amount of time from LH surge to LH peak to ovulation with greater precision. The ovum is expelled, to be sucked up by the ciliated, fingerlike fimbriae of the fallopian tube. The granulosa cells of the ruptured follicle enlarge, undergo "luteinization," and become the corpus luteum.

The corpus luteum produces progesterone and estrogen and reaches a peak of activity 8 days after ovulation. If there is a fertilized ovum, it is implanting in the endometrium of the uterus at about that time. If fertilization does not occur and there are no hormones from the developing embryo to increase its life span, the corpus luteum regresses. After 14–16 days, its life is over.

In the human, it is not yet known what causes the demise of the corpus luteum, or even precisely what sustains it before it declines. During the luteal phase, the FSH and LH in the blood are at their lowest levels at the same time estrogen and progesterone levels are at their highest, which would appear to be the consequence of negative feedback effects of the steroids on gonadotropin secretion. Although FSH and LH levels are low, however, they do not completely disappear, and there is always the continuous presence of small amounts of LH throughout the cycle. There is some evidence that the estrogen produced by the corpus luteum hastens its decline and that local concentrations of prostaglandin may be involved. Some researchers believe that the corpus luteum is autonomous and has an inherent life span of 2 weeks.

With the decline of the corpus luteum, the levels of estrogen and progesterone decrease rapidly, their negative feedback effect is diminished, and the FSH and LH can again increase to initiate a new cycle.

Endometrial Cycle

The endometrium consists of two layers: a superficial one that contains glands and occupies two-thirds of the endometrium, and a thin basal layer. The superficial layer is the one that responds to steroids, changes to a great extent during the cycle, and is almost completely lost during menstruation. The basal layer does not change in character during the cycle.

From days one to four, the endometrium is in the *menstrual phase*. After day four until a day or two after ovulation, the endometrium is in the follicular, proliferative, or regenerative phase. Even while menstruation is occurring, estrogen from the developing follicles in the ovaries causes cell division that repairs the denuded areas in the uterine lining. Regeneration of the sloughed-off superficial layer occurs rapidly. Within a few days of menstruation, the entire uterine cavity is covered with new epithelial cells. The gland cells of the superficial layer proliferate rapidly but do not accumulate much secretion at this time.

The second half of the endometrial cycle is called the *progestational phase* because the changes that occur in the superficial layer of the endometrium are the result of the action of progesterone. It is also called the *secretory phase* because the glands of the endometrium become dilated as they fill up with secretions of substances like glycogen and fats. The endometrium becomes twice as thick as it was in the previous phase, cushiony and nutritive, thereby forming a hospitable site for implantation of a fertilized ovum.

Toward the end of the progestational phase, if pregnancy has not occurred, the endometrium begins to regress. It is no longer supported by the high levels of progesterone and estrogen that are now undergoing rapid decline. The breakdown of the superficial layer of the endometrium is caused by lack of estrogen and progesterone stimulation, but the actual bleeding has to do with the special type of arteries present in the endometrium. The endometrium of the human female and the uteri of other female primates contain particular coiled spiral arteries, and only female mammals with this unique kind of spiral blood vessel are able to menstruate. As the endometrium regresses, these arteries become even more tortuously coiled in the rapidly thinning layer. The blood circulation

through them slows down. Some of the arteries intermittently constrict, cutting off the blood flow to the areas they supply, and the tissue dies from lack of blood. Prostaglandins, especially PGF_{2a}, may play a role in inducing arterial constriction. When the arteries dilate again, the blood escapes from the top where the tissue has already disintegrated. The little pools of blood rupture through the endometrial surface into the uterine cavity. The same sequence occurs in other arteries during the next few days. As more of the superficial layer disintegrates, small pieces of endometrium become detached, and glandular secretions and blood slowly ooze into the uterine cavity.

For four or five days, approximately 50 ml of blood, glandular secretions, and some tissue fragments flow from the uterine cavity through the cervical os and out the vagina. That amounts to $2\frac{1}{2}$ tablespoons of blood. Menstrual blood does not clot because some of the clotting factors ordinarily found in blood have been lysed by a proteolytic enzyme in the uterus. The clots that may appear to form with a heavier flow are a combination of red blood cells, mucus, glycogen, and glycoproteins, and form in the vagina rather than in the uterus, possibly because the amount of flow is too great for the amount of lysing enzyme.

MENSTRUAL HYGIENE

Menstrual blood is sterile and in no way unclean. Many women are overly concerned with hygiene and odor during the period of flow. Even the word "sanitary" in sanitary napkin implies a certain uncleanliness and reinforces the association of menstruation with impurity and sepsis—an association recognized by manufacturers who no longer use the term and now call their products mini- or maxi- "pads" or "shields." Actually, pads may be quite unsanitary, particularly if changed infrequently, because they form a haven and breeding ground for bacteria as well as a potential pathway for organisms to migrate from the rectum to the vagina or urethra. Since menstrual blood develops an odor only after being exposed to air, a more sensible means of protection against potential vaginitis, urethritis, and odor would appear to be internal absorption of the menstrual flow by tampons. The possible correlation between tampon usage and toxic shock syndrome, however, has created a quandary for women who are reluctant to give up the convenience of tampons but have qualms about the potential risk. A complete discussion of toxic shock and tampon usage follows in Chapter 4.

Any woman who chooses to use tampons can do so. There are no normal vaginas that cannot accommodate them. Difficulty in insertion usually turns out to be more emotional than physical. If necessary, lubrication with sterile jelly can assist in insertion. If the hymenal aperture is very small in adolescence, it is possible to accomplish painless stretching by using a tampon with a cardboard applicator and a little lubricating jelly. After a few months, there should be no problem in inserting tampons, and visible evidence of virginity will still exist, if that need is important.

MENSTRUAL SYNCHRONY

An interesting phenomenon, which mothers and daughters in the same house may have noticed but never really thought about, was reported first in 1971 and has had relatively little research effort since. Martha McClintock observed that females living in close proximity tend to menstruate at approximately the same time. When the time of onset of the menstrual flow was studied in 135 young women 17–22 years of age who were living together in a dormitory, the menstrual cycles were synchronized as to day of onset in a significant number of women who were close friends but not roommates. But when they were close friends and roommates, the simultaneous onset was even more highly significant. Stranger yet, when the young women were asked to keep track of their menstrual cycles and dating frequency, it was observed that those who dated most often had shorter cycles. Two Scottish researchers subsequently confirmed that menstrual synchronization occurred among women who were not roommates but close friends and who spent a lot of time together. In contrast to McClintock's data, however, they found no significant correlation between cycle length and the amount or type of interaction with men (Graham and McGrew, 1980.) Quadagno et al. (1981) reported that menstrual synchrony occurred among three and four women living together as well as between pairs, but also saw no effect of socialization with males.

The reasons for this curious synchrony are unknown, but it has been suggested that human odors may play a role. Russell et al. (1980) tried to find out whether underarm perspiration could be a stimulus for synchronous cycles. For a period of 4 months, they had five female volunteers daily rub on their upper lips cotton pads containing the perspiration odor from a donor female with regular cycles, while six controls rubbed plain cotton pads on their upper lips. At the end of the experiment, there was no change to a reduction in the onset dates of menstrual flow between the control subjects and the "sweat donor," but the average difference in date of onset of menstruation between the experimental group (the women who rubbed sweat on their lips) and the donor was reduced from 9.3 days to 3.4 days, suggesting that synchronization had occurred. This study, as yet unreplicated, may provide a clue to how menstrual synchrony can occur, but *why* it should occur at all is still a puzzle.

With some similarity to McClintock's data on shorter cycles and dating frequency, there is evidence that shorter menstrual cycles may be associated with sexual activity. Winnifred Cutler (1979a, 1979b) and her colleagues discovered that a small sample of college women who had regular weekly sexual intercourse had shorter menstrual cycles than women with less frequent sexual encounters. The same investigators have also provided data indicating that sporadic sexual behavior (defined as women who abstained for greater than 7 continuous nonmenstruating days) seemed to be one factor related to a type of infertility known as luteal phase defect. Such correlations are interesting and should stimulate additional studies, but to draw any inferences from them concerning the apparent benefit of coitus on menstrual regularity or fertility would be premature.

REFERENCES

Cutler, W. B., Garcia, C. R., and Krieger, A. M. Sexual behavior frequency and menstrual cycle length in mature premenopausal women. *Psychoneuroendocrinology* 4:297–309, 1979.

Cutler W. B., Garcia, C. R., and Krieger, A. M. Luteal phase defects: A possible relationship between short hyperthermic phase and sporadic sexual behavior in women. *Horm Behav* 13:214–218, 1979b.

Fritz, M. A. and Speroff, L. The endocrinology of the menstrual cycle: The interaction of folliculogenesis and neuroendocrine mechanisms. *Fertil Steril* 38(5):509–529, 1982.

Graham, C. A. and McGrew, W. C. Menstrual synchrony in female undergraduates living on a coeducational campus. *Psychoneuroendocrinology* 5:245–252, 1980.

McClintock, M. K. Menstrual synchrony and suppression. *Nature* (Lond) 229:244–245, 1971.

Quadagno, D. M., Shubeita, H. E., Deck, J., and Francoeur, D. Influence of male social contacts, exercise, and all-female living conditions on the menstrual cycle. *Psychoneuroendocrinology* 6(3):239–244, 1981.

Quigley, M. E. and Yen, S. S. C. The role of endogenous opiates on LH secretion during the menstrual cycle. *J Clin Endocrinol Metab* 51:179, 1980.

Robert, J. C., Quigley, M. E., and Yen, S. S., Endogenous opiates modulate pulsatile luteinizing hormone release in humans. J Clin *Endocrinol Metab* 52:583, 1981.

Russell, M. J., Switz, G. M., and Thompson, K. Olfactory influences on the human menstrual cycle. *Pharmacol Biochem Behav* 13:737–738, 1980.

Speroff, L. Getting high on running. *Fertil Steril* 36:149, 1981.

Vrbicky, K. W., Baumstark, J. S., Wells, I. C., et al. Evidence for the involvement of β-endorphin in the human menstrual cycle. *Fertil Steril* 38(6):701–704, 1982.

4. *Menstrual Problems: Causes and Treatments*

What is "normal"? What is "regular"? The one unvarying aspect of the menstrual cycle is how extremely variable it is. The studies that have produced data concerning length and duration of the menstrual cycle are as variable as the cycles themselves. They have been made on women of different nationalities, occupations, nutritional status, and age. The extent to which these and other social variables produce variations has not been determined. The methods of selecting samples of women who are included in the studies differ, and so do the methods of data analysis. Some investigations may have excluded women who do not "menstruate normally," whatever that means. In some studies, all the cycles may have been lumped together for analysis; in others, the individual cycle for each woman may have been analyzed separately. It may be that the women who agree to participate in a study do so because they menstruate more regularly than women who refuse to participate, and this results in data of questionable validity.

It is important to recognize that the statistics that follow are based on such investigations. The figures generally pertain to what is considered average and not necessarily typical. With that admonition, they can provide a basis for distinguishing between what is regular and what is irregular, and, within a broad range, what is considered normal and abnormal.

MENSTRUAL STATISTICS

The mean length of the menstrual cycle is 25–30 days. Only around 10%–15% are exactly 28 days, and anywhere from 20–40 days is still considered within normal range. When cycles are shorter than 3 weeks, the woman is said to have *polymenorrhea*. If the cycles are longer than 40 days, the condition is *oligomenorrhea*. These short or long cycles may indicate a disturbance that is not serious and is of little functional significance. Two-thirds of all cycles vary in length by 6 days, and one-third have greater variability. There is more regularity to the menstrual cycles that occur between the ages of 20 and 40. The preovulatory phase in a longer cycle is the one that is more variable. The postovulatory phase is generally constant and lasts 13–15 days. That means that a woman with a 21-day cycle ovulates on day seven and not halfway through the cycle.

The duration of menstrual flow is 4–6 days, although a flow for as few as 2 days or as many as 8 days is still considered normal. *Hypomenorrhea* refers to a normal interval for the cycle, but the duration of the flow is short and the amount is scant. In *hypermenorrhea*, again the interval is normal, but the bleeding is excessively heavy for a normal length of time. When the menstrual bleeding time is prolonged, it is called *menorrhagia*. If bleeding occurs between periods and is unrelated to the cycle, it is termed *metrorrhagia*.

The amount of blood that is lost at each period is variable as well. It makes up only about half of the total menstrual flow, and while the average is 50 ml, it may be as little as 20 ml or as much as 120 ml and still be considered within normal range.

MENSTRUAL DISORDERS

Many women experience deviations from the average in their menstrual cycles during their reproductive years. Since feelings about one's maturity, being a normal woman, and being able to bear children, for example, are likely to focus on the visible evidence that everything is all right, most women are understandably anxious about delayed periods, missed periods, periods that are too scant, too profuse, or too frequent. But since there is so much variation in frequency of cycles, amount of flow, and duration of flow, how is a woman to know when it is necessary to seek medical attention? Generally, a change in bleeding pattern from what is normal and usual for her that persists for several months is an indication that there may be a problem.

All women who are not of the regular-every-28-days-like-a-clock type should keep a menstrual chart (see Fig. 4.1). Even if a woman is very regular, a change in the frequency, duration, or amount of the menstrual flow can easily be determined from such a chart.

The cause of abnormal bleeding may be an endocrine disturbance because the hormones of the hypothalamus, the pituitary gland, the ovaries, the adrenal cortex, and the thyroid can all have an effect on the menstrual cycle. A change in bleeding pattern may also be caused by a physical disturbance of the tissues of the vagina, cervix, or uterus—what a physician calls an "organic lesion." Such a lesion could be a benign or malignant tumor of the reproductive tract and will be described later.

Abnormal bleeding may also be a result of acute infection, a blood or liver disease, hormone or drug administration, or the presence of an IUD. It may be caused by a complication of pregnancy—a spontaneous abortion (miscarriage) or an ectopic pregnancy. It may, as previously indicated, be caused by emotional factors mediated through the hypothalamus.

Dysfunctional Uterine Bleeding (DUB)

When no cause for the unpredictable, excessive, frequent, and/or prolonged bleeding can be determined diagnostically, it is called *dysfunctional uterine bleeding*. It is not to be inferred that there is no cause; it is simply that the cause cannot be determined, but it is known to be associated with a disturbance in hormonal mechanisms.

Possible Causes of Dysfunctional Bleeding
Some authorities think dysfunctional bleeding is always related to a cycle in which ovulation has not occurred. In such an anovulatory cycle, the endometrium is not stimulated by progesterone because a corpus luteum has not formed. The endometrium continues to grow in the proliferative phase. It becomes very thick but, lacking progesterone stimula-

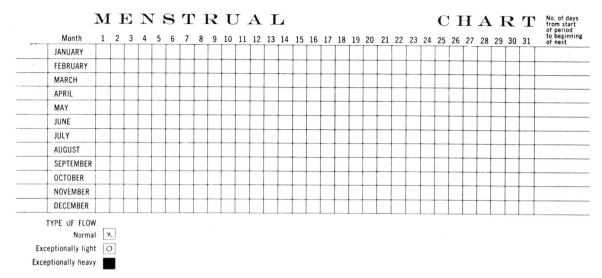

Month	1	2	3	4	5	6	7	8	9	10	11	12	13	14	15	16	17	18	19	20	21	22	23	24	25	26	27	28	29	30	31	No. of days from start of period to beginning of next
JANUARY																																
FEBRUARY																																
MARCH																																
APRIL																																
MAY																																
JUNE																																
JULY																																
AUGUST																																
SEPTEMBER																																
OCTOBER																																
NOVEMBER																																
DECEMBER																																

TYPE OF FLOW
Normal   X
Exceptionally light   O
Exceptionally heavy   ■

Figure 4.1. A menstrual record chart.

tion, never becomes secretory. When the ovarian theca and granulosa cells eventually disintegrate, the estrogen level decreases, causing menstruation. The bleeding from such an estrogen-only stimulated endometrium is prolonged and heavy, and parts of it are shed irregularly. The bleeding continues until a new crop of follicles in the ovary produce enough estrogen to stimulate new endometrial cell growth. Cell regeneration then stops the bleeding. The other type of DUB, accounting for perhaps 10%–15% of cases, is the premenstrual or postmenstrual spotting that occurs in an ovulatory cycle. It is due to a shortened or prolonged life span of the corpus luteum that probably disturbs the progesterone/estrogen ratio.

Anovulatory cycles are most frequent during the two transitional periods of a woman's reproductive life, just after menarche and just before menopause, and DUB also commonly occurs during these periods. Irregularity in an adolescent girl is almost considered normal, and it is treated only if the menstrual periods are distressingly heavy, frequent, and unpre-

dictable. Heavy and frequent menstruation also is not unusual in premenopausal women, but it is particularly important to be certain that no pathology exists. The incidence of uterine cancer increases with age, and any woman who has abnormal bleeding after she has completely stopped having menstrual periods should see a physician immediately for a complete diagnostic evaluation, especially if she is taking estrogen.

Treatment of DUB
This condition is probably one of the most frequently encountered in gynecological practice. The decision about how or whether to treat it is related to the age of the woman and the severity of the bleeding. There are two basic ways to control, stop, or at least regulate abnormal uterine bleeding—by surgery or with hormones. If the endometrium is removed surgically by D and C, abnormal bleeding does not recur for some reason, or at least does not return for several months. A nonsurgical or "medical curettage" may be effected with the administration of oral or injected

progestin. Given for approximately 3 days in large doses, a progestational agent will usually stop a hemorrhage within 24–48 hours. If it has been established that bleeding is a result of anovulatory cycles, it is possible to produce a secretory endometrium with progestin or to mimic a normal cycle by the sequential administration of estrogen and progestin. Usually, estrogen is given alone for 2 weeks, followed by estrogen and progestin for another week. When these are discontinued, withdrawal bleeding, which is similar to that of a normal cycle, occurs.

Of course it is possible to treat the symptoms without making a real attempt to get at the cause. To label an irregularity neatly as DUB and initiate hormonal therapy is a potentially dangerous medical practice if every effort has not been made to make sure that some underlying organic pathology has not been overlooked. Women should be reluctant to accept a diagnosis of "psychogenic bleeding," for example, until certain procedures have been carried out. Some of the diagnostic tools are use of the basal body temperature and vaginal smear to establish whether ovulation has occurred; determination of blood and urine levels of hormones or their metabolites; and/or procedures that require hospitalization, such as endometrial biopsy, diagnostic D and C, and laparoscopic examination. During their reproductive years, women should never agree to a hysterectomy for DUB except as a last resort after all other means of therapy have been tried.

Amenorrhea

Amenorrhea is the lack of menstruation during a woman's reproductive years. It is normal only during pregnancy. Amenorrhea itself is a symptom, not a disease. Once it has been established that there is no life-threatening underlying cause for the lack of menstruation, it is not medically serious. Psychologically,

however, it can certainly be a major cause of anxiety in a girl who has not started to menstruate when all of her friends have. And when periods stop occurring after several years of reasonable regularity, there are so many possible etiological factors that, while it may not be ultimately important to treat it medically, it is obviously necessary to try to establish some reason for it.

Causes of Amenorrhea

When menarche does not appear until the age of 16, it is called delayed menstruation and is most likely the result of a familial tendency to late maturation. Actual primary amenorrhea is the failure to menstruate by age 18. In more than 40% of cases, it is related to chromosomal abnormalities, malformation of the reproductive tract, or both. Less frequently, primary amenorrhea is caused by disorders of the thyroid, adrenal cortex, or (even more rarely) diseases of the pituitary gland. It can also result from extreme malnutrition.

Cryptomenorrhea (Greek: *kryptos*, hidden). This is a term for "silent" menstruation. Uterine bleeding occurs, but it is blocked from exiting from the cervix or vagina. The most common reason for the condition is an imperforate hymen, but an infection that causes subsequent scar tissue formation in the vagina or cervix may result in cryptomenorrhea. It is rarely the result of total absence of the vagina, a genetic malformation. Cryptomenorrhea is usually diagnosed several years after puberty, and it is treated surgically.

Secondary amenorrhea. This term is applied to the cessation of menstrual periods anytime between menarche and menopause when the cause is not pregnancy. In the same unknown way that emotional factors can cause abnormal bleeding, an abrupt failure to menstruate may also have a psychological basis. Secondary amenorrhea has frequently been traced to stress, fear, anxiety, or trauma. A great fear of pregnancy or a tremendous desire for pregnancy can cause it; so can a change in environment or a death in the family. Nutrition also has an effect on menstruation. Amenorrhea can result from a quick weight loss after a crash diet, or it may be related to obesity. Some of the reasons for secondary amenorrhea are listed in Table 4.1. A more complete list could fill several pages.

Many of the cases of psychogenic secondary amenorrhea are self-limiting. No one ever died from not menstruating; it is not in itself dangerous or life threatening. Since there is no evidence that administering hormones or inducing ovulation has any effect on the return of normal cycles if there is a psychological basis for amenorrhea, there is no reason to use hormone therapy unless the woman wants to become pregnant. Excessive bleed-

ing is always of greater concern than absence of bleeding.

One explanation of the amenorrhea that results from weight loss may be found in the "critical weight" theory of Frisch and Revelle (1970), who compared different samples of girls at the time of menarche. They discovered that although the girls' ages varied, their body weights did not significantly differ. The investigators postulated that there is a critical body weight (around 105–106 pounds) that is associated with menarche and that after the initiation of menstrual cycles, a weight loss of 10%–15% below the critical weight for maintaining cycles would stop them. Other investigators, claiming that the total body weight for girls at menarche varies over a large range, find little evidence to support the critical weight theory. Garn and LaVelle (1983) recently surveyed the reproductive histories of 79,000 girls and women of four racial and ethnic groups and

Table 4.1. SOME CAUSES OF SECONDARY AMENORRHEA

Physiologic (Normal)	Anatomic	Central Nervous System (Hypothalamic)
Pregnancy	Hysterectomy	Psychogenic
Lactation	Cryptomenorrhea	Environmental
Menopause	Destruction of endometrium	Nutritional
	Trauma (overly enthusiastic curettage)	Sudden weight loss
	Disease (T.B., e.g.)	Anorexia nervosa
	Irradiation	Obesity
	Ovarian	Iatrogenic (medically induced)
	Disease (cysts, tumors, etc.)	Oral contraceptives ("post-pill amenorrhea")
	Ovarian failure, premature menopause	Psychotropic drugs
	Destruction, through surgery, infection, irradiation	Thyroid dysfunction
	Pituitary	Adrenocortical dysfunction
	Tumors, disease	Chronic systemic disease
	Pituitary insufficiency	

found that body weight below 104 pounds was no deterrent to menarche, conception, pregnancy, or even repeated pregnancies. A currently more favored theory was proposed by Frisch and MacArthur in 1974. They suggest "critical fatness" or the body composition of fat as a percentage of body weight as the significant factor. The two investigators proposed that the minimal mean body fatness for the establishment of reproductive function at puberty had to be equivalent to 17% of body weight, increasing during adolescence to about 28% of body weight at age 18. The relative importance of body weight and body fat are still being debated, but it is also recognized that emotional, genetic, and as yet other unknown factors are important in the onset and maintenance of menstrual cycles.

That there is a relationship between body weight or composition and pituitary gonadotropin function is supported by studies that indicate that when a weight loss is self-imposed and severe, there is a decrease in the sensitivity of pituitary FSH and LH, and particularly of LH, to administration of LHRH. When doses of LHRH were given to women who had amenorrhea in conjunction with severe self-imposed weight loss, the FSH and LH responses were either completely absent or significantly lower than those of women of normal weight or women with secondary amenorrhea due to other causes than weight loss. When the women gained weight, their gonadotropic responsiveness to releasing factor returned, with a sudden increase in responsiveness occurring at 15% below ideal weight (Warren et al., 1975).

Some additional evidence for the critical weight theory is indicated by a study of women who had a history of regular periods before taking oral contraceptives but developed secondary amenorrhea after stopping oral contraceptives (post-pill amenorrhea). It was found that these women had significantly lower body weights than women who had had a history of menstrual and ovulatory irregularity before taking the pill and than a control group of women. Evidently, women of low body weight are more likely to develop post-pill amenorrhea even when there is no previous history of ovulatory dysfunction. Such women, it was suggested, should consider some other form of contraception (Hancock et al., 1976). On the basis of her data, Wentz (1980) goes even further and cautions that women who are on the pill and go on a diet and deliberately lose even a minor amount of weight or body fat are at a substantial risk for the development of secondary amenorrhea.

Women who literally exercise off most of their excess body fat have also been discovered to have irregular or absent menstrual periods, and the term "athletic amenorrhea" has been used. Surveys of women long distance runners participating in National Amateur Athletic Union cross-country championships showed that a significant number were found to have fewer than two cycles a year. Wakat and coworkers (1982) found that half of the cross-country runners studied had oligomenorrhea, and there have been similar findings of menstrual dysfunction in marathoners, gymnasts, and professional ballet dancers, who train as hard as football players and also worry more about their weight and appearance. Weight loss in itself, although common with intensive training, is apparently not the determining factor, because not all athletes or ballerinas have dysfunctional periods, and those that do are often no different in height and weight than those who menstruate normally. They may differ, however, in their specific body composition, having greatly decreased subcutaneous fat and greatly increased lean muscle mass. Female athletes have generally been found to have body fat ranging between 5% and 6% of their body weight compared with an average fat range in women of 24%–34%. At any rate, when the grueling exercise stops, so does menstrual irregularity. Warren's study (1980) of ballet dancers found that cessation of exercise (usually due to injury) brought a return of normal cycles, but that when the young women went back to training, the menstrual dysfunction returned without any change in body weight. It should be pointed out that only the intensive physical training that accompanies competitive sports or professional ballet has been associated with menstrual irregularity; there is no evidence linking menstrual dysfunction with aerobic dancing, jogging, or any other kind of recreational sport.

Since a decrease in body fat is apt to result in amenorrhea, the occurrence of absent or irregular cycles with obesity becomes even harder to explain. One theory suggests that since fat tissue can synthesize estrogen, the excess production may interfere with the hypothalamic-pituitary-gonadal regulation of menstruation. Another possible explanation is that the amenorrhea is stress induced, the stress either resulting from or being present as the initial cause of the obesity.

EATING DISORDERS AND AMENORRHEA

The 1983 death of pop star Karen Carpenter, evidently as a result of *anorexia nervosa*, caused renewed public concern about a disease with which many parents were already distressingly familiar. Anorexia nervosa, the "starvation disease," afflicts primarily white, middle- to upper-class females (90%), who are attractive, highly intelligent teenagers from affluent, well-educated families. The disorder is infrequent in women over 30, quite rare in males, and almost never occurs in poor, minority females. The actual prevalence is unknown, since the figures of one in 100 to one in 250 adolescent girls are based on clinical

cases treated and do not take into account the borderline or milder unrecognized anorexics who are at risk for developing the full-blown severe form of the disease.

Anorexia means loss of appetite and is a misnomer for the condition. It is not want of appetite, but an obsessive fear of fatness that underlies the self-imposed starvation. The hunger mechanism is experienced, at least well into the illness, but food is rejected, although it is not unusual for the anorexic to be preoccupied with elaborate preparation of food for family or friends. Further diagnostic characteristics are a weight loss of at least 25% below the original body weight, but a stubborn refusal on the part of the victim to recognize the skeletal thinness, no medical illness to account for the weight loss, no other obvious psychiatric disorder, and the cessation of the menstrual periods. Since the lack of menstruation very frequently *precedes* the emaciation, it is generally believed that psychological factors affecting the hypothalamus, as well as loss of body weight, are involved in the amenorrhea. Other symptoms of anorexia nervosa may include thin scalp hair but excess body hair, very low blood pressure, low heart and breathing rate, a dry, almost sandpaperlike skin texture, constipation, and a surprising ability for physical activity, given the low caloric intake.

There has been a rapid increase in the incidence of anorexia nervosa, and at least part of the reason for the increase may be our national preoccupation with dieting and slimness, which can promote a distorted body image. We are obsessed and oppressed by fat. The message that we obtain from magazines, movies, and TV is that "skinny is beautiful," and in the emotionally vulnerable, it may trigger the onset of the disease. We are all susceptible to such messages, some of us more than others. The seriousness of anorexia nervosa is not to be underestimated. It is a life-threatening disease that is not easy to treat and requires medical help, nutritional counseling, and supportive psychotherapy.

Obviously, starvation alone, whether self-imposed or not, can be fatal. There are some anorectics, however, perhaps as many as 30%, who are also self-induced vomiters in an attempt to cut caloric intake even further. This is the group whose lives are in greatest jeopardy. Anorexia nervosa and vomiting leads to loss of vital electrolyte ions (sodium, potassium, chloride) and disturbs the acid–base balance of the blood. Hypokalemia (low potassium) is particularly dangerous and potentially lethal because of effects on the heart. Data indicate that anorectics who both starve and vomit comprise almost the entire category of fatalities (Mars et al., 1983).

Binging on food followed by induced vomiting has always been the dark side of anorexia nervosa, but a more recent concern has emerged—young women in their teens or early twenties who suffer from a gorge/purge syndrome called *bulimia* or bulimarexia. Whether the disease is actually a disorder separate from anorexia nervosa is still controversial, since there is overlap in diagnostic characteristics of the two conditions. Although there are some common factors in bulimia, there is considerable variation in the frequency of the episodes, the mood state at the time of precipitation of binge eating, the kinds and amounts of food eaten, and the weight loss that accompanies the behavior. Sufferers may be of normal weight or even obese, so that the effects of this eating disorder are not as apparent, and it is possible for people to keep the problem secret perhaps for years. Some bulimics take laxatives and diuretics, some vomit, and some do both, but there are those who merely eat very little between binges or exercise strenuously to counteract the effects of excessive eating. Amenorrhea is often present, although not as consistently as in anorexia nervosa. One study

of 30 female bulimics at an eating disorder clinic in Australia found that 23 of them reported at least one episode of absent periods for 3 months or longer. The data appeared to indicate that menstrual dysfunction was associated more with weight loss behavior than with actual alteration in body weight (Abraham and Beumont, 1982).

The long-term effects of the problem for those who vomit or take laxatives can be devastating. Vomiters experience tooth decay as a result of the acid eating away at tooth enamel. A constant sore throat, esophagitis, and massive swelling of the salivary glands are also accompaniments of vomiting, while dehydration and loss of potassium can result in disturbance to heart and kidney function. Some bulimics are not unaware of the deleterious effects of decreased potassium levels in the blood and take potassium supplements, according to Abraham and Beumont. Liver damage is also probable, and sudden death from stomach rupture or heart failure is a possibility.

Bulimia has been linked to depression, and therapy with antidepressants has been successful in control, if not cure, of the disease. People who suffer an eating disorder like anorexia nervosa and/or bulimia frequently struggle with the disease for much of their lives; management of the problem, rather than cure, may be the only attainable goal. The actual prevalence of bulimia is unknown. Some investigators have estimated that an incredible one out of five American college women engage in at least occasional binging and purging—a figure that almost suggests the disease is catching. One might speculate about the influence of TV, magazine, and newspaper accounts on the incidence of bulimia. Anorexia nervosa has become a familiar household word; bulimia is catching up. At least one woman related that she started binging and vomiting after reading an article about it. Described as a typical bulimic, she recalled: "I got tired of dieting, and I thought, 'Oh, that's a really good way'" (Carl, 1982).

DYSMENORRHEA

The medical term for pelvic pain or cramps that occur during menstruation is dysmenorrhea. It is called primary dysmenorrhea when a woman has had painful menstrual periods since menarche, and there is no organic pathology. It is termed secondary dysmenorrhea when it first occurs months or years after menarche, when it could be the result of some organic pelvic pathology.

Symptoms of Dysmenorrhea

Many girls and women experience moderate to severe discomfort just before the menstrual period starts or on the first day of menstruation. The pain generally lasts for 24 hours but in some cases continues for several days. Cramping is felt in the lower pelvic area together with a drawing sensation in the thighs and a mild backache. The intensity of the pain varies and can be incapacitating in some women, although individual response to discomfort could be a factor in how seriously the pain is perceived.

The actual prevalence of dysmenorrhea is unknown. Statistical studies that have been attempted to determine the frequency of menstrual cramps among large groups of women have produced inconsistent results. A 1957 survey of adolescent girls indicated that 50% were affected, but it was 21.9% in a 1981 epidemiological study of American teenagers. It is a rare woman who has no idea of what menstrual cramps are, but the number of women who have pain severe enough to limit

normal activities or to stay home from school or work is probably small.

Severe, incapacitating menstrual cramps accompanied by headache, nausea, and vomiting may be "normal" in the sense that there is no underlying pathology but should not be tolerated by any woman as merely being her lot in life. This kind of symptomatology demands a complete history and physical examination to determine whether existing uterine or ovarian problems are responsible. If not, and the pain is from primary dysmenorrhea, the suffering may be needless in view of the recent development of more effective therapy.

Etiology of Dysmenorrhea

The causes of primary dysmenorrhea are not definitively known, although there are a number of theories that try to explain it. Psychological, anatomical, and hormonal reasons have been proposed, but none of these, singly or in combination, has as yet provided adequate explanation. There are still Ob/Gyn textbooks (even published in the 1980s) that state that menstrual pain is primarily psychogenic in origin, and there are, therefore, many practicing physicians who have been taught to believe that the pain is in the head and not in the lower abdomen. Since the pain is not then "real," it cannot be taken seriously, and they are not particularly concerned with its alleviation or treatment.

It may be possible that in some women severe dysmenorrhea *is* a sign of an underlying emotional disturbance, but then presumably menstrual cramps would not be its only manifestation. It is equally possible that many women respond to monthly episodes of sharp and painful cramps by developing real anxieties and fears about their menstrual periods, which could then increase the severity and persistence of the pain. To the person experiencing it, pain is pain, whether or not there is any organic basis for its origin.

The anatomical theory holds that an occluded cervical opening is a major cause of menstrual pain because for many women, having their first baby decreases or eliminates dysmenorrhea. But since as many women are cured by cesarean births as by vaginal delivery, it is not likely that it is the cervical dilatation alone that accounts for the alleviation of menstrual cramps. It may be that the greatly increased blood supply to the uterus during pregnancy and the subsequent vasculature that remains are factors.

Other theories that have been advanced about the causes of dysmenorrhea are too much estrogen, not enough estrogen, imbalance in estrogen/progesterone ratio, too much progesterone, food allergy, or a reaction to some chemical factor in menstrual blood. The idea that some sort of menstrual factor was involved in uterine contractions had been theorized for many years. The factor was later identified as being prostaglandin, and a recognition of the role of PGE_2 and PGF_{2a} has resulted in the currently accepted theory of why it often hurts to menstruate.

The nonpregnant uterus is a far from quiet organ, since the smooth muscle of the myometrium undergoes continuous contractions. During menstruation, the frequency of the contractions decreases, but the intensity increases, and they are described as "laborlike." In severely dysmenorrheic women, there is exaggerated uterine contractility and a significantly higher prostaglandin content in the menstrual blood with a twofold to tenfold increase when compared with women who do not have menstrual pain. The proposed physiological mechanism to explain the pain is as follows: During menstruation, prostaglandins are produced locally by the uterine endometrium. Their function is to mediate normal contractions to ease endometrial shedding. In women with dysmenorrhea,

for unknown reasons, prostaglandins are produced in excessive amounts. Increased uterine contractility and increased pain result. Pain may also be a consequence of a reduction of arterial blood flow to the uterus (uterine ischemia) mediated by prostaglandins.

One shortcoming of the prostaglandin-uterine contractility-ischemia-pain theory is that it may not apply in all cases. Some severely dysmenorrheic women have neither increased contractility nor increased PG levels. Moreover, studies of uterine PG content have included only adult women with severe dysmenorrhea, omitting adolescent girls in whom the greatest prevalence of dysmenorrhea presumably exists.

One other factor in dysmenorrhea, although the mechanism is obscure, is that it almost always occurs from a secretory endometrium; that is, when ovulation has occurred. That is not to say that absence of menstrual cramps means an anovulatory cycle, but just that when progesterone effect is absent, the cycles always terminate in painless flow.

Treatment of Dysmenorrhea

Since anovulatory cycles rarely result in painful menstruation, the most effective way to eliminate dysmenorrhea is to eliminate ovulation. Suppressing ovulation with oral contraceptives or whatever hormonal therapy is used results in almost uniform success in producing painless menstruation. If a woman wants to become pregnant or is concerned about the other effects of hormonal therapy, she must weigh the potential risk against the benefit to her. Sometimes, if normal periods are allowed to resume after several months of hormone treatment, they are more comfortable, and dysmenorrhea may never recur.

Several surgical techniques have been used to alleviate menstrual pain. One is cervical dilatation, but it produces relief only about 25%

of the time, and frequently the problem returns for that percentage. A very drastic surgical procedure is presacral neurectomy, in which all the autonomic nerves to the uterus, both sensory and motor, are severed. Uterine function and sexual response are said not to be impaired by this total denervation. Obviously, no woman or her doctor should consider this form of alleviation of dysmenorrhea when other methods are available.

In addition to hormones, many drugs have been recommended for the relief of dysmenorrhea, and these range from mild analgesics like aspirin to potentially addictive compounds containing codeine, morphine, or their derivatives. Combinations of tranquilizers and painkillers, or amphetamines and analgesics, have been used with varying results. For many women, however, the miracle drugs for menstrual cramps are the nonsteroidal anti-inflammatory agents or prostaglandin inhibitors. Their known effect is to block the enzyme cyclooxygenase, thus interfering with the transformation of arachidonic acid into the intermediate endoperoxides. The most commonly prescribed and studied PG inhibitors are the fenamates mefenamic acid (Ponstel) and flufenamic acid (Arlef), ibuprofen (Motrin), naproxen (Naprosyn), naproxen sodium (Anaprox), and aspirin. Aspirin has a weaker enzyme-inhibiting effect than the other compounds but has been shown in clinical trials to be effective for mild dysmenorrhea, although heavier menstrual bleeding has been observed in some women. All of the PG inhibitors have possible side effects, primarily of the gastrointestinal variety and should be taken with food and not on an empty stomach. The drugs have been studied extensively, but there is little published data comparing one type with the others so the choice of which to use is subjective. Mefenamic acid has been shown to substantially reduce excessive menstrual blood loss (Fraser et al., 1983), and naproxen sodium and zome-

pirac sodium (Zomax) have reduced many of the other symptoms, such as swelling, dizziness, vomiting, and depression (Chan et al., 1983; Budoff, 1982). Zomax, however, was recalled by the manufacturer in early 1983 because of apparent allergic reactions. In May of 1984, the Food and Drug Administration approved ibuprofen for over-the-counter sales. Under the trade names Advil (produced by American Home Products, the makers of Anacin) and Nuprin (manufactured by Bristol-Myers, the company that makes Exedrin and Bufferin), the non-prescription drug is available in 200 milligram tablets, half the dosage of the widely prescribed Motrin.

Prostaglandins are produced by nearly all body cells and are likely to influence their function in many as yet unknown ways. Prostaglandin inhibitors do not discriminate between blocking prostaglandins in the uterus and preventing prostaglandin synthesis throughout the body, so the long-term effect of these drugs is still unevaluated. Aspirin and acetaminophen (Tylenol) may have a greater potential for gastrointestinal side effects, but they also have a more proven safety record than the more potent PG inhibitors. Fortunately, treatment of dysmenorrhea and its accompanying symptoms requires only once-a-month therapy of short duration.

Home Remedies

Some women may be unable to tolerate the newer antiprostaglandins or be unwilling to take them. For centuries, women have been advocating remedies for menstrual pain to each other. Since there are so many factors that can be involved in the cause of dysmenorrhea, it should not be surprising that there are so many ways in which it can be treated without using drugs or hormones. But what works for one woman may not work for another; each may have her own particular problems. "The Monthly Extract, An Irregular

Periodical,"* was a publication that formed a communications network for gynecological self-help clinics, In it, women reported ways of dealing with menstrual cramps that produce good results.

Heat, in the form of a hot water bottle, a heating pad, or a soak in a hot tub, has been effective, Heat promotes an increase in blood flow and decreases muscle spasm. A hot drink, like spiced or herbal tea or soup, is soothing and relaxing and may help to break the pain–tension/more pain–further tension circle.

Exercise relieves menstrual cramps in many women. It has been noted that physically fit women generally suffer less menstrual distress. Few women would want to do push-ups or kneebends when having cramps, but a brisk walk is frequently more helpful than going to bed with the pain.

Dysmenorrhea produces tension and anxiety. Two mechanisms that are used to relieve the stress and tense muscles accompanying cramps are yoga and transcendental meditation (TM). Yoga is a method of mind and body control that uses physical and breathing exercises as well as meditation. *Hatha Yoga* is the most popular form and is concerned with *pranayamas*, or breath-control exercises, and *asanas*, the postures or physical exercises. Several of the postures work specifically on the abdominal and lower back muscles and are recommended for women experiencing dysmenorrhea, both as a preventive measure and at the time of the pain.

Transcendental meditation is a method of altering one's mental activities and autonomic functions. In TM, energy and concentration are directed inward rather than outward. Attention focuses on all the body muscles in a conscious effort to relax them. TM advocates claim less anxiety and a more relaxed attitude.

* Publication ceased in 1978.

During orgasm, the uterus undergoes spontaneous contractions beginning at the fundus and terminating at the cervix, and immediately after orgasmic response, the external os of the cervix dilates slightly, remaining somewhat opened for 5–10 minutes. This may be a physiological basis for the reports of some women that orgasm attained by intercourse or masturbation is very helpful in alleviating menstrual cramps.

TOXIC SHOCK SYNDROME

Toxic shock syndrome (TSS) is not a new disease, not necessarily a tampon disease, and not associated exclusively with menstruation. TSS has been recognized as a rare childhood disease since 1927, when it was described and called staphylococcal scarlet fever associated with *Staphylococcus aureus* instead of streptococcus organisms. The reason for its dramatic emergence in 1979 as a potentially fatal disease affecting menstruating women is unknown, but its recognition was the result of astute observations by a medical detective, epidemiologist Jeffrey P. Davis of the Wisconsin Division of Health. In 1978, Todd et al. published an article in *Lancet* that described TSS in seven children aged 8 to 17 years. All had sudden onset of high fever, sore throat, diarrhea, a sunburn-like skin rash, and associated kidney failure, liver abnormalities, and a rapid drop in low blood pressure leading to shock. One child died and all of the survivors had skin peeling of the hands and feet during convalescence. Little attention was paid to Todd's article until Davis (1980) noted a curious similarity—that between July, 1979 and January, 1980 seven patients with the same clinical symptoms were hospitalized in Madison, Wisconsin. All were women and six of the seven were menstruating at the time of the onset of the illness. Suspecting an association with menstruation, Davis initiated a surveillance system among physicians for identification and reporting of TSS in Wisconsin. A similar system was established in the neighboring state of Minnesota, and other states also began active surveillance for TSS.

In interviews with the seven Wisconsin women, Davis discovered that most of them had used tampons during the menstrual period corresponding to the onset of illness, and he suspected an association. As other reports came in, more data suggesting a relationship between tampon use and toxic shock accumulated, and in June, 1980, a report from the federal Centers for Disease Control (CDC) verifying the apparent link, was released and received national media attention. Almost daily publicity followed, and there was frequent mention of the possible association of TSS and the new type of highly absorbent tampon, such as Rely, manufactured by Procter and Gamble. After the issuance of a second CDC report showing a statistically significant association between tampons and TSS, with the highest risk occurring in users of Rely tampons, Procter and Gamble was persuaded to voluntarily withdraw Rely from the market in September of 1980.

Contrary to what much of the public believes, however, the disappearance of Rely did not result in the disappearance of toxic shock syndrome. Although statistics from the CDC in Atlanta indicated a significant decrease in the incidence of TSS after the Rely brand was removed, the disease continued to occur primarily in young women using all major tampon brands and styles, numbering about 35 cases a month.

Fifteen percent of TSS cases currently being reported to the CDC are unrelated to menstruation or tampon usage. Female nonmenstruating victims of TSS have contracted the disease after childbirth, either by vaginal delivery or cesarean section. Others, both men and women, came down with TSS in associa-tion with surgical wound infections, deep and superficial abscesses, infected burns, skin abrasions, insect bites, boils, and other types of "staph" infections. Many researchers believe that the actual incidence of TSS is underestimated as a result of underreporting by physicians and that the current frequency of the disease may be, in reality, as great as it was in the "epidemic" year of 1980.

Symptoms of Toxic Shock Syndrome

One reason why the true incidence of TSS is unknown and why TSS cases may be more common than believed may result from the strict case definitions of toxic shock set up by the CDC. There is no specific diagnostic laboratory test to confirm TSS, so in order to qualify for reporting as a case of toxic shock syndrome, the following criteria must be met:

1. Fever—temperature higher than 102°F
2. Rash—diffuse, sunburn-like rash
3. Skin-peeling—usually 0–2 weeks after onset of illness, primarily on palms and soles
4. Low blood pressure—systolic BP 90 mm Hg for adults or below age-related norms, including a drastic drop in blood pressure or fainting when getting up from a lying to a sitting position.
5. Involvement in three or more of the following systems—
   Gastrointestinal—vomiting or diarrhea at onset of illness
   Muscular—severe muscle aches or laboratory enzyme tests indicating muscle damage
   Mucous membranes—reddening of throat, conjunctiva of eye, or vaginal wall
   Urinary—blood urea nitrogen (BUN) levels twice normal or pus cells in the urine in the absence of urinary tract infection
   Hepatic (liver)—enzyme levels (SGOT, SGPT) twice normal
   Blood—platelet count below normal

Central nervous system—disorientation, confusion, or alterations in consciousness when fever and hypotension are absent

6. Laboratory tests that differentiate TSS from other infectious diseases, such as Rocky Mountain spotted fever, measles, or streptococcal scarlet fever

Milder cases of TSS may have some combination of the above findings but usually do not show severe low blood pressure or shock. The stringent criteria for reporting may have resulted in fewer mild, early-recognized cases being reported to state health departments with the consequence of an apparent decrease in the numbers of cases. Besides, symptoms of vomiting, fever, and muscle aches sound very much like "the flu" or "a virus," or even as "that time of the month" problems for some women. Although TSS is relatively uncommon, women should be aware that any sudden onset of fever, nausea, and vomiting during or just after a menstrual period requires immediate medical assistance and, if a tampon is being used, it should be immediately removed.

Treatment of Toxic Shock Syndrome

The treatment of toxic shock syndrome includes massive fluid replacement and all other supportive measures for shock and heart rhythm irregularities within an intensive care environment. Because the throat and vaginal cultures will frequently show a penicillin-resistant strain of *Staphylococcus aureus*, the penicillinase-resistant semisynthetic penicillins, such as oxacillin, nafcillin, methicillin, or a cephalosporin antibiotic are used. If the TSS has occurred in a menstruating woman, sometimes local vaginal disinfectants or antibiotics may be used to reduce the number of organisms or toxins in the vagina. With early diagnosis, proper treatment, and luck,

the patient is generally released from the hospital in about a week.

Causes of Toxic Shock Syndrome

There is much about the pathogenesis of TSS that is as yet unknown. It has been established that specific strains of *Staphylococcus aureus* are capable of producing a unique toxin associated with symptoms of TSS. Originally, it was believed that the toxin-producing strains produced two types, enterotoxin F and exotoxin C. These were subsequently shown to be identical to each other, and the single entity is now generally referred to as toxic shock toxin, or TST. It has not been definitely proved, however, that the toxin causes TSS. Injections of TST into a number of various animals have shown that toxic shock symptoms can be produced only in rabbits and baboons.

Furthermore, not everyone who harbors *Staphylococcus aureus*, even the TSS strains, develop the disease. "Staph" bacteria are normally found in the nasal passages and on the skin of perhaps 20%–50% of the population, and an estimated 5%–15% of women have the organisms as natural constituents of their vaginal bacteria. Why relatively few women get TSS when there are so many that are potentially vulnerable is not known. It is believed that most women over 30 years of age have developed antibodies to the toxin, which may be why those who appear to be at the greatest risk for TSS are younger women who have not as yet had an opportunity to make antibodies after exposure.

Tampon use is clearly not a necessary requirement for coming down with TSS, since menstruating non-tampon users contract TSS, and 15% of TSS victims are nonmenstrual. A number of studies strongly indicate, however, that if TSS is menstrually associated, it is linked to tampons, especially the superabsorbent variety. But a cause-and-effect relationship has not been established.

How are tampons related to the presence of *Staphylococcus aureus* or the toxins they produce? What was there about Rely that apparently carried a higher risk? For that matter, what is in a tampon anyway? There are only speculative answers to the first questions, and as for the actual composition of today's tampons, only the manufacturers know for certain since "grandfather clauses" and "trade secrets" provisions of the Food and Drug Administration regulations generally protect total disclosure (see Table 4.2).

Tampons came on the market in 1936, launched by Tampax as a disposable, flushable means of internal protection. They had been invented in 1933 by a Denver physician, Earle Haas, who was experimenting with wads of cotton because his wife told him that the traditional napkins were not good enough. Interviewed by the *Chicago Tribune* at the age of 96, Haas said, "I just got tired of women wearing those damned old rags and I got to thinking about it . . . it was designed so that it would absorb. It didn't block. It followed the natural contour of the vagina. Of course, after it was full, it would run over. . . ." He patented his device, called it a tampon and sold it to a company named Tampax. For 40 years, tampons remained wads of cotton and a string packed into a cardboard tube, but then synthetic polyester and rayon fibers and materials like carboxymethylcellulose were added to tampons to make them more absorbent and avoid "running over." They became *superabsorbent* and no longer merely soaked up menstrual flow, but expanded within the vagina to block it. Gilles Monif (1982), an expert on infectious diseases, has postulated that the superabsorbent tampons effectively occlude the vaginal canal and convert the posterior vaginal area into an anaerobic (without oxygen) environment that sup-

Table 4.2 TAMPONS CURRENTLY ON THE MARKET

Name	Manufacturer	Style	Composition[a]	Applicator
Tampax	Tampax, Inc.	Junior	Rayon	
		Regular	Cotton	
		Slender Regular	Cotton, rayon, and cross-linked carboxymethylcellulose	Spiral paper strips glued together
		Super	Cotton and rayon	
		Super-plus	Rayon polyacrylate	
Playtex	International Playtex	Regular Super Super-plus	All rayon polyacrylate	Polyethylene plastic
		Deodorant	Rayon polyacrylate and fragrance	
O.B.	Johnson & Johnson	Regular Super Super-plus	Cotton and rayon	None
Kotex	Kimberly-Clark	Regular Super Security	Cotton, rayon, and cross-linked carboxymethylcellulose	Paper stick inserted into tampon
Pursettes	Jeffrey Martin	Regular Super	Rayon	None

[a] Main ingredients only. Binders, lubricants, perfumes not revealed by manufacturers.

ports the growth of TSS bacteria. A group of Danish scientists, however, found that the insertion of a Tampax regular (the most popular brand in Denmark), OB normal, or Playtex regular actually introduced more oxygen into the vagina and changed the atmosphere into a more aerobic environment. They speculated that it was the increased oxygen that supported the growth of *Staphylococcus aureus* (Wagner et al., 1984).

Another clue to the role tampons may play in toxic shock came from evidence presented by the plaintiff in a lawsuit against Procter and Gamble, manufacturers of Rely, in 1983. Previously unreported data from an investigation funded by Proctor and Gamble revealed that toxin-producing strains of *Staphylococcus aureus* produce more toxic shock toxin when grown on a Rely Super tampon (the only tampon on the market that contained polyester foam cubes in addition to cross-linked carboxymethylcellulose) than on most other tampons. In another experiment, toxin was produced when the staph organisms were cultivated on cross-linked carboxymethylcellulose (found only in certain brands), and no toxin was able to be detected when *Staphylococcus aureus* was grown on plain cotton under similar conditions. Data also showed that although the staphylococcus organisms were able to grow equally well on all brands of tampons, there were differences in the amounts of toxin that they produced (Marwick, 1983).

Other suggested mechanisms for tampon association with TSS are also speculative but include such possibilities as trauma or laceration of the vagina caused by the plastic applicator tips, ulcer formation on the vaginal walls because the "supers" absorb everything, including the normal secretions within the vagina, and perhaps even by increasing the number of organisms in the vagina through introduction from the fingers when inserting a tampon.

Making an informed choice about tampon usage is obviously hampered by all the questions, hypotheses, speculations and theories, and no real answers. Perhaps the risk of TSS could be almost eliminated by avoiding tampons entirely, but women who do not want to give up the convenience but are concerned about the safety may want to take the advice of a 1982 report issued by an Institute of Medicine committee. The committee cautioned that women, especially young women between the ages of 15 and 24, should avoid the use of high absorbency tampons to lessen their risk of developing TSS, advised that women who have given birth should avoid tampons for 6–8 weeks, and also that women who have had toxic shock should never use tampons at all. It also makes sense to be careful not only about the *kind* of tampons worn, but the way in which they are used. Women who choose to use tampons should wear them only when the flow is heavy and not throughout the entire period (never throughout the entire month) and change frequently, stick to the "regular" rather than the "super" variety, and use a pad at night. One study suggested that women who used either oral contraceptives or spermacides appeared to be at lower risk for development of toxic shock (Shelton and Higgins, 1981). The researchers speculated that an antibacterial effect of spermicides may inhibit *Staphylococcus aureus* and that the protective effect of oral contraceptives may be related to reduced menstrual flow and a change in cervical mucus.

ENDOMETRIOSIS

Endometriosis is a condition in which bits of functioning endometrial tissue are aberrantly located outside of their normal site, the uterine cavity. These endometrial implants can occur deep in the uterine muscle, on the surface of the uterus, on the ovaries, on the broad ligaments, on the pouch of Douglas, or anywhere else in the pelvis. Sometimes, they are found in the vagina and on the cervix, infrequently on the vulva and perineum, and (exceedingly rarely) even on the arm, leg, and lung. The glands and arteries of these ectopic endometrial tissues respond to cyclic ovarian hormones just like the endometrial lining of the uterus; when the lining bleeds during menstruation, so may these implanted sites, right into the peritoneal cavity. Endometriosis is a common disease, but it is often without symptoms. It has been reported as being found unexpectedly in 30%–50% of women undergoing pelvic surgery for any reason.

Internal endometriosis, or adenomyosis, indicates that the endometrium of the uterus has dipped into or invaded the muscle or myometrium of the uterus. More commonly found in older women, its greatest incidence occurs between 40 and 50 years of age. External endometriosis, located anywhere in the pelvis, occurs in younger, nonparous women and is often associated with infertility, perhaps because it so frequently (50%) occurs on one or both ovaries. When the ovaries are involved, the endometrial implants become bloodfilled cysts that can easily rupture, spill their contents into the peritoneal cavity, and form adhesions. The larger cysts that form as a result are often called "chocolate cysts" because of the color and consistency of their contents.

Etiology

The cause of this inordinate ability of endometrial tissue to become transplanted and grow in sites other than its regular location is not definitely known, although there are several theories. One of the most widely accepted explanations of pelvic endometriosis was proposed in 1921 by Sampson, who suggested that menstrual blood containing little fragments of endometrium was regurgitated upward through the fallopian tubes into the peritoneal cavity during menstruation. The escaped endometrial particles implant and then grow on the peritoneal surfaces. Based on this supposition, it has also been proposed that endometriosis may be encouraged to develop as a result of douching during menstruation, by strong menstrual cramping, or even by sexual intercourse during menstruation. There is no real evidence to support any of these ideas, but Cohen has noted that in Israel, where there are strong religious prohibitions against intercourse during menstruation, doctors claim that they never see instances of endometriosis (1977).

A more inclusive theory than Sampson's is that of "coelomic metaplasia"—that the cells of the peritoneum are embryologically derived from the same cells that give rise to the reproductive organs, the coelomic epithelium of the genital ridge. It is thought possible that some of the cells of the peritoneum may remain undifferentiated into adult life and, under hormonal or inflammatory stimulation, retain the capacity to become endometrium. This theory can account for endometriosis occurring anywhere in the pelvis, and in sites inexplicable by the Sampson theory. There are other ideas as well; perhaps more than one mechanism functions in the development of endometriosis. Malinak et al. (1980) suggest a genetic basis and estimated that woman with an affected mother or sister has a greater risk of developing the problem. For the woman who has the disease, where it came from is less significant than what to do about it.

Symptoms and Diagnosis

Endometriosis is not easy to diagnose. Some of its manifestations are the same as in other conditions, such as pelvic inflammations, ovarian cysts, and even ovarian cancer. It is possible that there may be no symptoms at all, whether the endometriosis is minimal or is extensive. On the other hand, there may be a great deal of pain, either before menstruation, during menstruation, and/or during sexual intercourse. There also may be abnormal uterine bleeding. As Willson et al. point out, the disturbing features of endometriosis are that "all degrees of involvement are encountered without a single symptom and that severe dysmenorrhea often occurs in women with minimal endometriosis" (1983). Kistner (1979) estimated that 30%–40% of women with endometriosis are infertile, and an earlier study by Boutselis et al. (1975) claims that absolute or relative infertility is present in approximately 75% of all endometriosis patients, virtually making sterility a symptom. The cause of the infertility is hard to explain, since the fallopian tubes are almost always open and ovulation occurs.

The only certain method of diagnosing endometriosis is by seeing it. This can be done through laparoscopy, the direct visualization of the internal organs with an optical instrument and light inserted through abdominal incision. A tissue biopsy of the suspected nodule taken at the same time and examined microscopically can confirm the diagnosis.

The treatment of endometriosis obviously depends on the extent of the disease, the age of the woman, whether or not she wants to become pregnant and is having difficulty, and if there is any pain. The sites do not become cancerous, and they are completely dependent on hormonal cycles during the reproductive years. If there are no distressing symptoms, and particularly if a woman is approaching menopause, no treatment is the best treatment.

Oral contraceptives eliminate ovulation, prevent dysmenorrhea, and control the growth of endometriosis. Given continuously

throughout the month, they completely eliminate menstruation and produce amenorrhea for long periods. In many instances, such hormonal suppression of ovulation and menstruation produces relief of symptoms of pelvic endometriosis. The anovulation and amenorrhea can persist for a long time after hormones are discontinued, however, so this is not a method for a woman who wants to become pregnant. Women who are concerned about taking hormones or who fall into the group for whom oral contraceptives are definitely contraindicated should also not choose this form of therapy.

Danazol (Danocrine) is a synthetic steroid derived from testosterone. It acts to suppress pituitary gonadotropin production, thus inhibiting ovulation. In the standard dose of 800 mg daily, danazol causes regression and atrophy of the endometrium in the uterus and the ectopic sites, resulting in amenorrhea and pain relief. The pituitary suppression is usually reversible, ovulation and menstruation return, and, with luck, the pain does not return for a while. In one study, 90% of the women experienced pain relief after 6 months on danazol, but more than half of the women had recurrence of symptoms within 1 year (Moore et al., 1981).

Hormone therapy suppresses, it does not cure. Danazol can "buy time" for a previously infertile woman and provide her with an opportunity for pregnancy after a course of treatment. Approximately 50% of women treated with danazol can become pregnant. Danazol, however, is a very expensive drug, costing well over $100 for a 1-month supply, and has side effects that include acne, weight gain, irritability, and, occasionally, masculinization symptoms. There is evidence that danazol affects lipids in the blood and causes a decrease in plasma levels of high-density-lipoproteins (HDLs) (Malkonen, et al., 1980). Since low levels of HDLs correlate with the development of coronary heart disease, Luciano et al. (1983) cautions that danazol treatment could be associated with cardiovascular risk.

Surgical treatment is used, but generally it is very conservative unless a huge cyst has formed on the ovary or the disease is widespread enough to cause bowel or ureter obstruction and acute pain. When a woman is older than 40 or somewhat younger but does not want any more children, has very severe symptoms, and extensive endometriosis, a hysterectomy can be performed without removing the ovaries. Pain does not usually recur even though ovarian function remains. Some doctors, however, advocate hysterectomy plus bilateral salpingo-oophorectomy (removal of both ovaries and both fallopian tubes), believing that leaving a source of estrogen and progesterone will encourage growth of endometrial implants and recurrence of the symptoms.

Endometriosis is a chronic condition that produces severe pain in many women. It is difficult to diagnose, its cause is unknown, and its treatment is often frustratingly inadequate. Since 1980 the Endometriosis Association, a national self-help group with local chapters, has served as a clearing house for information about endometriosis and to offer support and help to affected women. Their resources include a report of their nationwide questionnaire survey of 365 women that chronicles the history of each woman's experience with endometriosis, her symptoms, diagnosis, and treatment. Further information and chapter locations can be obtained by writing: Endometriosis Association, P.O. Box 92181, Milwaukee, WI 53202. Another organization with endometriosis information is Resolve, P.O. Box 474, Belmont, MA 02178.

PREMENSTRUAL SYNDROME

Many women have no premonition at all that they are going to menstruate or they have lit-

tle reason to pay any attention to their impending menstrual flow. A look at the calendar or the beginning of bleeding is their only indication that the time has come again.

Other women are aware of breast tenderness, abdominal swelling, perhaps constipation, or an aggravation of acne just before menstruation or in the 10–14 days of the preceding luteal phase. Some definitely recognize increased nervousness or notice that they are less tranquil and somewhat short-tempered before menstruation. Such psychological changes could be related to the physical manifestations, since it would not be unusual if puffiness, swelling, and an increase in the number of pimples on a woman's face were to make her vaguely tense and irritable. Most women who have these changes are not incapacitated by them and are able to consider them more as a nuisance than as *symptoms* or *problems*. Other women, however, do have a much greater degree of predictable and recurrent physical, psychological, or behavioral distress that is more than just annoying. Each woman is unique, and each experiences her menstrual cycle individually. Although it is possible that some of the same things that are ignored by some are considered distressing symptoms by others, it is evident that some women experience dramatic premenstrual changes that cause a disruption in their personal and professional lives and result in limitation of their usual activities. In the last few years, these difficulties have collectively been known as the *premenstrual syndrome* or PMS. In medical parlance, a syndrome is a group of symptoms that collectively indicate a disease state or at least an abnormal condition.

The existence of PMS is real. It is not an imaginary disorder produced in the minds of "neurotic" women, the way in which many women's health problems have been viewed in the past by health professionals. But the extent to which the symptoms debilitate or incapacitate a woman is highly variable, and while there is every reason to take PMS seriously, there is little consensus in the medical literature and among researchers about exactly what constitutes the disorder, what causes it, and how to treat it. Even obtaining an accurate and consistent description of the condition has been problematic. The term *premenstrual tension* was first used in 1931 when American gynecologist Robert Frank described 15 women with a "syndrome" of irritability, anxiety, depression, and edema (swelling) in the days before menstruation or in the first 4 days of the flow. Frank attributed the syndrome to "excess circulating levels of female sex hormones." The nebulous clinical state has subsequently been referred to as *premenstrual distress, congestive dysmenorrhea, pelvic congestion syndrome,* the *toxemia of menstruation,* and now *premenstrual syndrome.*

In the 50-plus years that have elapsed since Frank's description, the original four symptoms have been expanded to include many more physical and emotional manifestations, such as nervousness, inability to concentrate, paranoid attitudes, suicidal thoughts, insomnia, tendency to drop things, acne, greasy hair, dry hair, fatigue, exhaustion (or alternatively greatly increased physical activity), heightened acuity in sight, smell, and hearing, increased thirst, increased appetite, craving for sweets, weight gain, breast tenderness, diarrhea, constipation, headache, nausea, vomiting, hand tremors, decreased motivation, decreased efficiency, lack of impulse control, increased sexual receptiveness, even asthma and epilepsy—the list goes on and on. More than 150 different cyclic changes have been included, and just a few of these have any positive aspects; all the rest are illness symptoms. The psychological symptoms have been interpreted as either instigating or resulting from the physical symptoms, depending on the investigator. The timing of the symptoms in relation to menstruation has also been ex-

panded. Originally designated as occurring during the 8 days of the paramenstruum (4 before flow, 4 during flow), PMS symptoms can now also start at ovulation and continue through the menstrual period. They would obviously have to disappear at some time during the cycle, however, or there would be no basis for a menstrually related link.

Since there is vagueness and considerable confusion as to what should be included in PMS—some have even added dysmenorrhea to the list—it is not unexpected that the reported incidence reflects a similar inconsistency. Estimates in the scientific literature as to what percentage of women experience these symptoms are reported to be between 5% and 95% of all women, with ages of occurrence being between 10 and 60 years. Much of the difficulty in determining the exact incidence has resulted not only from indecision about which symptoms to include, but also because little attention has been given to the question of severity of the symptoms. Studies have not distinguished between tolerable premenstrual changes of the nuisance type and the incapacitating variety that restricts the daily routine.

When a range of physical and emotional changes becomes that widespread, the plausibility of a single explanation, that is, PMS, becomes questionable. After all, if one were to construct a questionnaire and ask all people—men, women, and children—to assess which of the aforementioned signs they experienced at any time during a 30-day period, it would not be unrealistic to expect that 100% of them would report one or more of them. Much of the evidence for the incidence of premenstrual symptoms does come from retrospective questionnaires that ask women to report on their memory of various changes. One such comprehensive and widely used instrument has been the 47-item Moos Menstrual Distress Questionnaire devised by Rudolph Moos. As typified by the title, primarily negative effects are measured. Mary Jane Parlee (1973) pointed out the questionnaire's methodological weaknesses (half of Moos' samples were taking oral contraceptives and 10% were pregnant) and suggested that because the questionnaire may really have been measuring unwarranted assumptions and stereotyped beliefs about the menstrual cycle, its results should be interpreted with caution. Nevertheless, until recently this questionnaire remained the only symptom-rating scale available. Even better designed self-assessment questionnaires, however, could still have problems with a built-in bias on the part of the subjects because of the pervasive negative expectations society has concerning premenstrual symptoms. Diane Ruble's study (1977) of women undergraduates at Princeton University demonstrated that when women were persuaded through bogus "brain-wave" tests that they were premenstrual, they reported more negative symptoms of premenstrual tension and anxiety than women who were led to believe that they were in the middle phase of the cycle. In actuality, all the women were somewhere in between these two phases, suggesting that there is a strong tendency to attribute any bad days or negative symptoms as being menstrually related, even though such experiences could occur randomly throughout the cycle.

Etiology of PMS

Many of the physical symptoms are the ones that are inconsistently present, and it is the psychological and behavioral changes that are more frequently reported. The one physical symptom that has been classically described, however, and is most easily measured is *edema*, or swelling, as a result of water retention. Edema of the digestive tract leads to bowel and abdominal distension; edema of the breasts results in swelling and tenderness

of the breasts; and edema of the cerebrum may account for premenstrual headache and could also be related to mood changes. Salt and water retention is widely accepted as one possible reason for PMS.

There is no evidence, however, that edema is actually related to premenstrual symptoms. When diuretics (to cause elimination of body water) were tested in double-blind controlled experiments in which neither the physician nor the subject knew who was getting the diuretic, there was no apparent relationship between salt and water retention and mood changes. Furthermore, when sodium and water retention occur in those conditions associated with adrenal, heart, kidney, or liver disease, it produces no emotional mood swings or other changes that are said to occur in the premenstrual syndrome.

It has also been difficult to establish the reason for the salt and water retention. One popular theory to account for it is faulty estrogen metabolism, which causes excessive amounts of estrogen to be retained instead of metabolized relative to the amount of progesterone. Similarly, the also popular hypothesis of faulty luteinization could result in the same disordered estrogen-progesterone ratio. Estrogens have a weak action in causing water and salt retention, which might account for the edema. But although some studies have reported elevated estrogen levels or elevated estrogen/progesterone ratios in women with PMS, virtually an equal number have found normal estrogen values and no changes in the estrogen/progesterone ratios in the luteal phase of the cycle.

An abnormality or some change in the circulating levels of other hormones, such as prolactin from the anterior pituitary gland, vasopressin from the posterior pituitary gland, or aldosterone from the adrenal cortex, has also been proposed to account for water retention and subsequent premenstrual symptoms. These theories have stood up to further exploration no better than the other hypotheses. Not only is there no evidence that prolactin has water-retention effects in humans, but in a number of studies when prolactin levels were measured in women with PMS and controls, no demonstrable differences could be shown. Neither have aldosterone nor vasopressin levels been proven as yet to be elevated in women with PMS.

Many investigations have attempted to identify a relative progesterone insufficiency as a cause of PMS since symptoms tend to intensify late in the luteal phase as progesterone levels normally decline. But as pointed out in a review by Rubinow and Roy-Byrne (1984), out of 10 studies of progesterone in relation to PMS, five showed *lowered* levels in the luteal phase of women with PMS when compared with controls, one demonstrated that one-third of the PMS patients had *significantly* low progesterone, three studies found *no difference* in progesterone levels between PMS women and controls, and one revealed a *higher* level of progesterone in women with PMS. Clearly, changes in progesterone levels have not been shown to account for premenstrual symptoms.

Vitamin B_6 deficiency has also been suggested as a cause of PMS on a theoretical basis; B_6 may enhance estrogen clearance from the blood or may act as a coenzyme in the synthesis of certain brain neurotransmitters (dopamine and serotonin) that could be involved in the regulation of mood and behavior. Moreover, because some women experience symptoms premenstrually that are similar to those produced by hypoglycemia or low blood sugar (faintness, fatigue, shakiness), abnormalities in glucose metabolism have also been proposed as a reason. So have changes in endogenous opiate levels, prostaglandins, melanocyte-stimulating hormone, and adrenal corticoid hormones. None of these theories has been proven, and there is no evidence that levels of any of the hormones differ in women with or without premenstrual symptoms. The exact cause or causes of PMS are still unknown, but it is possible that interactions between some hormones and other of the biologically active substances mentioned may play a role.

Treatment of PMS

The wide variety of symptoms that fall under the umbrella of PMS are not likely to have a single cause. They are, therefore, not likely to be amenable to treatment with a single therapy. Since there are no diagnostic laboratory tests that can reliably determine the existence of PMS, it is a woman herself who must decide whether what she experiences cyclically is too mild in magnitude to qualify as a syndrome or disorder, is moderate and can be handled through self-help methods, or is so severe and debilitating that she requires medical treatment. But because a syndrome is an aggregate of signs and symptoms and, in the case of PMS, an unusually complex number of them, there are no quick and easy pill-or-potion solutions to the problem.

In addition to various combinations of steroids, both natural and synthetic, doctors have prescribed diuretics (water pills), painkillers, prostaglandin-inhibitors, high doses of vitamins, tranquilizers and other psychoactive drugs, oral contraceptives, bromocriptine (a drug capable of lowering plasma prolactin), lithium carbonate (a drug used therapeutically in psychotic or manic-depressive states), danazol, and placebo. A placebo is an inactive substance administered for suggestive effect or to placate the patient. (The ethics of requiring a patient to pay for a prescription with no medical effect is another problem.) The results of these therapies are variable and, in general, inconclusive. All (including placebo) have reportedly been effective for some women. The most publicized therapy, advocated originally by Dr. Katharina Dalton (1964) in Great Britain, is natural progesterone that is inactivated orally and must be administered by injection or vaginal or rectal suppository. It can also be absorbed in solution through a rectal enema—not everyone's favorite mode of administration—or in powdered form sublingually under the tongue. Most women taking progesterone use the suppository form that must be specially compounded by a pharmacist. But even that method can be unpleasant since suppositories leak vaginally or are spontaneously expelled rectally. The monthly cost can range from $35 to several hundred dollars, depending on the amount used and its source. Immediate side effects of progesterone can include changes in the menstrual flow (heavier, lighter, earlier, delayed, spotting), dizziness, faintness, libido changes, and weight gain or loss. For some women, progesterone has been the answer, but long-term side effects of progesterone are unknown. There have been no controlled studies of its safety in large numbers of women over long periods of time. And despite the popularity of progesterone for PMS, there has been no scientific proof in controlled double-blind studies that progesterone is superior to placebo or to any other kind of therapy. It is the treatment of choice, however, in many of the private proprietary PMS clinics that are operating in many parts of the country. The first such center, Premenstrual Syndrome Program, was founded in a Boston suburb in 1981 by Harvard neuroendocrinologist-psychiatrist, Ronald V. Norris. He estimates that at some point in their lives, 85% of all women (a mind-boggling 49 million) suffer from PMS, and he claims that progesterone therapy, reserved for the more serious cases of PMS, is the most effective treatment known (Norris and Sullivan, 1983).

Not everyone is equally convinced that PMS is a well-defined entity that affects such large segments of the population and requires therapy at a clinic. Women's health advocates are increasingly concerned about the "medicalization" of menstrually related symptoms and the labeling of PMS as another hormone-deficiency disease to be "cured" by doctors with drugs with the potential for unknown long-term adverse effects. At this point, no one has all or even some of the answers about premenstrual syndrome, and it is evident that what works for one woman may not be effective for another. If symptoms are severe enough to require therapy, there are a variety of nonmedical self-help approaches that could be tried first. They may not offer the quick-fix promise of a drug, but they may provide enough relief to avoid a possible long-term risk.

Before any treatment, either self-care or medical, is instituted, however, it is essential to establish definitely the presence of PMS and the severity of the symptoms. This can only be done by charting a daily diary of distressing symptoms for at least two complete cycles. The day of the cycle, the date when bleeding occurs, and all symptoms should be noted. Any unusual changes from the regular routine or external stresses should also be recorded. Qualifying PMS symptoms should begin not more than 14 days before menstruation and should occur at almost exactly the same day of the cycle each month, disappearing on the day of, or shortly after the onset of, menstruation. If the symptoms shift around to other times of the month they are unlikely to be related to menstruation and are probably associated with some other factor (perhaps physical or an external stress). Once it has been established that the symptoms are cyclic, premenstrual in nature, and defined, the major ones should be selected for control. If the major symptoms are water-retention and edema, salt restriction may solve the problem. If they are similar to those caused by low blood sugar, elimination of sweets and institution of frequent protein and complex carbohydrate meals (a hypoglycemia diet) could be effective. If nervousness, irritability, and short temper are prominent, caffeine should be eliminated. When depression is a major symptom, there are reports that vitamin B_6 or pyridoxine, in doses of 100–200 mg daily throughout the cycle for at least 6 weeks may alleviate it. Vitamin B_6 in higher doses (up to 800 mg) has been tried and is allegedly beneficial for depression, but, unfortunately, megadoses of pyridoxine can produce symptoms of overdose that include nausea, headache, dizziness, *and depression.* Exercise and other means of stress reduction, important for fitness and health at any time, can also be effective in alleviating PMS. The point is, since prescription drugs have not really been shown to be any more effective than some of the above self-care regimens for PMS, it is certainly worthwhile trying them for 3 months. Progesterone and other hormonal or antihormonal treatments should be reserved for the most severe situations that do not respond. Michelle Harrison's book, *Self-Help for Premenstrual Syndrome,* (Matrix Press, P.O. Box 740, Cambridge, MA 02238) contains monthly charts and guidelines to diagnosis and self-treatment of PMS.

The Politics of PMS

One hundred years ago, Victorian physicians warned that menstruation might cause temporary insanity, and that women could go berserk, attacking friend and family and even killing their infants. Such unfortunate women are so subject to their menstrual influence, said the doctors, that they should be locked up during their menstrual years for their own good and the good of society. In 1981, in a harrowing echo of the violent menstrual insanity described by nineteenth-century doctors, a precedent based on women's periodic derangement was established in the British courts. But instead of being locked up after

going "berserk," the two Englishwomen involved were sent for hormone injections to quell their monthly lack of control. Both women claimed that it was premenstrual syndrome that provoked them to violence and, in separate court cases, successfully used PMS as a defense. One woman, 37 years old, was charged with murder after she drove her car at her lover, pinning him to a telephone pole. The other, a woman of 29 and already on probation for stabbing a woman to death in a brawl, threatened to stab a police sergeant for insulting her. She became a "raging animal" each month unless treated for premenstrual syndrome, according to her attorney. After the trials, during which Dr. Katharina Dalton appeared as a defense witness, both women were placed on probation, and the one who used her auto as a weapon was barred from driving for 1 year.

That there is something in a woman's physiology that allows for diminished responsibility in a murder charge is a chilling notion, but PMS is now a legal defense in England and in France can be the grounds for a plea of temporary insanity. Thus far the one case in a U.S. criminal court in which premenstrual syndrome was claimed as a defense involved child battering. The claim was dropped by the defendant after she and her attorney pleaded guilty to a lesser charge in a plea-bargain, but the potential for acceptance of the legal status of PMS exists, and many are concerned. The long- and hard-fought gains in the status of women could hardly be advanced by the idea that women's nature is prone to uncontrollable rages and uncontrollable acts. Believing in a woman's monthly insanity could even justify the Victorian remedy—if she is out of control, it is necessary to control her.

In the wake of the British court cases, PMS became the media disease of the year. By spring of 1982, it was the subject of nearly every TV and radio talk show, and articles with titles like "PMS, One Woman's Nightmare" appeared in newspapers and magazines. Women with premenstrual symptoms were called PMS sufferers and PMS victims, and there was an evident prevailing belief that a biological basis, rooted in a progesterone deficiency, was the cause of the problem.

How much scientific evidence is there that even some women can be periodically deranged, controlled by, and at the mercy of their recurrent cycles? Women are certainly exposed to wide fluctuations of hormonal levels during the month but, as previously indicated, attempts to correlate those varying hormone levels with symptoms that occur cyclically have produced conflicting and inconclusive data.

Katharina Dalton, who directs the Premenstrual Syndrome Clinic in London, has done research for 35 years and has written extensively (50 articles, 5 books) on what she has called "the curse of Eve." She claims that menstruation can produce disastrous effects, having correlated with the premenstrual or the menstrual period an increase in sick days among factory workers, more emergency admissions to hospitals for accidents or psychiatric disorders, an increase in misbehavior and decrease of intellectual performance among schoolgirls, and an increased incidence of crimes of violence and convictions for alcoholism and prostitution. She has reported that more than half of the emergency admissions of children to a British hospital occurred when their mothers had or were just about to have their menstrual periods, and Dalton attributed the occurrence to the mother's inattentiveness, poor motor reactions, or general bad temper. Evidently believing that the behavioral influences of the menstrual period apply to all women, Dr. Dalton wrote a leaflet for the Royal Society for the Prevention of Accidents in Great Britain in which she advised that women should "understand, recognize and sensibly adjust their lives" around their menstrual cycles and not drive an automobile on long journeys during the 8 days before, during and after their menstrual periods. Stating that women are two and one-half times more likely to have an accident at this time, she further cautioned that women are at their "lowest ebb" and have "increased irritability and aggression, duller mental and physical ability, (are) tense, irrational, impatient and more easily tired" during the paramenstruum. Although evidence for all of these detriments may be specious and inconclusive, the benefits of such advice, if heeded, to the Royal Society is clear: getting half the population of Great Britain off the roads for 8 days a month is bound to prevent accidents.

In addition to Dalton's investigations, there are a number of other studies and large accumulations of data that apparently substantiate the assumption that women are more vulnerable to serious psychological problems premenstrually or that, at least in some women, an underlying emotional disorder may be triggered or precipitated at that time (MacKinnon et al., 1959; Wetzel and McClure, 1972; Belfer et al., 1971; Mandell and Mandell, 1967). Study after study links the time just before and during the menses with drastic negative behaviors and almost obscures the obvious: that the vast majority of women manage to live through their monthly cycles without any extreme variation in their behavior. Some of the claims that relate the paramenstruum to great emotional upsets have been challenged by other investigators as to the methodology that was used in the studies, and the interpretations that have been placed on the data (Parlee, 1973; Sommer, 1983). When the relationship between the menstrual cycle and child illness, accident-proneness, attempted suicide, or crime is studied, rarely has the relevance of external environmental situations been considered. Rare, too, is mention of the fact that men also have car accidents, have psychiatric admissions to hospitals, attempt suicide, and commit crime, and that their percentage is higher that it is in women. Ultimately, someone will undoubtedly attempt to correlate the incidence of great emotional upset in these men with the menstrual cycles of their wives, mothers, sisters, or lovers.

As Paula Weideger and others have noted, there has been an obvious and serious omission in the studies that have produced the speculations, suggestions, and claims concerning the effect of the menstrual cycle on behavior. *There are virtually no investigations linking positive behaviors like creativity, increased self-confidence, and optimism with the phase of the cycle.* The assumption that the menstrual cycle affects behavior in a negative fashion may indeed be a reflection of cultural bias and stereotyped beliefs in our society.

It could also be possible that the degree of difficulty that a woman has paramenstrually depends on her personality, the current environmental stresses to which she is subject, and her general physical state, which could include hormonal changes. But it is not possible at the present time to establish biological variables that would have a cause-and-effect relationship to her difficulties.

Estelle Ramey has pointed out that men, like women, are also subject to a changing internal environment (1976). Cyclic changes in gonadotropins and steroids occur in males and may well affect mood and behavior, but these fluctuations have been extensively explored and studied only in females. If there are any behavioral effects of their internal rhythms in men, they have learned to deny them. Women have learned to accept, expect, or even exploit them. With all the emphasis on the psychological and physiological correlates of hormonal cycles in women, with all the literature both scientific and popular that described premenstrual syndrome, any depression, hostility, or anxiety that may have greater relationship to her stressful life experiences than to her hormone levels can easily be blamed on the menstrual cycle. Emotional instability is part of our culturally defined female stereotype. Some women may find that it is permitted—even expected—in our society to exhibit negative behaviors. They are supposed to be once-a-month witches and bitches.

Scientific investigation is seriously hampered in an atmosphere of unwarranted assumptions, sexual stereotyping, and cultural bias. Why should there be so much emphasis on only the menstrual cycle? The validity of data concerning the effect of biorhythmic phenomena on such things as behavior, illness, effects of medication, job performance, competency, psychiatric hospitalizations, and accident rates is obscured when it is applied to only one-half the population. When research frees itself from cultural bias and explores human rhythms in their broadest aspects, the contributions to society at large will be of greater value.

REFERENCES

Abraham, S. and Beumont P. J. V. How patients describe bulimia or binge eating. *Psychol Med* 12:625–635, 1982.

Belfer, M. L., Shader, R. I., Carroll M., *et al.* Alcoholism in women. *Arch Gen Psychiatry* 25:540–544, 1971

Binge eating and its management, Comments. *Br J Psychiat* 1141:631–633, 1982.

Boutselis, J. G., Vorys, N., and Neri, A. S. Endometriosis, in Gold, J. J. (ed). *Gynecological Endocrinology*, 2nd ed. Hagerstown, Md., Harper & Row, 1975.

Budoff, P. W. Zomepirac sodium in the treatment of primary dysmenorrhea syndrome. *New Engl J Med* 307(12):714–719, 1982.

Carl, J. Bulimia. *Wisconsin State Journal*, Sunday, Nov. 7, 1982.

Chan, W. Y., Fuchs, F., and Powell, A. Effects of naproxen sodium on menstrual prostaglandins and primary dysmenorrhea. *Obstet Gynecol*, 61(3):285–291. 1983.

Cohen, M. E. Current concepts in endometriosis. *J Reprod Med* 19(5)(suppl):323, 1977.

Dalton, K. Childrens' hospital admission and mother's menstruation. *Br Med J* 2:27–28, 1970.

Dalton, K. The Curse of Eve. *Occup Health* 28(3):129–133, 1976.

Dalton, K. *The Menstrual Cycle.* New York, Warner, 1971.

Dalton, K. *The Premenstrual Syndrome.* Springfield, Illinois, Charles C. Thomas, 1964.

Dalton, K. Premenstrual Tension and Driving. Royal Society for the Prevention of Accidents, 1979.

Davis, J. P., Chesney, P. J., Wand., P. J., et al. Toxic shock syndrome. *N Engl J Med* 303(25):1429–1435, 1980.

Frank, R. T. The hormonal causes of premenstrual tension. *Arch Neurol Psych* 26:1053–1057, 1931.

Fraser, I. S., McCarron, G., Markham, R., et al. Long-term treatment of menorrhagia with mefenamic acid. *Obstet Gynecol* 61(1):109–112, 1983.

Frisch, R. E., and MacArthur, J. W. Menstrual cycles: Fatness as a determinant of minimum weight for height necessary for their maintenance or onset. *Science* 185:949–951, 1974.

Frisch, R. E. and Revelle, R. Height and weight at menarche and a hypothesis of critical body weights and adolescent events. *Science* 169:397–399, 1970.

Frisch, R.E., Wyshak, G., and Vincent, L. Delayed menarche and amenorrhea in ballet dancers. *N Engl J Med* 303(1):17–19, 1980.

Garn, S. and LaVelle, M. Reproductive histories of low weight girls and women. *Am J Clin Nutr* 37:862–866, 1983.

Hancock, K. W., Scott, J. S., Panigraphi, N. M., and Stitch, S. R. Significance of low body weight in ovulatory dysfunction after stopping oral contraceptives. *Br Med J* 2(6032):399–401, 1976.

Harvey M., Horwitz, R., and Feinstein, A. Toxic shock and tampons. *JAMA* 248(7):840–846, 1982.

Humphries, L. L., Wrobel, S., and Weigert, H. T. Anorexia nervosa. *Am Fam Physician* 26(5):199–204, 1982.

Kistner, R. W. Endometriosis and infertility. *Clin Obstet Gynecol* 22:101–119, 1979.

Luciano, A. A., Hauser, K. S., Chapler, F. K., et al. Effects of danazol on plasma lipid and lipoprotein levels in healthy women and in women with endometriosis. *Am J Obstet Gynecol* 145(4):422–426, 1983.

MacKinnon, I. L., MacKinnon, P. C. B., and Thomson A. P. Lethal hazards of the luteal phase of the menstrual cycle. *Br Med J* 1:1015, 1959.

Malinak, L. R., Buttram, V. C., Elias, S., Simpson, J. L. Heritage aspects of endometriosis: II. Clinical characteristics of familial endometriosis. *Am J Obstet Gynecol* 137(3):332–37, 1980.

Malkonen, M., Manninen, V., and Hirvonen, E. Effects of danazol and lynestrenol on serum lipoproteins in endometriosis. *Clin Pharmacol Ther* 28(5):602–604, 1980.

Mandell, A. and Mandell, M. Suicide and the menstrual cycle. *JAMA* 200:792, 1967.

Mars, D. R., Anderson, N. H., and Riggall, F. C. Anorexia nervosa: A disorder with severe acid–base derangements. *South Med J* 75(9):1038–1042, 1983.

Marwick C. Holdup of toxic shock data ends during trial in Texas. *JAMA* 250(24):3267–3269, 1983.

Monif, G. R. Tampons and toxic shock syndrome. *Female Patient* 7(12):42–48, 1982.

Moore E. E., Harger, J. H., Rock, J. A., and Archer, D. F. Management of pelvic endometriosis with low-dose danazol. *Fertil Steril* 36(1):15–19, 1981.

Norris, R. V., Sullivan, C. *PMS: Premenstrual Syndrome.* New York, Rawson Associates, 1983.

Parlee, M. B. Stereotypic beliefs about menstruation: A methodological note on the Moos menstrual distress questionnaire and some new data. *Psychosom Med* 36(3):229–240, 1974.

Parlee, M. B. The premenstrual syndrome. *Psych Bull* 80(6):454–465, 1973.

Ramey, E. Men's cycles (they have them too, you know), in Kaplan, A. and Bean, J. *Beyond Sex-Role Stereotypes.* Boston, Little, Brown, 1976.

Rubinow, D. R., and Roy-Byrne, P. Premenstrual syndromes: Overview from a methodologic perspective. *Am J Psychiatry* 141(2):163–172, 1984.

Ruble, D. M. Menstrual symptoms: A reinterpretation. *Science* 197:291–292, 1977.

Sampson, J. A. Peritoneal endometriosis due to menstrual dissemination and endometrial tissue into the peritoneal cavity. *Am J Obstet Gynecol* 14:442, 1921.

Shelton, J. D. and Higgins, J. E. Contraception and toxic shock syndrome. A reanalysis. *Contraception* 24:632–635, 1981.

Sommer, B. How does menstruation affect cognitive competence and psychophysiological response? *Women & Health* 8(2/3):53–89, 1983.

The tampon controversy. *Chicago Tribune,* May 3–8, 1981.

Todd, J., Fishaut, M., Kapral, F., and Welch, T. Toxic shock syndrome associated with phage-group-1 staphylococci. *Lancet* 2:1116–1118, 1978.

Wagner, G., Bohr, L., Wagner P., and Petersen, L. Tampon-induced changes in the vaginal oxygen and carbon dioxide tensions. *Am J Obstet Gynecol* 148(2):147–150, 1984.

Wakat, D. K., Sweeney, K. A., and Rogol, A. D. Reproductive system function in women cross-country runners. *Med Sci Sports* 14(4):263–269, 1982.

Warren, M. P. The effects of exercise on pubertal progression and reproductive function in girls. *J Clin Endocrinol Metab* 51:1150–1157, 1980.

Warren, M. P., Jewelwicz, R., Oyrenfurth, I., et al. Significance of weight loss in evaluation of pituitary response to LH-RH in women with secondary amenorrhea. *J Clin Endocrinol Metab* 40:601–611, 1975.

Weideger, P. *Menstruation and Menopause, The Physiology and Psychology, The Myth and the Reality.* New York, Alfred E. Knopf, 1976.

Wentz, A. C. Body weight and amenorrhea. *Obstet Gynecol* 56(4):482–487, 1980.

Wetzel, R. D., and McClure, J. N., Jr. Suicide and the menstrual cycle: A review. *Compr Psychiatry* 13:369–374, 1972.

Willson, J. R., Carrington, E. R., Ledger, W. *Obstetrics and Gynecology.* 7th Ed., St. Louis, C. V. Mosby Co., 1983.

5. *The Basis of Biological Differences*

Back in the sixties, when young men began to let their hair grow down below their collars, and length of hair was virtually considered a measure of how radically young people had strayed from what was considered appropriate social behavior and political thought, the older generation would say, sometimes plaintively, sometimes with amusement, but often with resentment, ". . . from the back, you can't tell 'em apart. The boys look just like the girls!" This was considered a devastating comment on the lamentable actions of young people and was voiced long before unisex hairstyles and clothing became the fashion trend for all ages.

Although the statement was not intended as an accurate morphological observation, it is quite true. Furthermore, when people are fully clothed and we are denied a look at the external genitalia, sometimes it's hard to tell 'em apart from the front as well. Physical *sexual dimorphism*, the differences in body configuration between sexually mature adult males and females, is far less marked in humans than it is in some animals, in which the differences between the sexes are so great that the males and females do not even seem to belong to the same species. Even in our close primate relatives, the orangutans, gorillas, and baboons, the sexual differences in such features as skull size, dentition, general stature, and overall physique are highly exaggerated. This is not so in humans, who are the most highly variable of all species. There are many women who are taller and larger than many men; some men have narrow shoulders and wide hips, and some women have narrow hips and are small breasted. The physical differences in size, shape, and stature that occur between two individuals of the same sex are often far greater than the degree of difference between two individuals of opposite sexes. Not too many people actually possess the cultural ideal of the typical male or the typical female form; those that do tend to be photographed a lot and become models.

Our lesser sexual dimorphism, as compared with other primates, may be associated with our human *paedomorphism*, our tendency to carry many childlike, or even fetal, characteristics into adulthood. Anyone who has ever made fatuous comments about a pretty little baby girl only to be informed that *she* is actually a *he* recognizes that the physical differences between male and female infants, or even children until puberty, are minimal indeed. This retention of fetal or infantile physical features into adulthood is variable, but it is particularly apparent in some women who have high foreheads, smooth skins, and very rounded contours.

Adult retention of childlike morphology is also true of males, for they also diverge far less from their own juvenile forms than do other primates. From any fossil evidence available, physical differences between the sexes in humans is now less pronounced than it once was. Millions of years of evolution appear to have produced in us lesser, rather than greater, somatic variation between males and females.

Internally as well, males and females are anatomically very much the same. No one could distinguish the human liver, kidney, heart, or other organ of a man from that of a woman. The size of these organs is more related to the size rather than the sex of the owner. The brain weight in an "average" female is about 100 g less than in the male. Much was made of this to prove the lesser intellectual capacity and ability of a woman,

until it was recognized that the range in brain weights of adults is very great, and that proportionally, males and females differ less in brain weight than in total body weight. Table 5.1 shows the average weights of various human organs in adult males and females. Of the 10 trillion cells, more or less, that make up the human body, only those that are specialized into the reproductive system result in the physical differences between males and females. The functioning of all the rest of the systems is dedicated to the survival of the individual; only the reproductive system and its hormones, dedicated to the survival of the species, account for the anatomical variation between the sexes.

But "Vive la difference!" says the old sexual joke. Of course, there are dissimilarities in the bodies of men and women. A male has a penis and a female has a vagina. These are the primary sex differences and, with their appropriate internal organs and ducts, are what make a male, *male*, and a female, *female.* But besides the gonads, ducts, and genitalia, there is little physical difference between boys and girls until sexual maturity. One exception is in the length of the forearm, which, from birth, is

Table 5.1 AVERAGE WEIGHTS OF VARIOUS HUMAN ORGANS (ADULT IN GRAMS)*

Organs	Females	Males	Total Adult Range
Brain and meninges	1258	1375	1100–1600
Thyroid	34	30	11–60
Thymus	14	14	1–25
Heart	250	300	240–360
Lung (right)	525	626	400–650
Lung (left)	470	600	350–625
Liver	1500	1600	1200–1700
Spleen	150	165	80–300
Kidney	140	160	120–180

* From various sources

generally greater in males with respect to the length of the upper arm or the total body height, than it is in females. Girls also have somewhat more subcutaneous fat, on the average, than boys.

With puberty, however, visible changes appear that are the result of the secretion of pituitary gonadotropins and the subsequent secretion of androgen by the testes in males and estrogens from the ovaries in females. As the ability to reproduce is attained, the testes in males are able to produce millions of spermatozoa, and the male is able to have erections, ejaculations, can impregnate a female, and, thus, can father a child. Females have menstrual cycles, generally produce one egg a month, can become pregnant and carry a fetus for 9 months, give birth to it, and nurse it. More succinctly, females menstruate, ovulate, gestate, parturate, and lactate.

The tranformation from childhood to adulthood involves not only the acquisition of reproductive capacity but, also, certain changes in physique and physiology. These differences are the result of the action of sex steroids not only on the penis and scrotum in the male and the labia and clitoris in the female, but on many target tissues that are common to both sexes. Developmental changes in the skeleton, breasts, hair follicles, and muscles are quantitative rather than qualitative, however, and the adult differences in these *secondary sex characteristics* are merely a matter of degree.

CHANGES IN MALES AND FEMALES AT PUBERTY

The key word for the physical changes that accompany sexual maturation in both sexes is *increase*. There is a tremendous increase in the rates of growth of various body parts, and this acceleration results in changes in the skeleton, changes in the relative amounts of bone, muscle, and fat, physiological changes in the circulatory and respiratory systems, and development of the reproductive system. The age ranges during which the various changes occur is wide and depends on which event is being considered. James Tanner and his colleagues at the University of London Institute of Child Health did longitudinal studies of hundreds of British children, observing them and making measurements from onset to completion of puberty (1973). Figure 5.1 shows the average ages at which certain significant stages in puberty begin and end. In girls, the first outward sign of the onset of puberty is the beginning of development, or budding, of the breasts, and it can occur anywhere from ages $8\frac{1}{2}$ to $13\frac{1}{2}$. The growth of pubic hair is usually the next sign; it appears approximately a year before underarm hair starts to develop. About a year after the beginning of breast development, a girl reaches the peak of her adolescent growth spurt, and menarche starts a year or a year and a half later.

The results of several recent studies in the United States indicate that the average age at which girls menstruate is 12.8 years and is not significantly different from the age at which their mothers first menstruated. The progressive decline in the age of onset of menarche that has been occurring since the mid-nineteenth century is apparently stabilizing here as well as in many other countries, and it is thought that better nutrition and improved public health for greater numbers of people contribute to the trend.

In boys, the first outward indication of the onset of sexual maturation appears at age 11 or 12, later than it does in girls. Testicular enlargement is usually first, followed by growth of pubic hair. The penis begins to grow rapidly a year or so later. Axillary and facial hair appear usually 2 years after the development of pubic hair.

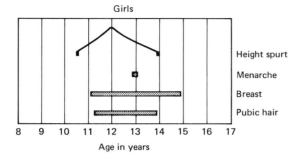

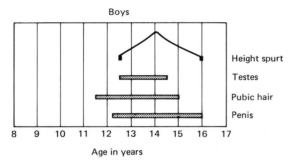

Figure 5.1. Significant events of puberty diagrammed as they occur in the average girl and boy. The highest point in the bar labeled "height spurt" is the peak velocity of the adolescent growth spurt. While the bars represent the beginning and completion of the pubertal events as they typically occur, any of them may start earlier and end later. For example, some girls begin breast development as early as 8 and have fully mature breasts at 13; others may begin at 13 and be completed at 18. Two-thirds of all girls show the first appearance of pubic hair after the beginning of breast development. Menarche, or the onset of menstrual periods, can take place anywhere from 9 to 17. (From "Growing Up," by J. M. Tanner. Copyright 1973 by Scientific American, Inc. All rights reserved.)

Girls start and complete their growth in height 2 years before boys. The increasing amounts of estrogen produced hasten the closing of the growth plates *(epiphyses)* in the bones. On the average, the long bones in females stop growing at 18. At that age, according to the National Center for Health Statistics, the average American young woman is 5 feet, $4\frac{1}{2}$ inches tall and weighs 123 pounds. Her male counterpart of the same age is 5 feet 9 inches tall and weighs 150 pounds, but he will grow another quarter to half inch until his early twenties, when his long bones, under the influence of testosterone, cease growing. Males end up about 10% taller because they have had 2 years longer in which to grow. In both sexes, the vertebral column continues to grow until age 30, but adds very minimally to the height. The velocity of growth in the shoulders and the hips increases in both sexes at about the time of the growth spurt in height. The rate of growth in the hips accelerates by approximately the same amount in males and females, so that hip width ends up about the same in both sexes. The shoulders and thoracic cage grow more rapidly in males, however, so that men have wider shoulders in relation to their hips, while women have wide hips compared with their shoulders.

The relative amounts of bone, muscle, and fat are different in males and females. As boys mature, they lose fat and gain in muscle mass and bone density. Females gain less in muscle and bone, but they continue to gain in fat, which is distributed differentially on their hips and buttocks and in their breasts.

There are more physical differences that develop at puberty, and these result from the high androgen output in males and the high estrogen output relative to androgen in females, and the effect of these steroids on what are essentially the same target organs. Males, therefore, develop a larger larynx and hence a deeper voice, greater facial bone growth (leading to more prominent features), a ruddier and coarser complexion, with more hair on the face and progressively less hair on the scalp, greater blood volume, more hemoglobin and more red blood cells, leading to a greater aerobic capacity and larger lungs and respiratory capacity. Although both sexes have the same numbers of sweat glands, males produce more sweat at a lower environmental temperature. Since sweating is a thermoregulatory mechanism that keeps body temperature down, men have a greater heat tolerance. Frye and Kamon (1983) found, however, that when men and women both exercise strenuously in a hot humid environment, the reduced sweat loss in women aids their conservation of body water and gives them the advantage.

After the first group of women cadets was admitted to the U.S. Military Academy at West Point, tests of their physical abilities compared with male cadets were performed both before and after basic training. Before training, there were some statistically significant differences between males and females in the physical abilities tested. The greater capacities of the male cadets in upper body strength and power, leg strength and power, and hand grip strength remained after training. That males still excelled in muscular strength and power both before and after training may be related to their superior size and prior greater muscle mass, but it is also possible that specific training exercises designed to develop arm and leg strength in women cadets could have narrowed the gap. The weeks of basic training did enable the women to attain the same cardiopulmonary (heart–lung) efficiency as the men.

In the final analysis, however, physical and physiological sex differences are relative. They only hold true in terms of an average difference that occurs in a large population. Individual variation can be enormous, and the overlap between the two sexes for any particular characteristic is usually considerable. People's genetic endowments have a lot to do with their appearance, and so does their nutrition (or lack of it), physical activity, social and economic status, geographical location, climate, and many other variables of environment and culture. Males and females do not

seem to have a great deal of difficulty in recognizing each other's gender, but many of those signals are artificially exaggerated and emphasized by dress and behavior. Ashton Barfield, in summarizing the factors responsible for sex differences, suggests, "You are what you eat, you are what you secrete, and you are whom you meet" (1976).

BASIC GENETIC MECHANISMS

Certainly for the first 8 years or so after birth, there is more similarity than difference in physique and physiology between boys and girls. Even after large amounts of the sex-appropriate hormones are secreted, which result in the dissimilarities that become defined at puberty, there is still a wide and almost infinite range of gradations in size, shape, and appearance of men and women. We cannot get any more specific than "on the average," or "for the most part," or "generally speaking," males tend to be taller, heavier, broader, and stronger than many females. For one basic difference, however, there is, at least 99% of the time, only an either/or possibility. For sex itself, for being born a male or being born a female, there is a determination at the moment of conception. When the sperm fertilizes the egg, *genetic sex* is established. This is the real sexual revolution—the biological phenomenon that originated several billion years ago to provide for the enormous hereditary variation on which natural selection has acted throughout evolution. Two parents, combining their genes to result in a new individual, provide the first embryonic event in a series that culminates in the birth of a baby boy or of a baby girl. But no biological process is ever 100% perfect, and some of the time there are errors in the sequence of developmental stages. Too many chromosomes, too few chromosomes, mutations of genes, hereditary metabolic defects, trauma, exposure to hormones or drugs, infections—any of these can alter the embryonic and fetal program that leads to normal sexual differentiation. An understanding of the basis of normal and abnormal sexual dimorphism requires a little medical genetics, so a discussion of cells, chromosomes, and genes is a logical place to begin.

Cells

Anyone who has ever taken a biology course has heard about how the seventeenth-century physicist Robert Hooke looked through the microscope at a thin piece of cork and coined the word "cells" for what he saw. The little empty squares reminded him of the small chambers in monasteries in which monks prayed. Hooke was looking at dead and dried-out plant tissue, but even now, 300 years later, when biology students examine thin, stained and fixed samples of animal tissue on a slide, they still, for the most part, merely see the same little squares. It is sometimes difficult for them to appreciate that these are the basic units of *life*, that these can respond to being poked or prodded; take in food and utilize it to get energy; synthesize new substances for their own use, or secrete the substances for functions elsewhere; in some cases, move under their own power; grow and reproduce— these things are not evident on a slide under the microscope. All those physiological properties of cells and more are based on the activities of subcellular components, or *organelles* within the cells, and of the macromolecules of which the organelles are composed.

Located in the cytoplasm of cells, all of the organelles are either made up of membranes or are closely associated with membranes. Surrounding the cell is the *cell or plasma membrane*, a highly invaginated complex structure, continuous with membranes inside the cell. Nutrients needed by the cell for its continued existence pass through the cell membrane, and all waste products and cell secretions or products also pass through it on their way out. Inside the cytoplasm of the cell, the *endoplasmic reticulum* (ER) forms a network of membranous, fluid-filled channels. Rough ER has tiny granules, or ribosomes, which are clustered along the channels and function in the synthesis of proteins. Smooth ER lacks ribosomes and is involved in the synthesis of lipids, steroids, and complex carbohydrates. *Peroxisomes* are small, discrete organelles that use oxygen to carry out their metabolic reactions and are thought to be responsible for detoxifying various molecules. *Mitochondria* are the larger, oval, but also membranous structures that are the site of the enzymes that catalyze the reactions called cellular respiration—the chemical reactions that derive an energy-rich molecule called ATP from food and oxygen. Cells store ATP as fuel for all cellular activities. Another system of membranous channels, sacs, and vacuoles, the *Golgi apparatus*, packages the products of the ER by enveloping them with membrane before their release from the cell. The Golgi body itself is involved in lipid and mucus manufacture. *Lysosomes* are organelles that contain enzymes that act as a digestive system in healthy cells and are able to break large molecules into smaller ones. They are also believed to be responsible for the changes that occur in tissues after death (see Fig. 5.2).

All the work that cells do both to maintain themselves and for the general good of the body is performed by the cytoplasm and its organelles, under the control of the nucleus. When a muscle cell contracts, when a nerve cell conducts, when a cell that lines the digestive tract secretes digestive enzymes or absorbs the products of digestion—whatever a cell does at any time in its life is the responsibility of its nucleus. Even when it wears out and dies a natural death, that event, too, is programmed from its nucleus. When cells have deteriorated or have been destroyed,

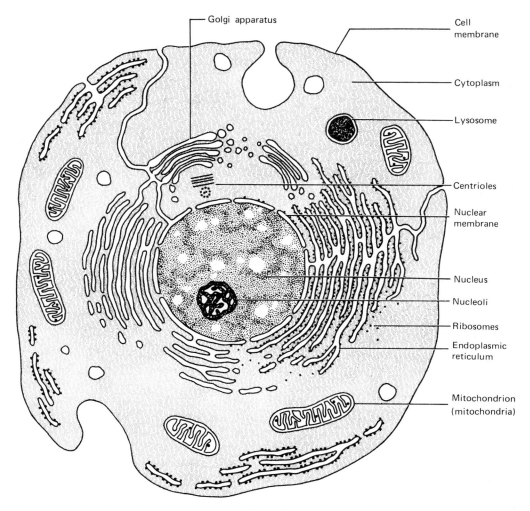

Figure 5.2. Drawing of a cell as it would appear when viewed with the electron microscope.

most of them, except for some that have become so highly specialized that they have lost the ability to reproduce, can be replaced by division of other cells. Reproduction of cells is also controlled by the nucleus.

Chromosomes and Genes

The reason that the nucleus is the controlling factor in all cellular activities, whether the cell is reproducing itself or breaking down nutrients to release ATP or synthesizing proteins to renew its enzymes, replace its organelles, or form its secretions, is because the nucleus contains *chromosomes*, and on these chromosomes are located the specific "how-to" units of instruction, the *genes*. When the cell is manufacturing a protein, the genes provide the blueprint for the synthesis of that particular molecule by transcribing the directions onto a messenger molecule that leaves the nucleus and takes the information to the cytoplasm. There, the message is translated and interpreted. When the cell is going to divide, the genes stop directing protein synthesis or their other metabolic activities and concentrate on duplicating themselves so that each of the two cells that is formed inherits exactly the same information as the original cell.

Chromosomes are composed of protein and an exceedingly long and thin filamentous molecule called *deoxyribonucleic acid,* or *DNA,* in the form of a double thread. It has been estimated that the human body contains somewhere between 10 and 20 billion miles of DNA distributed in its trillions of cells, so one can imagine the incredible way in which it is packed into the nuclei. A nucleus does not contain only one long filament of DNA; DNA is found in the form of a number of shorter pieces, or chromosomes. In a cell that is not about to divide, the pieces are highly coiled and folded so that they look like little masses or granules when they are stained and viewed under the microscope. They are then referred to as *chromatin.*

The number of chromosomes varies in different species of plants and animals, and human beings have 46 chromosomes in all the cells of the body, except for some cells that lack nuclei, or egg and sperm cells, which have only 23. Each human cell contains two of each distinguishable kind of chromosome in its total of 46, so there are actually 23 pairs. In males there are 22 matched pairs called *autosomes,* and two single, dissimilar chromosomes called the *sex chromosomes,* X and Y. Females do not have a Y chromosome; instead, the sex chromosomes are both X chromosomes. The *genotype,* or genetic constitution of normal males, then, is 44 autosomes plus X and Y, and is usually written as 46,XY, to designate the total number of chromosomes and to indicate that the male is normal for sex chromosomes. The genotype of normal females is 46,XX.

DNA Structure and the Genetic Code

The DNA molecule of a chromosome is in the form of a ladder that is twisted into a spiral staircase, the "double helix." * The side pieces of the ladder, or its backbone, are made of a sugar, deoxyribose, and of phosphate, occurring in alternating groups. The cross connections or rungs of the ladder are attached to the deoxyribose and are composed of nitrogen-containing compounds or nitrogenous bases called *purines* and *pyrimidines.* There are two kinds of purines, adenine (A) and guanine (G), and there are two kinds of pyrimidines present, cytosine (C) and thymine (T). Adenine always bonds to thymine, and cytosine is always attached to guanine, so that each rung of the ladder consists of a purine loosely linked to a pyrimidine. The term *nucleotide* refers to one of the purines or pyrimidines attached to the sugar unit, which is in turn attached to the phosphate group. The deoxyribonucleic acid molecule, then, is made up of two strands of nucleotides, strung along in various sequences, but always with a purine hooked up to a pyrimidine to form a nitrogenous *base pair.* The bonds that hold one member of a base pair to the other are very weak, and they can break apart easily (see Fig. 5.3).

The sequence of the base pairs as they form each rung of the ladder is referred to as the *genetic code.* Although there are only four variations of the linkage (i.e., thymine–adenine, adenine–thymine, cytosine–guanine, and guanine–cytosine), the order in which they occur can provide an almost infinite variety of combinations. A *gene* is a particular *segment of sequences* that gives the instructions for the

---

* James Watson and Francis Crick received the Nobel Prize in 1953 for delineating the structure of the DNA molecule. So did Maurice Wilkins, a crystallographer at Kings College in London. Students of biology should be aware that Rosalind Franklin, who worked with Wilkins and upon whose key observations Watson and Crick relied, did not receive the award. For further information, read A. Sayre's *Rosalind Franklin and DNA.* New York, Norton, 1975.

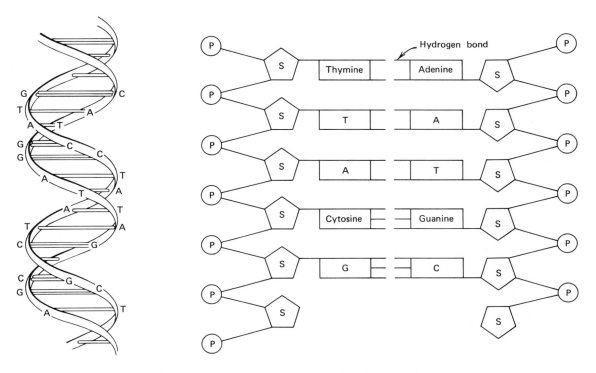

Figure 5.3. Diagrammatic representation of the DNA molecule. The twisted spiral (double-stranded helix) is very simply shown at the left. The drawing on the right illustrates how each sugar unit of a strand is attached to a nitrogenous base. The strands are held together by hydrogen bonds that link two bases in such a way that adenine can pair only with thymine and cytosine only with guanine. The genetic information is contained in a particular sequence of base pairs.

synthesis of a particular protein, which is in turn responsible for some other activity in the cytoplasm of the cell. Proteins are made up of long, linked chains of amino acids, and every sequence of three nucleotides, like AAT, or CTA, or CCG, and so forth, is a code word for a particular amino acid. Thus, genes are of different lengths, depending on the protein, and follow one another along the DNA strand. No one really knows how many genes are located on a chromosome—there may be tens or hundreds of thousands.

Reading the Code

Although the same genes are present on the same chromosomes of all the cells, not all cells perform the same work, synthesize the same proteins, or even look the same, for that matter. But since all the body cells do have the same DNA with the same information inscribed on it, it must be that different parts of the code are translated in different cells. That is, some genes are active in some cells, while other genes are active in other cells or at different times during development. FSH, for example, is synthesized in the cells of the anterior pituitary gland, and even though the genetic instructions for making FSH are present in the muscle cells, it is certainly not manufactured there. The kinds of genes that actually code for the synthesis of a specific substance like FSH or any other kind of protein that is concerned with the specific function of a cell are called "structural" genes. Other kinds, the "operator" genes, the "regulator" genes, and the "repressor" genes, dictate which genes in a cell are to function and which are to remain inactive.

Because chromosomes exist in matching or homologous pairs, genes also exist in pairs, called *alleles*. Each chromosome of a pair carries a sequence of nucleotides, the gene, that governs a particular activity, and the other chromosome carries the gene partner at the same location. While the two gene counterparts are concerned with the same function, their arrangement of nucleotides may not be exactly the same, and the individual is then said to be *heterozygous* for that particular gene. Then, one of these genes generally is *dominant* to its *recessive* allele. If the arrangement is exactly the same between the two members of the allele, the individual is *homozygous* for that pair.

If the arrangement of nucleotides in a gene is even minutely rearranged by mutation, a specific abnormality may result, a malfunction that is then transmitted as a hereditary disease to future generations. Tay-Sachs disease, cystic fibrosis, and sickle-cell anemia are examples of hereditary diseases caused by genetic mutation. Gene defects may also occur as a result of environmental factors, such as X-rays, chemicals, viruses, and infections. Classical human genetics is concerned with gene defects.

Cell Division

As cells go about their metabolic business, performing their own specialized work under the control of the genes in their chromosomes, they may have to reproduce themselves, or other cells have to reproduce to replace them, in the normal course of maintenance and repair. Some cells last longer than others. Nerve cells for example,

are not able to divide and must last virtually a lifetime, but every second, millions of blood cells die, and millions more must be there to take their place.

When cells do divide, it is tremendously important that all offspring cells have the identical chromosomal information as the parent cells, so that the genetic code can be passed to all future generations of cells. Therefore, before a cell divides, its chromosomes must replicate themselves. The process of chromosomal replication, followed by nuclear and cytoplasmic division that insures the preservation of the genetic code, unchanged from cell to cell, is called *mitosis*.

Mitosis

Mitosis of a cell takes place in four consecutive stages called *prophase*, *metaphase*, *anaphase*, and *telophase*, but the actual duplication of the chromosomes occurs before prophase, during *interphase*. Each ladder of the DNA double helix "unzips" its loose bonds between the purines and pyrimidines of its rungs, and a new strand of complementary nucleotides, made up from the free phosphates, sugars, and purines and pyrimidines that are available in the nucleus, zips onto each half of the ladder. Where there was formerly one DNA double-helix molecule, there are now two double strands that are formed. Each one has one-half from the old molecule, and one-half of each is new.

The doubled chromosomes are called sister *chromatids* at this stage in mitosis, and they lie next to each other in the nucleus, attached at a point called the *centromere*. Even though each of the 46 chromosomes consists now of two chromatids, it is still considered to be only one chromosome. Later in the process, when the centromere divides, the two chromatids separate from each other, and one of them from each chromosome moves toward the opposite end of the cell. Then each chromatid is considered a chromosome in its own right.

After the chromosomes move apart, the cytoplasm also pinches apart to divide. All of the organelles in the cytoplasm, the mitochondria, Golgi body, ER, and so forth, get roughly divided into equal parts, and mitosis is complete.

Meiosis

It was noted earlier that all of the body cells with nuclei have 46 chromosomes, except for the egg cells in the female and the spermatozoa cells in the male, which have only half that many, 23 chromosomes. It should be obvious why this is so. If a sperm had 46 chromosomes, and an egg had 46 chromosomes, when the sperm fertilized the egg, the resultant zygote would have 92 chromosomes, with further doubling of the number in each successive generation! Of course, this kind of doubling of chromosomes cannot occur, and the chromosome number of each given species must be kept constant. In the formation of the *gametes*, or eggs and sperm, a special kind of reduction division takes place, which insures that each gamete contains only *half* of the number of chromosomes. This special division is called *meiosis*. Then, when the sperm fertilizes the egg, the resultant zygote, from which all the cells of the new individual form, contains 46 chromosomes. Because the egg contains only 23 chromosomes and the sperm contains 23 chromosomes, meiosis also insures that in any body or *somatic* cell, half of those 46 chromosomes are of maternal origin and half of them of paternal origin. In both males and females, gametes are produced in the gonads and are derived from a cell with 46 chromosomes that undergoes two divisions to result in four cells. In males, all of the cells are viable and are called spermatozoa; in females, only one is a viable egg cell. All the rest become polar bodies and fail to develop.

The chromosomes replicate themselves in meiosis the same way they do in mitosis, but after duplication the two chromatids of each chromosome seek out the two chromatids of the homologous partner and arrange themselves next to each other, with all four chromatids stretched out along their entire lengths; that is, all the chromosomes of *maternal origin* lie next to their matching partners of *paternal origin*, with all their matching genes strung along, one after the other. The two homologous chromosomes, each consisting of two chromatids, are held together by their centromeres. At this point, a very important thing happens. In each pair of four chromatids, one of the maternal chromatids reciprocally exchanges material with one of the paternal chromatids. This is called *crossing over*, and the sites of the exchanges are called chiasmata. The only pair of chromosomes in which this does not happen is the XY in males. There is no particular pattern to the breakages and exchanges of fragments of DNA—it occurs absolutely at random—and the reshuffling of the genes, the rearrangement of the array of genes that results provides for the absolute uniqueness of every human being (see Fig. 5.4).

After crossing over is finished, each of the homologous chromosomes separates from the other and goes to the opposite pole of the cell. When the cytoplasm divides, each cell will contain 23 chromosomes, but each chromosome still consists of two chromatids. The second meiotic division resembles mitosis in that the chromatids separate, each to migrate to an opposite pole and then to become chromosomes in their own right. In females, the second meiotic division is completed only in an egg that has been fertilized by a sperm. Then the nucleus of the egg, containing 23 chromosomes, merges with the nucleus of the sperm, also containing 23 chromosomes, and the full set of 46 chromosomes is reformed.

That first cell of the new individual divides

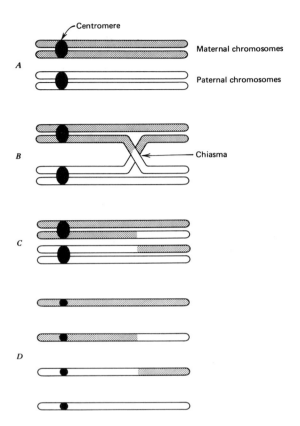

Figure 5.4. Crossing over and recombination of chromosomes. (A) Homologous chromosomes, held together by their centromeres, are paired together during metaphase of the first meiotic division. (B) Two of the chromatids undergo breakage at exactly the same point. A part of one maternal chromatid is exchanged with a corresponding part of a paternal chromatid in a process known as crossing over. The site of the exchange is a chiasma. (C) Appearance of the chromosomes following crossing over. Whatever genes were located on the original homologues are now in a new combination as a result of the exchange. (D) After separation, each gamete will contain one of the chromosomes. Note that two of the gametes will carry recombined chromosomes and two will not. In long chromosomes, several crossovers (breakages and recombinations) may occur.

to form the myriad numbers of cells of the body by mitosis. Within hours of fertilization, the first two cells form, then four then eight, 16, 32, 64, 128, all the while that the embryo is progressing down the fallopian tube. As the multiplications continue, the cells begin to develop along different lines, and from then on will no longer look alike. Some will become brain cells, some bone cells, some will be the eye, heart, or kidney cells, and some will even become segregated at that early stage to become future egg or sperm cells. That depends, of course, upon the sex of the embryo and was determined when the sperm fertilized the egg.

Sex Determination

Remember that the genotype of females is 22 pairs of autosomes and two sex chromosomes, XX, and that the genotype of males is 22 pairs of autosomes and the two sex chromosomes, XY. Since the sex chromosomes behave like members of an ordinary pair and separate during meiosis, it is apparent that while females can produce only eggs with an X chromosome, males can produce two kinds of sperm—those carrying an X, and those carrying a Y. Then, if a sperm bearing an X chromosome fertilizes an egg, the resulting XX *zygote* (or fertilized egg) will be female. If a Y-carrying sperm fertilizes an egg. The XY zygote will develop into a male. Because half the sperm will carry an X and the other half will carry a Y, the chances of a male or a female baby being conceived are 50/50 (see Fig. 5.5). The male and his sperm solely determine the sex of the offspring. What would all the wives throughout the ages who were blamed for not being able to produce sons, and all the queens in history who were discarded or beheaded for the same reason, have given for that information! (But who would have believed them anyway?)

Although the sex ratio of males and females

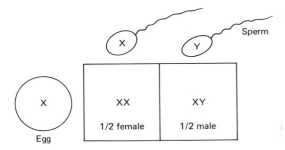

Figure 5.5. The male parent produces equal numbers of X-bearing sperm and Y-bearing sperm. It would then be expected that equal numbers of males and females will be conceived.

conceived is theoretically 50:50, in actuality, the actual birth sex ratio is 103 and 105 males born for every 100 females. The lower ratio occurs in non-whites (blacks and other ethnic groups) and the higher one in whites, both in the United States and in other countries that keep records. Greater numbers of males are born relative to female births, even though more male fetuses are spontaneously aborted or stillborn, so that there probably are more males than females conceived. The reasons why male conceptions have the edge on the numbers of females conceived is unknown, although several theories have been proposed. One idea is that the X-bearing sperm, larger and of greater mass, swims more slowly and does not get to the egg as successfully or as often as the Y. Another idea is that the Y-bearing sperm finds the environment in the female reproductive tract more favorable for survival. There is as yet no convincing evidence to really substantiate either of these speculations.

Genes on the Sex Chromosomes

The human Y chromosome carries a sequence of nucleotides, or a gene, that instructs certain cells of the embryonic gonads to form testes. Those testes will then secrete hormones that will direct further differentiation into a male reproductive tract and male genitalia. In the absence of the Y-chromosome, testes do not develop, and the embryo develops as a female. It is currently believed that a gene on the Y-chromosome induces the activation of an autosomal structural gene that codes for the production of H-Y antigen, a cell membrane surface protein. It is the H-Y antigen interacting with the embryonic cells that results in testes differentiation. In the presence of X-chromosome genes, the H-Y antigen is inhibited. As far as anyone knows, there are no other genes on the Y-chromosome—none for any other human activity or trait—except for that one that triggers the development of maleness in XY embryos.

In contrast to the Y chromosomes, the X does have genes that influence the expression of many characteristics other than those concerned solely with reproduction. These genes are called *sex-linked*, and among the most familiar are those affecting color vision, blood clotting, and muscle function. Abnormalities of these genes lead to color blindness, hemophilia, and a type of muscular dystrophy.

Barr Body

One of the X chromosomes in the nucleus from a cell of a female can actually be seen under the microscope, most easily in a stained smear of cells from the inner lining of her cheek, or in her white blood cells, but it is also visible in living cells with the use of phase-contrast microscopes. Originally called sex chromatin, or X-chromatin, its discovery in 1949 by two Canadian researchers, Murray Barr and Edward Bertram, led to a whole new field of study, human *cytogenetics*, the study of chromosome abnormalities. Barr and Bertram were working on stained sections of nerve cells in cats when they noticed a tiny, dark-staining blob of material that was always present only in the nucleus of the female cats and never in the males. It was subsequently determined that the little stained body was one of the X chromosomes, an inactive one that remained coiled and condensed so it could be seen, and that it was visible in human female cells as well. The *Barr body*, as it came to be known, is not seen in the nuclei of the cells of normal males. One of the two X chromosomes of a female embryo becomes inactive and, hence, visible at approximately 2 weeks of age, when the embryo consists of only a few hundred cells. From then on, it is the *invisible* X that has active functioning genes and goes about its business directing cellular activities. The X chromosome that appears as the Barr body is the inactive one, devoid of genetic duties. This inactivation of one X chromosome results in both sexes having an equal dosage of X chromosomal genes. Which of the two X chromosomes becomes a Barr body at the time of inactivation is a matter of chance. It could be the maternally derived X in some cells and the paternally derived X in other cells. From then on, interestingly enough, all those cells that are offspring of a cell in which the X chromosome from the father became active are then slightly different from all the descendents of the cell in the embryo in which the X chromosome from the mother's side became active. This is why geneticists say that a normal human female is a *mosaic*. She is composed of a mixture of two slightly different kinds of cells. The original description of the nature and significance of the physical inactivation of one of the X chromosomes was made by Mary Lyon, and the condensed X is also referred to as the Lyonized X (see Fig. 5.6).

Another way of identifying X chromosomes is by looking for the "drumstick," a little separate lobule that is apparent on approximately 0.5%–10% of the nuclei of white blood cells called *neutrophils* in a well-stained blood smear from females. So if large numbers of

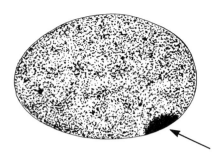

Figure 5.6. The Barr body, or one of the X chromosomes of a female as seen in the nucleus of a cell from a woman's cheek lining. A simple scraping made with a toothpick can be placed on a slide (buccal smear), stained, and examined. The percentage of Barr bodies in buccal smear surveys of normal populations varies from about 10% to 60% in females.

neutrophils are examined, it is possible to determine genetic sex in an XX individual (Fig. 5.7).

The identification of these sex-specific chromatin masses in cells has diagnostic value. The International Olympic Committee, in order to insure that only females compete in female athletic competitions, has made a "sex-test" mandatory for woman athletes. "Passing" the test provides the competitors with a certificate that validates their female-

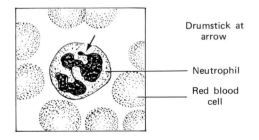

Drumstick at arrow

Neutrophil

Red blood cell

Figure 5.7. The drumstick appendage, equivalent to the sex chromatin of the Barr body, can be seen in a small percentage of certain white blood cells of a female.

ness. The presence of Barr bodies, however, does not always mean that an individual is female, and their absence is not always conclusive evidence that an individual is male. A Barr body only indicates that the cells contain at least *two* X chromosomes. In cases of certain kinds of chromosomal abnormalities, genetic sex can only be determined by establishing whether or not the somatic cells contain a Y chromosome. This is done by examining the chromosomes, arranging them in groups, and counting them. Viewed in this manner, the chromosome complement is called a *karyotype*.

Human Karyotype

The easiest way to examine chromosomes is to look at them during the metaphase stage of mitosis when they are condensed, separate from each other, and most visible. The easiest tissue to examine for cells is blood because it can be obtained from a vein with relatively little discomfort. In order to make a karyotype, a sample of blood is taken from an individual and incubated in a nutrient medium at body temperature for several days to encourage cell division. The culture is centrifugated to separate the red cells, which are then discarded. A nontoxic chemical derivative of colchicine, a drug that stops cell division in metaphase, is added to the white cells; after several hours a very weak salt solution is added to swell the cells and separate the chromosomes. When a drop of the cell suspension is spread out on a slide, stained, and then viewed under the highest power of a microscope, the chromosomes can be examined. This is called a metaphase spread. The problem is, how is it possible to tell one chromosome from another?

All of the chromosomes consist of two chromatids, held together by the centromere, so all of them will have an X configuration, but they come in different sizes. In order for a karyotype to be described in some standardized way, an international system of chromosome nomenclature is used to divide the chromosomes into groups based upon their length and the position of the centromere. When a photograph of the view under the microscope is enlarged several thousand times, the individual chromosomes can be cut out with a scissors, matched and arranged according to their length and shape, and divided into seven groups, A through G (Figs. 5.8A,B). The only difference in the karyotypes of a male and female is that the male has one X and one Y chromosome and the female has two X chromosomes.

Preparation of the human karyotype is obviously time consuming, and analysis of the chromosomes is certainly subject to human error. Newer methods of identifying chromosomes include the use of fluorescent dyes or special stains that preferentially bind to certain segments of DNA on specific chromosomes for more precise recognition. Such techniques are called *banding*. (see Fig. 5.9).

CHROMOSOMAL DISORDERS

The mechanisms underlying cell division are so complex that an occasional slipup in the orderly processes is not unexpected. Chromosomal abnormalities can occur during the early stages of meiosis in the development of the gametes, but can also arise after fertilization from faulty mitotic division of the zygote. There are two major kinds of chromosomal aberrations: either there are changes in the *number* of chromosomes or there are *structural* changes as a result of chromosomal breakage that may be due to environmental influences, such as drugs or infections. In *nondisjunction*, there is a failure of homologous chromosomes to be distributed normally when they pull apart, and the result is a cell with too many or too few chromosomes.

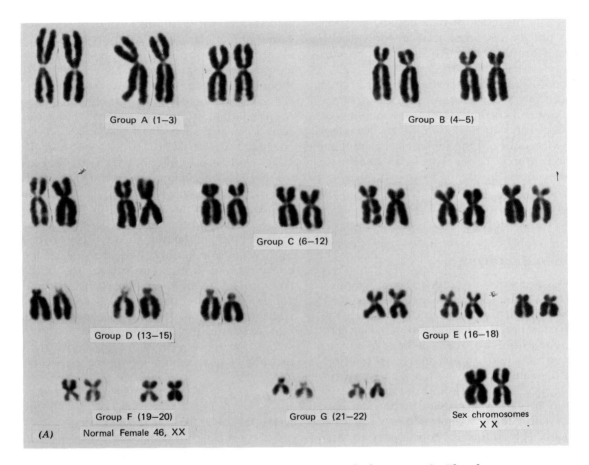

Figure 5.8. (A) Karotype of a human female. (B) Karotype of a human male. The chromosomes are cut out of an enlarged photograph of chromosomes in metaphase, arranged in pairs, and systematically grouped. The autosomes are numbered from 1–22 on the basis of decreasing size.

Trisomy indicates there is an extra chromosome, and *monosomy* means the absence of a chromosome. If a piece of a chromosome breaks off and is lost, it is termed a *deletion.* Sometimes the broken-off piece is not lost, but is inserted improperly on another chromosome. When broken segments are exchanged between two chromosomes, it is called *translocation.*

No one really knows the number of fertilized eggs that carry a chromosomal abnormality, but the number of children born with chromosomal disorder (at least one that is observable with a light microscope) is 0.5% or one in 200. Most of the chromosomal defects result in such major deformities in an embryo that they are incompatible with life and are spontaneously aborted, sometimes even before the woman realizes she is pregnant. When such spontaneously aborted fetuses

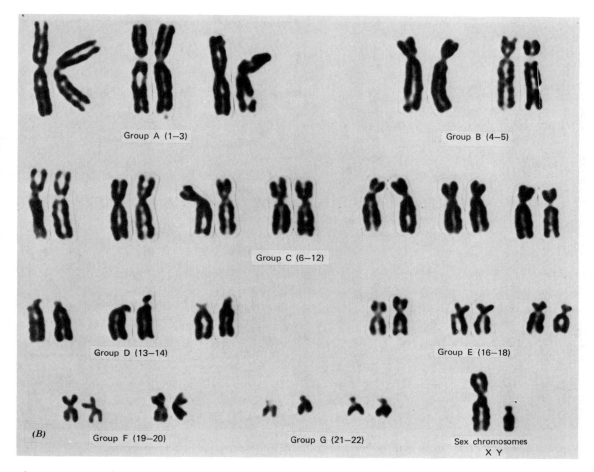

Group A (1–3) Group B (4–5)

Group C (6–12)

Group D (13–14) Group E (16–18)

(B) Group F (19–20) Group G (21–22) Sex chromosomes
X Y

Figure 5.8. (continued)

(called abortuses) have been analyzed for karyotype, chromosome abnormalities have been found in 20%–30%. Since there is no way of getting data, it is impossible to estimate the number of fertilized eggs that never implant at all and are lost because of chromosomal defects.

Down's Syndrome

All autosomal trisomies, in which the affected person has three rather that the normal two of a particular chromosome, for a total of 47 chromosomes instead of 46, cause severe mental and physical abnormality. The most frequent trisomy in live births, affecting perhaps one in 700 children, is trisomy 21, in which there are three copies of the smallest autosome, number 21. Children with this condition, also called Down's syndrome and formerly known as "mongolism," are almost invariably mentally retarded, small in size, and have poor muscle tone. There are about 10 or so major symptoms present, including the

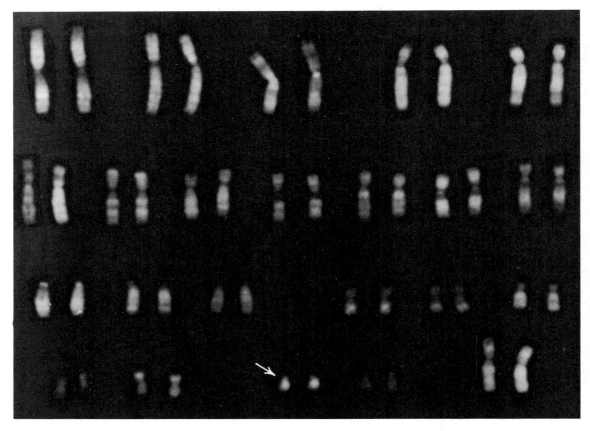

Figure 5.9. Female karyotype with quinicrine-induced fluorescence of chromosomes, believed to result from the stain's interaction with adenine–thymine-rich areas of DNA. Note that chromosome 21 (arrow) fluoresces more brightly than chromosome 22. The crosswise striations on chromosomes produced by quinicrine are called Q bands; other stains cause G bands, C bands, and R bands for detailed chromosome identification and spectacular effects.

"simian crease," a slight abnormality in palm and sole print patterns sufficiently different from the normal that it can be used for diagnosis.

Most cases of Down's syndrome result from meiotic nondisjunction of the number 21 autosome. At the stage of meiosis during which the homologous pair would normally separate, with each member of the pair going to an opposite pole of the cell before cytoplasmic division, the pair instead remains together. One cell will then get both chromosomes and the other cell will get none. An abnormal gamete with an extra chromosome is produced, which in the case of trisomy 21 is almost always the egg. If such an egg is fertilized by a normal sperm, autosome 21 will be in triplicate rather than paired (two chromosomes from the egg and one chromosome from the sperm), and the resultant embryo will have 47 chromosomes (see Fig. 5.10).

Although there is some evidence that paternal age, especially if the father is over 50, is also associated with an increase in Down's syndrome birth, the risk of having an affected child statistically increases with the age of the mother. The incidence is about one in 1,700 in women aged 15–19; one in 1,400 in mothers aged 20–30; one in 750 in those aged 30–35; thereafter, the risk becomes much greater, increasing about 15% with each additional year.

At the age of 40, the odds of having an afflicted child are one in 100; and when the maternal age is 45 or older, the chances of giving birth to a baby with Down's syndrome may be as high as one in 16, or 6%. If a woman has already had an affected child, the risk of recurrence also increases with maternal age. The higher incidence of meiotic nondisjunction in the ova of older women has generated a number of theories as explanation. It may be that the increased risk is related to the age of the ova in the older woman's ovary. The oocyte in a primary follicle that has resumed meiotic

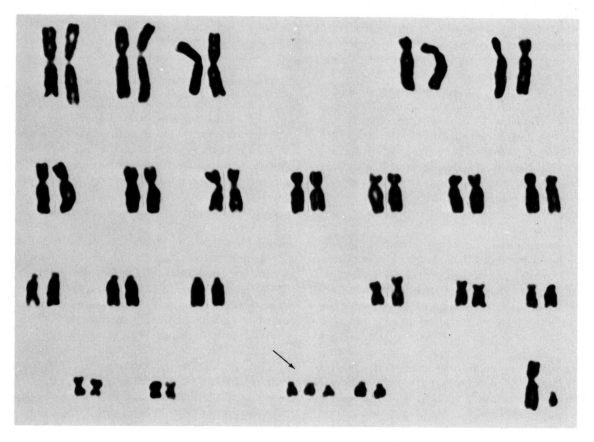

Figure 5.10. Karotype of a male with Trisomy 21, or Down's syndrome. This individual has three copies of chromosome 21, which results in mental retardation as well as a number of abnormalities affecting internal organs.

division on its way to ovulation in a woman of 45 has been dormant in the ovary for about 45 years and 3 months. Cells in an arrested state of meiotic division are known to be particularly susceptible to environmental influences—viruses, X-rays, toxic chemicals—that can interfere with normal cell division. The older a woman is, the longer these damaging agents have had to affect those oocytes. It has also been suggested that since older women are usually married to older men, and it is presumed that there is less frequency of sexual intercourse in older couples, the egg remains in the female tract longer before fertilization. In animal experiments, it has been found that such delayed fertilizations result in more chromosomal anomalies in the zygote. It has not as yet been determined whether (1) intercourse is really that much less frequent in people 40 years of age or older, and (2) whether couples aged 40 or older who have a child with trisomy 21 have had less frequency of intercourse than similarly aged couples who have produced a normal child. Thus, the delayed fertilization theory has not been substantiated. The reasons for the increased incidence of Down's syndrome in children of older women is really unknown.

Down's cases due to nondisjunction are not hereditary, but there is a small percentage of Down's syndrome births in which the cause is another kind of chromosomal aberration, called a translocation. In these cases, one of the parents, either male or female, carries the chromosomal rearrangement and may produce more than one affected child or pass the translocation to unaffected children who may subsequently have offspring with Down's syndrome. It is obviously important to identify such carriers, and all parents of a trisomy 21 child who shows a chromosomal translocation should have a chromosome analysis done.

There are several other kinds of autosomal abnormalities that result in multiple congeni-

tal (i.e., present at birth) defects that include both mental and physical retardation. Some are trisomies and some are deletions of a chromosome, but these occur much less frequently than Down's syndrome. Among them are trisomy 18, or Edwards' syndrome, which has an incidence of one in 6,500 live births but usually produces death within the first year. Lejeune syndrome, the *"cri-du-chat"* or "cat-cry" syndrome, is a very rare condition caused by a deletion of the shorter arm of one chromosome 5. Affected babies, who have physical deformities and are mentally retarded, have a characteristic high-pitched cry as a result of a defect in the larynx that sounds very much like the meowing of a kitten.

Sex Chromosome Abnormalities

The random accidents during meiotic development of the gametes that cause autosomal abnormalities can also, of course, produce trisomies and monosomies of the sex chromosomes. Sometimes the result is abnormal development of the gonads, but very frequently, there is no visible effect at all.

Nondisjunction of the sex chromosomes can occur during the development of the ovum *(oogenesis)* or during the development of the spermatozoa *(spermatogenesis)*. If such a number change has occurred, the result at fertilization can be an individual with any one of several abnormal sex chromosome complements: 45,XO or Turner's syndrome; 47,XXY or Klinefelter's syndrome; 47,XXX or trisomy X; 47,XYY or Double Y syndrome; also XXYY, XXXY, XXXX, and even XXXXY or XXXXX! People with a Y chromosome will be male, no matter how many X chromosomes they have. A cell, however, must have at least one X to survive. Combinations like YO or YY have never been observed, and it is assumed that they are lethal; that is, incompatible with life. Table 5.2 illustrates the mating possibilities between abnormal and normal sperm and ova that

**Table 5.2 VARIOUS MATING POSSIBILITIES BETWEEN
NORMAL AND ABNORMAL SPERM AND OVA AND
THE GENOTYPE PRODUCED**

Normal Ovum	Abnormal Sperm	Expected Genotype
X	O	XO
X	XX	XXX
X	XY	XXY
X	YY	XYY

Abnormal Ovum	Normal Sperm	Expected Genotype
XX	X	XXX
XX	Y	XXY
O	X	XO
O	Y	YO (inviable)

produce the most frequent clinical conditions. Table 5.3 shows the approximate frequency and the major clinical symptoms.

Generally, any physical or mental abnormalities produced by sex chromosome anomalies are very mild compared with those produced by autosomal trisomies and monosomies. There does seem to be a tendency for some mental retardation to be associated with extra X chromosomes, but even if there is some intellectual impairment, it is usually borderline and minimal. Surveys of patients in schools, hospitals, and institutions for the mentally handicapped indicate there is a higher incidence of XXX females and Klinefelter's males than in the general popula-

Table 5.3 FREQUENT SEX CHROMOSOME ABNORMALITIES

Genotype	Name of Syndrome	Frequency	Symptoms
XO	Turner's syndrome (gonadal dysgenesis)	1/3,500 females	Female, is short in stature, has typical webbed neck, poorly developed breasts, and immature external genitalia. Ovaries may be absent. The mental capacity is normal.
XXY	Klinefelter's syndrome	1/800 males	Male, is above average in height, with long arms and legs relative to the rest of the frame; has small testes and penis and is usually sterile. The breasts may be somewhat enlarged.
XYY	Double-Y syndrome	1/700 males	Male, is taller than average, normally fertile, but possibly of somewhat lower intelligence.
XXX	Trisomy X	1/1,000 females	Female, is normal physically, fertile, but possibly with greater tendency toward mental retardation.

tion. Perhaps because so many studies have been made on individuals in institutions, there may be a tendency to attribute too much of their behavior to their chromosome constitutions. A case in point is the concern that developed over XYY males when studies in Europe and the United States reported that they seemed to be overrepresented in penal and mental institutions, and the belief arose that these men were overly aggressive and had criminal tendencies. Newspapers and magazines have published lurid and sensational stories about the crime and violence committed by XYY males. It has subsequently been determined that the XYY defect was not at all as rare as was supposed, and that it was actually present in one of 700 births, making it the most common chromosomal disorder after Down's syndrome. Many unsuspecting XYY males must be living among us, leading quiet, nonviolent lives. Witkin and co-workers reported, after a large study in Denmark, that XYY males neither are overly aggressive nor exhibit particularly antisocial behavior, but may have somewhat lower intelligence than XY males (1976).

PRENATAL DETERMINATION OF GENETIC DEFECTS

It is possible to detect chromosomal abnormalities and a number of hereditary metabolic defects in the fetus before birth by *amniocentesis*, which involves piercing the abdomen of the mother to remove a sample of fetal cells for analysis.

Use of this procedure has become a controversial medical issue for several reasons, particularly for those who object to abortion under any and all circumstances. Despite the recent development of techniques that increase the potential for prenatal correction of some fetal defects, when an abnormality is de-termined by amniocentesis the only options available in most cases are to have an abortion or to give birth to a baby with the kinds of defects that may result in a lifetime of suffering for both child and parents. For some people, this would hardly be an agonizing decision; the abortion would be performed and another pregnancy would be attempted, if desired. For those who object to abortion, this type of prenatal diagnosis is not useful.

Actually, the notion that amniocentesis invariably leads to abortion is mistaken. Most women who undergo the procedure get good news, discover that they are not carrying a fetus with the suspected disorder, and deliver a normal baby. The March of Dimes Birth Defects Foundation claims that amniocentesis procedures are mostly lifesaving. They point out that fewer than 10% are for the purpose of determining specific genetic defects and that many amniocenteses aid in the determination of fetal maturity and detection of difficulties like Rh incompatibility that can be treated by intrauterine transfusion or induction of early labor to save the life of the fetus. There are indications, however, that some women elect to have amniocentesis as a method of sex selection; if the fetus they are carrying is not of the desired gender (and there is some evidence that there is a preference toward males), the pregnancy is terminated. Apart from the ethical and moral issues, the social implications of such a course have been questioned by even the most committed of those who believe in freedom of choice in abortion. Besides, any technique that is invasive (requires entering the body) is never entirely risk-free, and the potential benefits of such procedures should always be weighed against the potential risks. For the vast majority of pregnant women, "I don't care what it is, so long as it's healthy!" supersedes any actual preference for a boy or a girl baby.

The method itself has been known for more than a century, but has been used for prenatal diagnostic purposes only for approximately the last 25 years. During development in the uterus, the fetus floats in a fluid-filled sac, familiarly known as the "bag of waters," surrounded by two fetal membranes, the *chorion* and the *amnion*. Although some of the water and ions in this amniotic fluid may originate from the mother, the cells in the fluid have sloughed off from the fetus or from the amnion membrane and are genetically identical to those of the fetus. On an outpatient basis, and with or without local anesthetic, an amniotic fluid tap is performed at approximately the sixteenth week of pregnancy when the volume of fluid is about 175–225 ml. The location of the placenta is established by ultrasound to avoid puncture, and then a long needle attached to a syringe is inserted through the abdominal wall, through two layers of the peritoneum, through the uterine muscle, and into the amniotic sac. A sample of amniotic fluid is withdrawn and centrifuged. The sediment contains the cells, which can be incubated and cultured for Barr body or karyotype, and also for biochemical studies of various types. The supernatant can be used for biochemical enzyme analysis or for virus studies (see Fig. 5.11).

Amniocentesis in the second trimester of pregnancy makes it possible to diagnose specific chromosomal abnormalities, inborn errors of metabolism, and neural tube defects. It can replace a genetic probability with a diagnostic certainty and provide the option of therapeutic abortion and another pregnancy for particular high-risk couples whose former

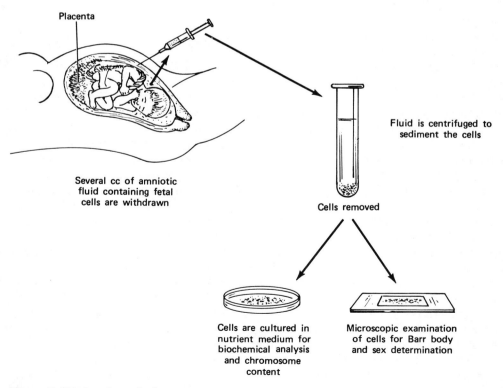

Placenta

Several cc of amniotic fluid containing fetal cells are withdrawn

Fluid is centrifuged to sediment the cells

Cells removed

Cells are cultured in nutrient medium for biochemical analysis and chromosome content

Microscopic examination of cells for Barr body and sex determination

Figure 5.11. Amniocentesis.

alternatives were refraining from childbearing or giving birth to an affected child. It cannot guarantee the absence of physical deformity or mental retardation; nutrition, drugs, and environmental pollutants are also suspect in causing birth defects. The major criterion for having the procedure would be when the probability of finding an abnormality is greater than the risk of having amniocentesis. Currently, this criterion is met in one or more of the following situations: pregnant women over age 35; any couple who has previously conceived a child with Down's syndrome or other chromosomal abnormality; when either parent has a chromosomal aberration; if there is a history of genetic disease in a close relative; when a previous pregnancy has resulted in a child with multiple structural malformations and no cytogenetic studies were performed; or when a couple has experienced three or more miscarriages, early infant deaths, or both. When the parents have indicated a willingness to accept the possible necessity of a therapeutic abortion, amniocentesis, performed competently by experienced physicians, is an invaluable and relatively safe diagnostic tool. A necessary adjunct to the procedure is skilled genetic counseling by trained professionals who can give accurate medical information and help couples interpret it within their own frame of reference.

In addition to the chromosomal aberrations that can be diagnosed by amniocentesis, there are more than 70 biochemical disorders for which prenatal diagnosis has been possible. These so-called "inborn errors of metabolism" are inherited defects due to a single mutant gene, generally resulting in the inability to produce a certain enzyme. Tay-Sachs disease, for example, is the failure to function or the complete absence, in nerve cells, of the enzyme hexosaminidase A. Lack of this enzyme leads to an accumulation of a carbohydrate-lipid molecule in nerve cells throughout the body, in turn causing blindness and men-

tal retardation. Table 5.4 lists some of the best-known examples of metabolic disorders diagnosed by amniocentesis. There are also certain structural deformities known as neural tube defects (NTD), such as spina bifida or anencephaly (in which the entire brain and spinal cord fail to form), that can be diagnosed prenatally. The test is based on greatly increased levels of the fetal protein, alpha-fetoprotein (AFP), that appear in the amniotic fluid when neural tube defects occur. The higher AFP levels are reflected in the maternal blood as well, and screening tests based on maternal blood levels around the fifteenth week of gestation have been developed. Since a raised AFP level in the mother's blood can also be due to other reasons than NTD, it is essential to make certain the result was not a "false positive" and to verify the elevated value through further examinations.

Risks of Amniocentesis

Although complications of the procedure are infrequent, amniocentesis is not without some danger to the fetus and to the mother. Fetal risks can include rupture of the membranes and a subsequent miscarriage, but a cause-and-effect association between amniocentesis at 16 weeks and miscarriage is difficult to analyze because a risk of spontaneous abortion is known at this gestational age anyway. The estimated increased risk of miscarriage as a result of amniocentesis is 0.08%–2.5%, based on data from a number of studies. The risk of miscarriage is greater in twin pregnancies.

Other consequences of amniocentesis for the fetus could be bleeding as a result of the needle hitting a fetal blood vessel or placenta or punctures of the fetus leading to scars, depressions, or dimples, but these injuries are not lethal and are generally believed to be not serious complications. One British study, however, suggested that amniocentesis sub-

Table 5.4 SOME HEREDITARY METABOLIC DISORDERS DIAGNOSED BY AMNIOCENTESIS

Disorder	Major Clinical Manifestations
Fabry's disease	Purple skin papules, renal failure, cardiac and ocular involvement.
Gaucher's disease	Mental retardation from birth; skeletal changes.
Tay-Sachs disease	Onset at 5-6 months, degenerative neurological disorder, profound mental and psychomotor retardation leading to death.
Niemann-Pick disease	Variable skeletal and neurologic involvement, four types possible.
Hunter's syndrome Hurler's syndrome	Gargoyle-like appearance, mental and physical retardation, liver and spleen enlargement, joint stiffness.
Maple syrup urine disease	Mental retardation, early death.
Galactosemia	Cirrhosis of liver, cataracts, mental retardation.
Adrenogenital syndrome	Virilization of female, adrenal insufficiency.
Lesch-Nyhan syndrome	Self-mutilation, spasticity, mental retardation.

Source: Modified from Dorfman, A., (ed). *Antenatal Diagnosis*. Chicago, University of Chicago Press, 1972.

stantially increased the risk of musculoskeletal deformity, congenital dislocation of the hips, and respiratory difficulties at birth (Turnbull et al., 1978). Other large-scale studies of amniocentesis hazards have not confirmed these findings.

Maternal risks can include puncture of the bladder, intestine, or a blood vessel, but the likelihood of serious injury is quite small, although it appears to be greater for the fetus and mother when amniocentesis is performed during the last 3 months of pregnancy.

Another method of prenatal detection of genetic defects that can be performed in the first trimester of pregnancy and yield results within a day of the test rather than the four-week wait required for amniocentesis is called *chorionic villi sampling*. The procedure was developed in medical schools in Milan and London and was first offered in the United States at Michael Reese Hospital in Chicago in 1983. Performed between the seventh and tenth weeks of pregnancy, the technique involves the insertion, under ultrasound guidance, of a small plastic catheter through the cervix into the uterus in order to withdraw about 30 mg of chorionic villi tissue, the fetal contribution to the not yet formed placenta. The tissue sample is processed for chromosome studies, and preliminary results of the chorionic cell analyses can be obtained within 24 hours with confirmation 3–5 days later. Should the tests indicate a defect, the couple may choose to have a first-trimester abortion, a safer and simpler technique than the second trimester pregnancy interruption that could follow amniocentesis. Candidates for the procedure would be those who fit the current criteria for amniocentesis—over 35 and/or a known carrier of chromosomal abnormalities or inborn errors of metabolism. The chorionic sampling procedure has the advantage of earlier detection over amniocentesis, but the risks have not been defined since the method has not as yet been performed on enough women. Several years of testing are necessary before the safety and risks are completely known.

NORMAL SEX DIFFERENTIATION

After the sex chromosomes have determined the sex in the normal course of events, the genetic constitution induces the differentiation of testes or ovaries in the embryo, and the testes and the ovaries produce hormones. In that hormonal environment, the result is the differentiation of the internal duct systems and the formation of the external genitalia and, just possibly, differentiation of the central nervous system to establish patterns of hormone secretions in the adult.

All of this happens long before birth. Actually, sexual differentiation (as are most of the major events in human development) is essentially completed by the twelfth week after fertilization.

During the fourth week of intrauterine life, only a few days after the first missed menstrual period and when the embryo consists of only a few hundred cells in all, there are large, round, primitive sex cells called *primordial germ cells* that become segregated from the other cells. These are visible in the wall of the yolk sac, a cavity that lies below the developing embryo. When the embryo folds during the fifth week, part of the yolk sac is incorporated into it, and during the sixth week, the primordial germ cells migrate to what will be

their permanent location, the gonads. Here they settle down, ready to be the progenitors or eggs or sperm when their time comes.

The gonads have become visible during the fifth week as a thickened area of epithelium called the gonadal ridge, which is located on the lower border of a temporary excretory organ, the mesonephros. The ridges contain numerous primordial germ cells rich in glycogen. Although the mesonephric kidney does not become the permanent organ (a role taken by the metanephros, which develops independently), the ducts of this transitory nephric system become incorporated into the genital duct system of the male.

In the seventh week of embryonic life, the gonads consist of an outer zone called the *cortex* and an inner area called the *medulla*, containing masses of cells, the primary sex cords. The primordial germ cells are incorporated into the sex cords. Two pairs of ducts are present next to the gonads at about this time. A paired *mesonephric duct* or *wolffian duct* forms to become a collecting duct that drains the mesonephric tubules of the primitive kidney. (Laterally, and a little lower down, the permanent kidneys and the ureters form; they open into the urinary bladder.) Alongside the mesonephric ducts, a pair of *paramesonephric* or *müllerian ducts* develop independently.

Even earlier, during the fourth week after fertilization, a small swelling, or bump, called the *genital tubercle* forms in the approximate position of the future external genitalia. Below it form the outer labioscrotal swellings, and inner urogenital folds form on either side of the cloacal opening, the common exit of the urogenital and gut tubes (see Fig. 5.12).

Now then, in the seventh week of life, the embryo is an "it," neither a "he" nor a "she." The only evidence of sexual organs is a bump and a swelling externally; internally, some noncommitted tissue masses and tubes exist, yet to differentiate. The embryo is 20 mm long

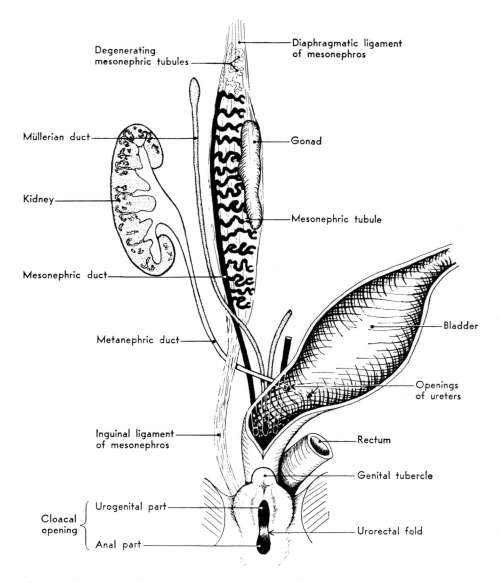

Figure 5.12. Schematic diagram showing the developing urogenital system before sexual differentiation has taken place. (From Human Embryology by Bradley M. Patten. Copyright © 1968 by McGraw-Hill, Inc. Used with permission of McGraw-Hill Book Company.)

(i.e., less than 1 inch) and looks unquestionably like a primate, although not necessarily human—that will take another week. Although the sex was determined at the moment of fertilization—XX or XY—those sex chromosomes have as yet had no visible effect.

Male Differentiation

The differentiation of the gonad into a testis takes place earlier than ovarian differentiation. If there is a Y chromosome in the cells, it has a strong testis-determining effect on the gonad mediated via the H-Y antigen, and the medulla portion develops while the cortex deteriorates. The primary sex cords become the seminiferous tubules of the testis, and the cells that lie between the seminiferous tubules increase in size and number. These interstitial cells secrete the androgen, testosterone. Testosterone and its metabolite, dihydrotestosterone, formed by the enzymatic reduction of testosterone by 5-alpha-reductase, are responsible for the stablization of the mesonephric or wolffian duct and causes the masculinization of the external genitalia. The mesonephric duct is incorporated into the male system as the epididymis, the vas deferens, and the ejaculatory duct, and it also gives rise to the seminal vesicle. The fetal testis also secretes a "müllerian duct inhibitor" that causes the regression of the müllerian duct. If both of these endocrine substances, the testosterone and the müllerian duct inhibitor, are secreted, and if the target tissues contain 5-alpha-reductase and have receptors for the hormones, a normal male will result (Fig. 5.13).

Female Differentiation

If there is no Y chromosome in the cells of the gonad, the medulla degenerates and the cortex develops into an ovary. The primary sex cords break up into clusters and become primordial ovarian follicles, each containing a germ cell surrounded by a layer of granulosa or follicular cells derived from the sex cords. Nothing has to be secreted, and no hormones are involved in the development of the female tract and the external genitalia. The mesonephric ducts spontaneously regress, and the upper parts of the paired müllerian ducts become the fallopian tubes, while the lower parts fuse. The upper portion of the fused müllerian ducts becomes the uterus, and the lower fused segment becomes part of the vagina. Fusion of the müllerian ducts also brings together two folds of the peritoneum, and the right and left broad ligaments as well as the pouch of Douglas and the uterovesical pouch are formed (Fig. 5.14).

During the conversion of the müllerian and the wolffian ducts into the adult structures, some remnants may persist in both males and females. Unless pathological changes develop in them, these vestiges are unimportant.

External Genitalia

At 7 weeks, the external genitalia are said to be in an "indifferent stage," although Figure 5.15 illustrates that they actually appear quite female-like. In females, the genital tubercle elongates rapidly at first, and then its rate of growth gradually slows; it becomes the clitoris. The urogenital folds become the labia minora, and the urogenital sinus outlet that they originally enclosed remains practically unchanged and constitutes the vestibule. The labioscrotal swellings become the labia majora, fusing anteriorly to form the mons pubis.

Under the influence of androgen in the male, the genital tubercle elongates to form the penis, and, as it gets longer, the penis pulls the urogenital folds forward to form the lateral walls of a urethral groove. The urogenital folds close over the urethral groove along the undersurface of the penis to form the pe-

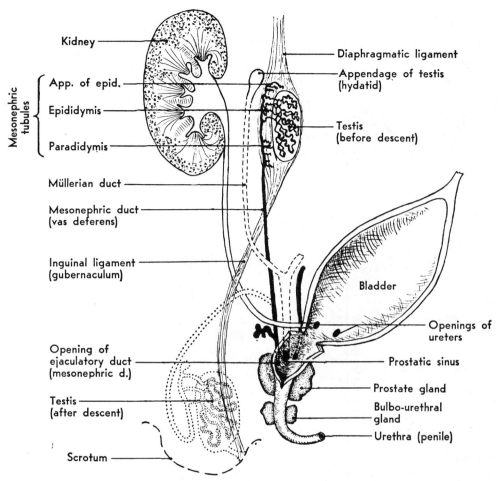

Figure 5.13. Schematic diagram showing the development of the male reproductive system. (From Human Embryology by Bradley M. Patten. Copyright © 1968 by McGraw-Hill, Inc. Used with permission of McGraw-Hill Book Company.)

nile urethra, and the external urethral opening in the male is caused to move from its original location at the root of the penis to its ultimate opening at the tip of the penis. The labioscrotal swellings develop and merge with each other in the midline; they become the scrotum, and later in fetal life the testes will descend from the body cavity to lie within the two scrotal sacs (Fig. 5.15, Table 5.5).

Who Really Came First—Adam or Eve?

It should be obvious that the actual state of affairs in the embryology of the reproductive systems cannot be an affirmation of the Adam and Eve myth. All embryos seem to be innately programmed to become females. In mammals, if the embryonic gonads are re-

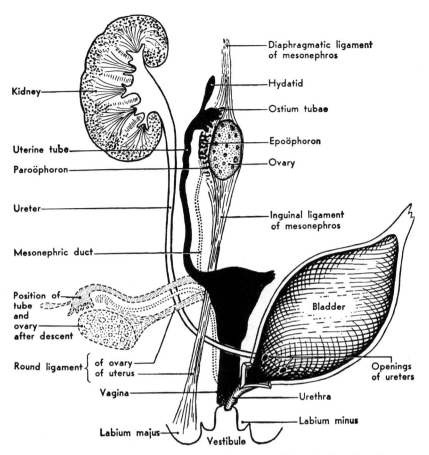

Kidney

Uterine tube

Paroöphoron

Ureter

Mesonephric duct

Position of
tube
and
ovary
after descent

Round ligament { of ovary
of uterus

Vagina

Labium majus

Vestibule

Diaphragmatic ligament
of mesonephros

Hydatid

Ostium tubae

Epoöphoron

Ovary

Inguinal ligament
of mesonephros

Bladder

Openings
of ureters

Urethra

Labium minus

Figure 5.14. Schematic diagram showing plan of developing female re-
productive system. The dotted lines indicate the position of the ovary
and fallopian tube after their descent into the pelvis. (From **Human**
Embryology *by Bradley M. Patten. Copyright © 1968 by McGraw-Hill,*
Inc. Used with permission of McGraw-Hill Book Company.)

moved before differentiation occurs, the embryo goes right ahead and develops as a female—but one lacking ovaries. In order for the basic female pattern to be changed, the Y chromosome must be present. The gonads have to become testes, and the testes must secrete testosterone to develop the male tract and müllerian duct inhibitor so that the fallopian tubes and uterus do not develop. In con-

trast, no hormones from the fetal ovary are necessary to induce the differentiation of femaleness. The developing ovaries form estradiol, and perhaps it plays some role locally in differentiating ovarian tissue, but the hormone is not essential in female development. Embryos of both sexes, of course, are exposed to levels of estrogen and progesterone from the mother, secreted both by her ovaries and

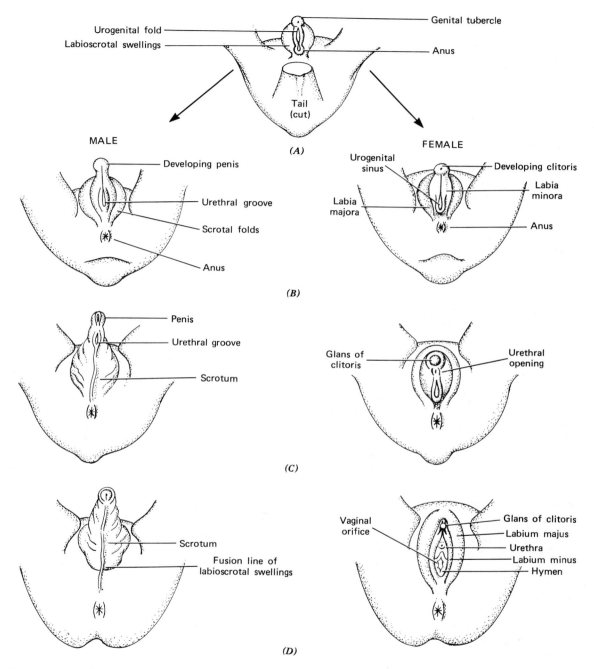

Figure 5.15. Diagrams illustrating development of male and female external genitalia. (A) At seventh week of gestation. (B) At tenth week of gestation. (C) At about the twelfth week of fetal life. (D) Shortly before birth.

141

Table 5.5 EMBRYONIC DIFFERENTIATION: GONADS, REPRODUCTIVE TRACT, AND EXTERNAL GENITALIA

Undifferentiated Structure	Male	Female
Cortex of gonad	Regresses	Differentiates into ovary
Medulla of gonad	Differentiates into testis	Regresses
Primary sex cords	Seminiferous tubules, tubuli recti, rete testis	Degenerate; follicles are derived from secondary (cortical) sex cords
Wolffian ducts	Epididymis, vas deferens, seminal vesicles	Small vestigial remnants
Müllerian ducts	Degenerate	Fallopian tubes, uterus, inner 4/5 of vagina
Inguinal ligament of mesonephros	Gubernaculum testis	Ovarian ligament, round ligament of uterus
Urogenital sinus	Urinary bladder, urethra, prostate gland, bulbo-urethral glands	Urinary bladder, urethra, lower vagina, Bartholin's glands
Genital tubercle	Penis	Clitoris
Urogenital folds	Underpart of penis	Labia minora
Labioscrotal swellings	Scrotum	Labia majora

the placenta, and to steroids from the fetal adrenals. Sufficient quantities of male hormones must be secreted by the male embryo's testes to overcome not only its own innate predisposition to femaleness, but also the high circulating levels of maternal hormones. As a result, mammalian males, even after sexual differentiation is complete, continue to show a very high resistance to experimentally injected estrogens, requiring very large amounts before any feminization appears.

Mammalian females, whose female anatomy is genetic and innate and not determined hormonally, are, in contrast, easily masculinized and responsive to even very small quantities of androgen. The presence of endogenous (from the mother) or exogenous (administered) male hormone during the critical differentiation period in a female fetus can interfere with normal female development of the internal organs and the external genitalia.

SEXUAL AMBIGUITY

Within moments after the birth of a baby, there is an immediate exclamation: "It's a boy!" or "It's a girl!" This instantaneous diagnosis is based on what is easily observed. This baby has a penis and a scrotum—he must be a male. This one has a vulva and vagina—she must be a female. More than 99% of the time, babies are born unambiguously male or female, but in approximately one in 500 births, the sex is doubtful because of the external genitalia. The baby is evidently not male, but neither is it clearly female. This apparent intersex state of the external genitalia is the result of an error of embryonic differentiation, and the extent of the abnormality depends on why or when the error took place.

True *hermaphroditism* (from Hermaphroditus, the son of Hermes and Aphrodite, half man and half woman) occurs very rarely and is the term for individuals who have both ovarian and testicular tissue internally. The genitalia are usually ambiguous, but the male configuration generally predominates. The sex assignment as a boy or a girl is made, as it is in all intersex states, on the basis of which appears to be the dominant sex and on the ease of surgical correction.

Pseudohermaphroditism implies that there is a difference between the external genitalia and the internal gonads. Male pseudohermaphrodites have testes but external genitalia that are feminine or of doubtful gender, and female pseudohermaphrodites have ovaries but masculinized or ambiguous external genitalia.

The reasons for these developmental mistakes are varied and sometimes completely unknown. Embryonic defects can sometimes occur spontaneously and without any obvious explanation. Some of them are the result of chromosomal abnormalities, but many of the XO, XXY, XYY, and XXY variants that have been described are not even recognizable as a difficulty until puberty. Even then, the chief problem in most of them is infertility. Sex chromosome abnormalities usually produce such minimal or subtle changes in the external genitalia at birth that they are likely to be overlooked.

Ambiguous external genitalia are more usually the result of hormone imbalance during embryogenesis, either inadvertently induced by hormone administration during pregnancy or because a genetic defect of metabolism was expressed as a hormonal dysfunction. The testicular-feminization syndrome that occurs in genetic males and the adrenogenital syndrome and progestin-induced virilization that can occur in genetic females are examples of abnormal sexual differentiation caused by hormonal imbalance.

Testicular-Feminization Syndrome

Testicular-feminization syndrome is also called the androgen-insensitivity syndrome. It is an X-chromosome-linked recessive gene defect in which women carriers transmit the disorder to half their male offspring. Gene expression results in the inability of the mesonephric ducts, the genital tubercle, or any somatic cells to respond to androgen, evidently because they lack androgen receptors in their cytoplasm. The karyotype of these individuals is male, 46,XY, but their external genitalia are female, usually with a blind-ending vagina. Internally there are testes, usually undescended, although they have been occasionally found in the labia, and there may be very rudimentary fallopian tubes and a uterus without a cavity. In the embryonic development of males affected with this condition, the testes, under the influence of the Y chromosome, produce both androgen and the müllerian duct inhibitor. But since the target cells, lacking receptors, cannot respond, the wolffian ducts do not develop to any extent. The müllerian ducts respond normally—they are inhibited. The external genitalia, also unresponsive to androgen, develop along female lines. The baby is born, is observed to be female, and is reared as a girl. Of course, the inability to respond to androgen continues throughout life. At puberty, stimulated by gonadotropic hormones, the testes release androgens and some estrogens as well. The target tissues respond only to the estrogens so there is breast development, and the body configuration is feminized. Menstruation does not occur, and this may be the first sign that something is wrong. A pelvic examination followed up by cytogenetic studies can establish that these females are actually chromosomal and gonadal males. If this knowledge is handled appropriately on both an intellectual and an emotional level (that is, if the parents are not abruptly apprised that their daughter is really their son) these girls do not necessarily have sexual identity difficulties. They have been raised as girls; they will grow up to be women and will marry and lead normal sex lives though unable, of course, to bear children. The testes of individuals affected with the testicular-feminizing syndrome are known to be more likely to become malignant, and because of this risk, they are frequently removed. If such removal is done before puberty, the breasts fail to develop, indicating that it is the testicular estrogens that cause the female development.

Adrenogenital Syndrome (Congenital Adrenal Hyperplasia)

In contrast to the above condition, where the body is female but the sex chromosomes, gonads, and ducts are male, there are disorders in which the external genitalia are masculinized to a varying degree, but the sex chromosomes and internal organs are female. Because the inherent program of development is female and androgen has to be added to produce maleness, any female embryo or fetus that is exposed to androgen during sexual differentiation may be born with ambiguous virilized genitalia. There may be an elongated clitoris with an external vaginal opening, or there may be a "penoclitoris" with enough midline fusion to obliterate any vaginal orifice. Only the external genitalia are affected because androgen alone is not enough to cause development of male duct structure; müllerian duct inhibitor also has to be secreted by male testes to prevent the uterus and fallopian tubes from forming so that the wolffian ducts can be organized into male structures. These fetuses are female, however, and lack testes so the internal ducts are not masculinized.

The most frequent cause of female fetal masculinization is an inherited defect (one in 20,000 live births) in the ability of the adrenal glands to synthesize glucocorticoids. The adrenals, two small endocrine glands that are located on top of the kidneys, produce three kinds of steroids: the mineralocorticoids, which control the amounts of sodium and potassium in the body; the glucocorticoids, which affect carbohydrate metabolism; and the sex steroids, estrogen and androgen, normally produced in insignificant amounts compared to gonad production. In the adrenogenital syndrome, also called congenital adrenal hyperplasia, the fetal adrenals fail to synthesize glucocorticoids because the necessary enzyme, C-21-hydroxylase, is lacking as a result of hereditary defect. Because there is less glucocorticoid hormone circulating in the fetus, there is less inhibitory feedback to the fetal pituitary gland. The pituitary attempts to compensate for the lack by putting out increased amounts of adrenocorticotropic hormone. The fetal adrenal glands are overstimulated, and the zone of the adrenals that produces androgens undergoes hyperplasia (more cells), and greater production of androgens results. Normally, the enzymes involved in glucocorticoid synthesis by the adrenals develop in the eighth to tenth week of fetal life; if they fail to develop, the timing is such that the excessive androgens are produced at the very time of the differentiation of the external genitalia, and they are masculinized. Most of the time, the deficiency of C-21 hydroxylase is incomplete, and the only result is virilization of the genitalia. In about one-third of the clinical cases, however, lack of the enzyme results in life-threatening deficiency of adrenal corticoids, and substitution therapy must be continued throughout the life of the individual. The masculinization effects of adrenogenital syndrome are limited to the external genitalia; once surgical alteration is made, the baby can be raised as a female, and further growth, development, and reproductive functions are normal.

Another source of virilization in a female fetus is an androgen-secreting ovarian or adrenal tumor in a pregnant woman, but this is rare since women with such preexisting tumors seldom become pregnant. More commonly, masculinization has resulted from the treatment of pregnant women with hormones to avert miscarriage. About 20 years ago, the accepted belief was that spontaneous abortion (miscarriage) was often due to insufficient production of progesterone from the placenta or corpus luteum of pregnancy. It seemed reasonable, therefore, to administer progesterone in the hope of maintaining the pregnancy. Natural progesterone is inactivated if it is taken orally, so synthetics that were orally effective were used in cases of threatened spontaneous abortion. The difficulty is that most synthetic progestins are androgenic to a varying degree. Although their masculinizing effects have been well known since 1958, and despite evidence that not only were progestins of doubtful value in preventing miscarriage but that these hormones caused other birth defects, they continued to be prescribed for pregnancy testing, prevention of miscarriage, and other complications of pregnancy well into the 1970s. The FDA withdrew approval of these drugs for use during pregnancy in 1974, but several years after the warning the drugs were still being administered, since only 10% fewer prescriptions were written in 1975 than in the year before the warning had been issued. Fortunately, the number of birth defects seen has been extraordinarily low considering the total number of women who have taken prescribed hormones, suggesting that only a small part of the population is susceptible to their effects. It is now recognized that there are few, if any, reasons to administer hormones during pregnancy, and all pregnant women should be aware of the potential hazard of taking sex steroids during the first 3 months of pregnancy.

IS THERE A MALE BRAIN AND A FEMALE BRAIN?

It is the male sex hormone that plays the critical role in the differentiation of the male sexual anatomy. Without androgen, all fetuses would develop into anatomic females. The question that has been the subject of a great deal of attention and research and has generated considerable controversy concerning the interpretation of such research is, does androgen also play a role in the differentiation of the central nervous system? That is, is it possible that there is also a critical or sensitive period during which the brain is susceptible to male sex hormone and is organized to result in subsequent masculine behavior, and perhaps not just reproductive behavior?

It is known that the hypothalamus contains an area of steroid-sensitive neurons that are responsible for the secretion of gonadotropin releasing hormone that, in turn, controls the release of FSH and LH from the anterior pituitary gland. According to Barraclough (1976) and others, this region, located above the optic chiasma and called the preoptic area, can be referred to as the cyclic center in mammalian females because the pattern of gonadotropic release is periodic and cyclic, rather than the continuous and basal secretion that occurs in males. There is ample evidence that in rodents, there are sex differences in the responsiveness of this area to steroids.

In rats, mice, hamsters, and guinea pigs it has been shown that the sensitive area is undifferentiated and potentially female until a certain critical period. If the brain is exposed to androgen during that period, the animal is defeminized, that is, the cyclic pattern of gonadotropin release in response to estrogen is permanently suppressed. If no exposure to androgen takes place, the hypothalamus of both male and female rodents will develop a female cyclic pattern. So the hypothalamus, like the genital anatomy, is potentially *female* in both male and female rodents, unless androgen is present. That critical period during which the hypothalamus is converted from a female type to a male type is postnatal in laboratory rats and takes place within the first 10 days after birth. During that time, it is possible to administer an injection of androgen to newborn female rats and change the hypothalamus from a cyclic pattern to a noncyclic pattern. When females like these reach sexual maturity, they will never ovulate and will remain sterile because FSH and LH are being continually (rather than periodically) released to produce cycles. Conversely, if the testes of a newborn male rat are surgically removed, the hypothalamus remains female and retains the ability to induce cyclic activity in an ovary. This can actually be observed if an ovary is grafted into the anterior chamber of the eye in such a castrated male rat.

When male and female rats are subjected to various experimental manipulations with their own gonadal hormones or with administration of injected steroids, they also undergo various changes in their mating behavior when they become sexually mature. If a female newborn rat is injected with androgens within 10 days of birth and is given no subsequent hormonal treatment, the typical female mating behavior toward male rats is inhibited, and she exhibits some male sexual behavior toward female rats. If the newborn female rat has been both androgenized and castrated—that is, had the ovaries surgically removed—her behavior at maturity will be sexually indifferent. She will exhibit disinterest to males and females alike. If such a postnatally androgenized and castrated female is treated with androgens as an adult, she displays some of the sexual behavior of a male.

Male rats castrated at birth with no hormone treatment will be sexually indifferent at adulthood—similar to the androgenized, castrated females. If such male rats are given doses of androgens during the first week after castration, a certain degree of masculine adult behavior is restored. If they are given estrogen and progesterone at maturity, they will display feminine sexual behavior.

Many of the same kinds of experiments were performed on other rodents, on other mammals such as guinea pigs, ferrets, dogs, sheep, and rhesus monkeys, and on several species of birds. In all animals studied, there is a developmental "critical period" for sexual differentiation during which the brain is sensitive to gonadal hormones. This time of increased sensitivity is postnatal in some species that are relatively less mature at birth and prenatal in those animals that are born more fully developed. The mechanism by which gonadal steroids are able to transform brain tissue to bring about particular behavioral and neuroendocrine functions remains undetermined, but there are several current hypotheses. Perhaps hormonal receptor sensitivity is altered, or maybe neurotransmitter function is modified. One possibility getting a lot of investigation is that hormones may induce a change in gene expression and lead to a local alteration in cellular growth (Naftolin and Butz, 1981). Steroid exposure could thereby affect brain circuitry, that is, change the way neurons are hooked up so that the wiring is rearranged. In substantiation of the growth theory, there has been increasing evidence for morphological changes in the brains of some species. In some of the animals studied, the anatomical sex differences are subtle and require ultramicroscopy to see; in others, they are clearly visible with a light microscope even on gross examination.

With the use of an electron microscope, Raisman and Field (1973) found that there were differences in the type and distribution of nerve fiber connections in male and female rat brains. Nottebohm and Arnold (1976) discovered that the song control centers, a chain of five discrete groups of neurons that controlled male singing patterns, were larger in male zebra finches than in females. William Greenough and his associates (1977) saw differences in the shape of the stimulus-receiving ends of the nerve fibers (dendrites) in the brains of male and female hamsters, and Roger Gorski and his co-workers (1979) found a very obvious sex difference in the brains of male and female rats that could be seen with an ordinary microscope. Located in the pre-optic area, a group of neurons they called the sexually-dimorphic-nucleus (SDN) is five times greater in size in male rats than in female rats. Although the volume of the SDN appears to be hormone dependent (the area shrinks with castration in male newborn rats and becomes larger if female newborn rats are given testosterone) the specific function of the nucleus is unknown.

In human brains, the only anatomical sex difference that has been reported is not in the hypothalamus, and not even in the gray matter of the brain. Two physical anthropologists, Lacoste-Utamsing and Holloway (1982), examined nine male brains and five female brains and discovered that a part of the corpus callosum, the bundle of white nerve fibers that connects the left and right cerebral hemispheres, was larger and differently shaped in female brains. The functional significance, if any, of the sex difference is unknown. It is possible that there simply may be a genetic difference between the male and female corpus callosum.

The studies on sexual differentiation of the central nervous system in various animals have led to the brain-organizing or brain-differentiating theory of the role of androgen: that fetal male hormone differentiates the embryonic bipotential brain into a "male" brain shortly after birth in rodents and during the prenatal period in other mammals. Since both gonadotropin release and mature mating and courtship behavior were evidently determined by this fundamental androgen influence on the developing brain, some workers further postulated a similar imprint operating for those characteristics thought to be uniquely male, such as aggressiveness, dominance, increased activity, rough-and-tumble play, and so forth.

The classic theory on the sex differentiating role of androgen on the brain has had to un-dergo some modification in the light of contradictory data. Different dosages of testosterone can produce different effects, and estrogen in animals is evidently just as effective in masculinizing the hypothalamus. Moreover, the biochemical pathways for testosterone action differ in various mammals. In rats, it is not testosterone that directly brings about sexual differentiation, but rather it is the brain's conversion of androgen to estrogen that causes the effects. Since estrogen rather than androgen does the differentiating, the male rat fetuses have to be protected from all the circulating maternal estrogen. This is accomplished in rats by an estrogen-binding protein called alphafetoprotein that selectively binds estrogen and not testosterone. There is no evidence that human alphafetoprotein has a similar protective effect. Brain tissue also contains an enzyme, 5-alpha-reductase, that converts testosterone to another metabolite, dihydrotestosterone (DHT). It is DHT that binds to the androgen receptors in the brains of some species. Nevertheless, it is well established that in rodents, gender-specific behavior patterns are due to neuro-organization of the brain by early androgenation, whether primary or secondary in nature.

It appears, however, that rat brains are not the same as monkey or human brains. Attempts to confirm the sexual differentiation of the hypothalamus in primates for cyclic versus noncyclic gonadotropin release, or for behavior, have not been particularly successful. When androgens were given to female rhesus monkeys throughout pregnancy, their female offspring were, unsurprisingly, born with masculinized genitalia (elongated clitoris, sometimes with the labia partially fused). There was no suppression of the hypothalamic cyclic gonadotropin release, however, because when these female monkeys matured, they had normal ovulatory menstrual cycles and were normally fertile. It was reported that such androgenized females did exhibit some masculinization of behavior patterns. When the variables of aggressiveness, rough-and-tumble play, energy expenditure, and mounting behaviors usually seen in young male monkeys were used as measures, these females tended to behave more like male juveniles.

In human females, the opportunities for observing the effect of prenatal exposure to androgen are obviously very limited. The two clinical conditions described earlier, adrenogenital syndrome and drug-induced virilization, are virtually the only sources of clinical data. When such genetic females are exposed to androgen and are born with masculinized genitalia, their brains, too, have presumably been androgenized. Females like these nevertheless menstruate and are fertile after a normal, but in some instances delayed, puberty. Like the female monkeys, the cyclic release of gonadotropins was evidently not affected. Their behavior, however, has been reported to have been influenced by male hormones acting on their developing brains.

Money and Ehrhardt studied a group of 25 such fetally androgenized females during a period of several years starting in 1967. Their work is the most frequently cited as prototype evidence for the ability of androgen to program the brain because of the increased incidence of "masculine" behavior found in the androgenized girls compared with a control group. These 25 young women displayed behavioral characteristics that they themselves, their mothers, and Money and Ehrhardt collectively called "tomboyism."

What kinds of behavior do "tomboys" exhibit that suggest that prenatal androgen left an imprint on the fetal brain?

1. These girls were athletic. They liked team games, such as neighborhood football and baseball, and, although some of them liked to play with girls as well as boys, some preferred boys exclusively as playmates.

2. They preferred clothing that was functional. Money and Ehrhardt say that these girls "chose to wear slacks or shorts" rather than "chic, pretty, or fashionably feminine dresses"—more practical, perhaps, for choice of activities. (Dressing for comfort rather than style, and pants on women were not as accepted in the late 1960s.)

3. The diagnostic group lacked interest in dolls and turned to the traditional boys' toys—cars, trucks, and guns—when they were younger. Later, they were said to be nonmaternal; they fantasized about having a career rather than, or in addition to, having babies and expressed little interest in wedding play. The researchers emphasized that even with all those indications of "tomboyism," these girls clearly had a female sexual identity. There was no evidence of homosexuality in their erotic interests as they became older.

4. There was a consistent tendency toward a higher IQ in the girls exposed to an excess of fetal androgen. But, as subsequently pointed out by Eleanor Maccoby, the intellectual levels of a group of prenatally androgenized females were elevated beyond that of the general population, but were not higher than the IQs of their own normal siblings. Ruth Bleier also notes that Money and Ehrhardt's patients formed a highly select group of children from families that had sought and received sophisticated medical treatment from the day of the girls' birth, with all the concomitant implications concerning their intellectual and socioeconomic status. The general impression has generally remained in the literature, nevertheless, that excessive prenatal androgen enhances intelligence in females.

Subsequent studies of girls affected with congenital adrenal hyperplasia confirmed the findings of the above investigation. The activity levels in the children were enhanced, and their interest in pregnancy, infant care, and wedding play was decreased. Genetic males with congenital adrenal hyperplasia were also evaluated; the only observed difference between them and unaffected male children was an increased energy expenditure at play (Baker, 1980).

It should be clear by now that these studies reflect a traditional view of masculinity and femininity. The researchers' interpretation is that children may show male behavior or female behavior, but not both. The behaviors characterized as tomboyism are only stereotypically masculine, however, and are not indicative that male programming of the embryonic brain has taken place. To want to overcome sexual stereotyping at an early age may not mean an androgenized brain in a girl, but her tomboyism may be one criterion for a successful professional career in later years. The August 1977 issue of *Women-Sport* Magazine contained interviews with a number of highly successful women, with the intent to compile a ''Who's Who'' of tomboys. In their list of politicians, college presidents, actresses, artists, and authors, there were many women who identified themselves as having been tomboys with the very same characteristics that were attributed to the diagnostic groups studied. These women, at the top of their professions, had played basketball, baseball, and football, had excelled at sports, had climbed trees, gotten themselves dirty, and generally had more *fun!* They were athletic, lively, and spirited, and they had a better time playing with the guys than with the girls. This diverse group of women included former First Lady Betty Ford, actresses Farrah Fawcett and Jane Fonda, the late Ella Grasso, former Governor of Connecticut, Congresswomen (at that time) Barbara Jordan, Margaret Chase Smith, and Shirley Chisholm, three of the four women mayors of major cities, nine of the twelve Women of the Year in 1976, and more. Such a listing does not indicate that success is based on taking on male characteristics in childhood, but more likely means that for these women, their world as children was less limited, that they enjoyed activities on their own terms and not according to some rigid stereotyped role. It also offers proof that females whose embryonic brains have not been exposed to the influence of male steroid do not differ markedly in attitudes, interests, behavior, or sexual preference from females who have been prenatally exposed to androgen.

In some mammals, at a particular period of maturation of the brain, exposure of a male fetus to his own circulating male sex hormones may result in the establishment of the adult male pattern of a constant and noncyclic secretion of gonadotropins. It is also possible that in some mammals, androgens present before birth act on some developmental processes in the brain, which are then programmed for subsequent male social and sexual behavior. Evidence that this occurs in male humans or subhuman primates is tenuous and based on inference, although it is undeniably true in rats, mice, hamsters, and guinea pigs. There is little, if any, justification for use of the terms ''male'' brain or ''female'' brain in humans, except in the sense that it is one found in a male or in a female.

GENDER IDENTITY

It has been seen that attempts to link gender-specific behavior or sexual preference in humans to a prenatal hormonal influence would at this time be purely speculative. There is more evidence that gender-identity differentiation of the brain as male or female occurs postnatally and depends not on hormones, but on the total environment in which a child is reared. When an error in embryogenesis has resulted in sexual ambiguity, corrective surgery can be performed to make the child unambiguously male or female to establish the morphological sex, and the gender can then be assigned to conform with the surgical alteration. When a child is called a female, looks female, and is treated as a female, she will grow up female, regardless of the genetic and gonadal sex. It is the sex of assignment and the sex of rearing, the entire postnatal psychosexual environment in which a child is raised, which is evidently the most critical in the establishment of gender identity as a man or as a woman, and not the chromosomes, not the ovaries or the testes, and not even the external genitalia. Usually, the chromosome constitution, the gonad structure, the morphology of the internal ducts, and the appearance of the external genitalia coincide; when they do not, the sex of assignment and rearing can apparently override both the genotype and the phenotype.

Money and Ehrhardt have described a situation that occurred in identical male twins. During circumcision on one of the boys, the surgeon accidentally and irrevocably damaged the tissue of the penis, which subsequently had to be amputated when the child was 7 months old. When sex reassignment to a female was suggested by a plastic surgeon, the parents had no other choice but to agree, and the required genital reconstruction was performed. At age 17 months, then, one of the twin sons became a daughter and was reared as a female; from then on, all clothing, hair styles, toys, and so forth were emphatically feminine. Six years after the reassignment of sex, the two children acted, dressed, and behaved as traditionally appropriate to their gender, even though they were genetically identical twins and both male.

Another case in point: Ewa Klobukowska was 21 years old and the co-holder of the world 100-meter dash record for women when doctors found she had "one chromosome too many" to be eligible as a woman for future competition. (The precise nature of the chromosomal anomaly was never made clear.) Because of this medical finding, the International Athletic Federation in 1968 withdrew ratification of all the records, victories, and medals she had won. Subsequently, the Polish former athlete lived alone in Warsaw, dated men, and was not averse to the idea of marriage, although she had been advised that her ovaries and tubes were atrophied and that she could never become pregnant.

It would appear that Ewa Klobukowska continued a belief in her femaleness despite her "failure" of the sex test; she seemingly had a solid sense of her identity as a woman—she dated, wanted to get married—even though her gender identity belied her genetic sex. The totality of gender, the *being, acting, feeling* like an adult male or female, is usually a coinciding of genetic sex, gonadal sex, hormonal sex, morphological sex, sex of assignment and rearing, and behavioral sex or sex role. Even if a discrepancy exists within the first four items cited, the last two can still produce a fulfilled and functional adult. It is only when the biological sex is out of alignment with the gender identity in an *adult* that the real psychological dilemmas result. Transsexualism is the intense feeling of identification with the sex opposite to one's chromosomal, gonadal, and morphological sex. It can occur in both sexes and is usually described as a "man born in a woman's body," or vice versa. Such incongruities, sometimes called *psychic hermaphroditism* or *gender dysphoria syndromes*, occur somewhat more frequently in males than in females. Hormonal and surgical sex reassignment, formerly acceptable as a method of treatment for those individuals for whom psychiatric therapy had been unsuccessful, has recently become more controversial. It is currently considered a last resort for a highly select group of gender dysphorics. Some reassigned transsexuals have received considerable publicity or have published their autobiographies.

The Huevodoce of the Dominican Republic

The well-established tenet that biological sex can be overridden by the sex of rearing—that a genetic male, raised *unambiguously* as a female will see herself unambiguously as a female—has been challenged. A 1979 study describing an apparent change in gender identity and gender role that occurred in genetic male pseudohermaphrodites has raised questions concerning the importance of early psychosocial environmental influences as the primary determinants of gender identity. The subjects of the research, originally discovered by a vacationing physician and rediscovered by a Cornell Medical School endocrinologist, Julianne Imperato-McGinley, were 38 males in the Santo Domingo area of the Dominican Republic, all of whom were afflicted with a rare genetic defect called *5-alpha-reductase deficiency*. Testicular feminization syndrome, in which the androgen receptors in target tissues are totally absent, was described earlier in this chapter. In 5-alpha-reductase deficiency, an autosomal recessive defect, the problem is not the absence of androgen receptors but rather a deficiency of the enzyme that converts testosterone to dihydrotestosterone (DHT), the form of androgen that binds to the cells of the fetal external genitalia in order for the male penis and scrotum to form. Male children with this inborn error of metabolism are pseudohermaphrodites, born with normal internal male ducts; testes in the abdomen, inguinal canal, or in the scrotum; a cleft scrotum that appears labialike; a blind-ending opening in the perineum that resembles a vagina; a phallus that is more clitoris than penis; and a perineal urethral opening under the phallus. In untreated individuals at puberty, conversion to DHT is no longer necessary for masculinizing changes to occur, and the increased levels of circulating testosterone result in the typical male secondary sex characteristics.

There were 38 subjects who were identified as being affected with the hereditary defect, and all came from 23 interrelated families in two rural villages. At the time of the study, 33 were still living. Of these, 19 had reportedly been unambiguously raised as girls. Seventeen of those evidently emerged from puberty with a reversed gender identity, and all but one not only felt, but acted like men, that is, they adopted a male gender role. Of those 16, 15 had lived or were currently living as males with women. Under the influence of testosterone, these individuals had experienced voice deepening, penis growth with erections and ejaculations that were emitted from the perineal urethra, enlargement and pigmentation of the scrotum, and descent of the testes if they were not already in the scrotum. They also developed an increase in muscle mass and strength. The only usual male characteristic missing was facial hair, and none of them developed acne. Beardless they may have been, but, nonetheless, here was evidence that females, raised and treated like girls, changed at puberty into men, taking on men's roles, jobs, and sexual identities.

On the basis of these findings, Imperato-McGinley and others concluded that in males, when the sex of assignment and rearing is contrary to the chromosomal sex, the Y chromosome prevails if the normal testosterone-induced activation of puberty is allowed to occur. They further reasoned that the prenatal exposure of the brain to male hormone is a stronger determinant of male gender identity than is the sex of assignment and the sex of rearing.

This report of sex role reversal taking place at puberty is necessarily controversial, and many doubts have been raised about its methodological validity and its conclusions. One criticism is that the affected children may not really have been unambiguously reared as girls. Individuals with the defect had been known in the society for four generations, and such children were referred to by other inhabitants of the villages as either "machihembra" (first woman, then man), or "huevodoce," which translates as eggs-at-twelve, but is actually slang for testicles or "balls" at twelve. The affected children were likely to have been confused about their identity as normal girls. Their genitalia before puberty were not completely masculine, but they were surely not normally feminine and must have looked peculiar to them. As John Money has said, " . . . if you're a girl who is not sure that you're supposed to be a girl because of your funny-looking genitals, you have an alternate choice . . . if you feel everything is wrong the way you are, maybe the correct way is the other way."

Perhaps Imperato-McGinley and her co-workers' study illustrated merely that in some societies, some psuedohermaphrodites who may be confused about their proper gender can undergo a psychological and social sex reversal after experiencing a masculinizing puberty, but the study has not demonstrated that a true gender reversal takes place. Neither has it presented conclusive evidence that prenatal androgen exposure is the primary determinant that programs the brain for gender identity in males.

THE OLD NATURE–NURTURE ARGUMENT

The very terms "gender identity" and "gender role" are confusing and unclear. What does it mean, to feel and act like a woman, or to feel and act like a man? Perhaps the feelings of identity as a male or as a female involve the personal view, the psychological awareness, the self-recognition, the intense conviction that one belongs to a particular gender. Then,

the gender role that one takes in society is the way that gender identity is acted out—the countless numbers of behaviors, habits, preferences, expectancies, and attitudes that are recognized as sex-specific to that particular gender in one's sociocultural environment.

The behavioral sex differences that are perceived to exist between males and females have been examined, investigated, explored, and studied as the subject of scientific writings for centuries. Psychologists, anthropologists, sociologists, biologists, economists, and political scientists all have attempted to understand what those sex differences are, how they originated, developed, and are manifested. A major question has been, how important is biology in the determination of one's behavior as a male or as a female? What are the sex differences in personality traits, abilities, interests, and values? Do these differences really exist, and if so, are they preordained by genes and hormones? Or are behavioral sex differences psychological and social in origin, and culturally determined?

The controversy surrounding these questions is very old: which contributes more to behavior, nature or biology, or nurture or environment? This kind of dispute could be dismissed as overly simplistic, irrelevant, and totally unproductive, except that it is still frequently resurrected to justify the alleged superiority of one group of humans of a different race or sex over another. If it is believed, for example, that one race or sex is genetically superior in intelligence to another, there is certainly nothing to be gained by providing equal educational opportunity to the inferior group. If it is considered that gender-appropriate stereotypes of feminine or masculine behavior are innate, natural, and are biologically based, it can justify discriminating against women in lower-paying, lower-status jobs. If a woman's place is meant to be in the home, why pass legislation for her equal employment opportunity?

A traditional view, unrelinquished by many scientists, is that physical, psychological, and behavioral sex differences are of biological origin. Behavior is then viewed as a natural consequence of biological differences in size, strength, and reproductive capacities. A woman's reproductive system gives her the ability to conceive, bear, and deliver children; her body, therefore, confines her to a maternal and nurturing role. It is her nature to behave in a "feminine" way—dependent, passive, and less active. Physically, a male is stronger—he is the hunter. He has always been characterized by aggressive behavior; it is in his biology—his nature—to be dominant. To quote British biologists Austin and Short (1972): "In all the systems that we have considered, maleness means mastery: the Y-chromosome over the X, the medulla over the cortex, androgen over estrogen. So physiologically speaking, there is no justification for believing in the equality of the sexes; vive la différence!" Or as Colley (1959) put it: "Persons do not exist; there are only male persons and female persons—biologically, sociologically, and psychologically."

There are many who have challenged the above opinions, believing that much of the research that attempts to prove a biological basis for the social roles of males and females from animal studies, observations on the newborn, hormone experiments, or from anthropological studies is frequently inconclusive and contradictory. It might be suggested that based on present knowledge, maleness means merely maleness and not mastery, and that people are human beings first and male persons and female persons second. The relevance of a research tradition that emphasizes the biological limitations and differences between the sexes rather than the similarities is debatable. We might also question the existence of only innate biological factors that determine the traits stereotyped as male qualities (those of aggressiveness, leadership, dominance, competence, independence) when those traits are highly valued and rewarded socioeconomically and directly opposite to those traits associated with women and femininity (submission, nurturance behavior, weakness, passivity, incompetence, emotionality).

Another approach to the nature and origin of behavioral sex differences is the assumption that biological and social forces both interact. Neither can be solely responsible in the determination of personality and behavior, but postnatal influences very likely have the greater importance. There is now a substantial body of research to uphold the contention that concepts of the "typical male" and the "typical female"—that is, behavior that is stereotypically masculine or feminine—is the result of socialization and is conditioned by cultural attitudes and expectancies.

In this view, acculturation to sex roles begins in infancy when parental attitudes toward male and female children differ to reflect society's perceptions of what is deemed appropriate for each sex. As the child grows, the sex role assignments continue to be reinforced by social pressure, and the gender identity, or awareness of self as male or female, develops within this context of gender-specific behaviors. There is ample evidence from the clinical cases of ambiguous sex that gender identity is firmly imprinted by between 18 months and 3 years of age. By that time, a child has a pretty strong awareness of "I am a boy so I will behave like a boy," or "I am a girl, so I will behave like a girl." The behaviors become quite rigidly polarized into stereotypes of masculine and feminine qualities, personality traits, and activities. Behaviors allowed to a particular sex are reinforced and rewarded, and those that overstep role boundaries are subject to social sanctions. Children are conditioned by their parents, their teachers, their textbooks, TV, the movies, and society in general to accept the traditional sex roles, and value is placed on strict adherence to the masculine or feminine image. Although little girls are generally allowed a somewhat greater leeway in displaying masculine-type behavior, at least until puberty, little boys are denied any behavior perceived as feminine, and their gender role is defined within much narrower limits. Even those parents, enlightened to a nonsexist childrearing that encourages giving trucks, erector sets, and tool kits to their daughters, have a little more trouble giving dolls to their sons.

Rigid polarization of sex roles into stereotypes of masculinity and femininity tends to further the notions of a natural male superiority, since only the "masculine" characteristics are those that are necessary for success in society. "Feminine" qualities are also valued, but they are seen as those characteristics necessary for successful homemaking and childrearing. If a woman works, it is at a job for which she is suited—being a nurse, teacher, librarian, or laboratory assistant. A young woman who wants to become a doctor, scientist, or engineer may then be regarded as deviating from her normal role, and she receives strong cultural messages that she is compromising her femininity by attempting to enter a man's world. Sex role stereotyping thus limits the full potential of all human beings by freezing them into traditional behaviors and attitudes.

The recognition that stereotyped generalizations about males and females are actually dysfunctional in our present world is a recent awareness, and it owes much to the women's movement. The roles of both men and women today have changed rapidly in response to the enormous social and economic pressures of a society that is overpopulated, inflationary, environmentally polluted, and rapidly running out of resources. A woman's place is not in the home and has not been for many years. Making up almost one-half of the workers now, women will continue to move into the labor force, many to occupy jobs that have always been perceived as male.

There are 55 longshorewomen in New York licensed to work on the male-dominated docks; they lift and stow boxes of bananas and 130-pound bags of coffee. And until they were laid off as a result of economic recession (last hired, first fired), there were a growing number of women coalminers in West Virginia, using a pickax alongside men and doing a "man's job." Not every woman (nor every man) has the strength or the inclination to be a dockworker, coalminer, truckdriver, or piano mover, but those who do would agree with women's activist, Florynce Kennedy: "There are only three jobs for which gender is a bona fide qualification: sperm donor, wet nurse and human incubator."

Not everyone is ready to believe that. There are many who see such challenges to our traditional attitudes about the roles of men and women in society as highly threatening and an attack on the home and family. They blame everything on the women's "libbers"—increasing divorce and crime rates, homosexuality, acceptance of abortion—and are convinced that a "reverse discrimination" now exists against white males. They see the traditional behavior of males and females as promoting stability and security, and they find it more comforting to cling to the stereotypes. Even if the result is the continued denial of equal opportunity to women, it is particularly appealing for some to still believe that behavioral sex differences are solely ordained by natural law, biologically determined, and that anatomy really is destiny.

While there are some scientists who contend that either nature or nurture alone is responsible for all sex differences, most contemporary researchers are more interested in assessing the relative importance of each. There have been thousands of studies of the differences in intellectual and cognitive abilities and in social and psychological behavior that exist between males and females. Current investigations continue to examine the importance of hormones on aggression or on nurturing behavior, to study the relationship between brain lateralization and sex differences, or to explore the possible existence of sex-linked genes for spatial ability and mathematical aptitude. Other studies are designed to demonstrate the greater importance of socialization processes in the development of sex differences. With all the research, *no measured behavior has as yet been found to be unique to either males or females.* Behavioral sex differences are never qualitative.

It is hardly realistic to raise children in an atmosphere totally free of generalizations, stereotypes, and expectancies about their behaviors, but only in this way can the relative importance of innate factors actually be determined. Perhaps biology does create a predisposition for the development of certain behaviors that are masculine or feminine, or perhaps it does not. And perhaps the answer is irrelevant. Of course, males and females exhibit differences in behavior, but so what? Our major concern should be that there are no sex differences, whatever their basis, that are able to justify sexist discrimination, that can imply male supremacy or female inferiority, or that can prevent individuals of both sexes from the realization of their full human potentials.

That sex differences exist is really not that important. It is that insistence—that demand—of society that they have to exist that can be stifling for both males and females. Psychologists Sandra Bem, Alexandra Kaplan, and others have proposed the concept of androgyny (*andro*, male and *gyne*, female), a model that acknowledges that valued feminine and masculine qualities can exist in the same individual to promote flexible and adaptive, rather than expected and stereotyped behavior. An inflexible categorization of masculine–feminine limits freedom of choice for women and cheats them of the opportunity to make their maximal contribution to all humanity. To set up limitations on some humans, to concentrate on the differences instead of the similarities between the sexes, not to integrate and prize the positive traits of both, is of little value. We are all people, capable of behaving in both masculine and feminine ways, whatever our sex. Such balance provides for the truly whole and integrated personality.

REFERENCES

Austin, C. R. and Short, R. V. *Reproduction in Mammals*, Book 2, *Embryonic and Fetal Development.* Cambridge University Press, 1972.

Baker, S. Psychosexual differentiation in the human. *Biol Reprod* 22:61–72, 1980.

Barfield, A. Biological influences on sex differences in behavior, in Teitelbaum, M. S. (ed). *Sex Differences: Social and Biological Perspectives.* New York, Anchor Press, 1976.

Barraclough, C. A. Modifications in reproductive function after exposure to hormones during the prenatal and early postnatal period, in Ganong, W. F. and Martini, L. (eds). *Neuroendocrinology*, vol 2. New York, Academic Press, 1976.

Bleier, R. Brain, body and behavior, in Roberts, J. (ed). *Women Scholars on Women.* New York, McKay, 1976.

Colley, T. The nature and origin of psychological sexual identity. *Psychol Rev* 66:165–177, 1959.

Dowling, Claudia. The tomboy who's who. *WomenSport* 4(8):33–40, August, 1977.

Frye, A. J., and Kamon, E. Sweating efficiency in acclimated men and women exercising in humid and dry heat. *J Appl Physiol* 54(4):972–977, 1983.

Gorski, R. The neuroendocrinology of reproduction: An overview. *Biol Reprod* 20:111–127, 1979.

Greenough, W. T., Carter, C. S., Steerman, C., and DeVoogd, T. J. Sex differences in dendritic patterns in hamster preoptic area. *Brain Res* 126:63–72, 1977.

Imperato-McGinley, J., Terson, R. E., Gautier, T., and Sturla, E. Androgen and the evolution of male-gender identity among male pseudohermaphrodites with 5a-reductase deficiency. *N Engl J Med* 300:1233–1237, 1979.

Kaplan, A. G. and Bean, J. P. *Beyond Sex-Role Stereotypes, Readings Toward a Psychology of Androgyny.* Boston, Little, Brown, 1976.

Lacoste-Utamsing, de C., Holloway, R. L. Sexual dimorphism in the human corpus callosum. *Science* 216:1431–1432, 1982.

Maccoby, E. E. and Jacklin, C. N. *The Psychology of Sex Differences.* Stanford, Ca., Stanford University Press, 1974.

Money, J. and Ehrhardt, A. *Man & Woman, Boy & Girl.* Baltimore, Johns Hopkins University Press, 1972.

Naftolin, F. and Butz, E. (eds). Sexual dimorphism. in *Science,* 211:1263–1324, 1981.

Nottebohm, T. and Arnold A. Sexual dimorphism in vocal control areas of the songbird brain. *Science* 194:211–213, 1976.

Polani, P. E. and Adinolfi, M. The H-Y antigen and its functions; a review and a hypothesis. *J Immunogenet* 10:85–102, 1983.

Raisman, G. and Field, P. M. Sexual dimorphism in the neuropid of the preoptic area of the rat and its dependence on neonatal androgen. *Brain Res* 54:1–29, 1973.

Tanner, J. M. Growing up. *Sci Am* 229(3):34–43, 1973.

Turnbull, A. C., Fairweather, D. V. I., Hibbard, B. M., et al. An assessment of the hazards of amniocentesis: Report to the M. R. C. *Br J Obstet Gynaecol* 85(Suppl 2), 1978.

Witkin, H. A., Mednick, S. A., Schulsinger, F., et al. Criminality in XYY and XXY men. *Science* 193:547–555, 1976.

6. *Female Sexuality*

Question: How often do you have sex?
He: Hardly ever—three times a week!
She: All the time—three times a week!

This bit of dialogue from a Woody Allen movie is hilarious because of our instant, if somewhat rueful, recognition of its underlying and universal assumption. Men like sex more than women do, they *need* more. There has always been this generalization—that males have only one thing on their minds and that females frequently say, "Not tonight, I have a headache." Stereotype it may be, but if the world were such that all couples basically agreed on the frequency of intercourse, then why are we laughing?

For centuries, women have been conditioned by society to suppress their sexual urges and told that their role was to satisfy the physical needs of their husbands without expecting any gratification for themselves. To bear the burden of sex was as much a duty for a woman as to bear infant after infant—it was accepted that she would bear all the obligations of her life. A modest woman, a "nice girl," "a lady" may have felt *something*, but it was separate and in no way equal to what

men felt and certainly not enjoyable without love. Currently, however, after several decades of research into the nature of human sexuality, women are told that they are in no way sexually inferior to males; that they have the anatomical equipment and physiological capacity for equal, if not greater sexual pleasure than males; and they are realizing that they are entitled to experience it. The research of Kinsey and Masters and Johnson should have laid to rest once and for all the myth that women obtain little satisfaction from sex. Women in today's society should feel free to express their feelings and desires more openly, without guilt or anxiety. Yet this is still a struggle. Why in this day of a supposed sexual revolution are there still so many women who merely endure; who submit, give, but do not receive, and fake it?

Perhaps some of the old myths are being dispelled only to be replaced by new ones. Some of the traditional and puritanical ideas about female sexuality have been at least partially discarded, but for many women, even those confident about their sex lives, the presumed sexual revolution has created additional confusion and conflict about the expression of their own sexuality and the expected sexual behavior of others. The publication of data that indicate what other people are doing, and the sexually explicit movies and magazines that show how they are doing it, may not be necessarily reassuring after years of social conditioning. Some of the recently changed standards of sexuality are seen by many women as goals that must be achieved: I must not be a virgin at the age of 15, or 19, or 21. I must want sex all the time, and in all ways. Today's woman, in the media image, is a superwoman, supersexual, who has been released, and now she's raring to go! She is seen in posters, hair wild, nipples erect, omnipresent, and bigger than life. And that is a most difficult burden to carry. Today's new sexuality sometimes offers women as few alternatives as yesterday's Victorian standards. The inability, unwillingness, or failure to be as "sexy" as women believe they should be can produce guilt and mental anguish in them. Has the revolution allowed women to make a choice about their sexuality, or has more openness about sex and more sex for more people, and more kinds of sex, only dictated a different kind of exploitation of womankind?

Word has it that the negative feelings toward premarital sex have changed for most people. The old idea, that males could and were expected to do anything they wanted but that females had to bring virginity to their marriage, is no longer tenable. The double standard is dead. Or is it? Perhaps women have achieved freedom in sexuality in the *philosophy* of our times, but in daily reality, many young women evidently do not feel increasingly free to decide for themselves the nature and the extent of their sexual behavior. They claim that while their mothers were told that they had to say "no," they are being told that they have to "yes," and if they are reluctant, there is something wrong with them. They feel that they are being pressured into sexual activity in order to maintain a close relation-

ship (". . . if you really loved me, you would. . . ."), only to discover that their expectations for the results of such relationships were unrealistic. They may find that casual and spontaneous sex can also be exploitive sex, but they are told by males that they are missing all the fun or that they are abnormal. A society that gives women little or no choice in their sexuality has not been very revolutionized. That kind of sex is as old as civilization.

The point is this: if singles' bars really are meeting places and proving grounds for sexually adventurous males and females, and if some women are having as much fun playing the field as men used to, then perhaps a sexual milestone *has* been reached. But no milestone for all womankind has been gained when today's sexuality is in danger of offering women as few options as before. True sexual liberation should include the right to say "no" to certain kinds of sexual relationships without guilt or anxiety. In particular, young people should not have to agonize over their supposed abnormality if they are still virgins past a certain arbitrary age.

That there is enormous peer pressure to narrow the gap between the onset of puberty and the onset of sexual activity is indicated by the increasing incidence of adolescent sexually transmitted diseases and more than 1 million teenage pregnancies each year. Whether sex education in the schools could alleviate the problem is an unanswered question, mainly because there is so little good sex education available. People stubbornly tend to believe that formal sexual education equals sexual permission, and the leading sex educators of children in the country continue to be other children. Sexual myths and misinformation are easily transmitted; attitudes of sexual rights and responsibilities and the fact that sexual activity is a choice are unlikely to be conveyed. Even when sex education is included in a school curriculum, the students learn a lot about reproductive physiology but do not get nearly enough about the necessity for respect and caring for other people in all relationships, and the reassurance that choosing not to enter into sexual relationships is just as healthy and appropriate as choosing to.

Another source of self-doubt and anxiety in women may be the result of the emphasis on the frequency and intensity of the orgasmic response. There is pressure for women to achieve multiple and wondrous orgasms. It is supposed to just happen. Books that have mass-market appeal are now allowing their heroines to have skyrocket orgasms in poetic language, and films show the same event, larger than life. But it may not just happen. Merely because women have been told that they *can* reach orgasm as easily as men does not mean that they do, at least in the way they believe they are supposed to—solely through vaginal intercourse. Many women, believing that the earth should move for them every time they have intercourse, think that they are "frigid" and incapable of response, either because the earth does *not* move for them in orgasm, or because they do not achieve orgasm at all. Women may be placing a huge burden on themselves and on their sexual partners by expecting "The Big O" to happen every time, on schedule, with the same intensity, from the same formula, and simultaneously with the male orgasm.

When a male ejaculates, he has had an orgasm. Male orgasm, producing spermatozoa-containing semen, is necessary to human reproduction. In contrast, the human race could continue to survive if no woman ever had an orgasm. Although the mechanism is the same in both males and females, a reflex consisting of sensory stimulation leading to motor muscular contractions, the focal point for female sexual stimulation is in the glans and shaft of the clitoris, an organ separate from the organs that function in conception, gestation, and delivery. In women, an orgasm

is a purely pleasurable experience, a natural physiological phenomenon that provides enhancement and satisfaction in sexual activity. There is nothing mysterious about an orgasm; it is a reflex, a physical manifestation, and it is a good feeling. So why is it so frequently missing in the sexual functioning of so many women? And why all the anxiety?

PHYSIOLOGY OF THE FEMALE SEXUAL RESPONSE

The actual nature of the female sexual response or any kind of sex research at all was largely a taboo subject in science until a zoologist at Indiana University, Alfred C. Kinsey, started to teach a marriage course in 1937. Dr. Kinsey was an entomologist, a specialist on insects, when he took over the course. According to Wardell Pomeroy, close associate and friend until Kinsey's death in 1956, the zoology professor discovered that so little was known about human sexual behavior that, in order to provide his students with the facts, he decided he would have to gather them himself. He approached his task with the same meticulous precision with which he studied his insects. With the publication of his observations in 1948 and 1953, popularly known as *The Kinsey Report*, he opened up the field of sex research as a valid scientific discipline, and foreshadowed the laboratory studies of human sexual responses observed by William Masters and Virginia Johnson at Washington University in St. Louis.

Masters and Johnson, under controlled laboratory conditions, reported the physiological responses of 382 women and 312 men between the ages of 18 and 89 during more than 10,000 "sexual response cycles," as they called them. They made observations during manual masturbation, during masturbation with a vibrator, during intercourse in several positions, while the breasts alone were stimulated without genital contact, and also during "artificial coitus" with a plastic penis containing a movie camera to record internal changes. The last method was the target for a number of attacks; the researchers were said to have mechanized and dehumanized sex. Sex in the laboratory was criticized and satirized, and many strongly objected to the approach. It was bad enough that Kinsey *asked* about what people did, but Masters and Johnson were *watching!*

While there may be some justification in the objections that Masters and Johnson's data missed the emotional aspects of sex by concentrating on the anatomy and physiology, until they published their research findings, there was little knowledge of the female sexual response because most of the genital response takes place *internally*. Vaginal lubrication, the distention of the inner two-thirds of the vagina, the gaping of the cervical os, the elevation of the uterus—all had been previously hidden from view. In contrast, no internal motion picture is needed to observe the male genital response; essentially, first the penis is erect, and then it is flaccid. Use of the artificial penis also made possible observations that had clinical value in infertility problems and contraceptive research.

For purposes of description, Masters and Johnson divided the female and male sexual response cycle into four phases: excitement, plateau, orgasm, and resolution. Those stages are successive; one follows the other along a continuum, and they are the same in both sexes.

During the four phases, there are two basic kinds of physiological mechanisms that cause the sexual response. One is *congestion*, or an increased flow of blood into the organs so that the tissues become engorged (swollen), and usually undergo a color change. The other main physiological phenomenon is *myotonia*, the increased muscular tension that occurs both in voluntary (skeletal) muscles and in involuntary smooth muscles. Vasocongestion and myotonia result in the physical manifestations that are visible in the genitalia, and also appear as extragenital responses, such as changes in heart rate, breathing rate, blood pressure, and perspiration.

THE EXCITEMENT PHASE OF THE SEXUAL CYCLE

Genital Responses

Vagina
The first sign of physiological response occurs in the vagina within 10–30 seconds of the initiation of sexual stimulation. The "sweating" of the vaginal walls as a result of the vasocongestion of the vaginal blood vessels results in the lubrication of the entire vagina. Almost simultaneously with lubrication, and as a result of vasocongestion, the entire vagina dilates in preparation for accommodation of the erect penis. The increased flow of blood also produces a color change in the vaginal wall, and it becomes darker, turning almost a deep purple.

Uterus
As the uterus, too, becomes engorged with blood, it moves up out of its regular position, rising in the body cavity into the false pelvis. The lifting up and back of the uterus and cervix, along with the expansion of the vaginal walls, produces a tenting or ballooning-out of the inner two-thirds of the vaginal barrel (see Fig. 6.1). This only happens if the uterus is in the typical anteverted position; it does not occur in a retroverted uterus.

Clitoris
Vasocongestion is the cause of the changes that occur in the clitoris as well, and the way in which the ischiocavernosus and bulbocav-

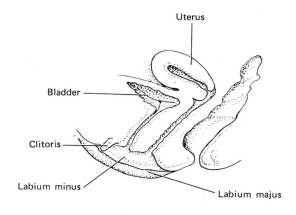

Uterus

Bladder

Clitoris

Labium minus

Labium majus

Unstimulated state

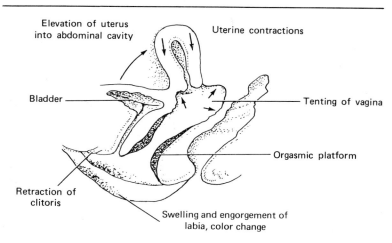

Elevation of uterus
into abdominal cavity

Uterine contractions

Bladder

Tenting of vagina

Orgasmic platform

Retraction of
clitoris

Swelling and engorgement of
labia, color change

During sexual response cycle

Figure 6.1. A diagram of the changes that occur in the pelvic area during the excitement and the orgasmic phases of the sexual response cycle.

ernous muscles contribute to clitoral enlargement was described in Chapter 2. In approximately 50% of the women observed by Masters and Johnson, the glans of the clitoris doubled in diameter; in the other half of the subjects, however, there was no visible difference in the size of the glans, although swelling could be seen if the glans were observed with a culdoscope. The blood pouring into

the corpora cavernosa of the shaft of the clitoris causes the shaft to increase in diameter, but only in about 10% of the women does the clitoral shaft visibly elongate.

Labia

As part of the whole vasocongestive reaction, the labia minora swell, and their color darkens. The labia majora also undergo congestive changes, and the extent depends on whether the woman has delivered children or not. Evidently, pregnancy results in a general increase in pelvic vascularity that results in the differences. In a nonexcited state, the major lips meet in the midline of the vaginal orifice. During the excitement phase in nulliparous women, the labia majora thin out and flatten themselves against the perineum, elevated away from and opening up the entrance to the vagina. In a multiparous woman, particularly if she has developed varicose veins, the major lips do not flatten, but become swollen and distended.

Bartholin's Glands
Very late in the excitement phase, or early in plateau, Bartholin's glands may secrete an insignificant amount of fluid, but this is after the vagina is well lubricated. They play, therefore, very little part in easing penetration.

Extragenital Responses

Breasts
Nipple erection occurs in many women early in the excitement phase. Sometimes one nipple becomes erect before the other. There may be an increase in nipple length of 0.5–1.5 cm and in diameter of the nipple base of 0.25–1.0 cm. The dark area surrounding the nipple (areola) enlarges and swells later in the excitement phase; by the end of this phase, the entire breasts have increased in size by 20%–25%—more noticeably in women who have not breast fed. The veins of the breasts become more apparent because of engorgement with blood.

Skin
Another vascular change that occurred in the laboratory in approximately 75% of the women and 25% of the men at least sometimes during the sexual cycle was the "sex flush." Beginning on the upper part of the ab-

domen and spreading up to the chest, throat, and neck, a pink mottling appears on the skin. It is especially obvious in fair-skinned people, and looks like a measles rash.

Muscle Tension
There is increased tension in the voluntary muscles of the arms and legs, the rectal muscles, some of the abdominal muscles, and the intercostal muscles of the ribs.

Cardiovascular, Respiratory, and Sweat Gland Responses

As sexual excitement increases, heart rate and blood pressure increase. There is no observable change in respiration or perspiration at this point.

PLATEAU PHASE

Genital Responses

Vagina
The bulbocavernosus muscles (so called because they arise on the *bulbs* of the vestibule and insert anteriorly on each corpus *cavernosum* of the clitoris) are the sphincter muscles that encircle the vaginal entrance. As they contract, they compress the veins in the bulbs of the vestibule, the highly vascular tissue that is equivalent to the corpus spongiosum in the male and which lie deep within the substance of the labia majora. The bulbs become turgid and swollen, producing a vulvar erection that reduces the diameter of the outer third of the vaginal barrel by approximately 50%. This is what Masters and Johnson called the *orgasmic platform*, and it actually produces a gripping clasp around the erect penis.

Uterus
More vasocongestion produces an increase in size, and the uterus elevates more fully into

the abdominal cavity and causes more tenting and ballooning of the inner two-thirds of the vagina. The complete elevation and vasocongestion produces an increasing uterine muscle irritability, and late in the plateau phase uterine contractions may start (see Fig. 6.1).

Clitoris

The clitoris also elevates in the plateau phase. Its normal position is overhanging the lower border of the pubic bone. In this stage of the cycle, the body and glans of the clitoris retract from that overhanging position and withdraw deep under the prepuce, carrying the labia minora and the suspensory ligament along to end up high on the anterior border of the pubic symphysis. It is no longer externally visible after this elevation, but it still responds to stimulation by pressure on the mons veneris over the pubic bone and/or by the penile thrusting that pulls on the labia minora, producing traction on the prepuce. A finger or a vibrator or any other kind of tactile friction will have the same effect. Unless clitoral stimulation continues, orgasm is not triggered.

Labia Majora and Minora

The labia majora continue to show more of the changes that started in the first stage. In a nullipara, they flatten out even more; in multiparous women, they become more swollen and engorged. Late in the plateau phase, the labia minora experience more of a vivid color change—a sure sign that orgasm will occur within a few minutes if clitoral stimulation is continued.

Extragenital Changes

Breasts

More turgidity of the nipples occurs, and there is further swelling of areola and entire breasts.

Skin

If a sex flush has occurred, it spreads over the shoulders, down the inner surface of the arms, and perhaps onto the abdomen, thighs, buttocks, and back.

Muscle Tension

There is a further increase in the tightening of both voluntary and involuntary muscles, and intercostals of the ribs, and there may also be voluntary tightening of the rectal sphincter and the buttock and thigh muscles.

Cardiovascular, Respiratory, Sweat Gland Responses

The heart rate increases from a normal 70 beats per minute to 100–175 per minute. There is an elevation in systolic blood pressure of 20–60 mm Hg above resting pressure and 10–20 mm above diastolic. Late in plateau, there is an increase in respiratory rate. The actual perspiratory reaction of the sexual cycle does not occur until after orgasm, but, of course, if a sexual encounter is taking place in a warm atmosphere, there will be sweating throughout the phases.

ORGASM

Genital Responses

Vagina

The major response is that the orgasmic platform—that is, the muscles that make up the outer third of the vaginal wall and those surrounding the turgid tissues of the vulva—undergo rhythmic contractions that occur at 0.8-second intervals. In a mild orgasm, there may be three or five such contractions; a more intense orgasm results in perhaps as many as 12.

Uterus

Along with the throbbing sensation in the vagina, the uterus also undergoes contractions; each begins at the top (fundus) of the uterus, works its way down to the middle, and then to the cervix. There are indications that these contractions are greater in intensity after self-stimulation than after intercourse. This may be why many women find masturbation helpful for dysmenorrhea, especially since the external os of the cervix remains dilated for about 30 minutes afterward.

Labia

There are no observable changes in the labia major or minora during orgasm.

Extragenital Responses

Breasts

There are no further noticeable changes in the breasts during orgasm.

Skin

If sex flush is present, it reaches a peak of intensity with the onset of orgasm and is most pronounced.

Muscle Tension

The entire body is so involved in the experience of orgasm that voluntary muscular control is mainly lost as involuntary contraction of many muscle groups takes place. The spontaneous involuntary reflex contractions of the perineal muscles and the pubococcygeus not only cause the throbbing of the orgasmic platform, but many also result in the same 0.8-second contractions of the external rectal sphincter. If they occur, the anal contractions do not last as long and usually involve only two to four spasms. Both superficial and deep muscles of the arms, legs, neck, abdomen, and buttocks are often contracted to cause stiffening and rigidity. The hands and feet are extended while the fingers and toes curl under,

appearing grasping or clawlike; this spastic posture is called carpopedal spasm. The amount of myotonia or muscle contraction that occurs with orgasm is an individual response. It may be so mild as to barely be visible, or it may be uncontrollably convulsive.

RESOLUTION

Within a few seconds of orgasm, the two physiological mechanisms that created the sexual responses, vasocongestion and myotonia, are rapidly dissipated.

Genital Responses

Vagina

The blood rapidly drains out of the engorged tissues, and the orgasmic platform disappears. The anterior wall of the vagina returns to meet the posterior wall, and the deep color change fades, although the complete return to normal coloring may take as long as 10–15 minutes.

Uterus

The elevated uterus returns to its normal unstimulated position. The cervix then dips into the "seminal pool" that has formed in the posterior fornix as a result of male ejaculation, if the woman has been on her back during intercourse. The external os of the cervix remains slightly dilated for approximately 20–30 minutes. If orgasm does not occur after a progression through the plateau phase, the vasocongestion that produced the enlargement of the uterus takes much longer to resolve, and engorgement may persist for as long as an hour—even longer in multiparous women.

Clitoris

Ten seconds after the last vaginal contraction, the clitoris returns to its normal position overhanging the pubic symphysis, but the engorgement with blood that produced the size increase takes longer to dissipate. The glans remains very sensitive for several minutes; in some women it is actually painful to the touch.

Labia

The labia minora return to their ordinary color within 10–15 seconds, and they return to their ordinary size. The labia majora return to their normal midline position and size more quickly in nulliparous women. In a woman who has had several children, the labia majora may retain their engorgement for several hours. The reason for the difference is the normal increase in the number and size of the blood vessels in the genitalia and reproductive organs that takes place with pregnancy.

All extragenital responses—that is, muscle tension, heart rate, blood pressure, and respiratory rate—show a rapid decline unless there is further sexual stimulation. The sex flush disappears in reverse order of its appearance, and in approximately one-third of the women subjects, there is the appearance of a widespread film of perspiration over the chest, back, and thighs. Heavier sweating occurs on the forehead, upper lip, and under the arms.

The swelling of the areolae of the breasts promptly disappears. The nipple erection remains longer than the areolar swelling; last to disappear is the swelling of the entire breast and the vascular tree formed by the vein outlines.

In males, the resolution phase after orgasm and ejaculation is followed by a refractory period, during which he cannot respond to sexual stimulation by another sexual cycle. The duration of the period is generally directly related to the age of the male; the younger he is, the sooner he is able to have another erection, ejaculation, and orgasm. In women,

there is no such refractory period, and they are said to be multiorgasmic, having the capacity to reach repeated orgasms should they want to. It is not unusual in some women to have multiple orgasms in rapid succession with no intervening resolution phase.

WHAT DOES ORGASM FEEL LIKE?

Physiologically, the term orgasm refers to the spasmodic and involuntary contractions of the muscles in the region of the vulva and the vagina. Subjectively, the experience can be perceived in the range from "O.K., nice . . ." all the way to ecstasy and "sky rockets," depending on the incredibly great individual variation that exists and the ease with which a woman can verbalize her sensations.

The Hite Report (Hite, 1976), a questionnaire survey of 3019 women aged 14–78, indicates that some words used in describing the subjective characteristics of arousal and orgasm are frequently repeated by many women: "tingling," "buzzing," "warmth," "pulsating" . . . with the focus of the sensations in the area of the clitoris. Other descriptions fairly consistently mentioned a prickling and tingling, heat, and throbbing sensations in the pelvic area. It has generally been assumed that a male's experience of orgasm—more sudden and explosive—is different from a female's experience. When a series of 48 descriptions of orgasm (24 male and 24 female) were submitted to 70 obstetrician-gynecologists, clinical psychologists, and medical students, however, they were unable to correctly identify the sex of the person describing the orgasm. There was also no ability to recognize any described characteristics of orgasm that would indicate a basis for sex differences in the subjective awareness of sensations (Vance and Wagner, 1976).

The findings in the above study suggest that the experience of orgasm for males and females is essentially the same, at least on the basis of their written descriptions of the event. But there are great differences in the way that women are able to interpret their feelings, and many psychological and cultural determinants of those differences. Some women may have orgasms and not be aware of them. The widely held assumption, "If you don't know if you've had an orgasm, you haven't!" is not necessarily true. Some women who think they never had an orgasm may recall having good feelings, fluttering or pulsations, waves, and then a sense of peacefulness afterwards, but never labeled their sensations as an orgasm. Such women may be suppressing many of the sensations, too inhibited, perhaps, to permit themselves to feel them. Alternatively, a good feeling about sex with a low intensity of response could be completely normal in a given individual. Usually, a feeling of warmth and some pelvic contractions are indications that orgasm has occurred.

Vaginal and Clitoral Orgasm

"What Freud did for female sexuality is equivalent to what Jack the Ripper did for surgery!" In this way a student summed up Sigmund Freud's influence on psychoanalytic doctrine concerning the role of the clitoris and the role of the vagina in female sexuality. In his *Three Essays on the Theory of Sexuality*, Freud hypothesized his ideas about the "leading zones" of eroticism in the bodies of men and women. Based on his observations in his psychiatric practice, he said that the main erogenous area in little girls was located at the clitoris, homologous to the glans of the penis, and was thus "masculine." In the transition to womanhood, girls had to transfer this childish masculinity in their sexual sensations to the vagina. With this transfer, a mature woman then abandoned her early clitoral eroticism and replaced it with a new and superior erotic zone. If this transition from the

clitoral zone to the vaginal zone is not made or is incomplete, women remain infantile and immature. Without any knowledge of the physiological mechanisms involved in orgasm, this theory is not illogical in Freud's context of women as imperfect men; that is, lacking and envying a penis. The clitoris in childhood is the substitute for the penis; it must be given up and transferred when at last the sexual act is permitted. Males, on the other hand, retain their leading erogenous zones, unchanged from childhood. The fact that women have to make the transfer, putting aside their "childish masculinity," makes them more prone to "neurosis and especially to hysteria" and creates the basis for their inevitable sexual inferiority.

Freud's opinions were enthusiastically codified by later psychoanalysts and psychiatrists. Like fundamentalists interpreting the Bible, orthodox Freudians and neo-Freudians proclaimed the Word—that vaginal orgasms were superior to clitoral orgasms. Women who had vaginal orgasms were feminine, mature, and normal; women who preferred clitoral stimulation were masculine, immature, and even frigid (although the latter term logically should be reserved for an inability to feel any sexual sensations at all).

The laboratory observations of Masters and Johnson proved that these assertions are nonsense. Orgasm is orgasm—the result, apparently, of primarily clitoral stimulation. Whether such stimulation takes place through direct friction of the clitoris by manual stroking during intercourse, by masturbation, or through homosexual activity with another woman, or whether it occurs indirectly by traction on the prepuce or clitoral hood when the male penis thrusts into the vagina, the mechanism of orgasm is the same. In some women, the quality of the orgasm may be enhanced by the repetitive pressure of the penis on the cervix, resulting in movements of the uterus and its associated ligaments. This may be the basis for the sensation of a deep, inner or "vaginal" orgasm. The percentage of women for whom such thrusting on the cervix is sexually important is unknown, but it may be related to evidence that some women who have had a hysterectomy experience lesser intensity of orgasm and hence greater sexual dissatisfaction after the uterus and cervix have been removed.

Compared with the male, there has been relatively little work done on the neurophysiological and neuroanatomical bases of the female orgasm, but available evidence indicates that innervation of the clitoris is similar to that of the penis. Sexual stimulation causes sensory nerve impulses from the clitoris especially and from the vulva to travel along the pudendal nerve. The sensory signals enter the spinal cord through the lumbosacral plexus into the spinal cord and are transmitted up to the brain. Reflexes in the spinal cord result in parasympathetic nerve impulses that cause arteries to dilate in the tissues of the genitalia, and the subsequent vasocongestion is then responsible for the other changes already described. If the sexual stimulation is continued, the continued sensory input to the spinal reflex center, which can be inhibited or facilitated by the higher brain centers, results in motor impulses to the pelvic area. The sympathetic discharge results in the changes in heart rate, respiration, blood pressure, perspiration, and so forth. The onset of approximately 10 rhythmic muscle contractions around and in the vagina causes most of the blood and fluid to then empty from the pelvis. Both the clitoris and the vagina form an integral part of the orgasmic reaction, and as psychoanalyst Helen Kaplan pointed out, to ascribe more prominence to one or to the other is as senseless as trying to distinguish between the sensory and the motor components of any reflex—blink, gag, startle, and so forth.

Put that way, orgasm sounds like a simple reflex: tactile stimulation causes sensory impulses to travel to a center in the spinal cord and results in a motor response, the contraction of muscles. Reflex it may be, but in most women it is far from simple. A reflex conditioned as much by psychology and sociology as by physiology is not really comparable to the knee jerk. The entire life experience of a woman is involved in her sexual response. If the brain says "go," then the appropriate reflexes are initiated from the spinal cord, and the throbbing sensations are transmitted back to the brain for the awareness of pleasure and for the peace and relaxation that follow. But if psychological inhibition prevents cerebral "letting go," the orgasm and its intensely pleasurable fulfilment is also inhibited, even in the presence of adequate local stimulation. Although every woman, unless there is some organic pathology or anatomical reason that prevents it, has the physiological capability to have an orgasm, she may not, for multiple reasons, have the psychological capability to function orgasmically.

SEXUAL PROBLEMS IN WOMEN

When men have problems with sex, it is related to performance. They are unable to get an erection (*impotence*), or they cannot exert enough voluntary control to delay ejaculation, and they reach orgasm too rapidly (*premature ejaculation*). What is called sexual dysfunction in males results in their inability to have intercourse. The sexual problems of women, on the other hand, assuming that vaginismus is not the trouble, are a matter of satisfaction rather than performance. All that women need to successfully engage in intercourse is a moderate amount of vaginal lubrication. Even if orgasms are lacking or infrequent, a woman still may have what is apparently a successful sexual relationship. If a woman wants to pretend to have sexual pleasure, only a minicamera and vaginal transducers could deduce the truth.

There are millions of women (estimates are 10%) who never consciously experience orgasm at all by any means—coitus, stimulation, or masturbation. Orgasm may not be the sole measure of sexual satisfaction or happiness in a sexual relationship, and an inability to achieve orgasm is certainly no criterion of inadequacy or abnormality. But there is no reason to suppose that it does not enrich and enhance the sexual act, or that sexual activity is better or to be preferred without orgasm.

There is evidence, however, that contrary to previously held opinions, a majority of women are unable to reach orgasms solely through intercourse without additional clitoral stimulation. Shere Hite's survey revealed that only 30% of the 3,000 women who replied to her questionnaire were regularly able to climax through coitus. For many of that group, penile thrusting alone was not sufficient, and they had to use specific positions to obtain clitoral contact. Because 82% of the women surveyed masturbated, and 95% of those women could always reach orgasm by self-stimulation, it would appear that the model described by Masters and Johnson for clitoral contact during intercourse may not always work. That is, for the vast majority of Hite's respondents, in-and-out movement of the penis did not necessarily produce labial traction on the clitoral hood for indirect clitoral stimulation. Women who are not able to achieve orgasm through intercourse but can through masturbation are, Hite believes, in no way dysfunctional, but they may have to take a more active responsibility toward their orgasm. The women surveyed were generally reluctant to express their need for manual stimulation of the clitoris to their partners, fearing to "put down" the male concerning the effectiveness of his penis. Even fewer were willing to stimulate themselves to orgasm while in

the presence of their male partner. Most of them did, however, express great concern that something was wrong with them because they were unable to have an orgasm in the "right" way. The pervasive influence of Freudian theory, even if demolished in the laboratory by Masters and Johnson, cannot be underestimated.

Hite's questionnaire was distributed to a particular segment of American women; that is, to those belonging to feminist groups and to readers of only certain magazines and newsletters. They tended to be relatively well educated and lived primarily in urban areas. Her findings, therefore, may not be representative of the feelings and experiences of all women, although they do tend to concur with the reports of other researchers, such as Kinsey and Kaplan. Bell and Bell (1972), who reported on sexual satisfaction among 2,372 married women, found that 42% had orgasms most of the time, but only 17% responded that they always have orgasms during intercourse.

The fact that many women can masturbate to orgasm but cannot achieve orgasm through coitus alone carries no implication that masturbation is "better" or represents a dire forewarning concerning the future of intercourse. Although the physiological orgasmic response is the same no matter how it is produced and brings with it the relief of sexual tensions, a climax reached through self-stimulation is essentially a solitary act. For a majority of women, orgasm that is reached with a loving and loved partner is likely to have more emotional meaning and greater satisfying significance than one that is self-induced. This is as true for women whose sexual preference is for other women as it is for heterosexuals. Desires and feelings based on mutual and warm understanding between two people are the primary factors for women in achieving sexual gratification.

Most women, however, are heterosexual. For them, given the knowledge concerning the anatomy and physiology of the female orgasm, it is unrealistic to always expect the male to "bring" a woman to climax without effective clitoral stimulation. The focus of male sexual excitement is penetration and thrusting movement; the basic mechanism for females is pressure and friction on the clitoris. Different couples are built differently. What is pleasurable for the male may not always be the specific technique for every female. Her own activities and movements, together with a communication of her needs for manual stimulation, may prove a more successful combination. In some couples, the touching and movement of the male pubic bone against the female mons pubis, for example, may provide the right contact for the clitoral area. Because of the individual variation in the prominence of the pubic bone, getting this contact may require a change from the usual man-on-top position, or perhaps some other accommodation.

Time is also a very important factor in female eroticism and ability to have an orgasm—not only the time involved in the actual sexual encounter, but the total time of a woman's accumulated sexual experience. As with most other activities, practice helps, and many women find that physical sex improves with age. In interviews with married women, Kinsey found that nearly 30% of women had not experienced an orgasm by any means when they were first married, but after 15 years of marriage, there were only 10% who were still nonorgasmic. It is possible to learn to have orgasms, and once learned, the response and letting go becomes easier as one becomes more adept at knowing what to do to produce it. One of the best ways to teach oneself to have an orgasm is by masturbating, but even knowing how to masturbate may not come naturally to women who have been taught since childhood that touching oneself is wrong and shameful. The very intense local stimulation that is delivered by an electric vibrator may help to achieve orgasm initially. Even with a vibrator, erotic sensations can be delayed by very strong cerebral inhibition, but persistence and a real effort to release the orgasmic reflex will result in success in most women.

Sex Therapy

Therapy is no way to generalize about something as subjective as sexual satisfaction, but that does not prevent a lot of people from trying. The only generalization that can be made about human sexuality without qualm is that is is incredibly, infinitely varied. There is no normal or abnormal in the capacity to respond to sexual activity or in the desire for the frequency of sexual activity. A woman could be labeled as "frigid" because her eagerness for sex is less than that of her partner. Another could be called a "nymphomaniac" because she experienced frequent desire or because her sexual excitement and need for a second coitus continued after her partner's orgasm. The persistence of such pejorative terms illustrates that many of us still have the illusion that there is some appropriate measure by which to gauge a woman's sexual appetite. But in common with every other human behavior, sex cannot always be 100% emotionally and physically satisfying, and the desire for or the pleasure derived from sex may change with time, opportunity, locale, sexual partners, and a variety of situational factors.

A woman is always being advised, however, on how she should be feeling about sex. Bombarded from all sides with information concerning statistical norms, orgasmic potential, multiorgasmic potential, sexual dysfunction, her total womanhood, that every woman can, and of all the joy she should be having in sex, it is easy to lose sight of the fact that *her* perceptions—what *she* finds satisfying, what *she* enjoys—are the only definitions of what is right for her. She should allow no one to decide these things for her. Neither her physician, her sexual partner, a magazine article, a book, nor a movie should tell her that she has a problem or is inadequate because she does not meet some presumed standard of sexual behavior.

But there are women for whom the expression of sexuality is tied up in knots because of the whole range of sociocultural attitudes and values they have had about sex. Their feelings of guilt or fear, their reluctance to lose control, their inability to communicate their need and desires to their partners, their possible low esteem and low expectations for themselves—some or all of these may result in patterns of behavior for them that limit or prevent their appreciation of sexual activity. Where there is total lack of interest in sex, or complete absence of response to erotic stimulation (or perhaps when vaginismus is a recurrent problem), or the woman perceives a difficulty that is serious enough to *her* to warrant treatment, formal sex therapy in a clinical atmosphere may be helpful.

Sex therapists believe that since attitudes are conditioned and learned, they can be unlearned. Their goal is to relieve the problem, usually without treating the underlying causes, although some do use psychotherapy as an adjunct. Most therapists employ a form of behavior modification, attempting to decrease sexual anxiety and encourage communication. They prefer to work with the couple if possible, rather than with the individual patient. When no partner is available, Masters and Johnson have used surrogate sex partners. Techniques among therapists vary: some use hypnosis; some may prescribe a series of specific instructions and exercises that the patient and her partner are to follow; some may perform surgery in order to improve body image, such as breast augmentation or facial surgery. Some doctors have performed a kind of clitoral circumcision to re-

move "adhesions" between the clitoral hood or prepuce and the glans of the clitoris. This is claimed to increase the sensitivity and provide for more direct contact with the clitoris, although efficacy of the procedure is still in question.*

Another therapeutic technique, used in conjunction with other methods, is "flooding" the patients with explicit graphic materials (most people would call it pornography) in order to desensitize their anxiety about the subject matter and permit themselves to be "turned on."

It should be recognized that along with the legitimate clinics that are usually associated with a medical school or hospital, there has been a virtual explosion of treatment centers, sex clinics, marriage counselors, and sex therapists throughout the country. Some of these centers are staffed by incompetents, charlatans, and quacks who charge exorbitant fees. At the very least, these self-styled "sex therapists" may dispense misinformation, but some may also cause reactions in their patients that compound the original problem.

Even accredited health professionals, lacking appropriate training and having their own anxieties about sex, may be no more sophisticated than the general public and no better able to deal with intimate sexual problems. One's family physician or obstetrician-gynecologist may not be the best person to seek out for counseling. With the growing awareness that total health care includes sexual health care, more medical and nursing schools are including sex education in their curricula and holding institutes and workshops for the postgraduate education of health professionals. Although this trains personnel to provide counseling and be better

* There is a very rare condition in some women in which the clitoral hood actually adheres to the clitoris. For them, clitoral stimulation and the resulting vasocongestion produces great pain, and the prepuce must be surgically separated, a much more complicated surgical procedure than freeing adhesions.

equipped to deal with sexual problems, it is necessary that such sex education extend beyond medical schools, hospitals, and universities to reach the general public through public schools and the mass media. The recognition that human sexuality is not just performance and satisfaction in intercourse, but is a lifelong human quality that encompasses love, warmth, and respect between human beings in their relationships, and that it can be expressed in a variety of ways, may eventually decrease the need for formal sex therapy. The realization that sex is important to *all* people as a basis for self-esteem and security will be an important step towards this goal.

LESBIAN SEXUALITY

What gay women "do," sexually, is frequently the source of curiosity (and even prurient interest) among nonhomosexuals, but what they do together is not substantially different from what heterosexual couples do or what a woman can do alone through self-stimulation. As indicated previously, the body responds to psychosexual stimulation with physiological effects, but the psychological part has as much or more to do with sexual arousal as the physical means of sexual expression. Lesbians may concentrate more on particular erogenous areas of the body and may have greater awareness of the role of the clitoris in achieving sexual gratification, but so may couples who are not of the same sex.

What gay women want in a sexual relationship is also no different than that of heterosexual couples—love, commitment, sharing, and meaning. But what homosexual couples need from society is taken for granted by heterosexual couples—public acceptance and better understanding of their sexual preference.

The usually quoted statistics indicate that 10% of individuals in the United States are or will become exclusively or predominantly homosexual. In the most comprehensive study of homosexuals to date, a Kinsey Institute report gave insights into the kinds of lives that those approximately 23 million Americans are likely to be leading. The findings of the research conducted by Alan P. Bell and Martin S. Weisberg (1978) were based upon interviews conducted with nearly 1,000 homosexual men and women in the San Francisco Bay Area. The major conclusion of the study was that "homosexual adults who have come to terms with their homosexuality, who do not regret their sexual orientation, and who can function effectively sexually and socially, are no more distressed psychologically than are heterosexual men and women." Contrary to stereotypical beliefs, most homosexuals do not exhibit bizarre behavior or engage in frenzied sexual activity. Three-quarters of the women were living with another woman in a monogamous relationship at the time of the study and differed very little from the heterosexual control sample in their psychological adjustment. Two-thirds of the men, who engaged in considerably more "sexual cruising," had contracted a venereal disease, but virtually none of the women had, and most of the lesbians had had fewer than 10 female partners. Only 5% of the women and 10% of the men were classified by the investigators as "dysfunctionals," those who came closest to the stereotypical tormented homosexual. Apparently, if the rest of the world allows it, most gay people can live satisfied, well-adjusted, fulfilled lives, and be as happy as their nonhomosexual counterparts. Fortunately, society in general seems to be moving toward a greater acceptance of alternative life-styles, including homosexuality. Rabid antigay sentiments still exist, of course, particularly in some segments of the population, but many communities have tolerant attitudes toward homosexuals, and a number of cities and states have passed specific legislation prohibiting discrimination against gays in jobs and housing.

SEX IN THE OLDER WOMAN

In 1959, Golde and Kogan conducted a study in which a group of university undergraduates was asked to complete this sentence: "Sex for most old people is. . . ." Almost all of them replied, "past," "negligible," or "unimportant," reflecting society's belief that after a certain age, people are, or should be, sexless. If not, they are dirty old men or weird old women.

That there may have been a change in attitude, at least among young people, is perhaps suggested by the continuing cultlike success of an early 1970s film, *Harold and Maude*, in which an 80-year-old woman has a loving sexual relationship with a 20-year-old man. It is not to be inferred that this innovative May–December romance was the only reason for the film's appeal to the younger generation, but evidently they were not entirely repelled by the idea.

There are many people, however, who do find the notion of sexuality in the aged distasteful, disgraceful, or just plain ridiculous. When elderly people marry, they do so for "companionship," and they are presumed to have no interest or capability for sexual activity. With these societal expectations, it is not surprising that a self-fulfilling prophecy takes place. Many older men and women do feel asexual and suppress any erotic interest they may have as being unsuitable or wrong.

All available evidence indicates that given reasonably good health, sexual function continues throughout life. Masters and Johnson, who studied women aged 40–73 and men aged 51–89, found that while there are decreases in the durability and intensity of the physical responses in the sexual cycle with increasing age, "the aging human female is fully capable of sexual performance at orgasmic response levels, particularly if she is exposed to regularity of effective sexual stimulation." The *Hite Report* quotes some older women who felt that sex was not that important to them any more, but the majority indicated that their sexual pleasure had actually increased with age, and that they were now enjoying more new and gratifying experiences. Postmenopausal women, free for the first time in their lives from the fear of an unwanted pregnancy and with fewer family responsibilities, may have an increase in their desire for sexual activity, particularly if they disregard the myths concerning how menopause is supposed to affect their sex lives.

Continuing to be sexually active is evidently the best insurance for the ability to remain sexually active. After menopause, there are sometimes vaginal changes that result from the decrease in estrogen levels. These changes may decrease the ability of the vagina to undergo adequate lubrication in preparation for intercourse. It is known that women who maintain regular intercourse or masturbation during their entire adult lives have much less difficulty in vaginal lubrication and expansion of the vagina than women who are sexually inactive. There are also fewer problems with atrophic vaginitis, an observation again substantiated in a recent study (Leiblum et al., 1983). The mechanism by which sexual activity helps to prevent vaginal changes secondary to estrogen decline is unknown. It may be that the principle of disuse atrophy, more commonly known as "use it or lose it," is playing a role.

Many older women, of course, lack the opportunity for the continuance of sexual involvement because they are widowed, divorced, or single, and many have never masturbated. It may be very difficult for those for whom self-stimulation has never been a part of their sexual value system to begin to masturbate in old age, even as a health measure. It is also curious that many people who have become accustomed to accepting childhood masturbation as almost mandatory for a functionally healthy sexuality in adulthood may still stigmatize masturbation in the elderly as a childish activity, or "sick," or sad.

SEXUALITY IN THE PHYSICALLY DISABLED

In the same manner that feelings of sexuality in the aged are sometimes regarded as inappropriate or even bizarre, society tends to regard those who are unfortunate enough to suffer from a congenital or acquired physical disability as nonsexual beings, lacking both desire and capability. We seem to harbor this universal fantasy, encouraged by the mass media—that ideal sexual activity takes place between young, beautiful people, all of whom are tall, slim, able-bodied, and active. When we see them on television commercials, it is obvious that any defects they may have are easily remedied by a different deodorant, a new shampoo, or a better toothpaste, and then they are again lovable and alluring. We have been literally programmed to think that vigor and physical appearance are of major importance in sexuality, and we have difficulty believing that people who are crippled by arthritis, who have cerebral palsy, who are confined to a wheelchair have any interest in sex. But sexuality is a human, lifelong characteristic, and this is equally true of an individual with chronic illness or a disabling injury as it is of any able-bodied person. The sex drive is rarely affected by disability, any more than physical handicap curtails hunger or thirst.

Approximately one out of 10 adult people has a physical impairment that was either present at birth, developed later as a result of illness, or was suddenly acquired through accident or injury. The sexuality of the physically disabled is a reality that has been largely ignored by the medical profession as well as by the general public. Many of the disabled, too, just like the elderly, have come to accept the prevailing belief that sex is unimportant to them. A few physicians, specialists in rehabilitation medicine, like Walsh in Great Britain and Cole in the United States, have written extensively on the need for sexual counseling, emphasizing that health professionals should expect to teach disabled patients about sexual behaviors in the same way they help them to deal with other activities in daily living. Cole very pragmatically pointed out that sexuality is so related to the self-esteem of the physically disabled, that sexual activity leads not only to a better adaptation to the disability, but also results in fewer medical complaints and less need for medical or social support, thus decreasing society's cost in dealing with the disabled. It does seem evident that those who have already lost their independence, their earning power, and their mobility should not have to cope with the additional loss of their sexuality.

Problems of sexuality, which certainly can exist in the physically able, are usually compounded in the physically disabled. One study of sexual problems in the disabled found that they generally fell into several broad categories (Stewart, 1976). Approximately 30% of the difficulties were in the realm of the psychological and social aspect of sexual relationships, in which feelings of hopelessness or depression lessened sexual drive, or where unfounded fears concerning the inheritance of the physical defect decreased sexual activity. The rest of the problems were physical or physiological, concerned with such things as physical comfort and safety during the sex act and with the loss of sexual capability as a result of paralysis or lack of sensation. People want to know what to do when stiff joints or muscles make sexual positions difficult or painful, or when dizziness or muscle spasms occur during intercourse. They are concerned about heart palpitations or breathlessness that may occur during sexual activity and whether it is safe for them to participate. They want advice on what to do about urinary catheters or other appliances or dressings that make sex uncomfortable or unesthetic. They want reassurance that alternatives to penile–vaginal intercourse are possible, and that other modes of sexual expression in a loving relationship exist that can be equally fulfilling. They also have a critical need for the compassionate and understanding attitudes of the people around them. Regrettably, their questions tend to remain unanswered. Sexual counseling is still taboo at many hospitals and rehabilitation centers, and too few health professionals or social workers have the training and expertise to deal with sexual problems in the disabled.

Part of the difficulty is that the subject of sexual functioning and dysfunctioning in the physically disabled is such a new area. There is relatively little in medical literature concerning patients with progressive chronic illnesses, such as heart, pulmonary, or kidney disease, or who have arthritis, surgical problems that impair sexual function, or a developmental disability. What information there is deals primarily with sexual function in spinal cord-injured males.

Spinal cord injury results in *paraplegia*, which is paralysis and the loss of most or all sensation in the lower part of the body and the legs; and if the injury is higher in the spinal cord, *quadriplegia*, the loss of voluntary muscular activity (paralysis) and some or all sensation from the neck down. The extent of the denervation depends on whether the spinal cord was completely or partially cut.

Because of war, athletic injuries, and their greater numbers of auto accidents and civilian gunshot wounds, four out of five spinal cord-injured adults are male. It is generally accepted that males have a much more difficult time in sexual readjustment after spinal cord injury because of their increased difficulty in erections and their loss of fertility. Although 50%–70% of spinal cord-injured men are capable of having an erection, almost all lose the ability to ejaculate, and the ability to produce spermatozoa is lost within a few weeks or months of injury. With the high value placed on male performance and activity in sex, it is evident that loss of some sexual function can be particularly devastating.

The medical literature pertaining to sexuality in spinal cord-injured females is scant, and what there is emphasizes that female sexuality is less affected. "In fact," as one physician cheerfully observed, "the picture is much brighter for the female paraplegic than for the male." Presumably, paralysis and loss of sensation below the waist are not viewed as being very different from the ordinary state of affairs in women; they can just lie there and be more passive. In contrast to males, women do retain full reproductive ability. They experience a temporary amenorrhea after injury, but normal cycles resume after 6–8 months, and they are completely capable of becoming pregnant and delivering a child. Contraception is as necessary for a paraplegic or quadriplegic woman as it is for any woman who does not want to become pregnant. Blood clots have a greater tendency to form in paralyzed limbs, however, and spinal cord-injured women are rarely given oral contraceptives because of their link with blood-clotting disorders.

As additional evidence that there is probably more sexuality that exists between the ears than between the legs, both men and women with spinal cord injuries have reported that, given the opportunity, they can experience full sexual satisfaction psychologically even if a physical sensation is lacking. For some, the areas of the body that are still innervated become new erogenous zones to compensate for those parts that no longer respond to touch, and many individuals have the ability to produce fantasized orgasms by concentrating on the sensations received from those areas. Others derive great satisfaction from the ability to give joy and pleasure to their partners. The actual physiological manifestations of the sexual response cycle in paraplegic and quadriplegic males and females differ little from those in neurologically intact individuals. According to Cole the vasocongestion responses are consistently present, but muscle responses in the genitalia may be absent. For example, in males, the only responses missing are those involved in emission and ejaculation. In females, the clitoris and labia become engorged with blood and swell, but the uterine and vaginal responses during the orgasmic platform are missing. All extragenital responses—heart, respiration, pulse, blood pressure, skin flush, and so on—are present in both men and women.

There is ample evidence that with a loving and caring partner and given the appropriate counseling, the reassurance concerning their sexual potential, and the technical advice about coping with physical problems during sexual activity, men and women who are physically disabled can accept themselves as total sexual human beings. We must not permit the narrow view that prizes youth and physical appearance as criteria for sexual activity to devalue this concept.

SEXUALITY DURING PREGNANCY

Masters and Johnson observed the sexual response cycles of six women throughout their pregnancies. They concluded that the physiological sexual responses during pregnancy were very similar to those of nonpregnant women, although some women complained of increased breast tenderness, especially in the nipples and areolae, during the excitement phase of sexual response. Four of the six had occasional cramping and aching in the pelvis after orgasm, and two of those four said they had subsequent lower back pain. All six women reported a heightening of sexual arousal in the second 3 months of pregnancy, which continued well into the final 3 months. Masters monitored the fetal heartbeat during orgasm and reported that, although it sometimes slowed down temporarily, it very rapidly resumed its normal rate. It was also discovered that the resolution phase after orgasm took longer and was less complete, in the sense that the women stated that orgasmic experience "did not relieve their sexual tensions for any significant length of time."

Masters and Johnson also supplemented their direct observation of the six with the subjective responses of 111 women who regularly reported their sexual feelings, behavior, and responses as their pregnancies progressed. Other investigators (Solberg et al., 1973; Kenny, 1973; Tolor and Digrazia, 1976; Keamy et al., 1982) used a similar technique of regular oral interviews or used a retrospective questionnaire to determine sexual attitudes of women during and after pregnancy. In general, the conclusions drawn from these studies indicate that the usual wide range of individual response exists, but that sexual interests, frequency, and the enjoyment derived from activity was essentially the same as it is in nonpregnant women until the last trimester of pregnancy, when a progressive decline in desire and frequency occurred.

Medical opinion concerning the advisability of sexual activity during pregnancy varies. There are still a few doctors who believe that intercourse during the first 3 months is contraindicated because of the fear of spontaneous abortion. There is the possibility that in a small group of women who have already miscarried several times and who obviously have difficulty in maintaining a pregnancy, it would be prudent to avoid all sexual activity including masturbation because of the uterine contractions. There is very little evidence, however, that intercourse causes spontaneous abortion. Women in the first trimester do have an increased pelvic awareness that accompanies the uterine changes; if they are concerned that the thrusting and movement during intercourse may be damaging to the fetus, it will tend to decrease the satisfaction derived from stimulation, and orgasm may not be as frequent. The sensations during intercourse may be different from those usually experienced, but they are in no way harmful.

By the second 3 months, a woman is usually used to the different kind of feeling in the pelvic area. Masters and Johnson report an increase in sexual functioning at this time, but other researchers have not confirmed the heightened arousal, reporting only approximately the same desire and frequency as before. There is no reason to abstain from intercourse during the second trimester.

The question of abstinence during the last 4–6 weeks of pregnancy is more "iffy." Many physicians are convinced that the uterine contractions during orgasm can trigger the onset of labor or that there is more possibility of infection. There is no proof of detrimental effects of intercourse right up until the time of delivery if the pregnancy is proceeding normally. Sometimes, the glans of the penis hitting the very vascular cervix may result in a little spotting; this need not discourage sexual relations if they are desired. The woman herself knows how she feels about sexual activity in the last weeks of pregnancy; should she want to continue intercourse, there is no reason not to. If all she is interested in is being held and loved, that is certainly another way of maintaining sexual activity.

After delivery, Masters and Johnson reported that sexual desire returned after anywhere between 2 weeks and 3 months, reappearing earlier in women who were breastfeeding. By the third week after delivery, the uterine discharge had stopped and the episiotomy incision, if present, had healed sufficiently to make resumption of sexual relations comfortable.

EFFECTS OF DRUGS ON SEXUALITY

Relatively few controlled, systematic studies of the effects of chemical agents on sexual activity ever appear in scientific journals. People take a very dim view of mind-control, and experiments that attempt to assess the effect of psychotropic or mood-altering drugs on human behavior are so fraught with ethical and moral problems that even the most respected and intrepid sex researchers have shied away. Many of the drugs that are presumed to have a relationship to sexual function are illicit, making direct laboratory evidence even more difficult to come by. Most of what is known is, therefore, primarily based on patients' reports of the side effects of medical drugs taken for other purposes, or on the subjective and anecdotal responses of individuals who have taken hallucinogens in order to enhance sexual pleasure—hardly the way to obtain solid objective data.

Another major difficulty in attempting to investigate the interaction of drugs and sex is in part due to the nature of drugs and the nature of sex. No given drug ever affects everyone in the same way, or even affects one person the same way at different times. The pharmacologic action depends on the dosage, the size and weight of the person taking the drug, and the length of time the drug is used. Other drugs, either taken deliberately as medication or inadvertently as food additives or environmental pollutants, may interact with the original agent, or perhaps there may be an individual sensitivity, allergy, or difference in metabolism that influences the reaction. Even the time of day the drug is taken is important. But probably the prime ingredients in the effect of a mood-altering or psychoactive drug effect are the mood, attitude, and expectations of the taker and the belief or faith one has in the prescriber of the drug.

By now, it should also be obvious that what one obtains or finds in a sexual experience also profoundly depends on what one brings to it. The entire cultural, sociological, ecological, and psychological aspects of the individual personality, the expectations, the conscious and unconscious needs, yearnings, fantasies, and attitudes of the moment are the determinant of the total emotional effect. There are so many psychological and biological variables in both drug action and sexual behavior that to attempt to measure the pharmacological effects of the one on the other is almost impossible. There is no known drug that has been found to have uniformly consistent sexual effects. Even with that caveat, it would still be interesting to look at some chemical agents for which there have been claims of either adverse effects or increase of sexual pleasure and capacity.

APHRODISIACS—MOSTLY WISHFUL THINKING

Aphrodisiacs are substances that arouse an individual to increased desire and ability to engage in sexual activity. For thousands of years, males have been looking for the perfect love potion to prop up a faltering phallus. Many of the earliest substances used to increase sex drive were plants or parts of plants that resembled the human form or, in particular, the genitalia. Mandrake and ginseng roots are still prized today for their presumed restorative properties, but Soviet physicians who frequently use ginseng therapeutically disclaim any aphrodisiac effect of the drug in humans (Baranov, 1982). Certain foods, like oysters or truffles, have also had aphrodisiac properties attributed to them, probably because of their presumed resemblance to testes. There is no scientifically documented evidence that any of these are effective, except as placebos. Believing that they work may make them work.

There are almost no historical references to women voluntarily taking substances that stimulated their desire for sexual activity, but only instances of their being given various agents that might improve their fertility or for enhancement of male pleasure with them. Even today, there are few reports of drug effects on female sexual responses. Several investigators have pointed out that although erection may be more visible and easily studied, the response is under the control of the same parasympathetic nerves that result in the labial swelling and vaginal lubrication in the female. Thus, penile erection is neurologically analogous to lubrication-swelling in the female, and it can be assumed that what affects the former will also affect the latter.

The supposed aphrodisiac with which most people are familiar is "Spanish fly," an extract from the pulverized wing parts of a beetle, *Cantharis vesicatoris*, which has been used for centuries. Cases of self-poisoning by men or poisoning of women by men in anticipation of producing sexual arousal still occur. The active ingredient, cantharides, is a toxic substance that produces bladder and urethral inflammation. The irritation may be severe enough to produce priapism, an extremely painful persistent penile erection with no associated sexual excitement. Permanent impotence, bloody and painful urination, or even death have resulted from its ingestion.

Amyl nitrite, popularly known as "poppers," has more recently been touted as an enhancer of sexual enjoyment. Medically, the drug has been prescribed for the heart pain that results from blocked coronary arteries; it causes vasodilation and increases the diameter of the arteries and results in a drop in peripheral blood pressure. The highly volatile substance is usually administered by breaking the capsule in which it is contained (hence the popping noise) and inhaling the drug, which then enters the bloodstream from the lungs. As a sexual stimulant, amyl nitrate is sniffed from an inhaler or "popped" just before orgasm. Subjectively, the drug produces a "flash" or "high" that is said to intensify or prolong orgasm. Not only does the drug have a very unpleasant odor, it can also have most unpleasant side effects that on occasion are dangerous or even life threatening. Some users have reported headache and aching eyes, but cardiovascular distress and even death have occurred. Amyl nitrite users who are simultaneously taking other drugs could experience potentially disastrous effects. Isobutyl nitrite, a chemical relative of amyl nitrite but a nonprescription drug, is an easier "popper" to obtain. Marketed as "liquid incense" under such trade names as Rush, Bolt, Locker Room, and Bullet, isobutyl nitrite has the same side effects and a similar capability for harm.

Levodopa, a therapeutic drug given to decrease the tremor, rigidity, and motor disturbances of patients with Parkinson's disease, has also been said to improve sexual function. Only a very small percentage of males have reported increased libido or slight increase in spontaneous or nocturnal erections following levodopa treatment, and it has been difficult to confirm or substantiate those effects. Indirect effects of general improvement in coordination of bodily movement may account for a

reawakening of sexual interest in some patients.

Methaqualone (Quaalude) is a sedative-hypnotic that is chemically unrelated to the barbiturates (Nembutal, Seconal, Tuinal) and glutehimide (Doriden) and is believed to act on a different brain site than the other sedatives. "Ludes" are called the "love drug" by recreational drug users, and there are claims that it causes great increase in sexual feelings and multiple orgasmic potential, but there is no known pharmacological reason for the alleged increase in desire and performance.

Marijuana has a strong reputation as a sexual stimulant, but its actual impact on eroticism is not clear. For many people, "pot" or "grass" decreases sexual inhibitions; it may also alter sensory awareness and distort time, and in that way enhance a sexual experience. Like many supposed aphrodisiacs, if marijuana is taken with the expectation of obtaining a desired effect, it is more likely to produce that effect.

Other, more powerful hallucinogens, such as LSD, STP, mescaline, and psilocybin, were more commonly taken in the 1960s and early 1970s, but there is a reported recent resurgence of LSD use, especially in California. These drugs are usually reported to be too "heavy," too "mind-blowing" to have an aphrodisiac effect.

As a recreational drug, cocaine is second only to marijuana in popularity, but the expense limits its usage. When the drug is inhaled or "snorted," users report that they experience an intense "rush" that some have compared to an orgasmic experience. The physiological effects include increased heart rate and blood pressure, dilated pupils, and anesthesia of the nasal passages. Cocaine may be at least partially effective in enhancing sexual pleasure, although the action may vary with the mode of administration. Gay and Sheppard (1972) reported that 15% of the male patients they saw in their clinic said they saved cocaine for sexual situations, and half of the respondents interviewed reported spontaneous erections upon intravenous injection of the drug. Some users like the anesthetic properties of the drug. When applied to the male genitalia, for example, cocaine allegedly delays ejaculation. Women have used the drug as a douche to obtain vaginal mucosal contractions and, as absorption slowly occurs, a systemic euphoria. Sniffing cocaine is said to provide pleasurable feelings, sensations of well-being, warmth, and energy (i.e., a "high"). Although it is physically nonaddicting, the desire for the rush and the associated effects eventually replace the desire for sex. In large doses with long-term usage, most street drugs decrease sexual desire and performance. It is well known that addiction to heroin, barbiturates, or morphine is usually accompanied by a loss of sexual interest.

Ogden Nash said, "Candy is dandy but liquor is quicker!" Alcohol has traditionally been thought of as a sexual stimulant—it relaxes inhibitions, decreases anxiety, and perhaps guilt. In many individuals, however, the release-of-inhibition-stage proceeds very rapidly to the central nervous system depressant or totally-zonked-out stage, resulting in impotence in the male and complete passivity in the female. Shakespeare recognized the antiaphrodisiac properties of alcohol when he said in *Macbeth*, "drink . . . provokes the desire, but it takes away the performance." The benzodiazepines, such as Valium and Librium, are antianxiety drugs and also sedatives, muscle relaxants, and anticonvulsants. Their effects on sexuality are similar to that of alcohol: decrease in inhibitions at low doses, but decreased libido at higher doses.

Drugs That Impair Sexuality

The vasocongestion and muscle responses that are a part of the sexual response cycle are under the control of the parasympathetic and sympathetic divisions of the autonomic nervous system. Anything that interferes with parasympathetic and sympathetic impulses may, therefore, inhibit some aspect of the sexual response.

Anticholinergic drugs are those that block the action of the parasympathetic fibers, and they are frequently used in the treatment of gastrointestinal disturbances because they decrease muscle spasm of the digestive tract and inhibit acid secretion in the stomach. It is possible that high doses of anticholinergics may result in erection difficulties in the male and, by extrapolation to the female, swelling and lubrication difficulties. Drugs that inhibit the sympathetic nerve impulses are antiadrenergic and are often used to lower high blood pressure. Men who take antiadrenergic drugs for hypertension have reported difficulty with ejaculation since the reflex muscular contractions that result in emission and ejaculation of semen are controlled by adrenergic or sympathetic fibers. Men have also claimed decreased libido while on antihypertensive medication. Women have no muscle response corresponding to emission and ejaculation, but the absence of specific reports from females should not be interpreted to mean that drugs taken to decrease high blood pressure could not affect their sexual function as well.

Some women have reported lessening of desire for sexual activity while on oral contraceptives, but some women have claimed an increased interest in sex occurs while on the pill. There have been indications from both men and women that tranquilizers and other kinds of antianxiety, antidepressant drugs result in decreases sexual interest, and some men have reported potency difficulties while on thiazide diuretics. Although effects of drugs on sexual function can occur at any age, they are more frequent and troublesome after age 50. Men obviously have a more visible and easier gauge by which to measure sexual dysfunction, but there is no reason why women should not also be alert to the possibility of disturbance while taking drugs with the potential of affecting some aspect of their sexuality.

REFERENCES

Baronov, A. L. *J Ethno Pharmacology,* 6:339–353, 1982.

Bell, R. R. and Bell, P. L. Sexual satisfaction among married women. *Med Aspects Human Sexuality* 6(12):136–146, 1972.

Bell, A. P. and Weisberg, M. S. *Homosexualities.* New York, Simon & Schuster, 1978.

Cole, T. M. Sexuality and physical disabilities. *Arch Sex Behav* 4(4):389–403, 1975.

Freud, S. *Three Essays on the Theory of Sexuality (1905).* New York, Avon Books, 1965.

Gay, G. R. and Sheppard, C. W. Sex in the drug culture. *Med Aspects Human Sexuality* 6(10):28–50, 1972.

Golde, P. and Kogan, N. A sentence completion procedure for assessing attitudes towards old people. *J Gerontol* 14:355, 1959.

Hite, S. *The Hite Report.* New York, Macmillan, 1976.

Kaplan, H. S. *The New Sex Therapy.* New York, New York Times Book Co., 1974.

Kenny, J. A. Sexuality of pregnant and breastfeeding women. *Arch Sex Behav* 2(3):201–203, 1973.

Kinsey, A., Pomeroy, W. B., Martin, C. E. and Gebhard, P. H., *Sexual Behavior in the Human Female.* New York, Saunders, 1953.

Leiblum, S., Bachmann, G., Kemmann, E., et al. Vaginal atrophy in the postmenopausal woman. *JAMA* 249(16):2195–2198, 1983.

Masters, W. H. and Johnson, V. E. *Human Sexual Response.* Boston, Little, Brown, 1966.

Pomeroy, W. in Brecher, R. and Brecher, E. (eds). *An Analysis of Human Sexual Response.* Boston, Little, Brown, 1966.

Reamy, K., White, S., Daniell, W. C., and Le Vine, E. Sexuality and pregnancy. *J Reprod Med* 27(6):321–327, 1982.

Solberg, D. A., Butler, J., and Wagner, N. N. Sexual behavior in pregnancy. *New Engl J Med* 288:1098–1103, 1973.

Stewart, W. F. R. Sexual rehabilitation—a gap in provision for the disabled. *Nurs Mirror* 142(5):48–49, February, 1976.

Tolor, A. and DiGrazia, P. V. Sexual attitudes and behavior patterns during and following pregnancy. *Arch Sex Behav* 5(6)539–551, 1976.

Vance, E. B. and Wagner, N. N. Written descriptions of orgasm: A study of sex differences. *Arh Sex Behav* 5(1):87–98, January, 1976.

Walsh, J. J. The spinal cord disabled. *Nurs Mirror* 142(5):53–54, February, 1976.

7. *The Mammary Glands*

Along with warm blood and skin covered with hair, the mammary glands are such a distinguishing characteristic of our taxonomic class that it is named Mammalia because of them. Mice, monkeys, whales, elephants, lions, tigers, and human beings are all mammals, and all have in common these specialized skin glands that secrete milk to nourish their offspring. Baby mammals are born in a relatively immature and highly dependent state, unable to forage for their own food or even to digest and assimilate an adult diet. Newborns completely subsist on the secretions of the mammary glands of their mothers. Sensibly enough, the number of pairs of glands has a general relationship to the number of young in the litter. An animal like the mouse, which regularly produces large families, may have six or seven pairs of breasts; humans, fortunately, normally have only a single pair. Considering the cultural fervor with which we humans view the female breasts, it is probably just as well we have only two of them about which to be concerned.

Sexual interest in breasts is peculiar to humans; in other species, the males are totally unimpressed by them. One has only to look at the paintings and sculptures in any art museum in the world to recognize that our preoccupation with breasts, our almost mystical veneration of the mammary glands, is not merely a current cultural phenomenon. The female breast as a source of eroticism, as a symbol of femininity, as a determinant of fashion, and as a measure of beauty has for centuries assumed an importance far out of proportion to the natural purpose of the glands—the nourishment of an infant. One can be certain that the comment, "Hey, get a load of those knockers!" is not referring to their sustenance-giving properties.

Whatever they are called—"boobs," "bazooms," or "tits,"—women know that by most male definitions, the breasts should be protuberant, conical, and large; the bigger the better. It is this *Playboy*-centerfold ideal of beauty, the large-breasted, long-legged, narrow-waisted standard, that women, too, have accepted, and which has led them in many instances to dislike their own bodies. Some of them, and they are not all Las Vegas showgirls, seek "remedies," such as breast implants and silicone injections for their presumed inadequacies, which can result in disfigurement and even death. Sometimes they delay or refuse to see a doctor when they discover a breast lump for fear of losing what they have been socialized to believe is their major badge of femininity.

Of course a woman is concerned about her breasts. As a major secondary sex characteristic that appears at puberty, the breasts are a symbol of feminine identity, forming a part of the body image and important to self-esteem. Breasts are a source of erotic stimulation, and they play a role in the expression of a woman's sexuality. Their size, shape, and appearance are unique to each woman, however, and there is no reason why they should conform to some idealized stereotype. They should not be viewed as a woman's most cherished assets, on which her "wholeness" as a real and complete woman depend. Perhaps under the influence of the woman's movement, and as women regain more control over their bodies and their lives, they will gain the self-confidence to increasingly reject the disproportionate glorification of breasts and of themselves as exclusively sexual objects. The strong pressure placed on women by our breast culture may diminish in time.

MORPHOLOGY OF THE BREASTS

Unlike other mammals, the breasts of humans, monkeys, and apes are located on the thorax and not the abdomen. Unlike other primates, who have flat breasts even during pregnancy and lactation, the human female has comparatively large and dome-shaped mammary glands. Generally speaking, the taller the woman, the higher up on the thorax the breasts are located. Differences in breast size and shape are completely determined by the relative amounts of fatty and connective tissue, dependent on genetic and endocrine factors, or as a result of how much body fat the

woman has (see Fig. 7.1). As women gain weight, their breasts become more pendulous and larger. All women have the same amount of actual mammary gland tissue—about a spoonful. The actual size of the breasts, whether 32AA or 38D, has, therefore, no relationship to a woman's ability to nurse a baby.

The breast is a modified skin (sweat) gland, lying over the pectoralis major muscles of the chest wall and attached to them by a layer of fascia, or connective tissue (see Fig. 7.2). Each breast extends approximately from the second to the sixth or seventh rib and from the lateral border of the sternum to the axilla, or armpit. The upper, outer portion of the breast is thicker, and primarily composed of glandular tissue. It extends to a variable degree as the "axillary tail" into the armpit—an anatomical feature significant in the spread of breast cancer. The longitudinal diameter of the breast, 10–12 cm on the average, is less than the transverse diameter, and the average weight of each breast is 150–200 g, but increases to 400–500 g during lactation. The right breast is usually somewhat smaller in size than the left breast.

The glandular tissue itself consists of 15–25 lobes, each with its own lactiferous duct, which all converge like the spokes of a wheel on the nipple. The lactiferous ducts enlarge slightly into a lactiferous sinus before emerging, and the nipple is then perforated by 15–20 tiny openings. The characteristic wrinkled and pigmented skin of the nipple extends out onto the breast for approximately 1–2 cm to form the areola. In prepubertal girls with light complexions and in blondes before pregnancy, the capillaries filled with blood under the areolar skin show through to impart a rosy pink color. The skin becomes more pigmented at puberty and darkens even more during pregnancy in fairskinned women. This pigmentation never completely disappears and, to a certain extent, can distinguish a nulliparous woman from a parous one. In darker

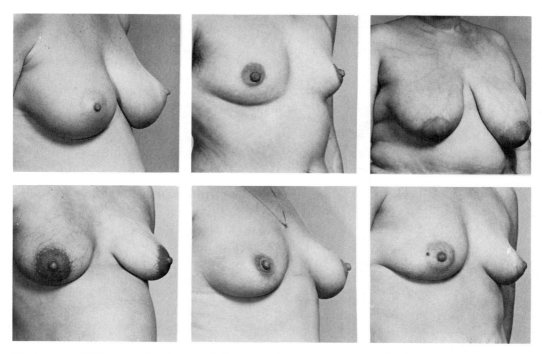

Figure 7.1. Differences in sizes and shapes of breasts in women who vary in age and weight.

skinned women, there is no noticeable change during pregnancy.

The areolae contain large modified sweat glands, Montgomery's glands, which increase in size and number during pregnancy and lactation and which may be important in lubrication of the nipple. Montgomery's glands become larger at puberty and also respond to monthly hormonal stimulation; they shrink and involute after menopause. Along the margin of the areolae are other large sweat and sebaceous glands, some associated with hair follicles, and many women are aware of the cyclic growth and recession of individual areolar hairs. The skin of the areolae and the nipples contain numerous longitudinally and circularly arranged smooth muscle fibers that are responsible for erection of the nipples

when they are stimulated tactilely or by exposure to cold.

The 15 or so major lobes of glandular tissue in each breast are further subdivided into 20–40 lobules. In a pregnant woman, each lobule, branching like a tree, sprouts little ducts that terminate in evaginated sacs called alveoli. The alveolar cells, under hormonal influence, are the actual glandular units that synthesize and secrete milk. The breast of a nonpregnant woman contains just a few alveoli that have budded off from the ends of the ducts. During pregnancy, the abundant ovarian and placental steroids produced by the mother cross over into the fetal circulation and are able to stimulate the fetal mammary tissue. Eighty to 90% of all newborn babies of either sex have slightly enlarged breasts that actually

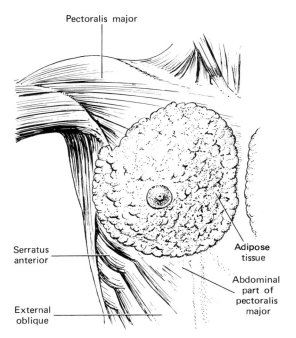

Pectoralis major

Serratus
anterior

External
oblique

Adipose
tissue

Abdominal
part of
pectoralis
major

Figure 7.2A. Position of the breast in relation to the superficial thoracic muscles.

secrete milky fluid ("witches' milk") for several days.

The lobes, lobules, and their ducts are surrounded and separated from each other by bundles of connective tissue. The largest of these partitions are called the suspensory ligaments of Cooper, and they extend from the layer of connective tissue or deep fascia over the muscles of the chest wall to the layer of superficial fascia just under the skin. Fat accumulates in these ligaments and in all the connective tissue of the breast, and so there is a fat layer overlying the breast just under the breast skin. Except for the smooth muscle of the areola-nipple complex, there is no muscle in the breast itself. It consists of the ducts, lobules, and alveoli of the gland tissue, a lot of fatty tissue, and connective tissue. Those amazing ads in the back pages of magazines

that promise to increase the size of the breasts by "bust-developers" describe an anatomical impossibility. Certain exercises are able to increase the size and strength of the pectoral muscles and can result in a small apparent elevation of the breasts, but no dramatic increase in size is possible. Better posture also raises the breasts, but there is nothing other than augmentation surgery that can change small breasts into large breasts.

Blood Supply of the Breasts

The major blood supply to the breasts is from a branch of the subclavian artery called the internal mammary, with additional contributions from the thoracic branch of the axillary artery. The veins that drain the blood from the breasts follow a pathway that quickly leads into the large vein, the superior vena cava, that enters the right side of the heart. From there the blood is pumped directly to the lungs to be oxygenated before it is returned to the left side of the heart for distribution throughout the body. The rather direct route from the breasts to the lung capillaries may be significant in the metastasis (spread) of breast cancer.

Lymphatic Drainage of the Breasts

All the cells of the body are bathed in a watery solution that contains nutrients and gases called tissue fluid. Tissue fluid, bringing essential food and oxygen to the cells, is produced from blood plasma at the arterial ends of the blood capillaries and is reabsorbed at the venous ends of the blood capillaries. Since a little more tissue fluid is produced than is absorbed, the excess passes into tiny lymphatic capillaries that are also found in the tissues. Called lymph, the tissue fluid is drained from these lymphatic capillaries into progressively larger lymphatic vessels, eventu-

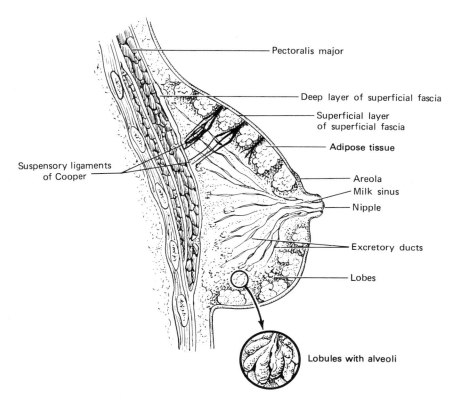

Pectoralis major

Deep layer of superficial fascia

Superficial layer
of superficial fascia

Adipose tissue

Areola

Milk sinus

Nipple

Excretory ducts

Lobes

Suspensory ligaments
of Cooper

Lobules with alveoli

Figure 7.2B. Sagittal section through the breast.

ally to empty into a major vein near the heart. If it were not for this lymphatic drainage, the tissues of the body would become swollen with excess tissue fluid, a condition called edema. There are several basic causes for edema, and one of them is obstruction of the lymphatic channels.

The lymphatic vessels are interrupted in their pathway by groups of small bean-shaped lymph nodes, an accumulation of cells called lymphocytes held together by connective tissue, sometimes called lymph glands. Clusters of lymph nodes in the neck, the armpit, and the groin are of particular importance clinically; they can be palpated, and the expression "swollen glands" really refers

to enlarged lymph nodes, which not only play a predominant role in immunity but also act as a filter, screening out harmful particles, such as microorganisms, debris, or cancer cells.

In the breast, most of the lymph from the central part of the mammary gland, the skin, nipple, and areola drains laterally toward the axilla. The first nodes encountered are the four to six anterior pectoral nodes, also called the low axillary nodes. From there the lymph passes to the central axillary nodes embedded in a fat pad in the center of the armpit. The flow then proceeds to the upper part of the axilla to the lateral nodes along the axillary vein and then to the deep axillary or subclavicular nodes. From the back of the breast, the lymph is drained to several interpectoral (subscapular) nodes located between the pectoralis major and pectoralis minor muscles. The internal mammary nodes that lie along the sternum lie in the pathway of the lymph channels that drain from the inside or medial part of the breast (see Fig. 7.3).

When breast cancer cells invade the lymph system, a process known as metastasis, they first reach lymph nodes, which retain the tumor cells and try to destroy them. The malignant cells grow rapidly, and, eventually, they overcome the nodes' capacity for destruction. Then the cancer travels on to other nodes. Because most of the breast lymph (75%) drains into the axillary nodes; it is believed that the removal of the primary tumor in the breast and the axillary nodes can frequently completely eradicate the cancer. This may not always be successful, because the secondary lymph channels that pass to the internal mammary nodes or, indirectly, to the subdiaphragmatic nodes that lead to the liver, or those channels that cross over to the opposite breast may play an equal role in the spread of breast cancer. When lymph nodes have been invaded by cancer cells, they are said to be "positive" nodes.

CYCLIC CHANGES IN THE BREASTS

Before menarche, the mammary glands of both males and females consist only of a few ducts surrounded by a lot of connective tissue. This is the way they remain in males. With the onset of puberty in females, however, the duct tissue, stimulated by estrogen, elongates and branches, forming the buds of the future lobules and alveoli. The connective tissue also proliferates and becomes infiltrated with fat, and the breast begins to assume the shape of the mature female gland. When ovulatory cycles begin to occur and the corpus luteum secretes progesterone, the characteristic duct–lobular–alveolar structure of the breast during the childbearing years is developed. Like the uterus, the breast prepares each month for the possibility of conception. During each menstrual cycle, there is a little increase in mammary development, but the major growth takes place during pregnancy, when the increase in size, volume, and density of the breasts is very great.

During the first half, or proliferative phase of the cycle, the high levels of ovarian estrogens in the blood stimulate the reproduction of cells in the glandular ducts and their terminal buds, the alveoli. This increase in the number of cells continues into the postovulatory or secretory phase of the cycle, when the increasing amounts of progesterone result in the dilation of the ducts and the differentiation of the alveolar cells into the actual secretory cells. After ovulation, there is also an increased blood flow to the breasts, and, in the week before menstruation, some women experience fluid retention. The resultant swelling, or edema, in the mammary connective tissue causes feelings of fullness and heaviness, and the breasts get larger. The enlarged duct diameters, the increased cellular growth, and possibly some newly formed alveoli all contribute to the engorgement and the increased lumpiness or nodularity of the breasts. When menstruation starts, there is actually some secretion of droplets by alveolar cells into the alveolar lumens, but toward the end of menstruation, all the proliferated tissue begins to regress and must be reabsorbed. The ducts become narrower, the lobules and alveoli become smaller, and the tissue edema disappears. But, before the tissue can totally regress, a new cycle has started and the estro-

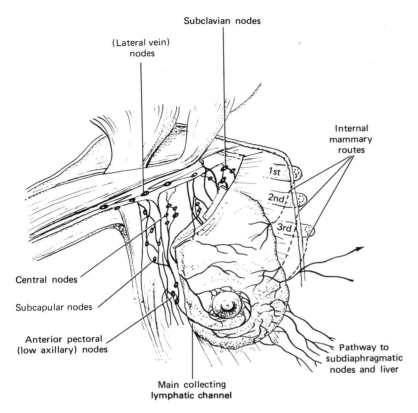

Figure 7.3. Pathways of lymphatic drainage from the breast.

gen again induces more proliferation. The ducts and alveoli never have a chance to completely return to the way they were before the preceding cycle, and each additional ovulatory cycle therefore results in a little more mammary development. This continues until approximately age 35.

Fibrocystic Disease

Most women are aware of some feelings of fullness and lumpiness in their breasts just before menstruation. This is so common that women are told not to examine their breasts for lumps until after their periods. Some women have very painfully engorged breasts, with a large number of tender nodules in the weeks before and during menstruation. This is physiological lumpiness and is not a disease, although the dividing line between such changes and the clinical entity known as "fibrocystic disease" is arbitrary and a matter of degree. The increased nodularity usually appears in both breasts in approximately the same location and disappears after menstruation. A woman who regularly performs monthly breast self-examination learns to distinguish her own normal thickenings and lumpiness from something new. Any one lump, however, that begins to grow out of line with the rest is suspicious.

There are several kinds of benign tumors that may be related to hormonal stimulation, although the cause is really unknown. One, the most frequent, is cystic disease of the breast. In cystic disease, small fluid-filled, palpable cysts, which may or may not be painful, develop rather quickly in one or both breasts. They are treated by withdrawing or aspirating the fluid with a needle and syringe or by surgical removal. Another kind of non-malignant growth in the breast is a solid tumor commonly called adenofibroma or fibro-adenoma. These tumors develop most frequently in women younger than 25 years of age and are removed surgically.

Both of these disorders fall under the catch-all term, fibrocystic disease, which is further complicated by being also named "chronic cystic mastitis," "mammary dysplasia" "cystic mastalgia," "benign breast disease," and a host of other designations. The number of variants of this benign condition is equaled only by the morass of terms used to describe all of them—around 40, at last count. The symptoms of cyclic pain, tenderness, a feeling of heaviness, and breasts that feel diffusely nodular, granular, or just generally "lumpy," occur in an estimated 50% of all women at one time or another in their reproductive lives, and microscopic examination of the breasts of women at autopsy has revealed that 90% have the histological signs. Such widespread occurrence has led Love and her co-workers (1982) to question defining the condition as a "disease" at all. They suggest "lumpy breasts" or "physiologic nodularity" as alternatives. By whatever name, however, benign growths of the mammary gland have clinical significance for several reasons. First, it is important to distinguish these from cancer. Not only can the

nodules imitate a malignant growth, but they can possibly hide a cancer present among all the lumpy changes. Second, there may be a relationship between lumpy breasts and cancer. There is some evidence that the risk of developing cancer is greater (between two to four times, depending on the study) for a woman with certain types of fibrocystic disease, and a large percentage of women who already have breast cancer have also had fibrocystic disease. The link has not been proved, however, and is still debatable. Henry Leis et al. (1983), Susan Love, and other investigators suggest that the potential of fibrocystic disease for subsequent malignancy has been overestimated and is, therefore, over-rated. The condition is so common that if it were really strongly related to the development of breast cancer, the incidence of cancer should be greater than the current one out of every 11 women.

But even if a woman could be reassured that the diagnosis of fibrocystic "disease" does not place her in a higher risk category for breast cancer, swollen lumpy breasts *hurt*, with discomfort that can range from merely being a nuisance helped by a brassiere with greater support to tenderness severe enough to make even turning over in bed a problem.

There is considerable controversy among physicians concerning the treatment of fibrocystic disease. Almost every kind of hormonal treatment has been tried with varying success, but since the reasons for benign breast changes are not completely understood, the reasons why various hormones or hormonal inhibitors alleviate the condition are also not clear. Estrogen causes mammary duct growth, and progesterone results in alveolar growth and duct dilation, but prolactin, growth hormone, ACTH, thyroxin, and androgen are also concerned with breast function at various times. One current theory hypothesizes that the primary factors in benign breast disease are absolute estrogen excess or absolute or relative progesterone deficiency resulting in an imbalanced estrogen/progesterone ratio. Another theory blames an inappropriate or exaggerated breast tissue response to the normal variations in ovarian hormones. An underfunctioning thyroid gland has been associated with fibrocystic disease, and, possibly, hypothyroidism should be investigated in women who suffer from breast lumps.

That an alteration in the estrogen/progesterone ratio may play a role appears to be substantiated by the reduced incidence of fibrocystic disease when a woman is on oral contraceptives, that provide a balanced source of estrogen and progesterone.

Oral progestin in the form of medroxyprogesterone acetate (Provera) has been used with some benefit, and in the past, injected androgens were given, but the masculinizing side effects of hair growth and acne outweighed the pain relief. One clinical study has reported lessening of symptoms when a progesterone cream was rubbed into the breasts daily (Lafaye et al., 1978). The big breakthrough in steroid treatment of fibrocystic disease, however, came with the approval in 1980 by the Food and Drug Administration of danazol for fibrocystic disease as well as for endometriosis. Danazol, the androgenlike antigonadotropin, is said to have response rates for severe fibrocystic disease in the 80%–90% range, depending on the dosage and the duration of treatment. The subjects of a 1983 clinical trial by Leis were 42 women with cystic disease severe enough to have required 10 to 28 needle aspirations in the previous year and a half. After 400 mg of oral danazol daily for 6 months, 18 women had no further cyst formation in the 2-year follow-up period, 21 developed between one and four cysts, and only three women had more than four cysts.

As previously indicated, the side effects of danazol are substantial, but, according to reports in the medical literature, they disappear rapidly upon discontinuation of the drug.

Several small clinical trials, with few women as participants and as yet unconfirmed by additional studies, have demonstrated that fibrocystic disease can be treated with dietary changes and/or vitamin supplementation. Women may want to try these measures before choosing hormonal or antihormonal therapies. John P. Minton, of Ohio State University Medical School, reported in 1979 that fibrocystic disease disappears when women with the condition avoid caffeine. Because studies have shown that cellular levels of the intracellular messenger cyclic AMP are elevated in fibrocystic disease but not in normal breast tissue, and that caffeine, theophylline, and theobromine, all known chemically as methylated xanthines, inhibit the enzyme that destroys cyclic AMP, Minton and associates (1975a, 1979b) reasoned that eliminating methylxanthines from the diet should improve the symptoms of fibrocystic disease. They persuaded 20 out of 47 women, who had long-standing benign breast nodules and who were consuming a daily average of methylxanthines equivalent to that in four cups of coffee, to abstain from coffee, tea, cola, cocoa, and caffeine-containing soft drinks, cold remedies, and painkillers. Thirteen of the 20 women had complete resolution of their breast lumps in 1–6 months. In the remaining group of 27 women who refused to avoid caffeine, only one experienced a spontaneous disappearance of her nodular condition, and the rest subsequently had to have surgical biopsies. Minton also advocates a diet high in fish, chicken, veal, and grains and low in red meat, salt, and fats.

The other trial was reported by London and colleagues in 1978. In 12 women with fibrocystic disease, nine obtained good results with no side effects by taking 600 units of vitamin E daily, and several additional studies by the same researchers over the next few years produced similar response rates (1982). Vitamin E, believed to act by altering adrenal steroid synthesis, could have adverse side effects when more than a gram is ingested daily (Roberts, 1981). It is recommended that daily doses beyond the 400–600 unit range not be exceeded.

Vitamin B complex in 100 mg daily doses may possible enhance liver deactivation of estrogen and is reportedly helpful. According to some women, vitamins C and B$_6$ have provided symptom relief. In mild fibrocystic disease, merely restricting intake of salt to avoid swelling could be sufficient to alleviate breast tenderness.

The former approach to fibrocystic disease was primarily surgical, with frequent breast biopsies and cyst removals. Current medical journals advocate treatment that attempts to circumvent a need for surgery: abstention from methylxanthines, other dietary changes, perhaps a diuretic taken premenstrually, and cyst aspiration for diagnosis rather than surgical removal. If the above measures are ineffective after an adequate trial, danazol is prescribed for severe fibrocystic disease. There are still, however, many doctors who recommend prophylactic surgery for women who they believe are at high risk for breast cancer because of their severe fibrocystic disease or who have intractable pain. According to Leis (1983), an increasing number of simple or subcutaneous mastectomies (removal of the breast tissue below the skin and replacement with an implant) are being performed throughout the country for fibrocystic disease. Although the logic of "if you don't have a breast, you can't get breast cancer" appears incontrovertible, there are other organs of the body equally subject to malignancy, and no one is advocating prophylactic removal of a lung or the prostate gland. Moreover, the reasoning is substantially flawed. Not only is the risk of cancer in the presence of fibrocystic disease probably overestimated, but anything short of a total mastectomy (also called a modified radical mastectomy) is unlikely to offer complete protection, since all mammary gland tissue is not removed by a subcutaneous mastectomy, and a cancer could develop in the remaining tissue.

BREAST CHANGES DURING PREGNANCY

After a number of Joe's organs had dramatically told their stories in *Reader's Digest* ("I Am Joe's Kidney," "I Am Joe's Liver," etc.) the magazine carried an article called, "I Am Jane's Breast." In it, Jane's breast rather smugly declares, ". . . I am capable of baffling, almost miraculous, chemical conversions. I change blood into milk."

Actually Jane's breast is guilty of a bit of overstatement. The mammary gland does not change blood into milk any more than the salivary gland changes blood into saliva, the stomach changes blood into gastric juice, or the kidney, blood to urine. The cells of the glandular alveoli of the breasts utilize the nutrients brought to the gland in the circulating blood—the plasma amino acids, fatty acids, and glucose. From those building blocks, the gland synthesizes milk protein, milk fat, and milk sugar, adds ions, vitamins, and water, and secretes it as milk. Throughout the 9 months of pregnancy, the breasts are preparing for nursing.

Within the first 3–4 weeks of gestation, the ducts of the mammary glands sprout branches, and more lobules and alveoli are formed. These changes are much greater than those that occur premenstrually, and the breasts become definitely larger and feel tender and heavy. Each breast gains nearly a pound in weight by the end of pregnancy; the glandular cells, partially filled with secretion, the increased number of blood vessels, and the increased amounts of connective tissue and fat result in the enhanced size.

The hormones that are involved in stimulating and preparing the breasts during pregnancy are the sex steroids from the ovaries and placenta. Prolactin and growth hormone from the anterior pituitary, human placental lactogen and human chorionic gonadotropin from the placenta, adrenal corticosteroids, insulin from the pancreas, and thyroid and parathyroid hormones also all contribute to the duct and alveolar growth and differentiation into secreting cells.

Anterior pituitary prolactin, secreted in increasing levels throughout pregnancy, triggers the actual synthesis and secretion of milk after the birth of the baby. During pregnancy, prolactin, in the presence of large amounts of estrogens and progesterone from the ovaries and placenta (and the other hormones mentioned), causes some synthesis and secretion of *colostrum*. Colostrum is a milklike fluid containing protein and carbohydrate but lacking milk fat. It is only after birth, however, when the high levels of sex steroids are abruptly withdrawn, that prolactin is able to cause the glandular cells to secrete large quantities of milk instead of colostrum. This takes place within 2–3 days of birth. Development of the glandular tissue, then, is stimulated mostly by sex steroids in the presence of the other placental, pituitary, and metabolic hormones mentioned. The actual milk synthesis and milk release into the lumens of the alveoli is caused by prolactin. The milk "comes in" after several days of colostrum secretion, but before the baby can get it, another hormone, *oxytocin*, must act so that the milk can be ejected from the alveoli to the nipples.

The stimulus of sucking on the nipple causes oxytocin to be released from the posterior pituitary gland.* Oxytocin gets into the bloodstream, travels to the mammary gland, and induces *myoepithelial cells* surrounding the alveoli to contract. The milk enters the ducts and begins to flow easily from the breast that was sucked, and also from the other breast. As long as the mother continues to nurse, the milk will continue to flow, even for several years. Prolactin and oxytocin result in milk production under the stimulus of sucking. When the baby is weaned, the prolactin and oxytocin secretion stops, and milk is no longer synthesized and released. (A further discussion of breastfeeding can be found in Chapter 10).

MAMMOPLASTY

Plastic surgery of the breast is termed mammoplasty and is not as well known as other forms of cosmetic surgery. We have become accustomed to the idea of having faces "fixed": nose jobs, face lifts, hair transplants, and removal of bags under the eyes. All are possible though expensive. Most health insurance plans do not cover surgery performed for cosmetic or esthetic reasons, and such procedures, hence, are mainly for the rich. Besides, plastic reconstruction, unless performed after an accident, is generally done on healthy, normal, if "imperfect" organs. One then runs a risk of complication or even death for an elective operation that was not really "needed" in the first place.

The definition of necessity varies, however, from one individual to another. What is abso-

* In a nursing mother, milk flow can also often be elicited merely be hearing another baby cry, or even by thinking about a baby.

lutely essential for the confidence, happiness, and self-esteem of one person may be really inconsequential to another. When plastic surgery is contemplated, it is very important, as it is with any other medical procedure, to understand what risks are involved, what to reasonably expect as results, and to make one's own decision on the basis of that information.

Breast Augmentation

No matter how it looks to the rest of the world, if an individual thinks that a body part is ugly or inadequate, it can become the focus of a sense of body-image deformity and produce anxiety and concern that interfere with his or her life. Small breasts (micromastia) in a woman do not have to impair her sex life, her activities, or her attractiveness. But if the small-breasted woman is convinced, and certainly our society helps her in her perception, that such smallness is at the root of all her difficulties, her self-image may be impaired, and she can never be truly comfortable with her body. Even though the usual method of enhancing the size of small breasts is by wearing "falsies," a woman may consider that it is more real to surgically implant the falsies. Essentially, this is the method that is used in augmentation mammoplasty.

Early operations to increase breast size used homografts, the woman's own subcutaneous fatty tissue transplanted from her buttock. This method is still used today in some countries of the world where the price of implants is too expensive, but the results are frequently unpredictable and usually asymmetrical because of the tendency of fat to be reabsorbed. The original types of alloplastic (nonhuman, man-made) implants were plastic sponges. They were unsatisfactory because they become infiltrated with scar tissue that hardened and flattened them out to pancakes. The widely used prosthetic device for breast augmentation today is the silicone or Silastic gel implant.

Contained within a soft and thin-walled envelope, the Silastic substance is solid but with a certain amount of elasticity and "give." There are many variations of the prosthesis, some reportedly more successful than others, but the solid gel implant is said to have a tendency to cause increased hardness. Collagen or scar tissue may become a problem, and some women have ended up with two baseballs instead of natural-looking and natural-feeling breasts. The fibrous capsule may then have to be split or released, which requires a second operation.

Other implants are the inflatables, and they rival the solid gels in popularity. The hollow inflatable implant of a suitable diameter is inserted, either through an incision made in the fold under the breast, or circularly around the areola. After placement, a fluid, usually saline solution, is injected until the desired volumetric breast size is reached. Inflatables feel softer than the gels and require a smaller incision for insertion. They have a distressing tendency to rupture or leak, but this is said to be less of a problem in the inflatables of most recent design. The skill and experience of the surgeon is very important to their success. Too much fluid injected can cause bulging; too little fluid can cause wrinkling. If all the air is not removed before inflation, the breasts may "squish" on palpation. Scar tissue will form around both the saline inflatable and the silicone gel implants, but many surgeons favor the saline type, believing they produce a better result.

The most inexpensive, but also the most dangerous method of breast-size enhancement is the injection of free silicone fluid, and no reputable and ethical surgeon in the United States will use it. The procedure should never be considered by any woman. Silicone fluid, injected in the large quantity required, generally disperses to migrate away from the injection site. At the very least, a woman could find herself with her breast augmentation located on her hip. More likely, an infection in the breast that does not readily respond to treatment may develop. The removal of the breast to stop the infection from spreading may be ultimately necessary.

Even when augmentation is performed by a competent plastic surgeon whose standards for quality medical and surgical procedures are high, the operation can still be unpredictable and subject to several kinds of complications. The contour and the feel of the breasts may be esthetically unsatisfactory, and the surgery may have to be performed again. Different women vary in the amount of scar tissue they produce, which can cause a problem with hardness of the breasts or with appearance of the scar. Some women have lost sensation in the nipples and areolae, but others have felt an enhancement of sensitivity. Infection is always a possibility, but it occurs less frequently than hematoma, the accumulation of a mass of partially clotted blood within the tissue. There is no evidence that breast implants cause cancer, but they may make early detection somewhat more difficult. Women who have had augmentation mammoplasty should, like all other women, perform regular monthly breast self-examination, and have frequent periodic clinical examinations by their physicians.

Breast Reconstruction After Mastectomy

Until very recently, the primary treatment for breast cancer has been mastectomy, complete removal of the breast. The response of a woman to mastectomy is intensely personal, and probably no woman really knows how she feels about losing a breast until it happens to her. To most of us, cancer is our greatest

health fear. To find out that one has breast cancer is enough to produce the initial emotions of shock, fear, resentment, denial, and perhaps panic in all women, but when that knowledge is accompanied by the surgical loss of a highly visible part of the body, the psychological effects can by tremendously magnified. Some women may be as devastated by the mastectomy as they are by the diagnosis of cancer. For others, survival is their main priority. They can say, "Take my one breast, take the other one, do whatever you have to, but leave me my life," and they mean every word, particularly if it has been determined that the cancer has already spread to the lymph nodes.

Although there are many factors that affect a woman's attitude toward mastectomy, probably the most significant is the status of her disease, Rose Kushner (1975) discusses the psychological aspects of breast removal in her excellent book, *Breast Cancer, A Personal History and an Investigative Report*. She discovered that women whose cancers have not been cured by mastectomy, and who are undergoing treatment to prevent even further spread of the disease, are likely to be far less concerned about the loss of their breasts than women who have already gone 5 or 10 years without a recurrence of the cancer.

Kushner also feels strongly that women should not be worried about being appealing and attractive to men after mastectomy. She believes that staying alive should mean more to a woman than how she looks to a man when he sees her naked.

Certainly, if a man truly loves a woman, his concern should be for her and not for her breasts. If a woman is surrounded by people who are loving and supportive, if her life has been secure and stable before mastectomy, she may be better able to reconcile herself psychologically to the loss of her breast and have little interest in undergoing additional surgery in order to replace it. But in Kushner's 1982 revision of her book, *Why Me?*, she reveals that although she had little difficulty in emotionally accepting the mastectomy, she chose to have her own breast reconstructed primarily for the physical convenience of being able to discard the prosthesis worn over the skin for one implanted under it. With the improvement of surgical techniques in the late 1970s, the procedure of breast reconstruction has gained in popularity. Women should discuss the possibility of reconstruction with their surgeons before surgery if mastectomy is to be the treatment for their cancer (there are now other available alternatives). If the medical opinion is that the reconstruction will not compromise the best chance for curing the cancer, the realization that something can be done to restore the breast may help a woman through the ordeal. Reconstruction is also an option for women who had their mastectomies 20–30 years ago, but since a radical mastectomy was more frequent then, the technique of creating a new breast contour is more technically complicated.

Postmastectomy breast reconstruction, however, is still relatively rarely performed. Of the estimated 1 million women with mastectomies in the United States, only an estimated 15,000–20,000 have undergone the procedure. In some parts of the country, reconstruction is virtually never sought by women, perhaps because they are unaware of its availability. But with the increasing interest on the part of both women and plastic surgeons in breast restoration, there will be more information available, more doctors performing the operation, and more patients having reconstructive surgery.*

The two breast reconstruction techniques used most frequently are the subpectoral prosthesis method, in which an implant is

placed under the chest muscle, and the musculocutaneous skin flap technique, using a section of skin and muscle transferred to the chest from another part of the body. Although there has been improvement and modification in the techniques, it is still generally agreed that the final result, however good, may be somewhat less than a cosmetic triumph. With current methods, the breast still cannot be restored to its preoperative normal appearance. The woman will look better in a brassiere than in the nude, and the two breasts will most likely be unequal in size. The opposite normal breast can be reduced to match the reconstructed breast, or an inflatable implant can be used in the reconstruction to achieve as much symmetry between the two breasts as possible. Sometimes, several operations may be necessary before the desirable result is achieved, but the less extensive the mastectomy to begin with, the simpler the reconstruction procedure afterward. Extensive postmastectomy irradiation treatment to the chest wall poses a particular difficulty and may make reconstruction almost impossible because of the tissue damage.

If a nipple and areola are to be rebuilt, they will require more surgery, so some women may prefer to do without them. Should reconstruction of the areola be desired, however, a graft can be taken from the other breast. Some doctors are against such areolar sharing because of the possibility of transplanting a present or potential malignancy from the normal breast. Another possible technique is to use a graft from the pigmented skin of the labia minora, which is removed under local anesthesia. A nipple can also be simulated by an implant of scar tissue, cartilage, Silastic, by nipple-sharing from the other breast, or by other surgical methods.

The most appropriate timing of the reconstruction is still a controversial issue, centering primarily on the question of whether the reconstruction may mask and thus delay the

---

* Information on breast reconstruction can be obtained from RENU, Reconstruction Education for National Understanding, 1148 Euclid Avenue, Cleveland, Ohio 44115.

discovery of a recurrent cancer. One group of physicians believes that reconstruction should be delayed for 2 years because a local recurrence generally develops within that period. Other surgeons believe in reconstruction at the time of the initial mastectomy. Some prefer to wait a few months to allow the resolution of trauma and scarring and to permit the completion of all postoperative therapy, such as radiation. One study at the University of Texas Cancer Center (1982) found that if a cancer recurred at the site of mastectomy, it developed, on the average, more than 2 years later. The Texas surgeons concluded that although there was little reason to delay reconstruction for 2 years, immediate reconstruction was inadvisable because until the axillary lymph nodes are stained, sectioned, and examined by a pathologist—a procedure that takes several days—there was always the possibility of a false-negative evaluation of the nodes.

Not only when, but also who should have breast reconstruction is also subject to controversy. The easiest candidate to operate on would be a young woman in good health with early cancer, no lymph node involvement, and a simple mastectomy. Women who have had a more extensive modified radical or the classical radical surgery are not excluded, but should recognize that the nature of the operation may be more extensive. Breast reconstruction post-mastectomy is necessarily subject to the same complications as breast augmentation—the possibility of implant hardness, hematoma, and infection.

If a woman is highly motivated, has made her own decision, and recognizes that there will be improvement but not perfection as a result of the reconstruction, the procedure can be of great value to her. She should be aware, however, of its shortcomings, and do it to please herself, not because anyone else wants her to have it done.

Reduction Mammoplasty

Perhaps in the opinion of some men, breasts can never be too large. But ask a woman who has macromastia—breasts that are disproportionately large in relation to her other body dimensions—whether she actually feels twice blessed. Excessive breast size can cause physical discomfort for a woman all her adult life. She may have neck and back pain, deep grooves in her shoulders from her bra straps. chafing, a skin rash and itching under her breasts, and have difficulty in the ordinary movements of walking and running. She would welcome a procedure to reduce mammary tissue and not merely for cosmetic reasons.

The surgical technique of breast reduction is called reduction mammoplasty. Some of the actual breast mass is removed, the skin over the breast is proportionately adjusted, and the nipple and the areola are relocated to a new appropriate position. There is some scarring, and the results are not always perfect and lasting. Most women, however, who have had the operation are very satisfied to get rid of the discomfort and excess weight of their breasts and are able to find and wear clothes that fit for the first time since they were children. In most instances, there is no loss of sensation in the nipples, and women are still able to lactate and nurse a child after surgery, but some doctors recommend that the operation be delayed until childbearing and breast-feeding are no longer desired. Postoperative complications of hematoma and infection are possible, but the skill and the experience of the surgeon are critical factors in making such complications relatively rare.

Ptosis

Ptosis, or sagging breasts, can occur without macromastia and may sometimes follow pregnancy. The surgical procedure which elevates ptotic breasts is called mastopexy. Sometimes, depending on the patient, it can be done on an outpatient basis under local anesthesia. The surgery is considerably simpler than reduction mammoplasty, although the scarring may be about the same, and there is a lesser incidence of postsurgical complications. When there has been a reduction in breast volume as a result of aging or after pregnancy, a woman may feel it is important to restore the fullness of the breast as well as its elevation. Augmentation with a Silastic implant can then be combined with mastopexy.

Opinions vary as to whether going bra-less contributes to sagging breasts. It seems logical that supporting the weight of the breasts against gravity with a brassiere diminishes the stress on the network of connective tissues that supports the fat and glandular elements of the breasts. Larger breasts would be more affected by gravity and are thought to be more subject to "Cooper's droop," the weakening of Cooper's ligaments. Women with larger and heavier breasts are usually more comfortable while wearing a bra, anyway.

BREAST CANCER

According to the American Cancer Society, one of every 11 American women will someday develop breast cancer, and one of every six girls born will have a biopsy at some point in her life. Breast cancer is not only the most frequent site of female cancer, but the greatest killer as well. Every 15 minutes, a woman somewhere in the United States dies as a result of it. In the late 1960s, more than 65,000 women developed breast cancer and 33,000 died. In 1984, an estimated 118,000 new cases were diagnosed, and the death rate remained essentially unchanged. With grim monotony, the incidence of breast cancer continues to increase while the same mortality rate per-

sists. Statistically, half the women found to have breast cancer this year will not be alive after 10 years. The various forms of treatment, the billions of dollars poured into research to find a cause, a control, a cure—none of these has thus far been able to significantly extend the lives of women with breast cancer.

With these scary statistics, is there any reason for optimism in the battle against breast cancer? Although it cannot be considered a silver lining, there are some encouraging aspects that have emerged during the campaign. First of all, the increased incidence and the unchanged death rate could mean the survival rate is increasing—a lukewarm victory, but heartening nonetheless. Besides, the overall survival rates mean virtually nothing for an individual woman. Not all breast cancer is the same, and there can be many variables, such as the virulence of the cancer or the resistance of the woman, that influence the course of the disease. Secondly, more localized breast cancer, caught before spread, is being found today than ever before because of increased public awareness and improved technological methods of detection. It is becoming evident that if it is detected early enough and treated promptly, the diagnosis of breast cancer is not an automatic death sentence. Early detection, the discovery of the cancer at a very early stage, before the disease has spread or when the spread is minimal, is the only approach that holds any promise at all of reducing mortality. Women themselves, if they practice monthly self-examination, can come to know their own breasts and develop more sensitive fingers than anyone else. They can detect a small change, an irregularity that could be serious, better than their doctors can. Even though only 9% of all women—and no more than that—will ever develop breast cancer, that presents enough of a lifetime risk to make it essential for all women to practice monthly breast self-examination.

If more were known about the causes of breast cancer, more efforts could be directed towards its prevention. There is no way to explain why some women do and other women do not develop the disease. As yet, there has only been the identification of some characteristic risk factors that appear to be associated with an increased frequency of breast cancer. Women who fall into those categories of high risk should, in addition to performing monthly self-examination, have more frequent clinical examinations by a physician that include other methods of diagnosis, such as routine low-dose mammography, or X-rays, of their breasts.

RISK FACTORS AND CAUSES OF BREAST CANCER

Breast cancer is rarely found in girls younger than age 15, and only 1%–3% of all mammary cancers occur in women younger than age 30. At 45, incidence increases and continues to rise rapidly with age. Although breast cancer occurs in men, it is 100 times more frequent in women.

Having had cancer in one breast increases the risk of developing it in the other breast by five to seven times. Benign fibrocystic disease may increase the risk 2.5 times beyond that of women with no evidence of the disease, but the issue is contentious. A history of a previous breast biopsy is a more recognized risk factor. A woman with a family history of breast cancer in a first-degree relative—that is, in a mother or a sister—has a doubled risk of getting breast cancer herself. If she has two affected sisters, her risk before age 40 is increased ninefold, compared with women of a similar age; if her sister and her mother both have breast cancer, her risk before age 40 becomes almost 50 times as great as other women her age. Even with affected relatives,

however, her own risk substantially decreases as she gets older, and by ages 60–79 her risk is the same as in the general population.

There are other, relatively slight influences on the occurrence of breast cancer that have to do with the reproductive life of a woman. Women whose menarche occurred before age 12 have a slightly increased risk over women who were age 15 or older at menarche, but the relative risk does not increase further with decreasing age at menarche. A late menopause also increases the risk. But if a woman has had an ovariectomy (and thus a surgical menopause) before age 35, the risk is lessened.

Never having had a baby puts a woman at greater risk. Pregnancy is protective, but the evidence presently available suggests that the degree of protection is associated with the age of the woman at the time of the first full-term delivery. The younger she is when she has her first child, the lower her risk for breast cancer. Having five or more children also decreases the risk, but pregnancies interrupted by miscarriage or abortion offer no protection. The risk of breast cancer actually increases with age at first delivery. If a woman has her first baby after 35, her likelihood of developing breast cancer is greater than if she had never had children at all.

For years, it was believed that breastfeeding protected against breast cancer. It seemed logical to correlate the increasing incidence of breast cancer with the greater numbers of women who were not nursing their babies. After many studies, conducted both in the United States and internationally, it has now been concluded that lactation has little effect either way.

There has also been the discovery that virus particles very similar to those of mouse mammary tumor virus were present in breast milk in both normal women and women with breast cancer. This led to the hypothesis that there was a viral agent responsible for breast cancer that could be transmitted in human breast milk. There has been no evidence, however of any association between being breastfed and the subsequent development of breast cancer. The incidence of breast cancer is the same in women whether or not they were breastfed as infants.

The relationships between such factors as the length of menstrual activity and childbearing and the development of breast cancer suggest that hormones are involved, but how and when they influence the disease are still only speculated. Estrogens are known to have carcinogenic potential in mice, but little is definitely known concerning the significance of a womans's own estrogens and ovarian activity relative to her development of breast cancer. During their reproductive lives, women produce estradiol, estrone, and estriol. Estradiol is the strongest of the estrogens in biological activity, but it has less carcinogenic potential in animals than estrone. Estriol is not only the weakest estrogen, but it has very little or no carcinogenic activity in mice. On this basis, some investigators have credited an increased risk of breast cancer to estradiol and estrone, whereas estriol is believed to have a protective effect.

Lemon (1973, 1980) theorized that estriol can block the carcinogenic action of estradiol and estrone on the breast and that a factor in breast cancer may be a disorder in the estriol to estrone/estradiol ratio. During pregnancy, the production of all three estrogens is greatly increased, but estradiol and estrone increase over pregnancy levels by only a hundredfold; estriol increases a thousandfold. On the possibility that it is the estriol of pregnancy that provides the protection of pregnancy against cancer, Lemon measured and found a subnormal urinary estriol excretion compared with estradiol and estrone in breast cancer patients, and speculated that such women have a genetically impaired ability to metabolize estradiol and estrone to estriol. But subsequent studies did not confirm that urinary estriol was lower in women with breast cancer, and, as pointed out by Vorheer (1980), many researchers have reported that urinary estriol was *increased* in breast cancer patients. The current view tends to discount a protective role for estriol.

More attention is now being directed toward the possible roles of (1) long-term exposure to normal levels of estrogen and progesterone or (2) the possibility of excess estrogen stimulation in the absence of cyclic progesterone being a factor. Women who have a later natural menopause (a risk factor) and an earlier menarche (another risk factor) are exposed to their own normal estrogens and progesterone for a longer period of their lives than women with an earlier menopause or later menarche. Lack of ovulation, with no corpus luteum formation and thus no progesterone secretion, has also been associated with an enhanced risk. Several studies have shown an increased risk of breast cancer in women with anovulatory cycles (Cowan et al., 1981; Gonzalez, 1983) and suggest that excessive estrogenic stimulation of the breast unopposed by progesterone can influence the development of the disease.

Different hormones than the sex steroids have also been suggested as factors in the development and progression of breast cancer. Other hormones that have been implicated include prolactin, thyroid hormone, or gonadotropins. Bulbrook and colleagues suggested that excreting abnormal amounts of androgen metabolites may precede the development of the disease (1971). Although there is little doubt that endocrine glands and their secretions are associated with breast cancer, just which hormones, in what combinations, and how and when they are involved still remain to be determined. When the answers are known, when the hypotheses become realities, it may be possible to measure hormone levels in order to identify women at higher or lower risk for the disease.

Other factors besides hormones appear to affect the development of breast cancer. High-fat diets, and obesity in particular, have been correlated with an increased risk of breast cancer. The relationship between obesity and breast cancer may actually have a hormonal basis, however. After menopause, it is known that estrogens are produced by conversion from androstenedione, a product of the adrenal cortex. This conversion takes place primarily in adipose tissue and, thus, more estrone is produced in obese women. Moreover, premenopausal obese women also have higher blood levels of estrogen and frequently have irregular menstrual periods and anovulatory cycles. As indicated above, anovulation may predispose to breast cancer. A further link between diet and estrogen metabolism is suggested by studies that compare vegetarian women with nonvegetarian women. For example, Goldin and co-workers (1982) found that vegetarian women had an increased fecal excretion of estrogen and a corresponding decreased level of estrogen in the blood, possibly due to the increased fiber in their diets. The amount of fat and fiber eaten could be a factor explaining the high risk of breast cancer in Western countries where a high fat and animal protein diet is consumed and the much lower risk of breast cancer in Third World countries where women eat vegetarian or semi-vegetarian low-protein, high-fiber diets.

There are some mysterious geographic and ethnic differences in the worldwide incidence of breast cancer. Incidence rates are five to six times higher in North America and Europe than in Asia and Africa, and Caucasians have a much higher rate than Orientals. Eskimo and Yemenite women almost never get breast cancer; the incidence is also very low in Japan and in the Chinese population of Singapore. But if Orientals migrate to the West, their incidence rate then becomes that of the local population in a few generations. One investigator estimated that when the Japanese dietary fat intake in Japan approaches the level of ingestion of fat in this country, the Japanese breast cancer death rate will increase to the present United States rate (Hirayama, 1978).

In contrast, incidence and mortality is very high in western Europe, particularly in the Netherlands; in the United States, it is highest of all in Connecticut, Nevada, and in the Alameda County and the San Francisco Bay areas in California. Although the curious cultural and sociological factors that appear to be involved in the incidence of breast cancer must be acting through some biological mechanisms, they have not as yet been identified. What underlies the relationship of breast cancer to diet, or to stress, or to lifestyle? There are no answers, but it is known that breast cancer occurs more frequently in affluent women than in poor women. In one group of Indian women, the Parsis of Bombay, the incidence of breast cancer is three times what it is in the Hindu, Muslim, Christian, and Jewish women in India. Parsi women are wealthier and better educated that the rest of the Indian population. Jewish women in New York City have more breast cancer than the average for the rest of New York and the United States. Even more strangely, the rate is higher for Jewish women in New York who have gone to college!

There is no reason to be overly concerned about risk factors; almost everyone is bound to have one or more of them. Even having all of them does not mean a woman is going to get breast cancer. It means only that her risk is greater than that of other women without the factors. But even with none of the risk factors, a woman still has a greater chance of developing breast cancer if she lives in the United States, and the reasons are unknown. Some American women, however, do have a statistically lower risk of developing breast cancer by the age of 75. These include, in decreasing order of risk, black women (one in

14); Chinese-American women (one in 16); Japanese-American women (one in 19); New Mexican-Hispanic women, (one in 21); and New Mexican-American Indian women (one in 40).

When certain groups, such as the Parsis, have a greatly increased incidence, or the American Indians have a much lesser incidence, it would appear that a genetic predisposition or innate resistance to the development of breast cancer may be operating. And when the incidence of breast cancer is higher in women relatives on both sides of a family, it would seem obvious that a genetic basis for the disease exists. That breast cancer is inherited, however, has been difficult to confirm as yet. Since other risk factors that include type of nutrition or age at first birth are also intertwined with a woman's background and heritage, it has been hard to separate heredity from culture. A cancer-prone or cancer-susceptible gene is suspected but has not been identified, and the inability to find any kind of reliable marker in the blood or cells of affected family members has not helped. It is commonly accepted that breast cancer can be *familial*—a kind of hedging word for hereditary. In the last decade, however, there has been accumulating evidence suggesting that some primary genetic factor may be operating when breast cancer clusters in families and that there may be an autosomal-dominant mode of transmission.

Even establishing beyond a doubt that hormones and/or heredity cause breast cancer is of limited value in preventing the disease since not much can be done to change those factors. In practical terms, there is much greater potential for prevention of breast cancer in knowing how nutrition and dietary fat intake are related to the development of breast cancer. There are epidemiological data that strongly correlate the geographical incidence of breast cancer to differences in fat intake, and ethnic or religious incidence to

differences in dietary practices. For example, Japan has a low breast cancer rate and a low average fat intake, and the United States has a high breast cancer rate and an average diet in which 40% of the calories are derived from fat. In addition, there are substantial laboratory data to show that a high-fat diet fed to rats and mice will produce mammary gland tumors.

A number of studies that compared the reported past dietary habits of breast cancer patients with those of matched healthy control women have provided data consistent with the hypothesis that increased consumption of fats increases the relative risk of breast cancer (Phillips, 1975; Miller et al., 1978; Lubin et al., 1981). The inherent weakness of such studies is that the current diet may not be the same as the diet during the lengthy period during which the breast cancer cells were becoming malignant, and most women would have difficulty in recalling how and what they ate 20–30 years ago. An investigation by Graham et al. (1982) of the dietary practices and food consumption habits of 2,024 breast cancer patients and 1,463 controls between 1952 and 1965 found no differences in fat intake between them, but, as one of the co-authors points out (Mettlin, 1984), both fat intake and the use of vegetable fats have markedly increased in the general population since 1957, and the same study today might produce far different results.

Most of the research attention has been focused on how nutrition and fat consumption may initiate or promote breast cancer, but there is recent interest in the possibility that natural inhibitors of breast cancer may also occur in the diet. Although the 1982 Graham study found no evidence of increased risk with fat intake, it did identify a small subgroup of older women in which a decreased consumption of foods containing vitamin A was related to an increased risk of breast cancer. This discovery supported earlier studies by the same group of epidemiologists that found decreasing levels of vitamin A in the diet to be associated with increasing risk of cancers of the respiratory and digestive tracts. There is also evidence, from mostly animal and some human studies, that the risk of cancer may be lessened when foods rich in selenium, vitamin C, and the cruciferous vegetables (brussels sprouts, cabbage, turnips, broccoli, and cauliflower) that contain carcinogenic inhibitors are consumed.

In *The Etiology of Human Breast Cancer*, published in 1974, Papaioannou considered all of the available evidence at the time and set up, in the order of their importance, the possible factors that interact to contribute to the development of breast cancer. He cited genetic predisposition or innate resistance, viruses, hormones, psychogenic stress, dietary fats, environmental carcinogens, and as yet unknown factors. He also emphasized the cardinal importance of immunological mechanisms that permit or mediate the action of all the other causative factors. A decade later, there has been no compelling evidence to indicate that the factors or their sequence in this list should be changed.

Some of the research in progress on the causes of breast cancer is highly promising, and eventually some of the questions will be answered. Currently, there are only suggestions and speculations and no definitive information about its etiology. Reducing fat consumption in the diet is prudent and sensible, not only as possible prevention of breast and other cancers, but to lessen the risk of the other killer, cardiovascular disease. The only other known method of preventing breast cancer is to bear a child at an early age and then have five more—hardly feasible as a protective measure! Until other means of prevention are known, every woman must recognize that while she may never develop breast cancer, she certainly should place herself in the best possible position for survival should it

happen to her. The greatest chance for cure of breast cancer lies in catching it early, and monthly breast self-examination is the best non-technological way of early detection.

Detection and Diagnosis

Self-Examination

Breast self-examination, or BSE, is not a difficult technique, and every woman can become as competent, if not more expert, than her physician. There is no need for a woman to doubt her ability to recognize a lump, if it is present. The doctor examines many breasts, but a woman is concerned only with her own two breasts and can become very familiar with what is normal for her. As she continues to practice monthly BSE she becomes so confident about the topography, the feel, and the appearance of her breasts that she is able to detect the smallest nuance of irregularity. She can easily perceive something that was not there before, the extraordinary from the ordinary.

The BSE method does not cost anything and is the simplest, quickest, most effective way women can take the responsibility for the well-being of a part of their bodies. Most things we can do as preventive health measures probably require some expense, giving something up, changing our diet, increasing our exercise, and are generally much more *involved*. But the painless procedure of BSE requires only a little time each month to provide the reassurance that everything is all right—certainly a more positive feeling than an ineffectual fear or anxiety about breast cancer. Even if symptoms should appear one month, it rarely means a malignancy. Eight out of 10 lumps are benign. If it is cancer, the chances are excellent that the woman who practices monthly BSE is not going to end up as a mortality statastic. More than likely, the earlier detection means that the cancer is still localized, and the earlier diagnosis and treatment mean cure and survival. What better motivation could a woman have?

When to Perform BSE

Between menarche and menopause, BSE should become a regular monthly health habit as soon as the menstrual period is over. At that time, levels of ovarian hormones are low, and the breasts are less likely to be painful and nodular. Postmenopausal women should select the first day of the month, the 15th of each month, their birthdates, or any other significant day that will help them to remember to perform self-examination. A premenopausal woman who has had a hysterectomy does not have a menstrual flow to remind her; if she still has an awareness of premenstrual breast tenderness, she should choose the day each month when those symptoms are no longer present.

The incidence of breast cancer is low until age 30, but the American Cancer Society has recently extended its educational efforts concerning BSE to girls of high-school age. They believe it is important to establish a lifetime routine of self-protection early. Many doctors agree, and they instruct their adolescent patients in self-examination when they come for a summer camp checkup or annual physical.

Technique of BSE

A woman may want to modify the *sequence* of the procedures to be described, depending on her schedule, her privacy, or her inclinations, as long as the three steps, feeling the breast while wet, inspection in front of a mirror, and palpation while lying down, are performed. "Self-examination" can also be performed for a woman by another person as an expression of love and concern. One preferred routine should be chosen and followed regularly, however, to establish total familiarity with the physical characteristics of the breasts. Only in this way can any irregularity be easily identified.

Examination While Wet

While taking a shower or a bath, move gently over every part of the wet breasts with a soapy hand. Raise the left arm behind the head, and with the flat fingers of the relaxed right hand, feel the entire left breast, beginning at the outermost top and moving in decreasing concentric circles, finishing at the nipple. Move high into the armpit and all the way over onto the breast bone. Then repeat the procedure for the right breast, using the flat of the left hand and raising the right arm behind the head.

Inspection of Breasts

Before getting dressed, sit or stand in front of a mirror with the breasts exposed. There should be a good, strong light, and the mirror should be large enough to allow inspection of the breasts while facing it directly. With arms at the sides, look at the breasts for bulges, asymmetry, or areas of surface flattening. Many women have unequal-sized breasts; they may be surprised at that discovery if this is the first time they are carefully inspecting them. As long as the contours of the breasts are symmetrical, inequality in size is perfectly normal (see Fig. 7.4).

Continue the inspection, looking for any deviation of a nipple, which could mean that it was being pulled toward an underlying tumor. Check for retraction of the nipple, the areola, or the skin. Examine the breasts for redness, any sore or ulceration of the skin, and for edema of the skin, which can exaggerate the pores so the skin looks like orange peel. Make certain there has been no change from the previous examination.

Next, raise the arms high above the head in order to expose the sides and undersurfaces of the breasts and repeat the same observa-

Figure 7.4. Breast self-examination. Arms at the sides.

Figure 7.5. Arms raised above the head.

tions (see Fig. 7.5). Cancer can produce retraction of the skin, manifested by anything from a small skin dimple or puckering to shrinkage of the entire breast. Raising the arms moves the muscles under the breast and may cause skin retraction or nipple deviation that may not have been apparent before.

Then, place the palms on the hips and press down in order to contract the pectoral muscles (Fig. 7.6). Again, changes that are not visible in other positions may become apparent. Bend forward to let the breasts hang free and check for anything unusual (see Fig. 7.7).

Lying Down

The rest of BSE consists of palpation of the breasts performed while lying down on a bed or a couch. To examine the right breast, a small pillow or a folded bath towel should be placed under the right shoulder. In a woman with anything but very small breasts, this maneuver flattens and spreads out the breast evenly over the chest wall (see Fig. 7.8). The right hand is placed behind the neck; this also helps to flatten the breast into a thin layer. otherwise palpation might be more difficult because of the lateral thickness of the breast.

With the left hand, fingers flat, palpate

Figure 7.6. Contraction of the pectoral muscles.

Figure 7.7. Breasts hanging forward.

gently in small circular motions around the entire breast as if it were an imaginary clock face (see Fig. 7.9). Beginning at the upper outermost top of the right breast for 12 o'clock, move to 1 o'clock, 2, 3, and so forth, all around the outermost part of the breast back to 12 o'clock. A ridge of dense nodular tissue at the lower curve of each breast is the *inframammary fold* and is quite normal (see Fig. 7.10). The ridge is frequently tender. Then move the fingers in about an inch toward the nipple, and repeat the circling. It may take three or more circles to palpate every part of the breast *including the nipple.*

The nipple is then gently "milked" between the thumb and forefinger. Unless a woman is pregnant, any discharge, clear or bloody, is suspicious, but a slight scaling on the nipple associated with the menstrual period is not unusual.

The same clock-circling procedure is repeated on the left breast with the right hand, a pillow under the left shoulder, and the left hand behind the neck.

What the Breast is Palpated For

The palpation is done to discover (1) a discrete new lump or mass of any kind; (2) an area of nodularity or thickening that was not

Figure 7.8. Breast self-examination while lying down. A small folded towel is under the shoulder of the side to be examined.

Figure 7.9. The breast is palpated in a circular clockwise direction until every part has been checked.

Figure 7.10. Location of the inframammary ridge. It is more prominent in some women.

present before; (3) any area of tenderness. Essentially, a woman is looking for a change in consistency of the breasts from the previous month's examination.

A good time to start practicing BSE is after a breast examination by the woman's physician, when one is not as likely to be as concerned about the dismaying variations in consistency that one may find. A professional examination is reassuring that all those little irregularities in the breasts are normal. As long as those nodules and thickenings remain the same size and are diffusely scattered through both breasts, there is nothing to worry about. It is the presence of a single dominant mass that was not there before that should arouse suspicion.

If a lump or skin dimple or nipple discharge is discovered during BSE, there is no reason to panic, but there is also no reason to delay in calling for an appointment to see a physician. These findings do not necessarily mean can-

cer. Actually, most of them turn out *not* to be cancer, but a woman should not put herself through the needless anxiety and sleepless nights worrying about what it *might be* when an immediate visit to her doctor can tell her what it *is*.

Mammography

The smallest mass or lump that is capable of being felt by palpation during BSE or clinical examination by a doctor is about 1 centimeter in diameter. At that small size, a cancer may have been present in the breast for many months or even for several years. It is thought that breast cancers are capable of metastasizing to the lymph nodes or to other organs very early in their development. Once this dissemination has occurred, a complete cure becomes virtually impossible, and efforts can only be directed at controlling the disease. But if the cancer can be detected before there are any clinical signs and symptoms and when there is very little probability of spread, if it can be found in an *extremely* early stage, the cure rate is greater than 90% and the death rate can be significantly reduced. Such minimal cancers can often be detected by mammography, or X-rays of the breasts (see Fig. 7.11).

The use of X-rays to identify breast diseases has been known for many years, but the technique has become widely used as a routine examination for breast cancer only since the late 1950s, when it was standardized by radi-

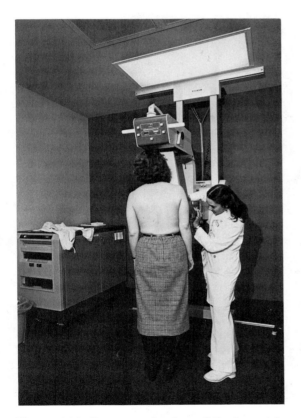

Figure 7.11. X-ray examination of the breast by xeroradiography (xeromammography) for early detection of breast cancer. The most recent procedures utilize low-dose techniques that minimize radiation exposure to the breast.

ologist Robert Egan. In mammography, a woman's breasts are alternatively placed on a metal plate and two X-rays of each, from the top and from the side, are taken and developed on photographic film. There are various radiographic abnormalities, such as tiny spots of calcification or areas of increased density and change in breast patterns, which a trained radiologist may recognize as very early cancer. The procedure is not 100% accurate, and much depends on the skill and experience of the interpreter of the films. Generally, the accuracy of mammography is greater in older women, whose breasts are not as glandular and dense.

Xeroradiography is a variation of mammography. A standard X-ray generator is used as the radiation source, but the image is recorded xerographically on paper rather than photographically on film. Some radiologists prefer the resolution with the xerox image, others like film better. Figure 7.12 is a photograph of the xeromammographic image of the breast of a woman with breast cancer.

Mammography has come to be recognized as a major diagnostic tool in the early detection of breast cancer and as a valuable adjunct in the treatment of breast cancer. Mammograms can verify a doubtful diagnosis when a mass is discovered by palpation, can determine the exact location of a mass for biopsy or for surgery, and have been shown to play a major role in the reduction of mortality by detecting a substantial proportion of breast cancers, perhaps as many as 45%, which are not as yet apparent by palpation. A 1963 study conducted by the Health Insurance Plan of Greater New York (HIP) had unequivocally shown that mass screening with mammography and physical examination of 31,000 asymptomatic women between ages 40 and 64 resulted in increased cancer detection and decreased mortality in women 50 years of age or older when compared with a matched control group. The study, however, did not show

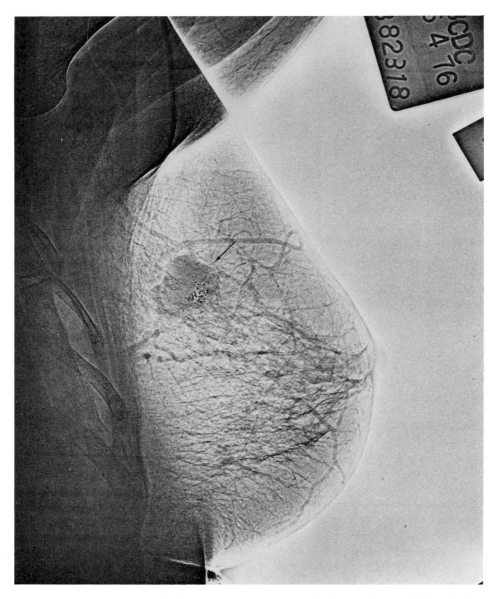

Figure 7.12. A photograph of a xeromammogram of a breast in mediolateral view. There is an obvious carcinoma (arrow) in the upper inner quadrant. The poorly defined mass in itself would be suspicious of cancer; the calcifications (black dots) are characteristic and verify the diagnosis. The large vessels are subcutaneous veins (Courtesy, Dr. John Milbrath, Breast Diagnostic Clinic, S.C., Milwaukee).

a similar decrease in mortality for women younger than age 50. In one-third of the women, the cancers were found only by mammography. In two-fifths of the women, however, the cancer was found on palpation alone, indicating that neither method is perfect, but the combination of mammography and physical examination can be of great value in finding women with cancer.

On the basis of the HIP study, the American Cancer Society and the National Cancer Institute (ACS-NCI) jointly launched a 10-year, nationwide Breast Cancer Detection Demonstration Project (BCDDP). By the mid-1970s, there were 280,000 women between the ages of 35 and 74 who were participating in the program at 28 geographically distributed detection centers. Along with personal interviews and health history, physical examination, and thermography (a means of recording heat patterns of the breasts), the volunteers were receiving five annual mammograms. They were to remain in the program for 5 additional years of follow-up study.

The initial results of the project appeared very encouraging. The cancers that were detected were considered minimal, and in nearly 80% of the women, there was no evidence of spread to other parts of the body. There began to be some misgivings, however, concerning the risk of annual X-rays of the breast. Although there was no doubt that the value of mammography for women aged 50 or older justified the possible hazard, in younger women it was claimed that repeated mammography was likely to cause as many deaths from breast cancer as it would prevent (Bailer, 1975; Bross, 1976). X-rays are ionizing radiation, a form of energy that is able to damage cells in animals and cell cultures. As shown by follow-up studies on women who had been treated for acute breast infections with X-rays and on women who received many chest fluoroscopies during treatment for tuberculosis, radiation can cause breast cancer

years after the original exposure. The incidence of breast cancer among women survivors of the bombing of Hiroshima and Nagasaki has also shown that the risk of developing the disease increases with the degree of exposure to radiation. Present-day mammography, however, subjects women to five hundred-fold lower levels of radiation than those reported to have caused breast cancer in the cited studies. It has been estimated that at the *highest* amount of exposure presently being used in mammograms, 1 rad (for radiation-absorbed-dose), a woman of age 35 would have to have consecutive annual irradiations until the age of 70 in order to increase her risk of developing breast cancer to 12% from her present 9% risk with no X-rays. Nevertheless, the carcinogenic potential of mammography is theoretically cumulative, and as yet there are no studies that have been performed to evaluate the effect of exposure at lower doses.

In the late 1970s, the use of mammography in women younger than age 50 became highly controversial, and the publicity that surrounded the medical debate created much misunderstanding and fear in many women who should be employing this diagnostic technique. Some women began to be reluctant about having any X-rays at all, even to detect gallstones, ulcers, tooth decay, or any other diseases totally unrelated to the breasts. Many doctors, particularly those associated with the BCDDP projects, defended the use of mammograms in younger women. Approximately half the women screened at the detection centers were between ages 35 and 50, and it was claimed that early detection was worth the minimal risk, since some cancers were picked up by mammography that were not detectable any other way. Further criticism was directed at the projects, however, after an unanticipated finding was revealed in late 1977. In the 1,850 women in whom breast cancer had been detected solely by mammography, 66 turned out not to have a malignancy, or at least the diagnosis was debatable, when their breast tissue was later reexamined by a team of pathologists. By then, of course, most of them had had their breasts removed.

Even the value of early detection was questioned. It was asked whether it is necessary to treat by surgery and follow-up a cancer so tiny that it might not cause a problem for years—or ever, for that matter. No one really knows very much about the early development of cancer. It could be possible that many people walk around all their lives with unknown minimal malignancies in various organs of their bodies. Under the perhaps reasonable assumption that "don't both it until it becomes big enough to bother you," it was suggested that treatment may actually be a disservice (Culliton, 1977).

Because of the mammography controversy, the NCI commissioned four studies of its breast-cancer screening program. On the basis of those investigations, new guidelines and limitations for the use of mammograms in asymptomatic, apparently healthy women younger than age 50 have been issued. The recommendations, released in September of 1977, were further assessed in 1978 by another panel of experts appointed by the National Institutes of Health and again revised in 1983 by the American Cancer Society. The consensus that has emerged is as follows:

Because younger women live longer and have more time to be exposed to the cumulative effects of radiation, no X-rays of the breasts are recommended for those younger than age 25. They are to be used between the ages of 25 and 34 only if there is a diagnostic problem, and never should be done periodically.

Between the ages of 35 and 40, a woman should have a mammogram with her annual physical examination at least once to establish a baseline breast X-ray. Further mammography should only occur when deemed necessary for diagnosis. Annual mammograms should be restricted to women who have already had breast cancer.

For women aged 40 to 49, a mammogram should be performed every 1 or 2 years. Annual mammograms would be for the woman at greater risk for the development of breast cancer: having had a previous mastectomy or with a mother or sister with the disease; either nulliparous or having had a first child after age 30; early menarche; a history of benign tumors; any lumps; nipple discharge; and severe pain in the breasts.

As a woman approaches age 50, is in her fifties, or is older, she should have annual screening for breast cancer, including a mammogram. *There has never been any doubt that at this age the potential benefits of mammography far outweigh any possible risk from radiation.*

All women, however, should make certain that the mammographic equipment being used on her is technologically state-of-the-art. It should be a "dedicated" mammographic unit that employs new, low-dosage techniques. One such improvement is the so-called screen-film system in which an X-ray intensifying screen is used in combination with a double-emulsion film. Substantially reduced exposure to all parts of the breast—about 0.06 rads for a two view examination—is possible. The screen-film technique requires that the breasts be firmly compressed; the tight squeeze during the procedure may produce momentary discomfort.

Thermography

Thermography is a harmless, inexpensive technique that can be used for the detection of breast cancer—a method that involves no risk at all to the woman. A thermographic examination involves transferring the infrared heat patterns of the skin of the breast to a visual display that can be photographed, producing a permanent record of the mammary heat pattern. Much like a fingerprint, each woman has her own distinctive thermal print. Since the skin temperature over a cancer is higher than that over normal tissue, it should be possible to use thermography in screening and diagnosis of breast cancer. Unfortunately, a positive thermogram is not specific for cancer. There are a variety of other nonmalignant reasons for an increase in heat, and they may be the cause of false-positive thermograms. Conversely, some cancers may not cause an increased temperature in the skin overlying them, and false-negative results are also possible.

It is generally agreed that thermography alone is far less effective than mammography or clinical examination for the detection of breast cancer and is useful only in conjunction with them.

Ultrasound

In the search for a safe, no-risk method of breast cancer detection, ultrasonic examination—the use of sonar to produce echo images of breast tissue—has been tried as a substitute for mammography.

In the breast ultrasound unit, the breasts are immersed in water in order to conduct the high-frequency sound waves. As the reflected waves bounce off the various fatty and fibrous tissues of the breast, they are converted electronically for display as images (Figs. 7.13 and 7.14). Advantages of ultrasonography over mammography include the ability to visualize a lump in the young, very dense breast and the ability to distinguish between a simple fluid-filled benign cyst and a solid tumorous lump—both are difficult with mammography. The disadvantage of ultrasonography as a screening tool is its inability to find subclinical cancers because the small-est mass detectable by the technique is still too large to be considered a minimal cancer. Some experts have demonstrated masses as small as 2 mm, but in general, ultrasound is not able to find a small and curable cancer. Future improvements in equipment may make ultrasonograms more useful as a screening method, as researchers in ultrasound claim that the technique could theoretically pick up very tiny lesions.

Other Techniques of Early Detection

In diaphanography, the breast is examined after being transilluminated with a high intensity light beam, and color photographs record the visualization. Breasts vary in translucency, depending on their size, amount of fat and fibrous tissue, and blood supply. Any benign cysts or neoplastic lumps, hemorrage, or inflammation in the breasts will also affect light absorption and change the appearance of the transilluminated breast.

When certain radioactive elements are injected into the body, they preferentially localize in an area of high metabolic activity, such as a tumor. In scintigraphy, such radionuclides are administered, and then the tissue is scanned with a scintillation counter to visualize greater radioactive emissions appearing in one area, as opposed to another. While bone, brain, and liver scans are effective in the determination of metastases from a primary breast cancer, breast scintigraphy used for early detection has not yet proved to be worthwhile. Moreover, no instrument that uses any other kind of radiation—light, sound, or heat—has, as yet, been found to compare with the efficacy of the mammogram.

Also in a preliminary stage is the use of biochemical markers as detection measures. It has been noted that blood serum levels of cer-

Figure 7.13. Breast ultrasound machine. The technician prepares for the ultrasono-gram by filling the tub and getting the bubbles out of the water (Courtesy, St. Francis Hospital, Milwaukee).

tain substances are elevated in breast cancer. With further investigation, such a biological assay may turn out to be the simplest, most sensitive, least expensive screening test for early breast cancer, but to date, no specific biomarker has been found. Right now there is no completely safe, economically feasible or logistically practical way to screen all women in the country for breast cancer. Even if there were, there are probably not enough medical professionals in the country who are expert enough to accurately interpret the results. The only hope of reducing the annual mortality from this disease is through monthly breast self-examination. Early diagnosis means early treatment; early treatment increases the chances of cure.

TREATMENT OF BREAST CANCER

Surgery

From all historical indications, cancer has been a human affliction probably for as long as we have been human. Breast cancer, being easily accessible, has always been treated by removal. Along with prayer, poultices, and potions, the primary approach in the treatment of breast cancer has been to cut it out, usually damaging or destroying much of the breast in the process. Surgery on the breast, in various forms and techniques, and with heroic endurance on the part of the patient, has been performed in an attempt to cure breast cancer for centuries.

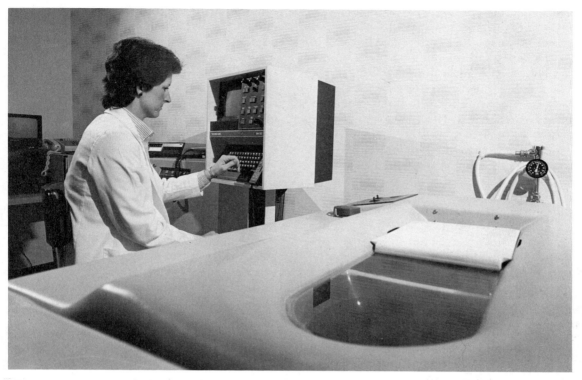

Figure 7.14. During the ultrasound procedure, the woman being screened will lie prone on the examining table with her breasts immersed in the water. The image will be displayed on the screen and photographed (Courtesy, St. Francis Hospital, Milwaukee).

Eighty years ago, a formalized surgical procedure that has come to be known as the Halsted radical mastectomy was described by Dr. William S. Halsted. He reported that with his operation, he was able to cure 50% of women with breast cancer; very few had ever been cured before his kind of surgery. Halsted's technique was enthusiastically adopted, and it persisted all these years with little challenge until the past decade. The operation removes the entire breast, a liberal amount of skin, all the subcutaneous fat overlying the breast tissue, the major muscles of the chest wall (pectoralis major and minor), and all the axillary lymph nodes and fat in the armpit. Complications as a result of the sur-

gery may include damage to the brachial plexus or network of nerves that supplies the upper arm, forearm, and hand. Another, not infrequent, functional defect is some limitation of shoulder motion; some women have difficulty in raising their arms on the affected side to even shoulder height. The anatomical deformity on the chest area can at least be covered by a breast prosthesis and clothing, but a disfiguring and highly visible result of radical mastectomy is often the enlargement or edema of the upper arm that occurs because of the interference with the lymphatic circulation.

The Halsted radical mastectomy became the traditional treatment of choice for breast cancer in this country. It was the preferred method of the great majority of American surgeons, the procedure with which they felt the most secure, and the one that many believed has consistently proven to be the most effective treatment. Even the preoperative procedure became routine. In the one-step procedure, a woman with a suspected breast cancer entered the hospital. There, under general anesthesia, the surgeon performed a biopsy on the tissue. While the woman remained asleep on the operating table, the specimen was rushed to the pathologist in the lab. A frozen section was examined, and if the pathologist said it was malignant, a Halsted radical mastectomy was performed. The woman had been made aware before biopsy that when she awoke, her breast may or may not be there. This routine was the same for everyone—as much for the woman down the block as for a President's wife, Betty Ford.

The Halsted radical remained the standard primary therapy for operable breast cancer for 80 years. Statistically, the number of women who were still living 5 years after surgery with no recurrence of cancer (in medical terms, the disease-free survival rate) was 75% if the cancer was in an early stage and 50% if the cancer was designated more advanced. The 10-year disease-free survival rate was about 70% for women with early cancers and around 40% for women with breast cancers designated as Stage II.

The Halsted radical is designed to remove the cancer mass and the local areas to which it already has or is more likely to spread. The principle on which it is based is consistent with the concept of cancer that prevailed in Halsted's day, and which is still accepted by many—that breast cancer is a rather orderly disease that starts at a primary site in the breast, remains there for a certain time, and then spreads through the lymphatic system to the axillary lymph nodes first. The nodes act as barriers for the cancer cells for a while,

but when the primary cancer gets large enough, the regional nodes can no longer contain the cancer. The cells metastasize themselves to the bone, liver, brain, and other distant sites.

It now appears evident that breast cancer is not just one disease. It probably has a variety of causes, appears in a variety of forms, and varies enormously in its progression and clinical course. The model just cited, while true for some forms of breast cancer, may not be valid for all types. It is now believed that not all cancers progress first to the axillary lymph nodes before spreading. Some cells from certain malignant tumors are capable of bypassing the regional lymph nodes and of metastasizing directly to distant tissues while the original cancer is still very small. No evidence of cancer cells in the lymph nodes at the time of diagnosis of breast cancer may still mean the presence of micrometastases in other tissues. Dr. Bernard Fisher, at the University of Pittsburgh Medical School, suspects that most, if not all, women with breast cancer have disseminated disease by the time the cancer is palpable. He believes that although women are apparently cured by an operation like the Halsted radical mastectomy, it does not necessarily mean that the surgery removed every last cancer cell in the body. It is likely that surgically removing the burden of the growing large tumor may have left sufficiently few cells, which could then be eradicated by the woman's own immunological mechanisms. The newer concepts of tumor biology challenged the continued use of the Halsted radical mastectomy.

Actually, there began to be misgivings concerning the routine use of one surgical procedure for all breast cancer patients approximately 25 years ago. Since the Halsted procedure ignored the fact that one of the main routes of lymphatic drainage is through the internal mammary nodes, some surgeons advocated even more extensive surgery, evidently on the premise that getting every last cancer cell in every lymph node produces better results. The "extended radical mastectomy" combined the standard Halsted technique with the additional excision of the internal mammary nodes, and the "super-radical mastectomy" excised the supraclavicular nodes as well. The latter operation requires removal of the lateral border of the breastbone and several rib attachments. It proved to do little for the long-term survival of the women on whom it was performed, and it increased morbidity and mortality from the surgery itself.

About the same time, other surgeons began to perform less extensive procedures, either alone or together with postsurgical radiotherapy or chemotherapy. Several alternatives to the orthodox radical operation began to be used. The one that has virtually replaced the Halsted radical in the United States and many other countries is the modified radical mastectomy.

Modified Radical Mastectomy

This procedure is also called conservative radical mastectomy or mastectomy with axillary dissection. The entire breast and the axillary lymph nodes are removed, but the pectoralis major muscle is left in place. One type of modified radical (Auchincloss-Madden) also preserves the pectoralis minor muscle; another form of the procedure (Patey) cuts the pectoralis minor.

By permitting the pectoralis major muscle to remain, however, the normal contour of the chest wall and shoulder is maintained, there is less frequent edema and limitation of motion of the arm, and breast reconstruction is easier to perform.

In 1979, the National Cancer Institute accepted the modified radical mastectomy as the standard treatment for breast cancer, and the modified version, usually of the Patey type, gradually supplanted the Halsted radi-

cal that is now infrequently performed. The only randomized clinical trial that was set up to compare the Halsted radical directly with the type of modified radical operation that left the pectoralis minor in place reported its results in 1981 (Turner et al.). No difference in survival rates between the two groups of women was distinguishable.

Although the modified radical is now the treatment of choice, other, more breast-sparing procedures are beginning to be used. Usually these more conservative operations are performed in conjunction with some type of axillary node removal and followed by radiation therapy.

Total, Simple, or Complete Mastectomy

The entire breast, including the nipple and skin over the breast is removed, but the pectoral muscles and the axillary nodes are left in place. Some of the nodes may be dissected, however, and when total mastectomy plus even more axillary node dissection is performed, the procedure becomes a modified radical mastectomy. One of the many controversies in breast cancer treatment is whether the axillary node removal should be performed for therapeutic purposes or merely to determine the stage or extent of the cancer progression. Those believing in the staging, rather than prophylactic value, of axillary node dissection remove only the low- and midaxillary nodes.

In a subcutaneous mastectomy, all the breast tissue is removed, but the nipple and skin overlying the breast are preserved. This procedure makes it easy for an implant to be inserted should reconstruction be desired. Frequently, however, some cosmetic surgery must be performed so that the two breasts will match.

Segmental Mastectomy

In this procedure, the primary tumor, some of the surrounding breast tissue, and usually a small amount of overlying skin are removed.

Depending on the varying amounts of breast tissue that are removed with the tumor mass, the operation is also called partial mastectomy, sector mastectomy, extended tylectomy, or quadrantectomy. Most patients undergo some axillary node dissection.

Lumpectomy

This is the most conservative, and also the most controversial, procedure. The operation involves the simple excision of only the breast lump and a narrow margin of normal tissue. If the cancer is not very large, there is no breast deformity. One of the objections to this procedure is that it may leave other microscopic areas of cancer (multicentric foci) in the breast. Radiation therapy is believed to take care of any such potential areas. As generally practiced in the United States, women having this surgery also have some axillary node dissection. Within 1–3 weeks after surgery, supplementary radiation therapy over a 4- to 5-week period is given both to the breasts and to the lymph nodes. A subsequent "booster" dose of radiation, either by external irradiation or by implantation of a radioactive source, is then directed to the quadrant of the breast where the cancer was removed. Dr. George Crile, Jr., of the Cleveland Clinic, has been one of the most vocal proponents of lumpectomy and the somewhat inelegant term is said to have originated with him. He believes that women would be less likely to delay treatment for breast cancer if they knew that breast-conserving alternatives were available.

Surgical Clinical Trials

Few surgeons would currently argue that the Halsted radical is superior to the modified radical mastectomy. There is enough evidence to show that the two procedures are equal in terms of overall survival and disease-free survival. Treatment by the alternative breast-preservation methods has become highly controversial, however, and the focus of great

medical debate. Although the less extensive procedures have not been shown to improve the survival rate, neither have they been shown to decrease it. But there are studies in the medical literature that strongly suggest that results in terms of survival from breast cancer are very similar regardless of what method of surgery is used. In his all-inclusive monograph on the history and treatment of early breast cancer, Mansfield (1976) evaluated the results of various modes of therapy: the extended radical without postoperative irradiation; the Halsted radical and the modified radical with or without postoperative irradiation; and the simple mastectomy or the local excision, both with postoperative irradiation. He concluded that no one method had as yet been shown to be superior to any other, and advocated that it would be logical for a woman with early breast cancer to choose the operation that would be least mutilating and least cosmetically deforming. The key words, of course, are "early breast cancer." Few doctors would agree that conserving the greater part of the breast is possible for all women with breast cancer, but it has been difficult for many of them to accept the idea that the less extensive procedures are feasible for even some women with breast cancer. Many doctors have claimed that the studies reported in the literature have not been controlled and the the patients who have had the different kinds of treatment are actually not comparable.

Some of the more vocal critics of the medical profession have claimed that physicians' motivations for performing the radical in lieu of the more conservative procedures stem less from concern with a woman's future and freedom from cancer and more from other reasons that are not as benevolent. Since insurance companies pay more for more extensive operations, some have hinted that profit may play a role. Sexism could also be a motive; it has been suggested that surgeons may not be particularly concerned with conserving breast tissue, believing that a woman should be so grateful for her life that only a neurotic would worry about the loss of a breast.

But surely not all doctors, any more than all women, are stereotypes. Not all surgeons are venal, sexist, or lacking in compassion and awareness of the psychological trauma of breast cancer. There are surgeons who sincerely believe that the standard modified radical surgery is safer until it is proven that the alternatives are equally effective. A number of controlled, scientific, clinical trials have been initiated that challenge the traditional forms of breast cancer surgery and will ultimately provide the answers on the alternative, breast-preservation treatments.

In 1971, investigators at 34 institutions in the United States and Canada began participation in the National Surgical Adjuvant Breast Project (NSABP) protocol B-04, designed to answer several of the questions. Women who entered the trial with no evidence of cancer in the lymph nodes (that is, clinically negative nodes) were randomly assigned to treatment either with radical mastectomy, total mastectomy followed by radiation of the chest wall and regional lymphatics, or total mastectomy alone. Women with clinically positive (cancerous) nodes were randomized to treatment with either radical mastectomy or total mastectomy and irradiation. A 6-year follow-up of the 1,665 women in the study showed that women with negative nodes fared equally well under all three treatment alternatives. The women who had clinically positive nodes did not survive as well but showed little difference in disease-free survival under the two treatment alternatives. Disease-free survival rates after 10 years were not substantially different from those observed at 5 years (Fisher et al., 1984).

Dr. Bernard Fisher, who conducted the first trial B-04, initiated NSABP protocol B-06 in 1976. Again, a randomized study is comparing

three groups of women, all of whom had small breast cancers and negative nodes. They received either a modified radical mastectomy, a segmental mastectomy plus staging axillary node dissection and no postoperative irradiation, or segmental mastectomy plus staging axillary node dissection and postoperative irradiation for 5 weeks. Women who undergo segmental mastectomy and are found to have positive nodes also receive a 2-year course of chemotherapy. A similar study is taking place at the Cancer Institute in Milan under the direction of Veronese and associates. The trial, involving 701 women who entered between 1973 and 1980, is comparing radical mastectomy with quadrantectomy, axillary dissection, and radiotherapy treatment. As reported in 1981, there has been little statistical difference in survival, disease-free survival, or local recurrence of the tumor among the groups.

A National Cancer Institute trial is comparing total mastectomy with axillary dissection and lumpectomy with axillary dissection followed by external irradiation to the breast and a subsequent iridium implant. Women with positive nodes receive chemotherapy. The data from these trials are being accumulated and analyzed. It could be years, perhaps as many as 10, before the results of these and other controlled trials provide the definitive information concerning the effectiveness of the less extensive surgical procedures.

Clinical Staging

Despite the debate over the best primary treatment for breast cancer, one trend does seem to be developing: the general agreement that no one kind of operation or postoperative therapy should be routinely applied to every woman with breast cancer. The problem is, of course, to decide which operation and which course of treatment should be selected. The classification of the breast cancer and the criteria for treatment are called "staging." Every woman with a suspected malignant tumor in her breast should be clinically staged before mastectomy. The whole point of mastectomy is to remove the local disease. If there are already distant metastases, performing even a modified radical mastectomy would be senseless, and other forms of therapy could be used. A few days of assessment in the hospital before surgery can determine the tumor size, extent, growth rate, the likelihood of lymph node involvement, and by X-ray and other scanning techniques, whether or not the cancer has already traveled to the bone or other tissues. There is no need for a woman to be put to sleep not knowing whether the mass in her breast is benign or malignant or whether she will have a small incision or no breast at all when she wakes up. The one-stage procedure and its accompanying routine are pointlessly archaic. A biopsy can be performed under local or general anesthesia, and then the specimen can be thoroughly examined from a permanent stained section, where there is less chance of error, instead of quickly from a frozen section. A second opinion from another pathologist, or even a third opinion, if necessary, can be obtained. A delay of several days, except in very rare cases, will not make any difference to the cancer, but can be of great value to the patient. Certainly, waiting it out while the diagnostic process is being completed can be incredibly difficult for a woman, but it will also give her the opportunity to be completely informed, understand exactly what her situation is, and allow her to participate in the decision as to why the surgery she will have is chosen as the one most appropriate for her condition.

Although the advantages of the two-stage procedure seem obvious, there are still advocates of the one-stage procedure who argue that waiting means that two operations, with all the surgical risks of anesthesia, infection,

or hemorrhage, are then necessary. They also maintain that there is greater psychological stress for a woman if she has to wait several days to a week before the second surgery and that no one really knows whether a prolonged delay between the biopsy and the ultimate treatment decision is harmful.

The identification of the anatomical extent of the disease is called clinical staging and it provides a method of assessing the status of the cancer. In this method, T refers to the tumor, N means the regional lymph nodes, and M refers to the distant metastases. Subscripts to these letters indicate the stages of cancer progression in each. Thus, a woman with no breast cancer would be staged at T_0 N_0 M_0.

A woman is staged at Clinical Stage I, II, III, or IV depending on the following factors: size of the tumor (T_1–T_4); whether the axillary lymph nodes on the same side are palpable but considered negative (N_{1a}) or palpable but considered to have metastatic growth (N_{1b}), whether the lymph nodes or the tumor are fixed to the underlying pectoral fascia, muscles, or chest wall; whether there are any other grave signs, such as edema, ulceration, or pitting of the skin of the breast; and whether there are any distant metastases (M). The overall 10-year survival rate for a woman with Stage I breast cancer is 80%; for a woman with Stage II, it is about 50%. The outlook is not as good for women with Stage III or IV disease, although wide variation owing to individual differences in the woman and in the cancer occur within the clinical stages.

Cancer staging can get very complex, but as an example, clinical Stage I can include a woman who is T_2 N_{1a} M_0; that is, her tumor is larger than 2 cm but smaller than 5 cm in its greatest dimension, but the lymph nodes are still considered negative. If the lymph nodes are movable and palpable, are suspected of containing metastases but still not larger than

2 cm in size or fixed to the underlying connective tissue or any other structures, the woman is considered at Stage II.

There is no doubt that determining the stage of the cancer is a most important factor in determining the kind and extent of the treatment. The real dilemma, and probably the reason that many surgeons feel justified in always taking the entire breast and the axillary nodes immediately, is the difficulty in identifying true Stage IIs. A certain number of Stage I patients with lymph nodes *clinically* assessed as negative may actually be Stage II patients whose lymph nodes are, positive. There is really no certain way of knowing whether the lymph nodes are involved until they are removed, sectioned, stained, and examined under the microscope by a pathologist. This fact was illustrated in the NSABP B-04 study in which 39% of the 350 patients thought to have negative nodes on physical examination turned out to have positive nodes after surgery, and of the 280 patients believed to have clinically positive nodes on examination, 27% proved to have negative nodes at surgery (Carter, 1984). This is why node "sampling," the dissection of some low- and midaxillary nodes, now more and more frequently accompanies the most limited surgery.

Radiation Therapy

Strong doses of X-rays, usually delivered by a cobalt beam or a supervoltage X-ray machine, damage cells. The irradiation is aimed directly at the abnormally dividing cancer cells and attempts to destroy them without damaging the normal tissue surrounding the tumor (Fig. 7.15).

Radiation has been used as the sole (primary) method of treatment of breast cancer in an attempt to replace surgery. Little data are available from the United States, where sur-

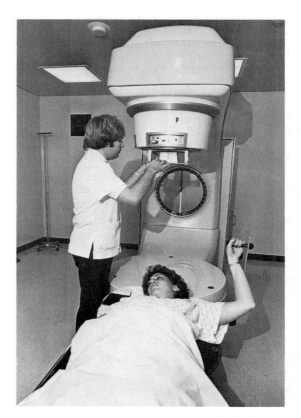

Figure 7.15. Simulation of radiation therapy. The Clinac 4MV linear accelerator delivers 4 million eV. The usual treatment dose to breast tissue is 25 rads.

gery dominates the treatment of breast cancer, but primarily evidence from other countries indicates that the survival rates following the use of only irradiation are similar to those after surgical treatment. As yet, information is too scanty and is without the benefit of years of follow-up to evaluate the role of irradiation as a replacement for mastectomy.

Strong radiation is also used to reduce the size of the tumor when it is so large that it is considered inoperable or to palliate (i.e., lessen) the pain in bone or other tissue metastases in advanced cancer. When it is used as an adjunct to surgery in early breast cancer, radiotherapy is used postoperatively to destroy cancer cells that may have been left in the breast or the lymph nodes after surgery. Although the use of postoperative irradiation has been shown to decrease the local recurrence of cancer in the breast, it has not been shown to increase the 5-year survival rate. There has also been the suggestion, as yet unproved, that postoperative radiotherapy is actually harmful: that it decreases the survival rate because it suppresses the immune response of the body against the cancer. Those who oppose radiation after surgery also cite complications, such as an increase in arm edema, the loss of skin on the chest wall, the difficulty of breast reconstruction afterward, and the carcinogenic potential of high-dose radiation. Doctors who favor postoperative radiotherapy maintain that it plays an important role in the treatment of breast cancer. They believe that reducing the incidence of local recurrence of the cancer is necessary even though there may not be an ultimate beneficial effect on survival. They claim that any complications are the result of poor irradiation technique and not the fault of the irradiation itself.

Although the use of postoperative irradiation is controversial, it has never become as acrimonious a medical debate as that which surrounds radical mastectomy and the lesser procedures. In actuality, the question of whether postoperative irradiation enhances survival is currently becoming irrelevant and is undergoing reevaluation within a new framework—whether chemotherapy, alone or in combination with radiation therapy, will provide better results in terms of overall or disease-free survival. There is a general agreement at present that irradiation has value in some cases but that it should not be used routinely. The trials currently in progress, which compare various surgical modes with and without radiotherapy, will provide additional information concerning its advantages and disadvantages.

Endocrine Therapy

Both surgery and irradiation are local treatments for breast cancer; they are directed at the primary cancer in the breast and at the axillary, internal mammary, and supraclavicular nodes. Neither breast surgery nor radiotherapy is able to affect any tumor cells circulating through the blood or lymph or any groups of cancer cells that have taken up residence in the liver, lung, bone, or brain. Once metastasis has occurred, systemic therapy is necessary to track down and deal with the cancer cells out of reach of the knife and the X-ray beam.

The first kind of systemic therapy employed in breast cancer was endocrine manipulation—attempts to change the hormone environment in the woman. It was recognized late in the nineteenth century that removal of the ovaries (oophorectomy) would sometimes produce relief of cancer symptoms and prolongation of life in certain premenopausal women. In the years that followed, oophorectomy and irradiation of the ovaries to destroy them became two accepted treatments for advanced breast cancer. Paradoxically, postmenopausal women would sometimes improve when they were given added estrogen, progesterone, and androgen as therapy, although the response rate was variable. In younger women, castration had been widely used preventively and was often performed immediately after mastectomy.

If a woman benefited from removal of her ovaries, the next procedure, if the cancer continued to progress, was to remove her adrenal glands (adrenalectomy), since these produce estrogen in place of the ovaries. If a relapse occurred, the adrenalectomy was followed by hypophysectomy, the removal of the pituitary gland to get rid of the source of all gonadotropins and adrenocorticotrophic hormones. Should a woman, whose initial surgery had very likely been a radical mastectomy, survive all these additional "-ectomies," she would have had to be maintained on a daily replacement of the now-missing hormones, some of which are essential to life. Additional morbidity and stress resulting from these successive surgeries could hardly add to the quality of life, but those women who did benefit from endocrine gland removal had relief from pain and prolonged survival, sometimes for as many as two years. If there was no improvement, the additional operations only added to their suffering.

Removing one gland after another is hardly the best way of selecting subjects who will benefit from hormonal manipulation, and fortunately there are better methods in use today. Normal breast cells contain receptors for estrogens, progesterone, and androgens. Certain breast cancers are also hormone dependent; that is, they contain protein estrogen and progesterone receptor sites in the cytoplasm of the tumor cells. If the primary tumor is analyzed for receptor sites when it is removed, it is possible to determine with a fair amount of accuracy whether removal of the estrogen source will be helpful. It is estimated that if positive estrogen receptors exist in the cancer, there is a 50% chance that the women will respond to endocrine-excision treatment. The presence of both estrogen and progesterone sites increases the probability to 75%. If no hormone receptors are found on biochemical assay, there is very little chance of response. Postmenopausal women less frequently have estrogen-dependent tumors.

The presence or absence of estrogen receptors in her tumor cells is currently used to predict a woman's response to endocrine therapy, and there has been the suggestion that the absence of estrogen receptors correlates well with a better response to cytotoxic (cell-killing) chemotherapy. There is also evidence that analysis of the tumor for estrogen receptors may have some bearing on the prognosis of the disease. A 3-year follow-up study of 286 British women who had mastectomies for early breast cancer showed that the rate of recurrence of the disease was significantly higher in those women who lacked estrogen receptors in their cancerous tissue (Cooke et al., 1979). The determination of estrogen- and progestin-receptor status requires that some of the tissue be saved for analysis at the time of its removal. If the cancerous tissue is prepared for permanent section and stain in a pathology lab, it ruins the opportunity to do a test for steroid receptors. The tissue should be frozen immediately or at least kept on ice for no longer than 30–45 minutes after surgery and then transported (or shipped) to a laboratory capable of doing receptor-site analyses.

Recently, in lieu of surgery, there have been some encouraging results in alleviation of symptoms and increased survival in advanced breast cancer by the administration of tamoxifen, a nonsteroid antiestrogen that competes with estrogen at the receptor sites. The majority of women taking tamoxifen have been post-menopausal, but although there are some clinical trials in progress in younger women, tamoxifen is not believed to be as efficient as ovary removal in the premenopausal woman. Aminoglutethimide, a compound that blocks steroid biosynthesis in the adrenal glands, is considered to be preferable to adrenal removal, but the drug produces the serious side effects associated with acute toxicity. Progestins, danazol, and luteinizing-hormone-releasing-hormone and its analogues are also being investigated for their effect.

Chemotherapy

Chemotherapy is the use of cytotoxic drugs to damage or kill malignant cells. The anticancer drugs function at the cellular level to inhibit the growth of, or actually destroy, the susceptible cells. Some of the agents are called antimetabolites, and they interfere with the synthesis of DNA and RNA, the necessary components of cellular function and multiplication. Others are alkylating agents, which crosslink the double helix of DNA and prevent replication; antibiotics, which react with DNA and interfere with its transcription into RNA; and some drugs that prevent cell division by disrupting the orderly processes of mitosis. Table 7.1 lists some of the chemotherapeutic drugs used to treat breast cancer.

The compounds that are used in therapy do not selectively seek out the cancer cells for disruption and injury, although sometimes the damage is more extensive to them because they may have altered metabolic processes. Generally, cytotoxic drugs affect all rapidly dividing cells, normal as well as malignant: in the skin, the hair follicles, the mouth, the gastrointestinal tract, and the bone marrow. Reproduction is frequent and rapid in such cells so they are also affected. Side effects, depending on the individual tolerance, can, therefore, include loss of hair, skin changes, sores in the mouth, nausea, and vomiting. Tiredness and lowered resistance to infection may occur as bone-marrow suppression results in fewer red and white blood cells. Some patients are fortunate and have mild or few side effects; others are unable to tolerate a particular drug that has to be stopped, perhaps to be later resumed.

In the past, the chemotherapeutic drugs were used only in advanced breast cancer. They were considered to be the "big guns," to be brought out as a last resort after all else had failed. Although the majority of drugs that are used in cancer therapy are in themselves, ironically, carcinogenic, it was considered justifiable to use them—the patient was going to die anyway, and there was no need to worry about long-term effects. By then a woman was usually so debilitated that her tolerance to a drug was limited. Even so, a single agent frequently resulted in a remission, or temporary abatement, of her symptoms and a short improvement in survival.

Chemotherapy is now used earlier in breast cancer and in combination form. Several drugs, each independently active but with no overlapping toxic effects, are combined. For the most part, such multiple chemotherapy has supplanted the use of a single drug. In addition, different ideas have developed in the past 10 years concerning the biology of breast cancer and the use of cytotoxic drugs. It is now believed that many women already have small micrometastases in other organs by the time their cancer is diagnosed, especially if the axillary nodes are involved. It is also now known that small tumors have a greater percentage of actively dividing cells than do large tumors. Since the drugs affect rapidly dividing cells, the cancer with the least mass would be the most susceptible to chemotherapy. Chemotherapists feel it makes more sense to use drugs early, as an adjuvant to surgery, rather than as a last ditch effort. When several clinical trials appeared to produce encouraging preliminary results, there was a recognition that combination chemotherapy in conjunction with surgery and/or radiation therapy could be used in early breast cancer to prevent the cancer from recurring and, it was hoped, to actually cure the disease.

At the present time, there are clinical trials in progress all over the world to evaluate the use of different drug combinations administered to women with early breast cancer and some node involvement. The two largest studies were initiated in the United States by Fisher and associates in 1972 and by Bonadonna and associates in Milan, Italy. Both trials compared surgery alone with surgery plus a chemotherapeutic agent or agents. The results thus far have appeared to be better for premenopausal women than for older

Table 7.1. SOME CHEMOTHERAPEUTIC DRUGS USED IN BREAST CANCER

Drug	Trade-Name	Mode of Action	Toxic Effects
Antimetabolites			
Amethopterin	Methotrexate	Blocks folic acid reductase, essential enzyme for nucleic acid synthesis.	Hair loss, nausea, vomiting, bone marrow depression, kidney and liver damage.
5-fluorouracil (5-FU)	5-Fluorouracil	Blocks thymidylate synthetase, essential enzyme for nucleic acid synthesis.	Nausea and vomiting, bone marrow depression, inflammation of mucous membranes of mouth and intestine, disturbance in muscle coordination.
Alkalating Agents			
Triethylenethiophosphoramide	Thiotepa	Exchanges alkyl groups with hydrogens of nucleic acids; modifies DNA structure.	Bone marrow depression
Cyclophosphamide	Cytoxan	Same as above	Bone marrow depression, hair loss, cystitis.
L-Phenylalanine mustard (L-PAM)	Alkeran	Same as above	Bone marrow depression, partial hair loss.
Vinca Alkaloids			
Vincristine sulfate	Oncovin	Destroys mitotic spindle, stops mitosis.	Neuritis, bone marrow depression, nausea and vomiting.
Vinblastine	Velban	Destroys mitotic spindle, stops mitosis.	Hair loss, bone marrow depression, nausea and vomiting.
Antibiotics			
Doxorubicin	Adriamycin	Modifies DNA function	Cardiac toxicity, hair loss, oral membrane inflammation, nausea and vomiting.
Actinomycin D	Dactinomycin	Modifies DNA function	Hair loss, nausea, diarrhea, tongue and oral membrane inflammation; marrow depression.
Steroid Hormones	Depo-provera, Megace and others	Stimulates or interferes with growth; may act at cell membrane.	No serious toxic effects.

women. The Mayo Clinic has compared L-PAM with Cytoxan, 5-FU, and prednisone, a steroid. Again, there was greater statistical improvement for the premenopausal women than for the postmenopausal group (Ahmann, 1984). The Southwest Oncology Group is matching L-PAM only against a five drug combination of Cytoxan, Oncovin, Methotrexate, prednisone, and 5-FU, and in Italy, Bonadonna is studying prolonged administration of Cytoxan plus Methotrexate plus 5-FU (CMF) as opposed to no therapy. In Birmingham, England, long-term 5-FU is being compared with Cytoxan. Thiotepa is being contrasted with Cytoxan in Moscow, and in Tokyo, investigators are giving some women 5-FU while others are receiving 5-FU, Cytoxan, and mitomycin C, an antibiotic.

Other agents with possible promise that are being tried at some centers include bleomycin, mitoxanthrene, bisanthrene, and elliptinum acetate.

At this point, it appears that multiple-drug therapy has proved to be superior when compared to a single agent in terms of the disease-free interval, but not enough time has elapsed to evaluate the effect of polychemotherapy on overall survival. Moreover, not everyone is equally enthusiastic about the early use of chemotherapy in younger women. Adjuvant chemotherapy produces remissions (the interval without recurrance of the cancer), but it is too soon to judge whether it aids survival. The drugs are toxic, produce acute side effects, and are carcinogenic in themselves, and those considerations are difficult to ignore. The long-term effect of cancer is death; the long-term effects of the drugs are as yet unknown. Only as a result of the controlled trials will it be possible to eventually know which patients should get the drugs, when to give them, for how long, and whether the benefit actually outweighs the potential toxicity of the chemotherapy.

Immunotherapy

The attempt to utilize the body's own defenses to control or cure cancer is still in the experimental stage, but many believe that manipulation of the immune response could be a major breakthrough in the treatment or even prevention of malignancies. The immune reaction is the specific response the body has to defend itself against foreign invaders or antigens. It is made possible by two types of cells, called the B lymphocytes and the T lymphocytes; the "natural killer" (NK) lymphocytes that are neither B cells nor T cells; and a number of other cells produced in the bone marrow plus chemical entities in the blood, such as complement and the antibodies produced by the B cells. Figure 7.16 illustrates the components of the immune response. Although much remains to be learned about the normal immune reaction to cancer, it is believed that when something foreign to the body tries to establish itself, the killer cells detect and destroy the foreign invader. These destructive cells are responsible for the rejection of organ transplants, and they are also able to recognize malignant cells as "not self" and attack them before the abnormal cells become established as life-threatening tumors. The body is under constant surveillance by the immune mechanism, and it is now recognized that cancer cells are antigenic, that is, sufficiently different from the normal to evoke the response of the T cells and NK cells. The T-helper cells are hypothesized to improve the immune response to cancer, and the T-suppressor cells may do the opposite. When more is understood about the immune response, it may be possible to immunize a person against malignant tumors or to utilize the immune response against tumors that are already actively growing.

There is a general belief that frequently the immune response of some breast cancer pa-

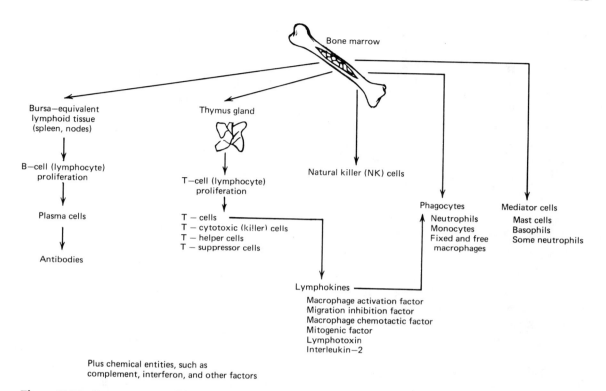

Figure 7.16. Components of the immune response. All blood cells originate in the bone marrow. White blood cells, or leukocytes, are involved in the body defense system. Those destined to become lymphocytes mature either in the thymus, and become known as T cells, or in lymphoid tissue like the spleen and lymph nodes, and become the B cells. B cells mature into plasma cells and produce antibodies. Certain T cells further differentiate and acquire special functions. Lymphokines are chemical messages released by sensitized T cells to recruit and activate the macrophages.

tients is impaired, and that this influences the prognosis; that is, the better the host defense mechanisms, the better the chances for the host. Some substances, such as bacillus Calmette-Guérin (BCG) or *Corynebacterium parvum*, are nonspecific potent stimulants to the immune system. Gutterman and his co-workers (1976) and Israel (1976), among others, have administered immunostimulants to breast cancer patients with and without accompanying chemotherapy. Although initial results of these and other studies suggested

that stimulating the immune response has some positive effects, a follow-up after 5 years made this approach to the treatment of breast cancer appear far less promising.

The most exciting advance in immunology that many believe holds the greatest potential for both diagnosis and treatment of breast cancer has been the development of monoclonal antibodies. The new technique, known only since 1975, is like an antibody factory—it allows a virtually unlimited number of identical antibodies with a predetermined specificity to be produced in a culture medium. It is based on cell hybridization, the fusion of two different cell lines, usually from different species, to produce a new cell that contains all or part of the genetic information from the parent cells. In the production of monoclonal antibodies, one of the cell lines is from an antibody-producing lymphocyte; the other is from a myeloma, a malignant tumor of the immune system that has the ability to survive indefinitely in cell culture. In 1975, Köhler and Milstein reported that by fusing a cultured mouse myeloma cell with spleen lymphocytes from a mouse that had been immunized against sheep red blood cells (the antigen), a "hybridoma" was produced. This hybrid cell had both the lymphocyte's ability to produce specific antibodies against the sheep red blood cells as well as the immortality characteristic of the myeloma cell. The hybridoma could then be cloned, and the clones could be maintained in culture for indefinite periods, continually producing their pure and identical antibodies to the original antigen—hence the term, monoclonal antibodies. For their achievement, in 1984 the two researchers were honored with both the Nobel Prize in medicine and the Albert Lasker Medical Research award.

The major use of monoclonal antibody technology thus far is in its application to biological assay techniques and the characterization of hormones, viruses, and other biologically active substances. Research on the clinical use of monoclonal antibodies in the detection and treatment of cancer is in a very early stage, but it is believed that there are several possibilities for their use in the near future. Radioactive tracers could be tagged with monoclonal antibodies to tumor cell antigens and enhance scanning techniques. It may also be possible to attach cell-killing drugs or other agents to monoclonal antibodies so that they can be specifically delivered to the tumor without injuring normal cells or,

perhaps, to develop monoclonal reagents that can do the job alone.

At the very least, the development of the monoclonal antibody technology is a significant breakthrough in immunology, genetics, and cell biology in general; it may also be the hoped-for crack in the enigma of cancer diagnosis and therapy.

The "Best" Strategy of Treatment

What, then, do all these possibilities for treatment mean? For doctors, it generally means being prejudiced in favor of their own particular speciality. Surgeons believe that the primary treatment for breast cancer is surgery. Some surgeons are still convinced that radical mastectomy is the only operation for all breast cancers. There are radiologists, however, who think irradiation after minimal surgery or irradiation alone is the true way. Medical oncologists have hopes that long-term remission and even cure are possible through early chemotherapy. Immunologists believe that immunotherapy may turn out to have the greatest potential for treatment or even prevention of cancer.

It is likely that the most effective treatment involves combinations of therapy, but there is only preliminary scientific evidence as yet to indicate which treatment or combination is better than any other. Obivously, no single therapeutic approach could possible be best for all women with breast cancer, and the most beneficial treatment for a particular woman must be selected on an individual basis. A woman who discovers that she has breast cancer is particularly vulnerable to the idea that the doctor knows best, but it is especially important for her at that time, scared and shocked as she may be, to know what the options are. She has to make certain that the status of her condition is appraised before surgery and that she gets a full explanation of the treatment that has been selected for her.

She must be assertive enough to let her doctor know that she, too, has some knowledge of breast cancer and its treatments and that she wants to participate in the decisions that are made concerning her.

A woman who has enough knowledge and money may be fortunate enough to get herself into the hands of an oncology team, a group in which the surgeon, radiologist, pathologist, and medical oncologist work together. Such teams are usually located in a cancer research center or in a teaching hospital affiliated with a medical school. Most women, regretably, do not have the opportunity for this kind of treatment and have to rely on a local doctor or a clinic. At least they should insist, difficult as that may be, that their doctors consult with a cancer specialist so that more than one opinion is available. Breast cancer statistics are still too grim for any doctor to feel satisfied about the results of any one treatment. Physicians must be willing to abandon their own prejudices and consider alternatives, particularly in view of the evidence accumulating from the controlled clinical trials that are in progress.

In the midst of all the unknowns concerning the cause and course of breast cancer and the controversy surrounding the most effective treatment, there is one unchallengable certainty—the earlier the diagnosis, the better the prognosis. When breast cancer is found early, the expectation for cure is greatest. Five to 10 minutes of self-responsibility, the monthly breast self-examination, must become a part of the lives of all women.

REFERENCES

Ahmann, D. L. Status of adjuvant therapy in patients with breast cancer. Cancer 53(3): 724–728, 1984.

Bailer, J., III. Presentation at the 66th Annual Meeting of the American Association for Cancer Research. Proceedings of the Eleventh Annual Meeting of the American Society of the Clinical Oncology. San Diego, California, May 7–11, 1975.

Bross, I. L. D. and Blumenson, L. E. Screening random asymptomatic women under 50 by annual mammographies: Does it make sense? J Surg Oncol 8(5):437–445, 1976.

Bulbrook, R. D., Hayward, J. L., and Spicer, C. C. Relation between urinary androgen and cortocoid excretion and subsequent breast cancer. Lancet 1:1076–1079, 1971.

Carter, S. K. Clinical trials and primary breast cancer: The therapeutic implications. Cancer 53(3):740–751, 1984.

Cooke, T., George D., Shields, R., et al. Oestrogen receptors and prognosis in early breast cancer. Lancet 1:995–997, 1979.

Cowan, L. D., Gordis, L., Tonascia, J. A., and Jones, G. S. Breast cancer incidence in women with a history of progesterone deficiency. Am J Epidemiol 114:209–217, 1981.

Culliton, B. J. How important is early detection? Science 198:171, 1977.

Fisher, B. Some thoughts concerning the primary therapy of breast cancer. Recent Results Cancer Res 57:150–163, 1976.

Fisher, E. R., Sass, H., and Fisher, B. Pathologic findings from the NSAPB for breast cancers (protocol No. 4). Cancer 53(3):712–723, 1984.

Goldin, B. R., Adlercreutz, H., Gorbach, S. L., et al. Estrogen excretion patterns and plasma levels in vegetarian and omnivorous women. New Engl J Med 307(25):1542–1547, 1982.

Gonzalez, E. R. Chronic anovulation may increase postmenopausal breast cancer risk. JAMA 249:445–446, 1983.

Graham, S., Marshall, J., Mettlin, C., et al. Diet in the epidemiology of breast cancer. Am J Epidemiol 116(1):68–75, 1982.

Gutterman, J. U., Cardenas, J. O., Blumenschein, G. R., et al. Chemiommunotherapy of advanced breast cancer; prolongation of remission and survival with BCG. Br Med J 2(6043):1222–1225, 1976.

Hirayama, T. Epidemiology of breast cancer with special reference to the role of diet. Preventive medicine 7(2):173–193, 1978.

Israel, L. The role of nonspecific immunotherapy in the treatment of breast cancer. Recent Results Cancer Res (57):189–195, 1976.

Kohler, G., and Milstein, C. Continuous cultures of fused cells secreting antibody of predefined specificity. Nature 256:495–497, 1975.

Kushner, R. *Breast Cancer, A Personal History and an Investigative Report.* New York, Harcourt Brace Jovanovich, 1975.

Kushner, R. *Why Me?* Philadelphia, W.B. Saunders, 1982.

Lafaye, C., Aubert., B. Action de la progesterone locate dans les mastopathies bénignes. Etude de 500 observations. *J Gynecol Obstet Biol Reprod* 7(6):1123–1139, 1978.

Leis, H. P., Cammarata, A., LaRaja, R., and Cruz, E. Fibrocystic breast disease. *The Female Patient* 8:56–77, May, 1983.

Lemon, H. M. Oestriol and prevention of breast cancer. *Lancet* 1:546–547, 1973.

Lemon, H. M. Pathophysiologic considerations in the treatment of menopausal patients with oestrogens; the role of oestriol in the prevention of mammary carcinoma. *Acta Endocrinol* (Suppl) 233:17–27, 1980.

London, R. S., Solomon, D. M., London, E. D. Mammary dysplasia: Clinical response and urinary excretion of 11-deoxy-17 ketosteroids and pregnanediol following alphatocopherol therapy. *Breast* 4:19, 1978.

London, R. S., Sundaram, G. S., and Goldstein, P. J. Medical management of mammary dysplasia. *Obstet Gynecol* 59(4):519–523, 1982.

Love, S. M., Gellman, R. S., and Silen, W. Fibrocystic "disease" of the breast—a nondisease? *N Engl J Med* 307:1010–1014, 1982.

Lubin, J. H., Burns, P. E., Blot, W. J., et al. Dietary factors and breast cancer risk. *Int J Cancer* 28:685–689, 1981.

Mansfield, C. M. *Early Breast Cancer, Its History and Results of Treatment.* Basel, S. Karger, 1976.

Mettlin, C. Diet and the epidemiology of human breast cancer. *Cancer* 53(3):605–611, 1984.

Miller, A. B., Kelly, A., Choi, N. W., et al. A study of diet and breast cancer. *Am J Epidemiol* 107:499–509, 1978.

Minton, J. P., Foecking, M. K., Webster, D. J., et al. Caffeine, cyclic nucleotides and breast disease. *Surgery* 86:105–109, 1979A.

Minton, J. P., Foecking, M. L., Webster, D. J. Response of fibrocystic disease to caffeine withdrawal and correlation of cyclic nucleotides with breast disease. *Am J Obstet Gynecol* 135:147–149, 1979B.

Montgomery, A. C., Palmer, E. V., Biswas, S., and Monteiro, J. C. Treatment of severe cyclical mastalgia. *J Roy Soc Med* 72:489–491, 1979.

Papaioannou, A. M. *The Etiology of Human Breast Cancer.* New York, Springer-Verlag, 1974.

Phillips, R. L. Role of lifestyle and dietary habits in risk of cancer among Seventh-Day Adventists. *Cancer Res* 35:3513–3522, 1975.

Roberts, H. J. Perspective on vitamin E as therapy. *JAMA* 246(2):129–131, 1981.

Staff members devise new guidelines for the timing of breast reconstruction. Newsletter, The University of Texas System Cancer Center, 27(5):5–6, 1982.

Stoll, B. A. (ed). *Risk Factors in Breast Cancer.* Chicago, Year Book Medical Publishers, 1976.

Turner, L., Swindell, R., Bell, W. G. T., et al. Radical versus modified radical mastectomy for breast cancer. *Ann R Coll Surg Engl* 63:239–243, 1981.

Veronese, U., Saccozzi R., Del Vecchio, M., et al. Comparing radical mastectomy with quandrantectomy, axillary dissection, and radiotherapy in patients with small cancers of the breast. *N Engl J Med* 305(1):6–11, 1981.

Vorheer, H. *Breast Cancer: Epidemiology, Endocrinology, Biochemistry and Pathobiology.* Baltimore, Urban and Schwarzenberg, 1980.

8. *The Gynecological Exam*

Ask any woman how she feels about having a pelvic exam. Chances are it will never make the top ten of her favorite things to do. Not that most other preventive or diagnostic procedures are that much fun, either. Who really enjoys having a complete physical, having blood drawn, getting X-rays taken, or going to the dentist? We are pricked, poked, prodded, and subjected to various indignities by health professionals, we generally accept them with equanimity. But few things make a woman feel as exposed, vulnerable, and downright humiliated as does the gynecological examination. A procedure that requires spread legs in stirrups and a bare bottom hanging over the examining table is at best only tolerable.

Necessity permits the gynecologist, who is almost always male, as much or greater intimacy with a woman's body than she ordinarily allows herself or her sexual partner. The uneasiness or at least mixed feelings that a woman is bound to have as a result of the exam are frequently compounded by the attitude of the physician. He may use condescending terms of endearment ("move down, honey," or "just relax, sweetie") or adopt a ho-ho heartiness that he thinks will make her

more comfortable. There are some doctors who never look at the woman at all during the examination. They talk about everything under the sun—the weather, politics, other superficialities—probably because this is the way in which they have come to deal with their own discomfort at the underlying sexual connotation. Sometimes a physician will actually exploit a woman sexually during the breast or pelvic examination, but many woman are disbelieving, even if it happens. After all, the doctor is not supposed to act that way, so perhaps she imagined it. Or she may blame herself, thinking she did something to provoke his actions.

Women, however, do not go to a gynecologist-obstetrician to be seductive, embarrassed, or patronized. They go because they are concerned about their bodies, and they know that a gynecological examination and a Pap smear are a necessary part of preventive health. They go because they may need a method of birth control; they may have a vaginal discharge that is irritating or painful; they may have irregular periods or dysmenorrhea; or perhaps they have missed one or more periods and suspect a pregnancy. If there is a specific problem, they want it handled effectively, and they want to know how to prevent the difficulty in the future.

Ideally, every Ob/Gyn would conduct the examination in a sensitive but matter-of-fact manner, would listen attentively to what the woman had to say, would answer questions unhurriedly, would explain what is being done, and, as one adult to another, educate a woman about how to take care of her body. Some are like that, and some are not, and it is not always possible to shop around. Most women have neither the time nor money to try doctor after doctor until they find one with all the right qualities. Many women are geographically limited to the medical care available in their area. There are some things that a woman herself can do, however, if she is not always able to see a doctor who is patient, supportive, communicative, and who respects her dignity as an individual.

Women must remember that the doctor who performs an examination is there to perform a service—the delivery of quality health care. This is the responsibility of the physician, and it is the consumer's right to demand it from the medical profession. But the entire burden for health care is not solely the physician's. Women, too, have a responsibility. They must change their own attitudes about the pelvic examination, rid themselves of their trepidations, and strip it of its mystique and fear. The sexual organs are a part of the body like any other, and there is no need to be embarrassed about their examination. Women must arm themselves with the knowledge to understand what an "internal" is, what is involves, what is taking place during the procedure, and how to recognize a good, thorough, gynecological exam. They must not permit themselves to fall into the subservient female role, to be handled, and told what to do. They must know what kinds of questions to ask, not be afraid to ask them, and they must persist until their questions are answered. Only when

women have the information about the medical procedure necessary for their own health maintenance will they be in the position to actively participate in decisions pertaining to their health.

Although some women already know exactly what is taking place during the exam, are medically sophisticated, very knowledgeable, and have no embarrassment concerning the procedure, they still leave the office dissatisfied. They may feel they have been rushed through, brushed off, or that the doctor was authoritarian or judgmental. The physician, male or female, that still treats patients in this way should be challenged. Certainly, it is difficult to sit in a doctor's office or lie on an examining table and criticize the attitude of the physician conducting the examination. But with knowledge, and the power that accompanies knowledge, it is possible for a woman to attempt to change attitudes for the doctor's good as well as her own. A woman can try to educate her doctor, one-to-one, to the partnership in her health care in which both of them can have responsibility. Confident assertiveness can only improve the relationship between a woman and her doctor. It may never make her feel anything more than just neutral about a pelvic, or any other necessary medical procedure, but at least the examination may become a joint enterprise, one in which she, as well as her pelvis, participates.

HOW TO KNOW A GOOD GYNECOLOGICAL EXAM

Every human being is a whole person and not a set of organs to be examined and diagnosed. Similarly, the health of the sexual and reproductive organs does not exist in a vacuum. It is part of, and should not be separated from, the total health of the woman. A complete medical history that includes both general medical and gynecological information is an essential part of the first visit to a doctor. If a woman continues to use the same physician, the detailed history does not have to be repeated each time because the information will have been noted on her record. But if on the second visit it appears to her that the doctor has not looked the information over, she should not be reticent about bringing up particular aspects she thinks are relevant. If a woman goes to a clinic, she may not be seeing the same doctor each time, and she should make certain that her record is read.

In the offices of some physicians, the medical history is taken by the receptionist or secretary, or the woman may be asked to fill out a form. These are obviously not desirable procedures. A woman who may be reluctant to discuss details of her medical, menstrual, or sexual function with a strange doctor will feel even more uneasy telling her history to a nonprofessional or confessing on paper that, for example, she has had an abortion. It is also possible that information given to someone other than the physician or written down on a form may not reach the doctor before the physical examination. It is better to be asked questions personally by the examining clinician, but this seems to happen less and less.

It should never be assumed that past or chronic problems are irrelevant to a gynecological difficulty and that they need not be mentioned. A vaginal infection may be associated with diabetes, for example, or with taking antibiotics. If the doctor does not ask about things that the woman may think are important, she should bring them up herself. The following list of questions provides the kind of information that should be obtained from the history. If the questions are not asked, the examination has not been complete, and its quality is questionable.

Family History. Are there any diseases that run in the family: diabetes, heart disease, cancer?

Previous Medical History. Have there been any serious illnesses, either physical or emotional? Any rheumatic fever, any infectious diseases, such as T.B. or V.D., any previous gynecological disease? What about operations? For what reason were the surgeries performed and under what kind of anesthesia? Is there any history of allergies in the family? Have there been any urinary difficulties, such as frequent bladder infections, stress-incontinence (dribbling while coughing, sneezing, or laughing but not at other times), or frequency or urgency not related to stress? Has there ever been blood in the urine?

Gynecological History—Menstrual. What was the age at menarche? Were the initial cycles regular? Is the cycle regular or irregular? How long of an interval between periods? What is the duration and amount of flow? Are external pads or tampons used? Are they "regular" or "super," and how often are they changed? How many are used for the duration of the period? (If the woman has kept a menstrual chart, the pattern of cycles is evident at a glance.) Are there menstrual cramps? When do they occur—before, during, or after the period? Are there any other discomforts besides cramps? Have there always been cramps with menstrual periods, or have the menses only recently become painful? How severe are the cramps, and what is done for them?

Obstetric History. Has there ever been a pregnancy? How many? Any miscarriages and in which month did they occur? Abortions or term deliveries? If a pregnancy was terminated in abortion, what method was used? What kind of labor and delivery was there? Were there any complications? What were the birth weight and sex of each child, and were the babies breastfed? Are there plans for future pregnancies?

Contraception History. What methods of birth control have been used? What methods, if any, are being used presently?

In taking the medical history, a sensitive

physician will not make automatic assumptions of heterosexuality. Doctors who respect women patients as individuals, regardless of their sexual preference, should be aware enough of alternative life-styles to ask a few courteous questions before attempting to provide birth control information, for example. Neither should lesbian women immediately be equated with sexually active gay men concerning the occurrence of sexually transmitted diseases. Lesbians, at least those who are not bisexual in preference, are at much lower risk for developing venereal diseases than are gay men, and gonorrhea and syphilis are virtually absent. On the average, lesbians also have fewer gynecological infections than sexually active straight women.

If the office visit has been for a checkup, breast examination, and a Pap smear, the physical examination should follow the taking of the history. If a woman has come with a specific complaint, even more detailed questions that deal with the problem will be asked. The most common gynecological difficulties involve irregularity in uterine bleeding, amennorhea, pelvic pain, vaginal discharge, the presence of a lump or mass in the abdomen, a protrusion from the vagina, urinary incontinence, hot flashes, infertility, or a problem related to sexual intercourse. A woman should tell the doctor all she can about the onset, severity, and what, if anything, she herself has done thus far about the specific complaint. Her symptoms can direct the physician's approach to her examination. If she suspects what the cause of her problem is, she should say so. She may be guessing incorrectly, but on the other hand, she could be right in her diagnosis.

External Examination

Before going to the doctor, there is no need to take any special cleansing measures at home other than taking a shower or a bath. Advertising to the contrary, a woman's normal secretions are not offensive. A douche should never be used before the examination. If there is a problem with a heavy or odorous vaginal discharge, trying to get rid of it for esthetic reasons may ruin the diagnosis.

Unless the doctor wants a urine specimen to check for cystitis, the pelvic exam will be more comfortable if the woman urinates before seeing the doctor. Besides, if the bladder is not emptied, its fullness could be mistaken for an abdominal mass or ovarian cyst.

If a woman has recently had a checkup, she will not need a complete physical. Some women, however, use their gynecologist as a primary care physician, and then they should not hesitate to ask to have their blood pressure taken, their eyes, ears, nose, and throat examined, and their heart and lungs listened to. If a blood sample is taken, a VDRL, the test for syphilis, should be performed on it. A woman should not feel insulted if the test is run on her blood; she should ask for it if it is not.

The breasts should be examined while the woman is seated and should be palpated while she is lying down. Any questions a woman may have concerning breast self-examination should be asked. If she is uncertain of the method, this is the time to find out how it should be done.

The actual gynecological examination is performed with the woman lying down on the examining table in lithotomy position; that is, her feet are up in stirrups, her buttocks are hanging over the end of the table, and a sheet is draped like a tent over the knees and the upper part of her body. This kind of positioning straightens the curvature in the lumbar region of the spine and relaxes the abdominal muscles. It is important for a woman to try not to tense up because it may make the procedure more uncomfortable for her. The usual steps in the procedure are as follows: palpation of the abdomen, breast examination, inspection of the external genitalia, and then, internally, speculum examination of the vagina and cervix, including cell smears for diagnostic analysis, bimanual pelvic examination, and rectal examination.

The doctor will look for any enlargement or tenderness in the abdomen. The vulva and the perineum are examined, and the labia majora and minora are spread apart to see the entrance to the vagina, the hymen, and the urethra. The doctor is looking for any inflammation, scarring, sores, or growths, such as warts, cysts, or tumors.

Pushing up the urethral opening against the pubic bone with the tip of the forefinger is called stripping or milking the urethra. In acute urethritis or in gonorrhea, a few drops of pus could be squeezed out from the paraurethral or Skene's glands. The thumb and forefinger are used on either side of the labia majora to palpate for Bartholin's glands, which normally cannot be felt but which may be enlarged and tender if infected.

Internal Examination

In order to see the cervix and the inside of the vagina, an instrument called the speculum must be used to separate and hold apart the vaginal walls (see Fig. 8.1). Specula come in various sizes, and the appropriate one to match the size of the introitus (vaginal opening) and the length of the vagina is chosen. When the speculum is inside, it may be slightly uncomfortable, but it should not be painful. If it hurts, the woman should say so. Perhaps its position may need to be adjusted.

The bivalve or duck-bill speculum may be made out of steel or plastic. It has two blades, and the posterior, or bottom, one is slightly longer than the anterior blade. It is designed so that it opens after insertion and can be fastened to remain open. The speculum cannot be lubricated with jelly because this

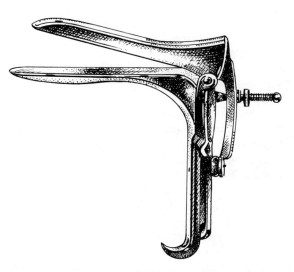

Figure 8.1. Graves vaginal speculum. Specula are available in different sizes.

would interfere with any analyses of vaginal secretions or cells. Usually, enough natural lubrication is present at the vaginal opening to ease the speculum into place. If not, water can be used.

The sudden entrance of a cold steel unlubricated speculum is a rather startling sensation—one that many women would, just once, wish their male physicians to experience. While one more discomfort added to the entire pelvic exam may not really be that important, it is considerate and more humane to warm the speculum before insertion. Some of the newer examining tables have warming drawers, some doctors keep their specula on a heating pad, and some will warm them by holding them under hot water. Plastic specula do not feel that cold.

The closed speculum is inserted, posterior blade first, into the vagina at about a 45° angle downward. In lithotomy position the angle of the vagina is down toward the sacrum. When the speculum is in up to the hilt, it is rotated and opened. The entire circumference of the cervix and the vaginal fornices can then be viewed. At this point, a sample of cervical or vaginal cells or discharge may be taken to test for cervical cancer, vaginal infections, gonorrhea, or to assess the endocrine status of the woman (see Fig. 8.2).

The speculum is then slowly withdrawn as the vaginal walls are inspected again to make certain that any redness, cyst, or other damage has not been missed because it was hidden by the blades. The doctor then puts on a disposable glove to perform the digital examination; that is, the insertion of the middle and the index finger into the vagina. As the examining fingers reach the full length of the vagina, the fornices are explored and palpated for masses or tenderness, and the cervix is palpated for size, shape, and consistency. The woman may be asked to hold her breath and "bear down." This increase in intra-abdominal pressure will reveal any weakness in the muscular supports of the bladder, rectum, or uterus.

The size, shape, position, mobility, and sensitivity of the uterus, ovaries, and fallopian tubes are ascertained by the *bimanual*, or two-handed, examination. The fingers of one hand inside the vagina are placed against the cervix to elevate it while the other hand presses downward on the lower abdomen. In this way, the body of the uterus can be outlined between the two hands. Then, the physician will try to identify the right and left ovary and tube by placing the vaginally placed fingers in each fornix while palpating abdominally with other hand. The normal ovary and fallopian tube cannot usually be felt as distinct structures, but the woman may feel a twinge of discomfort as the ovaries are pressed between the external and the internal hands (Fig. 8.3).

The pelvic examination should be concluded with the rectal or rectovaginal examination. There are some pelvic structures, such as the posterior surface of the uterus, the broad ligaments, the uterosacral ligaments, and the pouch of Douglas, that can be felt accurately only through the rectum. At the same time any abnormal growth that may be present in the rectum could also be located by the examining finger. An internal examina-

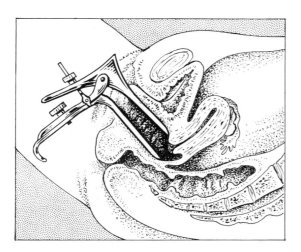

Figure 8.2. Lateral view of the speculum in position.

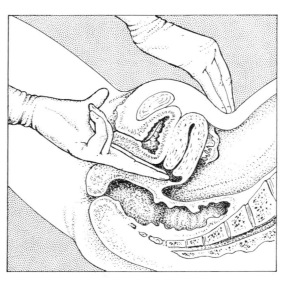

Figure 8.3. Position of the hands in a bimanual pelvic examination.

tion is not complete without a rectal examination (see Fig. 8.4).

When the physical examination is over, the doctor will tell the woman to get dressed. Some physicians will want to talk to her in the office after she puts her clothes on, but others will discuss the findings and explain any necessary treatment in the examining room. Even if the doctor has provided a running commentary throughout the examination, a woman may still have some unanswered questions. If medication is prescribed, she is entitled to know exactly what the drug is, for what condition it is being taken, and what the potential side effects may be. She should never feel that any question she may have is too "dumb" or trivial to be asked; if something is puzzling her she should persist until she is satisfied.

It should be noted that in many cities gynecological examinations are performed by nurse-practitioners or nurse-clinicians. Much of what women find distressing about their encounters with a male Ob/Gyn—the insensitivity, lack of communication, the "busy-doctor" attitude—disappear when the examination is conducted by a woman who knows from her own experience what a pelvic is like. Many nurse-practitioners routinely use a mirror and a light during the exam to let the woman see what is going on, and they make a special effort to educate her to understand what she can do to maintain her own gynecological health.

There is no mystery about the gynecological checkup—just a mystique and a vague uneasiness for the woman who cannot see what is happening to her, only feel it while she stares at the ceiling. There is no elaborate instrumentation used—only a speculum, a light, and a cotton swab or a wooden spatula for cell specimens. There is no reason to be humiliated by it or feel vulnerable and inferior and uneasy during the procedure—not if one has the knowledge and the confidence to be on equal footing with the doctor. An M.D. degree is certainly not necessary to be able to see what the doctor sees inside. Any woman with a tongue depressor, a mirror, and light can look at her own tonsils if she wants to; similarly, any woman with a speculum, a mirror, and light can look at her own cervix, if she wants to.

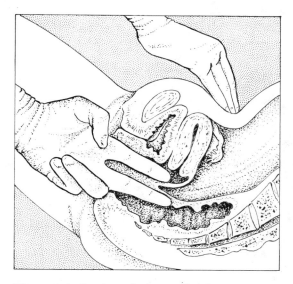

Figure 8.4. Rectovaginal examination.

VAGINAL SELF-EXAMINATION

A plastic, reusable, bivalve speculum can be purchased from a medical supply house, from a Planned Parenthood Center, or from a feminist self-help clinic, if there is one in the area. Depending on where they are purchased, specula cost from 50 cents to five dollars, and they come in three sizes—small, medium, and large. About 80% of all women would use the medium size.

To perform self-examination, a woman gets into whatever position she feels is most comfortable for insertion of the speculum. Some may want to squat over a mirror, others may stand with one leg resting on the bathtub or a stool. Many find that a semireclining position on a bed with the back supported by pillows works well. With the aid of a mirror and a high intensity lamp or a strong flashlight, the external genitalia can be examined. After checking the vulva (she will become accustomed rapidly to what is the normal appearance of her genitalia), the woman is ready to examine the vagina and cervix with the use of the speculum (Fig. 8.5A–E).

The tips of the speculum can be lubricated with a little K-Y Jelly or water before insertion. Vaseline should not be used because it can upset the natural balance of organisms in the vagina (besides, it is unpleasantly greasy). While the duckbills are held closed by the fingers of one hand placed between the

two sections of the speculum's handle, the labia are separated by the fingers of the other hand. The speculum is then gently inserted sideways at a slight angle into the vagina, aiming downward if the woman is on her back on a bed, or upward if she is standing—similar to the way in which a tampon would be inserted. The speculum should never be forced; if it is uncomfortable or painful, perhaps a smaller size is needed. The valves are inserted as far into the vagina as they will go, and the handle is then rotated up until it is perpendicular to the body. The fingers that are holding the duckbills closed are removed from between the handle sections, and the handles are squeezed together. This opens the valves and spreads the vaginal walls. The finger depression on the shorter handle is then pressed down while pulling up on the longer

Figure 8.5A. Sharing their knowledge in a comfortable atmosphere, women learn to use the tools and the techniques previously available only to their doctors.

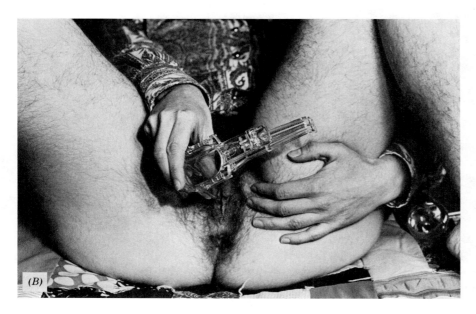

Figure 8.5B. After lubricating the tips of the blades with a sterile jelly or water, the speculum is inserted sideways at about a 45° angle aiming downward.

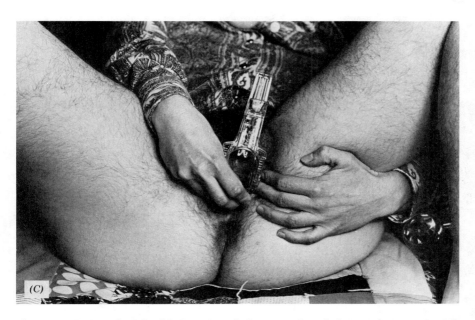

Figure 8.5C. Keeping the blades closed, the speculum is inserted up to the hilt and then rotated upward.

237

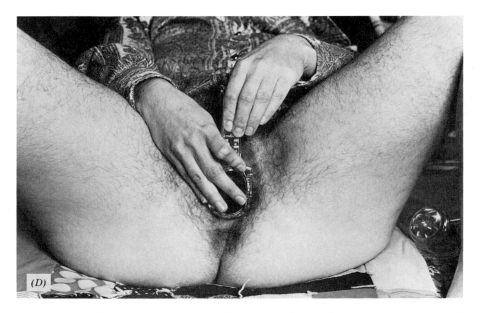

Figure 8.5D. The speculum is locked into place by pushing down on the small handle and pulling up on the long handle until a click is heard.

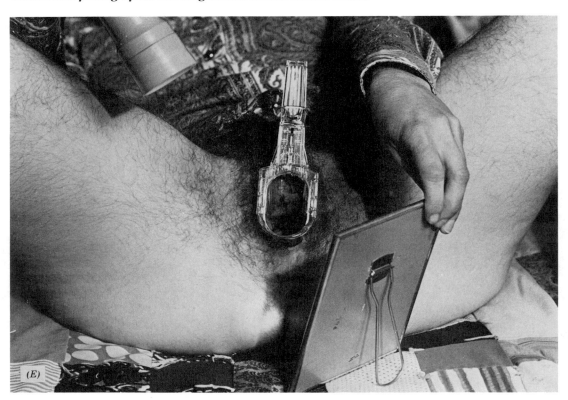

Figure 8.5E. The vaginal walls, the cervix, and the cervical os can be observed.

238

handle. This locks the speculum into one of three adjustable positions that can be heard clicking into place.

The mirror is then positioned between the legs while the light points into the mirror so that it reflects back a clear view of the vagina. If the cervix is not in view, the speculum may have to be pulled out again slightly and repositioned. Most cervices are not right in the center, and a little gentle searching may be necessary before they appear.

After viewing the cervix and the vagina, the speculum is removed by pulling it straight out without closing it first to avoid pinching the cervix or the vaginal walls. After the speculum is removed, it should be washed with soapy water. There is no need to sterilize it unless it will be used by another woman.

Why Self-Examination? Why Not?

There are many reasons for self-examination—there are health and financial benefits for women. Women who are acquainted with the normal physical appearance of their external genitalia and vaginal and cervical anatomy may be able to detect changes that indicate a developing pathology in its early stages. An infection caught early is easier, safer, and cheaper to treat than a full-blown case of itching vaginitis, for example. If a woman has reason to suspect a pregnancy and is familiar with her normal appearance, she may be able to detect its presence through changes in the color and consistency of the cervix and mucus—earlier than is possible through chemical tests.*

Or, a woman may have political reasons for self-examination. Women who band together to form self-help groups, sharing their knowl-

edge of self-care and preventive health, may find that mutual self-examination is another way of demystifying the practice of medicine and gaining more control over their bodies.

There may be some very personal, undefinable, reasons for a woman to examine herself. Just to see what she previously could only imagine—the pink, corrugated walls of the vagina, the smooth, shiny cervix, the moist cervical mucus oozing from the os, entry to the uterus. Now she knows the secret! And that results in there being no secret at all. Seeing the cervix is not the mystical revelation that

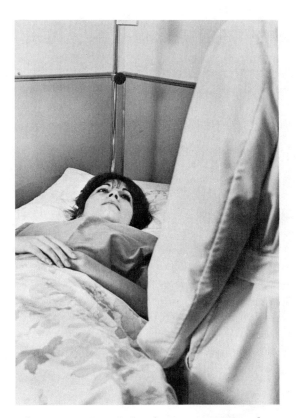

Figure 8.6. Knowledge is power. It can change the traditional doctor-patient relationship into a new and mutually beneficial partnership.

* Or course, while regular self-examination can be advantageous for the early recognition of changes, it is no *substitute* for annual or semi-annual complete clinical pelvic examinations with a cytological test for cancer. Competent professional backup is a necessary accompaniment to self-care.

can dispel all myths and provide all knowledge. It is an option, another choice that
women can make without depending on a
physician. It is not the solution, with far-
reaching implications, to all the problems
women have with the health care system.
Women can work toward greater self-determination and responsibility for their own health
care with or without a speculum, but for
many women it is a way to become more
comfortable and knowledgeable about their
bodies, to take themselves out of the hands of
men only and more into their own hands. It is
a tool that can be used to take the dominance
and mystery out of the doctor's role.

When the doctor stands less huge and all-
knowing, and the woman lies less confused
and troubled, the difference in power between them is reduced. Women must find the
confidence, however possible, for a more
equalized relationship between themselves
and their physicians (Fig. 8.6).

9. *Gynecological Difficulties*

Truly fortunate, and rare, is the woman who has never suffered an agonizing itch that she cannot scratch in public. And lucky is the woman who has never experienced the burning irritation of a bladder infection, who has never been told she has fibroids, an eroded cervix, or an ovarian cyst, who doesn't know what a menstrual cramp is, and who has never had any of the other "-rrheas." These kinds of gynecological disorders are so common in women and comprise so many of their medical complaints that all women should have a basic understanding of what they are, how to recognize them, and what can be done for them. Some of the more frequent menstrually related problems have been discussed in Chapter 4. This chapter will focus on the vaginal and venereal infections and infestations, and on some of the benign and malignant diseases of the reproductive tract.

VAGINITIS AND THE BIG THREE OF INFECTIONS

Vaginitis is the generic term that means inflammation and infection of the vagina. A more correct term is really "vulvovaginitis," because the symptoms are itching and inflammation of the vulva and the vaginal opening and an abnormal vaginal discharge. Pain on sexual intercourse, pain and burning on urination, and local swelling or edema may also occur to varying degrees.

There can be hundreds of causes for vaginitis. In some women it may result from an allergic reaction to soap, laundry detergent, bath oil, feminine deodorant sprays, commercial douche preparations, spermicidal jellies, or even colored toilet paper. Once the allergic contact is known, it can be eliminated, and thus the vaginitis will disappear. Sometimes an apparent vaginitis may be caused by a forgotten tampon or perhaps a diaphragm left too long in the vagina. Removal of the object will then remove the vaginitis as well. More often, however, and in more than three-quarters of the women with vaginitis, the cause is infection by one of three organisms—a fungus, *Candida*, a protozoan, *Trichomonas*, or a bacterium, *Gardnerella*.

The normal physiological environment of the vagina is hostile to harmful organisms that cause infection. The lactic acid-producing Döderlein's bacilli, acting on the glycogen stored in the estrogen-influenced vaginal epithelium, are responsible for the lowered pH that protects against invasion. But anything that disturbs the natural healthy balance between the Döderlein's bacilli and the potentially pathogenic microorganisms—anything that changes the hormonal stimulation, decreases the acidity, or directly damages or kills off Döderlein's bacilli—can result in the growth and flourishing of infectious organisms and a subsequent vaginitis.

The normal vaginal discharge may vary in amount depending on the stage of the menstrual cycle, but it does not itch, burn, or have an unpleasant odor. Infection and inflammation in the vagina, however, result in an excessive, abnormal discharge called *leukorrhea*. It may be heavy in consistency, white or greenish yellow, sometimes malodorous, and it does produce symptoms by irritating the vulva. The inner walls of the vagina, poorly supplied with nerve endings, are relatively insensitive. In contrast, the vulva and vaginal entrance, richly supplied with nerve endings from the pudendal nerve, respond with pruritus, a distressing and embarrassing itchiness . Leukorrhea and painful outside itching signal a vaginal infection.

Candidiasis

Candidiasis, moniliasis, thrush, fungus infection, and yeast infection are all names for vaginitis caused by the same organism, the genus *Candida*, primarily by the species, *Candida albicans.* These yeastlike organisms are commonly found on the skin, in the digestive tract, and may normally inhabit the vaginas of many women (10–15%), but they remain in a balanced relationship with the other organisms. When certain conditions predispose to the overgrowth of *Candida*, the result may be clinical symptoms of vaginitis. The infection can range in severity from a mild and hardly noticeable irritation to a dis-

seminated, highly dangerous, and potentially fatal disease. Fortunately, systemic candidiasis—affecting the entire body—is very rare.

Causes

Since the introduction of the so-called wonder drugs to conquer infection—the penicillins, streptomycins, sulfas, and tetracyclines—the incidence of candidiasis has increased dramatically. It is now the most common cause of vaginitis, reportedly outnumbering the other types by a ratio of three to one, or even higher. It is believed that antibiotics, given for some infection elsewhere in the body, suppress the normal bacteria of the vagina that ordinarily keep *Candida* in check, perhaps by competing with the fungus for the available nutrients. When the harmless and susceptible vaginal bacteria are partially eliminated, the opportunity for the antibiotic-resistant fungus to take over is present. Broad-spectrum antibiotics particularly, such as tetracycline, frequently prescribed for acne, have been implicated in vaginal candidiasis. (Sometimes the antibiotic does not help the acne, either).

The change in estrogen and progesterone levels during pregnancy also predisposes to the development of *Candida* infections.* The high hormone levels result in an abundance of glycogen in the vaginal epithelium. The subsequent ample supply of sugars favors a fungal overgrowth. There are data which indicate that for the same hormonal reasons, women on combination oral contraceptives also have a greater incidence of this kind of vaginitis. Of course, the pill may have an effect on the increased frequency of the disease through other than endocrinological mechanisms, since women on the pill are not protected from their partner's possible infections by

* Since the infection can be transmitted to the fetus while it is still in the uterus or as it passes through the birth canal, a pregnant woman with candidiasis should not delay treatment.

condoms, or contraceptive creams and jellies, which are believed to have an effect in controlling fungus.

Diabetes, because it too produces sugar in the urine and a "sweet" environment in the vagina and vulva, has always been suspected of increasing the growth of this yeastlike organism. There have been, however, several large-scale studies that do not support the contention that diabetic women have a greater tendency to *Candida* infection when compared with control populations (Cibley, 1977).

Other reasons for the overgrowth and increase of *Candida* in the vagina include the administration of corticosteroid therapy, malnutrition, a general rundown condition, and too-frequent douching, particularly with an alkaline or antiseptic solution.

Diagnosis and Treatment

The symptoms of candidiasis are intense itching of the vulva, which is usually worse at night. Because the vulva is frequently red and inflamed, burning, especially after urinating, is also a frequent complaint. The vaginal discharge may be light and watery, but it may also be quite heavy, white, and cottage cheese-like in consistency. Generally it is odorless. Speculum examination may reveal white plaques located on the vaginal walls and variable amounts of the cheesy discharge. Vaginal pH is in the range of 4.3–4.8. The definitive diagnosis is made by a wet smear technique. A cotton swab removes some of the discharge, which is mixed with a few drops of saline solution. One drop of this mixture is placed on a slide that is then viewed under a microscope to search for the filamentous hyphae and spores characteristic of a fungus (see Fig. 9.1). Sometimes there is so much cellular debris present that a drop of 10% or 20% potassium hydroxide (KOH) is added to the slide. The KOH dissolves the epithelial cells but leaves the *Candida* undisturbed for easier viewing. *Candida* species appear as oval bud-

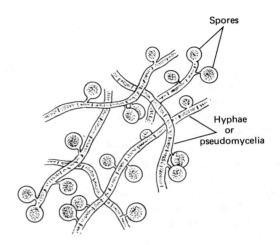

Figure 9.1. Appearance of Candida albicans in a smear prepared from vaginal discharge.

ding cells plus the filaments or pseudomycelia; other nonpathogenic yeast cells may be present as spores only.

When it has been determined that the vaginitis is the result of a *Candida* infection, it can be treated with antifungal preparations. These include both prescription and over-the-counter drugs. Nystatin (Mycostatin, Nilstat), miconazole nitrate (Monistat), and clotrimazole (Gyne-Lotrimin, Mycelex-G) are antibiotic, antifungal agents that can only be obtained by prescription. When inserted into the vagina in suppository or cream form, these drugs specifically attack *Candida* organisms by disrupting their cell membrane permeability to interfere with mitochondrial enzyme activity in the cell. The preparations are very poorly absorbed from skin or mucous membrane of the vagina and so do not get into the bloodstream. They apparently have no toxic effects, and few women are ever allergic to them. They are highly effective when they are used throughout the entire recommended course of treatment—anywhere from 7 days to 2 weeks—depending on which drug is prescribed. Stopping the use of the drug after a few days because the itchiness has disappeared is an invitation to recurrence.

Sometimes, stubborn cases that are resistant to therapy may be the result of reinfection. Because women with vaginal candidiasis frequently harbor the organisms in their gastrointestinal tracts, oral nystatin may be used concurrently with the vaginal therapy to insure that no "seeding" of the fungus from the rectum to the vagina takes place. Another source of reinfection is sexual activity, particularly if the sexual partner is an uncircumcised male. Candidiasis inflammation of the glans under the foreskin is not uncommon, but the infection is rare in circumcised men. It may be helpful for the man to apply nystatin or miconazole cream to the glans of the penis before intercourse. *Candida* species have also been found in the semen of males whose partners have frequent vulvovaginal candidiasis. If the fungus does inhabit the reproductive tract of the male, it is generally asymptomatic. This is fortunate because there is no effective way of treating it in the male. Oral nystatin has such a low rate of absorption into the bloodstream that it could never get to the site of infection. The best protection against ping-pong infections is for the man to wear a condom.

Gentian violet is an aniline dye that acts as a fungicide by blocking enzyme action within the *Candida* cells. There are tampons, creams, and suppository tablets that contain the dye (trade names, Genapax, Gentia-Jel, HYVA tablets), but they are not as effective as a 1% gentian violet in 10% alcohol solution swabbed all over the vaginal and vulval tissues. This is very effective for quick relief of painful itching and inflammation, but it is terribly messy. Not only does the dye stain underpants, towels, and pantyhose bright purple, but it can produce some startling effects on the sexual partner after intercourse. As a general rule, coitus should be avoided during treatment with any medication because it may cause additional irritation. Aside from its temporary esthetic drawbacks, another disadvantage of gentian violet is that it causes allergic reactions in about one out of 100 women.

When their patients are taking antibiotics for acne or other infections, some physicians advise taking Lactinex tablets or eating yogurt to restore the natural intestinal bacterial to the digestive tract. Lactinex contains live *Lactobacillus acidophilus* (the same as Döderlein's bacilli) and *Lactobacillus bulgaricus;* yogurt with live culture contains *Lactobacillus bulgaricus.* The tablets and yogurt are said to be useful in avoiding or treating the diarrhea, gas, and other symptoms produced by the antibiotic's destruction of the normal inhabitants of the tract. There is little scientific evidence to support the effectiveness of ingestion of either the tablets or yogurt, but many people claim benefit. Oral antibiotics will also kill off the lactobacilli of the vagina, possibly causing a *Candida* overgrowth with resultant vaginitis. Some women's self-health groups have, therefore, maintained that the restoring of healthy and friendly bacteria to the vagina early in a fungus infection can actually cut it short before it proceeds into a flaming vaginitis. Two or 3 tablespoons of natural (no fruit!) live-culture yogurt* may be used as a douche, or the yogurt may be directly placed into the vagina with a tampon tube or a vaginal jelly applicator. Most doctors are apt to view this kind of self-treatment as absurd, but there are some who believe the procedure has value. Physiologically, putting the lactobacilli directly into the vagina for candidal vaginitis makes as much, it not more,

---

* To determine whether yogurt contains live bacteria, put a couple of tablespoons of the yogurt into a cup of lukewarm milk and leave it overnight in a warm place. If the yogurt contained live culture, the milk will be thickened by morning. Dannon Plain Lowfat Yogurt (Dannon Milk Products Division, Beatrice Foods Co.) contains live lactobacilli; so does Knudsen Plain Yogurt (Knudsen Dairy Products).

sense than eating the bacteria. The method is easy, inexpensive, not harmful, and is worth trying—it may work. If after several days the self-treatment proves ineffective and the discharge persists, becomes worse, and the tissues more irritated, self-help should be discontinued and a physician seen for treatment.

The efficacy of boric acid in treating candidiasis has been known for years, but a 1981 double-blind controlled study showed that boric acid was a more effective cure for vaginal yeast infections than nystatin. Van Slyke and coworkers compared the use of daily intravaginal gelatin capsules containing 600 mg boric acid powder with identical capsules containing 100,000 units nystatin in nonpregnant college women with verified vaginitis due to *Candida* infection. The period of treatment was 2 weeks. The women were reexamined, and cultures were performed again 7–10 days and 30 days after treatment was discontinued. After 7–10 days, 92% of 56 women who used the boric acid capsules were cured, compared with 64% of 56 women treated with nystatin. The cure rate after 30 days was 72% after boric acid and 50% for nystatin. There were no adverse side effects, and blood analyses for boron showed little absorption from the vagina, an important consideration because of the known toxicity of ingested boric acid. The authors report that boric acid is now their preferred medication for candidiasis and have prescribed the inexpensive capsules, which they make up and dispense themselves, to more than 200 women who came to the student health center for treatment. Since in most areas it is necessary to have a physician's prescription to obtain size 0 gelatin capsules to fill with the boric acid powder, the opportunity for self-treatment is restricted, but women may want to ask their clinicians about the option. Van Slyke and colleagues prescribe one 600-mg capsule daily for 7 days followed by one capsule twice a week for 3 weeks.

"Trich" Infections

When an itching and burning sensation of the vulva and outer vagina is accompanied by a discharge that is greenish yellow or gray, frothy or bubbly, and possibly malodorous, the microscopic wet smear is likely to reveal the presence of *Trichomonas vaginalis*, a parasitic protozoan that moves by means of four whiplike flagellae (see Fig. 9.2). There are estimates that 15%–25% of women carry the organism in their vaginas, but many do not experience any symptoms. It is known that *Trichomonas* is present more frequently in sexually active women, particularly in those with multiple partners, and the organism is currently considered a venereal disease, transmitted through sexual contact. Preadolescent girls occasionally get trichomoniasis, however, and because the protozoa are known to survive 30 minutes in tap water and

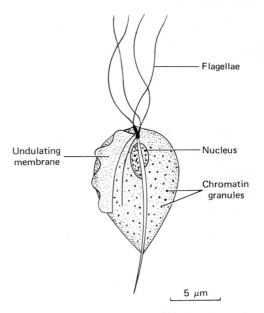

Flagellae

Undulating membrane

Nucleus

Chromatin granules

5 μm

Figure 9.2. Trichomonas vaginalis. Seeing motile trichomonads in a drop of vaginal discharge mixed with saline is diagnostic of a "trich" infection.

6 hours in warm saline, it is remotely possible that an infection may be picked up from swimming pools, toilet seats, and wet wash cloths. Since an estimated 3%–10% of men also harbor trichomonads, mostly in the urethra or prostate, it is generally recommended that when a woman is infected, both she and her steady sexual partner should be treated simultaneously, whether or not the organism has been shown to be present in the male.

The organisms can remain dormant in the vagina for years, although an asymptomatic woman could develop an active infection at any time. Unlike *Candida albicans*, which happily thrives in the acid environment of the vagina, *Trichomonas vaginalis* finds the normal vaginal pH unfavorable for growth, and circumstances that reduce the vaginal acidity thus encourage proliferation of the protozoa. Factors that may result in symptoms of trichomoniasis include pregnancy, trauma to the vaginal walls as a result of childbirth, or even sexual activity. Sometimes a generalized lowered resistance resulting from illness, a crash diet, or even going without enough sleep for a while can induce the growth of trichomonads. Some women experience symptoms after menstruation. In others, growth is triggered by some other factor that produces excessive cervical secretions. Candidiasis or other vaginal infections may activate an infestation with *Trichomonas*, and even an emotional upset will sometimes precipitate or aggravate the symptoms.

The usual treatment for protozoan vaginitis is metronidazole (trade names: Flagyl, by Searle; Metryl, by Lemmon; Protostat, by Ortho, or Satric, distributed by Savage Laboratories). Metronidazole is one of a group of related chemicals that are effective against anaerobic organisms like *Trichomonas vaginalis* that thrive best only in the absence of oxygen. The current method of treatment with Flagyl is a relatively high dose given over a short period of time. One popular schedule is 500 mg metronidazole (two 250-mg tablets) orally every 12 hours for 5 days for a total of 5 g. One 2-g dosage all at once is usually equally successful and probably better for people who forget or dislike taking medication. The 5-g amount produces a 98% cure rate in nonsexually active women but drops to 50%–85% if treated women are reinfected by sexual partners who harbor the organism. Concomitant treatment is thus recommended. Adverse side effects during treatment occur in 10% of people taking the drug, and these are mostly gastrointestinal—dry mouth, nausea, bitter aftertaste, and sometimes diarrhea and abdominal cramps. Headache, dizziness, and other central nervous system reactions also have been reported. Alcohol should be avoided, because drinking during, just before, or after treatment can induce nausea and vomiting.

Metronidazole is a trichomonicidal drug; it is unsurpassed in killing the organisms when taken orally. It also is effective against other organisms normally occurring in the intestinal or genital tract that can cause infection, given the opportunity and an appropriate set of circumstances. A recent issue of the medical journal *Surgery* attests to the drug's ability to substantially lower the risk of anaerobic infections when it is administered prophylactically before various types of gynecological and other surgery. But from the time the drug was released by the Food and Drug Administration for use in the United States, there has been concern that it may do more than eradicate trichomonads and produce unpleasant side effects. Studies indicated that metronidazole was carcinogenic (cancer causing) in rats, induced gene mutations in bacteria, resulted in birth defects in guinea pigs and mice, and caused a transient leukopenia (decrease in white blood cell count) in humans (Rustia and Shubik, 1972, 1979; Voogd et al., 1974, 1981). The evidence for carcinogenicity and mutagenicity of metronidazole in humans is inconclusive, and several follow-up studies of women treated for trichomoniasis with the drug found no evidence of potential for birth defects or increased risk of cancer (Roe, 1983).

Because trich infections never cause permanent ill effects and the disease is not severe or life threatening, there are some who believe that metronidazole should never be used at all. A substantial amount of the drug is ordinarily consumed in a single course of treatment, and it does not always prevent recurrence. Hence, many women are exposed to repeated courses of treatment. According to Sidney Wolfe and Anita Johnson of the Health Research Group, a Washington-based consumer organization, more than 2 million annual prescriptions for Flagyl are written, many times for conditions other than trichomoniasis, such as fungus infections and gonorrhea. In their opinion, such widespread prescribing of a potentially dangerous and strong agent for the nuisance of vaginitis is irrational and is not justified since the potential risks of Flagyl far outweigh the benefits. They requested that the Food and Drug Administration withdraw approval of the use of Flagyl as treatment for trichomoniasis. While such a ban began to be considered by the FDA in 1976, it was never implemented, although the FDA did call for a revised smaller total dosage and a labeling revision that included a warning about the potential hazards. There are currently several research projects, sponsored by the National Institutes of Health, that are underway to assess the risks of metronidazole exposure, but it will be some time before all the questions about the safety of the drug are resolved. It makes sense to avoid taking metronidazole in the first 3 months of pregnancy and to take it only if absolutely necessary during the second half of pregnancy. If not pregnant, it is also prudent, because of the side effects and even the slim possibility of carcinogenesis, to make certain that the vaginitis is actually due to *Trichomonas* organisms.

A trich infection may not have any effect on longevity or general health, but it is certainly not insignificant to the woman who has it. Getting rid of it may be, at the time, of greater importance than the presumed risk of the therapy. It may not be necessary to take Flagyl, however, because there are some alternative treatments. Women with trichomoniasis who are pregnant or those who have a blood or central nervous system disease cannot take Flagyl and have always had to be treated with a local topically applied antitrichomonal agent. Examples of prescription drugs that are used are AVC, Vagilia, or Vagitrol cream or suppositories, which contain sulfa drugs, Aci-Jel containing acetic and boric acids with oxyquinoline, and Aquacort Supprettes. For women whose symptoms are not severe, an over-the-counter preparation, Betadine solution (povidone-iodine) in a swab that is painted on the vaginal walls or as a Betadine douche may be helpful and is not harmful. A mild vinegar douche (1 or 2 tablespoonfuls in a quart of warm water) twice a week can also help to control the symptoms.

These preparations may reduce the itching and discharge temporarily, but they cannot possibly get to all the areas of the genitourinary tract that may be infested by the organisms. In some women, the vaginitis, unfortunately, is very likely to recur, but recurrence also is possible after Flagyl treatment as well.

While metronidazole is, in doctor's language, the "treatment of choice" in the United States, other chemically related compounds, nimorazole (Naxogin) and tinidazol (Fasigyn), are being investigated in Europe and Asia. Given in single oral doses of 2 g, they evidently are equally effective, but they, too, have been implicated as having the same mutagenic potential.

Only 10% of women have the acute classic *Trichomonas* infection—the profuse, frothy, gray malodorous discharge, the severe itching, redness, swelling of the vulva and vaginal opening, the pain on intercourse, the pain on urinating, and the urinary frequency. The other 90% have symptoms that are far less severe. In fact, according to Monif (1982), almost three-quarters of the women harboring trichomonads in their vaginas never have any symptoms at all. To attempt to totally eradicate every last protozoan in the vaginas of women with only occasional mild complaints on the possibility that at some future time they may become more symptomatic seems excessive and needless. Nevertheless, many doctors prescribe metronidazole, a powerful systemic drug with proved carcinogenic and mutagenic effects in animals, because they believe that, as potential disease transmitters, even those women with asymptomatic trichomoniasis should be treated. Since women and men both, however, are exposed to many carcinogenic environmental agents over which they have no control, they need not unnecessarily ingest another one.

No one should trivialize an extensive and severe case of trichomonal vaginitis. It can be sheer misery for the one out of 10 women who has it. The decision as to whether acute symptoms should be treated by metronidazole should be made by women themselves, after it has been determined that the infection is indeed trichomoniasis, and that other, alternative treatments for controlling the symptoms have proved ineffective.

When vaginitis has been diagnosed as a trich infection but the symptoms are relatively mild, a choice as to the mode of treatment can be made, but it should not be left entirely to the doctor. The Food and Drug Administration may also someday get around to making a judgment on the risk/benefit ratio of metronidazole. Informed women need not wait for a federal agency or their physicians to resolve the problem; they are in the best position to decide whether the benefits for them outweigh the possible potential hazards.

Gardnerella Vaginalis

A third common infection of the vagina is bacterial, caused by a gram-negative bacillus formerly called *Hemophilus vaginalis* now known as *Gardnerella vaginalis*. Formerly, when neither fungi nor protozoa could be identified in the discharge, and a vaginitis was present, the condition was called "nonspecific vaginitis." In 1955, Gardner and Dukes isolated the bacterium from the discharge of women with vaginitis and classified and named it *Hemophilus*. They successfully demonstrated the pathogenicity of the organisms by inoculating them into the vaginas of 15 asymptomatic women and causing vaginitis in 11 of the women.[*]

The primary symptoms of a *Gardnerella* infection are discharge and a characteristic "stale-fish" odor. Since the bacteria do not ordinarily cause an inflammatory response in the vagina, itching and burning are usually absent.

The discharge is thin, gray-white, and homogeneous in consistency. It may be scant, moderate, or profuse enough to soak through several layers of clothing and require the use of pads or tampons. If the secretion is mixed with a drop of 10% potassium hydroxide on a slide, a characteristic fishy odor caused by the release of two amines is produced. The same odor may be present after intercourse when the vaginal discharge comes in contact with the alkaline semen. The organisms can be diagnosed by saline wet smears because of the

[*] Prior to the regulations governing experimentation on human subjects, such procedures were not particularly uncommon and illustrate another instance of "woman-as-guinea-pig." According to the published reports, the 15 women were patients in a volunteer clinic, and one can only speculate whether they knew what they were getting into or paid for. Some were allowed to go untreated for as long as 4 months to illustrate that a spontaneous cure did not take place. It might be argued that the information gained was very beneficial and that no lasting harm resulted. But on the other hand, the doctors did not use their own wives as experimental subjects.

presence of characteristic "clue" cells—epithelial cells peppered with myriad small short bacilli. A vaginal pH higher than normal is usual but not diagnostic, and an infection is sometimes confirmed by culture. When it has been determined that the leukorrhea is caused by *Gardnerella*, the reportedly most successful treatment is oral ampicillin, a semisynthetic antibiotic derived from penicillin. An alternative antibiotic is tetracycline, although local intravaginal therapy can also be an option. Metronidazole has also been used with an 85% cure rate according to some gynecologists (Eschenbach, 1983). Apparently, it is not the drug itself, but its metabolites that are effective against *Gardnerella vaginalis*. Sulfonamide creams or suppositories (Sultrin by Ortho and Gantrisin by Roche) or painting the vagina with povidone-iodine are presumably not as successful in curing the infection as any oral therapy unless the local treatment is carried out for at least 14 days. Since *Gardnerella vaginalis* is spread through sexual contact, both partners must be treated, or the woman may become reinfected. Should reinfection occur, it is better to use topical therapy rather than repeated courses of antibiotics or the original problem could be compounded by an additional *Candida* infection.

When vaginitis is not caused by *Candida*, *Trichomonas*, or *Gardnerella* as revealed by wet smear, it may be the result of other bacteria, viral infections, cervicitis, or any number of other reasons, allergic or chemical. Unless a specific organism has been verified, there is no point in taking specific medications. Women should always make certain that an accurate diagnosis has been made before permitting treatment. Empirical treatments with various local and systemic medications that attempt to eradicate vaginitis can sometimes make the original condition worse through overtreatment.

How to Prevent Vaginitis

Perhaps, if they are compared to serious diseases of the reproductive tract, common vaginal infections appear insignificant, but women who have suffered with the discomfort and annoyance of a daily vaginal itch never want to have it again. There are several common-sense precautions that all women can take that may help to avoid vaginitis or to prevent its recurrence.

Daily soap and water washing of the vulva and vaginal area is important because all the little moist and dark crevices can provide a good home for organisms and should be kept clean. After defecation, the perineal area should always be wiped from front to back to avoid spreading any bacteria from the anus to the vagina. Of course, the vagina is not sterile, any more than the oral cavity is, but under the same line of reasoning, one should never put anything into the vagina that is too dirty to go into the mouth. If an unwrapped tampon falls on the bathroom floor, for example, it should be thrown away and another should be used.

Very young girls get vaginitis, and so do women who never have sexual intercourse, but since sexual activity plays some role in the transmission of several kinds of vaginitis, the cleanliness of the sexual partner is also important. Once an infection has become established in a woman, she should abstain from sexual intercourse, and her partner should wear a condom or be treated with the same medication to avoid spreading the organism back and forth between them. Condoms are also very helpful for avoiding infections in the first place. There is evidence from laboratory studies that vaginal birth control creams and jellies can be lethal to organisms other than sperm. A woman using spermacidal foam or jelly and a diaphragm may have less trouble with vaginitis.

Obviously, it is important to avoid anything

that upsets the natural organisms' balance or the natural pH of the vagina. Antibiotics are useful and necessary drugs but should never be taken without good reason. If they have to be used, eating a lot of live-culture yogurt during treatment may prevent digestive upsets, but whether it will maintain the lactobacilli in the vagina is conjectural. As the first symptom, it could be more effective to use yogurt directly in the vagina.

Too much douching can cause trouble. The normal vaginal discharge varies in amount in different women and from time to time in the same woman with hormonal changes during the menstrual cycle. If a woman wants to douche, she should probably not do it any more often than once every 2 weeks. As for the douche itself, there is some thought that no matter what the solution—vinegar and water, sodium bicarbonate, salt, plain water, or a commercial preparation—the vaginal pH restores itself to normal within 15 minutes. There is probably less chance of disturbing the vaginal chemistry, however, if the solution is kept acidic. The douche bag and apparatus should be kept scrupulously clean; they can harbor mold or bacteria. A woman should never douche while pregnant, and some doctors think that vigorous douching during menstruation may initiate endometriosis by driving the endometrial fragments upward. To douche at all is a personal decision, but if the procedure is used, the douche bag should never be higher than 2 feet above the hips, or the water pressure may be too great.

Mundane as it may sound, one of the greatest preventatives for vaginitis is the cotton crotch. Nylon or other synthetic underwear does not "breath"—the weave is not porous or absorbent enough—and the fabric retains moisture, an ideal environment for the growth of organisms. Pantyhose without a cotton crotch should only be worn with a cotton-lined panty underneath. But wearing cotton-lined underpants will not be as effective if tight jeans, knit pants, or a leotard are worn over them. The point is to cut down on the body heat and resulting dampness that can cause irritation. Loose clothing and the ventilation that cotton underwear provides help to prevent vaginitis.

Women who have difficulties with recurring infections have to be especially careful. They should never sit around in a wet bathing suit, for example. At night in bed even wearing pajamas can cause trouble, and they should wear nightgowns or nothing at all. No perfumed or dyed toilet paper, no "deodorizing" tampons, and no vaginal sprays should ever by used by women who appear to be prone to vaginitis. Bath salts or bath oils should be avoided. All possible sources of irritation should be eliminated.

Despite all the measures taken to prevent it, suppose that a vaginal infection does occur and treatment is necessary. For various reasons, sometimes getting to a doctor and getting that treatment is delayed. Sitting in a tub of warm water with the hips and buttocks immersed (sitz bath) for 15 to 20 minutes will help the miserable itching of the swollen and irritated tissues. Between baths, the area should be kept very dry. If a woman has not yet begun using any vaginal tablets or suppositories, wearing a tampon may help because it absorbs the leukorrea inside the vagina and prevents it from spreading onto the sensitive vulvar tissues.

The topical medications used in treating vaginitis have to be inserted high into the vagina. If a vaginal tablet is used, it should be moistened so that insertion is easier; if a cream is used, the applicator should be washed and dried after each use. Because the symptoms usually disappear dramatically after 1 or 2 days of medication and the treatment is sometimes messy or troublesome to use, there is a great temptation to be a therapy dropout. Unless the remedy is continued for the full prescribed time, however, recurrence is likely. It is especially important to carry on treatment right through the menstrual period, but tampons should not then be used since they may soak up all the medication. Even after it has cleared up and no organisms are apparently present in the discharge, a vaginal infection is likely to return after the menstrual period. Application of the jellies or cream that were used on a daily basis to cure the infection should be resumed occasionally or during the first few menstrual periods, even if there are no symptoms or discomfort. Possibly the failure of local treatments to provide long-term cures is due to women not using them for a long enough time in the first place.

OTHER SEXUALLY TRANSMITTED DISEASES (STDS)

Perhaps because vaginal infections are not spread exclusively through sexual intercourse, they are not generally seen for what they really are—venereal diseases. For many years, only gonorrhea and syphilis acquired the major social and psychological stigma associated with VD—that they are dirty diseases and that people are promiscuous if they get them. The fear, ignorance, guilt, and anxiety that surrounded gonorrhea and syphilis influenced the ability to effectively treat these diseases and prevent their spread, since the public seemed determined to remain unaware that these afflictions are illnesses like any other. Now the foregoing two infections have been joined by "new" sexually transmitted diseases that had never even been considered significant to public health until recently. Genital herpes, for example, is epidemic in adolescents and is an infection that provokes even more fear, anguish, and loss of self-esteem than gonorrhea or syphilis. But if all the publicity surrounding herpes has created panic and psychological trauma

way out of proportion to the physical consequences of the disease, there have been two positive aspects: first, the herpes scare has caused the general decline in the nationwide incidence of gonorrhea and syphilis that no amount of public education was able to produce; and second, the recognition that sexually transmitted diseases are mainstream, middle-class, and affect millions of Americans. Diseases like candidiasis, trichomoniasis, herpes, chlamydia, anogenital warts, gonorrhea, and syphilis are contagious illnesses, communicated by people, and the only differences between them and other communicable diseases are the parts of the body they affect, that they are not spread via air or water, and that they are transmitted through very intimate contact. Anyone who is sexually active is at risk for getting a sexually transmitted disease. The only sure method of avoiding infection is celibacy, and even that is uncertain since some vaginal infections become active as a result of other factors than direct transmission. When two virgins meet, marry, and remain forever monogamous, the probability of either getting a venereal disease is virtually nonexistent. There are many in our society like that, but there are also many who are not. Today there are more people interacting sexually with more other people, and the risk of infection has increased dramatically.

The chances of acquiring a sexually transmitted disease can be substantially reduced, however, by incorporating some changes in the patterns of sexual behavior. People who have active sex lives with several contacts must have a constant awareness of their responsibility for their own good health and for the health of others. There is no point in becoming paranoid about venereal disease, and it is difficult to regard the person with whom one is involved in a sexual relationship as a potential threat, but any casual (or anonymous) sexual contact must be viewed with suspicion. The use of barrier contraceptives (diaphragm, condom, spermicidal foam or jelly) is a good idea because they may be at least a partial barrier against transmission. Also, since the symptoms of infection are much more obvious on the exposed genitalia of a man, it is important for a woman to look for them. If any lesion, bump, lump, or sore on the penis or scrotum is seen, she will do herself and her partner a favor by reconsidering sexual contact. If, later, she has any reason to suspect that she has been exposed, and if there are any symptoms, no matter how vague, she must go for diagnosis and treatment. All sexually active women should request a blood test for syphilis and a culture for gonorrhea with the annual gynecological checkup and should certainly never feel resentful or that their privacy has been invaded if their physicians routinely check for these diseases. Should a woman have acquired gonorrhea or syphilis, she must be willing to have sufficient treatment, come back for follow-up examinations and identify her contact or contacts. These two diseases can be prevented and eradicated, and most other STDs can be prevented and controlled if all people behaved responsibly, safely, and with concern for others in their sexual relationships.

VIRAL INFECTIONS

Herpes Genitalis

Herpesvirus has an affinity for attacking the skin surfaces and mucous membranes of the body. An infection can be intensely painful, is apt to recur periodically, and, as yet, there is no cure. At the time of initial onset, compared with almost any other type of vaginal affliction on a misery scale, a herpes infection ranks high. But compared with potentially devastating diseases like gonorrhea, responsible for

most female infertility, and syphilis, a possible killer, herpes is not even in the running.

Obviously, there is not much good about getting herpes, but neither is it as bad as we have been led to believe. It is incurable but is not a significant health risk for healthy adults, and it is no more life threatening than the common cold, which is also incurable. There are plenty of other incurable diseases, such as arthritis, diabetes, and hypertension, with far more serious consequences for the affected individual than herpes. Herpesvirus is transmitted from person to person by intimate contact that brings mucous membrane surfaces together, but people who have contracted herpes can lead good and active sex lives and never spread the disease as long as they take a few precautions. Recurrence of herpes becomes less frequent with time, the manifestations becomes far less severe, and complications of herpes are preventable and manageable.

Herpes infections are caused by a virus. Viruses are minute infectious agents that are composed of a single molecule core of either deoxyribonucleic acid (DNA) or ribonucleic acid (RNA) and an outer lipid-protein coat or envelope that is largely derived from the components of the host cell's membrane. Viruses can only grow inside living cells. Once a virus has invaded, the virus-DNA or virus-RNA acts as a template to use the host cell's DNA-synthesizing system to make more virus. The foreign virus-DNA units then escape from the cells to infect other cells.

Herpes organisms are DNA viruses with a lipid coat. *Herpesvirus varicellae* are the agents that cause herpes zoster, the painful disease known as shingles; they also result in varicella or chickenpox. *Herpesvirus hominis* is the other group; it causes the herpes simplex virus diseases that can involve the skin, the mucous membranes, the eyes, and in newborn infants cause a disseminated generalized infection that is usually fatal. There are two types of the herpes simplex virus, which were identified in 1962. Type I, or HSV-1, causes the familiar "fever blisters" or "cold sores" on the lips and face. It can also invade the cornea of the eye. Type II, HSV-2, is the kind that involves the mucous membranes of the genital tract and is known as herpes genitalis. Today it is believed that approximately 15%–20% of genital herpes infections are caused by HSV-1, and 80% by HSV-2. Research investigation has shown that there are differences in the two strains in terms of recurrence of the disease. Evidently, 60% of type 2 infections recur after an initial episode, but only 14% of infections caused by the type 1 virus recur (Reeves et al, 1981). It also appears that women have fewer recurrences than men. The strain of the virus acquired is irrelevant, since both types produce genital herpes with the same clinical symptoms.

As with other foreign invaders, when viruses attack the body cells, specific antibodies are produced by the body to combat them. But unlike some other viruses, which are completely eliminated by the specific antibodies, the herpesviruses persist after an initial or primary infection, remaining latent for long periods of time. The resistance conferred by the antibodies is relative. In some people, the infection may never return; in others, a secondary or recurrent infection can be triggered by any number of factors—a cold, a fever, a gastrointestinal upset, a severe case of sunburn, sexual intercourse, menstruation, or an emotional stress. The antibodies, too, remain in the bloodstream from then on as evidence of a previous herpes infection.

The first episode or primary attack of herpes genitalis is usually the worst and causes a great deal of discomfort. There are some women, however, in whom an initial infection is very mild or completely asymptomatic. A number of adult women, estimates are 15%–20%, have antibodies to type II virus in their blood but never remember having had an in-

fection. But, more usually, symptoms begin to appear within 2 to 20 days following exposure to the infectious virus. First, there may be a tingling or burning sensation, then an intense itching, and, as the infection progresses, severe pain. Multiple little blisters appear on the cervix, the vagina, and the vulva or clitoris. They may also be present on the buttock or thigh. In a day or two they rupture, become secondarily infected, and result in large, shallow, exceedingly painful ulcers covered by a white membrane. There may be systemic symptoms as well—headache, achiness, fever, and tender, swollen inguinal lymph nodes. Painful urination is not uncommon, and there also may be a profuse vaginal discharge. After a week or 10 days, unless the sores have become secondarily infected by bacteria, they will heal spontaneously, leaving no scars. If a secondary bacterial or fungus infection has occurred and remains untreated, the lesions may persist for as long as 6 weeks.

After the sores have healed, the initial active phase of the infection is over, but the virus is still present in the body and enters a latent phase in the nervous system. In oral herpes, the virus lies dormant in a large group of sensory nerve cell bodies (trigeminal ganglion) located near the cheekbone. The genital herpes virus is apt to enter the nerve endings of the lumbosacral nerves and migrate up to the cell bodies that lie next the lower spinal cord (dorsal root ganglion). When the latent virus is activated, it takes the same nerve pathway back down to skin supplied by the sensory nerve endings. Because the body's defense system is familiar with the virus and has produced antibodies to combat it, the recurrent or second or third infection is of shorter duration and has much milder symptoms. There may be vulvar burning, discharge, and pain, but the severity is considerably less than that of the primary attack. (People who have been exposed previously to HSV but who never had any clinical manifestations will have a first-episode infection that is similar to recurrence—less extensive and less severe). Many have no warning that a recurrence is going to take place, but some men and women have a premonition or *prodrome* of a tingling, burning, or itching sensation that lasts 6 hours to 2 days before the lesions appear. The virus can be transmitted during the prodrome as well as during the active outbreak. This situation poses a problem for sexual partners of individuals with no prodromal symptoms, but the use of condoms or diaphragms may provide partial protection. More than half of people affected with herpes for the first time will never have a recurrence. Only a very small percentage will have frequent, perhaps once-a-month recurrences. The rest will fall somewhere in between, with rare occasional outbreaks.

Transmission
Contact with an active sore that is "shedding" infectious virus is the primary mode of transmission. The chances of a susceptible person contracting herpes from sexual intercourse with an individual who has active genital lesions is approximately 75%. Kissing a person with cold sores will transmit herpes, and the virus can be passed from the lips to the genitals or vice versa by the hands or by oral sex. Self-inoculation to other sites, particularly to the eyes, is possible, and affected individuals who are in an active stage of the disease should be especially careful to wash the hands before touching or rubbing the eyes. This is particularly essential for contact lens wearers.

Although it is well established that abstention from sexual contact during the prodromal period and until the sores are completely healed is a necessity, whether the virus can be transmitted during asymptomatic periods is controversial. Several studies have demonstrated that small amounts of virus particles can be periodically recovered from the cervices of a small percentage of asymptomatic women and the urethrae of an even lesser percentage of asymptomatic men, but the role of such intermittent viral shedding in the transmission of genital herpes is unclear. Spreading the infection from one person to another depends upon the amount of virus shed by the affected individual and the immune susceptibility of the partner. Someone who already has antibodies to oral or genital herpes may not contract the disease after exposure to a small quantity of virus, whereas a nonimmune person could acquire a severe primary infection from a similar exposure. As long as there is a possibility that people who have herpes may shed the virus in between recurrences, the use of condoms and a diaphragm with jelly (vaginal spermicides may also be virucides) may help prevent transmission to the sexual partner.

It has also been reported that the virus can survive for some hours on inanimate objects (towels, toilet seats, door knobs) but no one, as yet, has proved that herpes can be acquired from nonhuman contact. Besides, people would have to position themselves on a toilet seat in a highly unusual fashion in order for a lingering virus to make contact with their genital mucous membranes.

Diagnosis
It would seem that the presence of blisterlike sores on the skin and mucous membranes of the genitalia would be enough to suggest a herpes infection, but there are genital lesions or eruptions from other infections that resemble herpetic lesions, especially if they are not in the blister stage. The most reliable method of confirmation is viral culture on living tissue, but a more convenient and cheaper herpes test, although less accurate than culture, is the Tzanck smear test. A gentle scraping of fluid from the lesion is placed on a slide, allowed to air-dry, and then stained. The presence of giant, multinucleated Tzanck cells

that are two to five times as large as white blood cells is characteristic. A Tzanck smear will not be positive, however, after the third or fourth day of a recurrent lesion.

Treatment

As yet there is nothing that will cure a herpes infection. There is no drug that will rapidly eradicate the lesions, and no medication is known to prevent their recurrence. There are, however, ways of preventing secondary infection, lessening the pain, and accelerating the healing process.

Keeping the lesions clean and dry is very important in promoting healing and preventing a secondary infection. The sore or sores can be washed with soap and water and dried with a blowdryer set on cool. Ice-cold compresses can provide pain relief, during both the prodromal stage and the outbreak. A local anesthetic, such as lidocaine or 2% xylocaine jelly, and a painkiller taken systemically will help in an acutely painful first attack. Any heat to the area only increases inflammation, and some people cannot even tolerate warm water, but unless warmth has an adverse effect, bathing in a warm sitz bath to which a drying agent, such as Epsom salts, Burow's solution, or baking soda, has been added is soothing and relieves discomfort. If pain is very intense during urination, filling up a tub of water and voiding into it will prevent the pain and will also prevent any potential problem with urinary retention. During an outbreak, loose-fitting clothing and cotton underwear are recommended to avoid further irritation.

Since drying up the sores and keeping them clean aids healing and guards against secondary bacterial infection, various salves and ointments one might think would be helpful are not recommended since they could actually prolong the outbreak and cause new lesions to form. These include such self-help treatments as aloe vera gel, vitamin E oils and creams, yogurt, or other kinds of herbal or natural remedies.

There are some drugs that are known to act as antiviral agents, that is, they interfere with the steps leading toward DNA replication. Among them are adenine arabinoside (ARA-A, also known as vidarabine), and acycloguarosine or acyclovir.

Acyclovir, marketed as Zovirax by the Burroughs Wellcome Company, was approved by the FDA in 1982 in the form of a 5% topical ointment. When applied to the skin lesions in a first-episode primary herpes genitalis attack, acyclovir helps shorten the healing time, and reduces pain, itching, and viral shedding. Unfortunately, the drug has no effect on either the incidence or duration of recurrent herpes attacks. Studies have shown similar clinical effectiveness on an initial infection for oral acyclovir, but neither oral nor intravenous forms of the drug appear to be able to eradicate the virus from its dormant state in the nervous system and prevent recurrence (Bryson et al. 1983). Other virucidal drugs with similar biochemical activity are in preliminary stages of clinical evaluation. They include fluoroiodoaracytosine (FIAC), bromovinyldeoxyuridine (BVDU), and several others with unpronounceable names. (e.g., BIOLF-62, ABpp, and Ara-MP). Investigators are also looking at 2-deoxy-D-glucose.

Since there is no effective cure for herpes, prevention of the disease by vaccination is an alternative that is getting considerable research attention. Currently, there are several ongoing trials testing the ability of newly developed vaccines to confer immunity and thus protection against herpes. Data from one study, reported in *Medical News*, was the result of a collaborative effort between the University of Birmingham in England and Rush Medical College in Illinois. After 1 year, there were only two cases of infection in 200 vaccinated partners of herpes sufferers, when ordinarily about 100 cases would be expected.

In a second group of 40 volunteers who were vaccinated after the primary attack, the recurrence rate was 25% rather than the expected 90%. Of a third group of 200 individuals, vaccinated after a primary attack and one recurrence, 50% reportedly had less frequent recurrences and reduced symptoms. Another trial conducted by Merck Sharp & Dohme pharmaceutical company was initiated in 1983 with the administration of their vaccine to 500 spouses of herpes victims. Although the results look promising, these clinical trials will take several more years before there are definite answers on the safety and efficacy of the current vaccine, and the concept of protection via vaccination is not without practical problems. Since having the natural infection does not seem to result in immunity to recurrence or autoinoculation to another skin site on the body, at this time the development of a safe vaccine that can actually confer a lifelong immunity seems doubtful. Moreover, the cost of mass-producing a vaccine is currently being projected as economically unfeasible, and the other questions of how to and to whom this expensive vaccine is to be administered are also unanswered at present. There is little doubt, however, given the enormous advances of research in molecular biology in the past 10 years, that effective vaccination against genital herpes will be possible within the next decade.

Complications

An active herpes genitalis infection during pregnancy can be very dangerous for the baby. If it occurs during the first 3 months, there is a significantly increased likelihood of miscarriage. If it occurs late in pregnancy, the risk of premature delivery increases. Furthermore, a maternal primary infection can be passed on the baby by delivery through the infected vagina. Such infection in the newborn, who has an undeveloped immune system, can be fatal or result in severe systemic or nervous system damage. Apparently, even asymptomatic women can pass infection to the infant through vaginal delivery. One study showed that 39 of 56 women who delivered HSV-infected infants had no signs or symptoms of herpes at the time of birth (Whitely et al., 1980). Jarratt (1983) has provided some recommendations for pregnancy and delivery when there is a risk of herpes in a newborn infant. Any woman who has had herpes genitalis or whose sexual partner is affected is a high-risk mother. She should have weekly cervical and vaginal cultures for HSV starting with a month before delivery. If the cultures are positive for HSV within 2 weeks before delivery, or if she has genital herpes lesions at the time of delivery, a cesarean section should be performed. If vaginal delivery is performed, there should be no fetal monitoring with scalp electrodes to avoid direct inoculation.

The undesirability of a herpes infection is obvious, but anywhere from 5 to 20 million Americans have it, and that means a lot of people have learned to live with it. All of the publicity surrounding herpes has frightened, more than enlightened, people about the reality of having the disease. The Milwaukee police official who wanted the department's cardiopulmonary resuscitation program closed down because he worried about herpes transmitted from the training mannequin is only one example of the irrational anxiety caused by the magazine articles and television programs.

It is important to remember that the infection is acquired through physical contact with the shed virus, either from a visible sore, or just prior to the appearance of an active outbreak. In the case of labial cold sore, this means no kissing or oral–genital contact. In the case of genital herpes, it means no sexual intercourse until the lesions are *completely* healed. The routine use of condoms and spermicidal jellies and foams in between recurrences may reduce the risk of infection since asymptomatic viral shedding, if it occurs, is likely to be in relatively low amounts. These contraceptive methods cannot, however, prevent infection when lesions are present.

If a woman is confronted with a first-episode case of herpes, it is important to remember it is not the end of the world. It may never become reactivated again, and, even if recurrences take place, they reportedly tend to burn themselves out in frequency, usually after about 3 years. Since the virus tends to recur when the body is weakened, doing one's best to maintain good general health could be the best defense.

The national organization called HELP (Herpetics Engaged in Living Productively) has local chapters that offer information and group support to herpes sufferers. The address is 260 Sheridan Avenue, Palo Alto, CA, 94306. The telephone number is 1-800-227-8922, which is also the VD National Hotline number.

Hepatitis

There are three main viruses associated with hepatitis or inflammation of the liver. Hepatitis A, formerly known as infectious hepatitis, hepatitis B, previously designated as serum hepatitis, and a form not related to either A or B and called non-A, non-B hepatitis, may all cause serious and disabling disease. All three types are believed to be sexually transmissible, although they may also be spread through other means. Hepatitis A is spread via the gastrointestinal tract. It can be acquired by drinking polluted water, eating uncooked shellfish from sewage-contaminated waters, from food handled by a hepatitis carrier with poor hygiene, and from oral/anal sexual contact. Type B is transmissible from saliva, blood serum, semen, menstrual blood, and vaginal secretions. The incidence of hepatitis B has been shown to be higher among gay men, prostitutes, and patients attending venereal

disease clinics, providing further evidence for its spread through sexual contact. The infectious virus passes through tiny breaks in the skin or across membranes, and the number of asymptomatic chronic carriers is high.

Little is known about the sexual transmission of non-A and non-B hepatitis viruses. Since they appear to behave very much like hepatitis B virus, it is presumed they are transmissible by sexual contact.

A pregnant woman with an active case of hepatitis B or carrier status can transmit the virus to her fetus across the placenta, but this is relatively uncommon. More frequently, the infection of the baby occurs at birth via contact with body fluids of the mother, and the infected infants then tend to become chronic carriers with a subsequent higher risk of developing liver disease. Hepatitis A evidently is unable to cross the placenta or cause fetal or newborn infection.

In common with all virus diseases, there is no specific treatment for hepatitis. Rest, good nutrition, and a slow return to previous levels of activity are the main therapies. Prevention of hepatitis, as with any sexually transmitted disease, is related to avoidance of multiple and casual sexual encounters with anonymous sex partners.

After exposure to hepatitis A, protection or reduction of the impact of the disease can be obtained through a single dose of gamma globulin. Vaccination against hepatitis B has been possible since 1981, when the Food and Drug Administration released a vaccine derived from inactivated virus particles obtained from the plasma of chronic carriers. This vaccine has proved to be both safe and effective, although it still is very expensive. The cost of the full treatment of three injections is $100.00, and the vaccine must be refrigerated at all times. This obviously limits the use of the vaccine in underdeveloped countries where hepatitis is a grave public health problem.

Cytomegalovirus

Cytomegaloviruses (CMV) belong to the herpesvirus family, but unlike infection with HSV-1 or -2 or the varicella/zoster types, they rarely produce any overt clinical symptoms. Most of the population acquires CMV during childhood by respiratory spread, and 40%–80% of the population is infected by puberty. After the age of 40, between 80% and 100% of adults have antibodies to CMV. This indicates that the virus must be a universal parasite on human populations and, in the vast majority, causes asymptomatic infection. In a few rare instances, the virus results in illness. Occasionally, some people develop a syndrome similar to infectious mononucleosis with the abrupt onset of a high, irregular fever lasting several weeks, tiredness and weakness, and blood cell changes, but the sore throat and enlarged lymph nodes characteristic of infectious mononucleosis are missing. CMV can also become reactivated to result in life-threatening pneumonia or inflammation of the liver in chronically ill patients or people whose immune systems are suppressed. In pregnant women, CMV infection has been incriminated in miscarriage and is able to cross the placenta to infect the fetus, resulting in the birth of an infant with possible mental retardation. Although as many as 95% of all infants infected in this way are asymptomatic at birth, a few may actually be profoundly affected. It is estimated that significant mental and hearing impairment may occur in one out of every 1,000 live births as a result of congenital CMV infection (Monif, 1982).

The transmission of CMV is by means of contact with body fluids containing the infectious virus material, specifically saliva, urine, tears, and transfused blood. The presence of the virus has been shown to occur in cervical mucus and in semen, making sexual transmission a possibility, although not a certainty.

Condylomata Acuminata

Condylomata acuminata, commonly called genital or venereal warts, are caused by a papilloma virus similar to the one that causes skin warts anyplace on the body. In men, the genital warts occur most commonly on the glans and urethral opening on the tip of the penis, on the shaft of the penis, and on the scrotum. In women, the most common site is the perineum, but they may also be scattered over the vulva, the vaginal opening, the skin of the thighs, or occur within the vagina and on the cervix. For unknown reasons, the condylomata have a tendency to become very extensive during pregnancy, and sometimes the growths are so prominent they may even obstruct the vagina. The means of transmission of the virus is primarily by sexual intercourse, although there have been cases where the warts have appeared in children or in individuals whose only sexual partner does not have them.

A possible association between genital warts and cervical cancer is suspected, largely based upon the hypothesis that cervical cancer is caused by a venereally transmitted virus. The identification by Meisels et al. (1981) of atypical, flat, genital warts on the cervix, almost indistinguishable cytologically, histologically, and colposcopically from very early carcinoma of the cervix, has renewed interest in the human papilloma virus infection as a causative agent. Further epidemiological evidence for the connection between condyloma and cervical cancer has been presented by Reid et al. (1982).

The treatment of genital warts depends on their size and number. If they are not too large or extensive, they can be removed by application of podophyllin, a resin obtained from the mandrake plant. The podophyllin, dissolved variously in alcohol, benzoin, or mineral oil in solutions of 15% to 25%, is painted on the surface of the warts and allowed to remain for 4–6 hours. If it is not washed off thoroughly at that time to remove the residue, the podophyllin can cause severe slow-healing chemical burns. Most people experience considerable pain from treatment; in some it can be persistent and even incapacitating. Two to 4 days after podophyllin application, the warts dry up and slough off, but in some cases, several weekly treatments are necessary. Wiswell (1983) suggested a mixture of 10% podophyllin, 10% salicylic acid, and 10% trichloracetic acid in a colloidin base as a more effective, less messy preparation that does not need to be washed off. Alternatively, under local anesthesia, the warts can be burned off with electrocautery, frozen off with cryosurgery, or treated with CO_2 laser surgery. Sometimes, many warts coalesce to form a large cauliflowerlike growth, which then has to be removed surgically.

BACTERIAL INFECTIONS

Gonorrhea

By law, all cases of gonorrhea must be reported to local health officials, making gonococcal infection the most frequently *reported* communicable disease in the United States today. A million cases are noted annually, but it is estimated that there are actually three times that many because of all the infections that private family doctors "forget" to report to the authorities. Only the common cold is more common.

Gonorrhea is a serious and potentially very severe bacterial disease, and it is rapidly becoming more and more resistant to cure. In common with all other STDs, it is an equal opportunity infection—no one is immune to it regardless of race, creed, sex, or sexual pref-

erence. Also like other STDs, gonorrhea is not a poor-person disease; anyone at any social or educational level can get it. Of particular significance is the fact that anywhere from 50% to 90% of the women who get gonorrhea are totally symptom-free. Unlike men who are well aware that they have contracted "a dose of the clap"—they have clear-cut, painful symptoms of urethritis—women may have symptoms that are so vague and nonspecific that they never go to see a doctor with a complaint. There may be no complaint.

Because women are so frequently asymptomatic, they are regarded as the real "problem" in the spread of gonorrhea—an unwitting vast reservoir of infection for unsuspecting males. Of equal, if not greater, concern is the problem of the irresponsible infected man who neglects to inform a woman of her exposure. Because of the widespread prevalence of gonorrhea, a woman must realistically recognize that unless she is celibate or has a mutually monogamous sexual relationship, she is sooner or later likely to be exposed to gonorrhea. A thorough knowledge of all aspects of the disease is essential.

The organism in gonorrhea is the gram-negative bacterial diplococcus, *Neisseria gonorrhoeae.* It grows well only in the moist mucous membranes of the body. Away from the body the bacteria are very susceptible to drying and lowered temperatures and die within seconds. Although nonvenereal transmission of gonoccocal infections has been known to occur in infants through a mother's contaminated hands, it would be extremely difficult to catch the disease from toilet seats, towels, bedclothes, doorknobs, and so forth. One study published in the *New England Journal of Medicine* (Gilbaugh and Fuchs, 1979) demonstrated that when pus from the urethrae of male patients with gonorrhea was *inoculated* onto toilet seats and toilet paper, the organisms could survive in the pus for 2–3 hours. The authors emphasized, however, that this finding alone would not be enough to explain the acquisition of gonorrhea from toilet seats, and some other factor must be involved in transmission. They suggested that "contaminated toilet paper has greater potential as a direct source than do toilet seats." Since it is extremely difficult to imagine circumstances under which individuals would share contaminated toilet paper, nonsexual transmission of gonorrhea, though a possibility, remains unproved.

Gonococcal infection begins with the direct contact of the mucous membranes of an infected person with the mucous membranes of an uninfected person. The surfaces of the gonococci have hairlike projections called pili that enable the invading bacteria to anchor themselves to the mucosal surface and colonize, which they may or may not do successfully. People have varying individual resistance to the disease. It is estimated that in one penile–vaginal sexual contact with an infected partner, a man has a 20%–50% chance of contracting gonorrhea, and a woman has more than a 50% chance. Of course, repeated sexual activity with an infected person virtually ensures that the disease will be acquired. Rectal intercourse may result in inoculation of the bacteria into the anus and rectum, and oral–genital contact can result in gonorrheal pharyngitis or tonsillitis. Mouth-to-mouth kissing alone cannot transfer gonorrhea bacteria.

After an infecting exposure, almost all men, (an estimated 10% are asymptomatic) know that they have caught something. After only 1 day or as much as 2 weeks later, there will be a thick, pus-laden discharge from the urethra at the tip of the penis, and urination will be very painful. The urine is cloudy with pus and sometimes even a little bloody. Many men also have enlarged and tender lymph nodes in the groin. These signs are hard to miss or ignore. The severe discomfort, coupled with the campaigns that have made VD a household word, are motivation enough to send men off to the clinic, the private doctor, or the student health service for diagnosis and treatment.

In women, the characteristic symptoms, if there are any at all, are vaginal discharge and painful and perhaps frequent urination. The discharge is distinctively green or yellow in color with an unpleasant odor, but there is so little of it that it may be unnoticed. Besides, these kinds of symptoms are so common to women throughout their nonsexual and sexual lives that they are very likely to completely ignore them.

If the disease remains local, that is, confined to the lower genital tract, the bacteria may involve the urethra, vagina, and the cervix. If Skene's glands are invaded, pushing up on the urethra will result in a pus-filled discharge from their ducts. Bartholin's glands may also become infected and their ducts obstructed. In a small percentage of women, one gland may become abscessed and produce tenderness that makes walking or sitting extremely painful. (Not all Bartholin's gland abscesses, however, are caused by gonorrhea.) The organisms can also lodge in the glands of the cervix and cause the production of a profuse and irritating gonorrheal discharge. Gonorrhea bacteria also can remain in the cervix, silently and latently, and later proliferate and become the source of an upper reproductive tract infection. The cervix can look normal with no unusual discharge, but a culture of material from the endocervical canal will provide evidence of the presence of the organisms. The anus and rectum may also be sites of involvement. The condition, called gonoccoccal proctitis, develops in 40%–60% of women with gonorrhea and is usually the result of the bacteria travelling from the infectious vaginal discharge to the anus. Few women notice any symptoms, although there may be a little itching in the perineal area or a little mucus drainage from the anus.

Sometimes, a local gonorrhea infection is self-limiting. There is no further spread—the disease either undergoes a spontaneous regression, or a woman can become an asymptomatic carrier. Actually, the first recognition that a gonorrhea infection is present may occur after several months when, in about 50% of untreated women, the bacteria spread to the upper genital tract and cause major complications. Undiagnosed and untreated, or inadequately treated, the disease ascends upward through the endocervical canal to the endometrial cavity of the uterus, further on to the fallopian tubes, and out into the peritoneal cavity. Although the cervix usually acts as a barrier to the spread of disease organisms, the cervix is dilated at menstruation, and the gonococci rise into the uterus where they rapidly propagate on the dead cells, glandular secretions, and blood from the sloughing endometrium. The inflammation of the fallopian tubes, or *salpingitis*, can be acute, with pain in one or both sides of the lower abdomen, a temperature of 102° or higher, and nausea and vomiting. Usually, fever and nausea mean that the organisms and their inflammatory products have spilled out of the fimbriated ends of the tubes into the body cavity. When the peritoneum and the ovaries become involved, the condition is known as *acute gonorrheal, salpingo-oophoritis*, less accurately called PID, pelvic inflammatory disease. A significant further complication of gonorrheal PID is tubal and/or ovarian abscess. A ruptured tubo-ovarian abscess necessitates surgery with removal of the uterus and both ovaries and tubes. Fitz-Hugh–Curtis syndrome is another major complication in which the gonococci migrate to the liver. The severe pain is similar to, and has to be distinguished from, the pain of acute hepatitis, pneumonia, perforated ulcer, kidney stones, and gallstones.

Subacute salpingitis produces much milder symptoms so similar to those caused by other gynecological problems that it is difficult to diagnose. Even if cured by treatment, the damage done by acute or subacute salpingitis remains permanently. The scarring of the fallopian tubes that results is a major reason for infertility and is a possible contributing factor in ectopic pregnancy. PID can become chronic, flaring up at intervals with recurrences of acute inflammation.

More rarely, and only in about 1% of untreated gonorrhea cases, the gonococci enter the bloodstream, a condition called disseminated infection. Occurring predominantly in women, a disseminated gonococcal infection produces a characteristic skin rash, chills, fever, and arthritic joint pains in the wrist, ankles, knees, and feet. If disseminated infection is unrecognized and untreated, or if treatment is delayed, the joints become permanently damaged. Occasionally the gonorrhea bacteria can also invade the heart (endocarditis), the brain (meningitis), and the liver (toxic hepatitis).

If a pregnant woman has gonorrhea when she delivers her baby, the infection is transmitted to the infant's eyes as it travels through the cervix and vagina. The eye infection, called *gonococcal ophthalmia neonatorum*, can cause blindness. It is prevented by dropping a silver nitrate or penicillin solution into all newborns' eyes immediately after delivery. As better protection against neonatal eye infections, all pregnant women should be screened for possible gonorrhea early in pregnancy and also several times during the last months.

Diagnosis and Treatment of Gonorrhea

The presence of a gonorrhea infection can be diagnosed by a microscopic examination of a stained smear of secretions from the patient. The inaccuracy of this method has resulted in its replacement by a bacteriological culture test. Specimens obtained from the major anatomical sites the bacteria invade—the urethra, cervix, rectum, and pharynx—are streaked across a special nutrient jelly.

Neisseria gonorrhoeae are not easy to culture; they need special media that contain antibiotics to inhibit the growth of other organisms and incubation in an atmosphere with increased carbon dioxide. In a positive culture, colonies of gonorrhea bacteria appear on the incubated medium within 24–48 hours.

Penicillin is still considered the best treatment for gonorrhea, except, of course, in individuals who are allergic to it. Aqueous procaine penicillin G in a dosage of 4.8 million units is divided into two injections, one given in each buttock. The procaine is a local anesthetic to minimize the pain of injection. Thirty minutes before injection, 1 g of the drug probenecid is given orally to prevent too rapid excretion of the penicillin by the kidneys. This combination of penicillin and probenecid maintains a high blood level of the antibiotic that is toxic to the gonococcal bacteria. Because people with gonorrhea have become increasingly likely to have a coexisting infection with chlamydiae, bacterialike organisms insensitive to penicillin, the Centers for Disease Control current guidelines recommend an alternative treatment for gonorrhea: 50 mg tetracycline by mouth, four times a day for 7 days, total dose 14 g.

Follow-up cultures are absolutely necessary to make certain that the gonorrhea has actually been eradicated. There was a time when the gonococcus was very sensitive to penicillin, and 200,000 units of that drug would destroy the bacteria. Now, after widespread use of penicillin and its availability in many parts of the world without prescription, it takes 4.8 million units, and the gonococci are becoming more resistant all the time. In the late 1970s, reports began to appear of gonococcal strains that were not susceptible to penicillin no matter how large a dose was administered. Evidently, some of the *Neisseria gonorrhoeae* strains had acquired the abil-

ity to produce an enzyme, beta-lactamase, or penicillinase, that inactivated penicillin and made it valueless in treatment. The new penicillinase-producing *Neisseria gonorrhoeae* (PPNG) strains, first isolated in southeast Asia, have been found in more than 40 countries and 16 of the United States. The reported incidence in the United States was about 400 cases a month in the early part of 1983, compared with more than 80,000 monthly reported cases of non-PPNG. The gonorrhea that penicillin is powerless to eliminate can be cured by spectinomycin, sold only by the Upjohn Company under the brand name Trobicin. It is four times as expensive as penicillin, limiting its use in many areas of the world. The prevalence of the new gonorrhea in the United States is still relatively low, but researchers are beginning to have nightmares about the future. When the penicillin-resistant strain spreads, and when it develops a resistance to spectinomycin as well, what then? There have already been reports from Asia and the Pacific of a spectinomycin-resistant penicillinase-producing *Neisseria gonorrhoeae* strain, double-trouble organisms that are insensitive to both spectinomycin and penicillin. Such doubly resistant strains can currently be successfully treated with yet another antibiotic, cefoxitin plus probenecid, or with cefotaxime in a single injection without probenecid. But the probable next chapter in the antibiotic resistance story is obvious.

Equally worrisome about the penicillinase-producing gonococci is their possible ability to spread their talent for penicillin-resistance to other bacteria. The genetic information for producing the penicillin-deactivating enzyme is transmitted by gonococcal plasmids, small circular extrachromosomal DNA molecules found in the bacteria's cytoplasm. Plasmids may be transferred between microorganisms by bacterial conjugation, a process somewhat analogous to sexual reproduction. Such plasmid interchanges can take place between different bacterial species, although the more distant the relationship, the less efficient the transfer. But the possibility of PPNG organisms incorporating their trait into the bacteria that cause meningitis, for example, is another nightmare for public health officials.

As with other STDs, a vaccine against the organisms that cause infection would be a significant breakthrough. One gonorrhea vaccine trial was initiated in 1983 and is undergoing clinical testing on volunteer American servicemen stationed in the Pacific. Several other kinds of gonorrhea vaccines are also being tested in the laboratory.

Syphilis

The prevalence of infectious syphilis (an estimated 100,000 cases a year) is nowhere near that of gonorrhea, the documented epidemic. But accompanying the tremendous increase in the incidence of gonorrhea has been a parallel rise in the number of cases of syphilis, and the long-term seriousness of this disease is much greater. Syphilis can be a killer. Untreated, it can affect the central nervous system to produce blindness, deafness, insanity, and ultimately, death.

Part of the difficulty with the treatment of syphilis may result from its lower incidence. Not expecting to see it, private physicians may misdiagnose the condition or miss it completely. In women, the infection can be clinically unapparent in the primary and secondary stages, only becoming obvious in the tertiary stage. Even then, the visible lesions that are present are similar to those found in perhaps 20 other diseases, and the confirmation of syphilis must be made by expensive and time-consuming tests. Blood tests for syphilis are mandatory in most states for people obtaining a marriage license, giving blood, or joining the army, and they are performed on all pregnant women. A woman not in any

of those categories can easily remain undiagnosed and untreated.

The microorganism that causes syphilis is a spirochete, *Treponema pallidum.* Its length is about the diameter of the largest white blood cell, but it is so thin it is almost undetectable by light microscopy and must be viewed with a darkfield microscope. It is an extremely delicate organism, very sensitive to drying and temperature, so transmission via inanimate objects is virtually impossible. An open lesion is highly infectious, and there have been instances where doctors, dentists, or nurses in their professional work have contracted the disease through a break in the skin. The usual transfer of the spirochete, however, results from vaginal, anal, or oral–genital sexual intercourse.

The *Treponema* organisms invade any moist mucosal surface, although they can also enter through a minute break in intact skin, and they reach the bloodstream. Within 24 hours, they spread throughout the body. After an incubation period averaging about 3 weeks (it may be as short as 9 days or as long as 90 days), during which the infected individual has no symptoms, a primary lesion called a chancre appears at the site of contact. The chancre is a hard, painless ulcer that in women appears on the vulva, vagina, or cervix. Treated or not, the primary lesion will spontaneously heal, leaving no evidence, but the blood is still infectious. If the chancre is on the vagina or cervix and a woman has had no other reason to see a gynecologist, the disease will remain undetected and untreated.

The secondary stage appears 2–6 months after the initial exposure, sometimes occurring at the same time the primary chancre is subsiding. The only symptoms of the second phase may be flulike—headache, slight fever, loss of appetite, perhaps a general achiness. In addition, there may be a generalized non-itching skin rash on the body. In the genital areas, the syphilitic rash may form *condylo-*

mata lata, growths that are similar to genital warts. These lesions are swarming with spirochetes and are highly contagious. It is during the secondary stage in a pregnant woman that *Treponema pallidum* is able to cross the placenta and infect the developing fetus. There are innumerable other diseases that syphilis mimics in the secondary stage—anything from "mono" to allergies have many of the same symptoms. Less common manifestations are hair loss and the presence of greyish white lesions in the mucous membrane of the mouth and throat. If there are eruptions on the oral membrane, kissing could be contagious, especially if the person being kissed by the infected individual has broken skin in or around the mouth.

Left alone, untreated, the symptoms of the secondary stage will also go away, although they may recur during the following 2 years. The disease has now progressed into the latent stage, and after a few years have passed, it is no longer infectious. In two-thirds of the people with untreated syphilis, that is the end of it; they are no longer contagious to their contacts, they have no further symptoms, and they may live out the rest of their lives without ever knowing that they had syphilis. A blood test will always be positive, however, and in a untreated pregnant woman with latent syphilis, the spirochetes in maternal blood can pass to the unborn child. Syphilis during pregnancy contributes to spontaneous abortion (miscarriage) and stillbirths. Should the pregnancy go to completion, most full-term deliveries result in a baby with congenital syphilis, having marked and severe deformities. Fortunately, the incidence of congenital syphilis is far less than it used to be. Blood tests for syphilis are a routine part of prenatal examination, and if the infection is discovered early enough, treatment produces protection for the infant. Even if the fetus is already infected, control of the disease is possible, and further damage is prevented.

In the remaining one-third of the untreated

latent syphilitics, the disease progresses to the tertiary or late syphilis stage. The complications of late syphilis can affect any part of the body. They may be benign, producing only a skin lesion, called a *gumma*, with no further disability, or there may be a massive body reaction to the long-term presence of the spirochetes that involves the brain and spinal cord (neurosyphilis), the heart and lungs, and many other systems of the body. For 10% of individuals affected with tertiary syphilis, it is fatal. There is no way of knowing which cases of latent syphilis will get tertiary syphilis or how severe and extensive the complications will be.

Diagnosis and Treatment

Syphilis can be diagnosed by examining the fluid from a lesion for spirochetes under a darkfield microscope. All of the blood or serological tests are based on antigen–antibody reactions that occur between *Treponema* and the host. The VDRL (Venereal Disease Research Laboratory) test is widely used because of its low cost and simplicity, but it has its limitations. Evidently any number of conditions—chicken pox, various collagen diseases such as arthritis or systemic lupus erythematosus, measles, pneumonia, even drug abuse or pregnancy—may falsely give a positive VDRL. It can also give a false-negative result since it is positive in only about 70% of patients with primary syphilis. A negative VDRL, therefore, cannot rule out active disease. Another widely used method, sensitive enough to detect all stages of syphilis and specific enough to result in very few false positives, is the FTA-ABS (Fluorescent Treponemal Antibody-Absorption) test. The TPI, or Treponema Pallidum Immobilization, is also a highly specific test that is currently used only for confirmation in problem cases because of its expense and difficulty. Tests used in screening large populations are the TPHA (Treponema Pallidum Hemagglutinin Assy) and the RPR (Rapid Plasma Reagin). Any positives or doubtfuls showing up with these rapid and simple tests can be evaluated further with more sensitive procedures.

The treatment of syphilis in any stage is easy: penicillin, in amounts appropriate to maintain a continuous low blood level for a number of days. *Treponema pallidum* has fortunately not developed a resistance to penicillin and can be eliminated by it. An alternate drug for individuals with a penicillin allergy is tetracycline, but this antibiotic, if given during the first 3 months of pregnancy, may result in birth defects and causes fetal teeth and skeletal problems if given in later stages of pregnancy. The recommended drug for a pregnant woman sensitive to penicillin is erythromycin. Treatment for syphilis must always have follow-up examinations and repeated blood tests for at least a year to make certain that a cure has taken place.

Although gonorrhea and syphilis are two very different diseases, it is obvious that they have many things in common. Both, except in very rare instances, are transmitted through sexual contact. Both present a danger to the unborn child when a pregnant woman is infected. Having either one of the diseases once does not confer any kind of immunity from future infection; it is possible to contract either gonorrhea or syphilis immediately after cure. It is even possible to contract them both and have them exist simultaneously. Both diseases are found to a great extent in younger people. The majority of reported cases of gonorrhea and syphilis in the United States are found among individuals between the ages of 15 and 29. Finally, the social and behavioral factors in the incidence and spread of both diseases may be equal to, if not more important than, the medical aspects.

CHLAMYDIAL INFECTIONS

An ubiquitous little bacterialike organism, *Chlamydia trachomatis*, may be edging out herpes genitalis for the dubious title of agent that causes the most prevalent venereal disease in the United States. It is now responsible for more cases of pelvic inflammatory disease and its resultant infertility than gonorrhea. Like herpes, chlamydia infections are nonnotifiable diseases, but there is a general perception that they are more common than gonorrhea with upwards of 3 million cases in the United States annually.

Chlamydia infection is an inclusive term that describes three major groups of diseases caused by 15 recognized serotypes or strains of *Chlamydia trachomatis*. Serotypes A, B, Ba, and C are responsible for a chronic eye inflammation called trachoma. Trachoma has long been endemic, or continually prevalent, mainly in Africa, the Middle East, and Southeast Asia where it affects hundreds of millions of people. It is highly contagious in early stages and, untreated, results in scarring of the eyelids to blindness.

Serotypes L-1, L-2, and L-3 cause lymphogranuloma venereum, (LGV), a sexually transmitted condition that causes genital or rectal sores, possible inguinal lymph node involvement and, in the later stages, massive swelling (elephantiasis) of the penis or vulva as a result of inflammation of the genital lymphatic vessels. LGV is relatively rare in the United States and is more common in tropical and semitropical climates, although one manifestation of the disease, anal proctitis, has recently become a greater problem for homosexual men. The remaining serotypes D, E, F, G, H, I, J, and K are the sexually transmitted strains that cause genitourinary infections in men and women associated with a variety of complications and acute eye infections or chlamydial pneumonia in newborn infants who have acquired these diseases in passing through the infected mother's genital tract.

Nongonococcal urethritis (NGU), as the name implies, is urethritis that is not caused by *Neiserria gonorrhoeae*. NGU is estimated by the Centers for Disease Control as being twice as prevalent as gonorrhea, and about half of the cases of NGU are caused by chlamydia infection. Males show more symptoms of NGU than females. In men, the infection differs from the urethritis produced by gonorrhea in that the discharge is watery rather than thick and mucoid and is more profuse. The pain on urination, however, is less severe; men are likely to complain more of smarting or stinging and not of burning. The mildness of the symptoms frequently leads them to delay treatment and complications, such as epididymitis, can occur. Moreover, lack of treatment provides more opportunity for the infections to be transmitted to their sex partners. More than two-thirds of women whose male partners had chlamydia-caused NGU were found to have a chlamydia infection of the cervix. (Sweet et al., 1983), and another study found that one-fourth of women whose sex partners had chlamydial urethritis also had chlamydial infections in their urethras (Paavonen, 1979). Women are often asymptomatic carriers, unaware that they harbor the infection. They are, hence, diagnosed and treated less often than men. Unattended and untreated, however, the disease can lead to serious complications, which include pelvic inflammatory disease, acute salpingitis that can result in subsequent infertility, acute inflammation of the tissues surrounding the liver or, in pregnant women, the possibility of transmitting the infection to the baby during delivery.

Julius Schacter, director of the World Health Organization Reference Center For Chlamydia at the University of California, San Francisco, has estimated that an infant delivered vaginally through the cervix of a woman with chlamydial cervicitis has a 60%–70% risk of acquiring the infection (Schacter et al., 1979). More than 100,000 newborns are infected annually. About 30%–70% of these exposed children develop an acute eye infection called inclusion conjunctivitis within the first

2 weeks of life, and 20% develop pneumonia within 2–3 months. Conjunctivitis and pneumonia are the two well-established sequelae of neonatal infection, but there is some evidence that chlamydia is involved in middle ear infections, nasal passage obstruction, inflammation of the bronchioles and even in sudden infant death syndrome as a result of an inability to breathe.

Babies delivered by cesarean section do not acquire a chlamydial infection, but there has been a report of severe maternal pelvic infection from *C. trachomatis* following a cesarean (Cytryn et al., 1982). Chlamydia has also been implicated in other complications of pregnancy and postpartum endometritis.

Since the majority of women who harbor chlamydia are asymptomatic and, thus, untreated, the disease is a potential risk to their health and any children they may have. It would seem reasonable to screen pregnant women for chlamydial infections so that they may be treated to eradicate the disease. This would protect the infant and reduce the risk of maternal complications. Until recently, the barrier to maternal screening has been the lack of an easy, inexpensive diagnostic test for chlamydia infection. A simple stained smear test, used to confirm many kinds of infections, is only 40% accurate in identifying chlamydia, and the only certain method of verification is by special tissue culture techniques that are both expensive and time consuming. But two simpler and cheaper tests were approved for marketing by the Food and Drug Administration in late 1983. The first one, developed by Abbott Laboratories, is an enzyme-linked immunoassay (EIA) test; the other, devised by Syva Company, uses an antibody obtained through genetic cloning techniques. The monoclonal antibody, labeled with a fluorescent dye, will attach to chlamydia organisms swabbed on a slide. When viewed through a special microscope, the organisms show up as fluorescent green. The availability of quick and inexpensive tests that can detect chlamydia without culture should have a significant impact on maternal and infant infection as well as on the treatment of the 3–4 million nonpregnant women and men with the disease, even if asymptomatic. Chlamydia infection is curable with the right antibiotics; these include tetracycline, doxycycline, and erythromycin, but not penicillin. The disease is so prevalent in the United States that, as previously indicated, the latest guidelines from the Centers for Disease Control recommend that tetracycline or doxycycline be used as treatment for gonorrhea since as many as 45% of gonorrhea patients have coexisting culture-documented chlamydia infections.

Schacter maintains that many chlamydia cases can be cured in men and women without using any diagnostic tests at all if physicians would recognize that there are presumptive indications for giving antichlamydial therapy. He recommends that all NGU cases and their sex partners be treated with tetracycline, that all gonorrhea cases also get tetracycline routinely instead of penicillin, and that all women with endocervicitis, pelvic inflammatory disease, and nonbacterial urinary tract infections receive the appropriate therapy for chlamydia-caused disorders.

URINARY TRACT INFECTIONS

The term urinary tract infection (UTI) is a catch-all phrase used to describe conditions responsible for the symptoms of dysuria, or burning pain on urination, and urgency, the feeling of having to pass urine immediately and often, even right after having voided. Other symptoms of cystitis, or bladder infection, include dull, aching pain above the pubic bone and passing urine that is either pink, blood-streaked, or frankly bloody enough to

redden the toilet bowl. The initial dysuria/frequency symptoms of cystitis can be alleviated or sometimes even cured by drinking a lot of water, but once the infection really takes hold or there is any blood in the urine, it is not going to get better or go away by itself. It means the lining of the bladder is severely inflamed, and a doctor should be seen as soon as possible. Urethritis and cystitis are lower UTIs; they produce misery but are easily treated and cause no complications. Untreated bladder infections, however, can progress to a kidney infection, an upper UTI that is potentially a far more serious disease. Symptoms of a renal infection are chills followed by fever higher than 101°F, pain and tenderness in the flanks, and nausea and vomiting.

Sometimes painful urination is caused by infection unrelated to the urinary tract. Vaginitis that results from *Trichomonas*, *Candida*, or *Gardnerella* infections can cause burning on urination as the urine (especially the last few drops) passes over the tender, inflamed vulva. This kind of pain on urinating is described by women as feeling "external" and is usually unaccompanied by urgency or blood in the urine. It has not been established whether the three organisms that cause vaginitis cause urinary symptoms because they actually infect the urethra or whether the dysuria is the result of the labia-irritating discharge, but clearing up the vaginitis or the chronic cervicitis will often completely relieve the urinary symptoms.

In women without vaginitis, symptoms of cystitis result from two other groups of organisms. The sexually transmitted pathogens—chlamydial, gonorrheal, or herpes—can cause "internal" dysuria and frequency, but rarely cause blood in the urine. A urine sample will show either no pus cells in the urine or what is referred to as sterile pyuria, that is, pus cells in the absence of demonstrable bacterial organisms. This latter finding is often due to chlamydial or gonococcal infection that must be treated by the appropriate antibiotic.

The most common cause of non-sexually transmitted UTI in women is the result of their own bowel organisms, primarily *Escherichia coli*, but also *Staphylococcus saprophyticus* and *Klebsiella* species. These bacteria, found in the colon but also on the perineum and in the vagina, can ascend the short (2-inch), straight urethra in women and migrate up into the bladder, there to reproduce to millions of bacteria within 24 hours. This is likely to happen at least once during the lifetime of almost all women; in some unlucky ones, it returns again and again and again. Why some women experience recurrent bladder infections while others have one or none is unknown, but there are several theories. It is speculated that those with frequent infections are more susceptible because they (1) are unable to produce local antibodies or lack the antibacterial defense mechanisms to fight their own colon bacteria once they get into the urethra or bladder; (2) lack the specific competing organisms that would inhibit the growth of colon bacteria; (3) possess colon bacteria strains that have an enhanced ability, perhaps through specialized attachment organelles, perhaps by unique virulence properties, that enable them to infect the urinary tract; or (4) have unique factors about their own cells lining the urinary tract that allows them to be more hospitable to the attachment of colon bacteria.

A sudden increase in the frequency of sexual activity results in bladder infections (honeymoon cystitis) for some women. The rectal bacteria have inadvertently been thrust up into the urethra during foreplay or intercourse. Urinating them away immediately after coitus can help rid the urethra of the unwanted invaders before they have the opportunity to spread and multiply. It may be, however, that a poorly fit diaphragm used for birth control can be an obstruction to the necessary washing out of the bacteria. Dr.

Larrian Gillespie, a woman urologist with a practice in west Los Angeles, has found that a contraceptive diaphragm that is fit too large may press against the bladder neck and constrict the flow of urine by changing the urethrovesical angle. She recommends a slightly smaller size or a switch to another form of contraception (Gillespie and Margolis, 1983). A general rule in diaphragm fitting is that if it can be felt by the woman, she has been fit with too big a size.

Even in women who are not diaphragm users, recurrent UTIs can often be traced to some factor and thus prevented. It is very important to try to empty the bladder completely when voiding; a small amount of urine left at the bottom of the bladder pools to form a culture medium for avid bacterial growth. Drinking a lot of water daily is another significant method of prevention. It flushes the urine as well as any bacteria out of the bladder. Some have suggested drinking cranberry juice to make the urine more acid and less hospitable to bacteria, but the amount of fluid consumed is probably more important than its content. Taking vitamin C (ascorbic acid) also acidifies the urine, but it is best to keep the amount below 1,000 mg daily. There is some evidence that long-term megadose amounts greater than 1 g in some individuals may increase oxalic acid levels in the kidneys and promote kidney stone formation. Other possible factors that have been implicated in recurrent bouts of cystitis include allergy to spermicidal jelly or cream, too-tight jeans, wearing synthetics instead of cotton in the crotch area, bath salts, or colored toilet paper. For women with a truly "touchy" urethra, symptoms of dysuria/frequency can result from exercise, the vibrations of long auto rides or the pressure from the narrow seat of a 10-speed bicycle.

For a confirmed, culture-proven urinary tract infection, the conventional treatment is antibiotic treatment. Women with chlamydial infection should be treated with tetracycline.

It is least expensive in generic form, but must be taken on an empty stomach and is apt to cause gastrointestinal symptoms. Doxycycline is synthetic tetracycline, much more expensive but more easily tolerated. It is also frequently prescribed for nonchlamydial UTIs. Other antibiotics used to treat the common *Escherichia coli* bacterial form of bladder infection are amoxicillin, ampicillin, and cephalexin, the latter more frequently used during pregnancy. Nitrofurantoin (Macrodantin) should be taken with food or antacids to avoid gastrointestinal difficulties. Sulfamethoxazole (Gantanol) and sulfisoxazole (Gantrisin) are old standbys. The trouble with the standard forms of treatment is that having to take one to two pills four times a day, sometimes for 7–10 days, especially if they cause stomach upset, leads many women to forget about them once the initial symptoms disappear. Relapse or reinfection are then very likely to occur. After 1980, a number of reports in the medical literature described the success of single-dose treatment, as long as the UTI was confined to the lower urinary tract. Depending on the antibiotic used, cure rates of 44%–100% have been reported (Hocutt, 1982). The choices include amoxicillin in a single 3-g dose, sulfisoxazole in a 2-g dose, or double-strength trimethoprim/sulfamethoxazole (Bactrim or Septra), usually taken as three tablets. A woman should discuss the less expensive, more easily tolerated option of single-dose treatment with her clinician.

Multiple recurrences of UTI, or a chronic infection that appears to be intractable to the usual antibiotic therapy, may have to be investigated for the possibility of a structural abnormality of the urinary tract or for kidney involvement. Cystoscopy (an X-ray of the bladder after it is filled with opaque fluid) and an intravenous pyelogram (IVP), in which contrast medium is injected into a vein to be eventually excreted and show up on kidney X-rays, are diagnostic methods to identify con-genital obstructions, stones, or other anatomical abnormalities. When the urinary system has been determined to be structurally normal and the cystitis frequently recurs, the exact therapy is controversial. Some clinicians treat each recurrence with antibiotic as though it were a new episode. Others will prescribe a long-term antibiotic prophylactically. There is some evidence that antibiotic given as prevention for 6 months kept women free of cystitis during that time, but fewer than 10% remained free of infection once the prophylaxis was discontinued (Harding et al., 1979). A 2-year period of prophylaxis in a small sample of women has been, in the researcher's opinion, shown to be effective and well tolerated (Harding et al., 1982). Repeated dilations of the urethra, a painful procedure advocated and practiced by some urologists, in an attempt to stretch the urethra so that the bladder can more easily empty completely, is viewed as a futile and unnecessarily traumatizing treatment for recurrent cystitis by other urologists.

Women who get frequent bladder infections will appreciate the information and self-help techniques in former cystitis sufferer Angela Kilmartin's book, *Cystitis: The Complete Self-Help Guide*, published by Warner Books, New York, 1980.

INFESTATIONS THAT ITCH—"CRABS" AND SCABIES

The general term for infestation of the skin and hair by lice is called pediculosis. Closely related to the head louse and the body louse, the insect that attaches itself to pubic hairs to cause the condition commonly known as "crabs" is *Phthirus pubis*. When viewed under the microscope, the louse, between 1–4 mm in length, looks very much like a tiny crab (see Fig. 9.3). The blood sucking lice attach them-

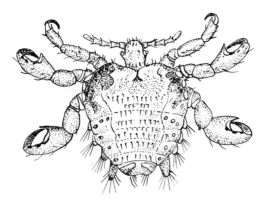

Figure 9.3. An adult female pubic louse.

selves to the skin at the base of pubic hairs and feed off the tiny capillaries in the skin. They are hard to see unless a hand magnifier is used. Lice do not move very much, and they look like little brown freckles on the vulva. The female lays her eggs or "nits" at the base of the pubic hair. In 7–9 days, the nits hatch and attach themselves to the skin.

Pubic lice are transmitted by sexual contact. Occasionally, infestation can occur from toilet seats, towels, or infected clothing or bedding, but the louse dies if it is separated from its human host for more than 24 hours. Most people with this infestation have intolerable itching, but others do not have any symptoms at all. Difficult as it may be to understand, some individuals have crab lice that they can pass on to others and evidently may never by aware of it. The treatment is easy—a prescription drug called gamma benzene hexachloride (brand name Kwell) that comes in a cream, lotion, or shampoo. Used as directed, Kwell gets rid of the organisms and brings rapid relief from the itching. It should never be used more frequently than the package insert recommends. Like other chlorinated hydrocarbon insecticides, Kwell is potentially toxic to the central nervous system if it gets into the blood stream at sufficiently

high-dose levels. Since it can be absorbed through the skin, exceeding the recommended usage by larger or more frequent doses can be hazardous, especially to children.

After treatment, all underwear, towels, washcloths and bed linen should be freshly laundered before using, but an outer clothing that has not been worn for 24 hours can be worn again. The lice are not able to survive for any longer than that away from the body.

Scabies, usually called "the itch," is caused by a mite (see Fig. 9.4). The female burrows into the epidermis of the skin and lays her eggs along the tunnel. After an incubation period of 4 weeks or more, the signs of infestation—a fine, wavy dark track with a small intensely itchy bump at the end—become evident. Scabies is transmitted by intimate sexual or close body contact with an infected person. It is also treated with Kwell or with benzyl benzoate preparations.

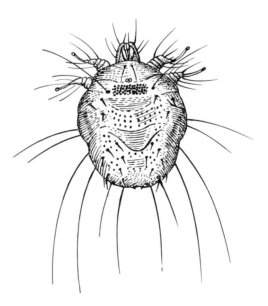

Figure 9.4. Sarcoptes scabei, the mite that causes scabies or "the itch."

ACQUIRED IMMUNE DEFICIENCY SYNDROME

The acquired immune deficiency syndrome (AIDS) is a recently recognized disease found primarily (73%) in young homosexual or bisexual men living in large cities in the United States. The next largest groups at risk for AIDS are intravenous drug users (17%), persons born in Haiti who have recently emigrated to the United States (3.5%), and patients with the hereditary blood-clotting disease, hemophilia (1%). About 6% of AIDS victims cannot be placed into any of the above three risk groups, are classified as "other," and include heterosexual partners of AIDS patients, recipients of blood transfusions, and patients for whom information on risk is either incomplete or unknown. In late 1984, the Centers for Disease Control (CDC) reported a total of nearly 7,000 cases of AIDS, and almost half of the patients had died. Ninety percent of the victims were 20–49 years old; almost half were between ages 30 and 39. Less than 10% of the cases were reported in women. The CDC has not included children in its surveillance of reported AIDS cases, but children who have what appears to be AIDS have been born to mothers who have either had AIDS themselves or were in a high-risk group for the disease.

The alarmist and hysterical publicity generated by AIDs would appear to be disproportionate, given the number of cases reported thus far. But the 40% mortality, and the rapid spread of the disease, with two new cases per day diagnosed, attest to the serious nature of the disease.

Victims of AIDS have a defective immune status. They have a lowered lymphocyte count; in particular, the lymphocyte population known as helper T cells is greatly depleted whereas the lymphocyte subpopulation called the suppressor T cells is not reduced. Helper T cells aid in maintaining immunity to infection and supressor T cells inhibit this activity. Severe reduction of the helpers while the number of suppressors remains the same can make AIDS patients highly vulnerable to "opportunistic" infections caused by organisms that ordinarily would be unable to produce disease in normal healthy individuals. Some AIDS victims develop *Pneumocystis carinii* pneumonia, herpes simplex, toxoplasmosis, cytomegalovirus, candidiasis, and tuberculosis, as well as an unusual form of cancer called Kaposi's sarcoma. Mortality in these individuals is associated with either the cancer or one of the other infections. Even if people with AIDS survive the infection, there has been no indication that they ever completely recover their normal immunity status.

The symptoms of AIDS are nonspecific: swollen lymph nodes, unexplained weight loss, bumps or a rash on the skin, chronic fatigue, fever, night sweats, diarrhea, cough, and shortness of breath. Having any of these signs does not mean an individual has AIDS, but their persistence, especially in a high-risk group, would be a good reason to get medical attention. Laboratory evidence of immune deficiency or the presence of Kaposi's sarcoma or any of the other serious opportunistic diseases would establish the diagnosis. The incubation period is currently suspected to be between 6 months and 2 years; whether a victim is infectious throughout this period is unknown at present.

The cause of AIDS is also not confirmed, and major efforts to pinpoint the agent or agents are underway. Based on epidemiological and research evidence, it is speculated that (1) AIDS is caused by a new transmissible entity, that (2) a member of the retrovirus family similar to the human T-cell lymphoma virus is a likely possibility, and that (3) the agent is spread by sexual contact or by injection in a fashion similar to hepatitis B. The transmission of AIDS appears to require direct exposure to the body fluids of an affected individual—blood, saliva, or semen. The common link between homosexual men, intravenous drug abusers, and hemophiliacs is suspected to be their exposure to such body fluids of affected individuals. For drug addicts the most likely mode of spread is through shared needles. Hemophilic patients require frequent injections of blood containing an antihemophilic clotting factor extracted from the pooled blood of many donors. Presumably, their total exposure to contaminated blood would thus be much greater than that of an individual who received transfusions of whole blood after surgery. But in order to reduce the risk to the public of contracting AIDS through a blood transfusion, the Public Health Service and the Food and Drug Administration issued guidelines to blood banks around the country instructing them to initiate procedures designed to limit the spread of AIDS. All persons at increased risk for AIDS should voluntarily abstain from donating plasma or blood. Further, special screening techniques that include laboratory tests, medical histories, and physical examinations should be used by blood banks to identify individuals with a high probability of transmitting AIDS. A method that checks donated blood for the presence of virus antibodies associated with expoure to AIDS has been developed, but the test's reliability as a mass screening device has been questioned. Until an accurate blood test for the detection of AIDS is confirmed, it was also recommended that any blood collected from known or suspected AIDS victims be used only for blood derivatives when the manufacturing process eliminates infectious agents. Although the risk for a nonhemophiliac of getting AIDS from a blood transfusion has been crudely estimated at about one in 100,000, it makes good sense to make use of autologous blood transfusions (use of one's own blood banked until needed after surgery) whenever possible.

To date, there has been no evidence that transmission of AIDS can be air- or waterborne, and casual contact, such as touching an AIDS victim, apparently presents no risk since friends and co-workers of AIDS patients have not contracted the disease. Of course, health care personnel and laboratory workers must protect themselves from direct contact with the body fluids of AIDS patients, but this can be accomplished by the use of standard hospital precautions for infectious diseases.

When a mystery disease with unknown cause, unknown transmission, and unknown potential for fatality suddenly appears, the serious concern for public health, both in the United States and internationally, is understandable. The research efforts to solve the enigma of acquired immune deficiency syndrome are a high priority, and the scientific community is relatively optimistic about the ability of advances in biomedical knowledge and sophisticated technology to come up with the answers, ultimately. But as one scientist observed, ". . . one must conclude realistically that the epidemic of AIDS will get worse before it gets better" (Dowdle, 1983).

BENIGN AND MALIGNANT CONDITIONS OF THE CERVIX

The cervix is the lower constricted part of the uterus. Half of the cervix projects into the vagina and is called the *portio vaginalis* or the vaginal cervix. It is covered with pale pink stratified squamous epithelium that is continuous with the rest of the epithelial lining of the vagina. The external os is the opening into the cervix from the vagina. It leads into the endocervical canal, which opens into the uterine cavity at the level of the internal os. The endocervical canal, about 3 centimeters long, is lined by a different kind of epithelium. Here, the cells, arranged in single rows, are tall, columnar shaped, and ciliated. There are about 100 infoldings of the endocervical epithelium that are called cervical glands. The cervical glands contain many mucus-secreting cells.

The junction of the red columnar epithelium of the endocervical canal with the pale pink stratified squamous epithelium of the vaginal cervix is called the squamocolumnar junction. The color change between the two different epithelia is obvious; it has been described as being the same as that between the skin around the mouth and red of the lips. In the majority of women, the union of the squamous epithelium and the columnar epithelium is not sharp and well defined. Rather, there is a zone of cells in which there is a gradual transition between the two types of linings, a transitional or transformation zone in which the cells can differentiate either into squamous or columnar cells. Most cervical cancers arise in this tranformation zone.

The most common benign conditions of the cervix are cervical erosions, cervicitis, and cervical polyps (see Fig. 9.5). Some of these conditions are so common that many authorities believe that the "normal" cervix—the perfectly healthy cervix—is the only ideal and is rarely seen. Most of the descriptive terms for the cervical appearance were originated when the cervix could only be examined with the naked eye. With the introduction of the colposcope, an instrument that allows direct stereoscopic visualization of the cervix with magni-

fications of six to 40 times, it became evident that much of what were formerly considered "conditions" were actually normal and physiological (see Fig. 9.6).

Erosions and Cervicitis

Cervical erosions are a misnomer because nothing is actually eroded or missing, and there is no loss of epithelium. In colposcopic terminology, an erosion is called an ectropion; that is, the endocervical lining is found outside of its normal boundaries. In a so-called erosion, patches of the red columnar epithelium of the endocervical canal have shifted position and have extended out beyond the original squamocolumnar junction to replace the paler squamous epithelium of the vaginal cervix. The patches, which have a soft and velvety appearance, contain the same glands as the regular lining of the endocervical canal, and they produce the same clear mucus. Why the endocervical mucosa moves from its ordinary location is not really understood, but it is assumed that the movements are influenced by a woman's hormonal status and, possibly, by local environmental factors in the vagina. In adult women, it has been postulated that cervical erosion may be associated with a previous inflammation of the cervix, with oral contraceptives, with the hormonal influence

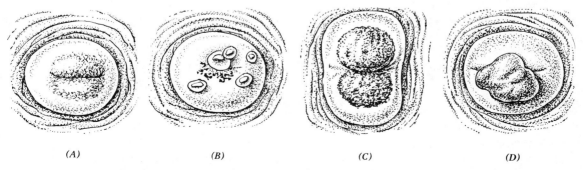

(A)   *(B)*   *(C)*   *(D)*

Figure 9.5. Benign conditions of the cervix. (A) An eversion, misnamed an erosion. (B) Nabothian cysts. (C) Laceration with eversion after pregnancy. (D) A cervical polyp.

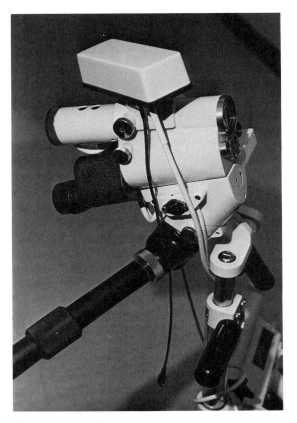

Figure 9.6. Colposcope.

of pregnancy, or with the trauma of child-birth.

Erosions commonly cause no symptoms, although in some women more than the usual amount of a clear nonirritating vaginal discharge may be produced. A normal sequel to a cervical erosion is the inward growth of the flat squamous epithelium to replace the endocervical epithelium in its normal limits. When this occurs it is called *squamous metaplasia.* If this ingrowth of squamous epithelium obstructs the openings of some of the endocervical glands, the cervical mucus continues to accumulate but has no way of exiting. It accumulates as little round bumps or nabothian cysts, named after an eighteenth century anatomist, Martin Naboth.

When speculum examination reveals a red patch and nabothian cysts on the cervix, and there are no symptoms, opinions vary as to what to do about such an "erosion." Some doctors regard all cervices that are not covered by squamous epithelium with suspicion, believing that any change in the normal growth pattern of cells in the transformation zone can be a forerunner of cervical cancer. They use electrocautery (burning) or cryosurgery (freezing) to destroy the area of displaced tissue. The epithelium regenerates and heals over, restoring the cervix to a more normal appearance. Since sensory nerve endings on the surface of the cervix are very sparse, extremes in temperature produce little or no discomfort. Electrocautery and cryosurgery can be performed in the doctor's office without anesthesia.

There is evidence that the destruction of cellular overgrowth does prevent the progression to cervical cancer, but, of course, not all cervices become cancerous. If a woman has no discomfort or distressing symptoms as a result of the condition, it may not be necessary to treat it at all. Even minor surgical procedures on the cervix carry some risk of complication and may be viewed as both medically and economically unwarranted. But the criteria of symptomatic or asymptomatic may not be enough to guarantee that erosions or ectropions should be treated or left alone. First, there must be a differentiation between a perfectly normal benign overgrowth of endocervical mucosa and a malignant process, and there is no way that the naked eye can do that. A Papanicolaou test (Pap smear) must be made. If the Pap smear is negative, no symptoms can be taken as an indication for no therapy.

Cervicitis

Chronic cervicitis is a catchall term that implies the presence of chronic inflammation or infection. It is used to describe everything from symptomless erosions, which are not caused by infection, to an obviously inflamed everted cervix, which bleeds on contact and produces quantities of a thick yellow-white purulent discharge containing organisms not ordinarily found in the vagina. Women with the latter kind of chronic cervicitis may also complain of backache, feelings of urgency and frequency of urination, and spotting or bleeding after intercourse. Symptomatic cervicitis is indistinguishable from cervical malignancy, and colposcopy or cytological smears and tissue biopsy must be performed. Once cervical cancer has been ruled out, the usual treatment for chronic cervicitis is electrocautery or cryosurgery.

Acute cervicitis is an acute infection, accompanied by vaginitis, and can be caused by any of the conditions already described—gonorrhea, syphilis, candidiasis, trichomoniasis—as well as almost any pathogenic bacterial agent and a number of viruses. The treatment of acute cervicitis is the appropriate therapy for the specific organism that has caused it.

Cervical Polyps

Cervical polyps are small extensions of the uterine endometrium or the endocervical canal that look like a little tail coming out of the external os. They may be single or multiple and are most common after age 40. If the extruded part is manipulated, even slightly, the polyp will bleed easily. Polyps very rarely become cancerous, but they also have little tendency to spontaneously disappear. Small polyps can be removed by a simple twist at their base and then cauterization of the base to prevent bleeding. Larger polyps may require a more extensive surgical procedure.

Cervical Incompetence

Cervical incompetence is a term used to describe a situation in which the internal os is too wide and is actually not competent to contain a fetus for the full 9 months. The result is spontaneous abortion in the second trimester (3 months) of pregnancy. Typically, when miscarriage is caused by an incompetent cervix, the cervix painlessly dilates, there are no labor contractions, there is very little bleeding, and the fetal membranes (bag of waters) rupture with a gush shortly before delivery. If a woman has an obstetrical history of repeated miscarriage after the first 3 months, it is likely that she has an incompetent cervix. If she is not pregnant, this diagnosis can be tested by determining whether it is possible to insert a dilator 6–8 mm in diameter through the external os into the endocervical canal without discomfort—a procedure that ordinarily would be very painful and require an anesthetic. A hysterogram, or X-ray of the uterus after an opaque dye has been injected through the cervix, may show widening at the level of the internal os and further confirm the diagnosis.

An incompetent cervix can be caused by a congenital anatomical defect, but many cases are caused by trauma. Dilation of the cervix to alleviate menstrual cramps, a procedure that has proved to be useless but may still be practiced, often causes damage resulting in later cervical incompetence. Sometimes a previous delivery caused cervical lacerations that were not recognized at the time and hence not adequately repaired, or perhaps a woman has undergone an overly enthusiastic dilation of the cervix when she had a therapeutic or diagnostic D and C.

If a woman wants to become pregnant after the diagnosis of incompetent cervix, a surgical treatment, known as the Shirodkar procedure or the Macdonald modification, is performed. A strip of suture material or nylon ribbon is stitched around the cervix in a pursestring fashion in order to keep it tight. The operation, said to have about a 75% success rate, does not guarantee that the pregnancy will be maintained. A woman who has had the procedure should be aware that any signs of impending miscarriage, such as contractions or bleeding, mean that she must *immediately* get to the hospital. There can be serious consequences during labor if the circular suture is not released. If the pregnancy has proceeded to term, either the pursestring is cut for vaginal delivery or a cesarean section is performed.

The Pap Test for Cervical Cancer

In 1928, George Papanicolaou, an American scientist, first described the use of the vaginal smear for the diagnosis of early genital cancer. His special staining technique for cervical and vaginal cells form the basis of the Pap smear, a test that should be performed at regular intervals on every woman. Because of this test, there has been a steady decrease in the number of deaths from cervical cancer; it is now one-third of the rate of 40 years ago.

To perform the test, cells from the endocervical canal are aspirated with a glass pipette or scraped off with a cotton-tipped applicator or wooden spatula and smeared onto a glass slide. The smear is immediately "fixed" to prevent deterioration of the cells by spraying with a commercial fixative or by dropping the slide into 95% ethyl alcohol solution. The cells are later stained and examined microscopically for malignant changes by a cytologist or a cytotechnologist, experts in cell analyses. The technique of the physician in collecting the sample and preparing the smear is as important as the interpretation of the cytology laboratory in obtaining accurate results.

The following is the original Papanicolaou classification used for reporting the results of the smears:

Class 1 Absence of atypical or abnormal cells

Class 2 Atypical cytologic picture, but no evidence of malignancy

Class 3 Cytologic picture suggestive of, but not conclusive, for malignancy

Class 4 Cytologic picture strongly suggestive of malignancy

Class 5 Cytologic picture conclusive for malignancy

According to the above system, Class 1 is negative, Class 2 is atypical but benign and probably caused by infection. Class 5 is definitely positive. Classes 3 and 4 are "maybe" and require more tissue study. Most cytologists today do not limit themselves only to this numerical classification, but also use a verbal description that indicates the precise nature of the abnormality.

Cervical Cancer

The report of an abnormal Pap test would understandably frighten any woman, but it does not necessarily mean that she has cervical cancer. It only indicates that some of the cells shed from the epithelium of the cervix or endocervical canal are atypical, and that there should be further investigation. If she has been having annual tests, a positive smear more than likely means that a possible premalignant condition has been uncovered. It can be treated and cured, making certain that further development to cancer cannot take place. The Pap test is the best kind of preventive medicine; if all women were screened, deaths from invasive cancer of the cervix could be completely prevented. For many reasons, based on social, political, and economic issues that make adequate health maintenance unavailable to large segments of the population, relatively few American women are being routinely screened by Pap tests. The incidence of cervical cancer is higher in low-income groups, and there is a higher rate among Puerto Rican women and black women than among whites. The disease is rare among nuns, virgins, and lesbians, and it is accepted that there is a direct relationship between sexual activity and cancer of the cervix.

Cytological screening could eliminate further mortality from cervical cancer since the disease in its invasive or life-threatening form does not just suddenly appear out of nowhere. There is every indication that true cancer of the cervix, a malignancy that has spread, may take 10 or more years to develop. The cancer first exists as a mild abnormality of atypical cells, which progresses through a series of intermediate stages, each more atypical than the last. Some of these early stages have the ability to regress to normal or, in a number of instances, may remain stable without further progression for an indefinite period. Even if untreated, an area of abnormality does not inevitably go ahead and become cancerous. The Pap test, however, picks up the early changes from the normal when they are preclinical and preinvasive, and the possible evolution to cancer can be stopped.

The question of how frequently Pap smears should be taken is currently controversial. For many years all women were told to have an annual Pap smear, but the most recent recommendation of the American Cancer Society is a screening every 3 years for low-risk asymptomatic women, primarily because yearly tests are presumably not needed for a cancer that develops as slowly as cervical cancer. Some doctors disagree with the longer interval, arguing that the incidence of invasive cervical cancer and subsequent deaths from the disease have declined in the United States, probably as a result of earlier diagnosis from annual screening. A National Institutes of Health consensus panel, convened to make recommendations concerning screening for cervical cancer, hedged somewhat by recommending that, if a woman has had two negative smears 1 year apart, rescreening should be repeated at regular intervals of 1–3 years. Given the differences of opinion, an individual woman's decision on the frequency of her Pap smears should take into account her own risk status for cervical cancer. The shorter interval is appropriate for a woman whose sexual behavior places her in the high-risk category—early age at first intercourse, currently sexually active and with multiple sex partners. This is consistent with the epidemiological evidence associating cervical cancer with a venereally transmitted virus. Of course, many women in a low-risk group will continue to have an annual relatively inexpensive Pap smear anyway in conjunction with their yearly breast and pelvic examination.

Most cancer of the cervix begins in the transformation zone of the squamocolumnar junction, where the cells have the ability to differentiate into squamous or columnar cells. More than 90% of cervical cancer is squamous cell carcinoma on the outside of the cervix; the rest are adenocarcinomas, malignancies of the glandular columnar cells of the endocervical canal. The natural progression of a squamous cell carcinoma is believed to be as follows:

Normal squamous epithelium $\rightleftharpoons$
　　　　　cervical intraepithelial neoplasia (CIN)

　　　dysplasia $\rightarrow$ carcinoma in situ
　　　　　　　　　　　　$\downarrow$
　　　　　　　microinvasive cancer
　　　　　　　　　　　$\downarrow$
　　　　　　　invasive cancer

Dysplasia is the first abnormality in the normal squamous epithelium of the vaginal cervix that may possibly evolve into invasive cancer. It means that there are atypical cells in the upper layers of the epithelium and that the cellular arrangement is no longer orderly. The nuclei are enlarged and stain darkly, and some cells are multinucleated. According to the numbers of atypical cells, it is categorized as slight dysplasia, moderate dysplasia, or severe dysplasia, but the abnormalities do not involve the entire thickness of the epithelium. Many cases of dysplasia regress and return to normal epithelium, but the likelihood of regression is decreased as the amount of atypical cells increases; that is, severe dysplasia has a greater malignant potential than mild to moderate dysplasia.

Carcinoma in situ, literally cancer-in-place, is the next stage in progression. While the malignant nuclear and cytoplasmic changes do not differ that much qualitatively from those in dysplasia, in carcinoma in situ, the entire thickness of the epithelial cell layer is involved. This stage is still a superficial condition, considered pre- or noninvasive because the abnormality has not broken through the epithelium into the connective tissue or stroma lying underneath. It has been widely accepted, although there is little direct evidence, that if cases of carcinoma in situ are left untreated, many of them will progress to invasive cancer within 10 years. Carcinoma in situ has, therefore, long been considered the "intraepithelial neoplasm." Once diagnosed, carcinoma in situ has generally been treated by hysterectomy with removal of the cervix, unless the woman still wants to have children, when a temporary lesser procedure called *conization* is used. In contrast, a diagnosis of dysplasia may be treated with cautery, conization, laser beam therapy, or sometimes ignored. Many clinicians, cytologists, and histopathologists have argued that overdiagnosis and overtreatment are as bad as underdiagnosis and undertreatment—that the standard practice of surgical removal of the uterus when atypical cell changes involve five or six cell layers of the epithelium, but leaving the uterus intact when the atypical occurs in only two or three cell layers should be reexamined. More currently thinking supports the concept that dysplasia and carcinoma in situ are part of a disease continuum, beginning with a mild abnormality and ending with invasive carcinoma and extending over a whole spectrum of changes. All degrees of dysplasia and carcinoma in situ are currently referred to as *cervical intraepithelial neoplasia* (CIN), with the implication that once CIN is diag-

nosed, there is the capability for progression to invasion if it is untreated. Each case, however, should be considered individually, and the most suitable treatment based on the severity of the abnormality determined by several diagnostic methods, the age of the woman, and whether or not she wants any or more children.

An abnormal Pap test, then, may mean any phase in the transition from normal cervical epithelium to a true cancer that has spread beyond the cervix. Further investigation by tissue biopsy is necessary.

Because it is not unusual for no visible difference to exist between a normal cervix and one with CIN, colposcopy or the Schiller test is utilized to aid in defining the suspicious area that should be biopsied after a positive Pap smear report. The Schiller test is based on the fact that normal cervical epithelium cells contain glycogen and will stain dark brown when painted with a 3% potassium iodide solution. Abnormal cells do not store glycogen and will, therefore, remain unstained. A pinching forceps is used to take a punch biopsy of the abnormal area—a "bite" of tissue that is then prepared for microscopic examination. If the histological or tissue examination indicates only a mild dysplasia, the area is treated by cautery or cryosurgery and there is frequent follow-up by Pap smears. When the site for biopsy has been directed by colposcopy, it is possible that the entire cervical lesion has been removed, and there is now no further treatment necessary.

If the pathologist's diagnosis is that the biopsy has revealed severe dysplasia, most doctors will then perform a conization, or cone biopsy. In this procedure, carried out in a hospital with the woman under general anesthesia, a cone-shaped piece of tissue is removed from the cervix with a scalpel.

The operation is apt to be a bloody one and is associated with a number of postoperative complications, unless the cone is shallow.

When the cone is deep and extends up into the endocervical canal, damage to the internal os and the canal itself is not uncommon. The subsequent scarring can threaten a woman's ability to conceive and deliver a child. The most common method of investigating an abnormal Pap smear, however, is by conization (see Fig 9.7).

Physicians who are trained in the use of the colposcope, which magnifies up to 20×, or colpomicroscope, which magnifies up to 180×, believe that conization is too extreme to be used for diagnosis purposes. They claim that in 90% of the cases where there are visible cervical lesions, the entire extent of the abnormality can be seen with the colposcope, and invasive cancer can be ruled out. The lesion can then be removed in the doctor's office by electrocautery, cryosurgery, or laser beam, painlessly without the use of anesthesia. Like a cone biopsy, these lesser approaches can also prevent the development of invasive disease, but without endangering a woman's fertility. Unfortunately, colposcopic

expertise and the sophistication needed to determine the circumstances needed to determine the circumstances for the use of the lesser procedures are not yet widely available. Unless physicians have the experience and knowledge to utilize colposcopy, or they send their patients to someone who does, there is likely to be a continued reliance on conization as the major diagnostic method used.

When carcinoma in situ has been diagnosed, there are the usual differences of opinion concerning the most appropriate way of treating it. Women should be aware that a body of evidence and a belief among many clinicians exists that hysterectomy is rarely indicated for carcinoma in situ. Several studies have found the results of conization to be as effective as hysterectomy in preventing invasive cancer from developing in the cervix or elsewhere. There are proponents of even more conservative outpatient procedures—electrocautery, cryosurgery, and laser therapy. Carbon dioxide laser surgery is a newer method of treating CIN that has many advo-

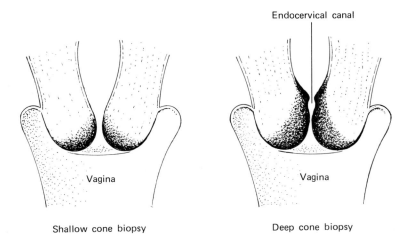

Endocervical canal

Vagina

Vagina

Shallow cone biopsy

Deep cone biopsy

Figure 9.7. Cone biopsy. The shaded area represents the extent of the cone. The procedure requires a general anesthetic and carries a 10% postoperative complication rate. Hemorrhage is a common problem after deep conizations.

cates. Laser is an acronym for Light Amplification by Stimulated Emission of Radiation. CO_2 laser refers to the production of energy in the form of an infrared light beam that literally vaporizes cells by turning their intracellular water to steam. Since the laser instrument is used in conjunction with an operating microscope, the beam of heat, highly accurate and controllable, is said to provide a preciseness of tissue excision greater than that of a surgical scalpel. Laser surgery is, as yet, too new a procedure for long-term follow-up data on its results, and it is considerably more expensive than cautery or cryosurgery.

If a woman gets a diagnosis of preinvasive carcinoma and a hysterectomy has been advised, there is plenty of time in which to get a second, or even a third, opinion. It takes an estimated 10 years for an untreated lesion, diagnosed as cervical carcinoma in situ, to progress to clinical invasion. There is some evidence that implies that only 10% of carcinoma in situ actually develop into invasive cancer (Green and Donovan, 1970; Coppleson and Brown, 1975). Other authorities are equally adamant in their belief that as many as 60%–70% of in situ cancers become invasive. At the present time, the preferred treatment in the United States for carcinoma in situ is vaginal or abdominal hysterectomy. Many clinicians are convinced that the recurrence rate for subsequent disease is lower when the uterus is removed.

Invasive Cancer of the Cervix

True cancer of the cervix which has gone further than the epithelium is staged as follows:

Stage 1 Strictly limited to the cervix (Stage Ia is for cases of microscopic invasion)

Stage 2 Spread to beyond the cervix, the upper two-thirds of the vagina, but not to the lateral pelvic walls

Stage 3 Has extended to either lateral pelvic wall and the lower third of the vagina

Stage 4 Involves the urinary bladder and/or the rectum or has metastasized beyond the true pelvis

Invasive cancer is most commonly treated by radiation therapy, either externally by a supervoltage X-ray or cobalt source and/or internally by way of a radium implant inserted into the cervical canal. In some cases, a radical hysterectomy, removing the uterus, upper part of the vagina, all the adjacent tissues, and the lymph nodes, is performed. Pelvic exenteration, an even more extensive operation, which removes the bladder and the rectum, is used in late stages of the disease or when irradiation has failed.

The choice between radical surgery or radiation therapy is usually made on the basis of the physician's preference; the 5-year survival rates from either are about the same—85% in stage 1 and 1%–5% in stage 4. As mentioned previously, the results of therapy for cervical intraepithelial neoplasia are virtually 100% effective. The best treatment for invasive cancer of the cervix is prevention—by regular Pap smears.

DES MOTHERS AND CHILDREN

Diethylstilbestrol (DES) is a nonsteroid synthetic estrogen that has estrogenic properties, is effective when taken orally, and is less expensive than the real thing. First synthesized in 1936, the drug was released for clinical use in the United States in 1941 and gained immediate popularity. It was used as replacement therapy in the postmenopause in the management of dysfunctional uterine bleeding, and frequently to "dry up" the breasts of women who did not want to nurse after delivery. In the late 1940s and for the following 2 decades, DES was widely prescribed for the treatment and prevention of habitual and threatened miscarriage and to prevent late complications of pregnancy that might be hazardous to the mother and fetus. Its use in pregnant women is a prime example of an increasingly frequent medical phenomenon—the therapeutic timebomb. A drug or a particular form of therapy, for example, X-rays, is generally used in accordance with accepted medical practice and believed to be safe and free of adverse effects. Many years later, the administration of the drug or therapy is discovered to have been a mistake—and often a tragic one.

In the late 1960s, eight cases of vaginal clear cell adenocarcinoma, a very rare type of cancer, were found in Boston among adolescent girls. Vaginal cancer is infrequent, making up less than 2% of all genital cancers. When it does occur, it is usually in women over 50, and most often of the squamous cell variety. Because of the unusual circumstances, further investigation took place. It was determined that the mothers of seven of these girls with vaginal cancer had received DES to prevent miscarriage during the first weeks of pregnancy, and the dose had been steadily increased during the remainder of the pregnancy.

Arthur L. Herbst, the physician at the Harvard Medical School who discovered the relationship between prenatal exposure to DES and vaginal cancer, set up a national clearing house, the Registry for Research on Hormonal Transplacental Carcinogenesis* in order to assess the extent of clear cell adenocarcinoma, as well as any other type of genital cancer. By 1983, more than 500 cases of cancer and about 80 deaths in young women had been reported. The youngest DES-exposed cancer victim was 7 years old; the oldest was born in 1948 and was 33 at the time of diagnosis. Not all the reported cancers are definitely

* Current address is Department of Obstetrics and Gynecology, University of Chicago, Chicago, Ill. 60637. For resource information, women can write or call: DES Registry, 5426 27 Street, NW, Washington, DC 20015, or DES Action, L.I. Jewish-Hillside Medical Center, New Hyde Park, NY 11040, 516-775-3450.

known to be associated with DES, but two-thirds have a history of maternal DES or similar compound ingestion. There may be more cases that have not yet been revealed to the registry or have not been detected.

On the basis of the data collected by the registry, Herbst has estimated that the risk of clear cell vaginal cancer developing in an exposed young woman up to the age of 24 is less than 1 in 1,000.

Fortunately, the most common abnormality related to DES exposure is not cancer but a benign condition of the vaginal epithelium called *adenosis*. In vaginal adenosis, the junction of the columnar glandular epithelium of the endocervical canal with the squamous epithelium of the cervix—the so-called transformation zone—occurs not just around the external os where it belongs but extends over the whole vaginal portion of the cervix and sometimes involves almost the entire vagina. The incidence of vaginal adenosis, as well as such other vaginal or cervical irregularities as abnormal transverse cervical or vaginal ridges (variously described as cockscomb cervix, cervical hood, pseudopolyp cervix), is about 35% according to Herbst, but over 90% according to Stafl and Mattingly (1974). These investigators, after studying 280 DES-daughters, were concerned about the abnormal colposcopic findings in the cervices and vaginas of more than 90% of the women. Pointing out that the incidence of squamous cell carcinoma of the vagina and cervix is known to be greatly increased in the presence of such atypical colposcopic lesions, they maintained that the major clinical danger from prenatal exposure to DES was not the development of clear cell adenocarcinoma, which infrequently occurs, but rather the subsequent progression of the abnormalities to squamous cell cervical and vaginal cancer when the exposed girls reached middle age. The researchers predicted an increase in the occurrence of cervical cancer in DES-daughters, placing their risk at 10 to 20 times greater than that for prenatally DES-unexposed women. Whether cervical cancer has become more prevalent in DES daughters still is controversial, however, primarily because not enough time has elapsed, and sufficient data to answer the questions have not yet been collected. Although several investigators have reported an increased incidence of the precancerous cervical intraepithelial neoplasia (CIN) in DES-exposed women (Fowler e al., 1981; Orr et al., 1981), others have found an equal or lower percentage of CIN in DES daughters when compared with nonexposed women (Robboy et al., 1981A, 1981B). More data addressing the issue were reported by Robboy and coworkers in 1984 from a study of nearly 4,000 DES-exposed women. In comparing the exposed women with unexposed matched control women, the researchers found that the incidence of dysplasia and carcinoma in situ was increased two to four times in the DES-exposed group. Virtually no cases of the precancerous condition occurred in DES-exposed virgins, providing additional evidence that cervical cancer is related to sexual activity. The authors suggested that DES exposure causes changes in the vaginal and cervical epithelium that may result in greater susceptibility to the risk factors of early coitus and multiple partners. The current attitude of researchers concerning the DES–cervical cancer link is that only time will tell, but that venereal transmission also plays a role in the cause of the disease. Another concern is the possibility of a relationship between prenatal exposure to DES and subsequent ovarian tumors. Schmidt and Fowler (1982A) observed the greater occurrence of benign ovarian cysts in DES-exposed women, and the same authors reported three cases of ovarian cystadenofibromas, one of which had borderline cancer components, in young DES-exposed patients (1982B). Whether these findings are significant will also be substantiated with time. What is clear, however, is that the more that DES-exposed progeny are studied, the more problems become apparent.

To date, the following physiological and structural abnormalities have been demonstrated in DES daughters: deformation of the uterus, distortions of the fallopian tubes, menstrual irregularities, infertility, a greater incidence of pelvic adhesions and benign ovarian cysts, abnormal healing response to cautery, cryosurgery, or other minor outpatient procedures, hirsutism or increased facial hair growth, higher male hormone levels, and a generalized dysfunction of the hypothalamic-pituitary-ovarian axis (Schmidt and Fowler, 1982A; Fowler et al., 1981; Peress et al., 1982).

The risks and the abnormalities as a result of DES exposure are not limited only to the female offspring. DES-exposed sons have been found to have lower ejaculate volume, more abnormal sperm, and a lower sperm count. Also reported are a significantly higher incidence of cysts in the epididymis, small penis (microphallus), undescended testes (cryptorchidism), underdeveloped testes (testicular hypoplasia), other urogenital abnormalities, and a possible greater incidence of testicular cancer (Bibbo et al., 1977; Gill et al., 1976; Beral and Colwell, 1981). Beral and her colleagues further found that a greater proportion of hormone-exposed men were either not married or not living with a woman, suggesting that interference with normal sexual function is a not uncommon aftermath to DES exposure in males.

The DES-mothers, the women who were given the drug to protect their pregnancies from miscarriage, have also been reported to suffer adverse effects. In 1978 Marluce Bibbo and her co-workers found that women who had been exposed to a maximum of 12 g of DES had a slight excess of breast cancer, 4.6%, compared with a control group of women who had not taken DES during pregnancy, in whom the occurrence of breast cancer was 3.1%. This was not considered to be statistically significant, although there were more deaths from breast cancer among the exposed group, due in part to the early onset of the disease. There was also slightly increased incidence of other endocrine-related cancers (ovary, endometrium, cervix, colon) in the exposed mothers, with an excess number of deaths occurring in the same order of magnitude as for breast cancer. Beral and Colwell's follow-up study of DES mothers (1980) found four cases of breast cancer in 80 exposed women and none in 76 nonexposed women. Brian and co-workers, however, found no association between DES exposure and breast cancer in their study (1980). Results of a more recent study (Greenberg et al., 1984) clearly point to an increased risk of breast cancer later in life when a woman is exposed to DES during pregnancy. The risk, however, is not as great as that associated with a family history of breast cancer.

The further discovery that the health risk spans three generations, that not only the women who took the drug and their sons and daughters, but also their grandchildren may be affected, is perhaps a not unexpected and tragic continuation of the DES story. The many anatomical abnormalities and distortions of the upper and lower reproductive tract in DES-daughters are necessarily associated with altered function. Women whose mothers took DES during pregnancy 20–30 years ago now have greater difficulty in delivering a healthy viable baby of their own. A number of studies have shown that DES-daughters have a higher proportion of spontaneous abortions, incompetent cervix, ectopic pregnancy, premature delivery, and greater infant mortality (Veridiano et al., 1981; Herbst et al., 1981; Ben-Baruch et al., 1981; Stillman, 1982).

The greatest use of DES during pregnancy was in the 1950s when an estimated 4%–7% of all pregnant women in the United States received it. A popular dose schedule was to administer 2.5 mg daily during or before the sixth week of pregnancy, increase to 5 mg by the seventh week, and to increase by 5 mg every other week until the fifteenth week when 25 mg was given. Every month thereafter, the dosage was increased by 25 mg until a maximum dose of 150 mg per day was given by the thirty-fifth week of pregnancy. During the sixties, and even earlier, there were published reports doubting the therapeutic value of DES during pregnancy. A decline in its use resulted, but it was still prescribed by many doctors, and in some areas of the country, the peak prescribing occurred during the middle 1960s. It was not banned by the Food and Drug Administration for used during pregnancy until November, 1971. By then an estimated total of four million American women had taken DES. No one born before 1941 could possibly have been exposed to DES, but it is obvious that there could still be hundreds of thousands of DES offspring coming into puberty.

It is evident that both the sons and the daughters of women who took DES, as well as the women themselves, should be alerted to the need for periodic and frequent examinations to detect possible abnormalities at as early a stage as possible. The problem is, who are all the DES offspring and the several million mothers? Many women who took DES do not remember taking it or were not aware or informed that it was given to them. Obviously, therefore, not every woman or man is in a position to know or to find out whether she or he has a history of exposure to DES in utero. Most of the knowledge concerning the DES hazards was originally provided through the activity of Dr. Sidney Wolfe, director of the Washington-based Health Research Group, virtually a storefront operation with a minuscule operating budget, and by cancer epidemiologist Kay Weiss. A number of widely read books by Gena Corea, Barbara and Gideon Seaman, Cynthia Orenberg, and others cover the complete DES story in detail. The entire population at risk, however, cannot be reached by books and magazine articles dealing with women's health. Wolfe called for governmental action to require follow-up studies by medical centers that supplied DES to pregnant women and to request that all private physicians search their files for DES-exposed women and notify them. The National Cancer Institute, by way of a Task Force Summary Report issued in 1978, recommended that all women who received DES in pregnancy, as well as their exposed offspring, be informed of their risk by a public information campaign and that individuals be notified by their private physicians. No federal funds were provided for implementation of these recommendations, however, and to date there have been no large-scale efforts by government agencies, by the doctors who prescribed the drug, or by the pharmaceutical companies who profited by its sales to inform individuals that they may be at increased risk for genital abnormalities and the development of cancer. The relatively few DES clinics that exist have been set up to do follow-up studies on DES-exposed women. One such program, the DESAD Project (from DES and ADenosis) is supported by the National Cancer Institute and currently involves a study group of almost 5,000 women at five cooperating institutions—Massachusetts General Hospital, Boston; the Gundersen Clinic, LaCrosse, WI; the Mayo Clinic, Rochester, MN; the Baylor College of Medicine, Houston, TX; and Cedars-Sinai Hospital, Los Angeles, CA. The DESAD Project, designed to study and observe daughters exposed in utero to DES, necessarily is aimed more at getting information than giving it, and more for collecting data on the prevalence and incidence of abnormalities than for informing women. In order to locate study participants, the DESAD Project has attempted to trace and notify individuals of their exposure by reviewing physician's records but ran into "monumental difficulty," as described by an investigative team at Massachusetts General (Nash et al., 1983). In addition to the problems of time and expense involved in searching through medical charts, the DESAD center workers encountered many doctors who refused to allow their records to be reviewed. Reasons cited by the doctors for denying access to their records were lack of adequate space in which to

conduct the chart review, infringement of their patients' privacy and rights to confidentiality, and the possibility of lawsuits by notified DES-exposed individuals. Some physicians refused to allow their patients to be contacted after they had been identified by the DESAD reviewers. The records were often unavailable because they had been destroyed, in some instances because the physicians had retired and no longer maintained them or because the doctors wanted to limit their personal liability. The physicians' concern about their liability may be unwarranted, since prevailing legal opinion on doctors' responsibility is that they would not be held liable for prescribing the drug prior to 1971 since they did so in good faith based upon the available information at that time. They could be held liable, however, if they do not make every reasonable effort to notify and inform the women to whom they administered DES (Saber, 1977). The DES issue is currently a medicolegal morass. There have been a number of successful class action and individual suits filed by women against the pharmaceutical firms that manufactured the drug for injuries arising from administration of DES, but to date there has been only one suit against private physicians.[*]

The health of millions of young men and

---

[*] Former Congresswoman Patsy Mink of Hawaii, along with two other women, in 1977 filed a class action suit against the University of Chicago, doctors on the staff, and Eli Lilly and Company. They alleged that as part of an experiment conducted by a doctor at the University from 1950–1952, they received, without their knowledge and consent, DES. They further claimed that none of the more than 1,081 women who were given DES knew that the pills contained a drug of any kind; Ms. Mink alleges that she was told that they were vitamin pills. The suit was settled out of court in February, 1982, for $225,000. The daughters of the three women are currently in litigation. In 1982, New York's Court of Appeals, after a series of trials in lower courts, granted $500,000 to a 25-year-old DES daughter whose hysterectomy and vaginectomy were the result of cancer. Lilly, the largest DES manufacturer, was liable, although it was not determined that company produced the drug her mother took. The legal principle developed was that of a "joint-enterprise liability" of all DES producers, regardless of the actual manufacturer.

women is at risk through no fault of their own but because their mothers were given DES. Routine physical examinations can detect and reduce the consequences of their exposure, but evidently the full responsibility and the major expense that accompany the avoidance of these medically induced abnormalities are to be borne by the DES sons and daughters and their mothers. Any woman who knows or suspects that she was given DES or a chemically related substance while pregnant with her son or daughter must have frequent physical examinations herself and must take her child for a pelvic examination soon after he or she has attained puberty. Unless their mothers have assured them that there was no pill-taking during pregnancy for a possible miscarriage or late complication of pregnancy, men and women born after 1941 can logically presume that they may have been prenatally exposed to diethylstilbestrol. For males, the preliminary data suggest that functional abnormalities that could interfere with fertility may be a problem, but future investigations are necessary. DES exposure before birth in young women is likely to be revealed by vaginal adenosis and cervical abnormalities. Gynecological examinations have to be more extensive than usual when there is a presumed or definite history of prenatal DES exposure. The Schiller test or a colposcopic examination has to be made. It is also recommended that a Pap smear from all quadrants of the vagina be taken separately from a smear of the cervix.

The biological effect of DES in creating an increased risk of cancer in both mothers and their daughters is not known. It may be that DES, in similar fashion to other estrogens, lays the groundwork for the possibility of future malignant changes. Perhaps DES reacts with other factors to result in the development of cancer at a later time or acts to speed up the onset of cancer. It may be possible that later exposure to more exogenous estrogen could be a critical and triggering factor. For this reason, the identification of DES-exposed women

and men is doubly important. They need to have this information, not only so that they can obtain what could be lifesaving examinations on a regular basis, but so that they can avoid any additional ingestion of estrogen. Because of the DES added to the food supply, this might be difficult.

The Hazardous Hamburger

When DES is added to farm animal and poultry feed, it accelerates growth. Without DES it takes 30 or more days and 500 more pounds of feed to produce a 1,000-pound steer. Obviously, meat producers can get livestock to market earlier and save feed by using DES. The animal feed incorporating DES is supposed to be withdrawn from use before slaughter. When proper procedures are followed, presumably no DES residues remain in any edible portion of the animal. But because more sensitive testing has revealed DES residues in meat, and because the Delaney clause of the Food, Drug and Cosmetic Act prohibits the use of any food additive known to be carcinogenic, the FDA moved to ban DES as a cattle feed additive or pellet implant in livestock in 1972 and 1973. The ban was not implemented, however, until 1979 when use of DES was to cease by November of that year. The next year, the federal government discovered rampant violations of the ban and determined that cattle on hundreds of feedlots in more than 20 states were still being illegally treated with DES. In 1982 and early 1983 the same story of widespread use of DES was repeated, but this time in veal calves. The law exists, but apparently not every meat producer abides by it. Since no one knows the possible effects of ingestion of even minimal amounts of DES or other estrogens, particularly when there has been an initial exposure, the implications for the public health are obvious.

The Morning-After Pill

Although it is presumed that DES is no longer being given to pregnant women, it should be noted that it is still being used to keep women from becoming pregnant. DES is the "morning-after pill," the postcoital contraceptive that has been approved by the FDA for use in "emergency" situations, such as rape or incest. When it is spread out in doses of 50 mg daily for 5 days, DES can often prevent implantation of a fertilized egg if it is taken within 72 hours of an unprotected sexual exposure. Not only does it frequently *fail* to prevent pregnancy, but the very unpleasant side effects of taking 250 mg of DES are similar to those produced by massive doses of any estrogen—nausea and vomiting. An antiemetic drug is usually prescribed along with the DES to help control the gastrointestinal symptoms.

Although the drugs may not be biologically equatable, the large dosage of DES in the morning-after pill is the equivalent of the amount of hormone provided in a 20-year supply of a commonly used oral contraceptive such as Ortho-Novum 1/50. All in all, and considering the known risk of exogenous estrogen, the exposure to DES should be of more concern than the exposure to pregnancy. Unless it is known that the sexual encounter occurred at the time of ovulation, the chances of pregnancy are not that great. No woman should take the DES morning-after pill; no responsible doctor should give it. DES is an abortifacient, and there are other and safer methods of terminating a pregnancy should one occur. An oral contraceptive morning-after pill is described in Chapter 12.

TUMORS OF THE UTERUS

The most common gynecological tumors are benign masses of muscle and connective tissue in the uterus called *fibroids*. Their true incidence is unknown, but there are estimates that they occur in anywhere from 20%–50% of women after the age of 35. They are very rare before menarche and after menopause, and estrogen is evidently important in their development. Estrogen definitely appears to stimulate their growth. Mattingly has noted a striking increase in the incidence of large fibroids in young women who have taken oral contraceptives. Genetic factors, too, may play a role in their incidence. According to DeAlvarez, fibroids are found somewhat more frequently in Jewish women, and it is well recognized that for no known physiological reason, they occur much more commonly, appear earlier, and grow more rapidly in black women.

Fibroids may be single or multiple; they may be microscopic in size, or they may grow to be enormous. The largest one ever reported in the medical literature was in 1888—it weighed 144 pounds! Although tumors of 40–50 pounds are still occasionally seen, earlier and better diagnosis has made excessively huge ones quite uncommon in the United States. Most of them average about 4–5 cm in diameter. They are round, smooth, quite firm in consistency, and when they are cut open, display pinkish white whorls and lines of muscle bundles.

The name, fibroids, implies that the tumors originate in fibrous connective tissue. Although they do contain varying amounts of fibrous elements, they are actually composed of smooth muscle cells and are more appropriately called myomas or leiomyomas (from *leios*, smooth, and *myos*, muscle). Depending on their location in the uterus, fibroids are classified as intramural, submucous, and subserous. Intramural are the most common and are found in the center of the uterine muscular wall. If they are small, they are frequently asymptomatic. Submucous fibroids grow between the endometrial lining and the uterine muscle. They may project into the uterine cavity and distort and stretch the endometrium. The subsequent increase in the surface of the endometrial lining may result in excessive menstrual bleeding. Submucosal fibroids may also be a cause of habitual miscarriage because they can interfere with the anchoring site of the placenta. Subserous tumors develop between the uterine wall and the external peritoneal covering or serosa. Because they have so much space in which to grow, these fibroids may attain sizes many times greater than that of the uterus itself. They also have a tendency to become pedunculated, that is, attached by a stalk or pedicle to the uterus. Occasionally, subserosal pedunculated fibroids can even become totally separated from the uterus and move away from it, becoming known as parasitic or wandering fibroids.

The most frequent reason for hysterectomy is a diagnosis of fibroids. Just because fibroids are discovered on bimanual pelvic examination, however, does not mean that they necessarily have to be treated. The vast majority of small fibroids are symptomless, grow very slowly, and only have to be watched from time to time. If they are not showing any evidence of rapid growth and are not causing any trouble, they are best left alone. Large fibroids may encroach on the bladder or rectum and cause pressure symptoms, such as urinary frequency, constipation, or sensations of abdominal fullness. Fibroids have also been shown to increase uterine contractility (Iosif and Akerlund, 1983) suggesting a reason for the menstrual pain and irregularity that some women have. But if a woman is close to menopause and the symptoms are those she can live with, the discomfort will disappear with the climacteric as the tumor or tumors stop growing and diminish in size (unless, of course, she is on estrogen replacement therapy). The risk of fibroids becoming malignant is very small—less than 0.3%—but the operative mortality risk of hysterectomy in the average American hospital is 1%.

In younger women, the presence of fibroids, even if they produce no symptoms, may be associated with infertility problems if they obstruct the opening of the fallopian tubes. If pregnancy does occur, small fibroids generally do not affect the course of the pregnancy. Conversely, neither does pregnancy affect the fibroids. If the fibroids are large, however, they may interfere with implantation and cause spontaneous abortion, or block the birth canal to prevent vaginal delivery. Myomectomy, the removal of only the fibroids and not the entire uterus, is performed when a woman is younger than her late thirties and wants to bear a child. The operation is technically more difficult than hysterectomy, and the fibroids may recur at a later date. The probability of pregnancy following myomectomy has been estimated as between 40%–50%, and it is generally necessary to deliver a baby by a cesarean section to avoid rupture of the uterus at the site of the incision for myomectomy. When fibroids are present in younger women, they should be aware that oral contraceptives may cause more rapid growth of the tumors.

When a woman is over 40, has a fibroid larger than a grapefruit with an excessively rapid growth rate, has severe pain, pressure, or bleeding, and nonsurgical treatment for the symptoms has been unsuccessful, hysterectomy, performed abdominally or vaginally, is the most specific form of treatment. Anyone who has virtually hemorrhaged her way through painful monthly periods that last for 9 days or longer or that occur more frequently than every 21 days and have made her anemic, is not going to be surprised to find out that a hysterectomy is indicated. She will be relieved to discover that the cause is benign and not malignant and recognizes that there is little chance that the procedure is being performed unnecessarily. If, on the other hand, a woman is advised that she needs a hysterectomy for asymptomatic or very mild symptoms of fibroids, a second opinion is mandatory. There have been so many abuses of hysterectomy, so many cases of what Gena Corea has called "remunerectomies," that some hospitals require consultation for hysterectomy. On some occasions the "tumor" was either nonexistent or has even turned out to be a pregnancy.

Total hysterectomy, which removes the cervix as well as the uterus has today almost completely replaced the subtotal or supracervical operation. Most physicians see little point in leaving the cervix to serve as the site of a potential 2% risk of cancer. Routine removal of the ovaries for the same prophylactic reason is more controversial. The majority of doctors advocate oophorectomy if the woman is 39 or older at the time of hysterectomy. Altchek typifies this opinion when he states,

> The ovaries will stop functioning in a few years anyway, and there is a small chance of an ovarian tumor developing in later life. The patient is made aware that the oophorectomy gives a surgical menopause, and she may suffer hot flashes and other symptoms of the climacteric. These can be controlled with hormone medication.

There is 1% risk of ovarian cancer in women over the age of 40, but there is also evidence that the ovary may continue to function for many years after menopause. There has also been the suggestion that surgical castration occurring 10 years or more before expected menopause contributes to the increased incidence of coronary heart disease, atherosclerosis, and osteoporosis. Other organs of the body, for example, the prostate gland, pose an even greater risk for the development of cancer, but are not routinely removed. Whether Altchek's statement would be as definitive if the word "testes" were substituted for the word "ovaries" is conjectural.

Cancer of the Uterus

More than 90% of uterine cancer is carcinoma of the endometrial lining. It is primarily a disease of older women, occurring postmenopausally at an average age of 55. Only a small percentage of cases occur in women under the age of 40, and there has been the suggestion that risk factors for the disease are related directly or indirectly to long-term or excessive estrogen stimulation. The incidence of endometrial cancer had been increasing since the 1950s, peaked in 1975, and has recently been declining for unknown reasons. It may not be entirely coincidental that the decrease in the number of cases has coincided with the decrease in the prescribing of estrogen replacement therapy for older women. Authorities are still cautious about making a case for estrogen as the cause of endometrial cancer, but many investigators believe it to be an associated factor. It is recognized that when estrogenic stimulation of the endometrium is prolonged and unopposed by progesterone, the risk of the disease increases. Women who, for various reasons, do not ovulate (no progesterone is produced) or who have estrogen-producing ovarian tumors have a greater incidence of uterine cancer. Also with an underlying hormonal basis, so do white women of high socioeconomic status who are more than 30% overweight, have not had any children, have had a late menopause, diabetes mellitus, high blood pressure, or prolonged estrogen-replacement therapy.

The major initial symptom is abnormal and painless vaginal bleeding. Any episode of bright red bleeding, either scant or moderate in amount, that occurs after menopause should immediately send a woman to her doctor. Just before menopause, irregularity is so common that it is sometimes difficult to distinguish normal from abnormal bleeding. At any age, however, a bloody discharge, intermittent spotting, or steady bleeding that takes place between menstrual periods should be investigated. Abnormal uterine bleeding is very rarely the result of uterine malignancy in a young woman. In the postmenopause, however, it should be regarded with suspicion. Depending on which authority is quoted, bleeding that occurs 6 months after menstruation has stopped indicates that a cancer is present someplace in the genital tract 20%–50% of the time. But since the symptom is such an early sign of the disease and an endometrial cancer tends to be such a slow-growing tumor, the chances of cure are very good if medical advice is sought quickly.

Diagnosis of the cause of abnormal bleeding is made by biopsy of the lining of the uterus. The sample can be taken in the doctor's ofice with only a local anesthesia to the cervix. In microcurettage, a small strip of endometrium is scraped away with a sharp spoon-shaped instrument called a Duncan or Kevorkian curette. Larger curettes are used for more generous samples. In suction biopsy, the curette tip is incorporated into a hollow tube inserted into the uterine cavity. When light suction is applied, the tissue sample is drawn out through the tube. Samples can also be taken by "jet-washing," brushing, or scraping cells from the endometrial surface, but these methods are seen as less effective. A Pap smear is also considered unreliable in ruling out endometrial cancer.

A positive diagnosis through endometrial biopsy is conclusive evidence of cancer, but a negative finding does not necessarily exclude it. The carcinoma may have been missed in an area that was not sampled. The most definitive diagnosis is made by the conventional D and C that includes a thorough systematic curettage.

Early endometrial cancer is treated by total hysterectomy plus removal of the fallopian tubes and the ovaries. More extensive carcinoma is generally considered an indication for preoperative radium treatment followed by surgery. When metastases to the vagina, pelvis, and lungs occur, chemotherapy in the form of massive doses of progestins has produced remissions. As in all cancers, increasing the rate of cures depends on early diagnoses. Vaginal bleeding in the postmenopause is a definite warning that could mean malignancy. It should not be ignored by a woman or her doctor.

OVARIAN TUMORS

The ovary is physiologically, normally, and naturally a "cyst-y" kind of organ. Every month between menarche and menopause, follicles develop and become little sacs filled with fluid—actually cysts. One of them, outdistancing all the others, grows to the size of 1.5–2 cm in diameter, bulges on the surface of the ovary, and bursts. The egg is released in a gush of follicular fluid, the large follicle is replaced by a corpus luteum, and all of the other follicular cysts disappear.

Quite commonly, and probably due to some alteration in the usual hypothalamic-pituitary-ovarian hormonal axis, a follicle does not ovulate but continues to grow, accumulate fluid, and produce estrogen. More rarely, the corpus luteum may not regress normally but continues to grow and produce progesterone. The hormones secreted by such a functional *follicular cyst* or *corpus luteum cyst* may produce dysfunctional bleeding—delayed onset of menstruation, prolonged menstruation, or perhaps a period every 2 weeks for a month or two. There may be some abdominal pain as a result of pressure from the cyst on the ovarian connective tissue capsule. These retention cysts will spontaneously disappear by fluid absorption or rupture within a month or two and do not have to be treated, by surgery or any other method, unless there is some complication.

Functional follicular and corpus luteum cysts occur frequently, and women are probably seldom aware of their existence. There is no reason to be upset if an enlarged ovary or ovarian cyst is detected on pelvic examination; it can be watched for several months. A true neoplastic or tumorous growth in the ovary will persist and progressively enlarge, but follicular and corpus luteum cysts will eventually vanish. Occasionally, a retention cyst will rupture through a small blood vessel on the surface of the ovary, producing intraperitoneal bleeding. If the bleeding is scant and spontaneously stops, there may be few symptoms. If the bleeding is extensive, the source can be diagnosed by laparoscopy (the examination of the pelvic cavity by an instrument inserted through a small incision in the abdominal wall). Surgery may then be necessary to control the bleeding and repair the ovary.

Unfortunately, as several authorities have recognized, too many doctors fail to appreciate the fact that the ovaries are normally cystic, and treat persistent follicular and corpus luteum cysts—physiological in origin—by surgical removal. In Max Bloom's words, the large number of ovaries that are "needlessly incised, resected, bisected, tinkered, punctured, wedged, suspended, and even ablated for no good reason at all . . ." bear silent testimony to that fact.

The unnecessary removal of wedges of ovarian tissue as treatment for Stein-Leventhal syndrome is an example. This condition, an endocrinological disorder also known as a type of polycystic ovarian disease, is manifested by a failure of ovulation, large numbers of follicular cysts, enlarged ovaries with thick capsules, excessive androgen production, and amenorrhea. Previously, the therapy for this disease was the removal of up to 50% of the ovarian tissue to decrease the production of androgens and presumably permit ovulation. Today it is believed that there is little indication for doing this type of operation. Any endocrine abnormalities can be corrected with hormonal treatment.

In actuality, cystic ovaries are not ovarian cysts; that is functional cysts in the ovary are not considered to be *neoplastic*, or new growths. True ovarian neoplasms, both benign and malignant, occur much less often but are far more significant. Growth that is new, a tumor, can increase to a very large size and must always be removed to exclude the possibility of cancer.

Neoplastic tumors of the ovary are classified according to their cellular origin; that is, whether the growth started in the surface epithelium of the ovary, in the germ cells themselves, in the connective tissue stroma, or other miscellanous cells. There are more than 60 kinds of ovarian tumors, but only the most common will be described.

Benign *serous cystadenomas* form about 20%–25% of all ovarian neoplasms occurring at any age but most often between the ages of 40 and 50. Most of the time a cyst will be found on only one ovary, but they also occur bilaterally in 20%–30% of the cases. A cyst may be very small and hardly palpable, but it may also be large enough to fill up the entire abdomen. Serous cystadenomas contain a clear watery protein-rich fluid, yellow in color or chocolate brown as a result of bleeding into the cyst. The type called *papillary serous cystadenomas* have the greatest likelihood for malignant change, and about half of that kind actually are, or potentially will be, cancerous.

Mucinous cystadenomas are filled with a thick, gelatinous viscous fluid. They can attain an enormous size, but become secondarily malignant much less frequently than the serous variety. They are usually unilateral.

Dermoid cysts, or benign cystic teratomas, make up about 20% of benign ovarian tumors. They can occur at any age but are most frequently found between the ages of 18 and 35. Dermoids are a strange kind of growth that contain multiple tissues, which are derivatives of all the three embryonic germ layers. Inside a dermoid are skin, sebaceous and sweat glands, and long strands of hair that bear no relationship to the hair color of the woman with the cyst. Cartilage, bone, and several well-developed teeth are not unusual, and even intestines and thyroid gland have been found. The tumors vary in size, but average 6–8 cm in diameter. They very rarely become malignant.

Dermoid cysts have been known since antiquity, and speculations concerning their origin have been many and disputed. One theory holds that a dermoid cyst develops by parthenogenesis, the reproduction of an ovum that has not been fertilized by a sperm. Another postulates that a dermoid is actually a womans' undeveloped twin. Neither idea is universally accepted.

Dermoid cysts have a tendency to be found in both ovaries. For this reason at the time of surgical removal of the cyst, the other ovary, even if it appears completely normal, has to be carefully inspected for a very small dermoid.

Malignant Neoplasms of the Ovary

Ovarian cancers are considered the worst of gynecological malignancies, primarily because they sometimes develop slowly and remain silent and symptomless until it is too late. There is no practical and certain way of detecting early cancer of the ovary—Pap smears are virtually useless for this—and it is usually found by chance. The highest incidence occurs in women between 55 and 65. While an ovarian mass could be observed for several months in a younger woman, in the postmenopause any enlargement of the ovaries is suspicious. One larger than 5 cm in diameter is an indication for immediate investigation. Diagnosis can be aided by laparoscopy, the tumor can be biopsied to determine its nature, ultrasound or computed tomography (CT) scans can be used, but most often abdominal surgery will be necessary to provide a definite diagnosis.

Malignancy of the ovary, when found, is always treated by aggressive and radical surgery. Depending on the clinical stage of the disease when detected, surgical treatment may be supplemented by radiation and chemotherapy. Some doctors do exploratory abdominal surgery or "second-look" operations on selected patients to determine the progress of the treatments, particularly after long-term chemotherapy.

The current overall 5-year survival rate, unchanged over the past several decades from all types of ovarian cancer is 15%–35%. The prognosis is still grim, but advances in the use of combination chemotherapy have recently made a diagnosis of cancer of the ovaries—although no reason for super-optimism—at least not as gloomy as before. The absolute cure rate does appear to be increasing.

British epidemiologist Valerie Beral and her co-workers discovered an interesting correlation between ovarian cancer and the number of children in a family (1978). Evidently the more pregnancies a woman has, the less her risk of mortality from ovarian cancer. Noting that ovarian cancer is infrequent in populations that do not practice birth control, Beral pointed that the death rate from this malignancy is lower in Catholic women, who until recently had an average family size greater than that of Protestant or Jewish women. She suggested that pregnancy, or some component of the childbearing process, in some way protects against ovarian cancer. Fathalla had previously (1971) suggested that "incessant ovulation" without respite from pregnancy or breastfeeding was traumatic for the ovarian surface and could, hence, predispose to cancer. Additional support for that possibility is provided by evidence from epidemiological studies that found having a late menarche, an early menopause, or using oral contraceptives that prevent ovulation also appear to have a protective effect against the development of ovarian cancer (Hildreth et al., 1981; Weiss et al., 1981).

REFERENCES

Altchek, A. Guidelines for hysterectomies, noncancerous conditions requiring surgery. *The Female Patient* 1(2):68–73, 1976.

Ben-Baruch, G., Menczer, J., Mashiach, S., and Serr, D. M. Uterine anomalies in diethylstilbestrol-exposed women with fertility disorders. *Acta Obstet Gynecol Scand* 60(14):395–397, 1981.

Beral, V. and Colwell, L. Randomized trial of high doses of stilboestrol and ethisterone in pregnancy: Long-term follow-up of the children. *J Epidemiol Community Health* 35(3):155–160, 1981.

Beral, V. and Colwell, L. Randomized trial of high doses of stilboestrol and ethisterone in pregnancy: Long-term follow-up of mothers. *Br Med J* 281(6248):1098–1101, 1980.

Beral, V., Fraser, P., and Chilvers, C. Does pregnancy protect against ovarian cancer? *Lancet* 8073:1083–1986, 1978.

Bibbo, M., Gill, W. B., Freidoon, et al. Follow-up study of male and female offspring of DES-exposed mothers. *J Obstet & Gyn* 49(1):1–8, 1977.

Bibbo, M., Haenszel, W, Wied, G., et al. A twenty-five year follow-up study of women exposed to diethylstilbestrol during pregnancy. *New Engl J Med* 298(14):763–767, 1978.

Bloom, M. L. and Van Dongen, L. *Clinical Gynaecology—Integration of Structure and Function.* London, William Heinemann Medical Books, Ltd., 1972.

Brian, D. D., Tilley, B. C., Labar, D. R., et al. Breast cancer in DES-exposed mothers: absence of association. *Mayo Clin Proc* 55(2):89–93, 1980.

Bryson, Y. J., Dillon, M., Lovett, M., et al. Treatment of first episodes of genital herpes simplex virus infection with oral acyclovir. *N Engl J Med* 308:916–921, 1983.

Cibley, L. J. Candidiasis: What to look for, how to treat. *The Female Patient* 2(10):14–20, 1977.

Coppleson, L. W. and Brown, B. Observations on a model of the biology of carcinoma of the cervix. A poor fit between observation and theory. *Am J Obstet Gynecol* 122:127–136, 1975.

Corea, Gena. *The Hidden Malpractice.* New York, Wm. Morrow and Co., 1977.

Cytryn, A., Sen, P., Chung, H. R., et al. Severe pelvic infection from *Chlamydia trachomatis* after cesarean section. *JAMA* 247:1732–1734, 1982.

DeAlvarez, R. R. *Textbook of Gynecology.* Philadelphia, Lea & Febiger, 1977.

Dowdle, W. The epidemiology of AIDS. *Public Health Rep* 98(4):308–312, 1983.

Eschenbach, D. A. Vaginal infection. *Clin Obstet Gynecol* 26(1):186–202, 1983.

Fathalla, M. F., "Incessant ovulation" and ovarian cancer. *Lancet* 2:163–168, 1971.

Fowler, W. C., Schmidt, G., Edelman, D. A., et al. Risks of cervical intraepithelial neoplasia among DES-exposed women. *Obstet Gynecol* 58(6):720–724, 1981.

Gardner, H. L. and Dukes, C. D. *Hemophilus vaginalis* vaginitis. *Am J Obstet Gynecol* 69(5):962–975, 1955.

Gilbaugh, J., and Fuchs, P. The gonococcus and the toilet seat. *N Engl J Med* 301(2):91–93, 1979.

Gill, W. B., Schumacher, G., Bibbo, M. Structural and functional abnormalities in the sex organs of male offspring of mothers treated with diethylstilbestrol (DES). *J Reprod Med* 16(4):147–153, 1976.

Gillespie, L., Margolis, A. Cystitis can be cured. *Ms.* 98–101, 1983.

Green, G. H. and Donovan, J. W. The natural history of cervical carcinoma in situ. *J Obs Gyn British Commonwealth* 77:1–9, 1979.

Green, J. and Staal, S. Questionable dermatologic use of iododeoxyuridine. *N Engl J Med* 295(2):111–112, 1976.

Greenberg, E. R., Barnes, A. B., Resseguie, L., et al. Breast cancer in mothers given diethylstilbestrol in pregnancy. *New Engl J Med* 311(22): 1393–1398, 1984.

Harding, G. K., Buckwold, F. J., and Macrie, T. J. Prophylaxis of recurrent urinary tract infection in female patients. Efficacy of low-dose, thrice-weekly therapy with trimethoprim-sulfamethoxazole. *JAMA* 242:1975–1977, 1979.

Harding, G. K., Ronald, A. R., Nicolle, L. E., et al. Long-term antimicrobial prophylaxis for recurrent urinary tract infections in women. *Rev Infect Dis* 4(2):438–443, 1982.

Herbst, A. L. et al. Reproductive and gynecologic surgical experience in diethylstilbestrol-exposed daughters. *Am J Obstet Gynecol* 141(8):1019–1028, 1981.

Hildreth, N. G., Kelsey, J. L., LiVolsi, V. A., et al. An epidemiologic study of epithelial carcinoma of the ovary. *Am J Epidemiol* 114:398–405, 1981.

Hocutt, J. E. Urinary tract infections; diagnosis and management. *The Female Patient* 7:40/15–40/23, 1982.

Iosif, C. S. and Akerlund, M. Fibromyomas and uterine activity. *Acta Obstet Gynecol Scand* 62:165–167, 1983.

Jarratt, M. Herpes and simplex infection. *Arch Dermatol* 119:99–103, 1983.

Mangan, E. E., Borow, L., Burtnett-Rubin, M. M., et al. Pregnancy outcome in 98 women exposed to diethylstilbestrol in utero, their mothers, and unexposed siblings. *Obstet Gynecol* 59(3):315–319, 1982.

Mattingly, R. F. *Te Linde's Operative Gynecology,* 5th ed. Philadelphia, J. B. Lippincott, 1977.

Mattingly, R. F. and Stafl, A. Cancer risk in diethylstilbestrol-exposed offspring. *Am J Obstet Gynecol* 126:543–548, 1976.

Meisels, A., Roy, M., Fortier, M., et al. Human papillomavirus virus infection of the cervix: the atypical condyloma. *Acta Cytol* 25(1):7–16, 1981.

Monif, G. *Infectious Diseases in Obstetrics and Gynecology,* ed 2. Maryland, Harper & Row, 1982.

Nash, S., Tilley, B. C., Kurland, L. T., et al. Identifying and tracing a population at risk: The DESAD project experience. *Am J Public Health* 73(3):253–259, 1983.

Orr, J. W., Shingleton, H. M., Gore, H., et al. Cervical intraepithelial neoplasia association with exposure to diethylstilbestrol in utero: A clinical and pathological study. *Obstet Gynecol* 58(1):75–82, 1981.

Paavonen, J. *Chlamydia trachomatis*-induced urethritis in female partners of men with nongonoccocal urethritis. *Sex Transm Dis* 6(2):69–71, 1979.

Peress, M. R., Tsai, C. C., Mathur, R. S., and Williamson, H. O. Hirsutism and menstrual patterns in women exposed to diethylstilbestrol in utero. *Am J Obstet Gynecol* 144(2):135–140, Sep., 1982.

Reeves, W. C., Corey, L., Adams, H. G., et al. Risk of recurrence after first episodes of genital herpes: Relation to HSV type and antibody response. *N Engl Med* 305:315–319, 1981.

Reid, R., Stanhope, R., Herschman, B. R., et al. Genital warts and cervical cancer. *Cancer* 50:377–383, 1982.

Robboy, S. J., Szyfelbein, W. M., Goellner, J. R., et al. Dysplasia in cytologic findings in 4,589 young women enrolled in diethylstilbestrol-adenosis (DESAD) project. *Am J Obstet Gynecol* 140(5):579–586, 1981A.

Robboy, S. J., Truslow, G. Y., Anton, J., and Richart, R. M. Role of hormones including diethylstilbestrol (DES) in the pathogenesis of cervical and vaginal intraepitheleal neoplasia. *Gynecol Oncol* 12(2 Pt 2):S98–110, Oct., 1981B.

Robboy, S. J., Noller, K. L., O'Brien, P., et al. Increased incidence of cervical and vaginal dysplasias in 3,980 diethylstilbestrol-exposed young women. *JAMA* 252(21):2979–2983, 1984.

Roe, J. C. R. Toxicologic evaluation of metronidazole with particular reference to carcinogenic, mutagenic, and teratogenic potential. *Surgery* 93(1):158–164, 1983.

Rustia, M. and Shubik, P. Induction of lung tumors and malignant lymphomas in mice by metronidazole. *J Natl Cancer Inst* 48:721–729, 1972.

Saber, F. A. The DES problem: Fashioning a physician's duty to warn. *J Legal Med* 5:21–30, 1977.

Schacter, J., Grossman, M., Holt, G., et al. Prospective study of chlamydial infection in neonates. *Lancet* 2(8139):377–380, 1979.

Schmidt, G. and Fowler, W. C. Cervical stenosis following minor gynecological procedures on DES-exposed women. *Obstet Gynecol* 56(3):333–335, 1980.

Schmidt, G. and Fowler, W. C. Gynecologic operative experience in women exposed to DES in utero. *South Med J* 75(3):260–263, 1982A.

Schmidt, G. and Fowler W. C. Ovarian cystadenofibromas in three women with antenatal exposure to diethylstilbestrol. *Gynecol Oncol* 14(2):175–184, Oct. 1982B.

Seaman, B. and Seaman G. *Women and the Crisis in Sex Hormones.* New York, Rawson Associates, 1977.

Stafl, A. and Mattingly, R. F. Vaginal adenosis: A precancerous lesion? *Am J Obstet Gynecol* 120(5):666–673, 1974.

Stillman, R. J. In utero exposure to diethylstilbestrol: Adverse effects on the reproductive tract and reproductive performance in male and female offspring. *Am J Obstet Gynecol* 142(7):905–1021, 1982.

Sweet, R. L., Schacter, J., and Landers, D. V. Chlamydial infections in obstetrics and gynecology. *Clin Obstet Gynecol* 26(1):143–164, 1983.

Van Slyke, K. K., Michel, V. P., and Rein, M. Treatment of vulvovaginal candidiasis with boric acid powder. *Am J Obstet Gynecol* 141(2):145–148, 1981.

Veridiano, N. P., Delke, I., Rogers, J., and Tancer, M. L. Reproductive performance of DES-exposed female progeny. *Obstet Gynecol* 58(1):58–61, 1981.

Voogd, C. E., VanderStel, J. J., and Jacobs, J. J. Mutagenic action of metronidazole: I. Metronidazole, nimorazole, dimetridazole, ionidazole. *Mutat Res* 26:483–490, 1974.

Weiss, N. S., Lyon, J. L., Liff, J. M., et al. Incidence of ovarian cancer in relation to the use of oral contraceptives. *Int J Cancer* 28(6):669–671, 1981.

Whitely, R. J. and Nahmias, A. J. The natural history of herpes simplex virus infection of mother and newborn. *Pediatrics* 66:489–494, 1980.

Wiswell, O. O. "Elegant" preparation. *The Female Patient* 8:14, Sept. 1983.

Wolfe, S. and Johnson, A. Letter to FDA on Flagyl, Oct. 22, 1974. Publ. 232, Health Research Group, 2000 P. Street, N. W., Washington, D. C. 20036, 1974.

Wolfe, S. and Johnson, A. Petition to FDA to ban the cancer-causing drug Flagyl. Publ. 163, Health Research Group, 2000 P. Street, N. W., Washington, D. C. 20036, 1974.

10. *Pregnancy, Labor, and Delivery*

A woman who had her children 25 or 30 years ago might reasonably ask what has happened to the whole process. She went to the hospital with her first pains, had a mask slapped on her face, had her hands tied down, and woke up with a baby. Today, even the vocabulary is new. People talk about "birthing," "parenting," "bonding," and "imprinting," about "choices in childbirth," about "psychoprophylaxis and Lamaze," about "Leboyer," and "birth without violence." Women have now openly acknowledged that there is a similarity between delivery of a baby and orgasm, and that breastfeeding might cause erotic pleasures and sensations—these were the unspokens in other generations. The great changes in childbirth have come because many women are no longer content to approach the process passively, letting the doctor take over. They want to be in charge, to make decisions about the manner of delivery, the procedure, the drugs, the instruments. They view pregnancy and delivery as a normal physiological process and reject the "sickness" view of traditional hospital obstetrics. Some women are deciding to have their babies at home, attended by a midwife rather than the traditional obstetrician.

These progressive and positive innovations represent a return of control to the mother and encourage the involvement of the father of the child. They emphasize the pleasures and the joys of pregnancy and delivery. The demands for alternatives have changed the attitudes of many hospital administrators and physicians and have stimulated research evaluating the benefits of different methods of childbirth.

The current emphasis on alternative methods for pregnant women—the media coverage, the many books, the classes, the articles in the women's magazines—is the result of the pregnant woman wanting choices for herself and her child and her awareness that she *should* be in control. It is, after all, her baby in her body. Unfortunately, the options are most available to the middle- and upper-class women who have planned pregnancy and have chosen to bear a child. A pregnant woman who already has more children than she wants, who is worried about how she is going to manage, who has neither the knowledge nor the money to get the best kind of prenatal nutrition or care, whose medical treatment is given in overcrowded county or city facilities, is not likely to place much value on a beautiful birthing experience. Neither is a pregnant teenager, who is physically immature, subsisting on an improper or perhaps even bizarre diet, and probably more concerned with hiding the pregnancy for as long as possible. Only when a pregnancy is planned, and the parent or parents have the money and time for options, can all the experiences and sensations of childbirth be truly understood and appreciated. The current agitation for options in maternity care is of great value and should not be underestimated, but it must be accompanied by a continued effort toward better health care for *all* women and include the right to choose whether to be pregnant at all.

Our discussion of childbirth, then, makes the assumption that the woman made a decision to have a baby and that she is ready to deal with all the joys, discomforts, highs, lows, and ambivalences of pregnancy. It is probably most relevant for the woman who has been fortunate enough to demand a certain standard of health care.

GETTING PREGNANT

All things considered, the act of conception is essentially inconceivable. To say merely that it occurs when the sperm unites with the egg is deceptively simple. Getting them both together at the right time involves an intricate interplay of hormonal preparation and an overwhelming number of natural barriers. That it happens at all seems amazing; that the world should be overpopulated as a result seems extraordinary. Obviously, there are a great many people who are overcoming the odds.

In order for a pregnancy to occur, a healthy ovum has to be liberated from the ovary, pass into a normal, open fallopian tube, and start being transported downward. The ovum must be mature but cannot get too ripe. If fertilization does not occur within less than a day after ovulation, the egg begins to deteriorate and die. If coitus does occur around the time of ovulation, between 250 and 400 million sperm in the ejaculate are deposited in the vaginal canal—for them, a very hostile environment. Not only can the acidity of the vagina be fatal, but sperm are subject to attack and ingestion by wandering leucocytes. It is essential for survival that they get out of the vagina and into the uterus in a hurry. So, with a limited amount of stored energy, they must swim upstream to a destination seven inches away—7,000 times their own length. (In humans this distance is the equivalent of 8 miles, or half the English Channel!) Many of the sperm start off in the wrong direction and get lost in the crevices of the vagina. Many more or them never make it up through the endocervical canal of the cervix. Even in the middle of the menstrual cycle when the cervical mucus is more watery, penetrating the cervix is still no small task. Presumably, with the aid of additional hormones—oxytocin is present in the female and prostaglandins are found in semen—and the muscular contractions of the uterus and the fallopian tubes, some of the sperm finally make it to the ampulla of the oviduct.

The sperm that actually get to the ovum, and only about .0001% do, find further formidable barriers. The egg is surrounded by a clear protein layer, the *zona pellucida*, and then outside of that, a densely arranged layer of cells, the *corona radiata*. The actual mechanisms of sperm penetration in the human are not completely known. It is generally believed that the sperm must undergo "capacitation," or the conditioning of a protective caplike coating from their heads. Only then are they able to secrete digestive enzymes and create a pathway through the corona radiata and the zona pellucida. Only one sperm actually penetrates the egg cytoplasm to fertilize it. Any others are blocked at the zona. The climax of all this effort is the union of the nuclei of the sperm and the egg. The two nuclei approach each other, make contact, lose their respective nuclear membranes, and combine their maternal and paternal chromosomes. The zygote, the beginning of a new individual, has been formed.

This event has been observed on human ova cultured in blood serum or follicular fluid and has even been filmed, using microcinematography. When scenes showing human fertilization are incorporated into television specials or other educational films, the approach of the sperm to the egg is generally accompanied by a rising crescendo of music. Then, at the very moment of conception, the drums roll and the cymbals crash—most appropriate for a happening this momentous. In real-life fallopian tubes, with no camera and no music, there is absolutely no awareness on the part of a woman that a potential pregnancy has been initiated. Whether she has been trying to get pregnant for over a year, or whether pregnancy is the last thing in the world that she wants, there is no physiological signal for any woman that says, this is it, now you are pregnant, and barring other circumstances, you will deliver a baby in 38 weeks! Several weeks will elapse before even one of the presumptive signs of pregnancy— missing the first menstrual period—will take place.

Influencing the Sex of the Child

Getting pregnant as soon as possible is generally the first priority when a couple decides to have a baby. Producing a daughter rather than a son, or vice versa, may not be of any particular concern unless the family already consists of two or more children of the same sex. It was explained previously (Chapter 5) that sex determination occurs at the time of fertilization and depends on whether the egg is fertilized by a Y-bearing sperm or an X-bearing sperm. An XX zygote will be a female, and XY will become a male. The chances of having a daughter or a son are 50:50. Statistically speaking, on any particular day throughout the world, about half of the babies conceived will be girls and half will be boys. If it were possible in some way to make certain that only X-sperm or only Y-sperm reached the egg first, it would then be possible to select for a particular sex. Landrum Shettles, former gynecology professor at Columbia University's medical school, believes he has such a method. Shettles claims that there are differences in sperm viewed with the phase-contrast microscope. Those with the larger elongate heads contain X chromosomes, and those with small round heads bear Y chromosomes. The lighter and more agile Y-sperm, he further suggests, reach the egg first when the environment is alkaline. The slower and hardier X-sperm can survive longer in an acid environment. Since the cervical mucus is most alkaline at the time of ovulation, intercourse at that time would produce a predominance of males. But intercourse 2 or 3 days before ovulation would assure that any sperm still alive when the egg arrives in the fallopian tube are the stronger female-producing sperm. His method requires that the precise time of ovulation be determined for several months by basal body temperature and cervical mucus tests (see Chapters 11 and 12) prior to attempting sex selection. The procedure also utilizes acid or alkaline douches before intercourse to change the vaginal environment.

Shettles claims that parents who have been instructed in his techniques successfully produced children of the desired sex 80% of the time. Considering that the couple has a 50% chance of getting what they want anyway, many may find the regimen not worth the effort. Shettles' procedures are listed for those who may want to try them.

In order to produce a male offspring:

1. Time intercourse as close to the time of ovulation as possible. The quicker Y-sperm can then get to the egg ahead of the slower X-sperm. Have the male drink strong coffee fifteen minutes before coitus in order to stimulate the sperm.

2. Precede intercourse by an alkaline douche consisting of 2 tablespoons of baking soda to a quart of water.

2. Make certain penetration is deep at the time of ejaculation and that the woman has orgasm.

4. Abstain from intercourse totally from day one of the cycle until ovulation.

Frequent intercourse lowers the sperm count, and Shettles believes that a maximum sperm count favors the Y-sperm. Tight clothing elevates scrotal temperature and also lowers the sperm count. It is suggested that the man avoid snug undershorts, jockstraps, or tight jeans for several months before attempting sex selection.

In order to produce a female offspring:

1. No abstinence is necessary. In fact, try to lower the sperm count by having frequent intercourse. But intercourse must cease 2–3 days before ovulation. This gives the stronger X-bearing sperm the best chance to survive until the egg arrives in the fallopian tube. Tight clothing is all right and even preferable, and a few hot baths may help, as long as the man does not overdo it.

2. Intercourse should be preceded by an acid douche consisting of 2 tablespoons of white vinegar in a quart of water. The woman should lie on her back during coitus, and Shettles advocates that if possible, she should avoid having an orgasm. It increases the alkaline secretions.

It should be emphasized that the Shettles method for producing a daughter should be used only by normally fertile couples. Lowering the sperm count, using an acid douche, and stopping all intercourse 2–3 days before ovulation so that the sperm have to lie in wait for the egg could, if there are any problems of infertility, be a way of not getting pregnant at all.

Moreover, the Shettles do-it-yourself technique has little support from other researchers who have found no experimental or statistical evidence to substantiate his claims. Few agree with Shettles that phase-contrast microscopy can readily distinguish X-bearing from Y-bearing sperm, and Diasio and Glass observed no difference in the ability of X or Y sperm to migrate through media of varying acidity in the laboratory. Glass concluded that it is highly unlikely that the sex of the child can be influenced by any douching technique. The other part of Shettles' theory, that the timing of intercourse at ovulation can produce males, is also controversial. Guerrero observed that the proportion of male babies conceived was greater both several days before and after ovulation, but fell to a minimum on the day of ovulation. Susan Harlap's study of almost 4,000 Israeli births further refutes Shettles' recommendations. In Israel, Orthodox Jewish women ritually engage in sexual separation each month. They abstain from intercourse until a week after menstruation, thus delaying intercourse to within several days of ovulation. Harlap showed that a significantly higher proportion of male babies were conceived when intercourse was resumed 2 days *after* ovulation. She warned, however, that couples should not try to conceive a boy baby by deliberately delaying intercourse until after ovulation because there is some evidence linking conception that occurs late in the cycle with birth defects, possibly because the egg has become "overripe."

If having a son is terribly important to a couple, there are some other methods of sex-selection that have been applied to humans that involve the techniques of sperm separation and artificial insemination used in livestock. But separating X- and Y-bearing sperm by sedimentation, centrifugation, or electrophoresis (running an electrical charge through the sperm in solution) has, thus far, been more successful in bull sperm than in human semen. One of the more publicized methods has been patented by Ronald Ericsson, a California biotechnician who founded a company called Gametrics, Ltd. Ericsson, whose company has 17 centers worldwide, has successfully used his procedure for artificially inseminating cattle with separated bull sperm. In a 1982 article, he and his co-worker, Ferdinand Beernink, described a similar technique for human sperm and claimed that it enhanced the chances of conceiving a male to 75%. The procedure starts with diluted human ejaculate layered onto a column of human serum albumin. The Y-bearing sperm, faster swimmers than the heavier X sperm, evidently move more quickly into the boundary between the semen solution and the albumin. After concentrating the sperm through two more albumin layers, the resultant sperm disc, now presumably containing 85% of the Y-bearing type, is resuspended in solution and inseminated into the woman's vagina near the cervix. The two workers reported the results from all Gametrics centers as 93 births, with 68 males and 25 females—a 75% ratio. The laboratory fees for Ericsson's sex-selection method are expensive even without including physician's costs for the insemination. Thus, only the relatively affluent who live near a Gametrics center and are willing to have sperm manipulation and artificial insemination can currently increase their chances of having a son by 25%.

At the present time, the only completely reliable method of ensuring an offspring of the desired sex is through selective abortion after amniocentesis—a procedure with physical risk and of questionable ethics when used for preselection. Most couples are primarily interested in having a healthy baby and are willing to take their 50:50 chances on its sex.

After Fertilization

When the egg is fertilized by a sperm, any symptoms of pregnancy are still weeks in the future, but for the fertilized egg, or zygote, a great deal of immediate activity takes place. The zygote somehow seals itself off so that no other sperm may penetrate it. The first cleavage takes place about 30 hours after fertilization, then the second and then the third, all the while the conceptus is slowly being transported into the uterine cavity as a result of the tubal muscular movements. After there are four cleavages, the sixteen cells appear as a solid ball or *morula*, meaning "little mulberry." The morula reaches the uterine cavity and cell division continues rapidly. Within the morula, an off-center fluid-filled space appears, transforming it into a hollow ball of cells, the *blastocyst*. On one aspect of the inner surface of the blastocyst is a mound of cells destined to become the embryo—the *inner cell mass*. The outer wall of the tiny ball becomes the *trophoblast*, containing the cells that will be responsible for the invasion of the uterine lining. By 6 days after ovulation, the pole of the blastocyst that contains the inner cell mass has attached itself to the endometrium of the uterus. Up until now, the developing conceptus has been nourished by the secretions of the tube and the uterus, but its further growth requires more food and oxygen. The blastocyst gains access to the maternal blood supply by burrowing into the uterine lining and becoming completely implanted in it. The trophoblast secretes enzymes that dissolve the cells of the endometrial blood vessels, glands, and connective tissue, and eventually the blastocyst comes to lie in a pool of maternal blood. Now the trophoblast cells proliferate to form branching sprouts, the primitive villi, which increase the surface area of the blastocyst and make it easier for the developing embryo to tap the maternal arteries for nourishment. The thick prepared endometrium is called the *decidua*, meaning, "to shed." The portion of the endometrium under the blastocyst is the *decidua basalis*; this will become the maternal portion of the placenta. Figure 10.1 is a diagrammatic summary of the developmental events during the first week after fertilization.

Concurrent with the development of the trophoblast and implantation there is further differentiation of the inner cell mass. Some of the cells will become the embryo itself, and others will give rise to the membranes that surround and protect it. Three embryonic layers of cells—the ectoderm, the endoderm, and the mesoderm—are formed (see Fig. 10.2A). The outer ectoderm will give rise to such structures as the entire nervous system and the special senses, the skin, and some of the endocrine glands. From the inner endoderm rises the digestive and respiratory systems and parts of the reproductive system. The middle layer, or mesoderm, differentiates into the future skeletal, urinary, circulatory, and reproductive systems. A cavity, the amniotic sac, develops from a cleft in the ectoderm. By about the thirteenth day, a second cavity—the yolk sac or primitive gut—appears in the endoderm (Fig. 10.2B). As the amniotic cavity enlarges, it becomes filled with a clear and watery fluid and will expand until its roof, the *amnion*, reaches the wall of the blastocyst. As the embryo develops, it is protected and cushioned by amniotic fluid within the amniotic sac. The second fetal membrane, or *chorion*, consists of the trophoblast cells and a mesoderm lining (see Fig. 10.2C). The trophoblast primitive villi, now with a mesoderm core, proliferate and branch to become the chorionic villi. Blood vessels develop in the mesoderm of the chorionic villi. They become attached to the body stalk connected to

the embryo, formed when the two edges of the amniotic sac came together. The body stalk is the future umbilical cord (Fig. 10.2D).

Originally, the chorionic villi are present over the whole surface of the blastocyst, but as the blastocyst enlarges, the surface that projects into the uterine cavity loses its villi and becomes smooth. This part of the chorion is known as the chorion laeve (smooth chorion). At the opposite side of the blastocyst the chorionic villi branch and enlarge. This area becomes known as the chorion frondosum (bushy chorion). The connecting stalk of the embryo is attached to the wall of the blastocyst at the chorion frondosum, which will become the fetal part of the special structure for fetal–maternal exchange, the placenta. A very complicated network develops in which the fetal villi are surrounded by and bathed in the maternal blood sinuses of the decidua basalis. Later, the fetal blood vessels, two umbilical arteries and one umbilical vein, are carried in the umbilical cord, which attaches to the fetal side of the placenta. These blood vessels continue on into the villi. Figure 10.3 is a diagrammatic section showing fetal and maternal tissues in the placenta.

At no time during the pregnancy is there any direct connection between the blood of the fetus and the blood of the mother, so that there can never be any mixing of blood. There are always layers of fetal tissues that separate the maternal and the fetal blood. This is the so-called placental barrier. Materials can be interchanged only through diffusion. The food and oxygen delivered by the maternal uterine arteries diffuse across the placental barrier into the bloodstream of the fetus to provide its life-support system. The waste products from the fetus travel in the opposite direction and are carried away by the mother's uterine veins. When the baby is born it is still connected to the placenta by the umbilical cord. The mature placenta, approximately 7 inches in diameter and 1 inch thick, is dark

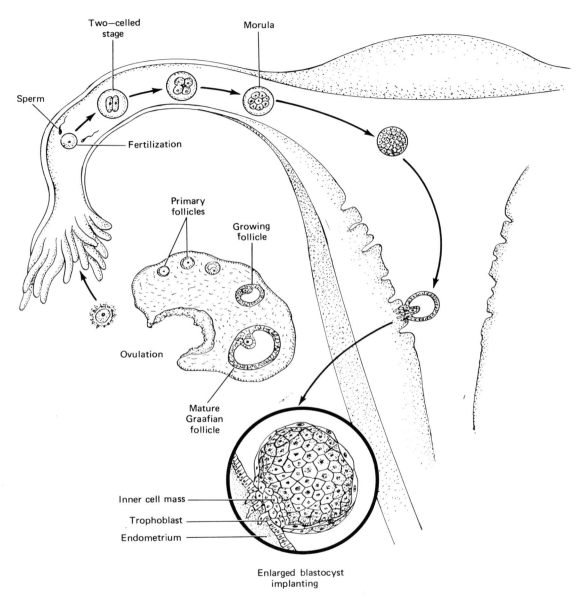

Figure 10.1. *Summary of the ovarian cycle, fertilization, and development of the zygote during the first week. During the second and third days after fertilization, the fertilized egg grows from two cells into the 16-cell morula stage. The next stage, the blastocyst, remains free in the uterine cavity during days four and five. By the sixth day, the blastocyst begins to attach itself to the endometrium of the uterus.*

red and weighs about a pound. It is delivered from the uterus after the baby, and the umbilical cord must be cut and tied off.

The placenta functions as a transfer vehicle for oxygen and nutritional substances passing from the maternal blood to the fetal circulation, and for carbon dioxide and waste materials from the fetus in the opposite direction. As the maternal blood enters the placenta, there is a fall in blood pressure that slows its passage and facilitates the respiratory, nutritional, and excretory transfer. The placenta is not only a transfer organ but a factory as well. It is capable of synthesizing enzymes and proteins, and it manufactures fats and carbohydrates that serve as a source of stored energy. The placenta is also a unique kind of endocrine gland; it has a feature possessed by no other endocrine organ—the ability to form both protein and steroid hormones. The hypothalamic-pituitary-ovarian hormones of the mother have set the stage for ovulation, conception, and implantation, but very early in pregnancy, the placenta begins to synthesize hormones. Placental *human chorionic gonadotropin*, HCG, is secreted very early by the cells of the trophoblast. In biological effects, HCG is similar to luteinizing hormone. Its function is to preserve the corpus luteum and its progesterone production so that the endometrial lining of the uterus, and hence pregnancy, is maintained. The production of HCG begins soon after implantation, rises to a peak production between the eighth and twelfth weeks of pregnancy, and then falls to a much lower level for the remainder of gestation. Its presence in blood or urine is used as the basis for pregnancy tests. Human placental *lactogen*, similar in function and chemical structure to both pituitary growth hormone and prolactin, is also known as human *somatomammotropin*. But there is some evidence that the placenta also produces growth hormone and prolactin, and possibly thyrotropin as well. After about the eighth week, the placenta is a

major source of progesterone, which it produces in large quantities to maintain the pregnancy. The placenta forms estrogen from androgen precursors, many of which are provided by the fetal endocrine glands.

The Period of Gestation

The first few days of human development after fertilization are spent in the fallopian tube, with the zygote busily cleaving itself into the eight-cell stage. In the next 2–3 days, the morula and then the blastocyst are free in the uterine cavity, but by the end of the first week the blastocyst is superficially imbedded in the endometrial lining of the uterus. It will not be completely implanted until $2\frac{1}{2}$ weeks have elapsed.

The third week of development coincides with the first indication for the woman that she may be pregnant—the missed period. It is not unusual, however, for vaginal bleeding as a result of hemorrhage from the implantation site to occur. Such bleeding, when it is misconstrued as menstruation, commonly results in a faulty calculation of the delivery date. Between the third and fourth week of pregnancy, the first organ system in the embryo becomes functional. The primitive heart tube is linked to the blood vessels in the connecting stalk and chorion, and there is now a cardiovascular system in the embryo, which can draw on the maternal blood supply in the uterus. With the formation and differentiation of a heart and blood circulation, the rest of the full development can take place. In the next few weeks, all of the organ systems of the embryo rapidly become established. By the end of the eighth week of pregnancy, the beginnings of all the major internal and external structures are present. Up until this time, the general body form was that of any mammalian embryo, but now it has taken on unquestionably human characteristics. With the change to a recognizable human, the develop-

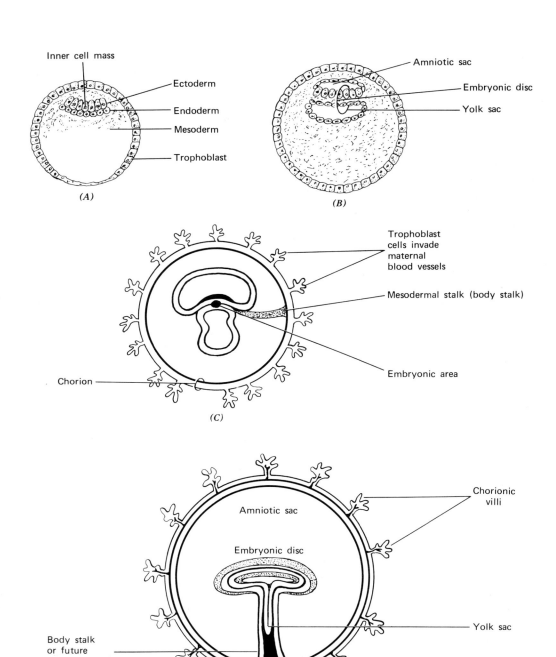

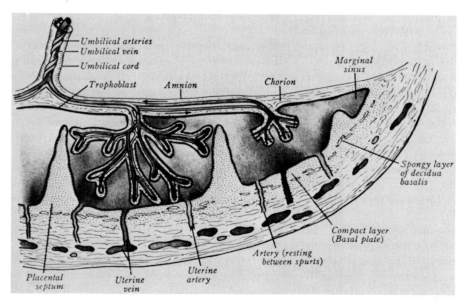

**Figure 10.3. Diagram of the maternal–fetal blood flow through the placenta.
Maternal blood carrying nutrients is delivered in spurts through the ends of the
eroded uterine arteries into the intervillus spaces to bathe the chorionic villi.
Nutrients and wastes are exchanged between the chorionic villi from the fetus
and the pools of maternal blood. (From Arey, Leslie Brainerd. Developmental
Anatomy. 7th ed. © 1974 by the W. B. Saunders Company, Philadelphia, Pa.)**

ing zygote is no longer called an embryo.
From the ninth week until birth, it is known as
a fetus.

The entire gestational period is commonly
divided into three parts or trimesters, each
consisting of three calendar months. The first
trimester is a critical period of development

because of the great sensitivity of the fetus to
teratogens, agents that can induce malformations. By the end of the first trimester, all of
the major organ systems are developed. The
rest of the fetal period is concerned with the
further growth and differentiation of the organs.

**Figure 10.2. Further development of the embryo and formation of the fetal mem-
branes. (A) Formation of the germ layers. (B) Two small cavities, the amniotic sac and
the yolk sac, appear. (C) Projections of trophoblast cells (primitive villi) invade blood
vessels in the endometrium, and a circulation between the uterine vessels and the
embryo begins to be established. The second fetal membrane or chorion forms. The
actual embryo will arise from the ectodem, mesoderm, and endoderm of the embryonic
disc. (D) The amniotic sac expands until it reaches the wall of the blastocyst. The two
edges of the sac come together to enfold the yolk sac and form the body stalk. The
blood vessels from the embryo extend along the stalk to become continuous with the
vessels in the chorionic villi, ultimately giving rise to the umbilical arteries and the
umbilical vein.**

The rate of body growth is astonishing. At the end of the first week of life, the embryo is smaller than the period at the end of this sentence. At the end of the first trimester, the fetus weighs a little over 1 ounce and its length from the top of its head to the end of its spine is the length of one's middle finger. In 9 months of fetal life, the ovum will average a daily increase of $1\frac{1}{2}$ millimeters in length and will increase its weight 6 billion times. If it kept growing like that after birth, a 10-year-old child would be 20 feet tall and weigh several trillion times as much as the earth!

By the end of the second trimester, 24 weeks after fertilization, the crown-rump length of the fetus is equal to the span of one's hand. It weighs about a pound and a half with very little subcutaneous fat stored under its skin. Consequently, the fetus looks wrinkled and wizened with more of the characteristics associated with old age than with infancy. Although all of its organs are well-developed, the respiratory system of a fetus born prematurely at this time is generally too immature to permit it to survive.

Beginning late in the seventh month the fetus accumulates rapidly increasing amounts of fat. As term approaches, it begins to take on the rosy plumpness of a newborn infant. Mortality rates are still high for fetuses born between 26 and 29 weeks, but some do survive independently. The last 3 weeks of fetal life, 35 to 38 weeks after fertilization, put on the finishing touches. Growth slows as birth approaches. On the average, fetuses reach a crown-rump length of 360 mm (14 inches), total length 20–22 inches, and weigh about $7\frac{1}{2}$ pounds. Male fetuses grow somewhat more rapidly than female fetuses, and baby boys tend to weigh about 3 ounces more than baby girls at birth. All newborns' eyes are the same color, blue-gray, and it is impossible to predict their later hue.

Multiple Pregnancy

Only one offspring at a time is the usual condition in humans and other large mammals. The most common type of human twinning occurs when two eggs are simultaneously ovulated from separate follicles, are fertilized by different sperms to implant in two separate, but closely apposed areas in the uterus. These are fraternal or *dizygotic* twins. They may be the same or of opposite sexes, and they resemble each other as much as any other siblings. *Monozygotic* twins originate from one fertilized egg. This kind of twinning usually starts at the end of the first week with the division of the inner cell mass into two embryos. Such twins are the same sex, genetically identical, and obviously are identical in physical appearance unless there has been some environmental factor during fetal life to cause a difference between them.

Twins occur about once every 87 births in the United States, and most of them are the fraternal type. The tendency toward dizygotic twinning is genetically influenced, being greater in blacks and less in Japanese, for example. The likelihood of twins increases with the age of the mother, and once a woman has given birth to twins, a multiple birth is about five times more likely to occur the next time. The chances subsequently decrease to three times. There are various opinions concerning the inheritability of monozygotic or identical twins, but it is agreed that age of the mother is not a factor. "Siamese" or joined monozygotic twins occur as a result of incomplete duplication about one in every 400 identical twin pregnancies.

Multiple births other than twins can be of the identical type, the fraternal type, or combinations of the two. Triplets are said to occur once in 7,600 births (87^2) and quadruplets once in 660,000 births (87^3). In recent years the use of drugs to induce ovulation has resulted in a greater frequency of quadruplets, quintuplets, and sextuplets, which are normally very rare.

DIAGNOSIS OF PREGNANCY

In the old movies, most of which are still seen on the Late, Late Show, pregnancy was always diagnosed by telephone. The phone would ring, the prospective mother would pick it up, listen, and then announce to whomever was around, "The rabbit died!" A whole generation of young moviegoers, only vaguely aware of how the daddy puts the seed in the mommy's tummy, grew up thinking that dead rabbits were somehow integral to being pregnant.

The rabbit did not just die, of course. It had to be killed so that its ovaries could be examined for ruptured follicles. Urine from a presumably pregnant woman was injected into the ear vein of a mature virgin female rabbit. The chorionic gonadotropin in the urine, if the woman was indeed pregnant, induced ovulation in the rabbit within 16 to 48 hours after injection. This constituted the Friedman test, one of the biologic methods for determining pregnancy. The test animals for the Aschheim-Zondek test were immature mice, and other biological tests were developed using immature rats, female toads, or male frogs. They were all based on the same principle—that chorionic gonadotropin present in the blood and urine of pregnant women will cause an effect in the reproductive tracts of test animals.

Although these biological tests were reliable, they required destroying test animals, took several days for determination, and gave the best results only after about 2 weeks past the missed period. After 1960, immunological tests to detect the presence of chorionic gonadotropin became available and have largely replaced the animal tests.

Immunological pregnancy tests are either slide tests or tube tests. In the slide tests, the woman's urine is mixed with the blood serum that contains antibodies to human chorionic gonadotropin. Latex particles coated with pure HCG are than added to the mixture. If the woman is pregnant, no clumping or agglutination of the latex particles takes place and the test is considered positive. If she is not pregnant, the agglutination of the particles can be observed, and the test is then negative. The basis for the test is that if there is HCG in the woman's urine, it neutralizes or inhibits the serum antibodies. The antibodies are then unavailable to react with the purified human chorionic gonadotropin that had been coated onto the latex particles; hence, no agglutination. Some slide tests are simplified to a quick one-step procedure. The latex particles, coated with anti-HCG serum, are mixed with the urine. If there is HCG in the urine, the particles agglutinate (positive test); if HCG is absent, there is no agglutination (negative test).

The tube tests are based on the same kind of agglutination inhibition, but the purified human chorionic gonadotropin is coated on sheep red blood cells. When anti-HCG serum, the coated blood cells, and pregnancy urine containing HCG are mixed in a tube, the HCG in the urine will neutralize the antiserum. The unagglutinated red blood cells sink to the bottom of the tube and form a ring of cells in a doughnut pattern, positive for pregnancy. The absence of the ring indicates a negative test; that is, no HCG in the urine.

There are a number of slide and tube immunoassay tests supplied to clinics, laboratories, and doctors in the form of pregnancy kits. No test is ideal, and all will occasionally produce false results. False-positive results, incorrectly diagnosing pregnancy in women who are not pregnant, are relatively infrequent, but can ocur when certain drugs, particularly tranquilizers, are taken. Protein or blood in the urine, common in bladder infections, also can result in false positives. False-negative results, which indicate nonpregnancy in women who are actually pregnant, are more common, particularly in early pregnancy. Some of the tube and slide tests have greater sensitivity for detecting low levels of HCG, but all of the routine immunoassays that are performed in a physician's office or at home are basically qualitative tests.

Many women would find a distinct advantage in being able to diagnose their own reproductive state at home, privately and confidentially. Whereas home pregnancy tests had been sold over-the-counter in Canada and Europe for many years, it was not until early 1978 that such a kit was licensed for sale in the United States. The "e.p.t. In-Home Early Pregnancy Test" was the first one available, and, like several others currently on the market, sells for about $10. The woman is told to add three drops of urine to a test tube that contains HCG on sheep red blood cells and HCG antiserum, to add a vial of water, and to place the tube within its plastic container undisturbed on a flat surface for 2 hours. If a dark ring has formed on the bottom of the tube, the test is positive for pregnancy. The instructions advise that if a test is negative, it should be repeated in a week. But since the test is not reusable, a woman would have to spend another $10. If the test is still negative and menstruation has not begun, the woman is advised to consult her physician.

The "e.p.t." kit is not to be used earlier than the ninth day after the first day of the expected menstrual period. The manufacturer claims that when the test is used as directed at home by consumers, a single test is 97% accurate if positive, and 80% accurate if a neg-

ative reading is obtained. Repeating the test in a week increases the accuracy of a negative test to 90% but doubles the expense. Since most cities have clinics, such as Planned Parenthood (check under Birth Control in the Yellow Pages), that will perform pregnancy tests free or for less money and certainly with equal or greater accuracy, the usefulness of a do-it-yourself kit, such as e.p.t., Predictor, Accutest, or Daisy II, is questionable and a matter of individual choice.

Radioimmunoassay tests (RIAs) are much more sensitive than the slide and tube tests, are capable of detecting very low levels of human chorionic gonadotropin in the blood serum instead of the urine even before the first missed period, and are usually quantitative tests. The earliest way of detecting HCG is by means of the RIA for the beta subunit of HCG, the portion of the molecule responsible for its biological properties. The beta subunit is produced by the trophoblast cells of the embryo 7–9 days after ovulation, and the RIA test can be performed by a laboratory that has the appropriate equipment in about 2 hours. Because the RIA beta-HCG test is able to pick up such very small amounts of the hormone in the blood, it is especially useful in the diagnosis of ectopic pregnancy, which is associated with very low levels of HCG. Another form of the RIA, the radio-receptor assay (RRA), was originally developed at Cornell University by Robert Landesman and Brij Saxena. It is commercially available as the "Biocept-G" from Walpole Laboratories. This test is reportedly 98% accurate and is capable of detecting pregnancy 10 days after conception. It costs about $15 and takes an hour to perform.

Another type of pregnancy test, the hormone-withdrawal, or clinical, test, was widely used until the early 1970s, when reports of serious birth defects associated with it appeared in the medical literature. In this test, a single injection of progesterone or a course of oral progestin was administered for several days. Within 2–5 days of taking progestin, 90% of nonpregnant women will have menstrual bleeding, and 90% of pregnant women will have no evidence of bleeding. The problem is that uterine bleeding after progestin administration is not conclusive proof that a woman is not pregnant. If she should actually be pregnant and then decides to retain the pregnancy, a powerful drug that carries the risk of birth defects has been taken at a time when the embryo is most vulnerable. Congenital heart defects (Levy, Cohen, and Fraser, 1973), as well as abnormalities of the esophagus (Oakley, Flynt, and Falek, 1973), have been ascribed to the use of hormone pregnancy tests. The Food and Drug Administration withdrew approval for the use of progestins for pregnancy testing, preventing miscarriages, or treatment of any abnormality of pregnancy in 1973. In the unlikely event that a physician should suggest such a test for pregnancy diagnosis, under no circumstances should a woman permit it. The urine immunological tests are reasonably accurate, and the blood serum tests are highly reliable.

Pregnancy tests are tools to establish the diagnosis of pregnancy when the physical signs are still inconclusive. By confirming a pregnancy they can reassure a woman who is very anxious to have a baby. They can allay anxiety by confirming nonpregnancy in a woman who is not. If a termination of the pregnancy is desired, a pregnancy test is useful since early abortions are safer. If a woman wants to be pregnant she can start a program of prenatal care and avoid needlessly exposing the fetus to drugs, smoking, or alcohol. Pregnancy tests, however, are rarely necessary for a wanted pregnancy. Within 2 weeks after a missed period, there are usually enough subjective symptoms so that a woman can be reasonably certain she is pregnant. At that time, a pelvic examination by an experienced practitioner can establish the diagnosis of pregnancy with accuracy.

Signs and Symptoms of Pregnancy

Traditionally, the portents of pregnancy have been grouped into those that are presumptive and experienced by the woman herself; those that are probable, and observed by the clinical examiner; and finally the positive, beyond-the-shadow-of-a-doubt signs. The latter are not evident until the sixteenth to eighteenth week of gestation.

The most obvious presumptive symptom is the absence of menstruation. Of course, just being late or even skipping a period is not a reliable sign of pregnancy, but when the cessation of menstruation is accompanied by morning nausea, fatigue, breast tenderness, and frequency of urination, pregnancy is very likely. Every woman is different, and these early complaints may not be present. But when several of them are associated with a missed period there is a $2:1$ chance of pregnancy.

A woman accustomed to vaginal self-examination would notice color changes from the usual pink to a dusky blue in the walls of the vagina and the cervix. This is caused by increased blood supply and venous congestion and is called Chadwick's sign. Such a color change can be present, however, in heart disease or with a pelvic tumor.

A rather reliable indication of pregnancy is an elevation of the basal body temperature. A woman who keeps a record of her BBT would know she is pregnant if a level of $98.8°–99.8°F$ remained for 3 weeks after ovulation.

Probable signs of pregnancy are those that appear on physical examination. Changes in the size, shape, and consistency of the uterus can be observed. Soon after implantation of the fertilized ovum, a soft bulge at the implantation site appears on one side of the uterus. Skilled hands can detect this softening of the uterus on one side while the other side remains firm (Piskacek's sign). Goodell's sign, a softening of the cervix, also appears early. Hegar's sign refers to the softening of the neck or isthmus of the uterus, the narrow part between the body and the cervix. First detectable at about 6 weeks, Hegar's sign disappears later when the entire cervix and the uterus have become soft and spongy in consistency.

The positive signs of pregnancy refer to proof that there is, beyond any doubt, a fetus growing in the uterus. Around 16–18 weeks a woman will begin to feel a sort of fluttering feeling (often mistaken for gas in a first pregnancy) that is known as quickening, or "feeling life." For the doctor or nurse-midwife, hearing the fetal heart sounds, a fetal electrocardiogram, or an ultrasound outline of the fetus within the fetal sac are all signs that make the diagnosis of pregnancy a certainty. By 14 weeks a procedure called internal ballottement can be performed gently during a vaginal examination. With the finger up against the cervix, the cervix is gently tapped. The fetus within the amniotic sac bounces up against the top of the uterus, and, as it sinks back down, the rebound is felt against the finger. By 24 weeks, external ballottement, or a push on one side of the abdomen with one hand in order to feel the fetus bounce to the other side and hit the other hand, can be performed.

Waiting to get these kinds of confirmation of pregnancy may seem to be unnecessary, but there are circumstances when demonstrating the presence of a fetus is the only way of distinguishing pregnancy from another condition, such as an ovarian cyst or multiple fibroids. Every now and then, a case is reported in which a woman, usually very obese, actually delivers a baby as her first indication of pregnancy.

Calculation of the Due Date

Should a woman be fortunate enough to know the date of conception, she can, based on statistical averages, expect to deliver a baby 266 days, or 38 weeks, later. Since the precise time of fertilization is ordinarily not known, the usual rule of thumb is to date the pregnancy from the onset of the first day of the last menstrual period that occurred—a total of 280 days, or 40 weeks, or 10 lunar months, or 9 calendar months. The estimated delivery date is obtained by taking the first day of the last menstruation, subtracting 3 months from that date, and then adding 1 week. If, for example, the first day of the last menstrual period was February 21, the termination of the pregnancy will be on or about November 28. This calculation assumes that ovulation, and hence fertilization, occurred 14 days after the onset of the menstrual period. This is only true in a 28-day cycle. Ovulation is known to occur 14 days *prior* to a menstrual period, but if a woman is pregnant, that is the menstrual period she will not have. The point is, the due date is only a time frame; delivery can normally occur 2–3 weeks before or after it.

MATERNAL CHANGES DURING PREGNANCY

Pregnancy is a physiologically normal condition and not an illness. This does not mean that a woman will feel terrific and tranquil for the entire 9 months. Who does, pregnant or not, for the better part of a year? Treating pregnant women as though they were ill and women thinking of themselves as such is one extreme. The other is denying that pregnancy can be anything but a healthy, happy, productive, problem-free period. Sometimes all the emphasis on the "naturalness" of pregnancy can produce unrealistic expectations about how a healthy and emotionally stable woman "should" feel and behave. But consider that there are profound and multiple physiological adjustments of all body systems that begin at the moment of conception, even before a woman is aware that anything has taken place. These changes continue throughout pregnancy, and some of them may not be reversed until 6 weeks after delivery of the baby. Such alterations are bound to be the basis for some symptoms. How a woman reacts and copes with them has a lot to do with her personality, life-situation, relationship with her partner and the rest of her family, and her feelings about having a baby. It would seem logical that there may be more ambivalence, regret, or conflict when the pregnancy was unplanned. But even a planned and wanted baby is no guarantee that the period of waiting for it will be gratifying and free from minor problems and complaints.

Everyone is different and unique. A great many women can sail through pregnancy experiencing no difficulties and enjoying every minute of it. No nausea, no backaches, no swelling mars their expectancy. They bloom, they glow; they apparently have no trouble with their skin or hair, and they don't get hemorrhoids, stretch marks, or varicose veins. Other women may have some or all of these symptoms, and they will obviously not get as much pleasure out of being pregnant. But whatever the ups and downs of pregnancy, and they do exist, a woman can have the underlying knowledge that she is participating in something exceptional—the process of creating another human being. The physical aspects of pregnancy are frequently unpredictable and sometimes uncomfortable, but an understanding of what is taking place and why, and an appreciation that other women have similar experiences, can provide a woman with the knowledge to work with her body and not against it.

Changes in the Reproductive System

Uterus

The uterus, which in a nonpregnant woman weighs 2 ounces, has a capacity of 2 ml, and measures 3 inches by 2 inches by 1 inch, undergoes a spectacular increase in size. At full term it weighs 2 pounds, is about five to six times larger, and has increased its capacity by 2,000 times to accommodate the developing fetus. This enormous enlargement of the pregnant uterus is accomplished by growth of the individual smooth muscle cells, or fibers, which become wider and longer. The blood vessels elongate, enlarge, dilate, and sprout new branches in order to support and nourish the growing muscle tissue, and the increase in uterine weight is accompanied by a large increase in uterine blood flow. Uterine contractility is evidently enhanced. Spontaneous, irregular and painless contractions, called Braxton Hicks' contractions, begin in the first trimester. They continue throughout pregnancy, becoming especially noticeable during the last month when they function in thinning out or effacing the cervix before delivery.

The uterus is still in the pelvic cavity for the first 3 months, after which it progressively ascends into the abdomen. As the uterus grows, it presses on the urinary bladder and causes the increased frequency of urination noticed in early pregnancy. By 20 weeks, the fundus, or top of the uterus, is at the level of the umbilicus. Thereafter, the growth of the fetus can be determined by measuring the distance from the pubic symphysis to the top of the uterine fundus. The expanding uterus begins to "show" around the fourth to fifth month, although the waistline seems to thicken earlier in subsequent pregnancies. Figure 10.4 shows the change in abdominal shape in a woman who is 7 months pregnant.

Cervix

Softening of the cervix by the sixth week (Goodell's sign) has been mentioned. Along with the softening, which is the result of vasocongestion, the endocervical glands increase in size and number and produce more cervi-

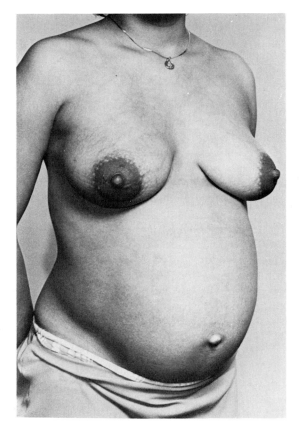

Figure 10.4. Pregnancy during the seventh month. The top of the uterine fundus is about three fingerbreadths above the umbilicus. By the ninth month, it will rise to about an inch and a half below the breastbone.

cal mucus. Under the influence of progesterone, a thick mucous plug is formed, which blocks the cervical os and protects the developing conceptus from bacterial invasion. The increase in cervical secretions results in leukorrhea, or increased vaginal secretion. The secretion is markedly acidic, which also helps in preventing bacterial infection.

The cervix is composed almost completely of connective tissue with very little smooth muscle. Page compares the cervix at term to a combination of rubber bands and salt water taffy. Under a slow steady pull, the cervix stretches but has very little rebound and can, therefore, progressively dilate.

Ovaries and Tubes
The increased blood supply to the ovaries and tubes causes them to become somewhat enlarged and elongated, and they tend to hang down alongside the uterus. The ovary that contains the corpus luteum, which reaches a maximum development during the third month of pregnancy, is even more pronounced in its enlargement. After the placenta takes over the major production of progesterone, the corpus luteum of pregnancy regresses. The anterior pituitary gland is inhibited by the large amounts of circulating steroid hormones, and no follicles mature or ovulate during pregnancy.

Vagina
Ordinarily a highly distensible organ, the vagina prepares for even greater distensibility during delivery by a thickening of the vaginal lining, a reduction in the density of the connective tissue, and an increased growth of the muscle tissue. The increased blood supply results in the color change previously mentioned.

Breasts
Many women experience a tenderness and enlargement of the breasts during the progestational phase of each menstrual cycle. This sensitivity remains and becomes intensified in early pregnancy. Tingling and soreness are common in the first 2 months; thereafter, the breasts become enlarged and even more tender and nodular. A woman accustomed to sleeping on her stomach may not be able to because her breasts hurt. Wearing a bra all the time, even at night, in order to support the breasts will help. The nipples enlarge, become more erectile, and deeply pigmented. As seen

in Fig. 10.5, the primary areolae become wider and deepen in color and Montgomery's glands are more prominent. Under the influence of steroids from the corpus luteum and the placenta, from placental lactogen, prolactin, and chorionic gonadotropin, the ducts, lobules, and alveolae of the gland tissue proliferate tremendously. By the tenth week of gestation, colostrum can be expressed from the nipples, but actual milk synthesis is inhibited by the high sex steroid hormone levels. If a woman is planning to nurse her baby, certain preparations can be made during the last 2 months of pregnancy. The colostrum should be massaged out of the breasts twice daily, and a wet washcloth should be used on the nipples and areolae to remove any crusting. Inverted nipples may have to be pulled out each day to encourage protraction. Some women's breasts become dry and chapped, and a petroleum jelly-based cream can be massaged into the skin.

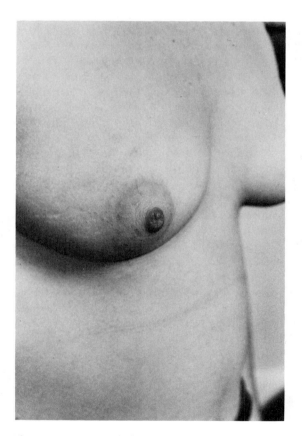 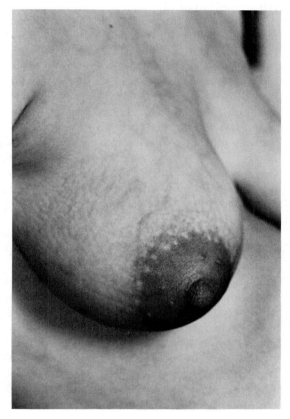

Figure 10.5. Comparison of the breasts of a nonpregnant women (left) with the breast of a woman who is 7 months pregnant (right). Note the size, the deeply pigmented areola, signs of increased vascularity, and prominent Montgomery's gland elevations in the breast during pregnancy. The second areola, a further pigmented mottling effect surrounding the primary areola, is evident.

Changes in the Digestive System

Morning Sickness

One of the most annoying symptoms of early pregnancy, experienced by 50%–80% of pregnant women, is the nausea and vomiting familiarly known as "morning sickness." While it occurs most often in the morning, the nauseated feeling can be present in the afternoon or evening, and some women have it all day. The nausea is not a dizzy nausea, the kind one gets with the flu, but is of sudden onset and seems to be associated with hunger pains, although it can be precipitated by odors from cooking, paint, cigarettes, and so forth. For many women, nausea can be eliminated by avoiding an empty-stomach situation. If the nausea is the mild prebreakfast type, munching a few crackers or dry toast in bed before arising may be the answer. For a particular woman, a period of trial and error may be necessary to find something that alleviates her nausea. With hot tea, fruit juices, candy drops, several small meals instead of three large ones, nonfat foods, and high protein foods, many women can solve this problem. Evidently, it is the frequency of eating rather than the quantity that is important.

The physiological basis for morning sickness is essentially unknown. There are various theories that attempt to link the nausea and vomiting with the high levels of human chorionic gonadotropin, high levels of circulating estrogens, reduced stomach acidity, and the lowered tone and motility of the digestive tract. The principal nutrient utilized by the fetus is glucose, and it derives its supply from the mother's blood. As a result of the placental transfer of glucose, the pregnant woman's fasting blood sugar level is about 10% lower than normal. This, and other changes in maternal carbohydrate metabolism may play a role in morning sickness since drinking a sweet beverage on arising often can prevent nausea and vomiting.

There are some women who try soda crackers, juice, hard candy, ice water—all the measures that apparently work for other people—without success and still experience multiple daily episodes of vomiting. Relatively rarely, the nausea and vomiting of pregnancy can develop into *hyperemesis gravidarum*, which as the name implies, is a severe and persistent vomiting that produces dehydration and disturbances in acid–base balance and requires hospitalization for intravenous feeding and other measures to relieve it. There is an analytic school of thought that associates hyperemesis with an unwanted pregnancy. "There is no doubt," states one Ob/Gyn text, "that most instances of excessive persistent vomiting (hyperemesis gravidarum) are due to deep-rooted psychological factors which result from a subconscious rejection of the pregnancy" (Page, Villee, and Villee). Another recent text on obstetrical practice, edited by Aladjem and presumably used to teach medical students, unequivocally states that hyperemesis is psychogenic vomiting and that a woman is unconsciously trying to "vomit up the baby" because she either rejects it or is frightened of having a baby with a serious birth defect.

Certainly, the involvement of the emotions with physical complaints is well recognized, and there may be cases in which excessive nausea and vomiting has a psychosomatic basis. Despite psychiatric conviction, however, there can be little actual evidence to prove or disprove this explanation since unconscious motivation is obviously difficult to validate. A woman whose morning sickness has become serious and persistent enough to put her in the hospital probably would have considerable anxiety as a result, and it would not be surprising if she very *consciously* rejected the whole idea of being pregnant at that point. In addition, labeling a woman with a preconceived psychological evaluation could easily lead physicians to ignore a consideration of other conditions, such as food poisoning or intestinal obstruction, that could cause hyperemesis.

The safety of any drug taken during pregnancy, including those to relieve nausea and vomiting, is questionable. Drugs should be avoided if at all possible, But women whose morning (or all day) sickness drastically interferes with their daily lives, is unresponsive to the usual nondrug remedies, generally makes them miserable for months on end, and can develop into hyperemesis with its accompanying metabolic disturbances, would have to weigh the benefits of taking an antiemetic drug against its potential risks. From the time of its introduction in 1956, an estimated 33 million women worldwide took such a proprietary antinauseant called Bendectin in the United States and Canada, Debendox in Great Britain, Lenotan in Germany, and Merbental in South America. When its manufacturer, Merrell Dow Pharmaceuticals, ceased production in June, 1983, and took it off the market, about 25% of the pregnant women in this country were taking Bendectin. Composed of an antihistamine (doxylamine succinate) and a vitamin (pyridoxine hydrochloride), Bendectin was the only medication for nausea and vomiting of early pregnancy that had specifically been approved by the Food and Drug Administration.

Since it has probably been the most commonly taken prescription drug during early pregnancy, it has also been the most thoroughly studied for the risk of birth defects. Its widespread use, however, has contributed to the inconsistency of the results. The incidence of birth defects in the general population—the risk that everyone faces—is 3%. Since one-quarter of all pregnant women in the United States (an estimated 400,000 annually) have taken Bendectin in the first trimester of pregnancy, there have been babies born with congenital abnormalities who have also been exposed to the drug, but this does not prove that Bendectin caused the defect. A cause-and-effect relationship between Bendectin and congenital malformations has never been clearly or consistently demonstrated, but there have been a number of studies that associated its use with birth defects that included digestive tract abnormalities, sternum concavities (funnel chest), heart defects, and cleft palates or lips.

A panel of experts convened by the Food and Drug Administration in 1980 examined the data from animal studies and 13 epidemiological studies and concluded that the incidence of malformations in the offspring of women who took Bendectin was no greater than the incidence of women who had not been exposed to Bendectin. Recognizing that it was impossible to state unequivocally that the drug was harmless for all fetuses, however, the government required that new labeling for Bendectin advise that use be restricted to those women for whom all alternative nondrug measures have failed, and they further recommended that the epidemiological studies be continued.

In 1982, the FDA, prompted by public concern and publicity about animal data that again raised questions about the safety of the drug, decided to reexamine all the Bendectin studies. An investigation had suggested an association between exposure to Bendectin in the first trimester of pregnancy and diaphragmatic hernias, a potentially fatal defect in which the abdominal organs herniate up into the lung cavity through a hole in the diaphragm, and a rat study in West Germany demonstrated a similar relationship. A monkey study in which the pregnant animals were treated with Bendectin at 10–20 times the human dose by weight linked the drug to a heart defect. Subsequent epidemiological studies, however, provided no evidence for a relationship between Bendectin use and such adverse effects as skeletal abnormalities, limb deformities, cleft lip or palate, or cardiac defects (Michaelis et al., 1983; Aselton and Jick, 1983; Golding et al., 1983). Only one retrospective investigation, by Yale researchers Eskenazi and Bracken in 1982, found that Bendectin appeared to be strongly associated with the occurrence of pyloric stenosis, a stricture or narrowing of the opening between the stomach and the small intestine. Even those findings were refuted in late 1983 by researchers at Boston University who reported that in their study of 325 infants with pyloric stenosis and 3,153 babies with other defects, Bendectin did not increase the likelihood of pyloric stenosis or any other congenital malformation (Mitchell et al., 1983).

But despite a lack of a statistical evidence that Bendectin causes birth defects, an anguished woman who has been exposed to Bendectin and delivers a child with a congenital abnormality is going to be convinced that Bendectin caused the birth defect, and juries may feel the same way. By early 1983, there were 300 lawsuits pending against Merrell Dow Pharmaceuticals, and the company's insurance rate was a million dollars a month. In March of 1983, claiming that costs of the pending malformation suits necessitated the action, Merrell Dow boosted the wholesale cost of Bendectin by 300%, resulting in a retail price of more than $90 for 100 pills. In June, 2 weeks after a jury recommended an award of $750,000 to the parents of a child with limb deformities, Merrell Dow announced that it was discontinuing production of Bendectin. It was not the FDA that removed the drug from the market, but the company itself, based upon their economic decision that they could not afford to pay the insurance and the costs arising from the suits.

Since Bendectin was the only antinausea drug approved for use during pregnancy, there are likely to be women who were aided by Bendectin that will be reluctant to have another pregnancy without its help. Mild or moderate morning sickness is bearable, but for many women, months of severe nausea and vomiting is not. There are also likely to be more women hospitalized with hyperemesis. Highly potent antiemetic drugs, such as the phenothiazines (Compazine, Tigan, Thorazine), have not routinely been used for morning sickness but have been reserved for the treatment of hyperemesis, where they produce a rapid and gratifying response. But the safety of phenothiazines during pregnancy has not been as well established as it has for Bendectin, and their potential for producing severe side effects in the mother (liver damage, blood difficulties) makes them even more controversial than Bendectin. Should phenothiazines be necessary, however, a pregnant woman can be reassured by the lack of consistent evidence for their teratogenicity, that is, ability to cause abnormalities.

Other Gastrointestinal Tract Difficulties

Digestive tract complaints may include excessive salivation, which occasionally occurs and then spontaneously disappears by the middle of the second trimester. Heartburn is somewhat more universal, especially during the last 3 months. It is caused by regurgitation of the stomach contents into the upper esophagus and may be associated with the general relaxed muscle tone of the entire gastrointestinal tract during pregnancy, a hormonal effect. The cardiac sphincter at the junction of the esophagus and the stomach that normally constricts to prevent reflux of the acid gastric juices into the esophagus is more relaxed. When this relative atony of the sphincter is coupled with the increased pressure from the growing uterus pressing on the stomach, heartburn results. It is not due to hyperacidity in the stomach, but antacids, such as calcium carbonate, magnesium trisilicate, or aluminum hydroxide preparations (Tums, Maalox, Gelusil), relieve the symptom. Baking soda or Alka-Seltzer, commonly used for heartburn, should not be taken because of their high sodium content. Other solutions for heartburn are avoiding spicy or greasy foods, small frequent meals, and sleeping on several pillows so that the head is elevated.

The decreased muscle tone in the large intestine results in decreased motility and increased water absorption, and leads to another common complaint, constipation and flatulence, or gas. Adequate roughage in the diet and exercise will usually minimize the problem.

A hormonal effect on the gums in the mouth may cause gingivitis—swollen, spongy gums that bleed easily. The condition may be improved by taking vitamin C; the difficulty will spontaneously disappear after delivery. There is no truth to the old wives' tale that pregnancy results in more tooth decay. The calcium in the teeth is fixed and is not withdrawn. Dental repair can be done at any stage, but because of exposure to the anesthetic, pregnancy is not the best time to have one's wisdom teeth extracted. Oral surgery should be postponed, if possible.

Changes in the Circulatory System

Cardiovascular alterations occur early in pregnancy. These changes are necessary to meet the demands of the enlarging uterus and the placenta for more blood and more oxygen. A pregnant woman is, after all, oxygenating the blood of the fetus as well as her own. The heart works harder and pumps more blood. By the end of the first trimester, cardiac output has increased by 25%–50%, and there is an accompanying increase in the blood volume. By the end of the pregnancy, blood volume will increase 40%–90%, depending on the size of the mother, the size of the fetus, and on which investigator is reporting the data. The number of red blood cells also increases, but there is a greater rise in the plasma volume as a result of hormonal factors and sodium and water retention. The hemoglobin concentration falls, the heart rate in-

creases and rises to a maximum of 15 beats above the nonpregnant state. Arterial blood pressure drops, reaches its low point at about 22 weeks of gestation, and thereafter slowly rises to prepregnant levels until term. The ability of the blood to clot increases significantly (hypercoagulabililty), but there is no increase in the incidence of thrombosis, or formation of blood clots, during pregnancy.

Varicose veins in the legs and the vulva are common during pregnancy. There may be a genetic predisposition to weakened venous walls and supporting tissues; women whose mothers or other family members have varicosities are more likely to have them. The developing uterus presses on the veins returning blood from the legs and impedes the flow of venous blood. This slowing of circulation, coupled with the engorgement of the pelvic veins, causes the blood to back up in the veins of the legs and exert increasing pressure on their walls. If the venous walls are not strong enough to withstand this increased pressure, they stretch and thin out to become the large tortuous varicose veins. They may not produce any symptoms, but some women have varying amounts of discomfort ranging from mild burning and itching to aching pain. Relief can be obtained by avoiding standing for long periods, elevating the legs whenever possible, and wearing support hose. In severe cases, putting on elastic stockings or bandages in the morning when the veins are collapsed and empty will help.

Hemorrhoids are varicosities of the rectal veins and can be very annoying during pregnancy. Constipation aggravates hemorrhoids, and straining during defecation may cause bleeding. There are various hemorrhoidal ointments that can provide relief.

Some women get occasional mild nosebleeds during pregnancy that may be related to the increased blood volume and resultant capillary pressure. These can usually be controlled by remaining upright, bending the head forward, and pinching the nostrils together.

Early in pregnancy, there is an increase in the minute respiratory volume—the total amount of air moved into the lungs each minute. The number of breaths per minute does not increase, but the amount of air breathed in with each breath does. Later in gestation there are anatomical changes as well. Even though the uterus in the abdominal cavity pushes up on the diaphragm and elevates it, the entire thoracic cavity compensates by an increase in its dimensions so that more air can be inspired. Shortness of breath, or dyspnea, develops in slightly more than half of pregnant women in the last month. Since the breathing capacity is not limited, the mechanism is unknown. It may be related to pressure on the diaphram and lungs. Dyspnea usually subsides when "lightening," the settling of the fetal head into the true pelvis, occurs. Lightening is experienced several weeks before labor begins with a first baby, but usually takes place during labor in a woman who has already borne a child.

Changes in the Excretory System

Urinary frequency occurs at the beginning of pregnancy and also near the end. Early in pregnancy, the enlarging uterus causes traction on the muscles of the bladder neck and results in the sensation of having to urinate often. As the uterus rises out of the pelvis, the frequency of urination subsides, but it returns in late pregnancy because of lack of room for the bladder to fill. There is little a woman can do about this symptom.

Probably as a result of hormonal influences, the ureters leading from the kidneys to the bladder become dilated by the tenth week of pregnancy. The right side is generally more affected than the left side. In some women, this enlargement may predispose to kidney infections.

Total body water and electrolytes are normally increased in pregnant women. Ankle swelling is very common and of little significance. A generalized body edema with swollen fingers and puffy eyelids occurs in about 25% of pregnant women. When it is accompanied by a rapid weight gain ($1\frac{1}{2}$ pounds in a week), an increase in blood pressure, and protein appearing in the urine, it may signal the beginnings of difficulties. In the absence of these signs, however, generalized swelling is of no consequence. There should be no attempts to reduce the edema with diuretics. Whether sodium restriction is helpful or harmful is uncertain. It is generally believed that the most prudent course is to neither increase nor decrease sodium intake, but to salt foods as usual and to rely on kidney mechanisms to maintain a normal salt balance.

Musculoskeletal Changes

By the tenth to twelfth week of pregnancy, and under hormonal influence, the ligaments that hold the sacroiliac joints and the pubic symphysis in place begin to soften and stretch, and the articulations between the joints widen and become more movable. The hormones involved are estrogen and progesterone and possibly relaxin, a peptide produced by the ovaries and placenta. Relaxin was originally identified in the blood serum of pregnant rabbits, rodents, and pigs. Weiss and co-workers (1976) have demonstrated the secretion of relaxin by the corpus luteum of pregnancy in women at term.

The relaxation of the joints is progressive and becomes maximal by the beginning of the third trimester. The purpose of the changes is to increase the size of the pelvic cavity and to make delivery easier. But the instability and looseness of the joints may cause discomfort in the pubic and lumbar regions. In an attempt to keep the joints "locked" during locomotion, a pregnant woman will assume the characteristic duck-waddling walk of late pregnancy.

The postural changes of pregnancy—an increased swayback and an upper spine extension in order to compensate for the enlarging abdomen—coupled with the loosening of the sacroiliac joints may result in lower back pain. The backaches can be relieved by heat, rest, a bedboard under the mattress, and by exercises that strengthen abdominal and back muscles.

Many women experience leg cramps during the last few months of pregnancy. A painful spasm of the gastrocnemius (calf) muscle is most likely to occur when lying in bed in a supine position. Immediate relief of the cramp can be obtained by flexing the foot upwards so that the toes point toward the face while simultaneously pushing down on the knee to straighten the leg. There are claims that high dietary phosphorus or insufficient calcium provokes leg cramps, but attempts to prevent them by adding calcium preparations or restricting dietary phosphorus are not always successful.

Some women notice occasional sensations in their hands, ranging from tingling or pins-and-needles (paresthesia) to actual pain that comes and goes. The symptoms are similar to those in carpal tunnel syndrome, pain and paresthesia of the hand and forearm as a result of compression of the hand's median nerve. Incidence of carpal tunnel-like symptoms in pregnancy has been reported in up to 50% of women. Although some have related the condition to relaxin, Voitk and his co-workers suggest that fluid retention in pregnant women is more likely responsible for the symptoms (1983).

Skin Changes

Many pregnant women are justifiably concerned about stretch marks, skin problems, and falling hair. Unfortunately, little is known about how to avoid them, and maintaining a healthy program of eating well and getting enough rest may be the best preventives.

Striae gravidarum, or stretch marks, are irregular reddish streaks that appear on the abdomen, breasts, and buttocks in about half of pregnant women after the fifth month (Fig. 10.6). After delivery, the red discoloration disappears but the striae persist indefinitely as whitish lines. The striae are the result of changes in the collagen and elastic fibers in the lower layers of the skin and have little or

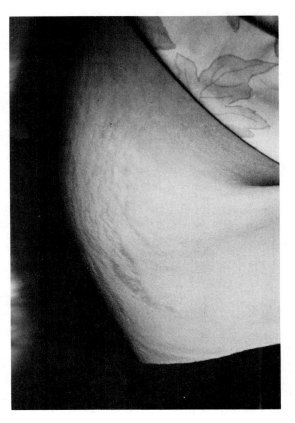

Figure 10.6. Stretch marks (striae gravidarum) on the hip and flank of a woman who had been pregnant with twins.

nothing to do with weight gain. They may be related to normal hyperactivity of the adrenal gland. Controlling weight within limits is sensible, but dieting to avoid stretch marks is useless and harmful to the fetus.

Complexion changes are not unusual. Clear skin sometimes develops acne, and problem skin sometimes clears up. The increased pigmentation that occurs on the breasts and genitalia may develop in other areas as well. Dark-haired women may notice a brownish-black line (linea nigra) appearing down the middle of the abdomen. Some women develop irregular patches of freckles on the forehead and cheeks to form the mask of pregnancy, or chloasma. (The same pigmentation may occur in women taking oral contraceptives.) Chloasma gradually disappears after delivery.

Another skin manifestation, believed to be secondary to the high estrogen concentrations, is the appearance of vascular "spiders," bright red circular areas of branching and dilated capillaries. These occur on the face, neck, thorax, and arms, and are especially obvious in white women but disappear after childbirth.

The blood flow to the extremities is increased. Women who have always had cold hands and feet will find them characteristically warm and even clammy during pregnancy.

Some women notice an acceleration in fingernail and hair growth during pregnancy. The hair follicles on the scalp and on the rest of the body normally undergo a growing and a resting phase. The resting phase is followed by a loss of hairs, which are then replaced by new ones. During pregnancy, fewer hair follicles go into the resting phase, and after delivery, there is a catching-up period with subsequent loss of hair until the follicles get back into their normal cycle. A similar hair loss sometimes occurs in women on oral contraceptives.

Staying Calm and Coping

The preceding list of physiological and anatomical alterations with their accompanying symptoms may sound discouraging, especially if one has never been pregnant. Almost all changes, however, disappear within a short time after birth, and a woman returns to her prepregnant state. Almost all the complaints are minor, rarely incapacitating, and more annoying than anything else. They have to be viewed with common sense and an attitude of acceptance. There are bound to be good days and bad days, and it would be unrealistic to expect only positive feelings for a period of 9-plus months. The point to remember is that pregnancy is not a test. No one passes or fails it. Neither is it a contest to see who can have the best or the worst one. Having the most "normal" pregnancy ever, and working right up until the moment of delivery is fine for some women, but it is just as normal not to, and it need not be a goal. The real goals are good health for the mother and the baby, and the foundations for health, as they are at any time in life, are good nutrition, exercise, and enough rest. A woman should listen to what her body tells her; it will indicate how to respond and what is right for her. If she feels like continuing her usual activities during pregnancy, then she should do so. But when she feels like resting, then she should rest. No woman has to prove anything to anyone with her pregnancy.

COMPLICATIONS OF PREGNANCY

Preeclampsia-Eclampsia

Healthy middle-class women, over 17 and under 34, with no serious physical illnesses, who are well nourished and have the opportunity to obtain good prenatal care, are considered low-risk pregnancy prospects. They approach term in good health and can expect to have a completely normal delivery. For them, the fifteen or so prenatal visits to the obstetrician's office appear superfluous, and even boring. They may sit around in the waiting room for several hours only to give up their morning urine specimen to the technologist, get weighed, then have their blood pressure taken, and finally have a brief abdominal examination by the physician. It may seem to them that they are getting very little health care, and less caring. With subsequent pregnancies, these women have such a lackluster feeling about seeing the doctor that there is a tendency to delay prenatal care until the fifth month or even later.

Those monthly, then bi-monthly, and later, weekly visits are more important than are generally realized, however. It takes just a few minutes for the examiner to establish that all is progressing normally and well.* For example, the evaluation of these three things— blood pressure, urine, and weight gain—is necessary to detect the onset of a real and major difficulty, a group of diseases called toxemias of pregnancy. Recognizing these complications early prevents their development into what could be life-threatening problems.

Toxemias of pregnancy occur in about 6 out of 100 pregnant women and are among the leading causes of maternal and fetal deaths. The term, toxemia, is a general one used to designate a number of conditions characterized by high blood pressure. About half of the toxemias constitute the disorder known as preeclampsia, a combination of

symptoms that includes hypertension, edema, and proteinuria (protein in the urine). A progression of preeclampsia into a more severe stage characterized by convulsions and coma is called eclampsia. Unless controlled, eclampsia may end in maternal death.

In early 1983, two reports suggested that preeclampsia was associated with the presence of a unknown flatworm parasite found in the blood of pregnant women with the condition (Lueck et al., 1983; Aladjem et al., 1983). These investigators named the worm *Hydatoxi lualba* and described its structure in blood stained with toluidine blue O (TBO) obtained from pregnant women with preeclampsia and eclampsia. Although the researchers received wide publicity as a result of their discovery, subsequent studies failed to substantiate their contention. Richards et al. (1983) found the "worms" in blood samples from all their subjects: women with preeclampsia, nonpregnant women, men, and five beagle dogs—but only when the blood was stained with TBO. They concluded that the structures represented artifacts that probably resulted from the technique of TBO-staining of blood smears. An electron microscopic study of the wormlike structures further supported that conclusion (Gau et al., 1983). Pregnant women, whose anxiety level tends to be heightened by worrisome newspaper and magazine articles, can at least scratch off their list of concerns a parasitic worm as a cause of preeclampsia.

Despite a century of investigation, the cause of preeclampsia is still unknown, but there are some recognized predisposing factors, including prior high blood pressure, prior chronic kidney disease, diabetes mellitus, obesity, severe anemia, twin pregnancy, and some types of malnutrition. The most favored current theory for the proximate cause of the disease is uterine ischemia, a reduction of uterine blood flow as a result of various physiological or mechanical factors that could in-

---

* A medical degree is not a prerequisite. Anyone can be trained to perform the examination, and in New York, the innovative Maternity Center Association initiated a pilot project of do-it-yourself prenatal care. Under the supervision of a nurse-midwife, women and their husbands do the prenatal examinations traditionally performed by an obstetrician—weight, blood pressure, urine, fundus height, and fetal heart tones.

clude a breakdown of immunological tolerance of the mother to the fetus, or a maternal blood coagulation disorder. The impaired blood supply to the placenta is believed to cause the release of protein substances that can affect blood pressure, kidney function, and capillary permeability.

The first in the triad of symptoms associated with preeclampsia is a rapid weight gain of at least 2 pounds in 1 week. Since the weekly gain normally should be less than a pound during the last 6 months of pregnancy, a greater gain indicates salt and water retention and causes a generalized swelling. This is the basis of the concern about an edema not limited to the ankles. The edema of preeclampsia is the deeply pitting type; an indentation or pit remains for a period of time if a finger is poked into the swollen tissue. The second symptom is a rise in systolic blood pressure to 30 mm Hg above the normal readings and a diastolic pressure of at least 15 mm Hg above normal. The third event is the appearance of protein excreted in the urine on 2 successive days. A woman who is not having her weight, blood pressure, and urine routinely checked, may not notice the symptoms. As the preeclampsia becomes more severe, symptoms of headache, dizziness, and blurred vision occur. Pain under the sternum, an indication of liver enlargement, is a serious sign. When it is accompanied by vomiting and scanty urination, it means that convulsions are going to occur.

The prognosis for a mild eclampsia is good if it is detected early enough. Since the only definitive cure is the delivery of the baby, treatment involves reducing the symptoms, preventing progression to the more severe stages, and trying to maintain the pregnancy until the fetus is able to live independently. For a mild form, if the blood pressure is not too high, bed rest at home and a high protein diet with vitamin supplements may result in improvement. More severe preeclampsia requires hospitalization and sedation. As soon as it is practical, labor is induced with drugs or a cesarean section is performed.

Since very little is known about the cause of preeclampsia, there is very little known about how it can be prevented. Good maternal nutrition, with a high protein diet and appropriate vitamin and mineral supplements when necessary, may reduce its frequency. Careful watching for early symptoms can reduce its progression to the more severe life-threatening form.

Spontaneous Abortion

Abortion means the termination of pregnancy before the fetus is able to live independently. Usually, a fetus is not deemed legally viable unless it has reached 500 grams and 20 gestational weeks, although few infants survive delivery before 24 weeks. Abortion may be spontaneous, that is, an involuntary expulsion from the uterus, or it may be induced intentionally by trained personnel in an appropriate facility. The familiar term for spontaneous abortion is miscarriage.

Of all recognized pregnancies, one out of six—an astounding number—terminates by spontaneous abortion. For all we know, there may be as many or even more that occur before the diagnosis of pregnancy and are considered to be only delayed menstrual periods. Estimates that take into account the number of fertilized eggs which for some reason fail to implant approximate that 50% of all pregnancies are aborted naturally.

Early spontaneous abortions are usually the results of defects in the fetus and may not actually be the tragedies they appear to be at the time. Although they may be disappointing, spontaneous abortions should be viewed as desirable in most instances. Studies on aborted fetuses have shown that 20%–60% of them have abnormal chromosomes. Since many miscarriages occur under circumstances where no scientist is there to study the abortus, it is reasonable to assume that other embryonic deformities exist as well. Later abortions, which occur in the middle trimester (only 2%), are more likely to be caused by an incompetent cervix, uterine abnormalities, toxemias, or preexisting chronic maternal disease.

Having a miscarriage during the first 3 months of pregnancy does not necessarily mean that a woman will have another one, although some authorities believe that her chances may increase. Habitual abortion, a term used when a woman has miscarried three or more consecutive times, is rare and occurs in only one out of 300 women. Even after three spontaneous abortions, the risk of another is only 25%.

When abortion is recurrent, it is an indication for genetic counseling and chromosome karyotyping of both parents since one may have a chromosomal abnormality. Sometimes it is not possible to determine a specific cause. Nutritional deficiencies, genital tract abnormalities, endocrine disturbances, and exposure to environmental agents have all been recognized as factors in spontaneous abortions. Evidence is accumulating that links chemicals and other hazards encountered by women working in hospitals and as industrial workers with the growing incidence of spontaneous abortions, stillbirths, and birth defects. Once detected, correction and elimination of a known condition in habitual aborters can often produce an uninterrupted pregnancy with a successful outcome.

The first sign of the possibility of abortion is vaginal bleeding, with or without cramping. Not all uterine bleeding, of course, indicates an abortion. Slight bleeding may be associated with implantation, or it may be originating from the cervix, the vagina, or the vulva. If the bleeding is occurring from the uterus without any dilatation of the cervical opening and, usually, without pain, it is termed a *threat-*

ened abortion. The treatment is to wait and see what happens. Sometimes the symptoms subside within several days and the pregnancy continues. Continued bleeding accompanied by uterine cramps and dilatation of the cervix means that progression to *inevitable abortion* has occurred. As the term implies, the abortion is going to take place, and any attempt to maintain the pregnancy is useless.

In most instances of abortion, the death of the fetus has taken place several weeks before the first symptoms. The stimulus of the growing conceptus is no longer there to cause the placental production of hormones that maintain the integrity of the endometrium and the placenta. Regression and placental separation follow. Obviously, if the bleeding in a threatened abortion is the result of fetal death, there is no way of preventing it. Going to bed or taking hormones is not going to change the ultimate course. If the early bleeding is not the result of a dead embryo and the pregnancy continues, the threatened abortion has little effect on later fetal development. It does not increase the probability that the baby will be born with a serious congenital defect.

The administration of progestational agents is still prescribed by some physicians as prevention or treatment for threatened abortion. Unless laboratory tests have confirmed that the miscarriage is due to insufficient progesterone production by the corpus luteum—a rare condition—hormone administration has no value and only delays expulsion of the fetus. Additional evidence of the futility of progestin therapy was provided by Smith and colleagues in a 4-year study of women with first trimester vaginal bleeding (1978). The purpose of the study was to clarify whether the site of implantation, which was determined by ultrasound scanning, had any relationship to spontaneous abortion or the subsequent development of placenta previa, another complication of pregnancy. While discovering that the implantation site had little affect on the cause or prognosis of abortion or on development of placenta previa, the study provided an additional opportunity to gain information concerning the effect of progestin administration. Forty-five of the 105 women in the study had been given progestins (Delalutin or Provera) in varying doses by their physicians because of prior abortion, habitual abortion, and prior infertility. They were compared with those women who had not received such hormonal support. Seventy-three percent of the women who received hormones aborted an average 19.7 days after the initial ultrasound scan. Nearly 68% of the women who had not received progestins aborted an average of 4.6 days after the scan. Indications were clear that progestins only prolong the period before spontaneous abortion but have no effect on the outcome. Of all the women in the study, 35 did not abort, continued in their pregnancies, and delivered infants with no birth defects. As it happened, 12 of those 35 women had taken progestins. It was fortunate that no visible effects occurred since there is a known increased risk of damage when sex hormones are taken during pregnancy.

In a *complete abortion*, the placenta and the embryo are totally evacuated, and the bleeding and cramping stops. More often, the abortion is *incomplete*. Some placental tissue remains in the uterus and has to be removed by curettage. Occasionally, long after fetal death, the fetal and placental tissue are retained in the uterus, and there are no signs of abortion. This is called *missed abortion*. The signs and symptoms of pregnancy gradually disappear, and curettage is usually necessary. Eventually the uterine contents will spontaneously be expelled, but it could take several months. By then, the amniotic fluid has been reabsorbed and the fetus is apt to be mummified or calcified.

If a fetus is meant to stay within the uterus for $9\frac{1}{2}$ months, it will. Tennis, horseback riding, or any other kind of physical activity will not result in its being expelled. The physical trauma of a fall or a blow, or even, in many instances, the crushing injuries to the pelvis in automobile accidents, have not resulted in interrupted pregnancy. Although it is generally believed that severe emotional shock may bring on an abortion, verification is difficult. As expected, the "rejection-of-motherhood-as-a-result-of-deep-seated-emotional-factors" theory has also been suggested by some psychiatrists as a cause of abortion, particularly in habitual aborters, but substantiation is obviously difficult. Presumably, blood flow to the placenta could be affected by autonomic factors as a result of emotional stimulation, and physiologically, the possibility for a psychological basis for pregnancy termination exists.

Placental Bleeding

Vaginal bleeding in the first 28 weeks of pregnancy may herald the onset of spontaneous abortion, but 15% of women who have bleeding retain the pregnancy with no adverse consequences. Ultrasound has become a valuable method for diagnosing the reason for bleeding in early pregnancy. Vaginal bleeding in the last trimester may be due to two other complications of late pregnancy—*placenta previa* or *abruptio placenta.*

Placenta previa, which occurs in one out of 200 pregnancies, is the result of implantation of the fertilized ovum in the lower part of the uterus instead of its more usual site higher up in the fundus. The placenta is then formed very close to, partially covering, or completely covering, the internal os of the cervix (see Fig. 10.7). Late in pregnancy or at the time of delivery a part of the lower edge of the placenta may separate from its attachment to cause characteristically painless bleeding that is bright red in color.

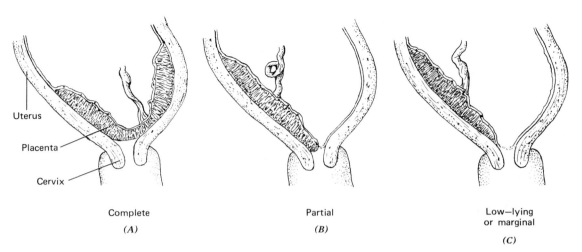

Uterus

Placenta

Cervix

Complete

(A)

Partial

(B)

Low—lying
or marginal

(C)

Figure 10.7. Types of placenta previa. (A). Complete, in which the placenta extends across the entire cervical opening. (B). Partial or lateral, in which the placenta covers only a part of the cervical opening. (C) Marginal, in which the margin of the placenta extends to the edge of the cervical opening.

If placenta previa is diagnosed at the time of labor, the baby is delivered by cesarean section. If the bleeding has occurred as early as 32 weeks, the pregnancy is allowed to continue with hospitalization, bed rest, and careful watching and waiting. As soon as there is no risk to the life of the fetus, a cesarean is performed.

When the placenta is normally located in the upper segment of the uterus but separates before the birth of the baby, the condition is known as abruptio placenta. Such premature separation occurs in only 1% of pregnant women and may be associated with hypertensive or chronic kidney disorders.

WEIGHT GAIN AND NUTRITION

For years it had been believed that excessive weight gain during pregnancy was associated with several complications, particularly preeclampsia, and was further related to mechani-

cal difficulties in labor and birth. Pregnant women were advised by their doctors to restrict caloric intake, sometimes drastically, in order to avoid excessive weight gain. The problem with this recommendation was that the definition of what constituted excessive varied considerably among different physicians. At one point it was generally agreed that weight increases should be limited to 20 pounds over the course of pregnancy, but many women were more severely restricted to as little as 16 pounds by their well-meaning but overzealous doctors. Some may even have been given diet pills (amphetamine appetite suppressants) to achieve that goal. It was also considered appropriate for overweight women to reduce while pregnant. Gains of nothing at all or just a few pounds were presumed safe for the fetus and for the mother if she was obese prior to pregnancy. Many women may remember how they dreaded that monthly visit to the office, reluctant to step on the scale and shamed to face the physician because a "good" patient was one who strictly adhered to the proposed weight control.

The idea that excessive weight gain predisposed to preeclampsia-eclampsia arose from rather specious evidence. During World War I, it was reported that there had been a startling reduction in the incidence of toxemia of pregnancy in central Europe. Since this decrease reverted to its former frequency as soon as the war was over, the change, without any further study, was attributed to the food blockade that had existed during the war. A theory that dietary restriction was protection against eclampsia developed, persisted, and was taught to several generations of physicians. There have been studies since that have attempted to prove a cause-and-effect relationship between diet and obstetric complications, but confirmation is lacking. Some investigations have noted an apparent relationship between the incidence of toxemia and prepregnancy obesity or overall maternal weight gain, but there have also been studies showing just the opposite. Underweight women seem to have an increased frequency of eclampsia, and there has been the suggestion that dietary deficiency and malnutrition pose greater risks for the development of toxemia of pregnancy. The report of the National Research Council's Committee on Maternal Nutrition concluded in 1970 that there was no evidence that caloric restriction to limit weight gain in pregnancy had any effect on the incidence of preeclampsia-eclampsia.

There is a great deal of evidence, however, that indicates that there is a relationship between weight gain during pregnancy and the birth weight of the infant. Babies who are smaller and lighter than normal, who are born at term weighing 2,500 grams (5½ pounds) or less, have an increased susceptibility to illness and an increased rate of mortality compared with heavier babies. A woman whose weight gain is limited and inadequate during pregnancy has a substantially increased risk of delivering a baby of low birth weight.

Of course, not all low birth weight infants are the result of maternal malnourishment only. Some babies are premature, but they are the right size for their fetal age. But when a baby is born at *term* and weighs disproportionately less than what is considered to be a healthy weight, it is said to be a small for gestational age, or SGA, infant. In SGA babies, some factor or combination of factors has retarded their intrauterine growth. Fetal growth retardation can be caused by the biological immaturity of the mother if she is younger than 17. It can be associated, not necessarily in order of importance, with pregnancy in women who have twins or triplets, who have had many prior pregnancies, who are short and small women themselves, or who may have been underweight and in a poor nutritional state before pregnancy. Additional known factors that influence fetal growth adversely are smoking, chronic maternal disease, certain infections, or some maternal vascular problem that interferes with the blood supply to the placenta. In study after study, however, one cause stands out as a major determinant of fetal weight, and that is maternal weight gain. Pitkin believes that maternal weight gain is second in importance only to the length of pregnancy in responsibility for the birth weight of the infant.

Until very recently, it was generally believed that the developing fetus could get everything it needed at the expense of the mother, as long as she received plenty of vitamins. This concept has been challenged by the investigational studies of the relatively new science of nutrition. The fetus is no longer seen as a complete parasite, miraculously able to draw on all the maternal tissues and remain unaffected even though the mother is malnourished. Animal studies have shown that maternal malnutrition decreases the ability of placenta to transfer nutrients from the blood of the mother to the blood of the fetus. Prenatal malnutrition has also been shown to cause a reduction in cell size and number of cells in fetal tissues, with disproportionate effects occurring in the brain (Winick, 1976). Furthermore, in animals, food deprivation during pregnancy always reduces the fetal weight proportionately more than it does the maternal weight, and there are data that suggest that the situation is similar in humans.

Two major episodes of human malnutrition that occurred during World War II have been studied extensively and have provided corroborating evidence that there is a strong positive correlation between maternal nutrition and infant birth weight. The seige of Leningrad in 1942 and the Nazi-induced famine in western Holland in the winter of 1944–1945 resulted in severe nutritional restriction, and hundreds of people died of starvation each month. The babies that were born during those periods of calorie deprivation were of significantly lower birth weight than those born before and after the famines. The infants also suffered excessive morbidity and mortality after birth.

The Leningrad and Dutch experiences of the effects of maternal undernutrition on the developing fetus were the result of profound deprivation. It is becoming evident, however, that less than adequate maternal nutrition, which is far short of actual starvation, also has an effect on the babies produced. For many years it has been known that birth weights were lower in developing countries, and that more smaller and lighter infants were delivered to poor women than to rich women in this country. Several nutritional intervention studies that took place in Formosa, Guatemala, and New York City provided additional evidence that calorie supplementation for the mother during pregnancy increases the birth weight and decreases the infant mortality rate. The experimental animal studies by Myron Winick and other investigators have pointed to an association between maternal malnutrition and brain development and function. Some human studies have suggested that malnutrition in early infancy, particularly when combined with low birth weight, can affect mental development. The full implications of maternal nutrition and malnutrition are not known, but it is obviously essential that the undesirability of imposing severe restrictions on weight gain during pregnancy be recognized. Unfortunately, however, there may still be some physicians who believe that weight gain during pregnancy should not exceed 15–20 pounds.

The current view, which has now begun to have significant impact on traditional thought, is that there is probably no one ideal value for the weight gain in pregnancy that would be appropriate for all women. A woman's age, prepregnancy nutritional status, and activity level during pregnancy are modifying factors, but it is generally agreed that she should gain at least 24 pounds to support the changes in her body and in the developing fetus. With that increase, the fetus and placenta account for about 11 pounds. The remaining 13 pounds are distributed as about 8 pounds in increased maternal blood volume and extracellular fluid, about 3 pounds for additional uterine and breast tissue, and approximately 2 pounds of extra fat tissue. Underweight women should gain more—about 30 pounds. Increases in the range of 22–30 pounds are seen as normal and commensurate with the fewest obstetrical complications and the most favorable outcome of the pregnancy. Less gain than that contributes to the likelihood of delivering a low birth weight infant. A gain of more than that, perhaps 40–50 pounds, will result in excessive fat deposition, be more difficult to lose after delivery, and contribute to future obesity. The total accumulation, however, is less significant than the *rate* of gain. By the end of the first 3 months of pregnancy, when the fetus still weighs only 1 ounce, the total gain should have been only 2–4 pounds. After that, during the next 6 months, the gain should be $\frac{1}{2}$–1 pound weekly. Thus, the weight is gained gradually, and only when there is a sudden and large increase should there be any reason for concern.

According to Roy Pitkin, who has written extensively on maternal nutrition, a weight gain of 1 kilogram (2.2 pounds) per month in the second and third trimesters is inadequate. Four pounds per month is optimal, but 6–7 pounds or more each month is excessive and will lead to postpartum obesity. Women who are overweight should not try to lose weight during pregnancy, but should gain at the appropriate rate. Pregnancy is no time to diet, and even obese women must put on the 22–30 pounds all pregnant women should gain. A woman who is reducing is likely to eliminate some of the essential nutrients from her diet. Moreover, dieting results in an increased fat metabolism, which elevates the level of ketone bodies in the blood. Not only is ketosis undesirable for a pregnant woman, it is potentially extremely dangerous for the fetus.

Thinking Nutritionally

The fact that a woman is encouraged to gain up to 30 pounds during pregnancy cannot be construed as permission to indulge in all the high-calorie, empty-calorie goodies she may have always denied herself. The quality of the food chosen—its nutrient properties—are extremely important so that the weight gained is not merely excess fat. Building an entire new person inside one's body while simultaneously increasing one's own uterine and breast tissue and blood volume is not possible on junk food. It requires the basic nutrients, proteins, fats, carbohydrates, vitamins, minerals,

and water, in the correct amounts. The problem with eating corn curls, potato chips, ice cream, or pastries is not that they are completely without nutrition, but that they are loaded with fat and sugar. They do have some nutrient value, but they can be so filling that there may be less appetite for the really necessary foods.

There is no necessity to eat everything in sight even if it is nutritious. A woman is not really eating for two—she is actually eating for about one (herself) and one-seventh. The calorie requirements, averaged out over the entire 9 months, are not as much as one thinks. The total energy needs, 75,000 calories, are about 300 extra calories per day, an amount equal to 15% over the prepregnant state. Those additional calories can be obtained from a peanut butter sandwich on whole wheat bread with a glass of milk, which would be excellent. A doughnut and coffee with cream and sugar provide the same calories, but little nutrition.

A woman who selects her diet carefully will assure herself of the appropriate essential nutrients. The physiological adjustments of the body to pregnancy and the optimal growth of the fetus do require, however, that special attention be paid to getting additional amounts of some nutrients.

Milk

Milk is very important for an expectant mother. It is a source of calcium, phosphorus, protein, and vitamins as well as some valuable minerals. The entire daily requirement for calcium during pregnancy is contained in 1 quart of milk. Three 8-ounce glasses of whole or skim milk are recommended daily during the first trimester, increasing to four glasses during the next 6 months. This would be an inordinate amount to drink for a woman with a small appetite, and would be virtually impossible for someone who actively dislikes milk. Fortunately, milk can be flavored, or it can be disguised in cream soups, custards, or puddings. When $\frac{1}{3}$ cup of nonfat dry milk is added to a glass of milk, the result is "double-milk"—the equivalent of two 8-ounce glasses. Buttermilk can also be substituted for regular milk. A $1\frac{1}{2}$-ounce chunk of hard cheese, $1\frac{1}{2}$ cups of creamed cottage cheese, or a carton of yogurt is the equivalent in nutrients (except for vitamin D) of a glass of milk. So is a cup and a half of ice cream or ice milk, but the extra fat and sugar are hardly bonuses.

Proteins

Protein needs double during pregnancy. Non-pregnant women need 0.8 g of protein for every 2.2 pounds (kilogram) of body weight. This increases to 1.3 g per kg during pregnancy. In most American diets, the amount of protein consumed daily, primarily obtained from meat, is already way above the requirement. Under most circumstances, women do not need to deliberately increase their protein intake since the recommended dietary allowance already includes a buffer against protein deficiency. Neither does the source of the protein have to be an expensive hunk of steak or filet of whitefish. Eggs, canned mackerel (cheaper than tuna), hot dogs, or several slices of luncheon meat (but watch the fat content), liver, tongue, or good old peanut butter on whole wheat can be used for the requirement. A half cup of cottage cheese contains 20 g of complete protein. Nuts, beans, lentils, or peas combined with grain or a small amount of animal protein become complete. A vegetarian who eats milk and eggs would have no problems in getting ample amounts of protein when pregnant, but a non-lacto-ovo-vegetarian may have to select her foods more carefully to insure that they are appropriately combined.

Iron

Iron is necessary for life. Approximately 70% of the body iron is incorporated into the "heme" part of the hemoglobin molecule, the oxygen-carrier present in the red blood cells. Smaller amounts of iron, about 5%–10%, are necessary for the synthesis of muscle protein, or myoglobin, and also form parts of certain cellular respiratory enzymes, essential in the extraction of energy from oxygen and food. The rest of the iron in the body, 20%–30%, is bound to proteins for iron-transport or is stored in the liver, spleen, and bone marrow. In adults, the levels of iron necessary for the body's needs are controlled by very efficient mechanisms that balance the daily intestinal absorption of iron in the diet with the daily loss of iron in body excretions and sloughed-off dead cells. Men absorb about 1 mg of dietary iron daily to replace the iron lost from the skin and from the intestinal tract. Women, who during their reproductive years lose an additional 15–60 mg each month by menstruation, need more and absorb more in order to compensate for the additional loss.

People who eat nutritious diets consume iron-rich foods. A woman who eats muscle and organ meats, fish and poultry, green vegetables, and enriched flour or breakfast cereals, has no problems with iron deficiency. Certain foods, such as prune juice, dried apricots, molasses, or dried beans and peas are very high in iron. During pregnancy, however, there are additional requirements for iron. The blood volume increases, and more iron is needed to provide hemoglobin for the increased number of maternal red blood cells for the placenta and for potential blood loss during childbirth. The iron is also needed for the fetal hemoglobin, the fetal myoglobin, and the fetal cellular enzymes. The fetus draws more heavily on the maternal iron in the last months of pregnancy. Since both breast and cow's milk are iron-deficient, and a newborn is able to absorb very little iron anyway, the fetus must accumulate enough storage iron to last for the first 3 months or so after birth.

The fetus will take what it needs from the iron stores of the mother whatever her state of iron sufficiency, and should there not be enough, it is the pregnant woman who will become anemic, not her offspring. It is generally presumed that woman's dietary intake is likely to be inadequate to meet the needs, and iron supplements in the form of 30–60 mg of ferrous iron salts are almost universally prescribed for pregnant women in the last half of pregnancy. Such routine supplementation is done to eliminate possible iron-deficiency anemia as a pregnancy complication.

The need for across-the-board administration of iron salts to all pregnant women is debatable, however. Women need a total 500–750 mg additional iron during pregnancy. Pregnant women absorb dietary iron from the intestine at twice the usual rate, and, according to Stevenson, an adult woman can store as much as 1,000 mg of iron, depending on her size and prepregnancy diet. Unless she has been living on a marginal diet, has had multiple past pregnancies within a short period of time, or has had excessive menstrual or other blood loss, her own reservoirs and her dietary intake will be sufficient for her needs and that of the fetus. The amount of supplemental iron in the daily dose would probably not be harmful, but some of the possible side effects of taking iron pills—nausea, heartburn, constipation, burping—are already experienced by many pregnant women. If iron is unnecessary, why supplement their gastrointestinal symptoms? It would seem obvious that iron salts should be administered only if a simple blood test has determined that a pregnant woman is actually iron-deficient.

Folic Acid

Folic acid, also known as folacin and folate, is a water-soluble B vitamin that is involved in the formation of DNA and in the maturation of red and white blood cells. Because cell division increases so greatly during pregnancy, folate requirements are increased as well and

are set at twice the recommended dietary allowance for the nonpregnant woman.

A number of foods are rich sources of folic acid and its derivatives: liver, dark green leafy vegetables, broccoli, asparagus, legumes, and, in particular, oranges and orange juice. Because surveys have indicated that a pregnant woman's diet may not be adequate to meet the additional requirements for folic acid, doctors tend to advocate a routine daily supplementation of 200–400 micrograms (μg).

Whether routine supplementation of folic acid during pregnancy is actually beneficial to a woman and the fetus is not established. Prolonged deficiency is known to result ultimately in maternal anemia, but there is little evidence to support the relationship between insufficient folic acid and other pregnancy complications or to impairment of fetal growth and development. While many authorities advocate prescribing additional folate to all pregnant women, others make a case for supplementation only to those selected women for whom it is indicated. It is not that difficult to obtain the required daily amount of folic acid from the diet. One hamburger patty and a large lettuce salad will do it. Folate is relatively unstable, however, and exposure of the food to air and light results in the loss of much of the vitamin. For this reason, and if fresh fruits or vegetables are not readily available, diets can easily become deficient in folic acid, and daily doses of 200–400 micrograms are advisable.

Other Nutrients

Besides additional protein, calcium, iron, and folate, a pregnant woman needs the daily amounts of the rest of the vitamins and other important nutrients she should have ordinarily. The recommended daily allowances for other vitamins and minerals are not significantly increased over that in the prepregnant state. Sodium, for example, should neither be increased or decreased. One can continue salting one's food in the usual way with no attempts at restriction. Common sense indicates that larger amounts than ordinary of salted foods should not be ingested.

In certain instances, an overabundance of a vitamin can be as harmful as an insufficiency. There is some evidence suggesting that there is an association between maternal hypervitaminosis for vitamin D and the development of severe fetal abnormalities. For this reason, a woman is advised not to increase vitamin D over prepregnancy levels.

Obstetricians, however, are very likely to prescribe multiple vitamin and mineral supplements during pregnancy in a shotgun approach. They recognize that such routine supplementation may not be helpful, but they presume it is not harmful and believe that it provides a margin of safety since women probably will disregard a recommended diet. This generally low opinion of the ability of women to understand or follow a food plan is almost universal and is held by some women themselves, who claim that they like to eat food and not servings of nutrients. Many women have learned to question their own capacity for maintaining a nutritional diet, perhaps as a result of the paucity of specific dietary advice that doctors give them. The nutritional counseling provided may be "eat a balanced diet" or "watch your weight" accompanied by a pamphlet. They are not told what to do. Few physicians utilize the services of a professional nutritionist in their obstetrical practice. It is not unusual for women who can afford a busy private obstetrician to get less nutrition advice than women with fewer socioeconomic advantages who are seen at federally funded clinics.

Women can certainly be presumed intelligent enough to make themselves nutritionally aware. A food plan during pregnancy or at any other time in life requires individualization

and extra time. A woman's life-style, family situation, food preferences, and a number of other ethnic, economic, psychological, and physical factors influence her eating patterns. Some of the dietary leaflets may sound patronizing or be written in a picture-book style, but the information is helpful and can easily be applied to one's own dietary plan.

Table 10.1 is a guide to the daily essentials. Whether they are consumed in three meals, five meals, or many meals and snacks is not important, as long as all the nutrients are included. If, after the basics are eaten, there is still room for junk food, that is the time to eat it. There is no need to curb calories, within common-sense limits, as long as the essentials are consumed.

If the total weight gained has not exceeded 22–30 pounds, a woman can count on losing 18–20 of them within a week after delivery. The rest of the pounds are likely to be completely gone by 3 months after delivery, espe-

Table 10.1 DAILY GUIDE TO ESSENTIAL FOODS

To be Consumed	Servings Each Day	
	Pregnant	Breast-feeding
Milk and milk products	4	4–5
Protein source (meat, fish, poultry, eggs, cheese, and at least one vegetable protein)	4	4–5
Leafy green vegetable	1–2	1–2
Source of vitamin C	1	1
Other fruits and vegetables	1	1
Whole grain breads and cereals	3	3–4
Water and other liquids (milk may be included in total)	6	6

cially if a woman is breastfeeding. If she is not lactating, she can speed up the weight loss by restricting calories (not nutrients). Under no circumstances should a woman embark on a fad diet, particularly the low-carbohydrate type. The best and only long-term solution to a weight problem, after a woman has learned to eat "right," is to continue in the right way, but with smaller portions.

PRENATAL CHILD ABUSE—EFFECTS OF DRUGS ON THE FETUS

Every year in the United States, around a quarter of a million children are born with a birth defect. Some have an obvious structural malformation; their limbs are crippled, they have a cleft palate, or a heart, gastrointestinal, or kidney defect. Others may be mentally retarded, be born blind or deaf, or have a metabolic disorder. Some may have problems that do not show up for years. Not too long ago, birth defects were attributed to a peculiarity of behavior of the mother, who could mark her child. Before that, deformities were blamed on witchcraft. Today, we are enlightened enough to know that structural and functional malformations are the result of genetic, viral, or chemical influences or, in many instances, are the result of interactions between genetic and environmental influences. Gene mutations or chromosomal aberrations are said to be the cause of about 25% of birth defects. Certain virus infections, drugs, and exposure to radiation are known to give rise to another small percentage. But at our present state of knowledge, the reason for 60%–70% of human abnormalities is essentially unknown and is presumed to arise from the interplay of several factors.

The recognition that environment influences prenatal development is astonishingly recent. The role of rubella, or German measles, infection in a pregnant woman causing fetal deformities was not established until 1941. And, incredibly enough, it was not until the thalidomide tragedy of the early 1960s that it became generally recognized that drugs taken by pregnant women may produce abnormalities in the developing children.

It had always been assumed that a placental barrier existed—that all the tissues through which materials must pass in maternal exchange constituted an effective obstacle that prevented an adverse effect on the fetus. What was safe for the mother to take was therefore considered safe for the fetus. When the new tranquilizer, thalidomide, became available in many countries it was believed to be so safe that it was sold without prescriptions. Besides, it had been extensively tested in pregnant rats and had produced no effects in the offspring. People took thalidomide for a variety of maladies, and pregnant women used it for morning sickness, anxiety, headache, and whatever else was bothering them. It is generally agreed that had it not been for the distinctive abnormalities caused by thalidomide—it particularly affected the long bones in the arms and the legs—the drug might not have been recognized as a *teratogen*, an agent that caused a high proportion of exposed embryos to develop malformations. When a virtual epidemic of deformed children appeared in 1961, the search narrowed to thalidomide as the causative agent. By the time the drug was withdrawn from the market, an estimated 7,000 children throughout the world had been born with phocomelia, abnormally short or absent arms and legs.

After that disaster it became mandatory for all drugs to be tested not only for their toxicity to humans, but for their teratogenicity on fetal development. In the past 25 years, a new discipline, fetal pharmacology, or the effects of drugs on the fetus, has generated a great deal of research. Symposia have been held, reviews have been written, central registers for birth defects have been established—much effort has been devoted to identifying those compounds that are able to cross the placenta. What is becoming increasingly recognized is the extent of the inadequacy of current knowledge. Relatively few chemical agents are positively known to be teratogenic, many are suspected, and probably all have the potential. Some drugs may be safe in animals (thalidomide is a case in point), but some drugs, considered harmless and not specifically associated with malformations in the fetus, have been known to induce deformities in several animal species.

The timing and the dosage of a chemical agent, that is, when it was taken during fetal development and how much, are very important to the possible occurrence of a congenital malformation. For the first 2 weeks after fertilization, the embryo is believed to be resistant to environment influences. If it is affected, presumably it undergoes such severe damage that implantation does not occur, and death and early abortion result. Animal studies, however, have shown that a number of chemical agents can actually penetrate the oviduct fluid to reach the fertilized egg. Fabro and Sieber were able to recover such substances as caffeine, nicotine, DDT, and phenobarbital in a rabbit blastocyst before implantation when those chemicals had been injected into pregnant rabbits (1969). Whether the ability of such substances to penetrate into the preimplantation embryo is of teratogenic significance is unknown, but the possibility exists.

The most critical period for disturbance of development of the embryo is from approximately day 15 to day 56, the time when all of the organs are forming. Each of the organs has its own particularly sensitive period, and interference at that time can result in malformations. Since the development time of many organ systems overlap, the outcome could be a whole group of abnormalities (see Fig. 10.8).

The effect of rubella, for example, is known to be most harmful if the disease occurs within the 4 or 5 weeks after fertilization when the brain, heart, eyes, and ears are undergoing rapid development. Although the risk to the child is greatest during the first trimester, susceptibility to environmental influences can continue. Brain development, for example, takes place throughout gestation and even extends into the postnatal period.

It has been determined that compounds that are positively or negatively charged, that are not soluble in lipids, are of high molecular weight, or are bound to large protein macromolecules, are not able to cross the placenta as rapidly as those drugs that are not ionized (charged), highly lipid-soluble, or of low molecular weight. The majority of drugs, however, is likely to be transmitted across to the fetus despite the physical properties if a high enough level persists in the maternal blood stream for an adequate amount of time. For all practical purposes, the placenta does not constitute much of a barrier.

The problem is further compounded by the diversity of ways in which a drug can affect the fetus. Even if agents are not able to cross the placenta, they are still potentially harmful. Some drugs pass across the membranes of the placenta and directly enter the bloodstream of the fetus to exert a toxic effect; others may be transformed by the liver of the mother into a toxic metabolite. Very little is known about the ability of the placenta itself to metabolize, detoxify, or in other ways affect the dose of the drug received by the fetus. There is the possibility that the placenta could convert an innocuous chemical into an active teratogen, mutagen, or carcinogen.

Two drugs, neither of which is particularly teratogenic, may interact to produce an adverse effect. It may be that a drug could be potentiated by genetic or other predisposing factors and is then able to cause a malformation. This might explain the puzzling fact that

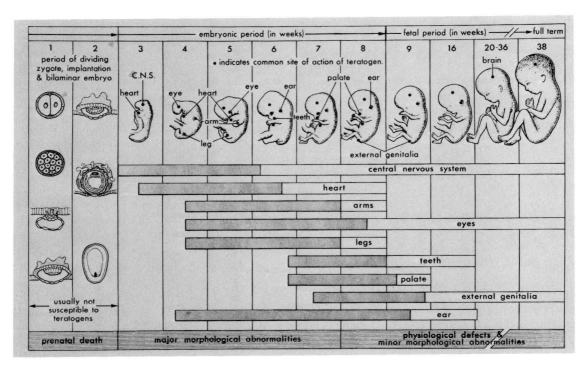

Figure 10.8. Schematic illustration of the sensitive periods in human development. The embryo is not believed to be susceptible to teratogens during the first 2 weeks of development. The dark part of the bar indicates highly critical periods; the light portion indicates stages less sensitive to teratogens. (From Moore, Keith. The Developing Human, 2nd ed. © 1977 by the W. H. Saunders Company, Philadelphia, Pa.)

not all women who are exposed to a particular chemical agent or infection will deliver a child with defects. A woman's overall health before pregnancy, her nutritional status during pregnancy, perhaps her own unique genetic susceptibility, the amount of alcohol she drinks or the number of cigarettes she smokes—all these can be factors.

A drug may indirectly produce abnormalities by interfering with the activity of another substance. The antitumor drug, aminopterin, is a folic acid antagonist that is known to induce abortion. Although it usually resulted in the deaths of the embryos when it was administered in early pregnancy, some of the fetuses that survived to term were grossly deformed. Methotrexate, another chemotherapeutic drug and an aminopterin derivative, has also been implicated as a teratogen.

The effects of progestational hormones for pregnancy tests and for the prevention of spontaneous abortion have already been discussed. Nora and Nora had reported in 1975 that some children who were inadvertently exposed to oral contraceptives prenatally were born with the VACTERL syndrome, a combination of vertebral, anal, cardiac, tracheal, esophageal, renal, and limb abnormalities, although some more recent studies found no link between oral contraceptives in pregnancy and birth defects (Harlap and Eldor, 1980; Savolainen, et al., 1981). The appearance of reproductive abnormalities in the sons and daughters of women given DES during pregnancy is a prime example of how unforseen, and even carcinogenic, effects can be produced 15–20 years after exposure.

A strong teratogen—one that produces a specific abnormality in virtually all fetuses exposed at any time in the first trimester to any dose—should be relatively easy to spot in the population because the sudden increase in the incidence of that unique defect would stimulate an intensive search for the cause. Unfortunately, most human teratogens rarely act that way. That a substance produces adverse effects in offspring when the mother takes it during pregnancy is difficult to verify, given the overall 2.7%–3% incidence of birth defects in the population. As described previously for Bendectin, published reports that associate a drug with congenital abnormalities do not constitute a cause-and-effect relationship. The association may be the result of the way the information was collected or analyzed. The sample size may have been very small, or the results may be of borderline statistical significance.

Information that appears in the medical literature may be in the form of a case report, in which an individual case of malformation and a history of drug exposure appeared so interesting or unusual that the physician decided to write it up and submit it to a journal. Although the drug may become suspect as a result of case reports, more information and further studies are needed to verify its potential teratogenicity. Epidemiological studies of the exposed populations compared with control populations can then be made to confirm or disprove the suspicion and to provide an evaluation of the risk of taking the drug. An estimate of risk can be obtained by a retrospective *case-control study*, in which the exposure to a drug is determined in a group of

women who delivered babies with a birth defect compared with a control group of women who are comparable in all respects except that they delivered normal babies. The extent of exposure may be determined by interviews with the women, by examination of their medical records, or both.

In a prospective *cohort study*, the frequency of occurrence of a birth defect in the offspring of a group of women who were exposed to a drug in question is compared with that in a group of control women who did not take the drug. Prospective studies are viewed as more reliable, since the ultimate proof of teratogenicity of a drug could come only from observations of the consequences of exposure to an agent in a well-controlled experiment. But since it is unthinkable to set up a study in which half of the women receive a suspected drug and half of them do not in order to see whether an adverse effect on their babies occurs, another way of assessing the teratogenicity of substances is to accumulate data from animal experimentation in the laboratory. Animal studies, however, also present problems. The differences in the metabolism and development of different species, the way in which drugs are absorbed and excreted in rodents and primates, and the basic difficulty in extrapolation of animal data to human populations, make estimations of human risk of a drug from its animal teratogenicity prone to under- or overestimation. Teratogenicity has been *established* for only a very few infectious or chemical agents, but some of the drugs that have been associated with the production of congenital malformations are listed in Table 10.2. There is no way of knowing how many drugs may be low-risk teratogens, that is, cause some adverse effects in some fetuses but no actual malformations. It is obviously important to determine which drugs result in major defects and which do not. Then if drug therapy during pregnancy is required, the least hazardous medication can

Table 10.2 HUMAN TERATOGENS, KNOWN AND SUSPECTED

Drug	Reported Abnormality	Drug	Reported Abnormality
Adrenocorticosteroids (prednisone)	Low birth weight, possible adrenocortical insufficiency in infants	Aspirin	Bleeding dysfunctions in newborn
Alcohol	Fetal alcohol syndrome—dose-dependent effects (?)	Benzodiazepines (Valium, Librium)	Reports of cleft lip and palate, no confirmed association from epidemiological studies
Anesthetic gases	Chronic exposure linked to spontaneous abortion, stillbirths, cardiovascular effects. Animal teratogenicity unconfirmed in humans	Caffeine	Teratogenic in animals at equivalent of 12–24 cups of coffee daily. No confirmation of teratogenicity in humans
Antibiotics		Cigarette smoking	Fetal growth retardation; low birth weight; stillbirths; dose-dependent effects
Streptomycin	Deafness		
Sulfonamides (Gantrisin, Gantanol)	Jaundice if taken late in pregnancy	Diuretics (thiazides)	Reports of bone marrow depression
Tetracycline	Tooth discoloration	Iodides, propylthiouracil	Thyroid enlargement, fetal goiter
Trimethoprim–sulfamethoxazole (Bactrim)	Teratogenic in animals (rodents)	Lithium	Cardiac defects
Nitrofurantoin (Macrodantin)	Hemolytic anemia	LSD (lysergic acid diethylamide)	Reports of chromosome abnormalities, limb reduction defects
Anticoagulents			
Coumadin	Hemorrhagic disorders, impaired psychomotor development, facial deformities	Marijuana	Low birth weight, fetal alcohol syndrome-like effects, adverse effects in animal studies
Wafarin	Craniofacial and limb defects	Mercury	Severe CNS damage (Minamata disease)
Anticonvulsive drugs (for epilepsy)		Morphine, heroin addiction (or methadone maintenance)	Low birth weight, withdrawal symptoms, later behavioral disturbances
Phenytoin, hydantoin	Cleft palate and lip; congenital heart disease, impaired digit and nail growth (fetal hydantoin syndrome)	Organic solvents (paint, thinners, lacquers)	Reports of CNS abnormalities; cleft palates; no conclusive epidemiological confirmation
Trimethadione	Growth and mental retardation		
Valproic acid	Spina bifida	Pesticides, herbicides DDT	Premature deliveries (?)

Table 10.2 *continued*

Drug	Reported Abnormality
Dioxin (2,4,5-T, Agent Orange)	Teratogenicity in rodents at high doses, unconfirmed in humans
Phenothiazines (Compazine, Tigan, Thorazine)	Reports of malformations, no conclusive evidence of associations
Polychlorinated biphenyls (PCBs)	Reports of low birth weight, stillbirths. Adverse effects in animals not documented in humans
Quinine	Deafness, CNS damage, blood disorders
Sex steroids	Female masculinization (progestins, androgens); benign vaginal adenosis, clear cell adenocarcinoma, etc. (DES); VACTERL (oral contraceptives)
Thalidomide	Reduction limb defects
Thiourea	Hypothyroidism
Tricyclic antidepressants (Tofranil, Elavil)	Reports of limb reduction defects, no epidemiological confirmation
Vaginal spermacides	Possible limb reduction defects, Down's syndrome, insufficient evidence for association at present

Sources: Yaffee and Stern, 1976; Sullivan, 1976; Cordero and Oakley, 1983; Kalter and Warkany, 1983.

be chosen. It is probable that any chemical agent, taken at a sensitive time of development and in a high enough dose, can affect an embryo or a fetus. At our present state of knowledge it is reasonable to assume that *drugs* are teratogenic, but some are more so than others. It would be prudent for a woman to avoid drugs unless she is certain that she is not pregnant.

This creates a dilemma for a pregnant woman who has heart disease, high blood pressure, toxemia of pregnancy, epilepsy, bacterial infections, or any other disorder for which drug therapy is indicated. Obviously, the risk of untreated maternal illness may be greater for the woman and for the fetus than the possible harmful effect of a drug. If a complete evaluation of the condition and all of its ramifications has been satisfactorily explained to the parents, if it has been determined that the drug is essential and that the least hazardous drug has been chosen, and if the beneficial effects of the therapy are greater than the potential harmful effects, then the decision to take the drug can be made by a woman and her doctor.

Over-the Counter Drugs

The drugs ingested during pregnancy are hardly limited to those prescribed by a physician. A woman who claims to have taken no drugs while pregnant has forgotten the times she sprayed the African violets, had her hair dyed, drank diet cola, painted the baby's room, or put saccharin in her coffee. People seem to forget that aspirins are drugs, that cigarettes and alcohol are drugs, that vitamins, cough drops, nose drops, antacids, laxatives, and antihistamines are drugs, and that coffee and tea are drugs when 10–15 cups are consumed daily.

Little is known about the potential hazards of over-the-counter drugs. Lynn Woodward and her colleagues (1982) interviewed 304 women who had recently delivered normal babies and found that they had consumed 93 different drug products, excluding alcohol, tobacco, and vitamins. The nonprescription drugs most frequently taken were, in order of decreasing consumption. Tylenol, Rolaids, aspirin, Maalox, Robitussin, Sudafed (an oral nasal decongestant), Alka-Seltzer, and Massengill Disposable Douche. Although some of the 93 preparations or their ingredients have been amply investigated for adverse or toxic effects during pregnancy, Woodward found no documentation on the potential teratogenicity or safety of fully two-thirds of the drugs or drug ingredients that the women consumed. Some nonprescription drugs, however, such as cigarettes, coffee, and alcohol, have been the subject of many studies.

Tobacco

Cigarette smoking in a pregnant woman is unquestionably related to low birth weight in the infant she delivers, and more than 50 studies have confirmed this finding. The relationship is dose-dependent, that is, the amount of growth retardation is directly correlated to the number of cigarettes smoked. The reduced birth weight with maternal smoking has been shown to be independent of all the other factors that may influence the size of the offspring, such as size of the mother, her race, socioeconomic status, birth order, or sex of the child. A 1977 study by Reba Michaels Hill and her co-workers discovered that women who smoked produced both premature and SGA (small for gestational age) infants. Manning et al. reported (1976) that when the mother smoked two cigarettes in succession, the fetuses of 32–38 weeks gestation had depressed breathing movements. Although it is possible that heavy smokers have lower maternal weight gain, which could concurrently retard fetal growth, the cause of the decrease in fetal weight (an average of 6 ounces) is probably related to the presence of excess carbon monoxide in the mother's blood. Repeated studies have shown that the level of fetal carboxyhemoglobin, the amount of hemoglobin combined with carbon monox-

ide, is up to 10 times higher in infants born of smoking mothers than in infants born of nonsmoking mothers. Hemoglobin that is tied up by carbon monoxide is not available to carry oxygen. Animal research has indicated that high levels of carbon monoxide can limit the oxygen supply to the fetus and affect the development of the brain cells. It has also been shown that sidestream smoke, that produced off the end of the cigarette, contains four times as much carbon monoxide as does mainstream smoke, that which is inhaled by the smoker. This means that pregnant women smokers who also spend long periods of time in a smoke-filled room or automobile could have alarming amounts of carbon monoxide buildup in their bloodstreams.

King and Fabro's review (1982) of the further effects of cigarette smoking on pregnancy and pregnancy outcome documents an association of smoking and an increased frequency of spontaneous abortions and a greater incidence of pregnancy complications, such as bleeding. The long-term growth and development of the offspring born to smoking mothers is also affected. There is even evidence that significant differences in height as well as reading and math abilities exist in children whose mothers smoked 10 or more cigarettes daily during pregnancy, and a number of studies have linked maternal smoking with a greater incidence of sudden infant death syndrome. Any woman who stops smoking for several months because of morning sickness only to resume smoking when she feels better is demonstrating an appalling disregard for her own health and that of her unborn child.

Caffeine

Animal studies, in which high doses of caffeine (equivalent to 12–24 cups of coffee daily) were fed to rats, produced skeletal abnormalities in the rat pups. On this basis, suspicion has been leveled at overindulgence in coffee or tea drinking during pregnancy, but it is evidently unwarranted. Two epidemiological studies, one by Rosenberg et al. (1982) and the other by Linn and colleagues (1982), found no association between caffeine consumption and birth defects. Linn's group studied over 12,000 pregnancies and discovered that women who reported that they drank more than four cups of coffee or tea daily had babies with a malformation rate of 2.0 per 1,000 live births, and women who drank no coffee or tea at all during pregnancy delivered infants with a birth defect rate of 2.5 per 1,000. On the basis of current evidence, it would appear that a pregnant women need not give up her morning cup of coffee or tea that contains between 100 and 200 mg of caffeine, but it would be prudent to keep the caffeine consumption in a reasonable range. A chocolate bar contains about 25 mg and a cup of cocoa or hot chocolate about half that much. Twelve ounces of noncaffeine-free Coca-Cola, Dr. Pepper, Tab, Pepsi, RC, or Mountain Dew, however, contain between 30 and 60 mg of caffeine.

Vaginal Spermicides

There have been several studies that have indicated that a very small increased risk of chromosomal abnormalities, limb reduction defects, penis and urethral abnormality (hypospadias), or Down's syndrome is associated with the use of vaginal spermicides (Jick et al., 1981; Huggins et al., 1982; Rothman, 1982). There are, virtually an equal number of investigations to refute these findings (Shapiro et al., 1982; Polednak et al., 1982; Mills et al., 1982; Cordero and Layde, 1983), and currently it is generally accepted that there is insufficient evidence to implicate spermicides as teratogenic.

Maternal Alcohol Consumption, or One For My Baby

Certainly, no mother in her right mind would give a glass of wine, beer, or hard liquor to her newborn infant. But alcohol, unlike some other drugs, passes with ease across the placenta. An embryo or fetus, therefore, is exposed to the same blood alcohol concentration as the mother, and when the mother drinks, so does her unborn child. The teratogenic effects of alcoholism or heavy maternal drinking have been recognized since 1973, when a fetal alcohol syndrome (FAS) was first described. FAS is a pattern of defects that may be completely or partially expressed and includes low birth weight and fetal growth retardation, especially microcephaly; facial abnormalities, such as small eye slits, short pug nose, underdeveloped jaws, and nonparallel low-set ears; a number of cardiac, urogenital, cutaneous, and musculoskeletal defects; and later developmental delay and mental retardation (Abel, 1982). Full-blown FAS has been observed to occur only in the offspring of chronic alcoholic mothers, for whom the risk may be as high as 30%. There is ample evidence, however, that a woman need not be an alcoholic to run a greater than normal risk of having a deformed or retarded child. For example, one study compared heavy drinkers, defined as women who consumed five or six drinks on some occasions with an average of $1\frac{1}{2}$ drinks daily, with moderate drinkers, women who drank more than once a month. Rare drinkers were those who used alcohol less than once a month and never consumed five or six drinks on any occasion. Of the 322 women in the study, the heavy drinkers had babies with twice as many congenital, growth, and functional abnormalities as did the moderate and rare drinkers. The risk was lowered when the heavy drinking women either abstained or reduced their alcohol use during pregnancy (Rosett et al., 1978).

Even moderate or light drinking has been linked to an increased risk of miscarriage and low birth weight (Kline et al., 1980; Olegard et al., 1979). Reba Hill's data showed that a slightly higher incidence of small for gestaional age infants were born to women who have only one drink nightly or once a week compared with infants born to nondrinkers. What is unknown at present is whether there is a minimal amount of alcohol safe to drink during pregnancy. Although there is no compelling evidence that a daily glass of wine, for example, is associated with any adverse effects, the United States Surgeon General is taking no chances and, via an *FDA Drug Bulletin*, has advised pregnant women, or women considering pregnancy, not to drink alcoholic beverages and to be aware of the alcoholic content of food and drugs. The Royal College of Psychiatrists in England has provided similar advice to British women.

Woodward, Hill, or Forfar and Nelson—researchers who investigated women's drug-taking during pregnancy—have discovered that 65%–95% of women self-medicate with over-the-counter preparations during the first trimester or throughout gestation. Women evidently do not view nonprescription drugs as drugs, even though they may be taking them for a physical complaint. Unnecessary drug ingestion is so much a part of our lives that it is equally common during pregnancy, so frequent that in most instances an association between a chemical agent consumed and a subtle congenital defect would never be recognized. Would anyone ever be able to relate aspirin, stomach preparations, or sleep aids with some subtle effect on the nervous system if it were manifested as a speech defect, a reading disability, a behavioral disorder, or a hyperactive child? Or, even more elusively, if it created the difference between a Phi Beta Kappa and a "C" student? With the present uncertainty concerning the effects of any environmental influence on the developing fetus, women should be conscious of everything that goes into their mouths or is in the air that they breathe from the time they miss their first menstrual periods until after delivery. They must take no drugs unnecessarily

and must make certain that any pharmacological agent ingested is not only beneficial but essential.

Birth Defects and the Father

The general consensus is that 3% of newborn babies are affected with a congenital abnormality of prenatal origin. The cause may be genetic, that is, a mutant gene or a chromosomal abnormality, or may be environmental, the result of maternal infection or other illness or maternal exposure to a teratogenic substance. There are hundreds of chemical agents that have demonstrated mutagenicity or teratogenicity when administered to pregnant animals; a logical question is, can a drug or toxic chemical be teratogenic through the male? In a review of the experimental evidence and the results of human clinical and epidemiological studies concerned with the exposure of men to toxic substances, even to those confirmed as teratogenic in pregnant women, Pearn (1983) concluded that there is not evidence for an increased rate of malformation in the offspring. Although workplace hazards for women have been linked to birth defects, many studies of male workers exposed to toxic substances indicate that the reproductive effect in men is on themselves in the form of reduced fertility, but does not directly affect the next generation in the form of congenital defects. Animal studies of drugs, such as narcotics, thalidomide, anticonvulsants, and alcohol, and toxins such as lead, pesticides, or other kinds of contaminants and pollutants, show that reproductive toxicity in the male adults, usually evidenced by a temporary infertility, occurred, but no malformations in the litters sired by the affected males resulted. Some human studies of men environmentally or occupationally exposed to various polychlorinated hydrocarbons (PCBs, PBBs, 2,4,5-T, dioxin, Agent Orange) have shown an increased rate of spontaneous abortion in their wives, but other studies have found no increase in miscarriages or stillbirths in their wives, and none has demonstrated an increased rate of congenital malformations in their offspring as a result of male exposure. A 1984 Air Force report of the health status of the 1,269 pilots and crew members involved in spraying Agent Orange over the Vietnam jungles revealed that the men had numerous medical problems but provided little additional information to what was already known. The study did reveal, however, that 14 deaths among infants fathered by the men had occurred compared with four infant deaths in an equal number of veterans who had not been exposed to the herbicide. The explanation for the difference was not known.

The hazards of environmental chemicals to which men, voluntarily or unwillingly, are exposed may be significant to their own health. They may develop liver problems or cancer or infertility, but there is no evidence as yet to suspect that any substance can act as a teratogen via the father.

LABOR AND DELIVERY

Around the fortieth week after the onset of the last menstrual period, pregnancy culminates in childbirth or *parturition*. By this time, most women have difficulty remembering what not being pregnant was like, and the big event is fervently anticipated. All of the changes that have taken place during pregnancy have been in preparation for expelling the products of conception—the fetus, placenta, and membranes—from the inside of the uterus out into the world. The mechanism by which this happens is labor, aptly named because it is muscular work.

The labor process is a series of rhythmic, involuntary, usually quite uncomfortable uterine muscle contractions that bring about a shortening (effacement) and opening (dilatation) of the cervix, and a bursting of the fetal membranes. Then, accompanied by both reflex and voluntary contractions of the abdominal muscles (pushing), the uterine contractions result in the expulsion of the baby. The first stage of labor begins with the first true labor contraction and ends with the complete dilatation of the cervix, large enough to permit the passage of the infant. The second stage of labor ends with the delivery of the baby. The third stage is the period from the birth of the baby through delivery of the placenta and the membranes, and the fourth stage, a very critical period, is the first hour or two after labor is ended.

Exactly what happens during labor is well known; why it happens is still a mystery. Proposed hypotheses to explain the onset of labor range from the rather unscientific—"the baby outgrows the uterus," or "when the apple is ripe it falls from the tree"—to a group of theories that explain the control of parturition on an endocrine basis. The withdrawal of systemic progesterone, the oxytocin effect, the progesterone/estrogen ratio, the secretion of the fetal adrenals, and the role of prostaglandins have all been offered as suggested mechanisms. Current thinking assumes that many, perhaps all, of these factors are implicated in causing labor; that labor is initiated by a sequence of interacting endocrine events involving hormones from both fetal and maternal endocrine organs. One possibility being explored is that altered hormone levels stimulate the release of some agent from the fetal and maternal tissues that causes the formation of prostaglandins, perhaps by activating enzymes that can liberate arachidonic acid (the prostaglandin building block) from its storage site in cell membranes.

The position of the fetus and its relationship to the placenta and membranes during late pregnancy is illustrated in Figure 10.9. This is the position assumed by the fetus 95%

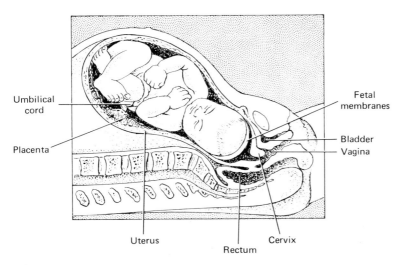

Umbilical cord

Placenta

Fetal membranes

Bladder

Vagina

Uterus

Rectum

Cervix

Figure 10.9. Cephalic presentation at the time of delivery.

of the time and is the one most favorable for normal labor and delivery. The head of the fetus is flexed so that the chin is down on the chest, and the presenting part, that which will descend first, is the top, occipital, or vertex part of the head. The presentation can be cephalic, that is, head first, but the presenting part may also be the face or the brow. A breech presentation means that the buttocks or the feet of the baby present first, and this occurs about once in 40 deliveries. Occasionally, the baby will lie in a transverse position, and the shoulder presents first.

Several changes take place in the uterus even before the start of actual labor. In most women who are having their first baby (primigravidas), lightening or dropping occurs. A few weeks before delivery, the presenting part of the fetus descends to settle into a position at the entrance of the true pelvis. Sometimes a first baby does not drop until labor begins, which is the usual situation in multiparas, women who have previously borne children. After lightening, the upper abdomen becomes flatter and the lower abdomen becomes more prominent. Women notice they are able to

breathe more easily again, but the urge to urinate frequently returns because of the increased pressure on the bladder. Pressure on the rectum and sacrum may also result in diarrhea and backache.

Another preliminary event is the taking up, shortening, or effacement of the cervix. During the first half of pregnancy, the uterine shape is the same as it is in the nonpregnant state. It consists of the upper corpus, or body, and the lower neck of the uterus or cervix, which extends into the vagina. As pregnancy advances, the corpus differentiates into two separate areas: an upper uterine segment that becomes the active muscular contracting part responsible for delivery, and the lower uterine segment—the thinner passive area located just above the internal os of the cervix. The demarcation between the upper and the lower segments is the physiologic retraction ring. During the last few weeks of pregnancy, the spontaneous painless Braxton Hicks' contractions that have been occurring all along, become stronger. Each time the muscle fibers in the upper uterine segment contract and shorten, a very remarkable thing happens.

Very unlike muscle in other parts of the body, these uterine muscle fibers never return to their former length when they relax. Instead they become progressively shorter and thicker, a property known as retraction or brachystasis. This continued shortening and thickening of the muscle fibers after contraction takes place in the lower uterine segment as well, but to a much lesser degree. As a result, the upper segment gradually increases in thickness, decreasing uterine volume, as the lower segment becomes thinner and elongated. The cervix softens, due to engorgement with blood, but remains closed, its canal plugged with mucus, until several weeks or even the last few days before labor begins. Eventually, however, the continual contraction and retraction of the upper segment causes the internal os to open and draws the entire cervix up from its position in the vagina. The cervix is thereby shortened, and the endocervical canal is obliterated as the cervix is pulled up around the fetal membranes and the presenting part and incorporated into the lower uterine segment (see Fig. 10.10). The external cervical os then remains as a circular opening with extremely thin edges. Now, the further contractions of the uterus will bring about progressive cervical dilatation, or enlargement, of the external os.

With a first baby, complete cervical effacement usually occurs before there is any cervical dilatation. In women who have previously delivered a baby, dilatation takes place when effacement is incomplete or concurrently with effacement. Cervical effacement is measured by percentage. A cervix one-half of its normal length is 50% effaced.

The effacement and dilatation of the cervix causes the fetal membranes that are attached to the uterine wall at the region of the internal os to become loosened. As they pull away from the uterine wall, the little mucous plug that corks the endocervical canal is set free. Loss of the plug is painless, but a little blood escapes with it, and it constitutes the "bloody show." Increased vaginal discharge during the period of cervical effacement is not unusual, but when pink-tinged or blood-streaked mucus is expelled from the vagina, it is a sign that active labor has started or is imminent (see Fig. 10.11).

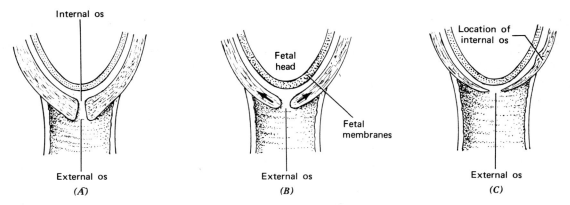

Figure 10.10. Effacement of the cervix in a first pregnancy. In a woman who has previously delivered a baby (multipara), effacement and dilatation occur simultaneously. (A) Before any effacement. (B) Internal os is being drawn upward around the fetal membranes. (C) Completely effaced cervix.

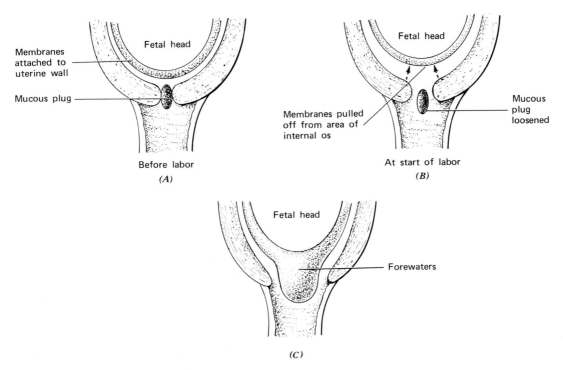

Figure 10.11. Cause of the bloody show. (A) Before the onset of labor the membranes are attached to the uterine wall. A mucous plug, or operculum, blocks the endocervical canal. (B) With effacement and dilatation of the cervix, the fetal membranes are pulled away from the internal os, and the mucous plug is set free along with a little bleeding. (C) After the plug is lost, the forewaters bulge in front of the fetal head.

First Stage

The criteria used to differentiate true labor contractions from the Braxton Hicks' prelabor contractions are regularity and discomfort. When the contractions become painful, less than 10 minutes apart, last 30–90 seconds, and are regular in frequency, labor has begun. For a first-time baby, the average duration of the first stage of labor is about 12 hours, but there are wide variations. The length may be as short as 3 hours if contractions are strong and frequent.

The fetal membranes, or bag of waters, usually rupture during the first stage, but they may have burst earlier or may even remain intact until delivery. Actually, the part that ruptures is only a portion of the amniotic and chorionic sac, a little pocket of fluid called the forewaters that lies in front of the fetal head. The successive uterine contractions keep compressing the forewaters and eventually it breaks, permitting a little gush of amniotic fluid to exit. The rest of the fluid remains behind the baby's head as the hindwaters. If the membranes have not broken by the end of the first stage or early in the second stage of labor, they may be artificially ruptured with a sharp instrument during a vaginal exam.

In 10% of pregnancies, the fetal membranes burst before labor begins, and premature rupture is said to have occurred. Labor will usually start within the next 24 hours. If labor fails to ensue spontaneously, there may be an increased risk of fetal and maternal infection, and it is generally believed that infant mortality rate increases if more than 24 hours elapse between rupture and delivery. Most physicians, therefore, are of the opinion that labor should be induced with an oxytocin solution if it does not begin after 6 hours. But in a review (1977) of a series of studies that evaluated the outcome of premature rupture, Mehl concluded that such a procedure may be an unnecessary intervention and of questionable value.

When the cervix is dilated to 10 centimeters in diameter, it is large enough to permit the passage of a fetal head of average size (the baby's largest dimension), and the first stage of labor is over. Cervical dilatation is gauged subjectively by vaginal or rectal examination and is expressed in centimeters or finger-widths. Throughout the first stage of labor there are regular vaginal examinations to determine the degree of dilatation. They are performed under sterile conditions, and the index and second finger are inserted into the vagina to feel the extent of the cervical rim that remains. A woman hearing that she is "five-fingers dilated" may envision the examiner's entire hand in the cervical os, but actually the subjective estimation is made by assessing the amount of spread between only the two fingers.

The last part of the first stage of labor, when the cervix is opening to complete dilatation, is called transition. It is the most difficult and, fortunately, the shortest phase for the woman, lasting approximately 1 hour in the first delivery and perhaps 15–30 minutes in successive births. At transition, the contractions are stronger, more painful, somewhat erratic, and last longer. Pressure on the rectum is great, and there is a strong desire to contract the abdominal muscles and push. Until the cervix is fully dilated, however, bearing down is not

helpful. It will only intensify the discomfort and may cause cervical lacerations.

The techniques taught in prepared childbirth education classes are especially valuable in coping with the difficult physical and emotional aspects of the transition period (Fig. 10.12).

Second Stage

The baby is completely passive throughout the entire progress of labor. With each succeeding uterine contraction it is pushed

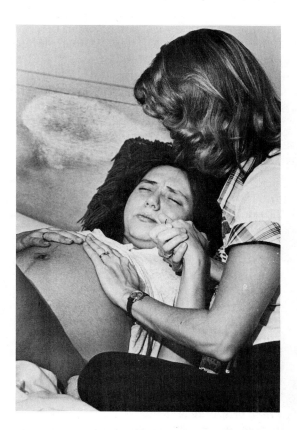

Figure 10.12. Transition stage during the first stage of labor in a birth at home. Coaching in the breathing techniques is being provided by a birthing attendant.

lower and lower, and its position alters as it accommodates itself to passage through the different parts of the pelvis. At the beginning of the second stage of labor, the birth canal through which the baby descends is formed by the completely dilated cervix, the distended vagina, and the stretched and distended muscles of the pelvic floor. When the fetal head meets the resistance of the pelvic floor, it rotates 45° from its former oblique position. The face is now directed posteriorly, facing the mother's sacrum. As further descent continues, the anus dilates and the vagina begins to open with each contraction. The fetal scalp becomes apparent at the vaginal opening but disappears between contractions. When the top of the head no longer regresses between contractions, it is said to have crowned. The perineum bulges and thins out with each contraction as the fetal head continues to enlarge the vaginal opening. If an *episiotomy*, a small incision in the perineum that is performed to prevent tearing, is indicated, it is done at this time.

Episiotomies are neither routinely necessary, nor routinely unnecessary. When it is apparent that a small surgical cut is going to prevent a jagged laceration in the perineal skin and underlying levator ani muscles, it has to be done. Sometimes the procedure is necessary to protect the fetal head, especially in a premature baby, from consistent battering against an unyielding or slowly dilating vaginal opening. Even if tearing does not occur, it is possible that excessive stretching could cause enough damage to the connective tissue and muscles of the pelvic floor to result in problems years later.

The episiotomy may be midline and extend from the vagina directly down toward the anus, or it may be extended in a lateral direction at about a 45° angle to the left or to the right. The timing of the cut is important: if it is done too early, the normal blood loss will increase; if it is performed too late, lacerations may have already taken place. The episiotomy has to be sutured after the baby and the placenta are born. The perineum is still extremely sensitive, and, even with a local anesthetic, the repair is usually quite painful. As the stitches heal and are absorbed in the postpartum period, there is also considerable discomfort. Obviously, it is undesirable for episiotomy to be performed routinely, but it is virtually routine in American obstetrical practice. Doris Haire has pointed out the lack of scientific evidence to support the contention that episiotomy decreases pelvic floor relaxation or reduces the possibility of neurological damage to the infant. There also has been a suggestion that the lithotomy position in which a woman is placed during a traditional hospital delivery necessitates a greater need for episiotomies because of the increased tension on the pelvic floor. Alternative birth positions, such as those used in home deliveries or nontraditional hospitals, appear to lessen the strain on the perineum. Perineal massage practiced at home for a month before delivery and then performed by the birth attendant during the second stage of labor is advocated by some nurse-midwives. It is believed to help prevent lacerations. There is no way, however, to categorically decide ahead of time that an episiotomy will be unnecessary and should not be performed. It depends on the situation. If an episiotomy forestalls an imminent serious injury to the posterior wall of the vagina or the rectum, it has to be accepted with equanimity.

When the head crowns, the neck of the baby is no longer flexed forward but is extended backward. First the top of the head emerges, followed by the forehead, and then the brow and the face. When the head is born, it drops down over the perineum and rotates sideways to restore its natural position relative to the shoulders. This is called restitution of the head. Then the shoulders rotate to present their narrowest diameter for passage and the head is turned farther to the side (external rotation). The posterior or left shoulder is born first. It falls backwards, the right shoulder slides out from underneath the pubic bone, and the rest of the body "squirts out" quickly and easily.

After delivery, the baby is kept below the level of the uterus. After a minute or so, when an adequate amount of placental blood has transfused to the infant, the umbilical cord is clamped and then cut about a half inch from the baby's abdomen. The other end of the cord still hangs from the vagina until the placenta is expelled.

Third Stage of Labor

Uterine contractions stop for a short time after delivery, but in a little while the relaxed uterus contracts again and causes the placenta, which has become partially separated away from the maternal endometrial surface during expulsion of the baby, to become completely detached. At the same time, the upper segment of the uterus changes shape to become smaller, firmer, and rounder. There is a small gush of blood from the vagina, and the umbilical cord appears to lengthen as the placenta is forced downward. Continued uterine contractions cause the placenta to be expelled within 5–30 minutes. Most physicians prefer not to wait for the woman to expel the placenta herself, and they hurry it along by applying a little external pressure to speed up the first contraction. Once the uterus is firm and globular, it is further massaged to aid in expulsion of the placenta. The rationale behind the assistance is that in some women a significant amount of bleeding occurs if the placenta is allowed to be extruded spontaneously. Massaging the fundus of the uterus through the lower abdomen will probably be distinctly uncomfortable right after delivery, however. It is all the more unpleasant because, presumably, the hardest part—birth of

the baby—is over. After the placenta is expelled, the uterus is still massaged to keep it contracting since contractions constrict the uterine blood vessels and minimize the possibility of hemorrhage.

The main cause of postpartum hemorrhaging is uterine atony, or lack of contractions. Customarily, a uterotonic drug is given to stimulate firm contractions and reduce blood loss. An oxytocic solution may be used, but more often ergonovine maleate (Ergotrate), the purified derivative of the rye fungus ergot, is administered intramuscularly or intravenously, after placental delivery. Both Ergotrate and the synthetic form methylergonovine maleate (Methergine), not only stimulate uterine muscle, but all smooth muscle as well and will elevate blood pressure. Oxytocin is, therefore, used in women with hypertension. Some doctors further prescribe oral tablets of Ergotrate every 6 hours until all danger of hemorrhage is past, about 48 hours. Obstetrician Joni Magee has cautioned that lysergic acid (LSD) is an ergot derivative, and that Ergotrate may be a mild hallucinogin. She notes that women believed to have symptoms of "postpartum psychosis" may be getting those symptoms only every 6 hours in a normal response to the drug.

If the newborn baby is permitted to nurse at the breast immediately after delivery and then at regular 3- to 4-hour intervals, pituitary oxytocin is released, which results in colostrum and later milk ejection from the breasts and which stimulates uterine contractions. Unless the mother has had little or no medication during delivery, however, the baby will generally be too sleepy to breastfeed right away.

Afterpains
Multiparas are frequently bothered by painful uterine contractions that may contine for as long as the first week after delivery. They get worse with each succeeding pregnancy, but almost never occur with a first baby. They last

for a half hour or so, but are really uncomfortable for only the first 2 or 3 days and can be relieved with painkilling drugs. Afterpains are usually stronger during breastfeeding because the oxytocin released by the sucking reflex results in uterine contractions.

Apgar Score

In 1960, American anesthesiologist Virginia Apgar introduced a method of assessing the well-being of the newborn withn 1 minute and 5 minutes after delivery. The baby is scored at 0, 1, or 2 in each of five categories: heart rate (absent, slow, or fast); respiratory effort (absent, weak cry, or good stong yell); muscle tone (limp, or lively and active); response to irritating stimulus, and color. A score of 7–10 means the baby is in the best possible condition, while 3–6 means moderately depressed; any score below 2 indicates problems. The rating essentially is evaluating the functioning of the central nervous system. Babies born to mothers who have been heavily drugged for pain relief generally score around 5 or 6. The extent to which a moderately depressed Apgar score as a result of medication influences the future neurological and behavioral development of the child is unknown, and it is assumed that there are no lasting effects.

A study by Harvard pediatrician T. Berry Brazelton indicated that medication during labor and delivery had an effect on milk production by the mother and sucking response by the infant. The combination affected the initial weight gain of the newborn babies when compared with a control, nonmedicated group (1970).

Childbirth Pain and Painless Childbirth

Only 8%–10% of all births are considered unusually difficult or abnormal. For the vast majority of women with normal labor, giving birth is not the acutely painful process we have been led to believe, but neither is it completely free from discomfort. To assert, as does one textbook, that ". . . the pain of delivery is the most intense that the human being can experience"[*] is certainly an overstatement. But to claim that even for confident, calm, and totally prepared women, all parts of labor and delivery are comfortable and easy is an equal misrepresentation. Undeniably, smooth muscle contractions can hurt, as anyone who has experienced menstrual cramps or the "green-apple runs" will recognize. Uterine contractions are similar to menstrual and intestinal cramps, but they are also unlike them. The peak discomfort of labor contractions, even when they are coming very close together, only lasts a few seconds and completely disappears between contractions. The action of the uterine muscle, however, is not the only component of labor discomfort. Another is the result of cervical dilatation. The traction on the cervix that results in the enlargement of the cervical os stimulates nerve impulses that are transmitted along sacral nerves to the spinal cord. Those sensations in the sacral region result in the intense backache associated with labor experienced by some women. A third reason for discomfort during labor and delivery is the stretching of the vagina and the perineum. When a full-term baby's head, 10 centimeters in its widest dimension, starts its descent into the birth canal, it creates pressure on the bladder, rectum, and all the surrounding tissues. The sensations, carried along the pudendal nerve to the spinal cord, have been likened to the feeling of having a bowling ball move slowly through the pelvis. All of these feelings do not have to be interpreted as excruciating, but the brain is certain to recognize them as different and uncomfortable.

Some women need, or think they need, a

---

[*] Niswander, K. *Obstetrics: Essentials of Clinical Practice.* Boston, Little, Brown & Co. 1976, p. 267.

great deal of pain relief. Others are able to manage with very minimal administration of medication, or none at all, depending to a certain extent on their attitude toward the event of giving birth. Unfortunately, the perfect painkiller, one that takes away all discomfort but does not hamper the progress of labor or present any danger to the mother or to the baby, does not exist. All drugs, whether they are given systemically, inhaled as gases, or injected for a local nerve block, have some risks and disadvantages. Any medication gets into the bloodstream, crosses the placenta, and, depending on the time of administration, the kind of drug, and the dosage, can affect the fetus and the newborn baby. Since a baby is unable to metabolize or excrete the drug as rapidly or effectively as an adult, the infant may be born with varying amount of respiratory and nervous system depression.

Heavy medication can result in a sleepy baby for several days or even a week or longer after birth. Light medication produces an alert and active baby. There are no meaningful scientific data yet to prove a long-term deleterious effect as a result of a lethargic entry into the world, but common sense indicates that the least depressing drug in the lowest possible dosage is best.

Tranquilizers and Sedatives

Tranquilizers and sedatives are used in early labor, if a woman is particularly tense and anxious. Barbiturates, such as Seconal, Nembutal, and Luminal, or tranquilizers, such as Valium, Vistaril, and Phenergan, do not relieve pain but they make one care less about having it. They are also thought to potentiate the effect of analgesics so that less pain-reliever is needed. Sedatives and tranquilizers sedate— they can slow labor, cause drowsiness or sleep, and occasionally, dizziness or nausea and vomiting. In some women, their use results in paradoxical effects, and mood elevation, excitement, restlessness, or even delerium may occur. The closer to delivery the drugs are given, the more likely is central nervous system depression in the newborn.

Intramuscular Demerol (meperidine) is a currently popular analgesic. Its action, the increasing of tolerance to pain, peaks at $1-1\frac{1}{2}$ hours after injection. It easily crosses the placenta, so the amount and timing of administration are very important. If labor takes less time than expected, the adverse effects of the drug on the fetus/newborn are evidenced by respiratory depression.

Inhalants

Inhaling gas through a mask to provide complete anesthesia or loss of consciousness is rare nowadays, but whiffs of low concentration of nitrous oxide–oxygen mixtures or trichloroethylene (Trilene) in a self-administered vaporizer may be used intermittently during the second stage of labor. Since there is a time lag of several seconds before the gas can get from the lungs into the bloodstream and to the brain, for maximum pain relief the inhalation has to be started before the contraction begins. Fetal depression is possible, but rare, with low concentrations of gas.

Maternal–fetal effects may not be the sole consideration in the use of inhalant analgesia. There is, or should be, increasing concern about the effects of long-term exposure to trace amounts of the inhalants on the health of delivery room personnel. Several studies have demonstrated that operating room nurses and anesthetists have a higher rate of miscarriage than the general population. Nitrous oxide, the most common anesthetic agent in a hand-held vaporizer, may be a particular health hazard. A study of dental assistants and wives of dentists exposed to nitrous oxide, among other anesthetic gases, revealed that their rate of miscarriage was greater than that in the general population (Cohen et al., 1980).

Regional Blocks

Regional pain relief means that the sensory nerve impulses from the pelvic area are blocked by the injection of a local anesthetic in the same way that the dentist "freezes" a tooth before drilling. The most common techniques used are the paracervical, pudendal, and the "spinals"—subarachnoid saddle, continuous caudal, and continuous epidural blocks.

A paracervical block is an injection that deadens the sensations arising from the dilating cervix during the first stage of labor. When the cervix has dilated to 4 or 5 cm, a local agent, such as Novocain, Xylocaine, or Carbocaine, is injected into the lateral fornix on either side. The anesthesia does not last very long and may have to be repeated after an hour or two. Transitory slowing of the fetal heart rate as a result of repeated injections has been known, so the total drug dosage must be kept at a minimum.

A pudendal block anesthetizes the nerve supply to the perineum, the vulva, and the vagina by an injection into the area surrounding the trunk of the pudendal nerve as it passes around the ischial spine. The injection can be made directly through the perineal tissue (transperineally) or, as is more frequent, through the vagina (transvaginally). The nerve block takes effect in 5 minutes and lasts about an hour. It is usually given late in the second stage of labor when the cervix is completely dilated, and it decreases the discomfort of the actual delivery. Pudendal block neither slows uterine contractions nor relieves their pain, so it is sometimes supplemented by another agent. With this kind of anesthesia of the pudendal and perineal nerves, a woman can be fully conscious during delivery but have a minimum of pain. It is believed to have no adverse effect on the baby.

The other regional anesthetics are called spinals because the injections are made into the coverings around the spinal cord itself. The brain and spinal cord are protected by three complete coverings or meninges: the outermost tough dura mater, the thin and delicate arachnoid mater, and the even more delicate pia mater. Between the arachnoid mater and the pia mater is the subarachnoid space filled with cerebrospinal fluid that cushions the brain and cord. The sensory nerves that carry pain impulses from the pelvic organs enter the spinal cord at the level of the eleventh and twelfth thoracic vertebrae. The motor nerves that cause uterine contractions exit from the cord higher up, at the level of the seventh and eighth thoracic vertebrae. Therefore, injections of an anesthetic that are made below the eighth thoracic vertebra will block all sensations but will not interfere with uterine contractions.

Subarachnoid spinal anesthesia is performed by the introduction of a needle into the interspace between the third and fourth lumbar vertebrae until it enters the subarachnoid space below the arachnoid layer of the spinal cord. The injection is frequently made with the woman in a sitting position with her back rounded. A lumbar puncture sounds intimidating, but is actually not that uncomfortable and does not take long to do.

The anesthetic solution is weighted with glucose or dextrose to make it heavier so that it settles in the lower spinal cord. The level of anesthesia up the back is controlled by positioning the woman or by tilting the table up or down following the injection. When the spinal block is kept very low, it is called a saddle block because the loss of sensation is in the part of the body that would sit on a saddle. A high spinal results in complete anesthesia from the waist down. Uterine contractions continue, but they may be slowed, and saddle block is performed when delivery is imminent during the second stage of labor. The effects last about an hour and alleviate the pain of episiotomy, delivery, and repair of episiotomy.

The administration of any type of spinal anesthetic requires the skill of an expert, and the subarachnoid injection is said to be safe and easy as long as it is properly performed. There is always the chance of nerve-root injury and possible neurological damage as a result, but the incidence is low. The major disadvantage to spinal injection is its tendency to cause a lowering of maternal blood pressure, which could impair the oxygen supply to the fetus. While there is a significant risk of such hypotension, it can be quickly alleviated by a change in the woman's position, either by elevating the legs or turning onto the left side.

Another drawback of this anesthetic is the chance of postspinal headache. Whenever cerebrospinal fluid is disturbed, either by injecting something into it or by removing some of it as in a spinal tap, a headache is likely to occur. Lying flat for 6–12 hours after a spinal may help to prevent a headache. Since the leakage of cerebrospinal fluid at the site of the injection may be a major factor in the cause of headaches, the use of smaller gauge spinal needles is said to reduce their incidence. If the spinal headache does occur, it can be very severe and incapacitating, last for several days, and can only be relieved by lying flat in bed. For mother and infant, this is a less than optimal way to start postpartum life.

Epidural block is achieved by an injection into the epidural space, which lies between the dura mater of the spinal cord and the ligaments that connect the dura and the vertebrae. Lumbar epidural analgesia, as it is called, has become the most popular anesthetic during labor and delivery. It gives complete relief from pain with fewer effects on the mother and infant than most other types of medication (Yurth, 1982). It may be given as a single dose just before delivery, but, more commonly, continuous administration is started during the first stage of labor. With the woman lying on her left side, a needle is introduced between the third and fourth lumbar vertebrae until it reaches the epidural space.

A plastic catheter is then threaded through the needle, the needle is withdrawn, and the catheter is taped into place on the skin so that appropriate doses of the local anesthetic drug, usually bupivacaine, can be given at intervals.

Continuous epidural block is popular because of the high degree of pain relief and the absence of fetal depression. There are, however, several disadvantages to its use. It requires an experienced anesthesiologist to time it properly and perform it safely and effectively. It slows labor in the first stage and causes the bearing down or pushing reflex in the second stage to be reduced or lost, so that the baby may have to be delivered by forceps. Maternal hypotension is an ever-present complication. Uncontrollable shaking of the legs or a temporary paralysis may occur. Neither is significant or lasts very long but can certainly be alarming if a woman is unprepared. And finally, if the dura mater is accidentally perforated, the result will be a subarachnoid anesthesia, which may not be desirable at the time and which increases the chance of a postspinal headache. For a cesarean section, a larger volume of drug is administered to produce greater blocking of pain higher up in the cord.

A caudal block is similar to an epidural in that it is made into a space outside the dura mater, but the site of needle insertion is different. The injection is made through an opening or foramen in the sacrum into the caudal space, an area below the level of the dura mater in the lowest part of the spinal canal. The procedure requires a larger dose of anesthetic and somewhat more technical skill than does an epidural injection, and the risk of infection is slightly greater. Both the pain-relief and the potential for complications are the same for both methods, and neither is thought to have much, if any, affect on the newborn unless maternal hypotension is prolonged and severe. Figure 10.13 shows the needle place-

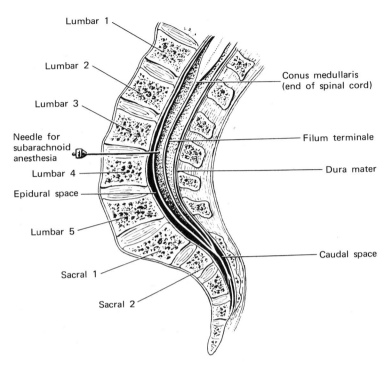

Lumbar 1

Lumbar 2

Lumbar 3

Needle for subarachnoid anesthesia

Lumbar 4

Epidural space

Lumbar 5

Sacral 1

Sacral 2

Conus medullaris (end of spinal cord)

Filum terminale

Dura mater

Caudal space

Figure 10.13. The spinal cord and its coverings in the lower portion of the vertebral column. Needle placement for a subarachnoid spinal block is shown. For an epidural block, the injection is made into the epidural space; in a caudal block the needle is placed in the caudal space.

ment for three types of spinal anesthesia.

Relief of Pain Without Medication

Childbirth educators, those who teach techniques of prepared childbirth, are reluctant to use the word "pain" at all. They refer only to "contractions" because of the negative connotation of the term, "labor pains." The fact that pain exists, however, is really not as significant as the way one feels about it and deals with it. There are mechanical and physical reasons for discomfort in childbirth; no one invents or imagines it. But there is a great deal

of evidence from laboratory studies and from the way people have been observed to behave during pain-producing situations that suggests that while everyone probably feels pain in the same way physiologically, there is enormous variation in the way that different individuals perceive, interpret, and respond to pain. The degree and quality of pain during labor and delivery can be greatly influenced by a woman's attitudes and preconceived notions toward the birth process.

For centuries, women have been exposed to what Dr. Pierre Vellay, an associate of Dr. Lamaze, has called "psychological pollution." The traditional and destructive conditioning concerning childbirth pain is still being reinforced today by the motion picture and television portrayals of the agonies of laboring women, writhing and screaming in apparently intolerable torment. Such images have great dramatic impact and are difficult to dispel, even with the recognition that they are only cultural myths that exist in the minds of the film directors. When a woman learns to expect pain in childbirth and to fear it, she is likely to experience what she anxiously anticipates. The goal of prepared or natural childbirth is to reeducate and condition women to a positive attitude toward childbirth so that labor and delivery can be, if not actually painless, at least comfortable enough so that they can experience the joy of the event. Prepared childbirth does not preclude the use of analgesic drugs, if a woman chooses to have them. Women who have been trained in its techniques, however, require far less medication. Not only can they actively participate in delivering the baby that they have carried for 9 months, but since lower doses of drugs can be administered, there is less risk for both mother and child.

The most widely used method of childbirth education in this country is the Lamaze method, also called psychoprophylaxis or prepared childbirth. Through the efforts of the American Society for Psychoprophylaxis in Obstetrics (ASPO) and the International Childbirth Education Association, increasing numbers of pregnant women are being prepared physically and mentally for the delivery of their children. First introduced in France in 1952 by obstetrician Fernand Lamaze, the method is based on the theories of the natural childbirth pioneer, Grantly Dick-Read of England. Dick-Read emphasized the relationship of fear and tension to pain and stressed education to removed the fear, exercises to prepare muscles and joints for delivery, and muscle relaxation and breathing techniques to relieve the tension. Lamaze utilized the

Dick-Read concepts, incorporating with them the Russian theory of psychoprophylactic preparation for birth.

The psychoprophylactic theory was based on the work of Russian physiologist Ivan Pavlov who discovered conditioned reflexes. Pavlov conducted experiments to prove that dogs could be conditioned to respond to painful stimuli by salivating, as though the stimuli were actually food, instead of responding with howls of pain. Russian doctors, who had experimented with various methods of suggestion and hypnotism in order to alleviate childbirth pain, began to use a similar kind of conditioning training for pregnant women. Through stimulus-response conditioning, it was discovered that women could teach themselves to develop new responses to the stimuli of uterine contractions and could ease the progression of labor. During the last 2 months of pregnancy women learned and practiced muscular and respiratory behaviors with which they would automatically respond during labor in order to eliminate or minimize the pain of childbirth.

The method, highly successful in Russia, was brought to France by Lamaze and spread to the United States in the early 1950s. As presently practiced, the technique is taught to pregnant women in the form of a series of lessons that begin at the end of the seventh month, although the orientation sessions may start as early as the second or third month. Various modifications of the method are taught, depending on the area of the country, or whether the classes are offered by a hospital, doctors for their patients, the Red Cross, the YWCA, or the Childbirth Education Association.

The basic principles of psychoprophylaxis are always the same:

1. Education about the anatomy and physiology of pregnancy, labor, and delivery, with

positive information to replace the myths and superstitions.

2. Physical exercises—squatting, tailor sits, pelvic rocks, various contraction and relaxation techniques—to strengthen, tone, and limber up the pelvic floor and abdominal muscles.

3. Relaxation and breathing techniques for use during labor and delivery (see Fig. 10.14A–C).

The woman learns neuromuscular control techniques, the ability to differentially contract isolated muscle groups while simultaneously relaxing other muscles. During labor she will be able to consciously "release" or relax muscles upon direction. A further component of psychoprophylaxis is what appears to some to be "gimmicks." These are the distracting stimuli that give the woman something to concentrate and focus on to take her mind off labor and pain. The distractions take the form of various breathing patterns and *effleurage*, a light rhythmic stroking on the abdomen that is performed in time to the breathing. The more absorbed a woman can become in these additional activities, it is believed, the less the brain will perceive pelvic sensations as being painful.

The childbirth education classes are taught by trained and qualified childbirth educators. Strong elements of group dynamics and group therapy are involved in the course; a woman has a feeling of community with the other participants and is not isolated in her pregnancy. A labor coach—the father of the child, the mother of the woman, or a willing friend—accompanies the pregnant woman to the classes. The coach learns the physical exercises along with the woman and is able to provide the necessary reminders, encouragement, and support throughout the labor and delivery process. While most hospitals are open to the idea of husbands in the delivery room, others have not progressed sufficiently to allow close friends or boyfriends to be present, particularly if the hospital has a religious affiliation. The presence of a labor coach is indispensible to psychoprophylaxis, and women may have to shop around to find both the obstetrician and a hospital that understand the necessity for the role of her collaborator.

Prepared childbirth works. Studies have shown that its techniques can effectively substitute for drugs in providing pain relief during labor and delivery. Scott and Rose (1976) compared 129 women who were having their first babies and had completed a Lamaze training course with an equal number of matched control women who had not. The prepared women received less frequent administration of drugs during labor, had fewer nerve blocks, and had a higher frequency of spontaneous vaginal deliveries than the control women. Richard Stevens, who studied the strategies of psychoprophylaxis in the laboratory under controlled conditions, found that the methods of systematic relaxation and purposeful attention focusing were successful in reducing pain perception. He concluded that prepared childbirth provided not merely a placebo effect by which highly motivated individuals could endure pain, but actually was a true psychoanalgesia (1977). Many aspects of yoga, meditation, autosuggestion, and hypnosis are similar to the mental focusing components of psychoprophylaxis; there is nothing mysterious about it.

With childbirth education, the pain in a normal labor and delivery is quite bearable for many women. When it does become fairly intense, the discomfort also may be completely overshadowed by the exhilaration and joy of participation. The fantastic, incredible, miraculous experience of giving birth only happens a few times in a woman's life; she should be there to enjoy it and not be totally out of it as a result of medication. High levels of medication also tend to hamper the possibilities for early mother–infant–father interaction. While the maternity-ward routine in many hospitals still plays the greater role in separation of newborn and parents, a sleepy baby is obviously more difficult for a mother and father to respond to.

The Lamaze method of psychological conditioning, while successful, cannot possibly be equally effective for all women. A woman should never feel that she has failed in some way, was not motivated enough, or have guilt feelings if she has to have some measure of chemical pain-relief. People differ, and childbirth should never be viewed as an endurance contest. Medication is an option, available according to individual need, but psychoprophylaxis will reduce the need. Prepared childbirth will enable the woman to redirect her emphasis. Pain may no longer be significant in an atmosphere of full participation and encouraging support.

Mother–Infant Bonding

The importance of the immediate period after birth for cementing the relationship between parent and child has recently been recognized. In a number of studies, (1972, 1976) Marshall Klaus and his co-workers emphasized the importance of contact, both visual and tactile, between mother and baby in the first minutes, hours, and days after birth. Bonds of affection can be particularly strengthened if eye-to-eye and skin-to-skin contact take place. Klaus and Kennel advocate the value of mother, father, and infant togetherness for the first hour after birth in the delivery or recovery room. De Chateau reported that when mothers were given their naked babies for 1 hour within the first 3 hours after birth, such extra contact resulted in differences in maternal attachment behavior—smelling, kissing, close body contact—1 month, 1 year, and even 2 years after delivery (1980).

The necessity of such bonding or imprinting during a "sensitive period" after childbirth may be overemphasized. Not all parents, nor all children, will need such early extra contact in order to establish a loving relationship. If it has been denied as a result of a premature or otherwise high-risk delivery, a mother's attachment to her baby and the child's subsequent development can be just as positive as if the skin-to-skin contact had occurred. Klaus and Kennel, themselves, have become concerned about a too literal acceptance of the word bonding that suggests that "the speed of this reaction resembles that of epoxy materials" and produces feelings of guilt and failure should it not take place (1983). The concept of the benefits of early contact, however, has helped to change the system that virtually excluded parents from the birthing process in most hospitals. The separation of mother and father from child during and after delivery is no longer routine and is unacceptable to many women who would choose to deliver at home if the hospital did not provide for parent–infant interaction.

Leboyer Method

Another French physician, Frederick Leboyer, has advocated what has become known as the Leboyer delivery. Leboyer appears less interested in the mother than in the baby. In his 1975 book, *Birth Without Violence,* he sees the process of birth as "torture of the innocent," a perception gleaned from his own psychoanalytic recollection of birth. His concept, rooted in the concept of primal pain, is that the uterus is a prison, that contractions crush, stifle, and assault, and that, among other horrors, as the "monster" bears down, twisting the baby "in a refinement of cruelty," the baby, "mad with agony and misery, alone, abandoned, fights with the strength of despair." To minimize this incredible trauma and pain associated with being born, Leboyer rec-

Figure 10.14. Prenatal conditioning exercises. (A) Tailor sits stretch the perineal muscles and increase flexibility. (B) Pelvic rocks strengthen abdominal muscles to aid delivery and to increase good posture during pregnancy. These exercises relieve abdominal pressure and backache. (C) Modified situps to tone and strengthen abdominal muscles.

Figure 10.14. (*continued*)

ommends a wide episiotomy for a controlled delivery, and gentle support of the head, neck, and sacrum as the baby emerges. The infant is then placed on the mother's abdomen, and its back is gently massaged by the obstetrician. While the cranial-sacral axis is continually supported, the baby is transferred to a warm waterbath for several minutes, dried off, and diapered. All this time the room is kept very quiet and dimly lit to avoid overstimulation of the newborn.

Leboyer's attitude toward women in the process of birth is highly questionable, but his attitude toward establishing a more humane, gentle introduction to the world than that provided by the usual hospital practices is valid. Women could consider aspects of the Leboyer-type delivery that would be applicable to their labor and delivery. Most of Leboyer's suggestions are not new; they have been traditionally practiced by women and midwives.

INTERVENTION

The whole purpose of health care during pregnancy, labor, and delivery is directed toward achieving the outcome everyone wants—the birth of a normally functioning, healthy baby to a healthy and happy mother. About 85%–90% of pregnancies would terminate in spontaneous normal delivery, an uncomplicated labor that produces a healthy infant, even if the only birth attendant were Mother Nature. Few women would choose to deliver their babies alone and unattended, but most may be getting far more assistance than they require. "Normal delivery," as depicted in a medical teaching film by that name distributed under the auspices of the American Medical Association, is childbirth that is helped along by the following procedures:

1. 100 mg of meperidine (Demerol) administered to the mother during the first stage of labor
2. Paracervical block during the first stage
3. Artificial rupture of the membranes to hasten delivery
4. Bilateral pudendal block early in the second stage
5. Low forceps delivery when the head has crowned
6. Midline episiotomy
7. Oxytocin injection in the second stage to assure uterine contractions in the third stage
8. Massage of the fundus of the uterus in the third stage to facilitate expulsion of the placenta
9. Manual removal of the placenta from the uterus if it is not spontaneously expelled within a few minutes
10. Repair of the episiotomy

One viewer, awed by all the injecting, cutting, sewing, and other maneuvers, commented that it certainly looks as though birth were impossible without that much help. The implication is clear; delivery is not a normal physiological process, but a hospital illness that must be managed and technologically assisted. The normal delivery film is several years old, but revision today might be likely to show additional, rather than fewer, interventions. The laboring woman, for example, would probably be hooked up to a machine that electronically monitors fetal heart rate, maternal blood pressure, and uterine contractions through the entire birth process, and the climax of the film would more likely be a cesarean section rather than a vaginal delivery.

Many women and some doctors have begun to question the need for and the value of such active obstetrical management of the normal spontaneous delivery. Birth is no longer the responsibility of the woman; it increasingly has become that of the attending physician. The substitution of a technically facilitated childbirth for letting nature take its course is being challenged, and there are doubts concerning the safety and the benefit of many of the interventions.

When it is indicated and necessary, an oxytocin infusion to start labor is a useful tool. How often is it done for convenience because the doctor wants to get back to an office full of patients, has a meeting to attend that night, or wants to leave on vacation the next day? Why does the incidence of use of forceps delivery vary from 1% to 2% in some hospitals to almost 100% in others? In discussing the reasons for forceps use, one Ob/Gyn textbook points out that a major factor is the maintenance of the skills of the obstetrician (Dilts, Greene, and Roddick, p. 117). It has been suggested that normal delivery is too routine, too boring, and presents little challenge to the highly trained obstetrician. Intervention and management of labor, with all the technological gadgetry that accompanies present hospital obstetrics, may make the process more interesting as well as educational for the obstetrical residents.

Some deliveries do require every medical advantage to make certain that no tragedy occurs. Certain pregnancies are designated as high-risk, and there is a significantly increased possibility of fetal or maternal mortality or morbidity. In that category are women with a history of previous obstetric difficulty, who have a major medical problem such as heart or kidney disease or diabetes, who have developed a hypertensive or a blood disorder, or who are having a multiple pregnancy. Along with the purely medical reasons, poverty, with its resulting social, nutritional, and emotional deprivation, is also seen as a major contributing factor to risky delivery.

Two-thirds of all complications in pregnancy and delivery occur in women who have been placed into a high-risk classification. These are the pregnancies for which all the tools of technology—the ultrasound scans, the oxytocin challenge tests, the amniocentesis for fetal maturity determination, the continuous fetal heart-rate monitoring, the inductions to initiate labor, the cesarean sections to replace vaginal delivery—play an essential role in assuring a favorable outcome. Although difficulty in delivery can occasionally occur without prior indication, the 10% of births that are abnormally complicated or unusual can almost always be predicted. But by some extrapolation of high risk to all risk, techniques formerly reserved for dangerous deliveries are now almost mandatory for all deliveries in most hospital-based obstetrical practice.

The usual medical controversy exists. An increasing number of doctors use maneuvers and techniques formerly reserved for the most exceptional of high-risk pregnancies as necessary and safe methods for all deliveries. Others view the liberal use of testing, monitoring, and management during labor and delivery, often culminating in cesarean section, as invasive and dangerous in themselves. No one could argue that now, when birth rates in the United States are low and presumably most pregnancies (except in teenagers) are planned and wanted, every effort to ensure the birth of a healthy, mentally and physically sound infant should be taken. Doctors who favor maximum interference imply that it is better to be safe than sorry and that new obstetrical techniques are responsible for a more favorable fetal outcome. Are they right?

After a decade of increasingly medically managed births and expanded reasons for cesarean sections, the infant mortality rate in the United States, although still one of the worst in developed nations, declined from 29 deaths per 1,000 births in 1970 to 11 per 1,000 in 1983. Whether a cause-and-effect relationship exists is highly questionable, however. Approximately 60% of infant mortality occurs in newborns weighing less than $5\frac{1}{2}$ pounds. Perhaps improved techniques in neonatal intensive care units for premature and low birth weight infants have contributed to the mortality decline. Maybe the statistics are better because more women with greater education and understanding of the roles of smoking, drinking, and weight gain on birth weight are having babies. Besides, the improved figure reflects the total rate of infant deaths; there is a 9.9 rate for white babies but still a 19.3 rate for black babies, who are twice as likely to be of low birth weight and twice as likely to die in the first year of life. There is no reason to feel proud of a decrease in infant mortality unless it applies to all infants. High-technology intervention in normal deliveries may be affecting the health status of some children, but may not be a significant component of the improved infant mortality statistics.

But a further question exists. If interventions during prenatal care and parturition may not provide greater benefit to maternal and fetal outcome, do they present some inherent risks? That is, are there adverse effects of medical intercession that go beyond the psychological ones—the possible feeling of dissatisfaction and regret of a woman with the uneasy sense that it was the doctor, the hospital, and the machines that delivered her baby, and not she.

Chicago obstetrician Frederick Ettner maintains that "hospital technology breeds pathology" and that women should refuse to be part of hospital protocols that consider the extraordinary as routine. From the time a woman enters the hospital until she leaves with her baby, there are surely some aspects of obstetrical care that have become so institutionalized that they are performed almost ritualistically, with no real evidence that they have any effect on the improvement of maternal or fetal well-being. There are indications that some techniques may be not only unnecessary, but hazardous as well.

Induction of Labor

Labor can be induced or a slow labor can be enhanced by the administration of a solution of oxytocin. Usually the cervix has to be "ripe," defined as one that is soft, dilated, and effaced in order for induction to be successful, but labor can also be frequently induced in a woman whose cervix is "unripe." Induction is medically indicated when an early delivery is necessary, as in severe toxemias and hypertension, Rh incompatibility, and diabetes mellitus, or for a pregnancy that is considerably prolonged past the due date, or when the membranes have ruptured prematurely and infection could occur. The use of oxytocin should rarely be necessary in a normal pregnancy and delivery, but the hormone has been used for a planned delivery, that is, induction of labor at a predetermined date for convenience or to ensure an on-time birth. The safety of purely elective induction is questionable, and the Food and Drug Administration in 1978 required new labeling for oxytocin, restricting its use in inductions for medical reasons and warning that the hormone should not be used merely for convenience. Presumably, oxytocin is used today in normal deliveries after premature rupture of membranes or in order to enhance or accelerate what could be a very prolonged labor. The drug is highly potent and difficult to control and, for that reason, should only be administered intravenously so that it can be immediately stopped should there by evidence of uterine hyperactivity or fetal distress. Overdosage can result in uterine contractions that are too strong or too frequent, and if the uterus is hypersensitive, the results could be disastrous. Fetal heart rate abnormalities during labor and a significant increase in the incidence of neonatal hyperbilirubinemia (excessive levels of serum bilirubin with jaundice after delivery), have been directly attributed to the use of oxytocin induction (Conner and Seaton, 1982). Women should never be left alone after induction; they have to be observed constantly.

Unfortunately, hormonal induction of labor can be one component of a cascading series of interventions that turn a normal birth into a full-blown medically managed delivery. For example, the membranes may be artificially ruptured to start or hasten labor, a procedure called amniotomy. When such surgical induction is unsuccessful in initiating contractions and infection becomes a concern, an intravenous oxytocin infusion may be started. The fetus then must be electronically monitored to make certain it is getting all the oxygen it needs during the strong contractions, and the monitoring is likely to be internal, with electrodes attached to the fetal scalp—a procedure that can lead to fetal complications. Also, few women are able to cope with the intense contractions produced by induction so the Lamaze techniques they have practiced become ineffectual. They require some kind of analgesia or perhaps epidural anesthesia for pain that then eliminates the ability to "push" so a forceps delivery becomes necessary. Or alternatively, indications of fetal distress on the monitor are interpreted as reason for a cesarean delivery.

Oxytocin Challenge Test

The oxytocin challenge test (OCT), also called the contraction stress test (CST), is a diagnostic procedure performed to determine the fetal heart response under stress, that is, when contractions are induced by oxytocin. The test tries to duplicate the stresses of labor in order to assess the condition of the placental-fetal circulation and the fetal ability to tolerate

labor. The indications for performing an OCT are in those pregnancies in which a placental insufficiency is suspected—preeclampsia, intrauterine growth retardation, a previous stillbirth—or when an irregularity of fetal heart rate has been observed. The test takes place 5–6 weeks before term, but it is also used when the baby is past due, at 42 or 44 weeks of pregnancy. Oxytocin is administered, and fetal heart rate patterns and uterine contractions are recorded. The tracings are classified as positive or negative. A positive OCT is one in which the fetal heart rate shows ominous changes in association with uterine contractions and is seen as an indication for cesarean rather than vaginal delivery. The test has serious limitations as is evidenced by the large percentage of false positives (25%–50%) and occasional false negatives (Hill, 1982). Originally prescribed solely for high-risk cases, there are indications that the OCT is now widely used as a general screening for a successful labor and delivery, despite the inherent risks of the test itself.

There has been the suggestion that the kind and quality of uterine contractions that are stimulated by oxytocin during the stress tests can themselves compromise the oxygen supply of the fetus sufficiently to produce a "positive" interpretation of the results. The result may be an unnecessary cesarean section.

Nonstress Test

A fetal nonstress test (NST) that can be performed in the doctor's office or an outpatient clinic is based on the premise that fetal well-being can be assessed by the increase in fetal heart rate in response to fetal movement. After a meal or a sweet drink, because the fetus becomes more active following glucose ingestion, a woman is placed in a semireclining position and an external monitor to record fetal heartbeat during spontaneous movements is attached. If the fetus is sleeping, pressure on the abdomen or sounds may be used to awaken it. Manipulation of the fetal head or buttocks will also stimulate movement. After 20–40 minutes of recording, the results are classified as "reactive" or "nonreactive." A reactive or normal result is a baseline heart rate between 120 and 150 beats per minute with fluctuations of 10 beats per minute or greater and accelerations of heart rate of 15 beats above the normal baseline that accompany at least two fetal movements during a 10-minute period. A nonreactive result is absence of the preceding; that is, no heart rate accelerations with fetal movements, no response of the fetus to manipulation or stimulation, or fewer than two accelerations in the 10-minute period. A nonreactive test has been correlated to a higher incidence of fetal distress during labor, fetal mortality, and intrauterine growth retardation and is seen as an indication for an oxytocin stress test as a follow-up procedure. The nonstress test is noninvasive, nondangerous, and a simple way to indicate fetal well-being in high-risk pregnancies. When fetal mortality from congenital abnormality, postnatal infection, or trauma at the time of delivery are excluded, a reactive NST is almost 100% predictive of fetal survival. The major disadvantage of the NST is the high rate of false-positive (nonreactive) and false-negative (reactive) results. In one series of tests, about half of the women showed an abnormal nonreactive result, but when they were retested, most of them turned out to be falsely abnormal (Hill, 1982). Decisions on the need for immediate delivery or other interventions should never be based on a nonstress test alone.

Fetal Heart Rate Monitoring

The use of continuous or intermittent electronic fetal heart rate monitoring (EFM) during labor is a part of the routine in almost all hospitals and has replaced the stethoscope as the primary method of observing changes that would indicate fetal distress. The normal fetal heart rate is between 120 and 140 beats per minute, although it is normal for the rate to slow during contractions. Prolonged slowing with irregularity, along with several other signs that include an abnormally low fetal blood pH, are indications that the fetus is suffering from lack of oxygen, or hypoxia. Deprivation of oxygen results in asphyxia or suffocation. All tissues are affected by oxygen lack, but most cells can regain their function and regenerate even if damaged. When brain cells are deprived of oxygen, however, they undergo irreversible injury, and brain damage or death can be the result. Fetal heart rate monitoring is seen as a more precise and accurate way of identifying the fetus in acute distress so that stillbirth or brain damage can be prevented.

There are two methods of monitoring fetal heart rate patterns, externally and internally. In external monitoring, two electrodes are strapped around the woman's abdomen. One picks up the uterine contractions, and the other records the fetal heart rate. The electrodes are attached to a monitoring machine that produces a paper tracing of the recordings or a nonfading oscilloscope display. Since the tracings produced by external monitoring are subject to distortions from maternal movements, fetal movements, and other kinds of vibrations, this form of recording EFM is frequently used as a screening of women when they start labor. Should there be any evidence of fetal distress, the internal monitoring technique is initiated. For internal EFM, two electrodes or leads are inserted through the cervix when it is 2–3 cm dilated and after the membranes are ruptured. One spiral electrode with a maximum penetration of 2 mm punctures the presenting part of the fetus, usually the scalp, to transmit the fetal electrocardiogram. The other lead is a catheter or very thin tube placed between the fetus and the uterus to record the changes in uterine pressure during contractions. Some machines permit simultaneous recordings of the fetal heart sounds (phonocardiogram or FPCG) and a Doppler cardiogram (or DCG), which records the heart beat signal obtained by ultrasound. Usually, an alarm device signals a deviation from the normal pattern, and, as the displays become more complex, greater skill is needed for their interpretation. Since it has been recognized that heart rate monitoring alone cannot provide an infallible guide to fetal distress, it is recommended that fetal scalp blood samples to analyze blood gas pressures and to determine the acid–base balance become a part of the procedure.

The criteria for the interpretation of the fetal heart rate were devised by one of its developers, Dr. Edward Hon of the University of Southern California Medical School. Hon described "innocuous" EFM patterns not associated with fetal asphyxia, and "ominous" patterns that may indicate distress, acidosis, and ultimately death. After a baseline heart rate has been established, a marked decrease or deceleration is of significance.

A type I, or early deceleration, is evidenced by a dip in the shape of the baseline heart rate that occurs at a very early stage of uterine contraction and recovers before the contraction ends. Early deceleration is considered benign and is associated with pressure on the fetal head. In a type II, or late deceleration, the dip comes toward the end of uterine contraction and may mean that the fetus is suffering from oxygen lack (hypoxia) or a reduced uterine–placental exchange. The variable deceleration, so called because it occurs in no uniform relationship with the uterine contractions, is associated with umbilical cord compression.

There is no evidence that variable decelerations indicate fetal oxygen lack when they occur in a moderate number. But frequent and prolonged variable decelerations are associated with fetal distress. Unless the cord compression is relieved, the fetus will suffer hypoxia and begin to suffocate. There will be a disturbance in the metabolic acid–base balance of the blood, and eventually, because neurons do not recover from damage, the possibility of brain injury. Most late and variable deceleration patterns can be restored to normal by changing the woman's position to relieve pressure on the cord or by giving oxygen. A sample of fetal blood obtained by scalp puncture should be taken to measure levels of oxygen in the blood and to determine the pH (acidity). If fetal hypoxia and acidosis are confirmed and all efforts to eliminate oxygen lack are unsuccessful, an urgent delivery by cesarean section is performed. But since fetal heart rate patterns that appear on the printout are not easy to interpret, an "innocuous" pattern may be misread as "ominous," and a deceleration pattern may result in overreaction. Variable deceleration patterns are known to appear in 80% of fetuses at the end of the first stage of labor, but only 30%–40% of them will actually be acidotic.

Yet, the benefits of continuous fetal monitoring seem obvious. Some studies have indicated a decrease in the number of stillbirths when all patients, both high risk and normal, are monitored. Other researchers maintain that fetal asphyxia during delivery, detectable by electronic fetal monitoring and fetal scalp blood sampling, is responsible for brain injury that results in mental retardation or cerebral palsy. Such contentions provide strong arguments for the use of EFM. Given that an ounce of prevention is worth a pound of cure, it would appear that routine application of monitoring to all pregnant women would be ideal. Or would it?

In a 1983 review of the possible relationship between fetal asphyxia and subsequent brain injury, Niswander points out that according to the existing literature, the cause of cerebral palsy is still not known, that a brain injury cannot be predicted by abnormal EFM patterns, most fetal asphyxia during delivery does not lead to cerebral palsy, that both EFM and scalp blood sampling have resulted in false-positive rates of 20%–80% for a diagnosis of fetal asphyxia, and that "more randomized clinical trials desperately are needed to determine the usefulness of EFM." Another review of EFM by Haverkamp and Orleans (1983) reported similar findings: no beneficial effect of EFM on neonatal deaths, neurological outcome, or subsequent health of the child. Both reviews emphatically affirmed, however, that increased cesarean section rate is associated with EFM.

If universally applied EFM is not really an ounce of prevention, it may still constitute considerably more than an ounce of risk. The position necessary for EFM, lying flat on the back, is the worst possible position for labor and delivery. It adversely affects comfort, lowers uterine activity, and causes maternal hypotension. In some women, blood pressure falls by more than a third, and fetal oxygen lack results. Furthermore, in order to attach the electrodes for internal monitoring, the fetal membranes have to be artificially ruptured (amniotomy) early in labor. Since the protection afforded by the amniotic fluid is now lost, the fetal head is more vulnerable. The uneven pressure on the head produced by uterine contractions results in a decreased cerebral blood flow and a decrease in fetal heart rate. Both the position and the procedure for EFM, designed to indicate fetal distress, can actually cause fetal distress.

Additional adverse effects of fetal monitoring include an increased incidence of uterine infections as a result of the leads introduced into the uterus and the fetal complication of scalp abscess and infection. For the woman in labor, the electronic fetal heart rate monitor can be disconcerting, if not frightening. It may be impossible to concentrate on Lamaze techniques while hooked up to a machine with flashing lights, heartbeat sounds, and "innocuous" decelerations to watch (Fig. 10.15).

Cesarean Sections

In the past decade, the substitution of cesarean section, an abdominal and uterine incision to remove the fetus, for vaginal delivery unquestionably increased in association with fetal heart rate monitoring. As reported in the March 27, 1978, issue of Time,

> Mrs. Lutz had been in labor less than an hour when monitors detected an irregular infant heartbeat and other signs of fetal distress. A difficult natural birth might have produced a brain-damaged or even stillborn baby. So her doctor performed a cesarean section, safely lifting out a six-pound girl. Says Mrs. Lutz, "A scar is a small price to pay for a healthy baby."

In 1970 only 5% of deliveries nationwide were cesareans; by 1981 the number had nearly quadrupled, to 17.9%. In some areas, one in three to one in four births is by a surgical delivery. A number of explanations have been suggested for the precipitous increase. Fetal distress, as detected by means of increased electronic heart rate monitoring, is seen as a valid reason for emergency cesarean surgery, although some authorities claim that there is little need for the C-section rate to rise if EFM is used correctly and fetal blood sampling is done to confirm fetal distress. Currently, virtually all breech presentations are delivered by cesarean before labor starts in order to avoid any risk of fetal asphyxia. Although most older physicians were trained in the delivery of breech babies, many younger doctors, having been instructed during their medical education or residency to handle breech cases by cesarean, lack experience in vaginal delivery. Forceps deliveries too, because of increased risk to the fetus, are often avoided in favor of surgical delivery.

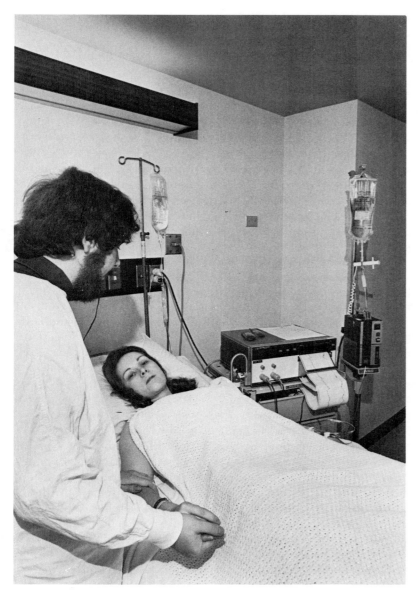

Figure 10.15. Internal fetal heart rate monitoring. One electrode from the machine is attached to the fetal scalp to record heart rate changes while the other electrode records the uterine pressure changes during contractions.

An additional factor contributing to greater incidence of cesarean sections is the older age of first-time mothers, those who have delayed childbearing and, hence, are seen as being at greater risk for complications after age 30. The cesarean rate has increased even more for women under 30, however, and may lead to even greater numbers in the future because of the dictum, "once a cesarean, always a cesarean." Fortunately, that prevailing policy has been challenged in the last few years, and vaginal delivery after cesarean section for selected cases is gaining in popularity. A review of studies from 1950 to 1980 (Lavin et al., 1982) presented data to show that vaginal delivery is a safe alternative to a C-section in 74.2 percent of women for whom there was no recurrent indication for a cesarean delivery after the first time and even a safe alternative in 33.3 percent of women whose previous cesarean was the result of dystocia (fetal pelvic disproportion, difficult and prolonged labor).

Some cesarean sections are performed because of the obstetrician's fear of malpractice suits, that is, as a precautionary measure if there is any indication of fetal distress. And finally, that any financial considerations play a role in higher C-section incidence is lamentable, but the facts are that the number of obstetricians has increased while the birth rate has decreased, the fee for surgical delivery is higher than that for vaginal delivery, and the hospital stay is twice as long for a cesarean. As further evidence that monetary incentives may be a factor, a recent study reported that the rate of cesarean section delivery is highest in the Northeastern part of the United States, in hospitals larger than 500 beds, also in private for-profit hospitals, when Blue Cross or other insurance is the source of payment, and for mothers aged 35 years or over. The C-section rate is lowest in the North Central area of the country, in hospitals with fewer than 100 beds, in government nonprofit hospitals, for patients who lack health insurance, and for

teenage mothers (Placek, Taffel, and Moien, 1983).

There are circumstances under which a cesarean section is unquestionably necessary, but there is also evidence that the escalating trend is the result of somewhat less justifiable reasons—a situation not in the best interests of mother and babies. A C-section is significantly more risky than a vaginal delivery. The incision may be a cleverly placed "bikini cut" just above the public hairline, but it is nevertheless an abdominal incision. While many of the hazards of other abdominal surgery are quite minimal in a cesarean birth, the dangers of anesthesia, hemorrhage, and infection are still present. Almost half of the 4,000 cesarean patients in one 16-year study had one or more complications, some severe enough to compromise future pregnancies (Hibbard, 1976). Kettering and Wolter's 5-year study (1977) of 547 cases showed an overall complication rate of 22%, including urinary tract infections, pneumonia, and pulmonary embolism. According to the report of a governmental Task Force on Cesarean Childbirth, the risk of maternal death is four times greater with surgical delivery than with vaginal delivery, with reported ranges from two to 27 times the risk.

A decline in the infant mortality rate has paralleled the rise in the cesarean birth rate in the United States, a statistic that could encourage complacency with regard to the increased surgeries. But, as indicated previously, there is little evidence that the decrease can be attributed to the C-section rate. Similar declines in fetal mortality have taken place in Europe but without the accompanying increase in cesareans. O'Driscoll and Foley's study (1983) showed that at the National Maternity Hospital in Dublin between 1965 and 1980, the cesarean delivery rate remained virtually unchanged at round 4.5%, while perinatal mortality fell from 42.1 per 1,000 infants to 16.8 per 1,000 infants.

Because the reasons for a delivery by cesarean section have evidently expanded beyond their previous indications, the surgery is conceivably a possibility for every pregnant woman. It should, therefore, be discussed with the physician well in advance of delivery. A woman should know her doctor's criteria for emergency cesarean birth and should be aware of all the ramifications. She should get an explanation of the procedure that would be used, the anesthetic, the aftereffects, whether she could have her husband or partner in the operating room with her, and whether subsequent pregnancies would also require abdominal surgery. It may be that women have little choice but to rely on the physician's professional judgment about the necessity for the operation, but they do have the right to participate in relevant decisions concerning the procedure (Fig. 10.16).

Diagnostic Ultrasound in Obstetrics

Ordinary sound, the kind that sets our eardrums to vibrating, travels through space in the form of energy waves (actually compressions and rarefactions of air molecules). The number of times, or cycles per second, a wave is repeated is the frequency of sound, and the greater the frequency, the higher the pitch of the sound. Audible sound ranges in frequency from 16,000 to 20,000 cycles, or hertz. Ultrasound is similar to audible sound except that its frequency is ultrahigh—in the millions of cycles or megahertz—and thus beyond the range of human hearing.

Ultrasound has had medical uses for many years. Ultrasound scalers are used by dentists to remove calculus from teeth, and, therapeutically, ultrasound waves of very high intensities (diathermy) produce a penetrating deep heat to treat muscle and joint injuries. But it is diagnostic ultrasound scanning for examination of all parts of the body that has become the greatest area of use in the past decade. The procedure can help detect abnormality

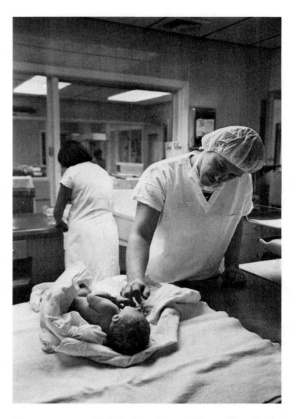

Figure 10.16. This infant was delivered by cesarean section, but the mother, under spinal anesthesia, was awake during the procedure. The father was present in the operating room, both parents were able to touch and hold the baby, and the father carried the infant to the nursery.

or injury from head to toe, and ultrasound scan has joined computerized axial tomography (CAT) and nuclear magnetic resonance (NMR) modality for "imaging" or diagnostic visualization of internal organs.

Actually, ultrasound scanners are not particularly expensive or very sophisticated instruments and have, thus, become available to any physician; they are especially popular with obstetrician-gynecologists for office use. The device consists of a small hand-held transducer or scanner that converts electrical energy into sound waves that travel into the body of the person being scanned. As the waves, emitted in the form of pulses or short bursts, encounter blood, bones, and organs of different densities, they are reflected back toward the source. These bounced-back echoes of sound waves are shown as a pattern on a viewing screen where they can be "read" immediately or stored and photographed for interpretation. Two kinds of imaging devices are used: the *static B-scan*, that produces a two-dimensional cross section image of body structures, and the *real-time* or dynamic ultrasound scanner, that made its first appearance in 1978 and can display moving parts like a motion picture. The Doppler ultrasound scanner, used to detect fetal heart rate, employs a transducer that emits continuous, rather than pulsed, sound waves. The echoes from moving parts are then recorded as audio signals.

For Ob/Gyns, ultrasound has dramatically increased diagnostic capabilities. It is a valuable adjunct to the detection of breast disease. In the pelvis, it reveals the presence or can confirm a suspected pelvic mass and can tell whether it is solid or cystic or how large it is. Ultrasonography can help detect ectopic pregnancy, show the number and extent of uterine myomas, localize an intrauterine device, evaluate the extent of inflammatory disease, and, by visualizing the follicles, determine the time of ovulation in an infertile woman who has been given ovulatory drugs. The applications to obstetrics are being perceived as even broader, and there is no doubt that when there are definite clinical indications for the use of ultrasound, the benefits are evident. A very significant advantage is that sonography, conducted at the appropriate time in gestation, can assist in accurately dating the age of the fetus. Knowledge of gestational age is important for amniocentesis, for determining whether preterm labor

should be stopped with drugs or allow the delivery to take place, and to assess the status of an alleged post-due-date fetus before induction of labor. Dating the fetus is absolutely essential when repeat cesarean is scheduled. Several studies have shown that a number of cases of respiratory distress syndrome, greater in premature infants, can be directly attributed to a too-early C-section, and that it is not unusual for babies delivered by surgical birth to weigh less than $5\frac{1}{2}$ pounds (Sabbagha, 1983). Ultrasound can also confirm suspected multiple pregnancy, evaluate the reason for bleeding in pregnancy, guide the needle during amniocentesis, follow fetal growth patterns if intrauterine growth retardation is suspected, and determine fetal abnormalities, such as brain and spinal cord defects, heart, gastrointestinal, or skeletal anomalies, or kidney or bladder problems when there is suspicion that such defects may exist. (Fig. 10.17).

Although there are few arguments by physicians against sonography when there are definite indications for its use, it is the routine application of ultrasound during pregnancy that has become problematic and controversial. Each year there are 3.5 million pregnant women in the United States; currently, about one-third of them receive at least one ultrasound examination, and the number is growing. Increasingly, obstetricians have a real-time ultrasound scanner in the office that is used not only for appropriate indications in pregnancy, but also defensively by the physician against malpractice, and indiscriminately to "show the baby moving" to all their pregnant patients. It has even been suggested that ultrasound visualization of the unborn child enhances bonding, presumably on the shaky premise that the earlier the "adhesion," the firmer the attachment.

There are several difficulties with progression toward a routine ultrasound screening of all pregnancies. One is that not all physicians are trained sonologists, and there are only a

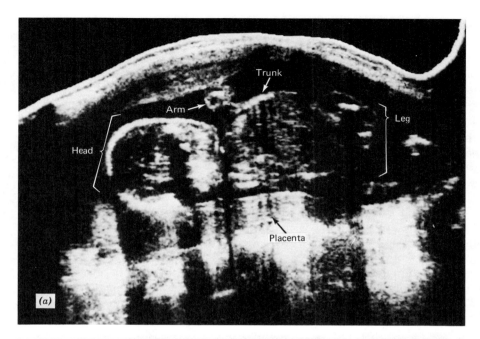

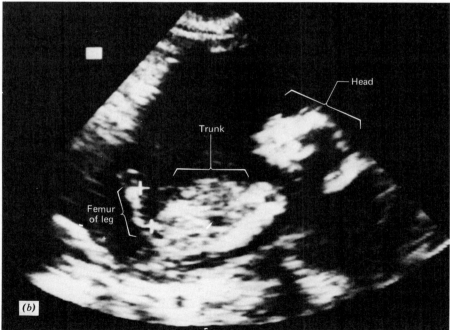

Figure 10.17. Ultrasonography of normal fetuses. (A) is a B-mode scan of a 25-week fetus. The mother's abdomen is outlined at the top of the picture. (B) is a real-time scan of a 15-week fetus. The distance between the two crosses at the left of the picture measures fetal femur length. The biparietal diameter of the head and the femur length are used to predict gestational age. Dating the fetus is most accurate between weeks 14 and 26. (Courtesy, Mt. Sinai Medical Center, Milwaukee.)

very small number of experts capable of detecting every possible fetal abnormality. Acquiring the skill to use the machine appropriately and interpret the results accurately requires time and experience (and, as some have admitted, a great deal of imagination). Moreover, the potential for false-positive and false-negative diagnoses is great. For example, at 20–24 weeks gestation, more than one-third of all fetuses lie in abnormal positions; at term, only 3%–4% are breech presentations. Similarly, at 20–24 weeks, one out of five placentas is low lying or appears to be placenta previa; at term, the incidence of placenta previa is only 0.5%. Even gestational dating, one of the most frequent reasons for ultrasound, is subject to pitfall. The correlation of the measurement of the fetal head, or the biparietal diameter (BPD), with fetal age is most precise only between 15 and 25 weeks of pregnancy, when the range of error is merely plus or minus 10 days. Before that time, between 8 and 13 weeks of pregnancy, a better dating can be obtained by measuring the fetal crown-rump length, which can pinpoint the age of the fetus back to the date of onset of the last menstrual period within 5 days. But beyond 26 weeks, a BPD measurement predicts fetal age within a range of error of about 3–5 weeks, with some small increase in accuracy with the addition of other measurements, such as long bone (femur) length and abdominal girth.

Another argument against routine ultrasound scans of all pregnancies is the lack of cost-effectiveness, except perhaps for the clinician who has paid for the machine and is effectively trying to recoup the cost.

As pointed out in a technical bulletin on ultrasound issued by the American College of Obstetricians and Gynecologists (ACOG), "no well-controlled study has yet proved that routine scanning of all prenatal patients will improve the outcome of pregnancy" (1981). Since some health insurance companies refuse to pay for routine screening procedures, ultra-sound performed for other than indicated reasons is not only ill-advised but expensive.

By far, the major argument against prenatal diagnostic ultrasound in *all* pregnancies is the matter of safety. As yet, there is no evidence of any dangerous effects of ultrasound to humans, but neither is there sufficient evidence to state without qualification that it is completely harmless. Ultrasound energy is not ionizing radiation, like X-rays, with known harmful effects. There are, however, at least two mechanisms of action by which ultrasound could cause biological damage to tissues. One is the production of heat, but this is apparently minimal in diagnostic ultrasound equipment that has high frequency levels and low intensity (in contrast to ultrasound diathermy machines where heat production is desirable and results from emission of low-frequency ultrasound waves with high intensity levels). A second mechanism is cavitation, a phenomenon that refers to the formation of gas microbubbles that form, increase in size, and burst in the tissue in response to sound waves. Although damage in insect eggs and some mammalian tissue has been produced by ultrasound induced cavitation, the implications for human tissue are unknown.

In cultured mammalian cells, there are reports that exposure to ultrasound induced abnormalities in growth and cell division, and there are other data reporting neurological, behavioral, and vascular changes in mice, rats, rabbits, dogs, and monkeys. In humans, one study showed that exposure to Doppler ultrasound induced increased fetal activity, and another study reported an increase in chromosome breaks when fetal cells cultured from amniotic fluid were exposed. Neither study has been considered statistically significant or confirmed by other investigators, but uncertainty about harmlessness remains. One could worry about elusive effects on the developing nervous system produced by sound waves bouncing against immature neurons, for example, or be concerned about the gonads, especially in a female fetus, receiving radiation energy that could have disruptive intracellular effects. No one has reported any evidence of obvious, short-term detrimental consequences of prenatal ultrasound exposure in children, but this should not allay concern about any cumulative or subtle effects that may potentially occur. One retrospective long-term follow-up study of 800 children, half of whom were exposed to diagnostic ultrasound between 1968 and 1972, is currently underway. An additional investigation, begun in 1977 in Canada, includes 10,000 women and their exposed offspring, 1,000 unexposed siblings, and 2,000 control children to be followed and compared for 20 years or more (Bolsen, 1982). Thus far, the results concerning miscarriage, congenital abnormalities, communicative disorders, and cancer in the exposed children are reassuring, but it will be some time before the elements of risk that surround the use of obstetrical ultrasound can be completely dissipated.

The Food and Drug Administration, the American College of Obstetricians and Gynecologists, the American Institute of Ultrasound in Medicine, the American College of Radiologists, and a panel of experts composed of 14 professionals from the fields of clinical medicine, scientific research, and law that was convened by the National Institutes of Health, have all taken a conservative attitude toward ultrasound in pregnancy and urge that it should be used only when medically indicated. The last-mentioned group, who had reviewed the literature and research on ultrasound for almost a year before their report at a conference in 1984, found many of the studies that had assessed the safety of ultrasound in humans to be inadequate and recommended further research to answer the unresolved questions. A woman may want to remind her Ob/Gyn of these positions should routine application of ultrasound be suggested.

Prenatal Therapy

A new kind of obstetrical practice is emerging, an aggressively technological kind of prenatal, natal, and antenatal practice. Like an interventionary chain reaction, one new obstetrical advance will generate another, and another, and yet another until some incredible ultimate in intervention will be reached. This pinnacle may already have been achieved with the introduction of fetal therapy. Now it is not only the pregnant woman who is a patient in a medically managed gestation, labor, and delivery, but the fetus is also a patient, to be diagnosed and treated while still in the uterus. After the development of real-time ultrasound "imaging" of the moving parts of the fetus, the two latest technological marvels are fetoscopy, or looking at the fetus through a special telescope, and fetal surgery, operating or in some way repairing the fetus before birth.

The first use of the term "fetoscopy" was in the early 1970s. It describes the technique of directly viewing the fetus within the uterus by means of a specially designed fiberoptic telescope. Although the field of vision through the fetoscope is not large, it is possible to see directly such anatomical abnormalities as limb and finger defects or facial clefts, undiagnosable by amniocentesis but suggested by an ultrasound examination. The procedure has also been used to provide fetal therapy, such as giving an intrauterine blood transfusion to a fetus with known erythroblastosis fetalis (severe Rh incompatability)—treatment that has generally been given after birth. The other major use of fetoscopy is for fetal biopsy, the obtaining of samples of skin or blood from the umbilical cord for diagnosis of suspected various types of genetic defects that cannot be detected by analysis of amniotic fluid cells.

Fetoscopy is a risky business; figures from the second International Congress on Fetoscopy in Athens in 1981 indicated that pregnancy loss through infection or miscarriage is 16% after the visualization procedure and 11% after tissue biopsy. The technique is currently performed to confirm or exclude a genetic defect found through amniocentesis or a congenital anomaly picked up by ultrasonography. It is rarely employed as the primary prenatal diagnostic method. According to Filkins and Benzie (1983), who reported that 580 fetoscopies were performed "for research and training purposes" on women having midtrimester abortions at Toronto General Hospital, the best time for viewing fetal parts is 18 weeks.

The expansion of the ability to diagnose prenatally certain kinds of fetal abnormalities through such techniques as ultrasonography and fetoscopy has also expanded the possibilities of what to do with the information. Until the early 1980s, the only options after detection by amniocentesis of chromosomal or metabolic defects were abortion or continuing with the pregnancy. Not that such decisions are ever easy, but with the appropriate and necessary genetic counseling, women who choose to have amniocentesis do so with a full understanding and recognition of the potential consequences of the procedure. After it became possible to diagnose some kinds of anatomical malformations while the fetus was still in the uterus, the choices have become more complicated. If the anatomical defect is of the type that does not appear to compromise the continued growth and development of the fetus, surgical correction could be delayed until after delivery. But, in the opinion of the pediatric surgeons who have pioneered in the prenatal treatment of the fetus, there are currently three correctable structural anomalies that encroach on subsequent development and do not allow it to proceed normally. These defects, potential candidates for surgical correction while the fetus is still in the uterus, are diaphragmatic hernia, in which the intestines push up through a hole in the diaphragm and compress the lungs; hydronephrosis, in which a blockage in the fetal urinary tract causes a backup of urine into the abdomen; and hydrocephalus, or a buildup of cerebrospinal fluid in the fetal brain as a result of obstruction to the normal flow, that compresses the developing nerve cells and can cause severe mental retardation. In 1981, after 20 years of research in perfecting the techniques in sheep and monkeys, the first surgical correction of hydronephrosis was performed on a 24-week-old fetus by making an incision into the uterus, operating on the partially removed fetus to alleviate the urinary tract obstruction, and then replacing the fetus who continued on to a nearly term delivery. Unfortunately, this medical milestone was a classic instance of "the surgery was successful, but the patient died." The baby died after birth because of other multiple abnormalities in addition to the kidney problem. The feasibility of prenatal treatment had been established, however, and a year later, Drs. Harrison, Filly, and Golbus of the University of California at San Francisco, who chaired a conference entitled Unborn: Management of the Fetus with a Correctable Congenital Defect, reported on experiences in fetal treatment derived from 13 centers in five countries. Correcting the obstructed urinary tract with a drainage catheter had been attempted in 21 fetuses, and placing a brain shunt to relieve hydrocephalus had been attempted in 8 fetuses, but the results were varied. Of the 21 fetuses with hydronephrosis, 10 survived with good kidney function. It has subsequently been determined that many fetuses with urinary tract blockages unexplicably undergo spontaneous correction on their own, so it is not known whether the interventionary treatment actually influenced the outcome. Of the eight fetuses with hydrocephalus, six survived, but knowing for certain whether they have any evidence of retardation requires long-term follow-up. Correction of a diaphragmatic hernia is the most difficult of the procedures and at that point had not as yet been tried on a human fetus.

Two years after the first surgery on an unborn human had been performed, it was evident that the initial expectations for prenatal treatment were overly optimistic. One of the leaders in developing the techniques cautioned that "merely because a procedure can be done does not mean that it should be done" (Harrison, 1983). Moreover, the moral implications of the procedure vis-à-vis maternal rights and fetal rights became a quagmire that was explored by both physicians and philosophers.* The future of fetal therapy is somewhat uncertain now, but if past history can be relied upon, we can anticipate that medical marvels, even those in the dubious achievement category, will continue to be advanced.

BIRTHING ALTERNATIVES

To escape what they believe is unnecessary intervention in a normal physiological event, some couples prefer to avoid hospital delivery completely. As an alternative to what they perceive as the dehumanizing, impersonalizing, unnatural hospital setting, and possibly because of the expense or because they may already have had one unsatisfactory experience in the hospital, many women are choosing to deliver their babies at home. The birth attendant may be a lay midwife, trained through apprenticeship and currently sanctioned in 13 states; a registered nurse-mid-

wife, one of the 2,000 nationwide and licensed in 47 states, or a physician. A woman who has a home birth is relaxed, in her own environment and in her own bed, in the birth position most comfortable for her, surrounded by friends, relatives, and, by family decision, older children who can see their brothers and sisters being born. Those who have experienced a home delivery are highly enthusiastic advocates; they believe that the participation of the entire family adds an invaluable psychological dimension to the birth process, which is ideal. (Figure 10.18A–L shows a birth at home, attended by midwives.)

No matter how psychologically satisfying and advantageous home births can be, if they are medically more dangerous, they should not electively take place. Most traditional obstetricians take a dim view of home deliveries. They consider them an irresponsible risk to the lives and health of women and their babies, believing that only in a hospital setting can the necessary standards of safety be met. The American College of Gynecologists and Surgeons reported that it had compiled information from 11 states on babies born at home and had determined that the mortality rate was substantially higher. The College flatly categorized home births and the use of midwives as a form of child abuse.

Of course, the ACOG study was based on statistics on all birth at home in those 11 states, including emergency deliveries and those not attended by any health professional at all. Owing to these inadequacies in data collection, it is probable that there is no real way of statistically determining the risks of home delivery or even how many out-of-hospital births are currently taking place. A more recent ACOG report claims that despite the vocal enthusiasm of home birth supporters and the publicity given to this alternative, the percentage of home deliveries has remained constant at around 1% in the United States since 1976. The recently formed Illinois-based

---

* See John C. Fletcher, "The Fetus as Patient, Ethical Issues" in *JAMA*, August 14, 1981, and William Ruddick and William Wilcox, "Operating on the Fetus, ," in the Hastings Center Report, October, 1982.

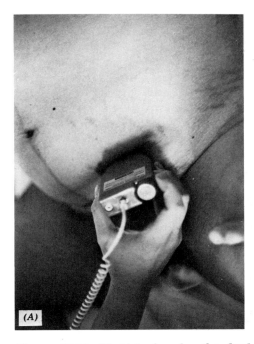

Figure 10.18. (A) Listening for the fetal heart tones.

American College of Home Obstetricians has denied that home deliveries are more dangerous than hospital deliveries when women are carefully selected for such births and a professional attendant is present. They claim that when women classified as high-risk are eliminated from the obstetrical population that participates in home deliveries and when a physician or a licensed midwife is present, home birth is just as safe as hospital birth, and probably safer.*

Other Options

Obviously, a birth at home, attended by a midwife, is not for every woman even if she has a normal labor and delivery. With the uneasy sense that even low-risk or no-risk deliveries

* Consumer resource groups on home delivery include National Association for Safe Alternatives in Childbirth, Box 1307, Chapel Hill, NC 27514 and Right to Home Birth, c/o Melodie Jarabek Rt. 2, Box 155, Denmark, WI 54208. Among professional groups on midwifery is the Midwives Alliance of North America, organized in 1982 to form a liaison between nurse-midwives and lay midwives. It has about 600 members.

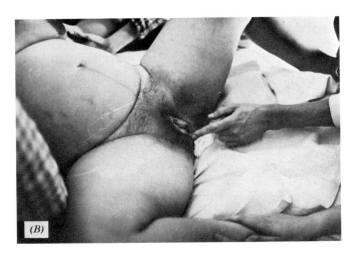

Figure 10.18. (B) Delivery takes place in a semireclining position, and the mother is supported by a birthing chair.

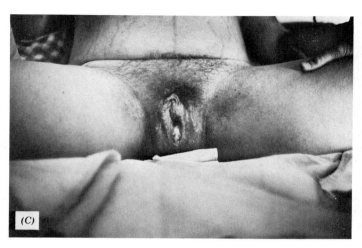

Figure 10.18. (C) Crowning of the head. The scalp is just visible.

Figure 10.18. (D) Easing out the head.

could suddenly become high-risk ones, some women would feel very uncomfortable unless they were in a hospital setting. Some women believe that they should have every advantage that modern medicine with its sophisticated technology can provide. There are some women, although the attitude is becoming more rare, who have little desire to participate or have more control in the childbirth process, and who are willing to have as much anesthesia as is necessary to have minimum discomfort, as long as it does not harm the baby.

For those who are dissatisfied with orthodox hospital-based obstetrics, but who may have some misgivings about delivering at home, some other options are beginning to be available. Not all obstetricians and hospitals have remained indifferent and inflexible to the not unreasonable requests that women have made. Some hospitals have changed to provide, if not the complete warmth and comfort of the home, at least a more humane, personalized, family-centered kind of maternity care. Obstetric departments have made efforts to change existing procedures and staff attitudes. Routine preparations for delivery that were standard for so long, such as shaving the

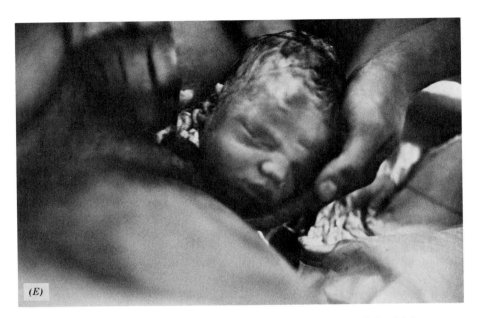

Figure 10.18. (E) *The head has rotated toward the mother's right thigh.*

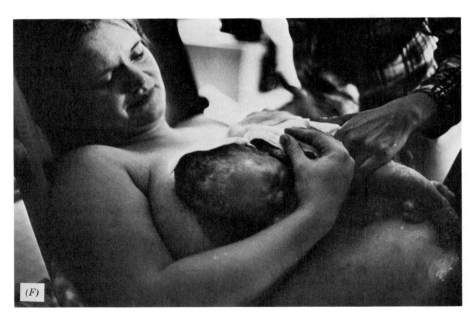

Figure 10.18. (F) *Immediately after delivery, the baby is placed on the mother's abdomen.*

374

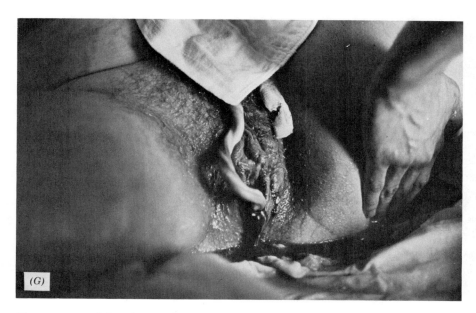

Figure 10.18. (G) Before delivery of the placenta.

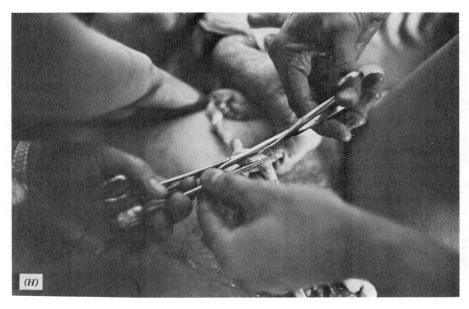

Figure 10.18. (H) The cord is clamped and cut by the father.

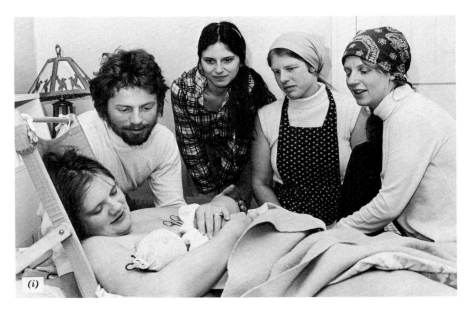

Figure 10.18. **(I) The three laymidwives look on as the father admires his son.**

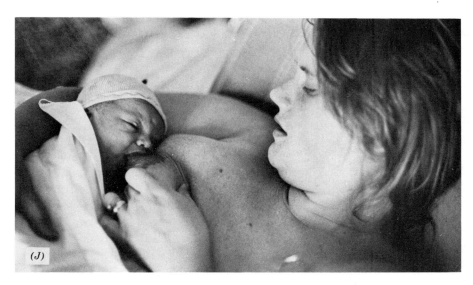

Figure 10.18. **(J) Suckling at the breast immediately after delivery stimulates uterine contractions, stimulates milk production, and is beneficial for infant–mother bonding.**

Figure 10.18. (K) *Placenta, fetal side.*

Figure 10.18. (L) *Placenta, maternal side.*

pubic area and giving enemas, have been eliminated. Physician interventions are kept to a minimum. Fetal monitoring may be performed only when required, and only intermittently even then. Labor rooms have been converted to "birthing rooms," pleasant and bedroom-like, where the woman, supported by her husband, partner, close friend, or relative, can labor, deliver, and recover without being transferred to a delivery room. When the baby is born, it is placed on the mother's abdomen, and the parents can touch and handle the infant. Several hours can elapse before the family transfers to the postpartum area, rooms that contain rocking chairs and other homey comforts. There is no rigid schedule to which mother and infant must conform, and for the most part, the woman may choose the type of care for herself and the baby that she wants, getting the advice of the staff when necessary.

Of course, pretty paint and rocking chairs are no guarantee that the traditional hospital and doctor obstetrical practices have really undergone any change to make them a more positive experience. But it is possible in some hospitals to have a midwife delivery. There are a few innovative hospitals that even encourage sibling visitation, not only to see Mama after delivery, but also to touch and hold the new baby.

There are some obstetricians who are willing to stand by and talk the prepared father through the delivery in much the same way that they would educate a first-year resident. And there are some doctors, although it may take some shopping around and word-of-mouth to find them, who are content to be birth attendants, not managers; who can be trusted to do, within the standards of medical safety, what the woman thinks is right for her; who see themselves not as central or authoritarian in the birth process, but as secondary to the main characters in the event, the mother and the child.

Right now there are still too few alternatives to what some women and health professionals see as the highly interfering, invasive, technical hospital deliveries, and those alternatives are found only in certain areas of the country. As more women make their demands known, however, more doctors and hospitals will change their present all-risk pathological approach to a normal and individualized view of childbirth. Perhaps the change in many instances will not be out of any sense of real commitment. But the economic pinch that results from the low birth rate and the renewed interest in midwifery and home delivery are likely to mandate a different concept of obstetrical care that considers the needs of the 90% of women who have normal deliveries.

BREASTFEEDING

For several generations, breastfeeding an infant has been seen as too restricting, inconvenient, lower-class, and suitable only for poor and uneducated women. New mothers who may have wanted to try it received very little instruction (the ability to nurse successfully does not always come naturally) and even less encouragement. It was assumed that few women wanted to breastfeed, and so they virtually routinely received an injection of an androgen-estrogen combination (usually, Deladumone OB, E. R. Squibb & Sons) to suppress lactation. Along with the recent movement toward nontechnological natural birth and with the recognition of its value for the newborn, however, there has been a resurgence of interest in breastfeeding. Now breasts are in, and bottles are out. Fewer than 25% of women were breastfeeding their babies in 1971; 10 years later, 45%–60% of women leaving the hospital after delivery said that they were breastfeeding, and the number is likely to be even greater by now (Figure 10.19).

Although breastfeeding has increased in the last decade in industrialized nations, it has declined in Third World countries with tragic public health effects. The illness of 10 million infants and the death of possibly a million babies a year have been attributed to inadequate formula feeding. There is nothing wrong with infant formula when there is money to purchase enough of it, sufficient education to prepare it properly, and appropriate hygiene to keep it uncontaminated. But when overdiluted formula is prepared from filthy water supplies, the result is starvation and sickness in babies. The aggressive marketing techniques of certain infant formula manufacturers in underdeveloped countries have generated enormous controversy, and some critics have called for a consumer boycott of all the products of one major infant formula company.

The American Academy of Pediatrics advocates breastfeeding for all full-term newborns, maintaining that ideally, breast milk should be the sole nutrient for the first 4–6 months of life. All available scientific evidence, and there have been scores of papers published in the last decade, overwhelmingly attests to the superiority of human milk for human babies. Cow's milk is undoubtedly ideal for calves, but as more is becoming known about the biochemical properties of human milk, it is becoming more evident that nature has designed it to be particularly suitable for the human species. Formulas made from cow's milk have been modified to make them more like breast milk, but even when cow's milk is made more "human," there are still some significant differences.

In comparison with cow's milk proteins, human milk proteins are completely digestible, provide all the essential amino acids, and are generally of greater biological value than the proteins in formula. Moreover, the proteins in cow's milk can cause allergic reactions, and circulating antibodies have been found in a majority of formula-fed babies even when they show no evidence of allergy, such as asthma, eczema, and diarrhea. Instances of unexplained crib deaths have been linked to anaphylactic, or very severe, allergic reactions to cow's milk (Parish et al., 1960).

The fat content in human milk is more digestible and allows greater uptake of fat-soluble vitamins A and D from the intestine. Human milk is unique in containing lactose as the principal carbohydrate. Lactose is the least sweet of sugars and is less likely to program the child for a future sweet tooth. The mineral content of human milk is less than that of cow's milk, which is not a disadvan-

Figure 10.19. Breastfeeding.

tage, since an infant's kidneys are not equipped to handle large loads of salt. Breast milk is always ready, at the right temperature, does not have to be pasteurized, and is not subject to the bacterial contamination highly probable in formulas. Breastfed newborns have a greater resistance to gastrointestinal disorders and to respiratory and ear infections than do formula-fed infants. The mechanisms that underlie the protective functions of human milk that result in the lowered infection rate are not completely known. Most antibodies to infectious diseases, such as scarlet fever, measles, and diphtheria, are transferable from the mother to the infant across the placenta before birth. How and to what extent the large molecule maternal antibodies contained in breast milk are able to survive digestion and pass through the infant's intestinal cells to get into the bloodstream to immunize the baby has not been determined. It is possible that maternal antibodies may function to inactivate organisms in the baby's intestinal tract so that they cannot cause infection. Whatever the mechanism, the anti-infective properties of human milk exist and are a major advantage that cannot be matched by cow's milk. The level of immune factors is highest in colostrum and milk produced in the first weeks of nursing.

Breastfed babies are not likely to be as fat as formula fed babies. The correlation between early obesity and overweight adults has not been proved, but there is some evidence that the seeds may be laid in infancy and childhood.

For the past 5 decades, researchers have attempted to measure the effects of breastfeeding on later intelligence, learning abilities, activity levels, personality, and adjustment of infants, but results have been inconclusive. Some data indicate an apparent behavioral advantage from breastfeeding, but there is equal evidence to indicate that no overall benefit exists. For many women, however, the

successful breastfeeding of their babies provides a secure, gratifying, and satisfying experience that contributes to the quality of mother–child interaction. There are some definite maternal physical benefits as well. The uterus involutes to its normal size more easily and rapidly, and there are fewer problems with uterine infection. It is much easier to reattain prepregnancy weight if a woman is nursing; the calories expended to produce milk help in shedding unwanted pounds. Nursing mothers are relatively infertile, since menstruation and ovulation return more slowly in women who lactate. The frequency and amount of nursing prolongs the amenorrhea, but is is also affected by body weight, being of shorter duration in heavier women. In many parts of the world where women are poorly nourished, breastfeeding is thus able to prevent a significant number of pregnancies. According to Vorheer (1974), half of the women who breastfeed are able to conceive within the first 6 months postpartum, while the other half seem to be protected, but, since there is no way of knowing in which 50% a woman may fall, lactation is no substitute for contraception.

The only possible health disadvantages to breastfeeding occur when breast milk contains substances that may adversely affect the infant. Although the levels of some substances smoked, drunk, breathed, or otherwise ingested by a woman may appear in breast milk in minuscule quantities, generally speaking, whatever she has in her blood stream can get to the infant through the milk supply. The baby, smaller in size and with immature kidneys and liver, is not as well able to handle or detoxify substances that are harmless to an adult. Some drugs are definitely known to cause health problems (Table 10.3); others, such as alcohol and caffeine, are considered safe in moderate amounts, but large quantities could result in responses in babies similar to those in adults. Few mothers would want

Table 10.3 SOME MATERNAL DRUGS KNOWN TO POSE POTENTIAL HEALTH PROBLEMS FOR THE BREASTFEEDING INFANT

Drug	Possible Symptoms or Effects on Infant	Drug	Possible Symptoms or Effects on Infant
Anticoagulant		Autonomic drugs	
Ethyl biscoumacetate	Bleeding problems	Atropine	Constipation
Phenindione	Bleeding problems	Laxative	
Anticonvulsant		Anthraquinone derivatives:	
Mysoline	Drowsiness	(Danthron, Dialose,	
Phenobarbital	Hypnotic effect	Plus, Dorbane,	
Phenytoin (diphenyl-hydantoin)	Methemoglobinemia	Dorbantyl, Doxidan,	
		Peri-Colace)	Bowel problems
Carbamazepine	Long-term effect unknown	Aloe	Bowel problems
		Calomel	Bowel problems
Antidepressant		Cascara	Bowel problems
Lithium	Loss of muscle tone, lowered body temperature, bluish skin	Narcotic	
		Heroin	Addiction
		Methadone	One death recorded
Antihypertension		Oral contraceptives	Gynecomastia, long-term effects unknown
Reserpine	Nasal congestion, weight loss, bluish skin		
Antimetabolite		Pain-killer	
Cyclophosphamide	Bone marrow depression	Propoxyphene (Darvon)	Addiction
Methotrexate	Bone marrow depression	Sedative	
		Barbiturates	Drowsiness
Antimicrobial		Bromides	Drowsiness
Chloramphenicol (Chloromycetin)	Refusal to nurse, sleepiness, vomiting	Chloral hydrate	Drowsiness
Metronidazole (Flagyl)	Causes cancer in test animals	Diazepam (Valium)	Lethargy, jaundice, weight loss
Nalidixic acid	Hemolytic anemia	Steroid	
Nitrofurantoin[a]		Prednisone	Poor growth
Sulfonamides[a]	Hemolytic anemia	Prednisolone	Poor growth
Antithyroid		Miscellaneous	
Iodide	Altered thyroid synthesis and release	Dihydrotachysterol	Renal calcification
		Ergot alkaloids	Ergotism
Thiouracil	Altered thyroid synthesis and release	Gold thioglucose	Rash, hepatitis, hematologic alteration
Radioactive iodine	Cancer, loss of thyroid activity		

Source: From Packard, V. S., *Human Milk and Infant Formula.* New York: Academic Press, Inc, 1982. Used with permission.

[a] This drug causes problems mainly in infants suffering the inherited enzyme deficiency glucose-6-phosphate dehydrogenase.

the possible nervousness, wakefulness, and irritability from excessive caffeine intake to be present in their infants. A lactating woman, like a pregnant woman, has to be careful of what she takes into her body.

One source of contamination of mother's milk over which women have little or no control is environmental pollutants. Pesticides, heavy metals, antibiotics, and all sorts of organic industrial wastes eventually find their way into food, and a woman has to eat. Industrial pollutants, such as polybrominated biphenyls (PBBs), polychlorinated biphenyls (PCBs), pentachlorophenols (PCPs), and dioxins, a contaminant of PCPs, are so ubiquitous in the food and water supply that it is impossible to avoid them no matter where one resides, although higher levels have been found in breast milk from women who regularly consume fish from contaminated waters. These organic toxic chemicals are stored in the body fat, resist breakdown by usual body detoxifying enzymes, and are excreted only through breast milk. That means that nursing mothers would have lower body levels of contaminants than other people, but their babies are exposed—a no-win situation. The role of such environmental contaminants in producing harmful effects in children is unknown, but that is primarily because it is virtually unstudied. Lactating women have been advised to have their breast milk analyzed for PCBs and PBBs if they work on a farm, have frequently eaten large quantities of fish from contaminated waters such as Lake Michigan and Lake Superior, or are otherwise exposed occupationally. Some states provide free testing; private laboratories may charge $60 and up for the analysis. Although some advisory "tolerance" levels for the contaminants in milk have been adopted, the results of breast milk analysis actually mean little because the long-term effects of "high," "moderate," or "low" levels of the chemicals, if any, are still unknown. Besides, it can take weeks before

the test analysis is returned, and what is a woman to do in the meantime? Lactation cannot be turned off and on again like a faucet. Breast milk analysis would appear to be of very limited value.

No women should avoid breastfeeding solely because our food and water supply contains toxic hydrocarbons. Cow's milk has them too, although in lesser amounts. It does make sense to avoid the only major dietary source of PCBs—sport fish from the Great Lakes—and to avoid making a concerted effort to lose a lot of weight quickly, which could release the chemicals suddenly from their storage site in fat tissue. What we all can and must do is persist in our efforts to have these potential poisons, as well as all other pollutants, eliminated from our environment.

All in all, there are economic, biochemical, psychological, and physiological advantages to breastfeeding, and a woman could at least try it. It is important to remember, however, that the emotional benefits are highly subjective, and that the physiological benefits are based on statistics. The decreased incidence of infection, disease, and allergy for babies fed with breast milk is not necessarily going to be evident for any individual infant on human milk or on formula. It should be remembered that much of the back-to-breastfeeding movement has been promoted by white, affluent, educated, and for the most part, nonworking mothers who are in a position to nurse their infants for the better part of a year. Women who, by choice or necessity, are unable to breastfeed—if, for example, they have to return to work right after delivery, or have a medical problem that requires taking large doses of drugs—should never feel that they have shortchanged their babies by bottlefeeding. Human milk is ideal for human infants, but if a baby must be fed with formula, there are commercial preparations that have been modified until they resemble human milk as closely as possible. Certainly a woman who chooses to or has to formulafeed her infant can develop a mother–child relationship that is just as close as if she breastfed.

Many women who are highly motivated to breastfeed can still have trouble with lactation unless they get the instruction, skill, and encouragement necessary for success. If they have initial difficulty in getting the baby started, or run into a problem with engorged breasts, leaking, or sore nipples, they may give up too easily and switch to formula. An inability to breastfeed is sometimes the result of anxiety and stress, which inhibit the milk-ejection reflex. Ordinarily, the stimulus of sucking at the breast causes the release of oxytocin, which results in the contraction of the alveoli of the glands and the ejection or "letting-down" of the milk into the milk ducts. Oxytocin, released from the neurohypophysis, is under direct neural control. Anxiety about breastfeeding, especially when accompanied by inadequate instruction and lack of encouragement, can readily inhibit the ejection reflex.

The support and information that many nursing mothers need can be provided by the La Leche League. Founded in 1956 by two women who had the idea of forming an organization where nursing mothers would inform and advise other nursing mothers, the League today has grown to more than 3,000 chapters throughout the world. In their manual, *The Womanly Art of Breastfeeding*, interested women can obtain the practical knowledge that is usually unavailable elsewhere. If there is a chapter in the area, La Leche League is listed in the telephone directory, and group meeting and individual counseling is provided.

Induced Lactation

Induced lactation, also called relactation, is the ability to produce breast milk in response to sucking stimulation when there has been no preceding pregnancy. Elizabeth Hormann, a pioneer of the relactation movement, has reported the results of a survey of 65 women who wanted to provide the psychological and physiological benefits of breastfeeding to their adopted babies. Eighteen of the women had never been pregnant, seven had been pregnant but had not nursed, and forty had been pregnant and had lactated before. The majority of the infants received by the adoptive mothers were under 1 month of age, but some of them were already receiving some type of solid food as well as formula when they arrived. All 65 women were successful in producing milk, although all but one had to supplement with formula since in most of the women the mammary gland production was not enough to fully sustain the baby. The act of nursing was evidently seen as having as much or greater value as the nourishment derived from breastfeeding (1977).

Producing breast milk without a pregnancy seems amazing to most people. It does require an inordinate amount of preparation and motivation, but is not that mysterious physiologically. It is known that it is possible to induce lactation in nonpregnant farm animals by milking. In humans, it has been shown that when the breasts and nipples are manually stimulated, or sometimes as a result of chest injury, surgery, or shingles (herpes zoster), lactation can result. Sensory nerve impulses from the breasts are relayed to the spinal cord and then up to the hypothalamus of the brain. The prolactin-inhibiting factor of the hypothalamus that ordinarily keeps lactation from occurring is suppressed, and the prolactin from the anterior pituitary gland can then be released to result in milk synthesis and production by the alveoli of the mammary glands (see Chapter 7). According to a study by Kolodny and co-workers (1972), manual breast and nipple stimulation in women resulted in an increase in serum prolactin levels within 5 minutes of self-stimulation or stimulation by the husband. Women who want to nurse adopted babies require a month or more of preparation through breast stimulation. Sometimes hormones can be taken to increase milk production. Khojandi and Tyson (1973) showed that 5 mg of oral estrogen plus nipple stimulation for 14 consecutive days resulted in prolactin hypersecretion and lactation in healthy nonpregnant women. Most of the women in Hormann's study produced milk without taking hormones, indicating that supplementation by drugs is not necessary.

Although nursing an adopted baby took a great deal of preparation beforehand and a great deal of time, since a large part of each day must be spent in nursing, the women in Hormann's study believed it was well worth the trouble.

As a further indication that with enough motivation, hormones, and help, *anyone* can nurse a baby, a Brooklyn doctor reported that his treatment of a 40-year-old male enabled that man to successfully breastfeed his infant daughter. The unidentified individual, a married transvestite who shared everything with his wife, including clothing and makeup, wanted to share equally in the raising of their child. After the man had received estrogen to develop his breasts before the birth of the baby, after the birth, the physician administered oxytocin to result in prolactin secretion. The man was then able to split the breastfeeding duties with his wife for 3 months. But probably not in public.

REFERENCES

Abel, E. L. Consumption of alcohol during pregnancy: A review of effects on growth and development of offspring. *Hum Biol* 54(3):421–453, 1982.

ACOG Technical Bulletin Number 63. Diagnostic ultrasound in obstetrics and gynecology, American College of Obstetricians and Gynecologists, October, 1981.

Aladjem, S. (ed.). *Obstetrical Practice.* St. Louis, C. V. Mosby, 1980.

Aladjem, S., Lueck, J., and Brewer, J. Experimental induction of a toxemia-like syndrome in the pregnant beagle. *Am J Obstet Gynecol* 145:27–38, 1983.

Allen, M. S., Schwingl, P. J., Rosenberg, L., et al. Birth defects in relation to Bendectin use in pregnancy. *Am J Obstet Gynecol* 147(7):737–742, 1983.

Aselton, P. J. and Jick, H. Additional follow-up of congenital limb disorders in relation to Bendectin use. *JAMA* 250:33–34, 1983.

Beernink, F. J. and Ericsson, R. J. Male sex preselection through sperm isolation. *Fertil Steril* 38(4):493–496, 1982.

Bolsen, B. Question of risk still hovers over routine prenatal use of ultrasound. *JAMA* 247(16):2195–2197, 1982.

Brazelton, T. B. Effect of prenatal drugs on the behavior of the neonate. *Am J Psychiatry* 126(9):1261–1263, 1970.

Cohen, E. N., Gift, H. C., Brown, B. W., et al. Occupational disease in dentistry and chronic exposure to trace anesthetic gases. *JADA* 101:21–31, 1980.

Conner, B. H. and Seaton, P. G. Birth weight and use of oxytocin and analgesic agents in labour in relation to neonatal jaundice. *Med J Aust* 2:466–469, 1982.

Cordero, J. F. and Layde, P. M. Vaginal spermicides, chromosomal abnormalities and limb reduction defects. *Fam Plann Perspect* 15(1):16–18, 1983.

Cordero, J. F. and Oakley, G. P., Jr. Drug exposure during pregnancy: Some epidemiologic considerations. *Clin Obstet Gynecol* 26(2):418–428, 1983.

De Chateau, P. Parent–neonate interaction and its long-term effects, in Simmel, E. G. (ed.). *Early Experience and Early Behavior.* New York, Academic Press, 1980.

Diasio, R. B. and Glass, R. H. Effects of pH on the migration of X and Y sperm. *Fertil Steril* 22:303–305, 1971.

Dilts, P. V., Greene, J. W., and Roddick J. W. *Core Studies in Obstetrics and Gynecology.* Baltimore, Williams & Wilkins, 1974, p 117.

Draft Report of the Task Force on Cesarean Childbirth, U.S. Department of Health and Human Services. PHS, National Institutes of Health, Washington, D.C., 1980.

Eskenazi, B. and Bracken, M. B. Bendectin (Debendox) as a risk factor for pyloric stenosis. *Am J. Obstet Gynecol* 144(8):919–924, 1982.

Ettner, F. M. Hospital technology breeds pathology. *Women & Health* 2(2):17–22, 1977.

Fabro, T. and Sieber, C. Caffeine and nicotine penetrate

the preimplantation blastocyst. *Nature* 223:410–411, 1969.

FDA Drug Bulletin, Warning on use of sex hormone in pregnancy. January–March, 1975.

Filkins, K. and Benzie, R. J. Fetoscopy. *Clin Obstet Gynecol* 26(1):339–345, 1983.

Forfar, J. and Nelson, M. M. Epidemiology of drugs taken by pregnant women: Drugs that may affect the fetus adversely. *Clin Pharmacol Ther* 14:632, 1973.

Gau, G., Bhundia, J., Napier, K., and Ryder, T. A. The worm that wasn't. *Lancet* 1:1160–1161, 1983.

Golding, J., Vivian, S., and Baldwin, J. A. Maternal antinauseants and clefts of lip and palate. *Hum Toxicol* 2:63–73, 1983.

Guerrero, R. Type and time of insemination within the menstrual cycle. *Stud Fam Plann* 6:367–371, 1975.

Haire, D. The cultural warping of childbirth. International Childbirth Education Association, revised. 1977.

Harlap, S. Gender of infants conceived on different days of the menstrual cycle. *N Engl J Med* 300(26):1445–1448, 1979.

Harlap, S. and Eldor, J. Births following oral contraceptive failures. *Obstet Gynecol* 55(4):447–452, 1980.

Harrison, M. R. Perinatal management of the fetus with a correctable defect, in Callen, P. W. (ed.). *Ultrasonography in Obstetrics and Gynecology.* Philadelphia, W. B. Saunders, 1983.

Harrison, M. R., Filly, R. A., Golbus, M. S., et al. Fetal treatment 1982. *N Engl J Med* 307(26):1651–1652, 1982.

Haverkamp, A. D. and Orleans, M. An assessment of electronic fetal monitoring. *Women's Health* 7(3):115–133, 1983.

Hibbard, L. T. Changing trends in cesarean section. *Am J Obstet Gynecol* 125:798–804, 1976.

Hill, R. M., Craig, J. P., and Chaney, M. D. Utilization of over-the-counter drugs during pregnancy. *Clin Obstet Gynecol* 20(2):381–394, 1977.

Hill, W. C. How to interpret antepartum monitoring. *The Female Patient* 7(1):32/1–32/13, 1982.

Hormann, E. Breast feeding the adopted baby. *Birth and the Fam J* 4(4):165–173, 1977.

Huggins, G., Vessey, M., Flavel, R., et al. Vaginal spermicides and outcome of pregnancy: Finding in a large cohort study. *Contraception* 25(3):219–230, 1982.

Jick, H., Walker, A. M., Rothman, K. F., et al. Vaginal spermicides and congenital disorders. *JAMA* 245:1329–1332, 1981.

Kalter, H. and Warkany, J. Congenital malformations. *N Engl J Med* 308(9):491–497, 1983.

Kettering, H. S. and Wolter, D. F. Complications of cesarean section. *Trans Pac Coast Obstet Gynecol Soc* 44:29–34, 1977.

Khojandi, M. and Tyson, J. E. Non-puerperal galactorrhea: Clinical and experimental. *Clin Res* 21:495, 1973.

King, J. C. and Fabro, S. Alcohol consumption and cigarette smoking: Effect on pregnancy. *Clin Obstet Gynecol* 26(2):437–448, 1982.

Klaus, M. and Kennell, J. Parent to infant bonding: Setting the record straight. *J Pediatr* 102(4):575–576, 1983.

Klaus, M. and Kennel, J. H. *Maternal-Infant Bonding.* St. Louis, C. V. Mosby, 1976.

Klaus, M. H., Jerauld, R., and Kreger, N. Maternal attachment. *New Engl J Med* 286(9):460–463, March 1972.

Kline, J., Shrout, P., Stein, Z., et al. Drinking during pregnancy and spontaneous abortion. *Lancet* 2:176–180, 1980.

Kolodny, R. C., Jacobs, L. S., and Daughaday, W. H. Mammary stimulation causes prolactin secretion in nonlactating women. *Nature* (London) 238:286, 1972.

Landesman, R. and Saxena, B. Results of first 1,000 radioreceptorassays for determination of human chorionic gonadotropin: New, rapid, reliable, and sensitive pregnancy test. *Fertil Steril* 27:357–368, 1976.

Lavin, J. P., Stephens, R. F., Miodovnik, M., and Barden, T. P. Vaginal delivery in patients with a prior cesarean section. *Obstet Gynecol* 59(2):135–148, 1982.

Leboyer, F. *Birth Without Violence.* New York, Alfred A. Knopf, 1975.

Levy, E. P., Cohen, A., and Fraser, F. C. Hormone treatment during pregnancy and congenital heart defects. *Lancet* 1:611, 1973.

Linn, S., Schoenbaum, S. C., Monson, R. R., et al. No association between coffee consumption and adverse outcomes of pregnancy. *N Engl J Med* 306(3):141–145, 1982.

Lueck, J., Brewer, J., Aladjem, S., and Novotny, M. Observation of an organism found in patients with gestational trophoblastic disease and in patients with toxemia of pregnancy. *Am J Obstet Gynecol* 145:15–26, 1983.

Magee, J. Labor: What the doctor learns as a patient. *The Female Patient* 1(11):27–29, 1976.

Man breast-fed baby, doctor says. *The Milwaukee Journal,* Feb. 20, 1980.

Manning, F. A. and Feyerbend, C. Cigarette smoking and fetal breathing movements. *Br J Obstet Gynecol* 83(4):262–270, April 1976.

Mehl, L. Options in maternity care. *Women and Health* 2(2):29–42, 1977.

Michaelis, J., Michaelis, H., Gluck, E., and Koller, S. Prospective study of suspected associations between certain drugs administered during early pregnancy and congenital malformations. *Teratology* 27:57–64, 1983.

Mitchell, A. A., Schwingl, P. J., Rosenberg, L., et al. Birth defects in relation to Bendectin use in pregnancy. *Am J Obstet Gynecol* 147(7):737–742, 1983.

Mills, J. L., Harley, E. E., Reed, G. F., and Berendes, H. W. Are spermicides teratogenic? *JAMA* 248:2148–2151, 1982.

Morelock, S., Hingson, R., Kayne, H., et al. Bendectin and fetal development. *Am J Obstet Gynecol* 142(2):209–213, 1982.

Niswander, K. R. Asphyxia in the fetus and cerebral palsy, in Pitkin, R. M. and Zlatnick F. J. (eds.). *Year Book of Obstetrics and Gynecology.* Chicago, Year Book Medical Publishers, 1983.

Nora, A. H. and Nora, J. J. A syndrome of multiple congenital anomalies associated with teratogenic exposure. *Arch Environ Health* 30(1):17–21, 1975.

Oakley, G. P., Flynt, J. W., and Falek, A. Hormonal pregnancy tests and congenital malformations. *Lancet* 1:256–257, 1973.

O'Driscoll, K. and Foley, M. Correlation of decrease in perinatal mortality and increase in cesarean section rates. *Obstet Gynecol* 61(1):1–5, 1983.

Olegard, R., Sabel, K. G., Aronsson, M., et al. Effects on the child of alcohol abuse during pregnancy. *Acta Paediatr Scand* 275(suppl):112–121, 1979.

Page, E. W., Villee, C.A., and Villee, D. B. *Human Reproduction,* 2nd ed. Philadelphia, W. B. Saunders, 1976.

Parish, W. E., Barrett, A. M., Coombs, R. R., et al. Hypersensitivity to milk and sudden death in infancy. *Lancet* 2:1106–1110, 1960.

Pearn, J. H., Teratogens and the male. *Med J Aust* 2(1):16–20, 1983.

Pitkin, R. M. Diet advice for the expectant mother. *The Female Patient* 2(1):38–41, 1977.

Pitkin, R. M. Nutritional support in obstetrics and gynecology. *Clin Obstet Gynecol* 19(3):489–513, 1976.

Placek, P. J., Taffel, S., and Moien, M. Cesarean section deliery rates: United States, 1981. *Am J Public Health* 73(8):861–862, 1983.

Polednak, A. P., Janerich, D. T., and Glebatis, D. M. Birth weight and birth defects in relation to maternal spermicide use. *Teratology* 26:27–38, 1982.

Richards, F., Grimes, D., and Wilson, M. The question of a helminthic cause of preeclampsia. *JAMA* 250(21):2970–2972, 1983.

Rorvik, D. M. and Shettles, L. B. *Choose your baby's sex.* New York, Dodd, Mead & Co., 1977.

Rosenberg, L., Mitchell, A. A., Shapiro, S., and Slone, D. Selected birth defects in relation to caffeine-containing beverages. *JAMA* 247:1429–1432, 1982.

Rosett, H., Quellette, E. M., Weiner, L., and Owens, E. Therapy of heavy drinking during pregnancy. *Obstet Gynecol* 51(1):41–46, 1978.

Rothman, K. J. Spermicide use and Down's syndrome. *Am J Public Health* 72:399–402, 1982.

Royal College of Psychiatrists. Alcohol and alcoholism. *Bull R Coll Psychiatrists* 6:69, 1982.

Sabbagha, R. E. Determining fetal age. *The Female Patient* 8:36/17–36/24, 1983.

Savolainen, E., Saxsela, E., and Saxen, L. Teratogenic hazards of oral contraceptives analyzed in a national malformation register. *Am J Obstet Gynecol* 140(5):521–524, 1981.

Scott, F. R. and Rose, H. B. Effect of psychoprophylaxis on labor and delivery in primiparas. *N Engl J Med* 294(22):1205–1207, 1976.

Shapiro, S., Slone, D., Heinonen, O. P., et al. Birth defects and vaginal spermicides. *JAMA* 247:2381–2384, 1982.

Shettles, L. B. Factors influencing sex ratios. *Int J Gynaecol Obstet* 8(5):634–647, 1970.

Smith, C., Gregori, C. A., Breen, J. J. Ultrasonography in threatened abortion. *Obstet Gynecol* 51(2):173–177, 1978.

Stevens, R. H. Psychological strategies for management of pain in prepared childbirth II: A study of psychoanalgesia in prepared childbirth. *Birth and the Family Journal* 4(1):4–9, 1977.

Sullivan, F. Effects of drugs on fetal development in Beard R. W. and Nathanielsz, P. W. (eds.). *Fetal Physiology and Medicine.* London, W. B. Saunders, 1976.

Tucker, R. E., Young, A. L., and Gray, A. P. (eds.). *Human and Environmental Risks of Chlorinated Dioxins and Related Compounds.* New York, Plenum Press, 1983.

U.S. Surgeon General. Surgeon General's advisory on alcohol and pregnancy. *FDA Drug Bull* 11:9–10, 1981.

Vellay, P. Psychoprophylaxis and its evolution. *Birth and the Family Journal* 2(1):19–22, 1975.

Voitk, A. J., Mueller, J. C., Farlinger, D. E., and Johnson, R. U., Carpal syndrome in pregnancy. *Can Med Assoc J* 128:277–280, 1983.

Vorheer, H. *The Breast, Morphology, Physiology, and Lactation.* New York, Academic Press, 1974.

Weiss, G., O'Byrne, E. M., and Steinetz, B. G. Relaxin: A product of the human corpus luteum of pregnancy. *Science* 194(4368):948–949, 1976.

Winick, M. *Malnutrition and Brain Development.* New York, Oxford University Press, 1976.

Woodward, L., Brackbill, Y., McManus, K., et al. Exposure to drugs with possible adverse effects during pregnancy and birth. *Birth* 9(3):165–171, 1982.

Yaffe, S. J. and Stern, L. Clinical implications of perinatal pharmacology, in Mirkin, B. L. (ed.). *Perinatal Pharmacology and Therapeutics.* New York, Academic Press, 1976.

Yurth, D. A. Placental transfer of local anesthetics. Clin Perinatol 9(1):13–27, 1982.

11. *Problems of Infertility*

Fertility or fecundity is the ability to conceive and produce a child. Most people take this capacity for granted. If anything, they are concerned about too much fertility, and this is a reasonable assumption. It is easy to make a baby.

Culturally, people have long regarded their own capability for babymaking as a measure of power in this world; it is a rare man who doubts his capacity to father a child, and just as rare is the adult woman who would ignore the possibility of pregnancy by thinking herself infertile. From menarche to menopause, a 40-year span, women live with the everpresent possibility of pregnancy. And it is obvious that there is much more effort and anxiety expended in trying to avoid conception than in trying to conceive. Parenthood is not necessarily being ruled out, but it is probably being delayed until the time is right. "Not now," says the single woman, "who needs the hassle?" "Not now," say the newly married couple, "we're not ready to make that commitment," or "we can't afford it yet," or "we want to get ahead professionally first," or "we want to buy a house," "travel," "get to know each other better first." For some couples who

want to replace "not now" with "now," however, an unsuspected and frustrating situation may emerge. Fertility cannot always be taken for granted. Until a man and a woman actually achieve a successful pregnancy, there is no way of knowing whether or not there is any problem. While most couples experience no difficulty in producing a baby, an estimated 15%–18% or one out of every seven couples in the United States, remains involuntarily childless. Couples, after practicing birth control for years, ironically may discover that they are infertile.

There are indications that the incidence of infertility is becoming greater, and several reasons have been suggested for the apparent increase in the problem. Amenorrhea and ovulatory difficulties are not uncommon after oral contraceptive use, and infections that result in an obstruction of the fallopian tubes have been associated with the use of intrauterine devices for birth control. The sexually transmitted disease epidemic has also increased sterility in both men and women. Moreover, while little is known about the long-range effect of environmental pollutants on fertility, there is a growing awareness that both male and female industrial workers are exposed to a variety of chemicals that may adversely affect their reproductive processes. Lead and other metals and inhalation of toxic fumes have been implicated, and Whorton and co-workers reported in 1977 that a number of cases of sterility in male workers resulted from their exposure to a common pesticide produced in a California chemical factory. Similar polychlorinated hydrocarbons are now known to be environmental contaminants to which all American couples are exposed. An additional reason for infertility may be the tendency in many couples to delay marriage and postpone childbearing until their thirties, after the age of maximum fertility is past.

Pointing to an already overpopulated world and the enormous cost, both financial and psychological, of raising a family, some would minimize infertility as a problem or might even consider it a blessing. Many couples are choosing to be childfree, guaranteeing it with a vasectomy or a tubal ligation. It is one thing to choose not to be parents as a result of mutual decision, but it is a very different and painful situation when a man and a woman have decided to have a baby and discover that their right to choice has been biologically denied. Just a few years ago there was little recourse for couples who did not want to remain childless. Today, advances in infertility research have made it possible with medical assistance for two-thirds of infertile couples to conceive and produce a baby.

WHAT IS INFERTILITY?

Infertility is defined as the inability to conceive a child during the course of 1 year of regular sexual intercourse unprotected by contraception. Data indicate that without birth control, 25% of couples conceive in 1 month, 60% within 6 months, 80% after a year, and that 90% have initiated a pregnancy by 18–24 months. Statistically, however, if a couple has not conceived by the end of a year, the chances of becoming pregnant decrease. The age of the couple is also a factor, and the older they are the more difficulty they have. A woman's peak fertility occurs between the ages of 20 and 25; it is less optimal after 30, and is at its lowest after 40. Similarly, male fertility is greatest in the mid-twenties, decreases in the thirties, declines markedly after forty, but never terminates completely as it does in a woman after menopause. Until the age of 30, men and women should probably wait a year or even longer before seeking medical help, but older couples should go for an evaluation if they have been unsuccessful after 6 months. If a couple has decided to postpone their childbearing until their thirties, they should at least start trying for a pregnancy early in that decade, giving themselves enough time to be helped, if necessary. Doctors are reluctant to attempt to help couples over 40, not only because the chances of success are so limited, but because the risk of producing a child with a genetic abnormality is so much greater with the increased age of the parents. With the availability of amniocentesis, however, that risk has been lessened.

That human infertility in men and women decreases with age is nothing new. Biological factors associated with the normal aging process are responsible, and there may be additional environmental reasons in the sense that the infectious agents or pollutants to which we are exposed throughout life have had a longer time to act. Most women have always recognized that their chances of getting pregnant decreased with age, but increased anxiety about possible infertility was generated by a report in 1982 by French researchers Schwartz and Mayaux of the Fédération CECOS. The study, evaluating the success rate for pregnancy in 2,193 women receiving artificial insemination because their husbands were sterile, reported that the percent of women conceiving after 1 year of attempted inseminations was 73% for women under 25, 74% between ages 26 and 30, dropped to 61% between ages 31 and 35, and went down another 5 percentage points between the ages of 36 and 40. What these findings mean to an individual woman *not* having artifical insemination is debatable, but the heightened concern about age and infertility was probably fueled by over-reaction to the study in the press and an editorial accompanying the report that actually suggested that, in light of the study, women may want to re-evaluate their priorities and devote their twenties to childbearing and their thirties to career development, reversing the current trend. Some women were justifiably skeptical, but others viewed these conclusions with apprehension, believing that their biological clocks were not only ticking away, but somehow running down.

The greater risk of infertility with age is real, but the outlook for getting pregnant in the fourth decade of life is obviously not that bleak. The number of first births to women in their thirties has risen dramatically, by 116%, in the last 10 years, surpassing by a wide margin the modest 7% increase in first births to women in their twenties. In fact, the largest rise in fertility in 1980 occurred among women in their thirties. These are women whose greater educational achievement and desire to get their careers established led them to put off having a family and resulted in fewer adverse effects on their offspring than in those of younger women. For example, there was less incidence of low birth weight infants born to women in their thirties than in any other age group. The French study was alarming to many women, but needlessly. Demographer John Bongaarts, who called it a "false alarm," questioned the validity of the conclusions and pointed out that the French findings of lower fertility with artificial insemination should present little cause for concern. The data from long-term studies of the general population indicate that the figures from France do not even approximate the considerably greater fertility of women over thirty when insemination occurs in the usual way, rather than artificially.

Traditionally, infertility has been seen as a "female problem." Even the words "barren" and "unfruitful" are feminine—who ever heard of a barren male? For this reason, all infertility research until recently focused on the female. Women, too, accepting the old beliefs and assuming responsibility in a childless marriage, have always been the first to seek help and become the more willing patients. Currently, the chances for correcting female infertility are still better than the possibilities for correcting male infertility, but now there is the general recognition that it takes both a man and a woman to make a baby, and it also takes the two to share responsibility for an inability to conceive. Some factor in the male is at fault 40% of the time, but because research on male infertility is in its infancy, treatment is not as successful. Actually, infertility is rarely the sole fault of either partner but is more usually the result of several minor dysfunctions in both. Together, they may be incapable of conceiving a child, but with another partner of greater fertility, there might be little difficulty.

Occasionally, there is no physiological reason for the infertility in either the man or the woman, but their inability to produce an offspring is the result of not knowing the best way or the best time to go about it. Not that there is anything wrong with their sexual technique—many people have read the manuals—but they may inadvertently be deterring rather than enhancing the possibility of conception. Perhaps the couple is having intercourse twice a day in an attempt to conceive, but increasing the frequency of coitus beyond four times a week may actually be lowering the sperm count to the point where pregnancy is unlikely. Even in a very fertile male, 12–24 hours should elapse between ejaculations in order for the sperm quality to have good fertilizing capacity. If a man is somewhat less than optimally fertile, it could take at least 48 hours to regain an appropriate sperm concentration. Sometimes, conception is hampered because the couple is using a commercial vaginal lubricant, such as K-Y Lubricating Jelly, that is lethal to sperm. Even saliva, usually the only lubricant recommended, has been found to be deleterious to sperm motility and activity (Tulandi et al, 1982). All such aids are somewhat spermicidal, but if couples really need the help of a lubricant, petroleum jelly, glycerine, and raw egg white have been tested and found to have minimal effects on sperm. Another mistake may be that the woman is getting out of bed too soon after coitus. She should lie on her back for an hour before arising, giving the sperm every opportunity to ascend through the cervix. And, since there is a strong likelihood of conception only 1, or perhaps 2 days, a month, a couple could theoretically be missing the right time for years if they attempt pregnancy in a kind of random, hit-or-miss fashion. The simplest way of knowing when ovulation—the optimal time for conception—occurs is to keep a daily basal body temperature chart.

Basal Body Temperature Curve (BBT)

If the body temperature is taken daily under basal conditions—that is, after a period of rest and before any regular activity, daily work, or meals—it will fluctuate daily over a very small range. This is true for all individuals, male or female, young or old. If the daily recordings are plotted on a chart in a healthy adult ovulating woman, however, the BBT pattern will show not only the daily variation in the period between two menstruations but also a defi-

nite elevation in temperature that occurs within 1–3 days of ovulation. This elevation of temperature remains until 3–0 days before the onset of the next menstruation, when the curve again deflects back to the postmenstrual level. The morning temperature generally varies from about 97.2° before ovulation to above 98.6–98.8° after ovulation. Women before menarche or after menopause, anovulatory women, and men do not show this biphasic basal body temperature curve.

If a woman is having menstrual periods at relatively regular monthly intervals and the periods produce some premenstrual symptoms, such as a little puffiness, breast tenderness, or complexion trouble, she can be reasonably certain that she is ovulatory. Another useful sign, although not every woman has them, is the presence of intermenstrual ovulatory pains. Regular examination of cervical mucus would also indicate the occurrence of ovulation. All of these criteria can be confirmed by keeping a basal body temperature chart for a number of months. While the actual day of ovulation cannot be accurately predicted, it can certainly be approximated after the chart has been kept through several cycles, and intercourse to produce a pregnancy can be timed around it. Waiting for the increase in temperature before trying to conceive is a mistake, however. The actual release of the ovum from the ovary probably occurs 24–38 hours before the temperature elevation. The most favorable schedule for conception would be coitus every other day for the 3–4 days before the rise, including the 2 days after it. After the basal body temperature has been elevated for 2 days, conception is biologically impossible. BBT charts are, therefore, useful for both infertility and fertility control.

Printed charts on which to record the BBT are available from a physician, but it is easy enough to make one using graph paper. Figure 11.1 illustrates the normal morning temperature curve and the optimal days for con-

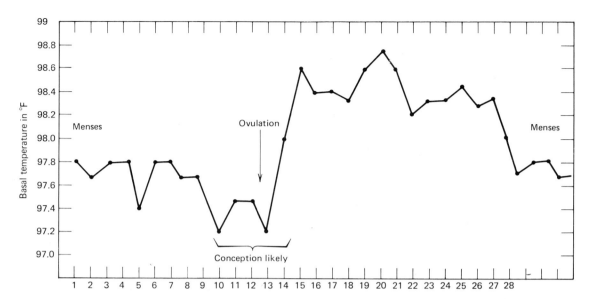

Figure 11.1. Basal Body Temperature (BBT) plotted during an ovulatory cycle is a biphasic curve. Daily temperatures tend to fluctuate in a pattern that is specific for every woman. Ovulation probably occurs the day before the elevation, which may be preceded by a drop to the lowest temperature of the cycle. Sperm retain fertilizing capacity for 24–48 hours, and eggs are able to be fertilized for 12–24 hours. Coitus every other day for about 3–4 days before and 2–3 days afterward would increase the probability of conception. Unfortunately, not all charts are as easily interpreted as this one.

ception. Each morning before getting out of bed, after at least 5 hours of sleep and before eating, drinking, or any extensive conversation, a thermometer is placed in the mouth for 3 minutes, read, and the temperature recorded on the chart. The rectal temperature has been determined to be more reliable and the most accurate measure of the actual basal body temperature, but the temperature taken orally or even under the armpit, as long as the same method is always used, will be accurate enough in most instances. Since the actual temperature in degrees is not as significant as the difference that occurs between successive recordings at the time of the elevation, it is possible to use a regular thermometer, but reading it may be more difficult. Special and

more expensive thermometers that show a range of only a few degrees can be purchased (see Fig. 11.2).

The temperature is marked on the chart for the appropriate day, and the mark is joined to the one for the previous day, thereby charting the curve for the month. The days of menstruation should be noted. Obviously any illness or infection that could cause a rise in temperature and be falsely interpreted as an ovulation elevation should also be noted.

A basal body temperature chart, recorded over five to six cycles, is easy to do and to interpret, requires only 5 minutes a day, and provides invaluable information to a woman interested in knowing her body's rhythms. It can tell her if she ovulates every month, when

Figure 11.2. A basal thermometer has increments in tenths of degrees between 96° and 100°F and, therefore, may be easier to read.

ovulation occurs, when is the best time to get pregnant, and when is the best time not to get pregnant, although its reliability in this regard is more suspect. If the temperature elevation remains, and no menstrual period occurs, she can assume that a pregnancy has occurred. If a couple is having difficulty in conceiving, they will certainly be ahead if the BBT has been charted when they consult a physician. The first thing the doctor is likely to do is hand the woman a basal body temperature chart to keep through several cycles.

When All Systems Are "Go"

Fertilization, and thus conception, occurs when a normal sperm unites with a normal ovum at just the right time. A man's testes must have produced a sufficient number of mature active sperm that move through unobstructed ducts so that they can be ejaculated in the appropriate concentration. He must be potent so that the sperm can be deposited in the woman's vagina, reach and penetrate the cervical mucus, and ascend into the uterus

and up to the fallopian tubes. The likelihood of sperm ascent is greatest when intercourse takes place in "missionary" position, with the woman on her back with her thighs flexed. Insemination is favored by this arrangement because the cervix will then lie in the pool of ejaculated semen that collects in the posterior fornix of the vagina.

The woman has to have produced a normal, fertilizable egg that has entered the fallopian tube so that it can be fertilized by a sperm within a period of several hours after ovulation. The resulting conceptus must move down the tube and implant in the previously prepared endometrium of the uterus so that it can continue its development. In both sexes, there must be appropriate levels of secretion of pituitary gonadotropins and gonadal sex steroids to sustain the reproductive processes. Any restriction or impairment of these basic events, even very slight, can cause a problem of infertility. Investigations of infertility are geared to systematic assessment of all the factors involved in the passage of the sperm and ova towards each other. They include five main tests aimed at diagnosing the possible defects that may be present and include an evaluation of the (1) male factor; (2) the vaginal and cervical factor; (3) the uterine factor; (4) the tubal factor; and (5) the ovarian factor.

EVALUATION OF INFERTILITY

Ideally, each couple seeking a diagnostic evaluation would be able to go together to an infertility clinic where a gynecologist, an andrologist (the counterpart for men to a gynecologist), a reproductive endocrinologist, and perhaps an immunologist and geneticist form an infertility team that uses the latest diagnostic tools and offers treatments unknown just a few years ago. Unfortunately these advantages are available only in major urban areas or at centers associated with university medical schools. But at least a doctor who specializes in infertility diagnosis and treatment should be sought. All gynecologists do not automatically qualify. It would also be wise, if possible, to find a physician whose practice is limited to infertility problems. The investigation will take 3 or 4 months, and some of the procedures would require immediate access to the doctor. Ovulation, for example, occurs any day of the week, including Saturday, Sunday, and holidays, and the availability of a physician who also maintains an obstetrical practice and performs gynecological surgery is obviously going to be limited.

It is logical to begin an evaluation of a couple's infertility by conducting the simplest, least complicated laboratory test—a semen analysis or sperm count. Knowing the extent to which the man is implicated in the couple's childlessness can direct the nature of the further tests to be performed on both partners. If he is completely normal, his part in the investigation may be finished. If he is completely sterile and the woman is subsequently determined to be normal, the couple may want to think about artificial insemination, adoption, or a childless existence. If the male is somewhat subfertile he may be contributing to, but not be solely responsible for, the infertility. The evaluation would then continue on the woman while further investigation and possible treatment could be initiated on the man. But obviously, a woman should not agree to having a complete diagnostic workup, which may involve invasive tests or even surgical procedures, until the role played by her husband has been determined.

The Male Factor

Male infertility can have many causes—nutritional, endocrinological, developmental, pharmacological, occupational, immunologi-

cal, anatomical—but they are generally expressed and recognizable in the semen analysis. For a semen examination, a man is asked to produce a specimen by ejaculation into a glass jar after 2–3 days of sexual abstinence. The ejaculate then has to be delivered to the laboratory for analysis within 1–2 hours. For religious or other reasons, some men find this method of providing a semen sample objectionable. Other methods of collecting sperm are by coitus interruptus (withdrawal before ejaculation during intercourse) or the semen can be aspirated from the woman's vagina immediately after coitus. It should not be collected in a condom, which may contain spermicidal chemicals.

When the specimen is brought to the laboratory, it is analyzed for volume, viscosity, number of sperm, sperm viability, motility, and sperm shape. The average volume of ejaculate is 3.5 ml, but it can normally vary from 2 to 10 ml. A low volume (1 ml or less) can be significant because it may not be enough to come in contact with the cervix when it is deposited into the vagina. For unknown reasons, a volume higher than 8 ml is also associated with infertility. After determining volume, the specimen is diluted and the numbers of sperm in a sample are counted under the microscope. To be considered normal, the specimen should contain at least 20 million sperm per ml, but experts in male fertility believe that the number of live and active sperm in the semen is a better correlate to fertility than their actual numbers. For there to be a strong probability of pregnancy, the semen should contain 50%–70% live sperm that move rapidly forward across the microscopic field and that retain their motility for several hours. A low sperm count with highly motile sperm usually represents greater fertility than a high sperm count with poor motility.

The shape and configuration of the sperm are also important. It is generally agreed that the chances for pregnancy are greatest when not more than 40% of the sperm in the specimen are considered to have abnormal morphology—too large, too small, too tapered, immature, or doubleheaded.

Upon analysis of their semen samples, some would-be fathers have sperm that clump together for no apparent reason. This condition, called nonspecific sperm agglutination, has been successfully treated with high doses of vitamin C. Why ascorbic acid restores sperm motility and activity is not completely known, but preliminary research by University of Texas investigators has shown that it may be possible for such men to regain fertility in less than a week of vitamin supplementation (Gonzalez, 1983).

There are so many variables that can affect seminal quality that several analyses have to be performed before a diagnosis of normal, subfertile, or sterile can be made. In 1976, Yanagimachi and co-workers discovered that when the zona pellucida of the hamster ovum is removed, the species-specificity for penetration by sperm is also lost and that any animal's sperm, including human, can then enter a hamster egg. A new technique, the zona-free hamster ova penetration test was developed to assess the actual fertilizing potential of a given male's semen. The hamster test is beginning to be used in conjunction with semen analyses to determine male infertility.

Some factors are known to have a temporary effect on male fertility. Virus disease, such as infectious mononucleosis or hepatitis, or actually any illness that produces a fever, can depress sperm production for several months. Certain tranquilizers or mood-elevating drugs may produce not only impotence but also suppress growth and production of sperm. Furadantin (Norwich Eaton Pharmaceuticals), an antibiotic used in urinary infection, is known to produce a temporary infertility. Testicular heating also adversely affects

sperm production. Scrotal temperature is ordinarily 2.2°C less than body temperature and optimum for sperm. Exposing the testes to heat by taking a sauna or steam baths, wearing tight pants, thermal underwear, or jockey shorts, or even having the kind of occupation (truck driver on cross-country hauls) that results in scrotal heating, can reduce fertility.

It has been found that about 25% of infertile males have a condition called *varicocele*, a varicosity of the left internal spermatic vein that drains the testes. A simple surgical procedure that ties off the vein has been found to improve fertility in the majority of men with the disorder.

When no reason for the poor semen quality can be determined, medical opinions vary as to whether there is any really effective way of treating the deficiency. Various hormones have been administered, but there is general consensus that their efficacy is questionable unless there is some definite indication for their use. Clomiphene citrate, a drug used to induce ovulation in women, has been tried in men in various dosage regimens. Some workers report improvement in sperm concentration and motility; others have given the same medication with little or no beneficial effect.

Artificial Insemination

When the sperm count has been determined to be only moderately deficient, artificial insemination using the male partner's own semen may be possible. One method is termed split ejaculate insemination. Only the first one or two drops of the ejaculate, containing a significantly higher sperm concentration than the rest, which is mostly glandular secretions, is inserted into the external os of the cervix. Another method that uses the subfertile male as donor is the freezing and storing of several ejaculates so that they can be combined to increase the numbers of sperm. Sperm can be successfully frozen and thawed without damaging them genetically, but their efficacy in producing pregnancy is not as great as that of fresh semen.

Artificial insemination using donor semen can be performed when the husband is considered clinically sterile. An estimated 10,000 babies are born annually as a result of this procedure. Artificial insemination, donor, or AID—an apt acronym—is an alternative to irreversible male infertility. It is becoming increasingly more acceptable, performed not only for infertile heterosexual couples, but also for single women who are homosexual or who expect to remain unmarried but want a child. Not all physicians are equally enthusiastic about performing AID. It involves ethical and legal considerations for the doctor, who must select the donor, preserve his anonymity, and worry about the threat of future litigation should anything go wrong. The latter possibility is circumvented by strongly worded consent forms. The donors are usually medical students, interns, and residents who provide sperm specimens for a fee. It is claimed that these men are preferred as donors because of their availability and their greater awareness of their own and their family's medical history. It can also be assumed that many doctors reason that the best source of sperm of high genetic quality should be from healthy stable men of superior intelligence—other doctors, naturally.

For the couple or the single woman, there are similar ethical and legal difficulties. There are unanswered questions about the legitimacy of the child and the rights and responsibilities of the parents. In some states, the husband must legally adopt the wife's child after birth. In others, some specific measures have to be taken to protect the child's right to the husband's estate. The considerable emotional and psychological ramifications of the procedure also must be confronted. Generally, it is recommended that the insemination be kept a completely private matter between the physician and the couple. Some have even suggested that the obstetrician who delivers the baby be different from the physician who did the insemination so that as few people as possible know how the pregnancy was established. Eventually, a single parent or couple has to deal with disclosure of AID status to the child, or risk the possibility of accidental discovery. The anonymous donors, who may have provided semen for a number of inseminations, are emotionally uninvolved, but where does that leave the child who wants to know the identity of the biological father? Some physicians routinely mix the donor's semen with the husband's sperm as a means of blurring paternity—a practice of better legal than medical value. For couples who are able to adjust to the idea of artificial insemination, it can be a method of achieving a pregnancy, but there are no guarantees. After a year of artificial inseminations, the reported national success rate averages 57%–70% with the majority of pregnancies occurring in less than 6 months of AID.

There has been some concern that overly generous sperm donors could be fathering so many children in a single geographic area that their offspring might meet someday and marry. The implications for the transmission of recessive genetic disorders could be considerable, given that there is no formal policy in most states that requires testing donors for evidence of genetic defects. Many physicians limit donations per donor for that reason. Others make use of sperm banks that are able to honor requests for sperm from men of specific racial or ethnic background. In 1979, a California businessman had an even "better" idea—to found a sperm bank and also contribute to the nation's intelligence by soliciting and accepting donor sperm only from Nobel Prize winners in science. The Herman Muller Repository for Germinal Choice, named after geneticist Muller whose death in 1967 eliminated *him* from the donors, provides highly intelligent (but aged, given the advanced years of Nobel Laureates) sperm for a price. The repository announced in 1982 that a woman gave birth to a daughter after being inseminated with the sperm of "an eminent mathematician." But the world will never know the true value of buying genius genes at a sperm bank since the parents of the girl declined to be identified.

Female Factors—Vaginal and Cervical

The uterine cervix is a gateway—it permits the entry of sperm to the uterus. The unique properties of the cervical mucus determine whether the gate is opened or closed. Around the time of ovulation, the cervical mucus is clear, watery and less viscous, more alkaline, and most receptive to sperm penetration. Within a day or two after ovulation, the cervical mucus, now influenced by progesterone, becomes thick, viscous, and opaque, and it impedes sperm passage. The cervix and its mucus have an obvious significance in problems of infertility.

Vaginitis is not known to cause infertility, but neither is it likely to encourage fertility. Semen is alkaline, the vaginal secretion is acid, and the combination is optimal for sperm survival. If a vaginal infection changes vaginal pH, sperm motility may be hampered. Even if there are no symptoms or clinical signs of vaginitis, a microscopic examination of vaginal discharge is a necessary part of an infertility investigation. Any foreign microorganisms present should be eliminated to ensure a normal vaginal pH.

There is some evidence that mycoplasma infections are associated with infertility. Mycoplasmas bear similarities to both viruses and bacteria. One of the organisms belonging to this group, T-mycoplasma or *Ureaplasma urealyticum*, is believed to account for about half of the cases of nongonococcal urethritis in men and is frequently found in the vaginas

of women where it produces no apparent symptoms. *Ureaplasma urealyticum* has been implicated by a number of investigators as the cause of repeated spontaneous abortion, stillbirths, and unexplained infertility, but its actual contribution is still unproved. When the infection has been discovered in the male partner of an involuntarily infertile couple, however, and the man was treated with a tetracycline derivative, pregnancy was achieved in a number of cases. One of the more recent reports, by Toth et al. (1983), noted that conception occurred in 60% of previously infertile couples once the *Ureaplasma urealyticum* infection was eradicated by doxycycline therapy for both partners. These authors suggest that this kind of hidden but prevalent infection may play a significant role in infertility.

The post-coital or Sims-Hühner test is an evaluation of both partners routinely performed in any infertility evaluation. The test, first reported by Sims over a hundred years ago, is made at the time of ovulation and consists of a microscopic examination of vaginal secretion and cervical mucus for the presence of motile sperm within some hours after sexual intercourse. It provides information concerning sperm quality and cervical mucus quality and is, therefore, an evaluation of both partners.

If a sample from the endocervix taken 4–6 hours after intercourse shows 10 or more motile and active sperm per high power microscopic field that are progressing forward in a clear and plentiful watery mucus, it is taken as evidence that semen is normal and that no block to fertility exists at the level of the vagina or cervix. If there are sperm in the vaginal secretion but not in the cervix, the assumption is that the cervical mucus is blocking their entry. If the semen analysis has been previously determined to be normal, but there are few or dead sperm in the vagina and cervix, it may mean that the cervical mucus is hostile or even lethal to the sperm. If there are no sperm or semen present in the vagina or the cervix, it may mean either that there was something wrong with the test and it should be repeated, or that there should be some tactful investigation of why there were no sperm deposited when intercourse presumably took place a few hours before. There could be some problem, possibly unrecognized by the couple, of incomplete penetration or even total lack of intravaginal penetration.

Mucus that is "poor," that is, thick and impenetrable, may have an endocrinological basis or be the result of infection and can frequently be successfully treated by endocrine or antibiotic therapy. If the cervical mucus remains a barrier despite treatment, one method that reportedly has resulted in pregnancy has been to inseminate a very small amount of previously collected semen directly into the endocervical canal, bypassing the barrier. Since bacteria, as well as sperm, get a direct introduction into the uterus by this procedure, the strong possibility of infection is a deterrent to its use.

Immunological Factors

There is little doubt that immunological factors play a role in infertility, but the actual mechanism and the site of action are poorly understood. Rare cases have been reported in which women have suffered strong allergic reactions to semen—hives, hayfever symptoms, vulvovaginal swelling—and, in isolated instances, even anaphylactic shock, an acute, severe response. Less dramatically, it has also been reported that a variety of sperm-immobilizing antibodies can be detected in the blood serum and/or the cervical-vaginal secretions of a proportion of infertile women. Their significance for otherwise unexplained infertility is unknown, however, because fertile women have also been found to have sperm antibodies circulating in their blood.

The presence of an unfavorable immune re-

action to sperm can be demonstrated by a simple agglutination test. A solution of diluted sperm is mixed with a sample of the woman's serum and observed for agglutination or clumping of sperm. If the test is positive, it indicates that sperm-agglutinating antibodies are present in her blood and that these may be interfering with fertility. It is assumed that once sensitization of the woman to her partner's sperm has occurred, continued and frequent intercourse will maintain a high level of antibodies in her blood. The condition is therefore treated by occlusion or "condom therapy." The woman is protected by contact with the sperm by the use of a condom for several months in order to cause enough of a drop in the antibody level so that sperm motility and, thus, pregnancy can take place. Some doctors have reported great success with this form of treatment; others believe the method to have doubtful value. It could take 9–12 months of protected intercourse before the sperm agglutination test becomes negative, and sometimes the antibodies to sperm persist even in the absence of exposure to sperm.

Sperm-agglutinating antibodies have also been demonstrated in the serum of infertile men. This kind of autoimmune response—a man is allergic to his own sperm—has not been shown in the serum of husbands of pregnant women. Shivers and Dunbar suggested that a similar autoimmune response—a woman developing an allergic reaction to her own ova—could be responsible for infertility in some instances. They reported the existence of antibodies in some of the serum samples from 22 infertile women that reacted positively with the covering around the ovum or zona pellucida of pig eggs. They speculate that antibodies to her own zona pellucida in an infertile woman can block fertilization by binding to the zona and making it impossible for the sperm to penetrate the ovum (1977). This hypothesis may hold true for some cases of unexplained infertility, but Sacco and

Moghissi (1979) found no difference between fertile and nonfertile women in their zona pellucida–antibody activity and also discovered that some women who had zona–serum–positive reactions subsequently achieved pregnancy.

Any kind of treatment for men or women that shows autoimmune responses to their own gametes is as yet unknown. Research on the immunological aspects of infertility is still in its earliest stages. The work that is exciting the most interest deals with the problem from another angle—the development of a vaccine to *prevent* pregnancy. When the exact nature of the antigens, the agents that cause the production of antibodies, can be identified and isolated, it may be possible to find a way of suppressing those antigens in the infertile and to prevent their effect on fertility reduction, and to develop a way to immunize women against sperm or against their own ova and thus reduce fertility.

Uterine and Tubal Factors

The uterus alone seldom interferes with fertility. A small uterus (the so-called "infantile" uterus) has successfully carried many pregnancies, and a "tipped" uterus, as long as it remains mobile and not fixed or bound down by adhesions, is rarely significant. Only if the retroversion and retroflexion (see Chapter 2) is very severe is it possible, but not likely, that the cervix is displaced enough to impair entry or passage of sperm through the endocervical canal. Fibroids are important only if they obstruct the fallopian tubes or interfere with the implantation of the fertilized egg. Endometriosis, even when the fallopian tubes and ovaries are unaffected, for elusive reasons does have an effect on infertility, however, and women who are aware that they have the disease should probably start to have their children early and space them close together before the endometriosis becomes further advanced.

A tubal disorder is the causal factor in 30% of couples with a failure to conceive. If the fallopian tubes are completely or even partially occluded, the egg may not be able to get into the tube from the ovary, the sperm may not be able to get to the egg to fertilize it, and, even if it does, the larger fertilized egg may not be able to pass down the tube to the uterus. Any bout with bacteria in the pelvis—either secondary to gonorrhea, abortion, miscarriage, normal delivery, an intrauterine contraceptive, a D and C, or a ruptured appendix—can cause scar tissue and result in tubal obstruction.

The patency, or openness, of the fallopian tubes can be determined by a relatively low-risk test that can be performed in the gynecologist's office. The Rubin, or uterotubal insufflation test, is a procedure by which carbon dioxide gas under pressure is injected through a cannula into the uterus so that it can flow out of the tubes into the peritoneal cavity. If one or both tubes are patent, only a normal amount of carbon dioxide pressure will be needed to produce the endpoint of the test—the experiencing of pain referred to the left shoulder when sitting up. The pain results from the gas passing into the peritoneal cavity to collect under the diaphragm.

When the carbon dioxide pressure has to be higher than normal to produce the shoulder pain, it probably means that the tubes are partially obstructed. If there is an absence of shoulder pain, the tubes are blocked. The pain is acute and unmistakable if it is present, but it is temporary and subsides as the woman lies down again and the gas is slowly absorbed. Usually, there is little or no discomfort after the test is over, but occasionally a diaphragmatic irritation will persist.

If the Rubin test is normal, that is, normal pressure to produce shoulder pain, some doctors will forego any further tubal evaluation at that point in order to proceed to investigate other factors. The test is uncomplicated, produces some discomfort but no other side effects, is not expensive, and occasionally is even therapeutic. It is evidently not unusual for a pregnancy to occur after the Rubin test has been used for diagnosis, perhaps because, as Rubin claimed, a clearing or flushing out of the tubes occurs.

The information provided by the Rubin test is limited, however, because all it can really determine is whether at least one fallopian tube is open. For this reason, and because technical problems with the Rubin apparatus can cause false-normal and false-abnormal test results, many physicians have replaced it with another test of tubal patency, hysterosalpingography. In this test, an opaque contrast medium, preferably water soluble, is injected through a catheter into the endocervical canal so that the uterus and tubes can be visualized during fluoroscopy and on X-ray film. If the fallopian tubes are patent, the dye will ascend upwards to distend the uterus and tubes and spill out into the peritoneal cavity. The disadvantages are that hysterosalpingography must be performed in the hospital or in a radiologist's office, and it exposes a woman to pelvic radiation. In its favor, the test can provide more information concerning the site of an obstruction or abnormality and seems to have even greater fertility-enhancing properties than the Rubin test. Although a hyterosalpingogram can pinpoint the level of obstruction in a tube, it cannot indicate anything about the nature of the problem. Further tests of tubal structure before treatment could include direct visualization of the pelvic anatomy by laparoscopy, hysteroscopy, or culdoscopic examination.

When it has been determined that the infertility is the result of some kind of tubal factor, the present prospects for successful treatment are still meager. Various kinds of surgical procedures have been developed, but oviduct tissue is extremely delicate, difficult to handle, and easily traumatized. Surgery on it is similar to operating on a thin short strand of spaghetti. Even in the hands of the most experienced infertility surgeons, using the smallest of instruments and sutures thinner than hairs, the results in terms of postoperative pregnancies are not that encouraging. A further reason for a general reluctance to attempt tube repair is its association with a high rate of subsequent ectopic pregnancies. Continued refinements of the tools and techniques of microsurgery may produce more positive results in the future.

The Ovarian Factor

The two basic functions of the ovary are to produce an ovum and to produce estrogens and progesterone. Both are equally important to pregnancy. It is estimated that 10%−15% of fertility problems are the result of some disorder of ovulation. Either the woman does not ovulate at all, ovulates very irregularly and infrequently, or ovulates and produces a corpus luteum that inadequately secretes progesterone and fails to sufficiently prepare the endometrium for implantation. The easiest profile of ovulatory function is provided by the basal body temperature chart. As previously described, a normal biphasic ovulatory pattern shows lower temperatures in the first half of the cycle, a dip to the lowest point before ovulation, and a sustained rise in temperature in the second half of the cycle until it falls again premenstrually. Another confirmatory test of ovulation that indicates if the secretion of progesterone is adequate is an endometrial biopsy, a strip of tissue removed from the lining of the uterus, just before menstruation. Since the histological appearance of the endometrium during the entire menstrual cycle is well known and standardized, the sample of endometrium is dated to see if it corresponds to the estimated date of ovulation that has been ascertained by the basal body temperature. An endometrium that does not conform to the normal histological pattern indicates a defect in the luteal phase. This relatively rare problem, estimated to occur in 3%–10% of infertile women, is treated with human chorionic gonadotropin or by progesterone administered through intramuscular injection or vaginal suppository. Progestational agents other than progesterone should not be used. Not only have they been proved ineffective for luteal phase defect, but they have been implicated in birth defects should pregnancy occur during their administration.

Endometrial biopsy is performed in the doctor's office, usually without anesthesia or with only a local numbing of the cervix. Thus, it may be painful for many women, rivaling, as Barbara Eck Menning (1977) observed, the hysterosalpingogram in discomfort.*

Other tests of ovulatory function include daily observations of cervical mucus and daily vaginal smears, which are stained and examined to indicate estrogen and progesterone secretion. Since ovulatory dysfunction may also be the result of hyperprolactinemia (elevated circulating levels of prolactin that could interfere with the synthesis or release of LHRH or progesterone) or a disturbance of the hypothalamic-pituitary-ovarian axis because of increased androgen production secondary to an adrenal disorder, daily measurements of the amounts of steroids and gonadotropins in the blood or of hormone metabolites in the urine can be made. These techniques are very expensive and time consuming and are generally more rarely used.

BYPASSING THE TUBAL FACTOR WITH TECHNOLOGY

The birth of a baby girl in mid-1978 to a woman whose tubes were irreparably obstructed raised the hopes of thousands of infertile women with similar problems. The method by which this child was conceived, however, while a remarkable scientific feat, cannot really be considered a feasible or easily available alternative to adoption or a childless marriage.

After 12 years of research and many unsuccessful attempts, British gynecologist Patrick Steptoe and Cambridge University physiologist Robert Edwards successfully established a pregnancy in 30-year-old Mrs. Lesley Brown through in vitro (outside the body) fertilization. After administration of a fertility drug to induce ovulation, a preovulatory ovum was removed from Mrs. Brown's ovary by laparoscopic surgery. The ripe ovum was combined for fertilization with husband Gilbert Brown's sperm in a "test tube"—actually a highly complex sterile laboratory apparatus. The fertilized egg was cultured for $2\frac{1}{2}$ days in appropriate nutrient media until it reached the 8-cell stage when it was then reintroduced for implantation into Mrs. Brown's hormonally prepared uterus. The embryo burrowed into the endometrial lining and survived. Nine months later, heralded by banner headlines, the world's first test-tube baby was delivered by cesarean section several weeks ahead of the anticipated due date.

Many more attempts at in vitro fertilization followed. By the time the first United States baby was born in December, 1981, at Norfolk General Hospital in Virginia, she was the nineteenth baby produced as a result of the procedure around the world. By now even the first test-tube baby has a test-tube baby sister. The technique advanced rapidly; 40 U.S. babies fertilized in vitro were born in 1983, including several sets of twins. By 1984, there were triplets delivered in Wisconsin, and programs offering the procedure were established in a number of states. Despite an estimated 500,000 women infertile as a result of blocked or missing fallopian tubes, however, the tech-

*Menning is the founder of Resolve, Inc., P.O. Box 474, Belmont, MA 02178, a support organization for the infertile.

nique is not likely to become a widely used or popular method of getting pregnant. The risks are substantial and the chances of producing a child per attempted ovum recovery and fertilization are slim. It could take as many as five or six attempts (at about $3,500 to $4,000 each) to achieve a pregnancy, and the overall probability of producing a live birth now is 15%–19%. Initially, widespread media coverage raised false expectations and misunderstandings about what the procedure could do, and even hysterectomized women tried to apply. Currently, more people have become aware of the limited success and the enormous cost (health insurance companies do not pay for in vitro fertilization) and are not standing in line waiting for service at the newly established fertility clinics across the country that offer the program.

The issue of in vitro fertilization is surrounded by ethical, moral, and legal controversy. There have been questions about the procedure causing visible abnormalities or psychological or intellectual adverse effects on the offspring. There is some unease and no explanation for the excess number of girl babies born relative to boys. Another objection has been to the routine practice of recovering as many ova from the woman's ovary as possible to be inseminated and incubated because of concern for the unused and possibly discarded embryos. A recent achievement that should allay such concern has been the successful storage of embryos through freezing so that they may be implanted at a later, perhaps more propitious time. After a number of unsuccessful attempts, an Australian research team reported on the birth, in 1984, of a healthy baby after the embryo had been frozen for 6 months, thawed, and then transferred to the mother's uterus for implantation. Two other Australian frozen embryos, however, became the focus of great controversy after the parents had been killed in a plane

crash since what to do with the stored "orphans" could not be resolved. It is obvious that the ability to store embryos for delayed implantation tends to increase the doubts rather than mitigate them since it opens the door to other possibilities with questionable social consequences—to frozen egg and embryo banks, for example, where couples could go to choose a fertilized egg of the "appropriate" sex and genetic heritage with both gametes provided by paid donors.

Wombs For Rent

Another highly controversial method of obtaining a child when the husband is fertile but the wife cannot conceive is through the use of a hired ovary and uterus or surrogate mother, in which a healthy fertile woman is paid by the couple to undergo a pregnancy achieved by artificial insemination with the husband's semen. The baby thus conceived would have half its genetic makeup from the husband of the couple and the other half from the woman who is donating her body for 9 months. The first known paid surrogate mother in this country responded to a newspaper ad placed by a California couple, was chosen over 160 applicants, and received 7,000 dollars. The price has gone up; most advertisements now promise at least $10,000 plus expenses. The practice is heartily condemned by those who find the notion of women as incubators unacceptable. It is also a legal dilemma. With little case law and no statutory guidance available on surrogate transactions, the answers to such questions as what happens if a surrogate mother wants to keep the child, or suppose the couple backs out and changes their mind because of divorce, illness, or death, or what about the birth of a child with physical or mental defects, are basically unknown. These problems have not deterred a number of physicians and attorneys from recruiting and referring surrogates or patients, and there have been some 20 surrogate parenting organizations established.

Another twist on surrogate motherhood is embryo transfer, which allows an infertile woman without functioning ovaries to bear a child fathered by her husband when another woman donates her ova. The ovum donor is artificially inseminated by the husband's sperm, and 4 days later, her uterus is flushed out with saline, the conceptus is recovered, and transferred by catheter into the uterus of the recipient. This nonsurgical technique of embryo transfer, exactly the same as the one extensively used by cattle breeders to allow ordinary cows to bear the calves of the most expensive purebred supercows, was employed in 1983 to allow several infertile women to become pregnant. The first healthy full-term infant to be born as a result was delivered in early 1984 by cesarean section (Bustillo et al., 1984). The Chicago-based Fertility and Genetics Research Inc., the developers of the technique, planned to patent the method with the profit-making aim of granting licenses to physicians allowing them to use it. The cost of embryo transfer, $4,000–$7,000, was viewed by the company as competitive with that charged for in vitro fertilization.

TREATMENT WITH FERTILITY DRUGS

Lack of ovulation (anovulation) or irregular ovulation (oligoovulation) has been treated in the past by various methods—thyroid medication, adrenocortical hormones, estrogen, progesterone, X-irradiation of the ovaries, surgical removal of a wedge of ovarian tissue— that were sometimes successful and sometimes not. In the past 20 years or so, the biggest breakthrough in the field of infertility research has been the development of the only really effective treatment available for ovulatory dysfunction. The use of fertility drugs to induce ovulation has made it possible for many women who would formerly have remained childless to become pregnant and bear children.

The three preparations used most often, either singly or in combination are (1) clomiphene citrate (Clomid), a nonsteroid synthetic; (2) human chorionic gonadotropin (HCG), obtained from the urine of pregnant women; and (3) human menopausal gonadotropin (HMG) or Pergonal, extracted from the urine of menopausal women.

The only indications for treatment with fertility drugs are absent or infrequent ovulation when other reasons for the couple's inability to conceive are lacking. Only after it has been determined that her tubes are patent, that she has ovaries capable of producing ova, and that her partner is fertile does a woman become a likely candidate for ovulation induction. Drugs have never been known to improve the chances for pregnancy in a woman who regularly ovulates.

Clomiphene citrate in low doses is the safest and least expensive method of ovulation induction and is generally used first, if conditions warrant it. Human menopausal gonadotropin is reserved for women who do not respond to this drug. Structurally, clomiphene is a distant relative to diethylstilbestrol, but it has no steroid hormonal effects in humans. Because the compound inhibited ovulation in laboratory rats, it was originally believed to have promise as a contraceptive and was tested on women in a clinical trial in 1961. An unexpected result, and probably even more surprising to the women subjects, was that clomiphene citrate turned out to induce, rather than suppress ovulation. Its action is still not completely understood, but it is now believed to be antiestrogenic. It appears to block the negative feedback effect of estrogen on the hypothalamus, perhaps by competing with estrogen for receptor sites. The hypothalamus, deceived into thinking that the endogenous estrogen level is low, produces releasing hormone that signals the pituitary production of FSH and LH to insure ovarian follicle production that ultimately results in ovulation.

The usual starting dose of clomiphene is 50 mg daily for 5 days, starting on the fifth day of the cycle. If ovulation does not occur that month within 5–12 days after the last pill, the dose is doubled in subsequent cycles to a maximum of 150–200 mg until the desired effect of ovulation, and happily, pregnancy takes place. Treatment plans vary. Some doctors will give human chorionic gonadotropin to increase the chances of successful ovulation after a few days of initial clomiphene therapy; others wait until the maximum dose proves ineffective. In some women, clomiphene's antiestrogenic properties tend to cause the production of a less receptive cervical mucus. Even if ovulation occurs, conception may be hampered. Small doses of estrogen started on day nine and continued through expected ovulation may be successful in counteracting poor cervical mucus.

There are a few major hazards and several side effects associated with clomiphene citrate therapy, but they are fewer with this drug than with the other ovulation inducers, and are relatively infrequent at the lower doses. When treatment is prolonged and large amounts are being taken, the ovaries may be overly stimulated to result in enlargement and cyst formation. If treatment is discontinued, spontaneous regression of ovarian enlargement will usually occur within several weeks. About 7%–10% of the pregnancies after clomiphene are multiple, primarily twins, but there have been at least one sextuplet and three quintuplet births. Since prematurity fre-

quently accompanies multiple birth, there is some fetal risk involved. No evidence of a greater incidence of birth defects following ovulation induction has been documented. There has been only one report, with no additional substantiation, of a greater number of chromosomal abnormalities in fetuses spontaneously aborted when conception occurred after ovulation was induced (Boue and Boue, 1973). Some side effects include hot flashes, breast soreness, nausea and vomiting, abdominal pain or soreness, and visual disturbances such as blurring or spots in front of the eyes, and temporary dryness or loss of hair. All of these symptoms occur infrequently and are reversible within a few days of discontinuation of the drug.

Clomiphene citrate functions by the domino effect; it acts on the hypothalamus, which acts on the pituitary, which produces gonadotropins, which act on the ovaries. Chorionic gonadotropins and human menopausal gonadotropins bypass the hypothalamus and pituitary and act directly on the ovaries to cause follicular maturation and ovum release. Human menopausal gonadotropin (or HMG) was originally obtained from the urine of Italian nuns, who are still a major source of the drug. Marketed in this country as Pergonal, HMG contains FSH and LH activity in a 1:1 ratio. Pituitary FSH and LH are chemically different from urinary gonadotropins, but they seem to have the same clinical effect on the ovaries. Pergonal is administered by daily muscular injection to cause follicular growth and development. When appropriate follicular maturation has been achieved, as measured by daily determination of blood or urinary estrogen levels, an injection of HCG (trade names Antuitrin-S, A. P. L., Pregnyl, or Follutein), which is biologically similar to LH, is given to create the LH surge that precedes ovulation.

The use of human gonadotropins to induce

ovulation has a high cost and high risk and is only used for women who do not respond to clomiphene citrate. The average multiple-birth rate is 25%, but more of the births are greater than twins after HMG, and this adds to increased fetal mortality. The incidence of ovarian enlargement is also considerably higher. The ovarian hyperstimulation syndrome, which includes enlarged tender and painful ovaries, distension of the abdomen by fluid, and weight gain, can be a life-threatening problem in its severe form. Serious hyperstimulation occurs in 1% of women treated with HMG.

HMG and HCG used sequentially are the most effective drugs available for the induction of ovulation. If the ovaries contain follicles and, hence, are competent, gonadotropins will cause ovulation in 90%–95% of women treated and will achieve a pregnancy rate of 40%–70% (Marshall, 1978; Gemzell, 1982).

Pergonal is a highly potent preparation, the last resort for the fewer than 5% of women who need gonadotropins because all other treatment plans for ovulatory dysfunction have failed. It should be administered only by a skilled and experienced fertility specialist who will do the intensive daily monitoring that is required. It should be taken only with full knowledge of the expense of the treatment, the willingness and time to undergo the daily therapy, and the full recognition of the potential hazards, both maternal and fetal, of the treatment.

Two other agents that are successfully able to induce ovulation are luteinizing hormone releasing hormone (LHRH) and an ergot alkaloid, bromocriptine (trade name, Parlodel). It has been found that LHRH is much more efficacious when it is administered in a pulsatile fashion, similar to the way in which it is secreted by the hypothalamus. When LHRH is delivered via a portable pump every 90–120 minutes, ovulation and pregnancy are very likely to occur within three or four treatment cycles (Miller et al., 1983). LHRH is not commercially available, and this kind of treatment is only possible as yet at major research centers. Bromocriptine is more effective in women whose amenorrhea, anovulation, and resultant infertility is the result of increased plasma prolactin levels. The drug acts by inhibiting the pituitary production of prolactin.

Psychological Factors

In 10% of the couples who are looking for an answer to their problem of infertility, no physical or physiological reasons for the difficulty can be detected. After all the tests are over, after the two people have willingly endured the indignities engendered by an infertility investigation—he, having sex several times with a clean wide-mouthed jar; she, having her reproductive organs insufflated, biopsied, dyed, fluoroscoped, and photographed, they, making love to produce a visible sample of sperm and cervical mucus, having sex on demand when a thermometer tells them to, answering innumerable questions about their private coital techniques—are now subjected to a final indignity. They are told they are "normally" infertile, and that there evidently are no answers.

As if coping and adjusting to the situation were not difficult enough, everyone, from well-meaning relatives and friends to members of the medical profession, believe they do have an answer. Since no physiological reason is apparent, the problem is obviously psychological. The couple is tense and anxious because they want a baby so badly, and it is hampering their fertility. All they really have to do is to stop worrying, to take it easy, and a pregnancy will occur. Some doctors are convinced that all these young people really need is a little calm reassurance from their friendly physician, and they relate stories about the couples that came just once to the office and then conceived with no further treatment. It is also commonly accepted that pregnancy frequently occurs after adoption, when the stress that precipitated the infertility is presumably relieved.

While there is no dearth of anecdotal folklore to bolster the hypothesis that there are psychogenic factors in infertility, scientific substantiation is absent. Even the generally accepted notion that adoption leads to pregnancy in apparently normal infertile parents has not been supported by statistical studies. Pregnancy rates for infertile couples who adopt children are the same as for infertile couples who do not adopt children (Weir and Weir, 1966; Banks, 1961; Tyler et al., 1960). A certain number of spontaneous pregnancies do occur in couples presumed sterile, but their association with a single visit to the doctor, with a Caribbean cruise, or with any other relaxing change in the environment is likely to be coincidental.

In the absence of any organic reason for the infertility—at least one discernible by present techniques—it is tempting to categorize it as emotionally induced. Not unexpectedly, the diagnosis of psychosmatic infertility is more frequently applied to women. It has been theorized that infertile women may consciously verbalize their wish to have a baby, but that they are masking their underlying rejection of pregnancy, childbirth, and motherhood. As previously noted, this subconscious-rejection-of-the-female-role theory has been used to explain many disorders—among them menstrual difficulties, spontaneous abortion, and the nausea and vomiting of pregnancy. Some recent psychiatric theories speculate that the women's movement has been responsible for a higher incidence of psychological infertility, that the newly liberated woman, career oriented, may suffer even greater psychic trauma than before as her bio-

logical role conflicts further with her personal goals. "Today's woman," writes a psychiatrist in a monthly journal aimed at the general practitioner, "has many ambitions and aspirations that compete with fertility." Today's physician, it might be observed, is obviously still being exposed to unsubstantiated generalizations about women. In a curious kind of reverse reasoning, women may even be blamed for infertility in men. It has been seriously suggested that the new sexual freedom in women has created demands that have resulted in a greater incidence of male impotence. Lack of performance in the male can be equated to lack of fertility.

Given that the hypothalamus is integral to the function of the reproductive tracts in both males and females, a psychological basis for infertility in men and women that could be mediated through some neuroendocrinological pathway is certainly a possibility that cannot be ruled out. But what are the mechanisms involved in the unconscious prevention of conception? If psychogenic infertility, rooted in conflict and anxiety, actually exists, it still would have to be manifested physically and physiologically on the cells, tissues, tubes, and secretions of the reproductive tract. In other conditions presumed to have a psychosomatic origin—stomach ulcers, colitis, asthma, tension headaches—there is also little real proof linking the disorder to the psyche, but at least there are visible signs of dysfunction. In the normally infertile woman, however, there are no visible abnormalities. It would appear, then, that there is even less reason to presume that the infertility is psychogenic.

For unknown reasons, a small percentage of evidently normal couples fail to conceive. Under such circumstances, any woman can be expected to react with feelings of frustration and depression. She should not also have to deal with the implication that no matter how much she may protest that she wants a baby, the reason that she cannot is her underlying subconscious wish not to be pregnant. Denbar's review of the medical literature concluded that there has been no meaningful documentation to support the hypothesis that psychogenic factors cause nonphysiological female infertility. Furthermore, there is no evidence, other than isolated case reports, that psychotherapy has ever been effective in causing previously infertile women to conceive.

REFERENCES

Aral, S. and Cates, W. The increasing concern with fertility: Why now? *JAMA* 250 (17):2327–2331, 1983.

Banks, A. L. Does adoption affect infertility? *Int J Fertil* 7(1):23–28, 1962.

Bongaarts, J. Infertility after age 30: A false alarm. *Fam Plann Perspect* 14(2):75–78, 1982.

Boue, J. G. and Boue, A. Increased frequency of chromosomal anomalies in abortions after induced ovulation. *Lancet* 1:679–680, 1973.

Bustillo, M., Buster, J. E., Cohen, S. W., et al. Nonsurgical ovum transfer as a treatment in infertile women. *JAMA* 251(9):1171–1173, 1984.

Denbar, S. P. Psychiatric aspects of infertility. *J. Reprod Med* 20(1):23–29, 1978.

Fédéfation CECOS, Schwartz, D., and Mayaux, M. Female fecundity as a function of age. *New Engl J Med* 306(7):404–406, 1982.

Gemzell, C. Gonadotropin treatment of infertility. *Female Patient* 7:42/60–42/64, 1982.

Gonzalez, E. Sperm swim singly after vitamin C therapy. *JAMA* 249(20):2747–2748, 1983.

Marshall, J. R. Induction of ovulation. *Clin Obstet & Gynecol* 21(1):147–162, 1978.

Menning B. E. *Infertility, A Guide for the Childless Couple.* Englewood Cliffs, N. J., Prentice-Hall, 1977.

Merz, B. Stock breeding technique applied to human infertility. *JAMA* 250(10):1257–1258, 1983.

Miller, D. S., Reid, R. R., Rebar, R. W., et al. Pulsatile administration of low-dose gonadotropin-releasing hormone. *JAMA* 250(21):2937–2741, 1983.

Sacco A. G. and Moghissi K. S.: Anti-zona pellucida activity in human sera *Fertil Steril* 31:503, 1979.

Shivers, C. A. and Dunbar, B. S. Auto-antibodies to zona pellucida: A possible cause for infertility in women. *Science* 197(4308):1082–1084, 1977.

Steptoe, P. C. and Edwards, R. G. Re-implantation of a human embryo with subsequent tubal pregnancy. *Lancet* 1:880–882, 1976.

Toth, A., Lesser, L., Brooks, C., and Labriola, D. Subsequent pregnancies among 161 couples treated for T-mycoplasma genital-tract infection. *New Engl J Med* 308(9):505–507, 1983.

Tulandi, T., Plouffe, L., and McInnes, R. Effect of saliva on sperm motility and activity. *Fertil Steril* 38(6):721–723, 1982.

Tyler, E. T., Bonapart, J., and Grant, J. Occurrence of pregnancy following adoption. *Fertil Steril* 11(6):581–589, 1960.

Weir, W. C. and Weir, D. R. Adoption and subsequent conceptions. *Fertil Steril* 17(2):283–288, 1966.

Whorton, D., Krauss, R. M., Marshall, S., et al. Infertility in male pesticide workers. *Lancet* 2:1259–1261, 1977.

Yanagimachi, R., Yanagimachi H., and Roberts, F. J. The use of zona-free animal ova as a test system for the assessment of the fertilizing capacity of the human spermatozoa. *Biol Reprod* 15:471–476, 1976.

12. *Problems of Fertility—Contraception*

As the morning paper or the evening news reports yet another complication of oral contraceptives or intrauterine devices, the old cliché seems inevitable. "Why can't a scientific technology that put a man on the moon develop a better method of birth control?" It probably could, given the same commitment, money, energy, and time that were devoted to launching satellites and the lunar landings. Finding new and improved ways of contraception is not a national priority, although the control of human fertility in terms of life on earth would probably be a greater boon for humankind than the conquest of space. Unless it suddenly became in the national interest to focus a similar kind of massive research effort on contraception, any significant improvements in contraceptive methods are not likely to be available until well into the next decade or beyond. Our present options in birth control remain limited, but it is possible to find a means of preventing pregnancy that is, if not ideal, at least fairly compatible with the different needs of different people.

We have learned to be skeptical about scientific "breakthroughs" in contraceptive technology, particularly concerning claims of increased effectiveness and safety. When oral contraceptives were developed in the early 1960s, it was believed that the ideal method of birth control had indeed been discovered. The pill was said to be totally effective, completely safe, easy to take, inexpensive, reversible, and did not interfere in any way with the sex act. All other methods suddenly became outmoded as millions of women went "on the pill." After a few years, the bubble burst. The oral contraceptive turned out to have only about half its presumed qualities. It is unquestionably effective, easy to use, and coitus unrelated, but its total safety and reversibility are highly suspect. And at a cost of around $175 to $200 annually, pills are not that inexpensive.

Then, the intrauterine device was described as the ideal contraceptive: it, too, was unrelated to intercourse, and it had an added advantage—once inserted it could be totally ignored. So easy—there was not even a daily pill to take. The side effects and complications associated with IUD usage, however, have made this method far less than ideal for many women.

The truly ideal contraceptive method would not seesaw between safety and effectiveness, but woud be both completely free of any present or future health hazard and completely effective in preventing an unwanted pregnancy. It would be completely reversible; fertility would be fully restored when the method was discontinued. It would be cheap enough for anyone to afford, distributed and marketed so that anyone could easily obtain it, and it would require no medical intervention or prescription. It would be so convenient to use that it would require a minimum of motivation, and it would be unrelated to the act of sexual intercourse—there would be nothing to put in or put on to interfere with lovemaking. The responsibility for using the method would be shared by both men and women, and it would meet everyone's cultural, religious, political, and philosophical requirements for controlling fertility. Not only does this paragon of pregnancy prevention not exist, but it should be obvious that no single method of contraception could biologically or sociologically ever have all the above attributes. What we really need, and could settle for, are a variety of new and improved methods of contraception that fulfill more of the criteria or at least strike a better balance between them. The choice between safety and effectiveness should not have to be made by any woman or man. At present, there are a rather limited number of contraceptive techniques, all with advantages and disadvantages, and none that is free of potential side effects or complications, either medical or emotional. No method except sexual abstention is 100% effective and reversible. No method that is highly effective, convenient, and unrelated to intercourse is completely safe. No technique that is completely safe is that easy to use. The choice of which method to use depends on the kind of tradeoffs an individual is prepared to make. Every woman has to decide, based on her life-style, her experiences, her personal preferences, and her right to do what she wants with her body,

which attributes of a particular method are most important. There is no best way of birth control. The only best method is the one that is chosen and is consistently used—the technique that a woman feels is right, natural, and comfortable for her and her partner.

A woman should not feel chained to a particular method throughout her reproductive lifetime. As circumstances and attitudes change, so will the reqirements for a contraceptive. When intercourse is infrequent or unexpected, it may not be necessary to use a method that provides constant daily protection. A married couple who is delaying or spacing their family may be more concerned about reversibility than effectiveness; a woman who does not ever want children or couples who have completed their families could decide on sterilization. Not infrequently, the choice of a contraceptive method depends on the availability. For adolescents or many unmarried young women, ambivalent and anxious about "planning" for sex, the psychological costs of obtaining and using birth control are already very high. Unless there is a family planning agency in the community, their access to medical contraception is limited, and getting drugstore methods can be embarrassing.

But the bottom line of birth control is the irrefutable fact that unprotected sexual intercourse, sooner or later, causes pregnancy. Engaging in sex carries the assumption of risk of pregnancy and the responsibility for preventing it. Each woman knows how she alone would be able to cope with the consequences of an unplanned pregnancy or the alternatives of abortion or childbirth. If the prevention of pregnancy is important she has to take the necessary steps to avoid it.

It is not possible to choose a preferred method in the absence of complete and accurate information about all existing techniques. Women who relied on their physicians for advice in the past have resentfully claimed that they were never told that anything more than nuisance side effects accompanied oral contraceptive or IUD usage. All of the emphasis was placed on the convenience and effectiveness of the two methods while considerations of safety were softpedaled. Some doctors rarely recommended use or provided information on any other types of contraception. Even today, when much more is known concerning the adverse physiological effects of pills and IUDs, women may still be steered away from safer methods because of the personal preferences or biases of their physicians or health professionals.

The two most highly prescribed methods—oral contraceptives and intrauterine devices—are undeniably highly effective in preventing pregnancy. What proves most surprising to many people is that *all major contraceptive methods* (that is, legitimate means of fertility control rather than hope and prayer) can be *approximately equally effective in avoiding an accidental pregnancy.* The major difference is that the effectiveness of methods other than pills and IUDs requires more *effort* on the part of the user to achieve the same success.

THEORETICAL EFFECTIVENESS AND USE-EFFECTIVENESS

The effectiveness of any method of birth control refers to its ability to prevent pregnancy. Oral contraceptives are known to be highly effective. All that is required of a woman to achieve this high efficacy is to swallow one every day. As long as she remembers to do that, the chances of her conceiving are extremely small. If she forgets to take her pills, however, or runs out of them without another pack handy, or for any other reason neglects taking them, she is not going to be as well protected by the pill. The *theoretical effective-*

ness of a method is its maximum success rate when it is used correctly, perfectly, and without error. The actual *use-effectiveness* of a method refers to all the possible ways it is used by everyone—not only properly, but also incorrectly, carelessly, and not according to instructions.

Any ranking of contraceptive effectiveness generally places oral contraceptives and IUDs up at the top of the list along with sterilization, abstinence, and abortion. Such methods as the diaphragm, the condom, and vaginal foam, while rated above rhythm, douching, and jumping up and down after intercourse, are usually considered well below the pill and IUD in effectiveness. Diaphragms have long been called "baby-makers," condoms are said to burst, leak, or otherwise prove untrustworthy; and spermicides are said to have a failure rate that makes them virtually useless for women who must not become pregnant. Such quantitative comparisons between common methods of birth control can be very misleading, however, because what is usually being quoted is the theoretical effectiveness of the pill and IUDs, and the use-effectiveness of all the other methods.

Measurements of contraceptive effectiveness are based on statistics gathered from major studies published in scientific literature over the last 35 years or so. These studies have been conducted at various times in many different countries on groups of people who varied in age, marital status, income, education, and in motivation to use the method. When all the results are lumped together, obviously a very wide range of effectiveness for a contraceptive technique or device is possible. The diaphragm, for example, is often said to be only about 80% effective. This could mean that if 100 fertile women get a diaphragm and use it for a year, at the end of that period, 20 of them may be pregnant—a statistic that would discourage almost anyone from using a diaphragm. But this takes into account all the ways in which a diaphragm is used—well fitted, properly, and consistently or poorly fitted, forgotten, and neglected.

Table 12.1 illustrates the theoretical effectiveness and the use-effectiveness of the various methods of birth control. Note the great latitude in effectiveness for most techniques. The ability of almost all contraceptives (omitting only the IUD) to prevent pregnancy is dependent on the person using the method. With motivation and consistency, the success

of a diaphragm or condom in avoiding conception compares very favorably with the pill or IUD. Even vaginal spermicides, long believed to be disastrously ineffective, have an equivalent success rate when used properly and have been used by many couples for long periods of time without a problem.

ORAL CONTRACEPTIVES

In June 1960, the Food and Drug Administration approved for safety and efficacy a little pink pill called Enovid-10, produced by Searle & Company. Enovid had actually been on the market for several years, prescribed for threatened spontaneous abortion and menstrual disorders, but its introduction as an oral contraceptive was the start of a new era in birth control—chemical contraception. Other brands soon appeared, and what generically became known as "The Pill" rapidly achieved enormous popularity. The social impact of what was hailed as a breakthrough in effective, easy, and safe contraception was so far-reaching that people began to speak of a "pill revolution" and then later of a "sexual revolution."

Never before in history has a drug treatment been so widely prescribed on a continuing basis to so many people who were normal and healthy. There were some scientists who had misgivings about the prolonged administration of sex steroids and questioned the wisdom of a daily hormonal onslaught on the basic female endocrine mechanisms. There were some doctors who were reluctant or refused to advise the use of oral contraceptives. Almost everyone else, however, was extraordinarily pleased with the pills—the drug companies that produced them at a substantial profit, the federal agencies that sanctioned their use, the doctors who prescribed them, and the women who took them as if

Table 12.1 CONTRACEPTIVE EFFECTIVENESS THEORETICAL AND ACTUAL USE RATES[a]

Method	Theoretical Effectiveness	Use Effectiveness (range)
Abortion	0	0
Abstinence	0	?
Tubal ligation	0.04	0.06
Vasectomy	0.15	0.15
Oral contraceptive (combined)	0.34	4–10
Progestin-only pill	1–15	5–10
Condom plus foam	less than 1	5
IUD	1–3	5
Condom	3	10
Diaphragm with jelly or cream	3	2–20
Cervical cap	3	2–20
Spermicidal foam or cream	3	2–30
Sponge	9–11[b]	13–16[b]
Coitus interruptus	9	20–25
Rhythm (calendar only)	13	21
Natural family planning (symptothermal)	2.5	5–40
Chance	90	90

Sources: Population Reports; Hatcher, et al.; DHEW Publication No. (SHA) 76-16030.

[a] Number of pregnancies per 100 women during the first year of use.

[b] According to manufacturer.

SWALLOW ME were written on every tablet. By 1962, 2 million American women were on the pill; by 1965, oral contraceptives were being ingested in this country alone at the annual rate of more than 2,000 tons. Pharmaceutical company stocks rose as "buy birth control" became the watchword on Wall Street. And society at large was gratified by continued progress toward a currently popular goal, the slowing of the population rate.

It is not surprising that women needed little encouragement to go on the pill. The rewards were many. For centuries the idea of swallowing something to prevent or eradicate pregnancy had caused women to dose themselves with ineffective and often dangerous potions and nostrums. Now the magic pill had really been found. For young single women it was such a nice, tidy, guilt-free way of avoiding pregnancy. The pill was completed dissociated from sexual activity, so there was no need to plan ahead for sex—no diaphragm and big tube of jelly in the purse, no other equipment to carry—to destroy the illusion of, or somehow mess up, a completely spontaneous encounter. It was true that there appeared to be some minor side effects associated with usage, but the biological advantage of almost perfect effectiveness plus the added benefit of the elimination of menstrual pain and bleeding for many users outweighed any disadvantages. Younger women, both single and married, took to the pill with alacrity, and a new kind of social ideology arose—that the entire responsibility for birth control belonged to women. But even that seemed to be liberating. A woman, after all, is the one who bears the burden of an unwanted pregnancy, and she should have complete control of her reproductive capacity without needing assurances that her partner would "take care of her."

Any anxieties about safety were relieved not only by women's physicians, many of whom became pill pushers, but also by numerous articles in popular magazines and newspapers. With such titles as "What You Should Know About Birth Control Pills," "How Safe are the Birth Control Pills?," and "Birth Control Pills: The Full Story," Vogue, Redbook, Ladies' Home Journal, and Good Housekeeping reassured readers that no adverse effects had as yet become apparent in the years of clinical trials since 1956. Later, the popular media were to sensationalize the risks of oral contraception; originally, the media were instrumental in allaying women's fears about them.

By the mid-1960s, it had become apparent that not only nuisance effects but also serious complications were possible for pill users. Reports that oral contraceptives were related to clotting disorders, strokes, and a number of other serious complications resulted in a decline in usage. Congressional hearings were held to determine the safety of oral contraceptives, and there were warnings and speculations concerning the possible risks to health. It also became evident that the adverse affects were estrogen related. When new formulations containing less estrogen came on the market, renewed confidence in the pill's safety resulted in a regaining of popularity. In time the prewarning levels of pill usage were again reached and exceeded (Fig. 12.1).

Sales of the pill in the United States climbed until 1976 when, in response to additional evidence of health hazards associated with usage, a sharp decrease occurred. Concern for safety was illustrated by a Gallup poll taken in July, 1977, when the pill was believed to be unsafe by 62% of the women and 43% of the men surveyed. But the drop in pill sales appeared to end in the late 1970s, and in recent years the percentage of women purchasing pills has remained approximately the same. Since trends in pill use appear to be linked to public dissemination of information about them, it is possible that the wide publicity given studies in the early 1980s that reported

Figure 12.1. Some of the various brands of birth control pills.

the noncontraceptive health benefits of pill taking and claimed that the risks were exaggerated will lead to increased usage. That public opinion had changed was shown by a poll of 10,000 women conducted by the research firm, Market Facts, in the spring of 1982. In contrast to the attitude exhibited five years earlier, 65% of the women respondents held a favorable opinion about the pill (Forrest and Henshaw, 1983). Currently, more than 50 million women throughout the world are estimated to rely on oral contraception, including anywhere from 8 to 10 million women in North America.

Mechanisms of Action of Oral Contraceptives

Although hormonal contraceptives are commonly referred to as "the pill," they are not all alike. They differ in their composition, their potency, and the side effects that they produce. Depending on the amounts and kinds of steroids they contain, the pills even differ in their contraceptive effectiveness. All combination pills are composed of a synthetic estrogen and one of five synthetic progestational agents, or progestins. Most are available in both 21- and 28-tablet packages. The progestin-only, or minipills, contain a low dose of one of the progestins, have a lower theoretical effectiveness rate than the combined pills, and are supplied in 28 and 42 tablet packages. There have been no real changes in oral contraceptives since their introduction in 1960, but most of the manufacturers believe there is a marketing advantage in being able to provide a variety of pills, so they keep introducing combinations of different strengths and potencies. The low-dose and ultra-low dose pills became available in the late 1970s, and recent entries in the product line are the "biphasic" and "triphasic" combinations that step up the dose of the progestin at intervals through the pill-taking cycle and provide a multicolor assortment of tablets during the month.

When the combined estrogen and progestin pills are taken daily, they are believed to function primarily by suppressing the production of the releasing hormone from the hypothalamus. Since pituitary production of FSH and LH is then inhibited, normal follicle growth, maturation, and ovulation of an ovum cannot occur. The synthetic steroids in the pills provide a constant amount of hormones, which are metabolized within 24 hours and which replace the normal cyclic production of a woman's own ovarian estrogen and progesterone. If for some reason the midcycle surge of LH is not completely inhibited, it is possible that an "escape" ovulation may occur. Secondary contraceptive mechanisms are then available from the progestational activity in the pills. Sperm penetration into the uterus is prevented by an alteration in cervical mucus, tubal transport of the ovum is slowed, and the endometrium undergoes changes that inhibit implantation. Progestin-only pills—the minipills—probably rarely prevent ovulation, and exert their function primarily through these secondary effects.

The monthly bleeding that occurs while taking oral contraceptives is a false menstruation produced by estrogen and progesterone stimulation of the endometrium, followed by withdrawal of the hormones 7 days before the onset of bleeding. On a 21-day pill, no hormones and no pills are taken for seven days, and on a 28-day regimen, the last seven pills are of a different color and contain no hormones. Menstruation is provided by the pills so that a woman feels more natural and comfortable about taking them. Should a woman want to avoid menstruating while on vacation or during some special occasion all she has to do is to take some extra pills from a different package—the menstruation offered by pills is so fabricated that it can be avoided completely.

Since ovulation and menstruation do not occur during pregnancy because of the inhibitory effect of the estrogen and progesterone from the placenta on the hypothalamus, it has been suggested that a woman who takes oral contraceptives and also does not ovulate is in a state of pseudopregnancy and can thus be physiologically equated with a pregnant woman. In this view, held by many including John Rock, a developer of oral contraceptives, the pills are the most "natural" form of birth control. After all, proponents of this notion suggest, early in human history, primitive women were likely to experience 10 or more pregnancies during their fertile years. In the absence of any contraception, all those pregnancies, plus the lack of menstruation as a result of breastfeeding afterwards, probably meant that early women had only rare intervals of regular menstrual cycles (Short, 1976). In contrast, this argument continues, contemporary women using any contraceptive other than the pill are having years of true menstrual cycles that could, in an evolutionary sense, be more unnatural. Biologically speaking, women may be intended to gestate and lactate more and menstruate less. The specious supposition is that oral contraceptives are more nearly in accord with Mother Nature's wishes because they suppress the many true menstruations that women were not meant to have anyway.

Even if one accepts the theory that under primitive conditions women spent the greater part of their reproductive lives in a state of amenorrhea, the major fallacy in this reasoning is that contraceptive steroids are not natural at all—they are synthetic. They do not produce the same effect as the hormones of pregnancy, and they function similarly, but not identically, to the estrogen and progesterone produced by a woman's own ovaries and adrenal glands. They have metabolic effects in addition to, and different from, their contraceptive effects. When oral contraceptives are taken in a constant daily dose they do not produce a physiologically natural state; they produce a pharmacologically hormonally-induced state. The endocrine balance in a woman on oral contraceptives is not the same as it is in a pregnant woman or in a woman not taking synthetic hormones, but this may or may not be harmful to an individual woman's health. Some women do develop serious complications associated with taking oral contraceptives. Unfortunately, it is difficult, and in many instances impossible, to predict which women will be affected.

Natural ovarian estrogens and progesterone are inactivated when they are taken orally. More than 40 years ago it was discovered that if estradiol were chemically changed by the addition of an ethinyl group at the 17 position (see Fig. 12.2), the estrogen became orally active. *Ethinyl estradiol* is, therefore, one of the two forms of estrogen in every combined oral contraceptive. The other form is *mestranol*, the 3-methyl ether of ethinyl estradiol. Mestranol has less estrogenic activity when tested in animals than ethinyl estradiol.

The five progestational agents, or progestins, in oral contraceptives are all synthesized from testosterone or hydroxyprogesterone. Some of them retain their androgenic activity in addition to their estrogenic and progestation potency. Some are also variously antiestrogenic and antiandrogenic. Norethynodrel is a highly estrogenic progestin, but it has no androgenic ability. Norgestrel has no estrogenic ability but has the strongest androgen effect of all the progestins as well as a strong antiestrogen effect. Norethindrone, norethindrone acetate, and ethynodiol diacetate are weakly estrogenic and androgenic at low doses but can become relatively antiestrogenic at higher doses. Thus, every pill on the market has its own profile of estrogenicity, androgenicity, and progestational potency (see Table 12.2). It is believed that the occurrence

of minor and major side effects is correlated to the differences in the various hormonal activities in different pills (see Table 12.3).

The tables and the charts that have been developed to express the hormonal potencies are based on data obtained through tests on animals and humans. In many instances, the potency differences are only relative, however, because biochemical interconversions of estrogens and progestins take place within the body. Although a knowledge of the hormonal activity of the various constituents of any one pill is useful, most doctors would admit that trying to initially tailor a particular formulation to suit a particular woman is mainly guesswork. According to some women's reports, an educated guess could take the form of a quick appraisal of the hairiness of a woman's arms and legs—certainly a less than adequate assessment of her hormonal status. Assuming there are no contraindications to the use of oral contraceptives, many physicians start a woman on a pill that contains 50 micrograms or less of estrogen. If side effects that can be correlated with the hormonal potencies of that initial pill develop, the dosage can be modified or a switch to another low-dose pill can be made. Some of the problems that develop, such as spotting, breakthrough bleeding that requires a pad or tampon, or failure to have a withdrawal menstruation, are due to hormonal excess or deficiency and can be solved by changing pills. Some of the other side effects, while presumed to be estrogen related, are less able to be pinpointed to specific greater or lesser hormonal activity. Switching around, while it may be effective, can also become arbitrary and empirical.

Complications Associated with Oral Contraceptive Use

A brochure that accompanies any package of oral contraceptives to be dispensed to a pro-

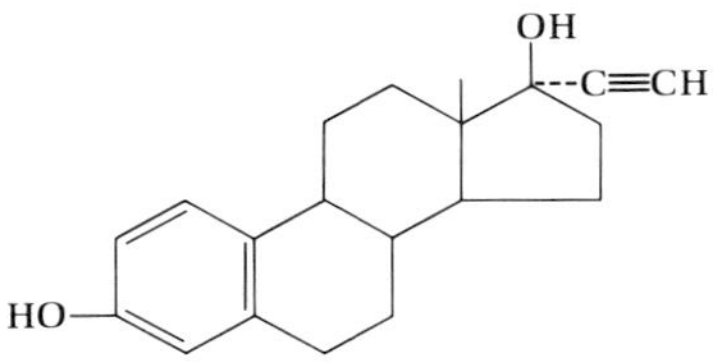

Ethinyl estradiol

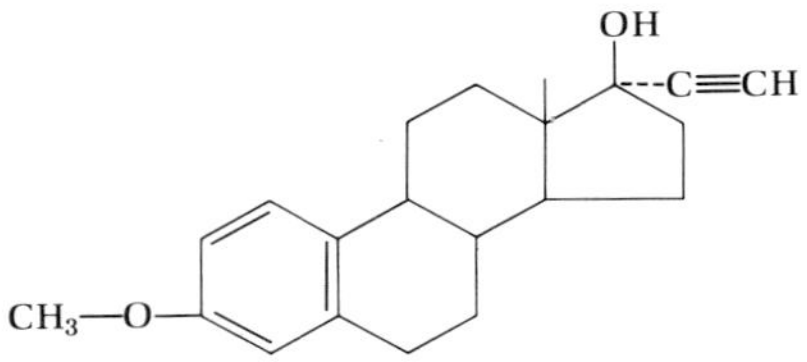

Mestranol

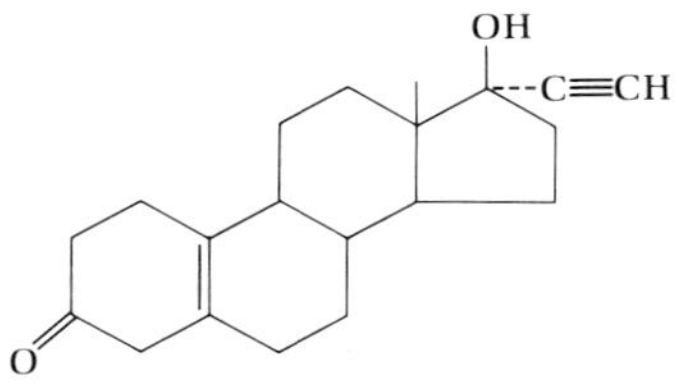

Norethynodrel
(Enovid)

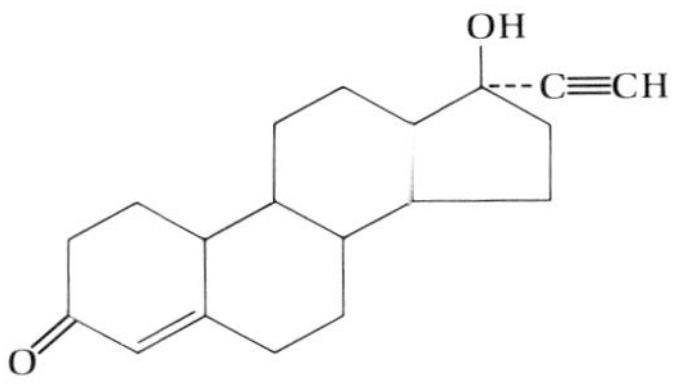

Norethindrone
(Ortho-Novum, Norinyl, Norlutin)

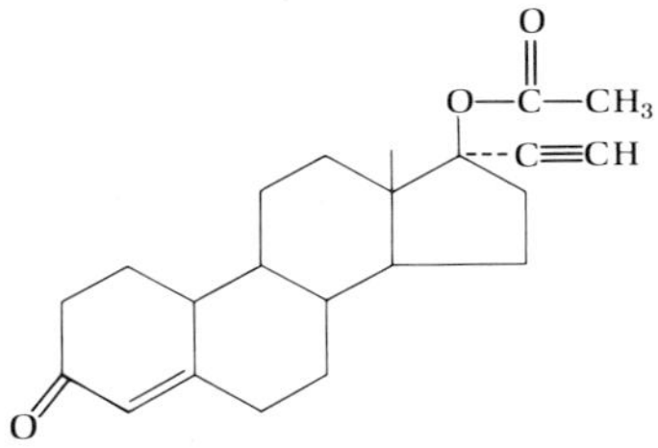

Norethindrone acetate
(Norlestrin, Norlutate)

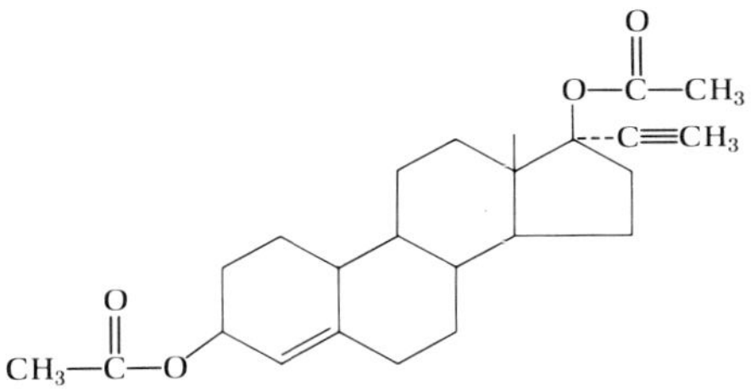

Ethynodiol diacetate (Ovulen, Demulen)

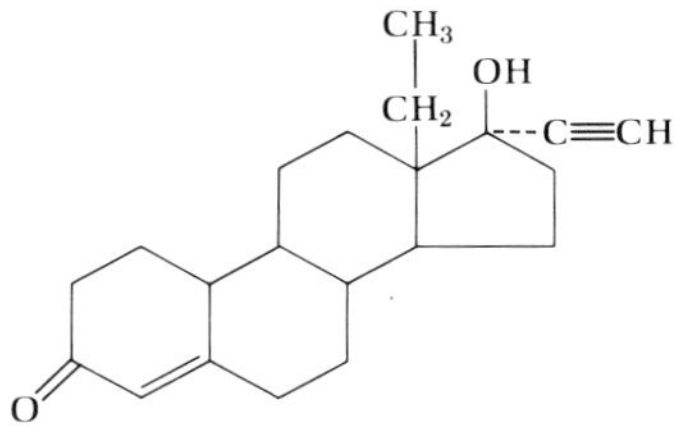

Norgestrel (Ovral)

Figure 12.2. Chemical structure of oral contraceptive agents.

416

Table 12.2 PROGESTIN AND ESTROGEN CONTENT AND CHARACTERISTICS OF ORAL CONTRACEPTIVES MANUFACTURED IN THE UNITED STATES

Name	Manufacturer	Progestin (mg/tablet)	Estrogen (μg/tablet)	Estrogen Potency[a]	Progestational Potency[b]	Androgenic Potency[c]
Brevicon	Syntex	Norethindrone 0.5	Ethinyl estradiol 35	42	.19	.17
Demulen 1/50	Searle	Ethynodiol diacetate 1.0	Ethinyl estradiol 50	26	.53	.21
Demulen 1/35	Searle	Ethynodiol diacetate 1.0	Ethinyl estradiol 35	19	.53	.21
Enovid E	Searle	Norethynodrel 2.5	Mestranol 100	80	.25	.00
Enovid 5	Searle	Norethynodrel 5.0	Mestranol 75	240	.48	.00
Enovid 10	Searle	Norethynodrel 9.85	Mestranol 150	438	.94	.00
Loestrin 1.5/30	Parke-Davis	Norethindrone acetate 1.5	Ethinyl estradiol 30	14	.65	.79
Loestrin 1/20	Parke-Davis	Norethindrone acetate 1.0	Ethinyl estradiol 20	13	.44	.52
Lo/Ovral	Wyeth	Norgestrel 0.3	Ethinyl estradiol 30	25	.30	.48
Micronor	Ortho	Norethindrone 0.35	—	1	.80	.16
Modicon	Ortho	Norethindrone 0.5	Ethinyl estradiol 35	42	.19	.17
Nordette	Wyeth	Levonorgestrel 0.15	Ethinyl estradiol 30	25	.30	.47
Norinyl 1/35	Syntex	Norethindrone 1.0	Ethinyl estradiol 35	38	.38	.34
Norinyl 1/50	Syntex	Norethindrone 1.0	Mestranol 50	32	.38	.34
Norinyl 1/80	Syntex	Norethindrone 1.0	Mestranol 80	42	.38	.34
Norinyl 2	Syntex	Norethindrone 2.0	Mestranol 100	46	.74	.67
Norlestrin 1/50	Parke-Davis	Norethindrone acetate 1.0	Ethinyl estradiol 50	39	.44	.52
Norlestrin 2.5/50	Parke-Davis	Norethindrone acetate 2.5	Ethinyl estradiol 50	16	1.02	1.31
Nor-Q.D.	Syntex	Norethindrone 0.35	—	1	/12	.12
Ortho-Novum 1/35	Ortho	Norethindrone 1.0	Ethinyl estradiol 35	38	.38	.34
Ortho-Novum 10/11	Ortho	Norethindrone 0.5 (10 pills) Norethindrone 1.0 (11 pills)	Ethinyl estradiol 35 Ethinyl estradiol 50	40	.29	.26
Ortho-Novum 1/50	Ortho	Norethindrone 1.0	Mestranol 50	32	.38	.34
Ortho-Novum 1/80	Ortho	Norethindrone 1.0	Mestranol 80	42	.38	.34
Ortho-Novum 2	Ortho	Norethindrone 2.0	Mestranol 100	46	.74	.67
Ovcon 35	Mead Johnson	Norethindrone 0.4	Ethinyl estradiol 35	40	.15	.14
Ovcon 50	Mead Johnson	Norethindrone 1.0	Ethinyl estradiol 50	50	.38	.34
Ovral	Wyeth	Norgestrel 0.5	Ethinyl estradiol 50	42	.50	.79
Ovrette	Wyeth	Norgestrel .075	—		.08	.12
Ovulen (21)	Searle	Ethynodiol diacetate 1.0	Mestranol 100	68	.53	.21

Source: Data on potency reprinted with permission from Richard P. Dickey, M.D., Ph.D., and Creative Informatics, Inc.

[a] μg ethinyl estradiol equivalents per day
[b] mg norgestrel equivalents per day
[c] mg methyl testosterone equivalents per 28 days

Table 12.3 PILL SIDE EFFECTS: HORMONE ETIOLOGY

Estrogen Excess	Progestin Excess	Androgen Excess	Estrogen Deficiency	Progestin Deficiency
Nausea, dizziness	Increased appetite and weight gain (noncyclic)	Increased appetite and weight gain	Irritability, nervousness	Late breakthrough bleeding
Edema and abdominal or leg pain with cyclic weight gain	Tiredness and fatigue and feeling weak	Hirsutism	Hot flashes	Heavy menstrual flow and clots
Leukorrhea	Depression and decrease in libido	Acne	Uterine prolapse	Delayed onset of menses following last pill
Increase in leiomyoma size	Oil scalp, acne	Oily skin, rash	Early and midcycle spotting	Dysmenorrhea
Chloasma	Loss of hair	Increased libido	Decreased amount of menstrual flow	Weight loss
Uterine cramps	Cholestatic jaundice	Cholestatic jaundice	No withdrawal bleeding	
Irritability	Decreased length of menstrual flow	Pruritus (itching)	Decreased libido	
Increase female fat deposition	Hypertension (?)		Diminished breast size	
Cervical ectropia	Headaches between pill packages		Dry vaginal mucosa and dyspareunia	
Contact lenses don't fit	Monilia vaginitis cervicitis		Headaches	
Telangiectasia (vascular "spiders")	Increase in breast size (alveolar tissue)		Depression	
Vascular type headaches	Breast tenderness without fluid retention			
Hypertension(?)	Decreased carbohydrate tolerance			
Lactation suppression	Dilated leg veins			
Headaches while taking pills	Pelvic congestion syndrome			
Cystic breast changes				
Breast tenderness with fluid retention				
Thrombophlebitis				
Cerebrovascular accidents				
Myocardial infarction				
Hepatic edenoma				

Source: *Contraceptive Technology*, 1984–1985 by Robert Hatcher, et al. Irvington Publishers, Inc., 1984. Reprinted by permission of the publishers.

spective user warns of more than 50 side effects, some merely unpleasant but minor and tolerable, others serious enough to be life threatening or fatal. Such detailed labeling was mandated by the Food and Drug Administration in 1978. At a news conference announcing the requirement, FDA commissioner Donald Kennedy, the first non-physician to head this government agency, commented that while his opinion was not the official recommendation of the FDA, his "personal advice to the people I love would be to find another method."

The package insert contains a lengthy description in very tiny print. Most women, once they have decided on the pill, go to the doctor for the prescription and fill the prescription at the pharmacy. At this point, having made the choice and paid for both pills and office visit, they are not likely to even read the labeling, much less be deterred by its content. But careful reading could prove a revelation to a woman who had not really been aware of the broad range of the risks associated with pill use. And while everyone may not want to "find another method" of birth control as a result of knowing all the dangers, those risks must be considered as another factor in addition to the convenience and effectiveness of oral contraceptives before coming to a decision to use them.

According to the information given to both physicians and women, the absolute contra-indications to pill usage are pregnancy, present or past history of blood clotting disorders, present or past history of cerebrovascular or coronary artery disease, known or suspected estrogen-dependent malignancies, any undiagnosed vaginal bleeding, or any present or past liver disease. This would imply that any woman who wants to rely on pills can take them, provided she does not have any of the above conditions. In actuality, the warnings and precautions that go along with those six absolute must-not-take-pills circum-

stances provide information that clearly indicates that there are a substantial number of women who also *should* not take oral contraceptives—women whose age, health, family history, heredity, and/or behavior is such that for them, taking the pill is too chancy. When, for example, a woman already has an underlying risk of developing heart disease or cancer, the superimposition of oral contraceptives on those predisposing factors is going to increase her odds of developing pill-associated problems.

When a woman has none of the conditions or predisposing factors that precludes oral contraceptive use, or no minor problems that might worsen as a result of taking pills, she would then appear to be a good candidate for this type of birth control at this time. For her, the risks of developing major or minor complications should be small. It should be kept in mind, however, that daily ingestion of synthetic steroidal hormones affects more than the reproductive tract. Oral contraceptives challenge the body's physiological resources, and some women by nature are probably better equipped to handle the challenge. Unfortunately, at the present time there are no physical clues and no laboratory tests that can accurately predict which presumably normal and healthy women on oral contraceptives will be less able to cope physiologically and could, therefore, develop the complications that can accompany pill use. And without any clues or predictions, the situation becomes a kind of roulette.

Thromboembolic and Other Cardiovascular Disorders

A *thrombus* is an abnormal blood clot that forms in an unbroken blood vessel. When a thrombus occurs in a vein *(phleb)* secondary to inflammation of the vein *(phlebitis)*, the conditions is called thrombophlebitis. Thrombophlebitis most frequently occurs in the superficial and deep veins of the pelvis and legs, sometimes for no apparent reason, sometimes because of injury or trauma, sometimes postoperatively as a result of immobilization of patients in bed. Once a thrombus or clot forms, it may spontaneously dissolve, but more often it remains intact and grows to interfere with the oxygen supply to the surrounding tissues, resulting in tissue damage and swelling. When a piece of thrombus breaks off from its attachment to be carried in the bloodstream it becomes a traveling clot, or *embolus*. Emboli are dangerous because they flow freely in the circulation, stopping only when they become stuck somewhere, and then they block the blood supply to a vital organ, such as the lungs, brain, or heart. A clot large enough to completely occlude the pulmonary arteries causes massive pulmonary embolism and instant death. A clot in one of the brain arteries could occlude circulation to a part of the brain and cause cerebral thrombotic stroke. Strokes can also occur as a result of cerebral aneurysm, a ballooning-out and bursting of an artery to cause hemorrhage.

It has been well established that oral contraceptives affect clotting factors in blood, and somehow produce alterations in blood vessel walls to create an increased risk of pulmonary embolism, cerebral thrombotic stroke, and cerebral hemorrhagic stroke. Women who use pills for birth control are four to 11 times more likely to develop these conditions than nonusers.

Any factor in a pill user that increases the likelihood of the development of thromboembolic disorders will obviously increase the risk. Women who have high blood pressure, diabetes, who are over 35–40 years old, or who have any past history of phlebitis, have predisposing conditions and should not take oral contraceptives. Because deep vein blood clots in the legs have a greater tendency to form after surgery, a woman must not initiate taking oral contraceptives or must discontinue their use if elective surgery is planned

within the next 4 weeks. There is some evidence, however, that changes in the clotting factors of the blood that could accompany pill use may persist for longer than a month after the contraceptives are discontinued. Some doctors, therefore, administer anti-clotting factors prophylactically before surgery. Pills also should neither be started nor taken if a woman has a limb fracture that requires long-term immobilization in a cast. For unknown reasons, women with type A blood have a greater propensity for developing clotting disorders when taking oral contraceptives.

The abnormal vascular changes induced by oral contraceptives can also increase the chances of a fatal heart attack (myocardial infarction). But while pill use alone increases the risk, the danger is much greater for women who already have some of the risk factors associated with heart disease—age over 40, high blood pressure, high cholesterol levels, diabetes, and of major importance, cigarette smoking. Smoking plus oral contraceptives can be a lethal combination not only because the risk of heart attack is increased, but also because of a greater incidence of other circulatory diseases. A 1978 report from the Kaiser-Permanente Contraceptive Drug Study in California showed that the risk of brain hemorrhage in pill users who smoked was 22 times greater than that of nonusers and nonsmokers (Petitti and Wingerd, 1978).

Cigarette smoking adds to the risks of oral contraceptives to such an extent that a woman must decide whether she wants to smoke or to take the pill; *she must not do both.*

While it had been known since 1967 that the pill had adverse effects on the heart and blood vessels to cause thromboembolic disease, cerebral hemorrhage, and myocardial infarction, a 1977 report from Great Britain clearly indicated that the use of oral contraceptives was associated with an increase in mortality from an expanded number of other cardiovascular disorders as well. The Royal

College of General Practitioners (RCGP) has been studying the effects of oral contraceptives on nearly 50,000 women in the United Kingdom, half of them pill users, since 1968. On the basis of data gathered through June 1976, Valerie Beral and Clifford Kay reported that the risk of dying from *all* circulatory diseases was four times greater in women who had ever taken the pill. The overall difference in mortality between ever-users of the pill and those who never used pills, or control subjects, was 20 or more deaths per 100,000 women per year, or one extra fatality as a result of pill taking for every 5,000 women.

The circulatory diseases implicated in the excess deaths included, in addition to the pulmonary emboli, strokes, and heart attacks that had been previously reported, also *subarachnoid hemorrhage* (beneath the arachnoid covering of the brain), *malignant hypertension* (a highly severe form of high blood pressure that literally bursts blood vessels), *acute myocardial insufficiency* (heart failure), and *cardiomyopathy* (disease and destruction of heart muscle). The findings of the RCGP study also suggested that the mortality risk was greater when oral contraceptives were used continuously for 5 or more years and that the danger may even remain for some time after pill use is discontinued.

A 1981 update report by the Royal College confirmed the risk of pill usage and cardiovascular disease but found that young nonsmoking women can use the pill with less chance of serious complications because the greatest risk of mortality concentrated in users who were 35 or older, especially if they smoked. In fact, smoking interacts with oral contraceptive use and age to the extent that a woman of 25 who takes OCs and smokes ages her heart and blood vessels by 10 years; that is, she faces the same risk of death from cardiovascular disease as nonsmoking pill-user who is 35 years old.

Another 1981 study, by epidemiologists

Slone et al., suggested, as did the RCGP study, that the risk of myocardial infarction persists among former pill users who had discontinued use 5–10 years previously. This report again underscored the harmful effects of smoking since the subgroup of users with the greatest risk were heavy smokers 40–49 years of age who had used the pill for 5 or more years and discontinued use within the past 10 years.

The risks of cardiovascular disease and death were suspected to be higher with a higher dose of estrogen, and the introduction and use of low-dose pills containing 50 micrograms or less of estrogen was believed to decrease the risk. A later finding, however, has been that the progestin component of the pill may also contribute to the risk, and data from the Royal College now shows a link between the incidence of arterial disease and progestin dose (Kay, 1982).

The development of *high blood pressure* is another cardiovascular effect that occurs for women on the pill. Abnormal elevations were estimated by Laragh (1976) to occur in about 5% of women users, which means that perhaps 500,000 of the possible 10 million women on the pill in the United States are walking around with induced hypertension—a condition that increases their risk of stroke and heart attack. Women who previously have developed hypertension, black women, or women with a family history of hypertension, are more likely to develop high blood pressure when taking oral contraceptives. Slight to moderate elevations of systolic blood pressure, still within the normal range, occur in almost all women after 3 years of continuous pill use (Ramcharan et al., 1974). Most of the women in the epidemiological studies above were taking the high-dose estrogen formulations, but there is more recent evidence to show that the low-estrogen combination and progestin-only pills also result in increased blood pressure readings (Khaw and Peart,

1982). In a study of women with oral contraceptive-induced hypertension, Weir (1982) found that changing from a high- to low-dose pill resulted in a marked decline in blood pressure, but the pressure never went back down to pre-pill-taking levels. The hypertension as a result of pill usage appears to be reversible when the contraceptives are discontinued, but the long-term effect of even "normal" induced high blood pressure is unknown.

The incidence of excessive bleeding and "dry sockets" is reportedly higher in women who have tooth extractions while on the pill. Obviously, not all surgery is scheduled in advance and there is often no opportunity to stop taking the pill. But whenever possible, oral contraceptive use should be discontinued and another form of contraception chosen for 1 month before any surgical procedure.

The vascular changes that accompany pill usage can also occur in the blood vessels that supply the eye. There have been some reports of visual loss as a result of retinal artery blockage, retinal swelling, and optic nerve damage, but the incidence is rare. But when a woman on oral contraceptives has blurring of vision, double vision, sees flashing lights, or has a sudden loss or diminishing of sight, it may mean a vascular spasm and an impending stroke. Such symptoms require an immediate cessation of pill taking and a medical consultation.

Since estrogen causes fluid retention, some visual changes may occur as a result of swelling or steepening of the cornea. Blurring or difficulty in focusing, the need for a new contact lens prescription, or the inability to wear contacts at all are side effects of pill taking that can sometimes be relieved by switching to a lower estrogen pill. Those who wear contacts may view this as more then merely a small annoyance; lenses are expensive to change, and most women would be reluctant to give them up completely.

There have been several reports in the medical literature that some women taking oral contraceptives acquire slight defects in color vision, particularly in distinguishing yellow-blue and green-blue hues. Two Canadian researchers confirmed an increased loss in color discrimination in diabetic women on the pill and have suggested that defective color vision in pill users may be related to an altered carbohydrate metabolism (Lakowski and Morton, 1977).

Migraine headaches are a vascular phenomenon. They may, therefore, be intensified by oral contraceptives, and women who are subject to migraines should not take the pill. A vascular headache that signals a cerebral thrombosis often has the same symptoms and may be indistinguishable from a severe migraine headache.

Carbohydrate and Fat Metabolism Effects

In a significant number of women, oral contraceptives produce an abnormal response to the glucose tolerance test, a measure of a fasting individual's ability to handle the ingestion of 1 gram of glucose per kilogram of body weight. Such a disturbance in carbohydrate metabolism is a pill-produced diabetes-like effect and makes it unwise for women with prediabetes as yet clinically undetected or women with a strong family history of diabetes to take oral contraceptives. Which hormonal component of the pill produces the altered glucose tolerance is unknown, but the change is reversible when pills are discontinued. Diabetic women on oral contraceptives do not get worse or need more insulin, but diabetics have an increased risk of developing cardiovascular disease and the pills may augment their danger.

The lipid content of the blood—neutral fats or triglycerides, phospholipids, cholesterol, high and low density lipoproteins—is influenced by oral contraceptive use. An elevation in blood triglyceride and cholesterol levels, reported in some pill users, may be a predisposing factor for heart disease. Studies have shown that women using oral contraceptives have reduced blood levels of high density lipoproteins (HDL) (Briggs and Briggs, 1982). High levels of high density lipoproteins are believed to be protective against heart disease; low or reduced HDL levels are considered a major risk factor. This alteration in the high-density lipoprotein fraction of the blood in older women pill users may be significant in the reported increase of mortality from circulatory disease.

The relationship between hyperlipidemia (high lipid levels in the blood) and heart attacks is well recognized. Even a young woman who has a family history of heart disease and high cholesterol and triglyceride levels should have a complete blood chemistry analysis before initiating oral contraceptives.

Liver and Gallbladder Disease

A number of sources, including the Boston Collaborative Drug Surveillance Program and the RCGP study, have reported that the incidence of gallstones (cholelithiasis) and inflammation of the gallbladder (cholecystitis) is higher in pill takers than it is in women who are not on oral contraceptives. This association appears to be greater only in the first few years of pill use, but women who have had previous gallbladder disease or who have had their gallbladders removed should choose another form of contraception.

Before the introduction of oral contraceptives, benign tumors of the liver (hepatic adenomas) were very rarely found in either men or women. Since 1973, however, more than 100 cases of benign liver tumors have been reported in the literature and linked to oral contraceptive use. A Registry for Liver Tumors Associated with Oral Contraceptives has been established at the University of California—Irvine since 1975.

For the 10 to 15 million women on the pill in the United States, this incidence of liver adenoma does not appear to be a large risk. But even though the tumors are benign, death as a result of rupture and hemorrhage can occur. Women taking the pill should always have an abdominal palpation to check for liver enlargement at their annual physical examination. Should they at any time experience pain under the right rib cage along with loss of appetite or nausea and vomiting, the pills should immediately be discontinued and medical treatment sought as soon as possible.

Edmondson and co-workers discovered that the incidence of liver tumors appears to be greater among women using pills containing mestranol than among users of ethinyl estradiol formulations (1976).

There is merely a trickle of reports in the literature that associate cancer of the liver with the use of oral contraceptives. No one has stated that a causal relationship exists. It has been suggested that the association may be more than chance, however, particularly since liver carcinoma usually appears in much older people or in conjunction with previous liver damage.

Oral Contraceptives and Cancer

After 1980, the twentieth anniversary of the pill, evaluations of the carcinogenic potential of oral contraceptives began to have greater credibility. For example, nearly all published epidemiological studies had failed to find an association between cancer and pill usage, but much of the data had been accumulated within 10–15 years of the introduction of oral contraception. If, as is believed by many researchers, some forms of breast cancer could take 20–30 years for induction by a carcinogen, it would take that long before anyone could say with certainty that pills do or do not cause breast cancer. Studies published in the late 1970s and early 1980s continued to provide reassurance that oral contraceptives were probably not carcinogenic, but there were enough other reports that made complete exoneration impossible.

One of the first suggestions that a woman's risk of breast cancer could be enhanced by taking oral contraceptives came from a group of researchers in the San Francisco Bay area. Paffenbarger and his co-workers (1977) interviewed 452 breast cancer patients and 872 women without breast cancer, matched for age, race, religion, and time of hospitalization. They discovered that not only was there an increased risk of breast cancer in women on the pill, but that women who had taken oral contraceptives for 2–4 years before having their first pregnancy had almost twice the chance of developing breast cancer than the matched control group. Similar observations were subsequently reported in three other studies (Pike et al., 1981; Harris et al., 1983; Pike et al., 1983). Clearly and consistently, these researchers demonstrated that there was an association of breast cancer with oral contraceptives when they were taken before ever giving birth. In the latter investigation, Pike and his colleagues specifically related the increased risk to long-term use of certain combination oral contraceptives with a high progestin content, such as Ovulen, Demulen, Ovral, Enovid 10, Norinyl 10, Ortho-Novum 10, Lo/Ovral, Enovid 5, and Norlestrin. This study, which appeared in the British journal *Lancet*, was widely publicized and widely criticized for its methodology but created considerable public concern. Another dimension had now been added to the question of whether to take the Pill—which pill to take?

In 1980, the Centers for Disease Control in Atlanta had begun the Cancer and Steroid Hormone Study, an investigation of the relationship of long-term oral contraceptive use to breast, endometrial, and ovarian cancer. Data were accumulated on study participants, women with cancer and their matched controls, aged 20–54 years, in eight geographic areas of the United States. The first report to emerge from this investigation, based on analyses of information collected after 6 months, was published in 1983. The findings of 689 women with breast cancer and 1,077 control women showed that neither short- nor long-term use of oral contraceptives appeared to increase significantly the risk of developing breast cancer. With respect to pill-usage before the first pregnancy, the risk of developing breast cancer appeared to be increased, and additional research on that issue, as well as on the influence of specific brands and doses of oral contraceptives, was planned.

There have been many reports indicating that women taking the pill have a decreased incidence of benign breast disease and that oral contraceptives are generally a protection against development of the disease. This is encouraging, since women who have had some kinds of fibrocystic disease may be more likely to develop breast cancer. Paffenbarger and his associates had found that women with a history of previously diagnosed benign breast disease who had taken oral contraceptives for 6 or more years had a risk of breast cancer 11 times greater than their matched controls. They suggested that perhaps the development of cancer is accelerated when transformed cells are already present in the breast during the period of pill use. The preliminary results from the Centers for Disease Control study above, however, found no increased risk of breast cancer from pill usage among women with benign breast disease or even with a family history of breast cancer.

Not everyone agrees with the apparent evidence that pills do not increase, and may actually decrease, the development of benign breast disease. Greenwald and his co-investigators in 1977 had pointed out an interesting possibility. Since women with existing fibrocystic disease are more likely to be cautioned by their doctors *against* taking oral contraceptives, a bias could have crept into the published reports that pills protect against the disease. It may be that the presence of benign breast disease influences the use of oral contraceptives and not that the use of oral contraceptives influences the risk of benign breast disease. There may actually be no protective effect. Nevertheless, one of the most frequently mentioned noncontraceptive benefits of oral contraceptives is the decreased incidence of benign breast disease. Epidemiologist Howard W. Ory estimated that 20,000 annual hospitalizations for benign breast disease are prevented by pill use (1982).

The government, however, appears to be taking few chances with the possible relationship of oral contraceptives to breast cancer. The FDA-mandated guidelines for the package insert accompanying oral contraceptives continued to warn in 1984 that "women with a strong family history of breast cancer or who have breast nodules, fibrocystic disease, or abnormal mammograms should be monitored with particular care if they elect to use oral contraceptives." A more prudent choice might be to elect another form of birth control.

The data concerning cervical cancer are also somewhat equivocal. One large study of 689 women with cervical carcinoma found that oral contraceptives did not appear to be associated with an increased risk, at least for women with normal cervices (Boyce et al., 1977). Similar studies produced conflicting results, but one common finding emerged in several of them: that long-term use of oral contraceptives appeared to be related to an increased incidence of cervical cancer. Still, investigators were reluctant to assume that the pill was causing cervical cancer, because sexual activity and a number of other factors are also believed to contribute to the development of cancer of the cervix. But in late 1983, a British 10-year follow-up study of more than 10,000 women provided evidence to strengthen the link between OC use and cervi-

cal cancer. The authors recommended that all long-term users (more than 4 years) have regular and more frequent Pap smears (Vessey et al., 1983).

The relationship between oral contraceptives and uterine cancer is more optimistic. Taking the combination pill appears to offer some protection against the development of uterine cancer. Although an increased risk of uterine cancer had previously been shown to be associated with the sequential pills that had been discontinued in 1975, in 1980 and thereafter, four major epidemiological studies, including the preliminary results from the Centers for Disease Control study, provided evidence that suggested that the risk of endometrial cancer developing in women who had ever taken combination oral contraceptives was less than that in women who had never used them. The mechanism for the protective effect was speculated to result from the progestin component of the pills. On the basis of the studies, Ory estimated that past and current use of OCs averts 2,000 cases and 100 deaths from endometrial cancer annually. As yet, however, the influence of other known risk factors for endometrial cancer, such as obesity and use of estrogen replacement therapy on the presumptive protective effect of the pill, have not been determined.

The other good news concerns the association between oral contraceptives and cancer of the ovaries. A number of studies published in the early 1980s demonstrated that taking oral contraceptives appeared to decrease the risk of developing ovarian cancer, and the Centers for Disease Control investigation confirmed that the small but significant decreased risk persisted even after cessation of pill usage.

But back on the negative side of the ledger, long-term use of OCs appears to be correlated with a quick-spreading type of skin cancer. Whereas some earlier studies had suggested a very weak or no link at all between pill use and malignant melanoma, a study by epidemiologist Elizabeth Holly and her colleagues (1983) of 87 women with malignant melanoma and 863 control women residing in the same county in the Seattle area did find an association. The researchers showed that pill usage for under 4 years or less resulted in no increase in risk; 5–9 years of use produced a 2.4 times increased risk of developing the cancer; and more than 10 years on the pill caused the risk to increase almost fourfold.

Obviously, the whole issue of the effect of contraceptives on cancer is far from resolved. OCs have statistically been shown to prevent against cancer of the endometrium—score "one" for pill benefit. They have also been shown to provide some protection against ovarian cancer—another "one" for the pill. But long-term use of the pill before the first pregnancy may increase the risk of breast cancer, and pill users may also increase their risk of cervical and skin cancer. The score on the pill's affect on cancer is evidently benefit 2: risk 3. How are women to interpret these studies and make an informed decision on oral contraceptive use? Although it is reassuring that after 2 decades of research on oral contraceptives, there appears to be no *conclusive* evidence that the pill actually initiates any form of cancer, it may be premature to conclude, as does *Population Reports*, the bulletin published by the population information program at Johns Hopkins University, that "in fact, taking the pill may be one of the few positive steps a woman can take that reduces the risk of certain types of neoplasia" (1982B). According to the statistical evidence, a woman who follows that advice and takes the pill to reduce her risk of endometrial or ovarian cancer could be increasing her chances of getting breast, cervical, or skin cancer—a remarkably bizarre tradeoff. What women really need are the results of long-term epidemiological studies of the total effects of oral contraceptives on *cancer*.

The Pill and Pregnancy

The danger to the developing embryo of prenatal exposure to sex hormones used for pregnancy tests or to ward off threatened abortion is recognized and has been discussed previously. The inadvertent taking of oral contraceptives in early pregnancy is a possibility, and a number of years after the introduction of the pill, Nora and Nora (1975) reported that exposure to OCs during pregnancy resulted in a combination of abnormalities called VACTERL syndrome—vertebral, anal, cardiac, tracheal, esophageal, renal, and limb defects. Other large-scale studies published shortly thereafter provided additional evidence that the risk of congenital heart defects appears to be greater when hormones, including oral contraceptives, were taken during the first months of pregnancy. But by now, a number of additional surveys, reanalyses of data, and reviews of the medical literature have minimized the connection between oral contraceptives and certain nongenital malformations. The current view is that the potential of inadvertent OC usage in early pregnancy for causing heart, limb, or central nervous system defects is small. The possibility of genital abnormalities—sexual ambiguities of DES-type defects—as a result of continued contraceptive use when pregnant has not been discounted, however, and should still be a concern.

If one or two pills are accidentally skipped during a cycle and the expected withdrawal bleeding does not occur at the end of that cycle, pregnancy is a possibility. If the pregnancy is to be retained rather than terminated, it is important to get a pregnancy test before starting a new package of pills to avoid any teratogenic potential of the pills during the next cycle. If no combination of pills have been missed and the period is skipped, there is no reason to worry; the chances of pregnancy are slim. If one pill is forgotten, two pills should be taken the next day. If two pills in a row are missed, two pills each day for the next 2 days should be taken. When three or more pills in a row are missed, there is a significantly increased risk of pregnancy, and another form of birth control should be used for the rest of the month if sexual intercourse takes place. Unfortunately, the effect of missing pills at several different times of the cycle is unknown.

Also of concern is the suggestion that there is a greater incidence of abortuses showing chromosomal defects when conception occurred soon after discontinuation of oral contraceptives. There is no evidence that more miscarriages occur after a period of pill taking, but only that when they did take place, chromosomal abnormalities were greater. For this reason, women are advised to discontinue use of the pill and use another method of contraception until they have had two spontaneous cycles before trying to conceive.

According to evidence from the Royal College of General Practitioners, many women experience a delay in the return of ovulation and fertility for a varying period of time after discontinuation of the pill. This is another good reason to stop taking the pills for 3–6 months before an anticipated pregnancy. A small percentage of pill users, about two to three out of every 100 women, develop a postpill amenorrhea, and fertility may be delayed for a prolonged period or even permanently impaired. Postpill amenorrhea evidently is unrelated to duration of pill use, type of pill, or dosage but is strongly correlated with a previous history of menstrual irregularity, late onset of menarche, and low body weight. A young, very slender woman who has always had infrequent, irregular, and scanty periods

is a prime target for postpill amenorrhea and should not take oral contraceptives if she wants to become pregnant at some future date.

Nutritional Effects of Oral Contraceptives

The steroid hormones in oral contraceptives affect not only carbohydrate and lipid metabolism but produce alterations in the metabolism and/or absorption of certain vitamins and minerals as well. Thus, women on the pill appear to have an increased need for vitamin B_6 and folic acid and, to a lesser extent, B_1, B_2, vitamin C and possibly zinc. In contrast, the blood serum levels of some nutrients increase, and women appear to have a lesser need for vitamin A, iron, and copper.

Table 12.4 illustrates the nutritional aberrations that can occur when oral contraceptives are used. These changes may not be induced in all women taking oral contraceptives since much depends on their prior and current nutritional status. Certain women, because of an inadequate diet, may be more vulnerable to a vitamin depletion, but the symptoms of actual deficiency have occurred rarely.

The most markedly increased requirement appears to be vitamin B_6. As measured by an observable alteration in the metabolism of the amino acid trypotophan, vitamin B_6 has been said by Theuer and Vitale to be depleted in 80% of pill users while the other 20% have an absolute deficiency (1977). The clinical significance of the altered trypotophan metabolism is unclear, but this biochemical change is reversed when oral contraceptives are discontinued or when a vitamin supplement is taken.

While Theuer has estimated that 20–30% of oral contraceptive users have abnormally low folic acid levels, clinical signs rarely appear. Fewer than 30 cases of megaloblastic anemia (abnormal red blood cell production) pro-

**Table 12.4 NUTRITIONAL ALTERATIONS ATTRIBUTED TO ORAL
CONTRACEPTIVES**

Nutrient	Function	Effect of OCs
Vitamin B$_6$ (pyridoxine)	Enzymatic reactions necessary for protein metabolism	Reduced serum levels Depression (?)
Folic acid	Metabolism of nucleic acids; normal blood cell production	Reduced serum levels Reduced erythrocyte levels Megaloblastic anemia (Very rarely)
Vitamin B$_{12}$	Metabolism of nucleic acids; normal blood cell production	Reduced serum levels
Vitamin B$_1$ (Thiamine)	General metabolism; cellular respiration	Reduced serum levels
Vitamin B$_2$ (Riboflavin)	General metabolism; cellular respiration	Reduced serum levels Glossitis (when diet is inadequate)
Vitamin C (Ascorbic acid)	General metabolism; especially in connective tissue	Reduced serum levels
Vitamin A	Normal vision, growth, reproduction; normal epithelial tissue	Increased serum levels
Iron	Oxygen transport; cellular respiration	Increased serum levels
Copper	Enzyme component in cellular respiration	Increased serum levels
Zinc	Enzyme component in cellular respiration	Reduced serum levels

duced by folate deficiency have been recorded in the medical literature. On the basis of several studies, nutritionist Daphne Roe, of Cornell University, concluded that the risk of a serious folic acid depletion developing in women taking oral contraceptives is related to their dietary intake of the vitamin. Raw leafy vegetables, whole grain cereals, organ meats, nuts, legumes, and yeast are excellent sources of folic acid. A woman who has a bowl of fortified cereal for breakfast and a salad during the day is not likely to develop a folic acid deficiency while taking the pill. Roe suggests that

a vitamin supplement of 35 μg of folic acid daily would more than compensate for any difference in folic acid status in women taking oral contraceptives. Should a state of vitamin depletion exist, she advocates 100-μg supplementation.

There are several multivitamin-mineral preparations on the market—Feminins by Mead Johnson Laboratories is one of them—that contain greater amounts of B vitamins and C than the usual cheaper multivitamin tablet. They also contain vitamin A. As Roe pointed out, vitamin A levels are increased in women on the pill; additional supplementation is not helpful and could even be hazardous with possible hypervitaminosis developing in women consuming large amounts of dietary vitamin A.

Much of the evidence that oral contraceptives have an adverse effect on a woman's nutritional status is incomplete, and many of the studies are inconclusive and contradictory. There is no real proof that supplemental vitamin and mineral tablets are necessary or useful for women on the pill who eat a balanced nutritious diet. Any supplementation should be confined to the B vitamins, especially B_6 and folic acid, and possibly to vitamin C. A moderate extra intake of water-soluble vitamins is not going to be harmful.

Minipills

Because so many adverse effects appeared to be associated with the estrogen content of the combined oral contraceptives, attempts were made during the mid-1960s to develop a progestin-only pill that could prevent pregnancy but would have less effect on the body's metabolism. The minipill, or small-dose progestin pill, has been available since 1973. Nor-QD and Micronor contain 0.35 mg of norethindrone, and Ovrette provides 0.75 mg of norgestrel. None of these contains any estrogen, although assays have shown that Nor-QD and Micronor both have some weak estrogenic potency.

Minipills are taken every single day with no interruption right through the menstrual period. They may prevent ovulation during some cycles, but that is not the primary basis for their contraceptive action. It is believed that one effect of microdose progestin-only pills is to slow the movement of the ovum down the fallopian tube, delaying it to the point where implantation is not likely to occur. The endometrium is also altered by the progestin to further inhibit implantation.

Sperm motility and penetration through the cervix is impaired or prevented because minipills cause the formation of a thick and hostile cervical mucus.

Minipills are said to have a somewhat lower theoretical effectiveness rate than combined pills, in the area of 98.5%–99%, or as low as 97% according to some sources. The additional number of pregnancies per 100 woman years of use may be attributed to the fact that in contrast to the combined pills, missing one day of minipills can result in pregnancy.

When a pill is missed, two should be taken the next day and another contraceptive method used until the next menstrual period. Since pregnancies are more likely to occur during the first 6 months of use, it is recommended that a second method be used during that period.

Because minipills contain no estrogen, they tend to produce less frequent menstrual bleeding, cause spotting or breakthrough bleeding, or result in changes in the volume of menstrual flow. When regular 23–30 days menstruation does occur, it probably means that ovulation is taking place each month. Many women, however, experience some regular cycles but can also have intervals between bleeding that may last up to 45 days or more. There are women taking minipills who menstruate very infrequently—perhaps once or twice a year. The known lesser effectiveness of minipills along with the irregular bleeding can make women on the longer cycles very nervous about a possible pregnancy. Manufacturer's instructions with Ovrette indicate that if no tablets have been missed and the period is delayed up to 60 days, the pills should be immediately discontinued. If withdrawal bleeding does not occur within a few days, a pregnancy test is necessary. Emory University's "bible" of family planning information, *Contraceptive Technology*, recommends that a woman see her doctor if a period has not occurred within 45 days of the last one. Further instructions to a woman on minipills recommend the use of a second method during midcycle (days 10–18) if bleeding is regular because ovulation is occurring. When bleeding is irregular, it is suggested that a backup method of birth control be used throughout the entire cycle since there is no way of knowing when or if ovulation has occurred.

It seems reasonable to question whether taking an oral contraceptive, even one containing only a microdose of progestin, is worthwhile if menstrual irregularities cause anxiety about pregnancy and the necessity for a second method of contraception. A substantial number of women on progestin-only pills do have some kind of disturbance of duration or amount of menstrual flow and many of them discontinue use as a result.

If one is willing to accept the possibility of irregular withdrawal bleeding and the probability of slightly lower effectiveness, are there any other advantages of minipills over a low estrogen combined pill? Progestin-only pills have been shown in laboratory and clinical studies to produce fewer metabolic effects than combined low estrogen oral contraceptives. The side effects associated with estrogen excess would not be expected to occur with minipills, and many of the progestin-excess miseries—because minipills contain less progestin than any of the combined oral contraceptives—also should not take place. It has not been proven that minipills increase the risk of blood-clotting disorders as much as combined oral contraceptives, but neither has it been shown that they do not. Actual rates of thromboembolic disease have not yet been established for minipill users. The FDA mandates that manufacturers of the minipills include a package insert that warns of the same health hazards that are associated with the combined orals. There is no present evidence to indicate that minipills, while theoretically safer, are actually less harmful or different in their adverse effects than other oral contraceptives.

The use of the minipill is not particularly widespread, and it accounts for only 0.2% of the total oral contraceptive market in the United States. Many pharmacies do not even stock them and have to special-order the pills when they are prescribed. Even the manufacturers—Syntex, Ortho and Wyeth—apparently prefer to promote the use of low-dose combined pills and seemingly never advertise the minipills in medical journals. Minipills might be used as the initial contraceptive for women over age 35 who insist on birth control pills, for women whose medical history warns against an estrogen-containing pill, or for those women who have experienced estrogen-related side effects on the combined oral contraceptives. Unlike combined pills, minipills evidently do not affect milk production in nursing mothers and may be recommended as birth control for women who are breastfeeding. There is some evidence that minipills may also prolong the time during which successful breastfeeding can take place. Progestin-only preparations may not affect milk production, but there are no long-term large-scale studies to show that the small amounts of progestin that get into the milk are harmless to the nursing infant. It makes more sense to choose another method of contraception until the baby is weaned.

Pills—Pro and Con

To be able to take one pill a day and then completely ignore the possibility of pregnancy is a very positive benefit of oral contraception, probably the main one for most women. There are, in addition, a number of noncontraceptive benefits that should be considered in making a decision about oral contraceptive usage. They include a decrease in menstrual discomfort and a decrease in the possibility of ovarian retention cysts, since follicular development does not take place. A woman who uses the pill for at least a year has only half the risk of developing pelvic inflammatory disease as women who use no contraception at all and has more protection against PID than that offered by barrier contraception (Rubin, Ory, and Layde, 1982). Less incidence of pelvic inflammatory disease would necessarily affect the incidence of ectopic pregnancy, and current pill-users are, therefore, protected against ectopics, as well.

A report from the Royal College of General Practitioners said that current users were only one-half as likely to develop rheumatoid arthritis as nonusers. A woman taking pills is not likely to develop iron-deficiency anemia since her volume of bleeding is less, and there is evidence that the possibility of developing uterine fibroid tumors is decreased. Her menses will be regular, and there may be reduced premenstrual tension. And, for what it is worth, there is a 25% reduction in the amount of ear wax formed.

On the other side of the scale are the intimidating number of possible side effects associated with pill use. Table 12.5 summarizes them as they tend to occur over a period of time. Assuming that a woman meets the medical criteria for oral contraceptive use—under age 35, neither underweight nor obese, regular menstrual cycles, no high blood pressure, no family history of cardiovascular problems, no diabetes or family history of diabetes, no breast disease or family history of breast cancer, no migraines, no previous gallbladder disease, a nonsmoker, and is willing to put up with minor side effects that may or may not become worse with continued pill use, she still has to answer the question, "Am I going to be better off or worse off if I use birth control pills?"

Only a woman herself knows what the potential benefits of using this form of contraception mean to her. If pregnancy at this time in her life would be an absolute disaster and abortion as a backup is out of the question,

the peace of mind concerning the effectiveness of pill use may outweigh the annoyance of minor side effects and the back-of-the-mind anxiety concerning the possible hazards. The risk of dying as a result of taking the pill is one in 5,000. A fatalist could point out that life itself is a risk, that the risk of walking across the street and getting hit by a car is likely to be greater than the probability of mortality from oral contraceptives. Moreover, it is recognized that more women die from pregnancy and childbirth than from pill usage (unless they are over 40 and smokers). The pill optimist could also point out that statistics can be juggled to mean almost anything; as the old saw says, "statistics lie, and we've got the statistics to prove it."

A pill pessimist is apt to look at the data from a different perspective. Twenty additional deaths per 100,000 women as a result of pill usage is admittedly not that high, unless one happens to fall among the 20. Concerning the argument that other causes of mortality are greater than those associated with taking oral contraceptives, Beral and Kay's 1977 study demonstrated that the mortality rate among pill users was more than *double* the rate from all accidents in the control population and much higher than the rate of 1.8 deaths per 100,000 women per year associated with childbearing. In younger women, the dangers of the pill are usually presumed to be very small compared with the much higher mortality associated with pregnancy in the absence of birth control. Other methods of contraception are available, however, making that reasoning irrelevant. Table 12.6 shows the overall lifetime risk of dying from using one type of fertility control continuously throughout the reproductive years compared with the risk of death if no contraceptive at all is used. Except for the older pill-user who smokes, the risk from use of *any* method is considerably less than the risk of childbirth and ectopic pregnancy when no method is ever used. A

Table 12.5 PILL SIDE EFFECTS: A TIME FRAMEWORK

Worse in First 3 Months	Over Time: Steady-Constant	Worse Over Time	Worse Post-Discontinuation
Nausea plus dizziness	Headaches during 3 weeks that pills are being taken	Headaches during week pills are not taken	[b] Infertility, amenorrhea; hypothalamic and endometrial suppression, and miscalculation of the expected date of confinement
Thrombophlebitis (venous)		Weight gain	
Leg veins	[a] Arterial thromboembolic events, blurred vision, stroke	Candidial vaginitis	
[a] Pulmonary emboli	Anxiety, fatigue, depression	Periodic missed menses while on oral contraceptives	One form of acne
[a] Pelvic vein thrombosis	Thyroid function studies		Hair loss—alopecia
[a] Retinal vein thrombosis	Elevated PBI	[a] Chloasma	
Cyclic weight gain, edema	Depressed T3 resin uptake	[a] Myocardial infarction	
Breast fullness, tenderness	Susceptibility to amenorrhea post-Pill discontinuation	Spider angiomata	
Breakthrough bleeding		Growth of myoma	
[a] Elevated serum lipid levels even to the extent of pancreatitis	Change in cervical secretions —mucorrhea	Predisposition to gallbladder disease	
	Decrease in libido	Hirsutism	
[a] Abnormal glucose tolerance test	Autophonia, chronic dilatation of Eustachian tubes rather than cyclic opening & closing	Decreased menstrual flow	
Contact lenses fail to fit because of fluid retention		Small uterus, pelvic relaxation, cystocele, rectocele, atropic vaginitis	
Abdominal cramping	Acne	Cystic breast changes	
Suppression of lactation		Photodermatitis —sunlight sensitivity with hypopigmentation	
Failure to understand correct use of oral contraceptives; pregnancy		One form of hair loss—alopecia	
		Hypertension	
		Focal hyperplasia of liver and hepatic adenomas	

Source: Contraceptive Technology, 1984–1985 by Robert Hatcher, et al. Irvington Publishers, Inc., 1984. Reprinted by permission of the publishers.

[a] May be irreversible or produce permanent damage.

[b] N.B. To avoid this complication in many patients, advice women desiring to become pregnant to discontinue pills 3–6 months prior to desired pregnancy.

432

Table 12.6 CUMULATIVE RISK OF MORTALITY PER 100,000 NONSTERILE WOMEN, BY FERTILITY CONTROL METHOD ACCORDING TO AGE GROUP

Regimen	15–44	15–32	15–19	20–24	25–29	30–34	35–39	40–44
No control	462	192	35	37	46	74	129	141
Abortion	41	26	3	6	7	10	9	6
Pill/nonsmoker	251	21	3	3	5	10	70	160
Pill/smoker	977	132	12	18	34	68	257	588
IUD	45	25	6	6	7	10	10	
Condom	23	19	6	8	4	1	2	2
Diaphragm/spermicide	53	28	10	6	6	6	11	14
Condom and abortion	1	1	a	a	a	a	a	a
Rhythm	68	36	12	8	8	8	14	18

Source: Reprinted with permission from *Family Planning Perspectives*, Volume 15, Number 2, 1983.

[a] Less than 1.0.

greater choice exists than between the pill and pregnancy-related mortality. And while no one can deny that life is full of risks, many of which we cannot control, it may be more prudent to avoid those over which we do have some control.

Oral contraceptives provide an effective and convenient, but potentially hazardous, method of birth control. Should a woman elect to use them, she must recognize that there are certain responsibilities associated with their use. It is essential that she have an initial examination by a private physician or a nurse-practitioner that includes a medical history, a physical examination that includes a pelvic exam, a Pap smear, a VD culture for gonorrhea and a blood test for syphilis, a urinalysis, blood pressure determination, blood chemistry analysis, and a breast examination. She should be aware that possible drug interactions with other medications may occur while she is taking oral contraceptives, and she must be certain to tell any other physician she consults that she is using birth control pills. She must return annually for a similar examination and must be constantly alert for indications of trouble. *Virtually none of the major complications of pill use occurs without*

some prior warning. The authors of *Contraceptive Technology* have listed five symptoms with the acronym ACHES: Abdominal pain that is severe; Chest pain or shortness of breath; Headaches, Eye problems, Severe leg pain in the calf or thigh. These are symptoms that must not be ignored nor accepted as a matter of course, nor expected to just go away. Delay could be fatal. As would be true with any potent drug, it is possible to minimize the serious complications of usage when informed and aware users are alert to the warning signs.

THE DEPO-PROVERA DISPUTE

Even the controversy surrounding the safety of oral contraceptives has paled beside the great debate on Depo-Provera, the injectable contraceptive manufactured by the Upjohn Company. Depo-Provera is the trade name for the synthetic progestin medroxyprogesterone in depot form and is known internationally as "the shot" or "the jab" because a single injection of 150 mg into the buttocks or upper arm prevents ovulation or acts like other progestin-only preparations to prevent pregnancy for 3 months. The drug is currently used in nearly 80 developing and developed countries by an estimated 2 million women, but was denied approval for use as a contraceptive in the United States by the Food and Drug Administration in 1978, primarily because animal studies conducted by Upjohn's own scientists had suggested a link between Depo-Provera and benign and malignant breast tumors in beagle dogs and endometrial cancer in monkeys. Subsequently, and at Upjohn's request, a special board of inquiry, consisting of three scientists, was convened to reconsider the agency's decision. In January of 1983, the board held a public hearing for testimony concerning the issues surrounding Depo-Provera's use. Supporters of the drug include the World Health Organization (WHO), International Planned Parenthood Federation and many other family planning organizations and population groups, and the American College of Obstetrics and Gynecology. A longtime foe of approval for the injectable contraceptive in this country is Sidney Wolfe of the Health Research Group, joined by a number of other opponents including the National Women's Health Network and, in an unusual alliance, highly conservative right-to-life groups who are against Upjohn because the company sells products that cause abortion.

The proponents of Depo-Provera claim that its benefits outweigh any potential and as yet unproven risks. They allege that epidemiological studies and 20 years of use have demonstrated no serious effects of medroxyprogesterone; that it is 99% effective, better than pills or IUDs; that it lasts 3 months without needing anything on a daily basis and thus offers a high degree of privacy to women who want to keep their birth control practices to themselves; and that it is particularly appropriate for use in Third World countries where other established means of contraception are often used improperly. Depo-Provera advocates have also challenged the validity of the ani-

mal toxicity studies that raised cancer possibilities, and these include the Upjohn researchers who have had to repudiate their own tests. In their opinion, the findings of breast cancer in beagles and endometrial cancer in monkeys should not be viewed as significant because of physiological differences between monkeys, beagles, and humans, and because of methodological flaws in the experiments.

The opponents, however, criticized the existing epidemiological studies that showed no harmful effects as flawed because of small sample size, insufficient exposure and follow-up, lack of appropriate controls, and other methodological problems. They remained unconvinced that these studies in humans have provided evidence for the long-term safety of the drug.

There are other concerns about Depo-Provera that extend beyond its possible carcinogenicity. The drug, unlike oral contraceptives, does not suppress lactation and can be used during breastfeeding, but there is little information about the effects of depo medroxyprogesterone acetate transmitted to the nursing infant. Moreover, almost all women on Depo-Provera have menstrual irregularities ranging from spotting and staining to occasional episodes of very heavy vaginal bleeding. Fifty percent become amenorrheic. Women who lose their periods do not know if they are pregnant or not, and possible prenatal exposure to the drug poses a risk of congenital abnormalities or masculinization of female fetuses. Also, the amenorrhea is apt to continue after cessation of Depo-Provera injections, so the return of fertility is frequently delayed, sometimes by as much as 2 years. Other side effects associated with Depo-Provera use are unsurprisingly those associated with hormonal contraception—headaches, weight gain, abdominal bloating, depression, dizziness, and fatigue.

Although the British government in mid-1984 approved its use as an injectable contraceptive, the board of inquiry in October, 1984 recommended against the release of Depo-Provera for marketing in the United States. The Upjohn Company evidently failed to demonstrate that the benefits of the drug exceeded the risks, but still believes an individual woman should not be denied her choices of contraceptive, and claims that "Depo-Provera is at least as safe as other hormonal contraceptives. . . ." With more than 50 serious side effects listed in the package insert for oral contraceptives, that degree of safety could be questioned. It should also be noted that between 1967 and 1976, many thousands of women in the United States received Depo-Provera for contraceptive use as part of an Investigational New Drug study approved by the FDA. Major administration occurred at Grady Memorial Hospital Family Planning Clinic in Atlanta that services primarily black women and in Los Angeles at the University of Southern California, where the recipients were mostly Hispanic women.*

Although the use of medroxyprogesterone is banned as a contraceptive, it is approved for other therapeutic purposes and is often used in the treatment of dysfunctional bleeding, habitual miscarriage, hormonally sensitive malignancies, and for endometriosis. It is generally considered both safe and effective by gynecologists when used for those health problems. The drug has also been sanctioned by the government for use on sex offenders to create "chemical castration." Several treatment clinics and some state prisons in the country are empowered, under strict guidelines, to administer 500 mg weekly injections of Depo-Provera to sex crime offenders who volunteer, sometimes in lieu of conviction and sentence, for the program. In those doses, the hormone inhibits the production of testosterone, thus sometimes, but not always, reducing the sex drive and causing testicular atrophy and impotence. But when the massive doses of hormones are stopped, so are the effects, and there is reversion back to the original sexual urges that resulted in the criminal acts. The question of how it can be guaranteed that the rapist who has been parolled or the child molester that has been released will continue to come back for the weekly expensive injections has not been answered. Obviously, any expanded use of the chemical castration program has become highly controversial. It is opposed for different reasons by such groups as the American Civil Liberties Union and Women Against Violence Against Women.

INTRAUTERINE DEVICES

That the presence of a small object in the uterus prevents pregnancy may have been known since ancient times. One frequently told story concerns the Arabian practice of inserting a small pebble into the uteri of camels to make certain that pregnancy was avoided on long caravan journeys—a procedure, considering the nature of camels, that must have required extraordinary tact on the part of the camel drivers. The writings of Hippocrates, the Greek physician who is called the father of medicine, describe the use of a hollow tube passed into a human uterus through which medication or a small device would be inserted. Whether the method was actually used for contraception is not clear.

For several thousand years after Hippocrates and ancient camel drivers there was no further historical mention of intrauterine devices, although intravaginal suppositories or *pessaries* were widely used as a method of birth control. The forerunners of the modern IUDs were most likely the stem pessaries of the late nineteenth and early twentieth centu-

* Another investigational contraceptive is being tested at Grady Memorial since 1975. Under-the-skin implants of estradiol pellets that provide 6 months of contraception are being used also to treat women "with premenstrual syndrome, low sex drive, and excessive hair growth." (*Contraceptive Technology Update*, 4(2):19–20, February, 1983.)

ries. Placed into the vagina ostensibly to correct a displaced uterus or other "female disorders," stem pessaries were small caps or buttons made of wood, ivory, pewter, or even precious metals. They fitted over the cervix and were attached to little stems that led into the cervical canal or even further into the uterine cavity. Little reference was made to their contraceptive ability, but they were probably quite effective. In 1909, a German doctor, Richard Richter, published his description of the first completely intrauterine device, a ring made of dried silkworm gut. In the 1920s, another German doctor named Pust combined Richter's silkworm ring with the older pessary idea of a glass disc that fit over the external os. He inserted this device into hundreds of women, claiming that no pregnancies or serious complications occurred. The idea never caught on with other physicians, however, who feared that the use of the device would cause pelvic infections. More attention was paid to the work of Ernst Grafenberg, still another German doctor, who began using completely intrauterine silkworm gut rings for contraception and later developed a silver ring device. The *Grafenberg ring* was very popular in Germany and elsewhere for several years, but enthusiasm waned in the middle 1930s as the devices were almost universally condemned by the medical profession as dangerous and ineffective. Although Grafenberg's most vociferous critics had little or no practical experience with his ring, they opposed it on the theoretical supposition that it might cause serious complications from pelvic infections. The Grafenberg ring fell into complete disrepute, and no more was heard about IUDs for the next 25 years.

In the early 1960s, two independent workers, Oppenheimer in Israel and Ishihama in Japan, who had earlier rediscovered the Grafenberg rings, published their reports of the successful use of intrauterine devices at approximately the same time. There was a sudden and renewed interest in IUDs, occurring, perhaps not coincidentally, with the recognition of a worldwide population explosion. Many clinical trials in various countries were initiated, and when the successful and favorable results of these trials were reported at two international conferences on IUDs held in New York in 1962 and 1964, it became evident that medical opinion had undergone a reversal. The IUDs became established as a medically acceptable contraceptive.

Today there are loops, shields, coils, and bows; wings, rings, T's, and 7's; the classic Antigon-F, Ypsilon, and Om-ga; the poetic Flower of Canton and the Shamrock, and the prosaic Massouras Duck's Foot. Although the Lippes Loop is evidently the most popular IUD, a remarkable array of devices differing in shape, size, and composition are presently in the uteri of an estimated 60 million women throughout the world. While it is impossible to know the exact number of women relying on an IUD, there are estimates that anywhere from 3 to 6 million have chosen this method in the United States.

Intrauterine devices were originally believed to be closest to the "ideal" contraceptive. Although they had the drawback of requiring a clinical procedure for insertion, once in place IUDs could be ignored except for periodic checking by the woman, and they provided long-term protection against pregnancy that was completely reversible once the device was removed. There were believed to be no serious systemic side effects associated with their use. After a decade and a half, it became apparent that IUDs did not fulfill their expectations. Problems of accidental pregnancy, spontaneous abortion, involuntary expulsion, perforation, pelvic infection, pain, and excessive menstrual bleeding have made IUDs far less than ideal for many women.

Attempts to improve the safety, effectiveness, and acceptance of IUDs have been only relatively successful. The contraceptive action has been enhanced, and some of the side effects have apparently been reduced by the addition of bioactive substances, such as copper or progesterone. The most commonly prescribed and commercially available devices in the United States today are the so-called inert Lippes Loop and the Saf-T-Coil, and the medicated Copper 7, Copper T, and the Progestasert. Currently, a woman requesting the insertion of an intrauterine device must receive, on order of the FDA, a lengthy and detailed information sheet prepared by the manufacturer that lists all the possible adverse effects. Some women may choose to forego the assets of the IUD after being informed of its disadvantages and select another contraceptive method. But even with full knowledge of what they may be getting into, or more accurately, what is getting into them, women may find that after weighing risk against benefit, that this form of birth control is still the one most suited to their needs (Fig. 12.3).

Mode of Action

IUDs function to prevent pregnancy, but exactly how they accomplish this contraceptive action is not clearly understood. The most widely accepted theory is that the IUD makes the endometrium of the uterus hostile to implantation of the fertilized egg, probably by causing a nonspecific inflammatory reaction—which sounds much worse than it is. Inflammation is a normal body defense against foreign material; it does not mean infection, although a similar kind of foreign body reaction would occur against bacteria. The presence of an IUD in the uterus evokes a kind of "sterile" inflammation—the release of large numbers of phagocytic white blood cells

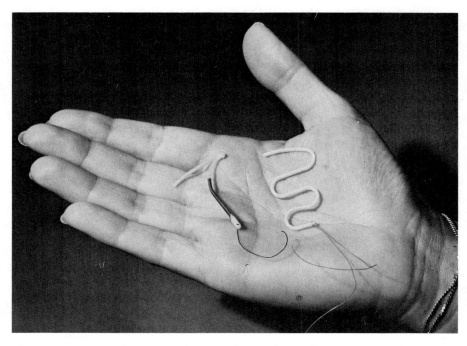

Figure 12.3. Two widely used intrauterine devices. The copper-7 (left) and the Lippes Loop (right).

that have the ability to engulf and devour cells or to disrupt them through release of toxic products. These phagocytic leukocytes may kill some of the sperm, may attack the embryo, or may change the endometrium so that it is no longer an ideal medium for implantation. Other mechanisms thought to contribute to the IUD's action may be an increased intrauterine prostaglandin level, an alteration in tubal motility so that fertilized ova reach the uterine cavity prematurely, or possibly an immunological change. IUDs evidently have no effect on ovulation or the function of the corpus luteum. Their antifertility effect is confined to the uterus, and it presumably disappears when the device is removed.

The addition of copper to an IUD increases its contraceptive ability. How this works probably involves a series of intracellular biochemical reactions, which are not clearly understood. In some fashion, copper ions enhance the prevention of implantation, perhaps by interfering with estrogen-receptors on endometrial cells or in other ways inhibiting intracellular enzyme activities, but they may also be toxic to sperm or inhibit their transport. It is known that the contraceptive effect is related to the amount of copper that is released into the endometrial cavity. The Copper T and Copper 7 are plastic devices with an amount of copper wire coiled around the lower stem to provide a surface area of 200 sq mm of copper with a total copper content of 95 mg. This amount of copper releases 50 μg daily. After 3 years, the contraceptive expectancy of the copper is exhausted, and the IUD must be replaced. Research efforts to improve the design of the copper-releasing mechanism are presently underway. It is projected that future models of the copper IUD will have a much longer duration of use, possible 10–25 years, before removal and reinsertion is necessary.

Copper IUDs have been in use for a relatively short period of time, and the long-term effects of constant exposure to copper are ob-

viously unknown. Most of the copper released from the IUD remains in the uterus and is expelled with the menstrual fluid, but the rest of it is presumed to be absorbed into the blood stream. The amount of copper that gets into the general systemic circulation has been calculated as being only 5% of that which is normally absorbed from the diet (Hasson, 1978). This small amount would ordinarily be expected to have no effect, but there have been several reported allergic reactions. Skin rash developed in one woman, and another got hives as a result of wearing a copper IUD (Barkhoff, 1976).

As yet there has been no evidence of a greater risk of cervical or endometrial cancer as a result of wearing a copper IUD. Should a woman become pregnant while using one, there is also thus far no evidence to indicate that exposure to intrauterine copper until the device is removed has any teratogenic affect on the fetus.

Another type of "medicated" IUD that released progesterone became available in 1976. Called the Progestasert (Alza Corporation), the device releases a continuous amount of hormone that is believed to add to the implantation-preventing effect of an IUD. Like the copper IUD, the Progestasert could then be reduced in size without losing its effectiveness or increasing the possibility of expulsion. The quantity of progesterone released daily is 65 μg. The manufacturer claims that this amount, a total of 35 mg in a year, is less than the amount of progesterone produced in 1 day by a woman's own postovulatory corpus luteum. Theoretically, then, the dose levels of progesterone contained in Progestasert should not interfere with any normal body function, suppress ovulation, or have any damaging effect on the fetus should pregnancy occur. Subsequent studies of hormone levels of FSH, LH, progesterone, and estrogen in women wearing the Progestasert have confirmed that this IUD does not appear to affect hypothalamic, pituitary, or ovarian function and, like nonmedicated types, exerts its contraceptive effect only locally. The current version of the progesterone-IUD requires removal and reinsertion of a new device every year, which tends to provide a deterrent to its use. Other models with an anticipated life-expectancy of 2 years were discontinued when pregnancy rates increased after they were left in that long. Research on medicated IUDs that are coated with synthetic progestins such as d-norgestrel, l-norgestrel or medroxyprogesterone acetate is underway. These types do have a systemic effect and inhibit ovulation (Cohen and Gibor, 1983). None is currently available for marketing in the United States.

Adverse Effects of the IUD

Pregnancy

A paramount factor in choosing a contraceptive method is its reliability. *Population Reports* has summarized the rates of failure to prevent pregnancy of the most commonly used intrauterine devices. For all IUDs, net pregnancy rates in the first year after insertion range from 0.0 to 5.6 per 100 women. The variation depends on which studies and followups were included and can also be attributed to different IUD characteristics in size and shape. The rate or pregnancy is similar to that of oral contraceptives and proves that, unquestionably, IUDs are highly effective. Failure rates have been found to be highest in the first few months after insertion, and some have recommended the use of additional protection, such as foam or cream, at least during ovulation in the first few cycles.

Ectopic Pregnancy

While IUDs prevent uterine pregnancies quite well, they have little or no effect in preventing ectopic pregnancies. The IUD wearer, therefore, unprotected from ovarian conception or tubal implantation, is going to exhibit non-uterine pregnancy in greater numbers than the general population. The diagnosis of ectopic pregnancy, never easy anyway, is more difficult in IUD wearers. Some of the side effects commonly associated with IUD use—episodes of bleeding and pelvic pain—are similar to the symptoms of an ectopic pregnancy.

Spontaneous Abortion

Approximately half of the accidental pregnancies that occur when a woman is wearing an IUD will terminate in a spontaneous abortion should she decide to continue the pregnancy. And approximately half of those miscarriages will be accompanied by infection and be preceded by a fever, a combination that is very rare when spontaneous abortion occurs in non-IUD wearers. Some of the infections, called septic abortions, have been known to spread with incredible rapidity and were severe enough to be fatal. A greater incidence of septic abortions and associated mortality has occurred when the IUD was a Dalkon Shield. This device, crab-shaped with little "feet" that clung to the uterine lining, was especially favored in the early 70s because it appeared to tenaciously resist expulsion, a particular problem in young nulliparous women.

After it had been implicated in 14 deaths and 223 septic abortions, the Dalkon Shield was voluntarily withdrawn from the market by its manufacturer, A. H. Robins Company in 1974.[*] There was subsequent disagreement among doctors as to whether the nearly 3 million women who were already wearing the Dalkon Shields should have them removed.

---

[*] Robins' altruism cost them little at the time of the Dalkon Shield's withdrawal. A press release indicated that their earnings for the year would not be adversely affected by more than 2 cents per share by the subsequent loss of sales. By 1984, however, their financial problems had increased. Robins told stockholders that it and its insurer had paid out $245 million and still had 3,768 pending cases and claims remaining.

While a kind of recall for removal did take place as a result of newspaper and television publicity, there were unknown numbers of women, many of whom had no idea what kind of device had been inserted, wearing a Dalkon Shield. Moreover, since Robins had only removed the IUD from domestic distribution, about 2 million more Shields were inserted abroad before sales were completely halted. Finally, in 1980, the Robins Company which by then had thousands of lawsuits filed against it, wrote to physicians in the United States recommending removal of the device because of the risk of infection. How many women were "recalled" by their doctors is unknown, but an estimated 80,000 women in this country (and probably hundreds of thousands worldwide) could still have had them in place. This is of particular concern in view of evidence that these women have a five-fold increase in risk over any other IUD for pelvic inflammatory disease (Lee et al., 1983). Any woman who had an IUD inserted before 1975 and does not know the specific type used should immediately see her physician at A. H. Robins expense. In late 1984, after a further $38 million settlement of 200 cases, the company offered to pay full costs of medical examinations to determine the kind of device and for removal if a Dalkon Shield was found.

The problem of spontaneous abortion associated with infection and fatality appeared to be greater with the Dalkon Shield, but it was not really determined that this IUD was any worse than the others in causing septic abortion. It is now recommended that if a woman decides to continue a pregnancy that accidentally occurs when any IUD is in the uterus, the device should be removed as soon as possible. The chances of spontaneous abortion as the result of removal, however, are 25%.

Even those pregnancies with an IUD present that proceed to term without miscarriage may not be entirely uneventful at the time of delivery. Problems of premature labor,

excessive bleeding, and stillbirths have been reported (Vessey et al., 1979).

Pregnancy in the presence of an IUD is rare because this method is an effective way of preventing it. When other contraceptives fail, however, a woman has to cope only with the pregnancy—not with life-threatening complications caused by the contraceptive. *Any woman wearing and IUD whose period is delayed by 1 week or more must have an immediate evaluation for pregnancy.* If she is pregnant, her IUD must be removed.

Expulsion, Pain, and Bleeding

A woman may choose to rely on an intrauterine device as a contraceptive, but her uterus may not agree with her decision. In the first year after insertion, expulsion rates range from 0.7%–19.3%, and removal rates for pain or bleeding range from 4%–14.7% (*Population Reports*, 1982A). This means that one out of every four to five women will either spontaneously expel the IUD or need to have it removed for intolerable side effects within a year after insertion.

Retaining the device is a real problem for some women and may be particularly difficult if they are young and nulliparous. Older women who have delivered one or more children are less likely to expel the IUD. Age may actually be a more important factor than parity.

Spontaneous expulsion is also most frequent in the first few months after insertion and usually occurs during menstruation. If a woman notices that the IUD has been ejected she might fare better upon reinsertion. More than half of the women who have experienced one expulsion are ultimately able to retain an IUD (Tietze, 1973). A woman could have a more serious problem—pregnancy—if she has not recognized the expulsion. Mishell estimated that one-third of the pregnancies in IUD wearers were the result of unsuspected loss of the device (1978). It is important to always look for the IUD on tampons or pads during menstruation and to check for its presence after each period by palpating for the end of the string that hangs out of the cervix. If it is not there, or it if appears to be longer than it was originally, an office or clinic visit to check its location is necessary.

A missing string could mean that the IUD has been expelled, but it could also mean that it has migrated in the other direction. Embedding of the IUD into the uterine wall or perforation through it to end in the peritoneal cavity are infrequent but possible occurrences. An examiner must then probe the endocervical canal to look for the string or must use an IUD "hook" to fish around in the uterine cavity for the device. If it still cannot be found, X-ray hysterography or ultrasound can be used to localize the IUD.

The case of the missing string is obviously easier to solve when a woman is certain that the IUD has been spontaneously ejected. Instances have been reported in which abnormal bleeding and pelvic pain were caused by the presence of *two* intrauterine devices. The second one had been inserted by a physician because it was assumed, when the string was no longer visible, that the first one had been expelled (Millen and Bernstein, 1976).

Pain on Insertion

Some discomfort may occur when the IUD is inserted, but most women find it tolerable and an anesthetic or analgesic are rarely necessary. Women can expect to feel a pinch from the tenaculum, the instrument used to grasp and steady the cervix, then perhaps a sudden stinging sensation as the cervix is dilated by the introduction of the IUD inserter barrel, and finally a slight cramping when the device is placed into the uterine cavity. In a few women, there will be a severe pain on insertion. A paracervical block can be used to numb the cervix to eliminate discomfort on insertion, but the cramps that most women feel afterwards will not be relieved by this form of anesthesia. The paracervical block cannot eliminate uterine pain. Very infrequently, the procedure of insertion causes women to experience changes in heart rate, cold sweats, nausea, and even fainting. The overall incidence of one or more of these kinds of responses has been reported as ranging from 1%–10%.

Problems Persisting After Insertion

Virtually all women can expect some cramping pain after an IUD is inserted. In women who have had children, the pain is usually minimal, lasts a few minutes to an hour at most, and requires only a couple of aspirin for relief. Very young or nulliparous women usually have moderate to severe menstrual-type cramps that last for hours or for several days. They need something stronger than aspirin in order to feel comfortable.

After insertion almost all women will have some vaginal bleeding accompanying the pelvic pain. The bleeding, too, lasts several hours to several days. Most women will also experience a change in their menstrual periods. The period is likely to start earlier, last longer, and be characterized by a heavier flow. In many instances, the presence of an IUD will also increase the incidence and severity of menstrual cramps, or at least change their character. Dysmenorrhea may appear throughout the entire period instead of merely on the first day, for example.

It has been claimed that the use of the Progestasert device results in less blood loss and less dysmenorrhea. According to Trobaugh (1978), women who formerly had severe menstrual cramps will have very mild or no dysmenorrhea after insertion of a Progestasert. It is hypothesized that the progesterone released by the device may block the production of prostaglandin, a factor in the cause of dysmenorrhea.

The reason for increased pain and blood loss during menstruation in women using IUDs is speculative. Perhaps there are increased levels of prostaglandins that trigger muscle irritability. It may also be possible that the pelvic pain is caused by the uterus attempting to adjust its size and shape to the size and shape of the foreign object within it.

Mishell and Moyer showed (1974) that the endometrial lining of the uterus of an IUD user has indentations that correspond to any part of the IUD in contact with it. When the muscular wall of the uterus periodically contracts, the movement of the IUD over the endometrial lining may cause abrasions to result in pain and bleeding. Howard Tatum, inventor of the T-shaped model, concluded that the pain and bleeding were the result of compression of the endometrial lining and distension of the muscle by the large stiff devices. His smaller T-shaped IUD was shown to decrease pain and bleeding, but it was less effective in preventing pregnancy. Addition of copper or progesterone to the T and 7 enhanced the contraceptive ability and resulted in fewer problems of pain and blood loss.

Excessive blood loss during menstruation, even in the absence of pelvic pain, can be uncomfortable and restricting. Depending on a woman's diet, it could even increase the risk of iron-deficiency anemia and require iron supplementation if blood tests indicate a low hemoglobin concentration. A woman has to decide for herself how much discomfort and incovenience she is willing to tolerate for the sake of the IUD. A menstrual period that lasts longer, is twice as heavy as before, and is accompanied by dysmenorrhea that may not have been a prior problem, could make many woman ask themselves, "Do I really need this grief?" and request removal.

Because a certain amount of additional monthly bleeding and pain is so commonplace in IUD wearers, there is a tendency for doctors and women themselves to ascribe these symptoms only to the IUD. The excessive blood loss has been found to persist with time, but *excessive* (this is the key word) pain that continues past the first couple of months after insertion should not be accepted and requires an examination and a reevaluation. In one series of 199 removals for pain and/or bleeding, it was later discovered that three women had pelvic infections, two had unsuspected tubal pregnancies, two had acute appendicitis and one had a ruptured corpus luteum cyst (Trobaugh, 1978).

Decreasing the Problems with IUDs

It has been recognized by authorities that one of the most important factors affecting failure and expulsion of IUDs is the skill and experience of the person who does the insertion. The only correct placement of the device is high up in the uterine fundus, but proper positioning is not always achieved. A 1973 study undertaken at Cairo University showed that within 3 months after insertion of Copper T's, they were actually where they were supposed to be in the uterus only 30% of the time. (Kamal et al.) If the IUD is placed insufficiently high to begin with, the device is more likely to be displaced, dislodged, or spontaneously expelled. What has not been indicated by the experts is how women can evaluate the competence of the person doing the insertion. Other than asking around to determine who does a lot of them, one might assume that a gynecologist who is cautious, considerate, and takes time with all procedures would be careful and meticulous with insertions. It would be wise, if possible, to go to a family planning clinic or a university medical center where the IUD insertion is regularly done by trained personnel.

Placing the IUD high up in the fundus of the uterus does not mean *through* it, but perforation of the uterus that unknowingly places the IUD into the pelvic cavity is a frequent complication of IUD insertion. Perforations that occur at a later date may also be related to original faulty insertion techniques. It has been claimed, although proof is lacking, that an IUD in the uterine cavity is not capable of penetrating through the uterine wall by itself. It has to have been aided by a trauma to the uterine wall at the time of insertion. Absolute necessities for proper placement and to rule out the possibility of perforation when an IUD is inserted are:

1. Bimanual pelvic examination of the uterus to determine its size, shape, and position. Perforations are more common in a retroflexed uterus.

2. Use of a tenaculum, an instrument that grasps the lips of the cervix to apply traction in order to align the body of the uterus in a straight axis with the cervix.

3. Sounding the uterus with an instrument to determine its depth.

4. Inserting the plunger in a very slow and careful manner without using force.

A uterine cavity less than the average 6.5 cm in length is a contraindication for IUD insertion. According to the recommendations of the American College of Obstetricians and Gynecologists, placing an IUD into a smaller than average uterine cavity is an invitation to complications.

Any history of excessive menstrual bleeding or significant dysmenorrhea would also exclude a woman as a candidate for IUD insertion. Another major reason to rule out the use of an IUD is the presence of any pelvic infection or even having once had a pelvic infection.

Pelvic Infection

An infection of the upper reproductive tract is generally referred to as pelvic inflammatory disease (PID). The term PID can be used to describe any bacterial infection involving the uterus (endometritis), the fallopian tubes (salpingitis), and the ovaries (oophoritis).

Wide-spread, acute PID usually refers to a salpingitis that has spilled over into the peritoneum to infect the adjacent pelvic structures. As noted in Chapter 9, an untreated gonorrhea can progress to the upper genital tract as gonococcal salpingo-oophoritis, but other bacteria, such as staphylococci and streptococci, can also invade the tubes, ovaries, and peritoneum and cause acute PID. One symptom of an acute pelvic infection is lower abdominal tenderness or pain, usually localized in an area above the pubic bone. Originally the pain may come and go, but it can also progress to a constant discomfort. It is accompanied by fever, sometimes by nausea and vomiting, and a general feeling of really being sick. There may be symptoms of urinary tract infection—frequency and burning. PID tends to occur more frequently after a menstrual period because the sloughed-off endometrial tissue forms a good growth medium for bacteria.

In 1968, the report from an Advisory Committee on Obstetrics and Gynecology of the FDA conceded that it was probably impossible to insert an IUD without some bacterial contamination. No matter how careful and aseptic the technique, several species of organisms normally found in cervical mucus are likely to ride along into the uterine cavity with the IUD insertion device. In most instances, the normal body defenses in the uterus are able to counteract this minimal infection. Mishell, Moyer, and others have shown in separate studies that uterine cultures taken within the first 24 hours after insertion were positive for bacteria, but that the infection was completely gone and the uterus was again sterile a month later. Depending on a variety of factors, however, (the technique of insertion, the number of bacteria introduced, their virulence, possibly the kind of IUD, the resistance of the woman, and most importantly, the presence of a preexisting infection that is aggravated by the insertion) the bacteria are not destroyed by uterine mechanisms. Instead, they survive, multiply, and cause pelvic inflammatory disease. A study by Tietze found that 2%–3% of women had symptoms of PID in the first year following insertion, but that the incidence was substantially higher (7.7%) in the first 15 days than during later periods. According to Tietze (1973), it is questionable whether insertion of an IUD can cause infection unless a minimal chronic or subchronic infection is present. Any woman with past episodes of PID is not a good candidate for IUD insertion.

Pelvic infection that occurs months or even years after insertion has been recognized as a serious complication in IUD wearers, but there has been a general reluctance to admit to the causal relationship between the IUD and an increased risk of PID. It seemed reasonable to suggest that at least part of the incidence of PID in IUD users could be related to the loss of protection against infection provided by condoms, diaphragms, and spermicides. It also was evident that it was difficult to attribute an increased risk of PID to the IUD because there were so few valid statistics concerning the incidence of PID in the general population. The exact figures were hard to obtain because different doctors tend to apply different criteria to the diagnosis of the disease. Furthermore, the majority of women who received IUDs from clinics during the 1960s were from the group referred to as "lower socioeconomic status"—poor, black, and other minority women. It is stereotypically accepted that the incidence of venereal or other intercourse-related PID is much higher in this group of IUD users.

Recently, however, there have been enough large-scale controlled epidemiological studies to indicate that while a factor such as socioeconomic level or sexual activity could contribute to the increased incidence of pelvic inflammatory disease in IUD users, the major reason for a high rate of PID in IUD wearers is the very fact that they are IUD wearers. They have, in fact, a relative risk of PID that is 1.5 to 10 times greater than it is in non-IUD users (Westrom, 1980; Vessey et al., 1981; Eschenbach et al., 1977; Ory, 1978). Vessey and coworkers showed that the higher rate of PID in IUD users was not the result of differences in age, parity, social class, or smoking, and provided convincing evidence that the link between PID and the IUD is one of cause-and-effect.

The actual reported incidence of acute PID in the United States is low. Eschenbach estimated that approximately 500,000 cases occur annually. Epidemiologist Howard Ory pointed out that certain women are initially at high risk for PID. Once a woman has had one or more episodes of the disease, she is more likely to get it again. Studies have shown that women under 30 have an increased risk as compared with women over 30. Finally, there are also statistics to indicate that the chances of contracting venereal disease, which may lead to gonococcal PID, are greater for women who tend to be sexually active with many partners or who change partners frequently. IUD use is not recommended for women at risk.

Even more disquieting is the evidence that IUDs in some manner may increase the incidence of nongonococcal pelvic infections. A tubo-ovarian abscess that occurs unrelated to gonorrhea is uncommon, but it has been reported in women wearing Dalkon Shields or a copper device (Golde et al., 1977). Actinomycosis pelvic infections, the result of organisms ordinarily found in the anus and mouth but never in the vagina, have also been linked to IUD use (Valicenti et al., 1982). References to nongonococcal salpingo-oophoritis and abscess formation in IUD wearers are beginning to be very prevalent in the medical literature. The reason for the increased incidence associated with the IUD is not known, but there have been suggestions that tissue damage and trauma to the uterine lining from the device may favor growth of organisms.

Pelvic infection can be dangerous and even life-threatening without treatment, but with antibiotic therapy, the infection clears up and the fever and abdominal tenderness will disappear. The real problem with IUD-related PID is the effect it may have on a woman's ultimate fertility. Even a mild infection in the fallopian tubes can contribute to infertility by causing the development of scar tissue or adhesions that block tubal transport of ova or sperm. This is a factor to consider in choosing the IUD as a contraceptive. Women who want to avoid pregnancy now but have thoughts of becoming pregnant in the future may want to think about the risk, no matter how small, of jeopardizing their fertility.

BARRIER CONTRACEPTIVES

Barrier contraceptives are forms of birth control that prevent pregnancy be preventing the sperm from getting to the egg. They include the diaphragm and the condom, which physically obstruct the passage of sperm throughout the cervix, and the contraceptive foams, jellies, creams, and suppositories, which chemically destroy the sperm in the vagina. Barrier contraceptives were very popular and widely used until the 1960s. After the introduction of oral contraceptives and IUDs, the use of diaphragms, condoms, and vaginal contraceptives declined, and they acquired a poor image. These methods were seen as esthetically inferior and far less effective, and they were considered old-fashioned, somehow suitable only for older couples. But with the disillusioning recognition of the shortcomings of the pill and IUD, there has been a recent new respect for the old reliables because they have no adverse physiological effects.

Barrier contraceptives, unlike oral contraceptives, and IUDs, are used only when they are needed. This is a major advantage since the body is not subjected to a daily dose of hormones or the constant presence of a foreign object. It can also be viewed as a major disadvantage, since the barrier must be present each and every time there is sexual intercourse and obviously requires more effort to use than methods nonrelated to coitus. But, except for an occasional allergy to a particular product, the barrier contraceptives present no health risk; they may, in fact, provide some health benefits to be mentioned later. Their drawbacks include less reliability when compared with sterilization or oral contraception and what some perceive as their nuisance or inconvenience factors.

Condoms

Condoms, probably the most widely used contraceptives in the world, are also the only kind of birth control mechanical device used by the male. A condom, also known as a "rubber," "safety," or "prophylactic," is a thin, usually transparent, and flexible sheath made of latex or animal membrane that is closed on one end and open at the other. The condom is rolled over an erect penis to fit tightly during intercourse in order to trap the ejaculate and prevent semen from entering the vagina (see Fig. 12.4).

The origin of the condom is unknown. As soon as people recognized that intercourse was related to pregnancy 9 months later, there must have been some clever individuals who figured out that covering the penis might help to prevent pregnancy. The first recorded account of a condom appeared in 1564 in the writings of Italian anatomist, Gabrielli Fallopius, who claimed that he invented it. Fallopius described a linen sheath to cover the penis in order to protect males from syphillis, and no mention of its use as a contraceptive

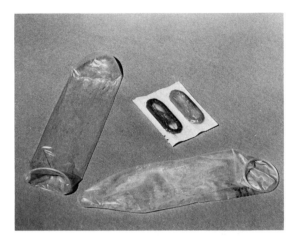

Figure 12.4. Condoms.

was made. Condoms of fabric, fish skin, and intestinal membranes were used as prophylactics (the term is still in use today) against venereal disease for many years thereafter and were common in the brothels of the eighteenth century. Their value as contraceptives was always secondary to their protective function and only came to be appreciated gradually. It could be that because of the connotations of condoms to VD, prostitution, and illicit sex—a throwback to war-training films and furloughs—older couples had a bias against their use. According to the 1975 National Fertility Study, the condom was the preferred method of contraception for only 11% of married couples. But a upswing in condom use may currently be taking place and for the very reasons they had previously been disdained—the recognition of their protective, as well as contraceptive, benefits. Evidence that condoms can exert a preventive effect not only on gonorrhea and syphilis, but also on herpes, chlamydia, PID, amniotic fluid infections during pregnancy, and cervical cancer has revived interest in their use. Planned Parenthood reports that currently, about 4½ million U.S. women choose condoms

as their preferred method, and they rank third in preference among all women, after sterilization and oral contraceptives. Apparently more people are willing to accept a little loss of spontaneity in lovemaking to obtain the protection afforded by condom use.

Condoms today are of two basic kinds—those made of latex rubber and the skin condoms. Skin condoms, actually made from the intestine of a young lamb, are now manufactured only in the United States and cost more than double the price of the most expensive latex condoms. It is claimed that the skin type stimulates the "feel" of the vaginal mucosa better and permits greater sensitivity during intercourse. Whether or not this is true, skin condoms are status items perceived as luxury or "classier" condoms.

Modern condoms are quality controlled, free of defects, prerolled, and sterilely packed in aluminum foil or paper envelopes. It is possible to by them one at a time, but they are usually sold in a pack of three or in boxes of a dozen in drug stores and family planning centers. American condoms come in only one size, 7 inches long, but they can be unrolled to a greater or lesser extent, until the entire penis is covered with the rim of the condom at the base. They can be purchased with a plain tip or reservoir end, smooth or textured, dry, or prelubricated with a water-soluble or a silicone substance. Most condoms are transparent, but they may also be opaque and colored red, green, yellow, black, or any other color. The most common brands are Trojans, Sheik, Ramses, and Shields. Newer brands carry names that reflect an emphasis on sexual enjoyment and sensuality—Excita, Stimula. Some of the varieties, openly aimed at increasing pleasurable sensations, are rippled or flocked with a rough rubber surface (Fig. 12.5).

A condom that is used correctly is put on over an erect penis before it enters the vagina and is worn throughout sexual intercourse. An objection to condom use has been that it

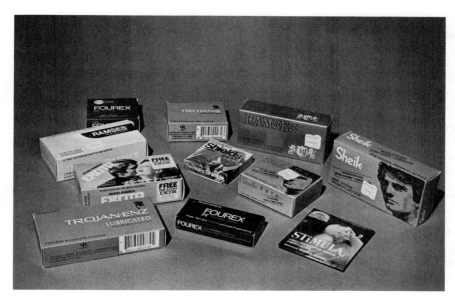

Figure 12.5. The variety in types of condoms should make it possible for couples to find a brand that is mutually pleasing.

requires having to interrupt loving sex play in order to roll it on. Some couples have found that a condom need not detract from, but can actually enhance, sexual enjoyment by incorporating the activity of putting on the condom as part of their foreplay. If the condom has no reservoir to hold the semen, the first half inch should be pinched with the fingers while unrolling in order to keep out the air and leave a space to catch the ejaculate. Breakage or tearing of a condom is rare, but when it happens it may be the result of friction or lack of lubrication, even in a prelubricated condom. Contraceptive foam is an excellent lubricant, and when used with a condom, the *combination is virtually 100% effective in preventing pregnancy.*

Soon after ejaculation and orgasm, the condom should be held tightly by the rim at the base while the penis is withdrawn from the vagina. If the penis becomes flaccid in the vagina it might be possible for the sperm to spill

over or swim out of the open end. Should the condom tear or come off in the vagina, contraceptive foam or jelly should immediately be inserted.

Condoms can be reused if they are washed in soapy water, dried, and powdered with cornstarch only. (Talcum or baby powder deteriorates the rubber latex.) Most people tend to feel more confident about effectiveness when the condom is not used again.

The major argument against the use of the condom, aside from its presumed failure rate, is that it interferes with sexual satisfaction. This disadvantage may be more psychological than physiological. Most women who have sufficient lubrication for an unsheathed penis would not be able to notice any difference in a sheathed one. An unlubricated vagina will make penetration difficult or painful whether or not a condom is used. Contraceptive jellies or foams, K-Y Jelly, saliva—anything but petroleum jelly or oil that affects latex—can be used to ease entrance. Men have been known to complain that using a condom thwarts their enjoyment of the sex act and is indeed a hindrance, somewhat equivalent to "taking a shower with a raincoat on." But unless a man is as exquisitely sensitive as the princess with the pea, it is physiologically unlikely that any dulling of pleasure can occur with a condom in place. Condoms are very thin—they range from 0.4 to 0.9 mm in thickness—and they transmit body heat and touch sensations with no problem. Men who claim loss of sensitivity are usually those who do not want to accept any responsibility for conception. Some men who have difficulty with premature ejaculation may find that their belief in the reduction of sensation provided by a condom may actually help them to retain their erections for a longer period of time. The rim of the condom may also help by providing a slight tourniquet effect.

Condoms have a remarkably underestimated value and potential as a highly effective method of birth control. Doctors seldom recommend them, possibly because most doctors are male, or because the condom requires no medical intervention in its use. With a prevailing expectation that the woman should be solely responsible for contraception, both men and women have learned to be biased against condoms without trying them. The simple condom is reliable, relatively cheap or at least comparable in price to other methods, is compact and easily disposable, completely safe, and needs no medical examination, supervision, or prescription to obtain. It provides excellent protection against pregnancy when used properly, approaching 100% when used with contraceptive foam. It gives the male a chance to participate in the responsibility for birth control, and it has the added advantage of protecting both partners against sexually transmitted disease and the woman against most types of vaginitis. With

so much going for them, condoms should no longer be ignored.

But aside from their poor image, a major obstacle to condom use in some areas is their limited availability. While condoms are supposedly easily accessible, in practice they may be neither easy nor convenient to obtain. In the misguided notion that ready availability of contraceptives leads to increased adolescent sexual activity, some states still have legislation regulating the display, advertising, or distribution of condoms. Even when it is not prohibited, nearly 50% of pharmacists refuse to place condoms on the shelves (*Population Reports*, 1982C). Sales are hardly encouraged when condoms are hidden behind the counters of drugstores to be brought out on request as if they were "girlie" magazines. Women who want to buy condoms and teenagers who have to face a pharmacist's disapproval are likely to be made uncomfortable by the manner in which condoms must be purchased. Even adult men may be embarrassed by having to ask for, rather than merely picking up, articles that are so obviously associated with sexual activity.

Fortunately, there are many areas where condoms are not considered indecent and are available for sale to anyone. There is also a thriving mail order business in condoms, and advertisements for their sale appear with increasing frequency in certain magazines and in campus newspapers. Women who are interested in sharing the burden of birth control might want to purchase them and keep them available for use by their partners. Condoms can also be obtained from family planning associations and are usually less expensive when purchased there.

It is interesting that the United States is virtually the only country in which innovative marketing techniques are not used to expand the availability of nonmedical contraception. On television, sexual images are used to sell almost everything, but contraceptive commercials can never appear. In other parts of the world, condoms are highly respectable items, and they are advertised and sold accordingly. In Japan, where the pill is illegal for contraception and condoms are the preferred contraceptive method for 70% of married couples, they are sold door-to-door by midwives, the "condom Avon ladies." In Sweden, impetus is provided by condom-advertising T-shirts and a national condom month (Fig. 12.6). In Great Britain, as in Japan, condoms are widely used contraceptives and are sold by mail order, in vending machines, an in barber shops as well as in pharmacies and the "chemists." In Egypt, free condoms are handed out at festivals and sporting events.

Diaphragm

A vaginal diaphragm is another highly effective and safe method of contraception that has suffered from distrust fostered by misleading statistics on the use-effectiveness rates. Two large-scale studies have demonstrated that with thorough education about the diaphragm, careful fitting, motivation, and encouragement in an atmosphere free of negative bias, the use-effectiveness rate of the diaphragm can actually be less than the theoretical effectiveness rate quoted in the beginning of this chapter! In a 2-year study of more than 2,000 women at Planned Parenthood Margaret Sanger Research Bureau, the overall failure rate was two pregnancies per 100 users after a year (Lane et al., 1976). The other 1976 report from British workers Vessey and Doll showed an overall rate of 2.4% in 4,223 women using diaphragms for an average of nearly 3 years each. There is every evidence from family planning associations across the country that there has been a well-deserved reversal in the trend of diaphragm usage. The diaphragm is being rediscovered by the older married

Figure 12.6. Advertising the Black Jack Condom in Sweden. (Courtesy, Population Information Program.)

couples and being chosen as the initial method by young unmarried women.

A vaginal diaphragm is a soft rubber dome surrounded by a metal spring. Used in conjunction with a spermicidal jelly or cream, it is inserted into the vagina to fit between the nooks of the anterior and posterior fornices to cover the cervix. The diaphragm itself is a good mechanical barrier to sperm, but it alone cannot completely bar the passage of sperm that might be able to get around its rim. The diaphragm must always be used with *liberal* amounts of a spermicide. It has been suggested that the diaphragm be viewed as a cup to hold the real contraceptive—the jelly, cream, or cream-gel—against the cervix (Fig. 12.7).

Diaphragms vary in both type and size. A woman must be properly fitted with the specific type and size suited especially for her.

Figure 12.7. Diaphragm kit with applicator for insertion of additional contraceptive jelly or cream.

They are available in sizes 50 mm–110 mm rim size, in increasing increments of 5 mm. Most women need a 70 mm–80 mm diaphragm. There are four basic rim types:

1. The coil-spring diaphragm, with a rim that is flexible all the way around, is particularly appropriate for a woman with strong vaginal muscles and no relaxation of the pelvic floor, a uterus that is not tipped and a vagina that is average in size and contour.

2. An arcing-spring diaphragm, which has a firmer rim and forms an arc when compressed, makes insertion easier. It is designed for women who have less than normal vaginal muscular support and those with a mild or moderate cystocele (bladder bulges into anterior vaginal wall) or rectocele (rectum bulges into posterior vaginal wall), or a tipped uterus. The firm ring may

be uncomfortable for some women, however.

3. The flat-spring diaphragm, less frequently used, is soft and flexible with a flat metal band in the anterior rim. It is especially suited for women whose anterior fornix forms a shallow niche in front of the pubic bone and who, therefore, need a thinner rim in that location,

4. The Matrisalus or bowbend rim diaphragm is not completely round, and its irregular shape makes it appropriate for women who need additional support for the anterior vaginal wall.

There should be a diaphragm from among these types to fit every woman, but there are some contraindications to use. Occasionally, a woman is sensitive to the spermicidal agent, but switching to another one usually solves that problem. Infrequently, there is an allergy to the latex rubber that prevents use. Rarely, there is an anatomical problem that makes it impossible to achieve a proper fitting. Sometimes a combination of short fingers and a deep vagina prevent a woman from inserting the diaphragm herself—a difficulty that can be circumvented by a properly instructed partner or the purchase of an inserter.

Proper fitting, in an unhurried and unharried positive manner, by someone who knows how and who is willing to take the time to instruct the woman in the techniques of insertion, removal, and use is the answer to successful use of the diaphragm. A properly fitted diaphragm is the largest size that can be tolerated without being noticed by the woman or her partner. Fitted too small, a diaphragm may become dislodged during intercourse and leave the cervix uncovered. A too large diaphragm may cause problems with vaginal or abdominal discomfort, recurrent bladder infections, or slip out of place to lie lengthwise in the vaginal canal. An inexperienced diaphragm fitter may not recognize that the size originally selected for a woman who is nervous about the fitting—and almost everyone feels somewhat awkward with the device at first—may be too small when she and her vaginal muscles are more relaxed. Other factors that affect diaphragm size besides the degree of tenseness at the time of fitting are those that occur during sexual activity. Since the vaginal barrel is known to expand during the sexual excitement phase of the female response cycle, the presumed correct size in the sexually unstimulated state may be too small and slip out of position during intercourse. The frequency and intensity of intercourse and the position during intercourse can also affect the diaphragm if it does not fit snugly enough. All of these possibilities must be considered in the selection of the appropriate size.

Fitting should be done with sample diaphragms rather than with a set of fitting rims. Otherwise, a woman has no opportunity to practice putting in and taking out the diaphragm. Aside from getting the proper size, a most important part of the procedure is the chance for a woman to learn to comfortably insert and remove her diaphragm herself without aggravation or frustration. No woman should have to leave the office after a fitting with only a prescription for the pharmacist and a printed set of instructions. She must have been shown proper techniques of insertion and placement and have practiced until she feels confident.

Because the fitting procedure takes time, patience, and education, it may be that a busy and harassed private Ob/Gyn, with an office full of waiting patients, is not the best person to see for diaphragm selection and instruction. Successful use of the diaphragm as a contraceptive requires *use* of the diaphragm. Trouble getting it in, trouble getting it out, and negative attitudes about using the diaphragm are reasons why it ends up in a drawer instead of over the cervix. The diaphragm's undeserved reputation as a baby-maker may have its origins in the kind of climate prevalent in the initial fitting procedure. If a woman does not feel good and positive about using a diaphragm after the fitting, it may be well worth it to merely write off that office visit and try again with another physician or a nurse-practitioner at a family planning association.

Using the Diaphragm

Inserting and wearing the diaphragm will be more comfortable if both the bladder and bowel are empty before putting it in. Step one in insertion is the application of the spermicide to the diaphragm. Jelly, cream, or cream-gel preparations are made for use with the diaphragm. Any of them can be used, depending on personal preference, but foam is not recommended, and vaseline, cold cream, or any other makeshift should never be used.

Some women find that jelly leaks, and that they like the consistency of the more viscous cream or cream-gel better. A teaspoon or more (more is better) of spermicide is placed in the bottom of the cup of the diaphragm and spread around the inside and the rim with the fingers. Using one hand, the rims of the diaphragm are squeezed tightly together, jellied-side up, allowing half or two-thirds of the diaphragm to protrude from the grasping hand.* While standing with one leg elevated, squatting, sitting, or lying down—actually any position that is convenient—the labia are spread apart with the other hand and the diaphragm is pushed into the vagina in the same way that a tampon is inserted—upwards and backwards—as far as it will go. It should then be checked for position by feeling for the cervix through the diaphragm and the rim just behind the pubic bone. If the finger cannot get by under the rim, neither can the penis. (see Fig. 12.8A–D).

The diaphragm, unlike the other barrier contraceptives, does not have to be coitus related. It can routinely be inserted daily or nightly and left in place for hours. Intercourse can take place immediately after insertion or within 4 hours without needing the protection of additional jelly or cream. Should more than 4 hours elapse, an additional applicatorful of spermicide should be inserted. After intercourse the diaphragm *must not* be removed for a minimum of 6 hours. Normal activities—urinating, bowel movements, bathing, working—can all take place, but the position of the diaphragm should be checked after a bowel movement. Should intercourse take

* Diaphragms with spermicide are slippery and can zing out of the hand if not held tightly. Diaphragms are not and need not be messy to use, but there may be occasions when lack of privacy for insertion and cleaning present problems that may temporarily favor the use of another barrier method. One summer camper, going behind the bushes to insert her diaphragm, temporarily lost control after applying the jelly. When she picked it up off the ground, the diaphragm was covered with bark, leaves, pine needles, and sadly, little insects.

place again during the waiting period, more jelly or cream must be applied to the outer part of the diaphragm without disturbing it. Some spermicides come with applicators for inserting the cream or jelly into the vagina. It may be that the additional loads of jelly provide too much lubrication, and some women may prefer the use of a condom.

Most women like to take the diaphragm out for washing every 24–36 hours to avoid the development of any discharge or odor. For removal, the finger is hooked under the rim of the diaphragm, and it is pulled down and out. Should removal ever be difficult because of the suction that forms under the diaphragm, squatting and bearing down with the abdominal muscles will dislodge it enough so it can be taken out.

After removal, the diaphragm should be washed with warm soapy water, rinsed and dried with a towel. It may be dusted with cornstarch to help keep the rubber from corroding, but this is not necessary. It should be stored in its plastic container away from heat and not in the same drawer with nail-polish remover or perfume because the fumes can cause deterioration of the latex rubber. A diaphragm may discolor with time, but it should last for several years with proper care. It should be checked frequently for tears or holes, particularly around the rim. Refitting may be necessary after pregnancy, pelvic surgery, or any weight loss or gain of 10–20 pounds.

A well-fitting diaphragm should not produce any discomfort, back pain, or difficulty in urination. Any irritation or itching may mean a mild allergy to the spermicidal perfume, and changing brands will eliminate the reaction.

The jellies and creams used with a diaphragm are not only spermicidal, but also toxic to bacteria. Use of a diaphragm therefore may offer some protection against venereal disease and some forms of vaginitis. It has

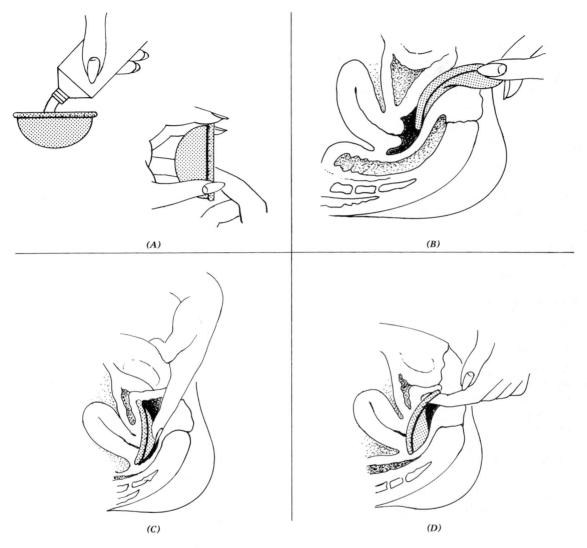

Figure 12.8. Proper procedure for use of the diaphragm. *(A) Jelly or cream specific for diaphragm use is applied to the cup and the rim. (B) Diaphragm is inserted into place. (C) Diaphragm is checked for position by feeling for the cervix through the dome. The rimm of the diaphragm should be just behind the pubic bone. (D) To remove, the finger is hooked under or over the rim of the diaphragm, and it can then be pulled out.*

also been claimed that the diaphragm may be a protection against the development of cervical cancer. Another advantage to diaphragm use is that it can be used during menstruation to hold the menstrual flow. This may be convenient for sexual intercourse during the period.

Cervical Cap

In contrast to the diaphragm, which covers the entire upper part of the vaginal canal between the pubic bone and the posterior fornix, a cervical cap, as indicated by its name, is thimble shaped and covers only the cervix. The caps are made of rubber and are used with spermicide in the same way diaphragms are. They must be fitted to an individual woman so that the rim of the cap surrounds the base of the cervix while the dome of the cap does not actually touch the cervical os. The device remains in place through suction and is removed by tilting the rim away from the cervix with the index finger to break the suction.

Forty to 50 years ago, cervical caps were considered an excellent method of contraception and rivaled the diaphragm in popularity both in the United States and in Europe, but their appeal waned with the introduction of oral contraceptives and IUDs. The only United States manufacturer of the caps ceased production in the early 1960s because there was little demand for the devices. Fifteen years later, when increased awareness of the adverse effects of hormonal and intrauterine contraception revived interest in barrier methods of birth control, the cervical cap was rediscovered. Feminist health centers imported the caps from Lamberts, Ltd., a British firm that currently is the only company manufacturing them, and distributed them to thousands of American women. In 1980, however, the Food and Drug Administration limited distribution and use of the cervical caps

by classifying them as medical devices that may not be marketed until sufficient data have been accumulated to establish their safety and efficacy. Now caps can be obtained only through a number of women's health centers, college health services, public clinics, and private physicians who have obtained an exemption from the FDA in order to conduct a monitored study of safety and effectiveness. The available cap styles are shown in Fig. 12.9.

The Prentif Cavity-Rim type, available in four inside diameter sizes, has a firm rim with a groove running around the circumference to enhance the suction at the base of the cervix. The Vimule style is rimless and has a shallower dome; its flaring sides adhere with strong suction to the vaginal vault rather than to the cervix, and it is useful when the cervix is short. Another type, the Dumas or vault cap, in infrequently fitted. It is very shallow with a wide diameter and fits like a small diaphragm, suitable for women with irregularly shaped or lacerated cervices.

One disadvantage of the cervical cap is that it may be more difficult to learn the proper technique of self-insertion and removal since the cap must be placed deep in the vaginal canal. Since the cap can stick by suction to any part of the vaginal walls or to the side of the cervix, it must be checked with the finger to make certain the cervix is covered. Another problem may be dislodgement during intercourse, although it is less likely than with a diaphragm. Difficulty with dislodgement may result from cervical changes in size or position that normally occur during the month or take place during sexual excitement. The cap could also be knocked off position by a different sex partner. Use of an alternate cap of a different size will frequently solve problems of periodic dislodgement, and women who are fit with caps are told to use condoms as a backup for the first month of use and check for cap position after each intercourse.

Although there are mechanical drawbacks

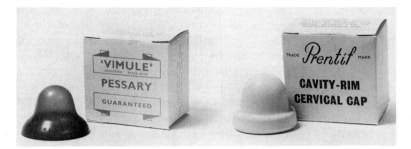

Figure 12.9. Two types of cervical caps imported from England.

to cap use, there are many women who prefer this method to a diaphragm. For one thing, the tube of spermicide appears to last forever. The amount of cream or jelly required to fill one-third to one-half way up the cap is a tiny fraction of that needed for proper diaphragm use. Like the diaphragm, the cap must remain in place for at least 6 hours after the last intercourse. Unlike the diaphragm, additional spermicide does not have to be added to the vagina if multiple intercourse occurs unless the cap is accidentally dislodged. The cap is thus less messy, and for many, more convenient.

The effectiveness of the cap appears to be about the same as the diaphragm, which seems reasonable, given the similarities between the two devices. Relatively few studies of efficacy have been performed on caps, and some of the earlier published data reflect evaluations of cap types no longer used. Statistics on the number of pregnancies per 100 women in the first year of use have been quoted as two to 20, the same use-effectiveness range as for the diaphragm. As previously indicated, such wide discrepancies in contraceptive effectiveness studies are usual, and probably result from differences in sample sizes, populations studied, the methodology employed, and the type of statistical analyses used. Koch's more recent study of 373 women found an 8.42% failure rate in women using the Prentif Cavity-Rim cap (1982). At a Cervical

Cap Symposium held in 1983, Koch reported similar findings on the experiences of more than 3,000 women fitted by him and his staff, and concluded that the Prentif caps were as safe and effective as diaphragms. Those figures were in the same ranges as the failure rates provided in the preliminary reports from the feminist health facilities conducting the FDA authorized studies.

Although several cases of toxic shock syndrome have been reported in diaphragm users, there has been no occurrence of TSS and only one report of adverse effect as a result of using a cap. In 1982, Bernstein and co-workers reported that vaginal lesions ranging from reddening of the vaginal tissue to frank lacerations had appeared in 10 out of 12 women upon their initial wearing of a Vimule cap for a period of 2–7 days. The degree of trauma to the vaginal mucosa appeared to be increased with an increased size of the cap, but the lesions were asymptomatic, caused no bleeding or pain in the women users, and spontaneously healed and disappeared with no treatment. Because of these findings the authors abandoned their plans to test the use of the Vimule cap (but not the Prentif) and speculated that the abrasions may be inflicted in some women by the rigid rim of that style. No such damage was evident as a result of wearing the Prentif Cavity-Rim cap. Examination of two women who had been using the Vimule caps for almost 3 years revealed unusual "redundant" formations of tissue at one site of the cap rim/vaginal mucosa junction, suggesting to the researchers that the vaginal epithelium may have responded to repeated trauma by additional cellular proliferation.

After the report of Bernstein et al., the FDA directed that approved cap providers notify the women who had been fitted with Vimules to return for examination and to report any further adverse effects noted to the agency. To date there has been no published corroboration of the link between vaginal lesions and the Vimule caps. On the possibility of adverse effects, however, some of the health facilities that have enrolled study participants, such as Bread and Roses Women's Health Center in Milwaukee, discontinued fitting of the Vimules.

An additional safety issue concerns the amount of time the cap can stay in place. In the past and when caps were made of rigid materials such as silver, ivory, aluminum, or plastic, women were advised that they could safely leave them inserted for the entire month, removing them only at the menstrual period. Current recommendations are far more conservative and suggest that caps be inserted only for 24–36 hours or less. This would minimize any possibility of irritation and also reduce the chances for developing an unpleasant odor. Until more data are accumulated by the studies currently underway, the cervical cap should be viewed as an improved, less messy, miniature diaphragm that provides greater comfort but is perhaps a little trickier to insert and remove.

Still in the trial stage is another version of the cervical cap designed to eliminate problems with dislodgement and odor. Invented by a dentist, Robert Goepp, familiar with techniques of dental impressions and plaster molds, and an Ob/Gyn, Uwe Freese, the cap is custom fitted to a woman's cervix and is held in place by surface tension rather than suction. The custom cap contains a valve for the escape of cervical mucus and menstrual flow and could theoretically be self-cleaning and be worn for indefinite periods of time. The initial clinical trials of the custom valve cap began in November of 1981 but ended abruptly 6 months later with a disastrous failure rate—nearly 50% of participants reported pregnancies from some study sites. New clinical trials with a different version of the caps made of other materials and cast from plaster molds from a modified impression-taking process are underway.

Contraceptive Foam

Foam preparations look, feel, and smell something like shaving cream. Foam consists of two ingredients; an oil-in-water emulsion medium to provide a mechanical cervical barrier, and a chemical spermicide, usually nonylphenoxypolyethoxyethanol (nonoxynol-9) that immobilizes and kills sperm. The foam is packaged in a pressurized container propelled by chlorofluorocarbons. It may be released into an applicator from the can or may be purchased in packets of disposable applicators for direct discharge into the vagina. Although contraceptive creams and jellies for use without diaphragms are also available, they are not recommended because they are less likely to spread out rapidly and evenly. Foam preparations are the most effective of the spermicidal contraceptives (Fig. 12.10).

With spermicidal foam, as with other barrier contraceptives, consistent use, each and every time there is sexual intercourse, is the main factor in effectiveness. Many couples have very successfully used foam as the only method of contraception for years. The effectiveness indicated by clinical trials shows such a wide variation that it is virtually impossible to assess the actual success of foam in preventing pregnancy. *Population Reports* reported failure rates of 1.75–29.25 pregnancies per 100 woman years of use occurring in trials conducted between 1961 and 1974. It is more than likely that belief and confidence in foam,

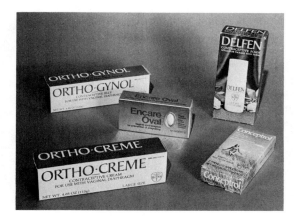

Figure 12.10. Spermicides. Packages tend to be close together on the pharmacy shelves, and care must be taken to choose the right preparation. The creams and jellies are specifically designated for use with the diaphragm and should not be used alone. Foam preparations are not meant to be used inside the diaphragm cup but may be used after the diaphragm is in place for repeated intercourse.

consistent use according to directions, and use with every intercourse can be highly effective. A fatalistic it's-not-going-to-work-anyway attitude will cause careless use and a high failure rate. Of course, foam and a condom used together result in supersafe and supereffective contraception.

The authors of *Contraceptive Technology* point out that even a conscientious foam user can make mistakes in use of the method. Errors that can be made include not using enough of the foam in the right place. They recommend the insertion of one full applicator of Emko, Emko Pre-fil, Because, or Dalkon, or two full applicators of Delfen or Koromex well back into the vagina as far as it will go near the cervix. The manufacturers indicate that foam can be inserted up to 1 hour before intercourse, but vaginal secretions may interfere with dispersal after a while. It is recommended that application take place just be-

fore intercourse or not longer than one-half hour before intercourse.

Other considerations important to effective use include making certain that the foam container still contains the material. With most of the brands there is no way to recognize when it is almost empty. Also, the foam container must be shaken vigorously at least 20 times to insure mixing of the spermicide and the vehicle. It is also important not to douche after intercourse. Foam does not usually drip out but a tampon can be inserted if an annoying wetness becomes a problem.

Other than an occasional allergy there are no adverse side effects from using a foam. A noncontraceptive benefit is the decreased incidence of vaginal infections and sexually transmitted disease.

Spermicidal Suppositories

Encare Oval was the first product of its type—a vaginal suppository that melts and then effervesces into a foam for dispersal—and was introduced in late 1977, amid claims of very high efficacy (99%). The claims turned out to be based on studies of rather questionable and unconventional design in West Germany. The manufacturer revised its original promotional advertising, and the Ovals, along with the newer market entries, such as S'Positive, Semicid, and Intercept suppositories, are to be considered no more or less effective than any other foam preparation—around 97% if used properly and consistently, 85% in typical use.

The various tablets and suppositories have to be inserted 10–30 minutes before intercourse to give them time to dissolve, melt, or fizz to form a barrier. If intercourse does not take place within 2 hours, another one must be inserted. Some women have found that the suppository or tablet does not dissolve within the appropriate time; others have complained, perhaps because they have more natural lubrication, that the preparations are messy and drippy. For some women, Encare's effervescence releases a small amount of heat that burns and feels unpleasant to the vagina and penis. Others find the sensation unnoticeable or pleasant. Trying the method out is the only way to find out if it is satisfactory. In concept, suppositories are attractive, providing the efficacy of foam in small, convenient, and for many, more esthetic form.

Vaginal Sponge

In April, 1983, the FDA approved a polyurethane foam vaginal sponge that can be worn for up to 24 hours as a nonprescription contraceptive. The sponge, marketed as "Today" by the VLI Corporation of Costa Mesa, California, is approximately 2 inches in diameter and $1\frac{1}{2}$ inches thick. It is saturated with 1 g of nonoxynol-9 that has to be activated with water before inserting and also contains small amounts of benzoic, citric, and sorbic acids to adjust it to the vaginal pH. The sponge has a small depression on one surface to aid in positioning it over the cervix, and a string loop on the other surface to help in its removal (Fig. 12.11).

Like other barrier contraceptive devices, the sponge has to remain in for 6 hours after the last intercourse and can be worn for up to 24 hours. Its advantages include its ease of insertion (although some women have had difficulty in removal); no waiting period after insertion; no additional applications of spermicide for additional acts of intercourse since it releases an average of 125–150 mg of spermicide over a 24-hour period. The sponge is not reusable and nonbiodegradable—it cannot be flushed down the toilet.

According to VLI, the theoretical effectiveness of Today in preventing pregnancy is 89%–91%, while its user-effectiveness is 84%–87%. In studies conducted by the manufacturer in the United States that compared 720 sponge users with 717 diaphragm plus jelly users, the pregnancy rate per 100 women was 15.8 for the sponge and 11.6 for the diaphragm. In those trials, the only adverse effects of sponge use were development of aller-

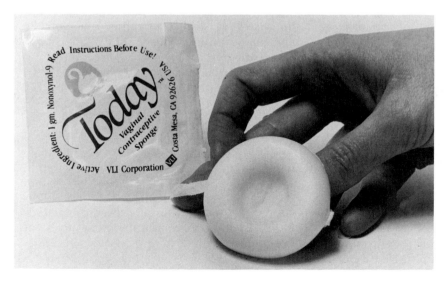

Figure 12.11. The vaginal sponge.

gic reactions, including irritation, itching, and rash in about 4% of the users.

Safety questions related to the development of toxic shock syndrome began to plague the developers of the sponge shortly after its introduction to the market. Six months after release by the FDA, one $5 million claim against VLI was already pending, and by 1984, the FDA had confirmed six additional cases of TSS in women who had used the sponge. The role of the device in TSS is controversial, because as yet the factors involved in development of the syndrome are not completely understood. A polyurethane sponge, like the polyester foam cubes in a Rely tampon, could potentially be a culture medium for *Staphylococcus aureus* organisms, but the risk of developing TSS with sponge use should be less than with a tampon since spermicide is also a bactericide. Moreover, the acidity of Today is hostile to *Staphylococcus* proliferation. The manufacturer has conceded that clinical trials with the sponge were not large enough to rule out the risk of TSS and recommends that the contraceptive not be used during menstruation.

A second concern is the risk of exposure to large amounts of nonoxynol-9 and the possibility of spermicide's absorption through the walls of the vagina. Little data on absorption are available for evaluation. As previously mentioned, data on the incidence of congenital defects in infants born to spermicide users is highly equivocal, and on the basis of present information, there is no reason to avoid spermicides to avoid pregnancy. If there is any reason to suspect a pregnancy, of course, it would be prudent to stop using spermicide and use a condom.

According to VLI, the vaginal sponge works by killing sperm, blocking the entrance to the cervix, and absorbing semen and trapping sperm. It is more realistic to view the sponge as another kind of vehicle to deliver spermicide to the vagina, and for some women, Today may offer improvements over previous forms of delivery. For many, the greatest deterrent to use may be the expense—a package of three sponges retails for $3–$4.

It may appear by now that the use of a barrier contraceptive is bound to be a nuisance. Just when lovemaking is moving right along in ecstatic fashion, one or the other of the partners has to say "stop," "just a minute," and proceed to roll on a condom, slide in a diaphragm, squirt in some foam, or push in a sponge or suppository—injecting what seems to be a mood-killing, jarring moment in their pleasure. It need not be that way at all. Two people who care about each other, who have mutual consideration and a respect for the need to avoid pregnancy, are not going to feel that their sexual experience is irrevocably marred by such a minor interruption. There can be additional emotional satisfaction that comes from knowing that an unwanted pregnancy is being prevented with complete safety to health. It is not that difficult to have a matter-of-fact acceptance of the need to halt the proceedings for protection. It may even make a couple feel closer—more sharing—to help or to completely perform the necessary procedure for the other partner.

CONTRACEPTION WITHOUT CONTRACEPTIVES—RHYTHM

Rhythm is an older term for a method of birth control based on periodic abstinence from sexual intercourse. It is also called variously "natural birth control," "natural family planning," "BBT,' "Billings Method," or "sympto-thermal method." No contraceptive devices are used. Instead, certain observations, techniques, and calculations are used to determine the "fertile" and the "safe" periods of the menstrual cycle.

Paul Ehrlich, professor of biology at Stanford University and founder of Zero Population Growth, has said that people who practice rhythm are called parents. This assessment of the method's effectiveness is probably unjustifiably severe. Currently, in its most sophisticated form, using a combination of several procedures along with the skills necessary for their proper interpretation and with motivation and cooperation of the man and the woman, certain couples are very successfully able to plan, delay, or avoid pregnancy.

Techniques for the use of periodic abstinence for contraception are based on several assumptions:

1. That ovulation occurs once per cycle around 14 days before the onset of the next menstrual period;
2. That the ovum is viable and capable of being fertilized for about 24 hours;
3. That sperm are able to survive in the female reproductive tract to fertilize an ovum for about 72 hours maximum.

Calendar Rhythm

The calendar method of calculating the fertile period during which sexual intercourse cannot take place requires keeping track of the length of each of the menstrual cycles over an 8-month span. With the first day of bleeding as day one, the number of days until the first day of bleeding of the next menstrual period is recorded. When she gets to the ninth cycle, a woman can calculate her fertile or unsafe days as follows: subtract 18 days from the length of the shortest cycle, and subtract 11 days from the length of the longest cycle. For example, if during the eight cycles the shortest period was 24 days and the longest was 33 days, the first fertile day during which intercourse must not take place is the sixth day after the onset of her period (24—18) and the last fertile day is the 22nd day (33—11). This method of calculation presumes that ovula-

tion occurred 14 days before menstruation in both the shortest and the longest cycles, but takes into account a possibly earlier or later ovulation, the lifespan of the egg, and the lifespan of the sperm.

Calendar rhythm works better, psychologically and physiologically, if a woman has regular 28-day cycles. With that cycle length, the period of abstinence can be shorter, and the likelihood of accuracy in estimating the period of greatest fertility is better. Cycle lengths in young women, in premenopausal women, and in women after childbirth, miscarriage, and abortion are likely to be irregular. Therefore, calendar rhythm alone is really a nonmethod of contraception and is now considered obsolete. Even with luck, its effectiveness is only about 60%–80%.

Basal Body Temperature

The use of the basal body temperature (BBT) chart is a way to enhance the effectiveness of periodic abstinence. Unlike calendar rhythm, which is mostly dependent on a regular cycle, BBT relies on a visible, measurable indication of ovulation—the elevation of the body temperature under basal conditions that occurs at the time of or shortly before the egg is released from the ovary. The use of the BBT chart has been described in Chapter 11 as a technique for couples who want to become pregnant. To use BBT to avoid pregnancy, a woman must have cycles in which an obvious rise in temperature occurs—only then can this method be used to determine ovulation. When sexual intercourse is restricted until three consecutive days after the elevation of temperature has occurred—that is, until the postovulatory phase when conception is biologically impossible—the BBT method can be a virtually infallible means of birth control.

The BBT of an adult, healthy, ovulatory woman throughout the month characteristically describes a biphasic curve between two menstruations. If the body temperature is taken orally or rectally under basal conditions (after at least 3 hours sleep, at rest, and before engaging in any activity) starting from day one of the cycle, the daily variations are in the range of 0.1°–0.2°F, until just before ovulation. The BBT then shows a slight drop—small and sometimes unnoticeable—followed by a sustained rise called the *thermal shift*. An elevation in temperature of at least 0.4°F that continues for at least 3 days means that ovulation has occurred. After that shift, unprotected sexual intercourse can take place until the onset of the next menstrual period, when the preovulatory phase of the next cycle is entered. Strict adherence to the BBT method in this fashion is highly effective.

In practice, a woman may have shifts in body temperature that are difficult to recognize. The elevation may not occur during 1 day but may take place gradually over 5 or 6 days, may rise stepwise with several plateaus in between, or zigzag up and down but with a progressive rise that can take as long as a week to level off. Illness, a night of insomnia, or even sleeping under an electric blanket may make interpretation of the chart puzzling. Although the same variation may not appear each month, a woman tends to have a consistent pattern that is unique and distinctive to her, and she has to learn to recognize her own graph (see Fig. 12).

During an anovulatory cycle there will be no temperature rise. Under those circumstances, waiting until it is safe to have intercourse will frustratingly take the entire month, but there is no way of knowing for certain whether ovulation is merely being delayed or is completely absent. In some instances, ovulation may be occurring even though the temperature elevation is absent. Moghissi reported (1976) on one study of 30 healthy menstruating women who recorded their BBT daily while their levels of estrogen, progesterone, and luteinizing hormone were

also being measured. As indicated by changes in hormone levels, ovulation occurred in 27 of the 30 women, but a rise in BBT occurred at the time of ovulation in only 21 women. Evidently, the absence of a rise in BBT may not necessarily mean an anovulatory cycle.

A major drawback to the BBT method is that it does not predict ovulation. It merely indicates when it has occurred. For complete safety, sexual intercourse is limited to the 10 or 12 days before the next menstrual period. In order to extend the range of days for unprotected intercourse other techniques must be used in addition to BBT to try to define the safe or infertile days that occur before ovulation.

Sympto-Thermal Method

The sympto-thermal method is the most recent refinement of the rhythm method. It relies on a combination of techniques to recognize ovulation, including basal body temperature, changes in the quality of cervical mucus, alterations in the position and firmness of the cervix, the openness of the cervical os, and other symptoms of ovulation, such as ovulatory pains, breast tenderness, and edema, that can be subjectively interpreted by the woman.

The cervical mucus produced by the endocervical glands undergoes changes during the menstrual cycle. Reliance solely on these mucus changes has been promoted as a separate method of contraception by John and Evelyn Billings in Australia and is called the ovulation method or the Billings method. Presumably, the Billings method eliminates thermometers, charts, and calendars, but unfortunately not all women experience easily discernible changes in cervical mucus or may have them in some cycles and not in others.

The cervical mucus method is not difficult to practice. A woman has to learn to recognize

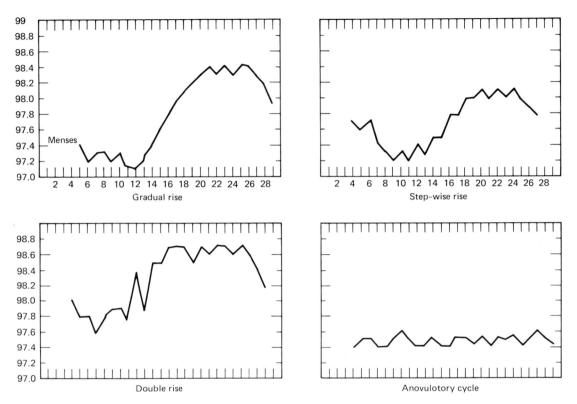

Figure 12.12. Not all BBT curves follow the usual pattern of an abrupt elevation around the time of ovulation. Some curves show a gradual rise, some a stepwise elevation, and others a double thermal shift or rise. Abstinence is always necessary until there have been three consecutive days of temperature elevation recorded at the highest point.

sensations of dryness and wetness of the vaginal discharge and be able to distinguish between feelings of stickiness or tackiness and of slipperiness or lubrication.

It may be possible to make some of these differentiations on the cervical mucus merely by noticing what appears on the underwear, but in order to gather the information properly, it is necessary to check the mucus as it appears at the vaginal opening, or even better, by a two-finger digital examination inside the vagina. The following patterns of mucus production in response to changing levels of hormone secretion occur:

1. Dry days, also called the early safe days, that occur immediately after menstruation. The vagina is moist inside, but the vulva is definitely dry. There is no sensation of wetness or lubrication or slipperiness and no staining in the panties.

2. Early preovulatory mucus. Mucus is

opaque white or yellow, and its consistency is described as tacky, gummy, pasty. A woman has a sensation of stickiness, an awareness of mucus in contrast to the dry days. This less fertile mucus continues unchanged from day to day until it becomes wet mucus.

3. Wet mucus—thin, watery, more profuse. Estrogen levels are high, and the cervical discharge increases in amount, becomes clear, and has a high salt content. At the time of ovulation, the mucus becomes very profuse, has the consistency of egg white, and at the peak production during the time of ovulation, has *spinnbarkeit*. This is the ability of the mucus to be stretched between the thumb and forefinger into a long clear, thin, strand. If this clear mucus is smeared onto a glass slide and permitted to dry, it can be seen under the microscope to fern, forming a highly branched pattern because of the salt content. These signs indicate very fertile mucus—a very unsafe time—and persist for several days after ovulation.

4. Postovulatory mucus—again sticky and thick under the influence of progesterone. This viscous mucus may again become clear and watery just before menstruation, but this stage occurs inconsistently.

Approaching ovulation can also be gauged by changes in the cervix itself. Right after menstruation, the cervix is easy to reach and feels firm and almost dry. As the early preovulatory mucus develops, the cervix remains easy to reach in the vagina and is still firm. When the wet fertile mucus days appear, the cervix begins to feel slippery, softer, and is harder to reach. The os gets larger and begins to open. On the peak mucus day when spinnbarkeit and ferning is greatest, the cervix is virtually out of reach of the fingers, and the os is definitely open. In the dry, postovulatory

days, the cervix again descends, is easy to reach, and the os is closed.

The sympto-thermal method of birth control requires a daily charting of the BBT, the cervical mucus, the cervical os, and any other subjective signs of ovulation—pain, spotting, pelvic pressure, and breast tenderness. The first 5 days of the cycle are considered safe, as long as cycles are never less than 26 days. Then the rule is no intercourse on dry *days* until mucus begins because semen may obscure the signs a woman is looking for during the day. Nighttime coitus is permissible. After the wet mucus begins, no intercourse or any genital contact is allowed until 3 days after BBT rise has occurred, the cervix has closed and lowered, and the mucus is again reduced and "dry."

Many people might justifiably feel that practicing this method is too involved and complicated considering that a barrier method can be used with equal safety, less psychological difficulty from abstinence, and probably greater effectiveness in preventing pregnancy. For couples whose religious beliefs prevent the use of other contraceptives, however, the sympto-thermal method can be acceptable and successful, especially when forms of sexual expression other than intercourse are used during the fertile period. In its favor, the method is a cooperative form of birth control; both partners must be equally committed to its use. If a woman has had little experience in exploring her own body, use of the sympto-thermal method creates a new and healthy body awareness.

There are few studies that have evaluated the success of the sympto-thermal method. In the largest trial to date, of 1,022 couples, conducted between 1970 and 1973 in five different countries, there were 7.2 pregnancies per 100 women in the first year of use; most of the women were over 30 and were well experienced in the method (Rice et al., 1977). The study of 590 U.S. users by Wade et al. (1980) reported 16.6 pregnancies per 100 woman-years. Another 1980 report of a clinical trial sponsored by the World Health Organization among new users in Colombia found a much higher pregnancy rate—34.4 per 100 women-years in the first year of use (Medina et al.).

Coitus Interruptus

The withdrawal of the penis from the vagina before ejaculation occurs is called by the refined Latin term, *coitus interruptus*. It is better known colloquially as "pulling out in time" or "don't *worry*, I'll be *careful!*" As a method of contraception, it is the oldest, best known, and most widely used means of preventing pregnancy in the world. It requires no device or preparation, it is cheap and always available, safe and reversible, but it does have one obvious disadvantage—for most couples it is less effective than other methods.

The method requires that the male withdraw his penis completely from the vagina before orgasm and ejaculate well away from the vaginal orifice. The first few drops of the true ejaculate contain the greatest concentration of sperm, and should some preejaculatory fluid escape from the urethra before orgasm, conception may result. (As Margaret Nofziger has commented. "A man is like a basketball player—he dribbles a little before he shoots.")

Obviously, this technique of contraception is not as desirable as some others, but it is probably underrated for its effectiveness and overrated concerning the potential physical and psychological problems. Psychiatrists, urologists, and sex therapists have warned of dire consequences as a result of habitual practice of withdrawal. There is no substantiated proof, however, that it causes premature ejaculation, pelvic congestion in the female, or any other inevitable difficulties.

When a couple uses coitus interruptus consistently, they are likely to have developed their own techniques to make the method mutually satisfactory. It does require rather rigid self-control during an activity in which loss of control is usually conceded to be more pleasurable. A woman may find it more difficult to relax and let go when she is concerned about the man's ability to withdraw in time. If a couple prefers that the male have complete charge of birth control, as he does with this method, it would seem that less "interruptus" and more peace of mind would occur when the man uses a condom.

STERILIZATION

Sterilization is a very attractive method of contraception for those who are certain that they do not want children. It is a onetime process, providing almost perfect protection against pregnancy (nothing is 100%), and after it is over, it never requires any further action, thought, or worry about which method to use and the possible side effects. This can be a tremendous emotional relief. Sex is bound to be a lot more fun when the constant burden of birth control has been removed.

Sterilization is the most popular method of contraception in the United States for couples married 10 years of more. Nearly one-third of all married couples, and nearly 12 million adults, married and single, have undergone voluntary sterilization. While the number of male sterilizations, or vasectomies, exceeded the female sterilizations, or tubal ligations, in the early 1970s, the popularity of the female procedure increased rapidly after 1973, partly because of the introduction of newer and simpler techniques, partly because the Supreme Court's ruling on abortion, giving women the right to control their own bodies, extended to voluntary sterilization and made the dictatorial formulas and criteria for sterilization used by doctors and hospitals obsolete.*

For the married couple who has all the children they want or who choose to be childless, the decision about which of them is to be sterilized can be influenced by many factors unique to their particular relationship. Vasectomy is the simpler operation, a 10-minute office procedure that ordinarily requires no general anesthetic. It is not expensive, and although both male and female surgeries are beginning to be covered by many group health insurance plans, some policies may not provide complete coverage and expense will be a factor. The present physical condition and the medical history of both partners have to be considered. Of equal importance is their emotional status—their feelings about each other, about their future together, and about the surgical procedure itself. A woman may have a very happy monogamous relationship but still choose a tubal ligation because she wants to retain control over her reproduction. The final decision about which one is sterilized, or whether sterilization should take place at all, should be a mutual decision. There is no legal requirement for the consent of the partner, but if one has to sell the other one on the idea, there may be later recriminations.

Obviously, voluntary sterilization is not only for the married. Single men and women who are certain that they will never want any children in the future have also chosen sterilization for contraception. Those using Medicaid funds for sterilization may find that there are

some federal and state regulations governing age and competence and some requirements for waiting periods or counseling. But anyone, married or single, who makes the decision must recognize that the result of the surgery is permanent infertility. Although various claims are made for the reversibility of vasectomy and tubal ligation, the operation has to be considered irreversible. What has been removed or cut cannot be expected to be replaced or put back together. Men or women who believe there is any possibility that they may change their minds should continue to use another form of contraception.

Vasectomy

The vasa deferentia are paired muscular tubes, about 35 cm long, that lead from the testes to the prostate gland where each joins with the ducts of the seminal vesicles to form the ejaculatory ducts. During orgasm, muscular contractions in each vas deferens propel the sperm from the epididymis of the testis to the urethra where they are expelled from the body along with the seminal fluid. The portion of the vas deferens that is located in the scrotum is surrounded by a sheath of connective tissue called the spermatic cord that contains pain nerves and blood vessels. In a vasectomy, local anesthetic is injected into the scrotum and a small incision, either single in the midline or on either side, is made. The spermatic cord is drawn out, opened, and the vas deferens is exposed. A small segment of the vas deferens is removed on both sides, and the two ends are tied or clipped and coagulated by cautery with an electric needle. The incision is then sutured and after a short recovery period, the man walks out of the office—rather carefully—and goes home.

There is some minimal postoperative discomfort. An ice pack to the scrotum helps to reduce swelling and pain. A scrotal support is

* The "120 Rule" was standardly used by the medical profession. A woman's age times the number of her children had to equal 120 before she could have a sterilization unless she could come up with a psychiatric evaluation or enough symptoms of pain and bleeding to justify a hysterectomy. Part of the alarming increase in "unnecessary hysterectomies" during the 1960s may have been initiated by women themselves, who had little choice about sterilization except by removal of the uterus.

worn for several days, and strenuous physical exertion is avoided until there is no further discomfort. Because some viable sperm remain in the tubes of the reproductive tract, sterility is not immediate, and contraceptive precautions must be taken until about 10 ejaculations have taken place. The man then brings a sperm specimen to the laboratory to see if it is negative. After two negative sperm counts, sterility is complete.

There is no statistical evidence for any long-term clinical effects resulting from vasectomy, although considerable publicity has been given to scare stories concerning the relationship of the surgery to virility as well as all kinds of pathological conditions from arteriosclerosis to thrombophlebitis. More than 11 studies have now reconfirmed the lack of association to any illness, and the largest epidemiological study to date, a federally funded look at 10,590 vasectomized men and 10,590 controls followed for 5–10 years, showed that the incidence of heart disease, cancer, or immune system diseases in vasectomized men was the same or lower than the incidence in nonvasectomized men (Massey et al., 1983). About one-half to two-thirds of vasectomized men develop antibodies to their own sperm, a condition that has been found, but less commonly, in both normally fertile and infertile men as well. No association between such sperm antibodies and any systemic condition has yet been established.

But despite overwhelming confirmation that vasectomy is one of the safest, easiest, and most effective, albeit permanent, methods of contraception known, occasionally reports of adverse effects continue to receive inordinate dissemination. One small study, described in *Contraceptive Technology Update* (1983) found what was described as a "shocking" link between vasectomy and impotence in 14 vasectomized men over 50 years of age. The research team at the University of California–Los Angeles was evaluating 42 men with

impotence in order to see if there were any organic reasons for the erectile dysfunction. The men had problems that included high blood pressure for which they took medication, coronary artery disease, peripheral artery disease, cerebrovascular disease, high blood fat levels, back injuries, and four of them were alcoholics. What shocked the researchers, however, was the discovery that one-third of them, or 14, had been sterilized, some as much as 20 years previously. Despite the tiny sample size and the numerous other factors that could have contributed to the impotence, the report generated, unsurprisingly, the usual publicity. There has been no prior or subsequent corroboration for the link, and it is unlikely that any will be forthcoming, but for some men who have considered vasectomy, the damage has probably been done.

Although half a million U.S. males already elect this procedure annually, reversible, rather than permanent, sterilization probably would have greater appeal to younger men. The procedure, called *vas anastomosis*, uses microcurgical techniques to reconnect the two stumps of the cut vas deferens. The surgery is very delicate and time consuming and has been successful only by the efforts of highly skilled and experienced microsurgeons. The reversibility of vasectomy is too unpredictable to make the possibility of changing one's mind a factor in the decision to be sterilized. Besides, the presence of sperm antibodies in the blood plasma of many vasectomized males makes the restoration of full fertility through reversal surgery even more chancy. Preliminary results in dogs using a *vas valve*, a device that is inserted into the vas deferens and can be turned on or off, permitting or restricting sperm passage, look promising. Before human trials can begin, however, there are apparently many bioengineering problems with the device that must be resolved.

Tubal Ligation

Ligation means tying, and tubal ligation has long been the classic operation for female sterilization and is the origin of the expression, "having her tubes tied." Simple ligation, or tying a suture around the loop of the fallopian tube without cutting it, is rarely performed today because of its high failure rate. The tie had a great tendency to slip off the loop. Today surgeons doing tubal ligation may electrically cauterize the tube in several places (fulguration), use a plastic clip or a silicone rubber band (the Falope Ring, see Fig. 12.13), a Silastic plug to occlude the tube, or simply remove a piece of the tube (resection). A popular, frequently performed operation uses the Pomeroy technique. In this method, a loop of the fallopian tube is elevated, the base of the loop is tied off with absorbable suture, and the loop is cut off or resected. When the suture absorbs several weeks later, the ends of the tube pull apart and leave a gap in between (see Fig. 12.14).

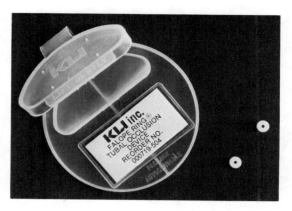

Figure 12.13. The Falope ring, a silicone plastic band that fits over the fallopian tube for sterilization.

Colpotomy

Several methods may be used to expose the fallopian tubes. In the vaginal approach, performed by colpotomy, there is no visible scar. The cul-de-sac, or pouch of Douglas, located between the front of the rectum and the back of the uterus, is entered via a small incision into the posterior fornix of the vagina. Each oviduct is brought into view, tied and cut, and the incision is closed. This form of tubal ligation is very popular in India but not performed that much in Europe or the United States because surgeons have not been trained in the procedure. In skilled hands, the vaginal approach is safe, effective, quick, and easy. The incidence of complications from infection and hemorrhage are known to be higher, however, when the physician is inexperienced.

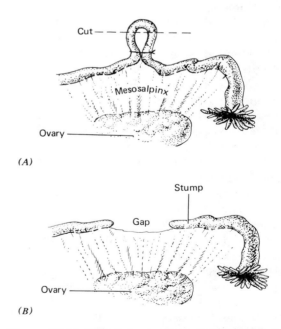

Figure 12.14. The Pomeroy technique for tubal ligation. (A) The fallopian tube is drawn up into a loop, tied with a dissolving suture, and cut. (B) Appearance of the fallopian tube a short while after surgery.

Laparoscopic Tubal Ligation

Laparoscopic sterilization, also known as the Band-Aid or the belly button operation, is the preferred technique of tubal ligation most frequently used in this country. A laparoscope is a form of endoscope or internal telescope that is used to visualize the abdominal organs. It consists of a long slender tube that has a fiberoptic light source—cold, nonburning light conducted through a bundle of glass or plastic fibers—and a lens system that permits a viewer to look at the internal structures. The laparoscope is inserted through a tiny incision in the lower edge of the navel, which is why the procedure has been called belly button surgery. Since the abdominal viscera are very close together and difficult to see, before introducing the laparoscope, several liters of gas, usually carbon dioxide or nitrous oxide, are pumped into the abdominal cavity via a Verres needle inserted through the same small incision (Fig. 12.15). When the abdomen is distended to the point where it appears that the patient is in the second trimester of pregnancy, the organs have been spread far enough apart so that the instruments can be placed without doing any damage to them. During the procedure, the operating table is tilted so that the head of the woman is lowered (Trendelenburg position), which allows the larger intestine to fall away from the reproductive organs. A trocar, a sharp instrument encased in a metal sleeve or sheath, is placed into the abdominal cavity through the umbilical incision. The trocar is then removed, leaving the sleeve in place, and the laparoscope is inserted through the sleeve. A second small incision along the pubic hairline is made for the cauterizing forceps, or the forceps may be introduced right alongside the laparoscope (Fig. 12.16). About 2 cm of each tube is then electrically cauterized to coagulate the tissue and seal it. Some surgeons also cut or remove a piece of tube. Newer techniques that avoid electrocautery use plastic

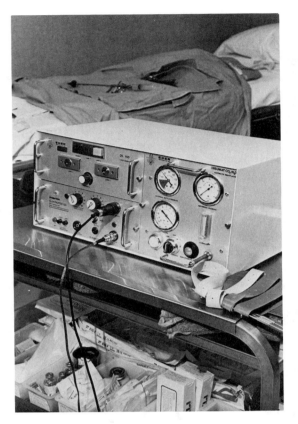

Figure 12.15. The automatic insufflator for distending the abdomen with gas and electrocoagulator for tubal sterilization.

clips or rings to ligate the tubes. When the operation is completed, the laparoscope and the instruments are removed, the gas is allowed to escape, and the incision is closed by a stitch or two covered with a Band-Aid.

A woman who has had a tubal ligation by laparoscopic surgery can expect to feel a little discomfort around the incision, but the greater source of pain may be in the shoulders and chest as a result of the gas distension. This may last a few hours or a day or two until all the gas is absorbed. The amount of

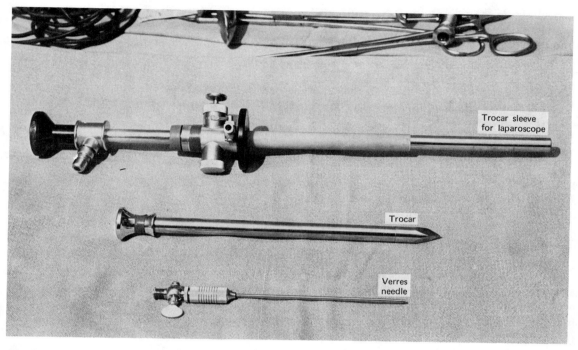

Trocar sleeve for laparoscope

Trocar

Verres needle

Figure 12.16. Instruments used in laparoscopic tubal sterilization.

pain, the aftereffects of the surgery, and the resumption of normal activities vary in different women; some recover much more rapidly than others. If a local anesthetic has been used, the whole procedure from admittance to dismissal will take just a few hours in the hospital. A general anesthetic usually requires an overnight stay. The vast majority of laparoscopic sterilizations are performed under general anesthetic, despite the risks associated with the anesthesia itself. Use of a general anesthetic may result in some postoperative nausea and probably a sore throat because of the tube that has been placed in the trachea as an airway. Most doctors who perform laparoscopy appear to prefer that the patient be completely "out" during surgery, perhaps because they are not accustomed or have not been trained to do the procedure

under local anesthetic. When local anesthesia is used, a tranquilizing or sedative premedication is injected into a vein, and the area of the incision is numbed with procaine or a derivative. Because the woman is sedated and feels nothing, there is no danger of her moving at the critical moment of cautery, although some doctors claim this as a reason for general anesthesia. Howard Shapiro, a physician and author of *The Birth Control Book*, contends that for him, an advantage of general anesthetic is the avoidance of a "distracting conversation with an inquisitive or apprehensive patient while laparoscopy is being performed." One would hardly disagree that nothing should be permitted to detract from the surgeon's utmost concentration to the task at hand, but, in practice, most operating rooms rarely lack irrelevant conversation. There is a great deal of casual, jocular, and presumably distracting conversation in the OR, frequently emanating from the surgeon, especially when a patient is totally anesthetized.

Obviously, no one kind of anesthetic is going to be appropriate for all women undergoing laparoscopic sterilization, but it is worth looking for a physician who is experienced at doing them under local anesthesia. Since the cost of the procedure itself from the surgeon may range from $300 to $1,000 or more, the added expense of the anesthesiologist and the overnight hospitalization can price the surgery beyond the capability of a woman whose insurance does not cover it.

The Minilaparotomy for Sterilization

A laparotomy means an incision into the abdominal cavity. A minilaparotomy is the term applied to a laparotomy through a very minimal incision in order to ligate the fallopian tubes. In the "minilap," a 1-inch incision is made just above the public hairline, the skin and muscle layers are retracted, and the fallopian tubes are brought to the surface and ligated, usually by means of the Pomeroy technique. Electrocautery may also be used, but since the oviducts can be brought out through the incision and directly ligated, the possible hazard associated with cautery can be avoided. No gas is needed or used to distend the abdomen, and the entire procedure usually takes 20 minutes or less.

Minilaparotomy is a safe, simple, and inexpensive procedure, but it is more common abroad. In the United States most gynecologists appear to prefer the laparoscopic method. Minilaparotomy can be performed under local anesthesia in an outpatient clinic, or it may also be done in a hospital under local, general, or spinal anesthesia. A physician who performed 200 minilaps in a period of 18 months accomplished them all in an out-of-hospital setting using the customary operating room procedures with a backup

hospital no more than 15 minutes away. He reported that none of his patients needed general anesthesia or transfer to the nearby hospital and stressed, interestingly enough, that the surgeon and the assistant "should be prepared to engage in casual conversation with the patient" throughout the operation (Penfield, 1978).

The complications of the laparoscopic tubal ligation are infection, hemorrhage, or cautery burns to the bowel or surrounding organs. A survey of 1,452 respondent members of the American Association of Gynecological Laparoscopists revealed that of the 89,492 sterilizations performed, the major complication rate was 1.8 per 1,000 women, and the death rate was two per 100,000 women. A pregnancy rate of six per 1,000 women was also shown, indicating that even sterilization is not perfect contraception (Phillips et al., 1981). As surgeons gain experience with the techniques of sterilization, the incidence of complications becomes more rare, but when adverse consequences of the procedure occur they appear to be related to the skill and experience of the laparoscopist. A previous AAGL survey provided evidence that surgeons who performed fewer than 100 sterilizations a year had a complication rate four times that of the more experienced doctors. (Keith et al., 1976.) A woman considering a tubal ligation by laparoscopy must make certain that her doctor is thoroughly familiar with the technique and regularly performs sterilizations by that method. This appears to present the usual difficulty—how is a prospective patient to evaluate the expertise of the physician? In this instance the best way is to conduct a telephone survey—to call the physician's office, to speak to the doctor, and to *ask*. The gynecologist should be asked what kind of training in the procedure has been taken and where; how many laparoscopic sterilizations have been performed annually; what kind of anesthetic is preferred by the doctor, and has there been experience in using a local anesthetic; and would the physician consider using a nonelectrical form of sterilization? These are reasonable and valid questions. A doctor who refuses to answer them on the telephone or makes a woman uncomfortable about asking is not the right doctor to use.

Obviously, physicians disagree concerning methods of female sterilization, and there are lengthy debates in the medical literature on laparoscopic tubal ligation versus minilaparotomy, epidural spinal anesthesia versus general anesthesia versus local anesthesia, and whether the gas infusion into the abdominal cavity is essential. In the same way that any one method of anesthesia cannot be appropriate for all operations, no one method of sterilization is suitable for all women. If a woman has uterine fibroids or any other condition of the reproductive organs that would make it advisable for the surgeon to be able to look in and around the pelvic cavity, then a laparoscopy is the more sensible operation. Under ordinary circumstances, when there are no abnormalities, a minilaparotomy under local anesthetic is less traumatic for a woman because it is simpler and safer. Neither of the surgical sterilization procedures is recommended for women who have a lot of subcutaneous fat under the abdominal skin. Obesity prolongs the operating time, creates a problem in getting to the oviducts, and makes the operation more hazardous.

Theoretically, one form of sterilization may be better than another; pragmatically, the best and safest operation in any local community is the one performed by a skilled and experienced surgeon who has done a lot of them. There is no point in demanding a vaginal approach or the minilap in preference to the Band-Aid operation, if the doctor is unfamiliar with the technique.

Since even sterilization is not 100% effective, a woman who misses a period or has any reason to suspect she is pregnant should check out the possibility without delay.

MORNING-AFTER CONTRACEPTION

No one really knows for certain what the risk of pregnancy is from one instance of unprotected intercourse. Some researchers have estimated that the chances are 2%–4%; others think that it may be as high as 20%–30%. In theory, this means that one can get away with playing the just-this-once game of unpremeditated sex at least 70% of the time, or, possibly 98% of the time. Even those kinds of odds are not likely to be comforting to a worried woman waiting for her menstrual period. But, considering the expense, the availability, and the possible health hazards inherent in the present methods of postcoital contraception, it might be more sensible, although more anxiety-provoking, for a woman who has had unprotected intercourse or a reason to believe she is pregnant, to wait and see whether pregnancy has actually occurred before doing something about it.

The dangers of diethylstilbestrol, or DES, the only "morning-after pill" approved by the FDA, have already been discussed. The unknown carcinogenic potential, the truly unpleasant side effects, and its questionable efficacy in preventing pregnancy make this method of protection a very undesirable choice. Even when a woman has been raped and has a great emotional need to be assured that pregnancy will not take place, a 5-day course of DES is not the answer.

Fewer side effects and similar effectiveness in preventing pregnancy after unprotected intercourse has been claimed for Ovral, the only pill containing the combination of 50 μg ethinyl estradiol and 0.5 mg of norgestrel. Two Ovral tablets taken immediately or not later than 72 hours after coitus, followed by two more Ovral pills 12 hours after the first dose had a failure rate of .88% when tested on 1,500 women (Yuzpe, 1979). Schilling (1979) reported treating 115 college women with the method with 100% success in preventing pregnancy. Whether such doubling up of Ovral pills has any harmful effects on a woman or on her fetus should she be pregnant and decide to continue the pegnancy is unknown. The only morning-after method currently approved in Great Britain, however, is the Ovral procedure. Obviously, this approach to pregnancy prevention is for emergency, one-time use only and is *not* a birth control method.

The morning-after insertion of a copper IUD has been used as a postcoital contraceptive. The copper IUD is effective probably because the copper begins to act almost immediately to interfere with implantation, while other inert IUDs could take days or weeks to reach effectiveness. Future protection against pregnancy is an added advantage to IUD insertion, but there is the increased risk of pelvic infection to consider. For a rape victim whose unprotected intercourse may also have exposed her to the risk of sexually transmitted disease, an IUD is not the answer.

There is another alternative for a woman unwilling to wait for a positive diagnosis of pregnancy. It is not a procedure performed the morning after, but it can be done within a day or two of a delayed or missed menstrual period. Called *preemptive endometrial aspiration*, the method is also known as menstrual regulation, menstrual extraction, minisuction, miniabortion, or even "lunch-hour abortion." A transparent flexible plastic cannula of very small diameter, eliminating the need for cervical dilation, is passed through the cervical os into the uterine cavity usually without using a local anesthetic. The cannula is attached to a 50-cc syringe that provides minimal but enough suction pressure to evacuate endometrial tissue and blood from the uterine lining. The procedure takes only a few minutes,

is relatively painless, and results in a low rate of complications. Such preempting of the uterine cavity will terminate a pregnancy if the woman has conceived and it relieves her from dealing with emotional and ethical considerations of abortion, if she finds that necessary. If she is not pregnant, an unnecessary procedure has been performed.

Once there has been a positive diagnosis of pregnancy, the syringe aspiration of the uterus can be successfully used to terminate pregnancy up to 8 weeks of gestation. Now called minisuction or miniabortion, this method is less expensive, takes less time, and is usually less painful than any other method of abortion. If necessary, a paracervical block is used to reduce discomfort when the cannula is inserted through the cervix. The woman is likely to experience only a few menstrual cramps, if anything, during the suction. The aspirated material is carefully examined for evidence of the pregnancy, and a repeat pregnancy test and follow-up pelvic examination are performed in 2 weeks. Some practitioners are reluctant to use the syringe method as a very early abortion technique because of the chance that the cannula may miss the implanted conceptus and fail to terminate the pregnancy. They prefer to wait until the seventh or eighth week after the last menstrual period and to use the conventional vacuum suction method to avoid the possible subjection of the woman to two surgical procedures.

The same technique of endometrial suction aspiration can be used diagnostically to obtain a sample of the uterine lining for biopsy. It can be performed on an outpatient basis and can replace, when appropriate, the traditional D and C that requires general anesthesia and a hospital stay.

The many names for the procedure are an indication of the various uses for which the method is employed. The terms, "menstrual extraction" or "regulation," and the technique were originally used by feminist self-help groups in California who developed the procedure in order to evacuate the menses at the beginning of a period, thus condensing 5 days into 5 minutes. In proving that there were medical techniques that could be performed for women by other *women*, and not necessarily by doctors, the concept of menstrual extraction was a great value. In practice, while there could be situations to justify its occasional use, using regular menstrual extraction for contraception or to rid oneself of the presumed nuisance or inconvenience of menstruation is too risky. Even when performed by the most skilled hands, putting an instrument from the germ-laden atmosphere into the sterile uterus too frequently is asking for trouble. Menstrual extraction is an invasive procedure. It always carries the possibility, although slim, of hemorrhage, perforation, and infection no matter how expertly it is performed.

ABORTION

No presently available method of contraception is infallible 100% of the time. One myth circulated by those opposed to legal abortion asserts that the women who decide to terminate their pregnancies are promiscuous women, that they do not contracept, and that they blithely use abortion as a form of birth control. The reality, based on common sense, indicates that since even the best form of contraception has a failure rate of 1%, multiplying that percentage times the number of women in the world that are faithful but fertile users of the pill can result in about a million of them finding themselves pregnant when they do not want to be.

People are not perfect even when they practice birth control conscientiously and consistently. Extenuating circumstances can cause a slipup, even in the most careful person. Statistics indicate that one-third of all couples who use birth control will still have a pregnancy that is unwanted when it occurs (Weller and Hobbs, 1978). The surveyed population did not include the unmarried woman, for whom an unplanned pregnancy is generally much more significant.

It is well established that legal abortion in an accredited facility, performed by a skilled practitioner, and done before the sixteenth week of pregnancy, is safer than pregnancy and delivery. As Cates et al., (1982) have shown, if a woman is faced with an unwanted pregnancy, her safest choice with respect to her life is abortion. This is in contrast to the status of abortion before the January 22, 1973 decision of the U.S. Supreme Court. Then kitchen-table abortionists or self-induced abortions using self-destructive methods with broomstraws, coat hangers, and chemicals were the only options, and resulted in the death or illness of thousands of American women.

The induced termination of a pregnancy is legal, safe, and is a woman's constitutional right. There is a small but vocal and powerful minority of Americans that would like to see that right abolished. Since the legalization of abortion, there have been constant campaigns that threaten women's decisions governing their bodies, and they have been partially successful. In June 1977, the Supreme Court ruled that the states need not pay for nontherapeutic abortions for poor women although they pay for childbirth. Congress then voted to bar federal funds for abortions for Medicaid-eligible women unless they furnish "proof" of rape, incest, or life-threatening danger from the pregnancy. Several states have now passed laws that strongly restrict abortion, and although the constitutionality of such legislation continues to be challenged, often successfully up to the U.S. Supreme Court, it is evident that legislators will continue in their attempts to provide more stringent criteria. Women may have gained a victory in 1973, but there is obviously a strong move to turn back the clock to the days when wealthy women could obtain illegal but safe abortions because they could afford them, while poor women suffered, died, or gave birth to unwanted children. Women watch legislation taking away rights they thought they had won; there is a helpless feeling as events move backward. It is not enough to say that women must not let this happen, because it does happen, and gains are being lost. Perhaps we can only recognize that all goals for women will involve massive efforts for tiny victories, and that every achievement, at least for now, is shaky even when it appears settled.

The decision to abort is a matter of personal choice, but it is not an easily or casually made decision. Abortion is not considered an everyday kind of occurrence for most women; it may be an unpleasant and frequently painful experience. Few women view an abortion as anything but a difficult and necessary alternative to an unintended pregnancy. It is not a method of contraception. The issue is not that abortion may take the place of sensible contraception, but that it is a backup method to contraceptive failure that provides a way out of an untenable situation.

Obtaining an Abortion

According to the Alan Guttmacher Institute (AGI), the research arm of the Planned Parenthood Federation, the number of abortions performed in the United States increased by 73% between 1974 and 1980. Currently, there are well over a million legal abortions occurring each year. Pregnancy termination is now the most frequently performed surgical procedure in the United States.

Most legal abortions take place in free standing nonhospital abortion clinics, which are located all over the country, but concentrated mainly in major urban areas. Only a relatively small percentage of abortions are performed in doctor's offices, and only 22% of general hospitals provided abortion services in 1980. Although virtually all other surgery and almost every childbirth occurs in private and public hospitals, many hospitals, with and without religious affiliations, refuse to offer abortions to women who want them. In the few states that still pay for Medicaid abortions, poor women in urban areas are more likely to utilize hospital services if available. Where Medicaid excludes abortion or the hospital does not provide them, low-income women have to get their abortions elsewhere or do not obtain them at all (Henshaw and O'Reilly, 1983). The paucity of hospital abortions present a major problem for a woman seeking an end to an unwanted pregnancy. The AGI survey indicated that a majority of counties in the United States did not have a single facility in which an abortion could be obtained. Women in rural or less populated areas of the country may have to travel to another county, sometimes to another state, in order to get a health service to which they are constitutionally entitled. Such women, if they are young, may be also the least likely to have the money, time, knowledge, and ability to cope with the difficulties of having to go outside their local area for an abortion. The likely delay in obtaining the abortion adds to the costs as well as the risks of the procedure, which rise for each additional week of pregnancy.

Even in metropolitan areas with several abortion providers, choosing the appropriate facility may present problems. The independent abortion clinics vary in their equipment, ambience, attitude, costs, and quality of the services. Abortion can be a highly lucrative, even multimillion dollar business. Abortion mills, concerned more with making a profit than with meeting needs, do exist. Women must be wary of the clinic with a hello-good-bye, in-and-out atmosphere, devoid of respect for a woman's dignity and concerns. The abortion itself must be performed by a licensed qualified gynecologist who has hospital affiliations. There must be appropriate equipment and personnel for the preoperative tests, the procedure, and emergency backup, if necessary. Since for most women abortion is more than merely a medical procedure, counseling and information provided by a professionally trained staff with a supportive nonjudgmental attitude must be an integral part of abortion services.

If a local community has a woman's crisis line, woman's center, or a chapter of the National Organization for Women, a phone call will provide assistance in finding the best abortion facility in a particular area. The nearest Planned Parenthood affiliate, which has a chapter in almost every state, or the national office in New York City, can also help to locate a highly rated abortion provider.

Methods of Abortion

The earlier in pregnancy an abortion is performed, the less complex and safer is the procedure. Primarily, the method chosen is determined by the length of the pregnancy which is to be terminated. First trimester abortion refers to pregnancies that are interrupted during the first 12 weeks dated from the first day of the last menstrual period. The main technique of first trimester abortion is through vacuum aspiration, also called vacuum curettage. This is very similar to the previously described menstrual extraction except that the cervix must be dilated to accommodate a larger suction cannula, and more powerful negative pressure is obtained through the use of an electrical vacuum pump. The optimum time and way to terminate a pregnancy is at 8 or less weeks of gestation by means of a vacuum aspiration method, either menstrual extraction or suction curettage. After 8 weeks, suction curettage is the appropriate technique.

Before the suction abortion is done, a medical history is taken, and certain routine laboratory tests are performed. Obviously, there must be a positive pregnancy test. Blood tests administered should include a hemoglobin and hematocrit determination to check for anemia, a blood clotting time, an ABO and Rh typing, and a sickle-cell anemia test when indicated. Rh negative women should be given medication to protect against antibody buildup in future pregnancies. A Pap smear, a gonorrhea culture, a VDRL, and a urinalysis are also standard parts of preabortion testing.

The medical procedure begins with a bimanual pelvic examination to determine the size of the uterus and its position in the pelvis. Then, a speculum is inserted to visualize the cervix. The cervix and the vagina are swabbed with an antiseptic solution and the upper portion of the cervix is grasped with a tenaculum to hold it steady during the rest of the procedure. A local anesthetic—usually a 1% lidocaine solution—is injected into either side of the cervix to cause a paracervical block. Some operators use a uterine sound at this point to determine the direction of insertion of the dilators; others eliminate this step and introduce the first of a series of tapered metal dilators. Progressively larger sizes are used until the cervix is dilated enough to permit the entry of the right sized cannula or vacurette. The entire dilating process takes just a few minutes.

Suction cannulas are numbered to correspond to their diameters in millimeters. After 8 weeks, the general rule is to use a size 2 mm less than the number of weeks from the last menstrual period. A No. 8 suction curette, for example, would be used for a pregnancy of 10 weeks duration, and a No. 10 would be used for a 12-week pregnancy. The larger the diameter of the cannula, the stronger the amount of suction.

The cannula is attached to the hose of the electric suction machine. When the machine is turned on, the cannula is gently rotated while moving from the fundus to the cervix of the uterine cavity. In this way the entire product of conception, including the placenta or chorionic villi, the fetal material, and some of the decidual tissue or remainder of the endometrium is removed (Fig. 12.17).

During and after evacuation, the uterine muscles contract to prevent hemorrhage and reduce the uterus to its original size. These contractions may cause rather strong cramps for a while, but they usually subside within 10–20 minutes after the abortion. A period of rest and recovery follows, during which the woman usually receives instructions on aftercare and contraception. She may then leave

and return to her normal routine, but most women prefer to rest for the remainder of that day and to avoid particularly strenuous activities for a few days. It is usual to experience cramps and bleeding for the first 2 weeks after an abortion, and it is not unusual for spotting to occur for another 2 weeks after that. The next normal menstrual period is likely to start 4–6 weeks after surgery, and most women ovulate within the first 3 weeks. Intercourse is not permitted for a week after abortion, and to avoid getting pregnant again immediately, a method of contraception must be available and used after that. It is possible to conceive even before the first menstruation.

The D and C for Abortion

Before the introduction of the vacuum aspiration technique, the usual method of pregnancy termination was the D and C, the dilat-

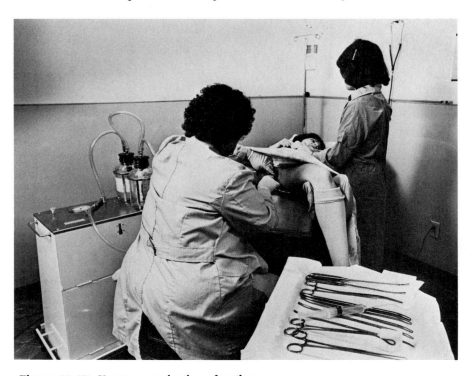

Figure 12.17. Vacuum aspiration abortion.

ing of the cervix so that a sharp metal curette could be used to scrape the uterine cavity free of the conceptus and remove the parts with a forceps. This procedure needs more cervical dilatation than the suction method, is a far bloodier operation, is more painful and requires general anesthetic and a hospital stay, and often does not evacuate the uterus as completely as the other method. It is, however, a technique with which every doctor trained in Ob/Gyn has some familiarity. A woman whose physician wants to perform a D and C for a first trimester abortion should find another doctor, even if it means traveling to another community.

Midtrimester Abortion—The D and E

Some abortion facilities will perform a simple suction curettage up to the fourteenth or fifteenth week of gestation using a No. 12 cannula and "twilight sleep" for analgesia—an intravenous Valium and Demerol injection. In general, however, a procedure that is an extension of both the conventional D and C and the vacuum aspiration is particularly appropriate for use during the thirteenth to sixteenth weeks of pregnancy, although some advocates will perform it through 20 weeks. Called a D and E, *dilatation* and *evacuation*, the method requires a greater dilatation of the cervix and a cannula of greater diameter. Because the fetus is larger, withdrawal is aided by an instrument called the ovum forceps, which fragments the conceptus. Suction is then used to complete the procedure, and some physicians will administer drugs to minimize blood loss.

The D and E requires more cervical dilatation than the traditional D and C. Since rapid dilatation of the cervix by the usual metal dilators can not only be very painful and traumatic but can also result in cervical injury or laceration, the cervix is usually enlarged gradually by the use of *laminaria tents*. Laminaria tents are made from two species of seaweed,

Laminaria digitata or *Laminaria japonica*, that grow in the North Atlantic and North Pacific oceans. The stems of the plants are dried, cut, and shaped into cylindrical smooth sticks about 6 cm long, and of varying diameter, from 3 mm to 10 mm. A string is looped through one end just beyond a plastic disc that prevents the stick from migrating up into the uterine cavity (see Fig. 12.18). The sticks

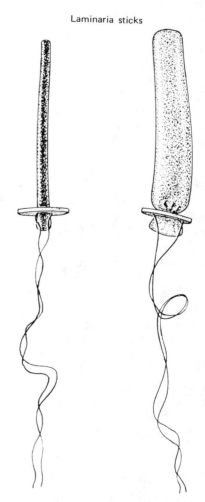

Laminaria sticks

Figure 12.18. Laminaria stick, before insertion and after swelling.

are sterilized by high energy gamma irradiation to prevent any possibility of infection.

Laminaria is enormously hygroscopic—it has the ability to absorb and retain water in a moist environment. When a dry laminaria stick is placed in the moist environment of the cervical canal, it gradually swells to three to five times its original diameter, slowly, progressively, and painlessly dilating the cervix. Sufficient dilatation is accomplished within 6 hours. In practice, a woman who will have an abortion with the use of laminaria will require an initial insertion of the stick and is then asked to return the next day for removal and for the abortion. If the laminaria tent were not removed for several days, it would be likely to induce abortion anyway because of the cervical dilatation. The danger of infection becomes so great after 24 hours, however, that the stick should never be permitted to remain in place for that long a period.

Intra-amniotic Instillation

The other technique of abortion after the thirteenth week is the inducing of a spontaneous abortion by the instillation of a solution into the amniotic cavity around the fetus. One of three solutions may be injected. The most commonly used is a 20%–25% salt solution, hypertonic saline, but prostaglandin or urea may also be the agent.

The method is really *amniocentesis* followed by *amnioinfusion*. A needle is inserted through the abdominal cavity and the uterus into the amniotic sac. If hypertonic saline is to be instilled, about 150–250 ml of amniotic fluid is withdrawn and an equal amount of saline is injected. If prostaglandin is used, a very small amount of amniotic fluid is removed, and 40 mg in 8 ml of PGF_{2a}, the only prostaglandin currently approved by the FDA, is injected. Then the woman is returned to her hospital bed to wait for the onset of labor, which may take anywhere from 12 hours to 2 days. There is no predicting the length of time

from intra-amniotic instillation of saline or prostaglandin until labor and expulsion because it varies in different women. So does the amount of discomfort that accompanies the abortion. The process is the same as in childbirth; the uterus contracts, the cervix dilates and effaces, and the fetus is expelled. Fortunate women have a few cramps and it is all over; others may have to experience many hours of painful contractions.

One advantage of prostaglandin is that the time from instillation to delivery is shorter than it is with hypertonic saline. It may not always be effective initially, however, and may require a second intra-amniotic injection. Other disadvantages of prostaglandin are that it commonly causes diarrhea and vomiting, and, because of the more powerful uterine contractions induced, an increased incidence of cervical trauma. Even more traumatically, there is the real risk that the fetus may still show some fleeting signs of life when delivered.

The mortality of the fetus is assured when hypertonic saline is used, but this method has the greater potential for severe complications. Should the solution inadvertently be injected into a uterine blood vessel and get into the general circulation, the rapid influx of such high salt levels into the bloodstream can result in death.

The intra-amniotic instillation of urea is used in Europe and England but has not been employed to any extent in the United States. Urea is safer, produces fewer side effects or major complications, and is always feticidal. When used alone, however, it frequently fails to induce abortion. Some doctors use urea as an adjunct to prostaglandin, reducing the need for that drug.

Safety of Midtrimester Abortions

Abortion through 16 weeks of pregnancy is safer than continuing the pregnancy and delivering a child. When abortion is delayed beyond the twelfth week, however, the relative risk of the procedure increases. The risk of both major and minor complications is affected not only by the delay in obtaining the abortion, but also by the choice of abortion method—a factor determined by the physician.

Willard Cates and his co-workers have investigated the morbidity risks of more than 80,000 legal abortions that took place between 1971 and 1975. They determined that the safest procedure at the safest time was suction abortion at 8 weeks or less of gestation. The complication rate increased at 2-week intervals, becoming 91% greater with suction curettage when the abortion was postponed until the twelfth week. After 12 weeks of pregnancy, a D and E, the extension of the suction procedure, produced another increase in complication rate that was higher than that occurring at any time during the first trimester. The risk associated with using D and E as a method of abortion during the midtrimester, however, was never as high as when intra-amniotic instillation of either saline or prostaglandin was used. (see Fig. 12.19).

The researchers concluded that the relative safety of a D and E for midtrimester abortions should make it the procedure of choice between the thirteenth and the fifteenth week and that specialists should be trained in its use so that this method can be made more widely available in abortion facilities. Those recommendations have been largely realized; between 1977 and 1979, dilatation and evacuation was the most common method of abortion at 13–15 weeks gestation, and by 1980, D and E replaced saline instillation as the most common method during the 16- to 20-week interval. There are still physicians, however, who lack experience with this method, and some doctors refuse to perform abortions at all between the thirteenth and the sixteenth weeks. This is the interval when the uterus is

deemed too small and the amniotic fluid too insufficient to perform the instillation procedure. In geographic areas where a D and E is difficult to obtain, a woman who has postponed her abortion beyond the twelfth week may have to wait another month before having her pregnancy terminated—and by a method with a higher risk of complications.

Although some women procrastinate obtaining an abortion because of ambivalence about the decision or perhaps in a futile hope that the symptoms of pregnancy will just go away, a delay may occur for a variety of reasons for which the woman is not responsible. A woman who normally has very irregular periods or who menstruates during the first month or two of gestation may not even know she is pregnant until after the third month. Or, a woman who starts out wanting her pregnancy may find herself by the third month in the midst of a separation or divorce or some other circumstance that would lead her to decide on termination. Many women may decide on an early abortion, but several months may elapse before they can accumulate the $175–$250 necessary to pay for it. Health insurance covers pregnancy but infrequently covers ending the pregnancy.

When amniocentesis detects a fetal abnormality, a woman is obliged to have an abortion, if she wants one, very late in the second trimester—perhaps at 20 weeks or more. It takes that long before all the results of the chromosomal and biochemical tests are in. But it is sensible, since each week of delay increases the relative risk of complications of abortion, that women who do seek an abortion soon after the first trimester have them performed by the D and E procedure.

Rooks and Cates (1977) suggested that while the D and E procedure is safer, cheaper, far less painful, and less emotionally traumatic for women, the instillation procedure is much easier for the physician. The infusion of hypertonic saline or prostaglandin takes 15 min-

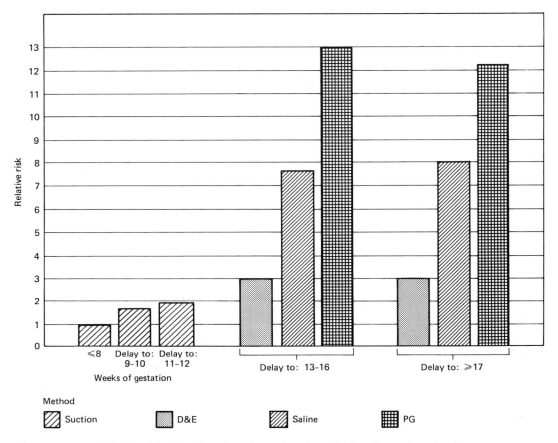

Figure 12.19. Relative risk of major abortion-related morbidity due to length of gestation and choice of method compared with the risk associated with suction at 8-weeks gestation. (Reprinted with permission from Family Planning Perspectives, Volume 9, Number 6, 1977.)

utes, and the doctor goes away, never seeing the woman in labor or the fetus she expells. It is a much "cleaner" way of performing an abortion than is the D and E method, which requires extracting fetal parts with a forceps— a procedure that may be distasteful and disturbing. This could explain the reluctance of many American physicians to perform D and E. But abortion is never pleasant. It can never really be free of emotional challenge either to a woman or her doctor, even in the first 12 weeks when what goes through the suction

tube is not identifiable. The physical disadvantages and potential risks to women of the instillation procedure should be of greater importance to physicians than their own discomfort in performing the D and E procedure.

The Sooner, The Better

To some degree, abortion is a stressful and problematic experience for all women, and hard to deal with even in the first trimester. Abortion after the twelfth week becomes an even greater difficulty, physiologically and psychologically. Many physicians do not perform the D and E procedure, and a saline or prostaglandin abortion is a particularly unpleasant, painful, and emotionally upsetting process that should be avoided, it at all possible. If pregnancy termination is the decision, getting it done simply and safely well before the twelfth week is the best and wisest choice, since there are far fewer complications and little chance of any deleterious effect, either psychological or physical, as a result.

A 1978 issue of *Family Planning Perspectives* summarized the results of five major studies conducted to determine whether prior abortion has any effects on subsequent pregnancies. The findings of investigations in Seattle, Boston, Hawaii, and New York indicated that there was no link between a previously induced abortion and a later complication or unfavorable outcome of pregnancy. The only exception, and the figures were not considered statistically significant, occurred in the Singapore study, where preliminary results showed a slightly increased incidence of miscarriage after prior D and C but not after suction abortion. Similar conclusions were drawn by Hogue, Cates, and Tietze after their exhaustive 1982 review of the medical literature to assess effects of induced abortion on subsequent reproduction. The researchers found that abortion had no association with infertility, ectopic pregnancy, miscarriage, early delivery, low birth weight babies, or greater infant morbidity or mortality. Even multiple abortions produced no increase in adverse pregnancy outcomes, except when D and C procedures with general anesthesia had been performed.

Nonsurgical Abortion

There is considerable research interest in the development of a nonsurgical procedure as an alternative to vacuum aspiration. Both naturally occurring prostaglandins and various analogues of the F and E prostaglandins have been used in intramuscular and suppository form to induce uterine contractions and subsequent abortion in early pregnancy. In a Stockholm study (Bygdeman et al., 1983) of 198 women pregnant less than 49 days, each received either a series of intramuscular prostaglandin injections or prostaglandin analogue suppositories to be inserted on a particular dose interval. Some of the women agreed to self-administer the vaginal suppositories at home. The success rate varied between 92% and 98%, which is not good enough, of course, and 40%–50% of the women experienced vomiting and diarrhea, the usual prostaglandin side effect. Some of the women had enough cramping to require analgesic injections, and many ran a slight fever, more pronounced in those who received the injected prostaglandin. It would appear that as yet, the frequency and probability of such side effects as vomiting, diarrhea, and cramping restricts this procedure from strongly competing with vacuum aspiration, but newer forms of prostaglandin analogues may produce better results. Being able to terminate a pregnancy at home with a self-administered safe and effective treatment would be an attractive option for many women.

THE FUTURE IN BIRTH CONTROL

The question of what can be expected in future alternatives to current methods of contraception can be rather simply answered— not much. Even with the spectacular explosion of knowledge in cellular and molecular biology, no method that really differs substantially from the present techniques of birth control is around the corner, or even around the decade. The major deterrent to progress in fertility regulation and the reason that contraceptive research is just limping along rather than forging ahead is lack of money.

Support for research on contraceptive development on a worldwide basis has been shrinking. In the United States, federal allocation of funds for reproductive research, never great, has actually declined since 1973. Roy Greep and the other authors of the *1976 Ford Foundation Report* on reproduction and contraceptive research estimated that it would take 15 years and $10 million to develop and produce a new contraceptive drug. By 1981, Djerassi claimed it would take $17\frac{1}{2}$ years and $46 million before a new female contraceptive agent would get onto the market. Currently it could reasonably be assumed that the introduction of any new contraceptive would take very long and cost an enormous amount. Most of the research support is now coming from agencies in the public sector, such as International Planned Parenthood Federation, the Ford Foundation, The Rockefeller Foundation, the Population Council, the World Health Organization, the U.S. Agency for International Development, and the National Institute of Child Health and Human Development. The drug industry, which extensively supported the development of oral contraceptives, evidently now has little interest in pouring the kind of time and money it would take into contraceptive research. With 50 million women in the world on the pill, and 60 million using intrauterine devices, the incentive is lacking.

Immunological Methods

Unfortunately, support for contraceptive research has lessened at the very time that major advances are being made in reproductive endocrinology and immunology. Among the newest and most promising possibilities are offered in the area of immunological control of fertility. Recombinant DNA techniques that allow large-scale production by synthetic means of certain antigens have revolutionized vaccine research. Currently, animal tests and preliminary clinical trials are underway to study the safety and efficacy of vaccines that can cause the production of antibodies to human chorionic gonadotropin (HCG). Antibodies to HCG interfere with that hormone's role in implantation and ability to sustain pregnancy. Preliminary investigations seem to indicate that when the antibody level declines, the contraceptive ability is reversed to permit childbearing. Another possibility being considered is to provide passive, rather than active, immunity by giving a woman a dose of antibodies produced by monoclonal antibody techniques.

Other immunological methods under investigation include the production of antibodies to pituitary gonadotropins or to hypothalamic releasing factors, and the development of a way to immunize women against sperm or against their own ova, or against the blastocyst so that implantation is prevented. Any commercial availability of such vaccines to pregnancy are not likely until the mid-1990s or later.

Abortifacients

There are also research efforts to develop a once-a-month pill to induce menstruation whether or not conception has occurred. Several substances are being clinically tested for their potential in suppressing the corpus luteum or in some other way blocking progesterone production. Other lines of investigation involve plants. For thousands of years, a large number of plants have had abortifacient properties attributed to them, and native doctors and herbalists in various countries still prescribe them. The World Health Organization is taking such folk medicine seriously and initiated a study of plants used worldwide in fertility regulation. After selecting the most promising 400 candidate herbs by the end of 1984, WHO will continue to fund testing in primates for toxicology and effectiveness of the five plants with the greatest apparent potential. For example, one extract that may turn out to be worthwhile is from the Mexican plant, zoapatle, the common name for *Montonoa teomentosa* (Gallegos, 1983). Another herb found in the southern part of Nigeria is *Momordica angustisepala*, used by Igbo native doctors to induce abortion in women and confirmed to have abortifacient properties in rodents by Nigerian researchers (Aguwa and Mittal, 1983).

Other agents that are being studied are those that are capable of blocking the progesterone-receptor sites in the uterine endometrium, thereby interfering with implantation, or drugs that inhibit cellular enzymes to prevent the synthesis of progesterone. Some of these substances appear to have undesirable side effects as well as a high rate of failure, and their safety is still questionable. While it is possible that one may emerge from the laboratory within the next few years, much work remains to be done.

Toward Male Contraceptives

Progress toward the development of a safe and effective pill for males is proceeding, but at a slow and careful pace. It may be, as has been suggested, that despite an apparent attitude change and an accompanying more equitable distribution of research funds, the male-dominated scientific community is really reluctant to meddle with the male reproductive tract. More charitably, it may also be that a lesson has been learned in the past 20 years—that hormonal contraceptives have hazardous side effects. At any rate, no one is rushing into production with the male birth control pill.

Sperm are produced in the seminiferous tubules of the testis under the influence of gonadotropins from the pituitary gland. FSH initiates the proliferation of sperm, and LH, acting on the interstitial cells between the seminiferous tubules, results in the production of androgen necessary for the completion of sperm development. Sperm in the tubules are nonmotile and incapable of fertilizing an egg, but after they are stored in the epididymis of the testis for a while, they acquire maturity and fertilizability. There are two main ways to regulate male fertility with a chemical agent: (1) inhibit sperm production by blocking FSH, LH, or the hypothalamic releasing factors and (2) interfere with sperm maturation in the epididymis. The trouble with using a steroid to suppress gonadotropins in order to interfere with sperm production is that androgen, responsible for sex drive and potency, is also inhibited. A synthetic testosterone derivative called danazol, which has been used as therapy for endometriosis in women, is an orally active antigonadotropin. Danazol, when used daily in combination with testosterone enanthate, an orally active androgen, decreased the sperm count without decreasing the libido. Other androgens, antiandrogens, and combinations of progestins and androgens have been used in clinical trials administered orally, through implants, or by intramuscular injection. Because they either suppress spermatogenesis or interfere with maturation or motility, they have promise as contraceptive agents. One such hormone is *inhibin*, known to exist in bovine and human semen and also in ovarian extracts and follicular fluid of cows, pigs, and humans. Porcine follicular fluid has been found to inhibit FSH secretion in rhesus monkeys and suggests that inhibin may thus be able to inhibit spermatogenesis, assuming that sperm production in human males is highly dependent on FSH. As yet, however, the source of inhibin in the testes is unknown, and the hormone has not been isolated, purified, or characterized. Even once that is accomplished, its potential for the suppression of spermatogenesis or ovulation in humans would still have to be assessed.

In 1972, Chinese scientists reported the development of a pill for men that is extracted from the seeds and other parts of the cotton plant. Called gossypol, the pill has been tested in nearly 9,000 men in China and reportedly has relatively few major side effects while reducing sperm counts below 4 million per ml of semen (Liu et al., 1981). Dry mouth, dizziness, fatigue, gastrointestinal symptoms, and decreased libido and potency were found in 3%–13%, but fewer than 1% developed a serious side effect of potassium deficiency. The major problem with gossypol, however, may be its possible lack of reversibility. Restoration of full fertility after gossypol ingestion ceases can take 12 months or more, and in 10% of the men, no living sperm were evident in the semen after several years.

A Finnish study (Ratsula et al., 1983) has tested gossypol gel used as a vaginal barrier contraceptive in 15 women and found it to be effective in immobilizing sperm. The same researchers had previously reported that gossypol inhibited herpes simplex virus and gonococci in in vitro (out of the body) experiments and suggested a future role for gossypol as a vaginal spermicide.

It is unlikely that any form of drug contraceptive for men will be any less free of adverse effects than a similar agent taken by women. While women are justifiably resentful that all the risks associated with pill use have been theirs alone for so many years, most would not feel, unless revenge is their aim, that exposing their husbands or lovers to similar health hazards is a way to solve the problem. The ultimate goal should be the elimination of risk for everyone.

LHRH Analogues for Contraception

Another group of compounds believed to have great potential as birth control agents are the luteinizing hormone releasing hormone analogues. Since 1971, when LHRH was finally isolated, purified, characterized, and synthesized—an undertaking that required the analysis of hundreds of thousands of sheep and pig hypothalami—more than 1,000 LHRH analogues, structurally different but functionally similar compounds, have been generated. The chemical modification of the parent LHRH molecule has made it possible to create two kinds of peptide derivatives, the antagonists and the agonists. The LHRH antagonists do what their name implies—they interfere with LHRH activity and are able to block production of FSH and LH, thus inhibiting ovulation or sperm production. The agonists, which require less structural tinkering with LHRH to derive, were expected to be valuable for the fertility-enhancing ability in anovulatory women because they act similarly to LHRH. The profertility applications of the agonists turned out to be disappointing, however, and their effectiveness in inducing ovulation or treating such conditions as amenorrhea or delayed puberty, has not been as successful as was anticipated. But in contradiction to their presumed actions, the agonists were shown to have highly active antifertility effects in higher doses. Tests on animals and clinical trials in humans revealed that paradoxically, some of the compounds were very potent or "super" agonists that could inhibit ovulation or corpus luteum activity and, thus, provide contraceptive effects. Taken orally, these agonists were inactivated, but it also became apparent that the compounds could exert their pharmacological activity after being administered via other than oral routes—nasally, vaginally, and even rectally. In a study in Sweden, LHRH analogues were administered to 27 women via nasal spray daily for 3–6 months and resulted in the inhibition of ovulation and menstruation with no pregnancies and no apparent side effects, at least in the short term. Upon discontinuation, ovulation and menstruation returned (Bergquist et al., 1979). A number of other investigations along similar lines have demonstrated that super agonists can clearly inhibit ovulation and thus constitute a promising lead in the search for a new female contraceptive. The most appropriate timing, dosage, and methods of administration and possible adverse long-term effects are all unknown as yet, although LHRH analogues should theoretically be safer and pose fewer problems than synthetic steroid-based oral contraceptives. We are still, however, a long way from "a sniff a day keeps the baby away."

Nevertheless, the promise of super agonists as contraceptives is greater at this point in women than in men. In males, agonists can inhibit or virtually prevent sperm development, but treatment is accompanied by the usual unacceptable side effects of testosterone inhibition and impotence. Men have to take androgen along with the analogue to prevent these problems. Compared with the number of clinical trials in women, there has been relatively little testing of LHRH analogues in men, and as yet, the best agonist to use, what dose, and how it should be administered have not been established. It is anticipated that an LHRH analogue contraceptive pill for males will take much longer to develop.

So What Else is New and Improved

In addition to searching for new methods of birth control, a major focus of contraceptive research is aimed at the improvement of the existing methods. Most of the advances in birth control technologies of the past decade, such as lower dose oral contraceptives, medicated IUDs, and better ways of barrier contraception, are really improvements on, or rectifying the problems of, known methods. A congressional report released in 1982 identified 31 new or significantly improved contraceptive technologies that are expected to be available by the year 2000. Assuming that the funds for intensified research and testing are available, the candidates most likely to appear first are:

1. Safer oral contraceptives and new ways of delivering hormonal steroids other than oral injection. These include improved long-acting injectibles, the intravaginal Silastic ring that fits around the cervix, and the Silastic rod implant under the skin.

2. Progesterone-medicated and copper IUDs that can remain in the uterus for 5–10 years and postpartum IUDs that can be safely inserted without fear of expulsion immediately after delivery.

3. Better barrier contraceptives for women: disposable diaphragms, one-size-fits-all diaphragms, spermicide-infiltrated diaphragms, cervical caps that can remain in place for months without removal, vaginal or cervical spermicide-releasing rings.

4. Improved methods of ovulation detection so that "natural family planning" becomes a more effective means of birth control.

5. Prostaglandin-analogues to be self-administered via vaginal suppositories that induce menstruation whether pregnant or not.

Other prospects in view by the end of the century are listed in Table 12.7.

Table 12.7 FUTURE FERTILITY PLANNING TECHNOLOGIES

Highly likely before 1990

1. Safer oral contraceptives
2. Improved IUDs
3. Improved barrier contraceptives for women
4. Improved long-acting steroid injections
5. Improved ovulation-detection methods for use with periodic abstinence
6. Steroid implants
7. Steroid vaginal rings
8. LRRH-analogue contraceptives for women
9. Prostaglandin analogues for self-administered induction of menses

Table 12.7 *Continued*

Possible by 1990 but prospects doubtful

1. Monthly steroid-based contraceptive pill
2. Improved monthly steroid injection
3. New types of drug-releasing IUDs
4. Minidose vaginal rings
5. Antipregnancy vaccine for women
6. Improved barrier contraceptives for men
7. Sperm suppression contraceptives for men
8. Reversible female sterilization
9. Simplified female sterilization techniques
10. Simplified male sterilization techniques
11. LHRH analogues for self-administered induction of menses

Unlikely by 1990 but possible by 2000

1. Antifertility vaccine for men
2. Antisperm drugs for men
3. Antisperm maturation drugs for men
4. Lactaton-linked oral contraceptives for women
5. Ovulation prediction methods for use with periodic abstinence
6. New types of antiovulation contraceptive drugs for women
7. Contraceptive drugs for women that disrupt ovum transport
8. Reversible male sterilization
9. Pharmacological or immunological sterilization for women
10. Pharmacological or immunological sterilization for men
11. Agents other than LHRH analogues for self-administered induction of menses

Source: U.S. Congress, Office of Technology Assessment. *World Population and Fertility Planning Technologies: The Next 20 Years.* Washington, D.C., U.S. Government Printing Office, 1982.

REFERENCES

Aguwa, C. N. and Mittal, G. C. Abortifacient effects of the roots of *Momordica angustisepala.* *J. Ethnopharmacology* 7:169–173, 1983.

Anderson, D. J. and Alexander, N. J. A new look at antifertility vaccines. *Fertil Steril* 40(5):557–571, 1983.

Barkhoff, J. R. Urticaria secondary to a copper intrauterine device. *Int J Dermatol* 15(8):594–595, October, 1976.

Beral, V. and Kay, C. Mortality among oral contraceptive users *Lancet* 2:727–730, October, 1977.

Bergquist, C., Nilius, S. J., Wide, L. Intranasal gonadotropin-releasing agonist as a contraceptive agent. *Lancet* 2(8136):215–217, 1979.

Bernstein, G. S., Kilzer, L. H., Coulson, A. H., et al. Studies of cervical caps: I. Vaginal lesions associated with use of the vimule cap. *Contraception* 26(5):444–456, 1982.

Boyce, J. G., Lu. T., Nelson, J. H., et al. Oral contraceptives and cervical carcinoma. *Am J Obstet Gynecol* 128(7):761–766, 1977.

Briggs, M. H. and Briggs, M. Randomized prospective studies on metabolic effects of oral contraceptives. *Acta Obstet Gynecol Scand* 105 (suppl):25–31, 1982.

Bygdeman, M., Christensen, N. J., Green, K., et al. Termination of early pregnancy: Future development. *Acta Obstet Gynecol Scand* 113(suppl):125–129, 1983.

Cates, W., Smith, J. C., Rochat, R. W., and Grimes, D. A. Mortality from abortion and childbirth, are the statistics biased? *JAMA* 249(2):192–195, 1982.

Centers for Disease Control Cancer and Steroid Hormone Study. Long-term oral contraceptive use and the risk of breast cancer. *JAMA* 249(12):1591–1595, 1983.

Cohen, B. J. and Gibor, Y. Metabolic and endocrine effects of medicated intrauterine devices, in Benagiano, G. and Diczfalusy, E. (eds.). *Endocrine Mechanisms in Fertility Regulation.* New York, Raven Press, 1983.

Dickey, R. P. *Managing Contraceptive Pill Patients,* ed 3. Durant, Okla, Creative Informatics, Inc., 1983.

Djerassi, C. *The Politics of Contraception.* New York, W. H. Freeman and Company, 1981.

Edmondson, H. A., Henderson, B., and Benton, B. Liver-cell adenomas associated with use of oral contraceptives. *N Engl J Med* 294(9):470–472, 1976.

Eschenbach, D. A., Harnish, J. P., and Holmes, K. K. Pathogenesis of acute pelvic inflammatory disease: A role of contraception and other risk factors. *Am J Obstet Gynecol* 128(8):838–850, 1977.

Five studies: No apparent harmful effect from legal abortion on subsequent pregnancies; D & C is possible exception. *Fam Plann Persp* 10(1):34–38, 1978.

Forrest, J. D. and Henshaw, S. K. What U. S. women think and do about contraception. *Fam Plann Persp* 15(4):157–165, 1983.

Gallegos, A. J. The zoapatle—a traditional remedy from Mexico emerges to modern times. *Contraception* 27(3):211–221, 1983.

Golde, S. H., Israel, R., and Ledger, W. J. Unilateral tuboovarian abscess: A distinct entity. *Am J Obstet Gynecol* 127(8):807–810, 1977.

Greenwald, P., Nasca, P. D., Caputo, T. A., et al. Cancer risks from estrogen intake. *NY State J Med* 77(7):1069–1074, 1977.

Greep, R. O., Koblinsky, M. A., and Jaffe, F. S. *Reproduction and Human Welfare: A Challenge to Research.* Cambridge, Mass, MIT Press, 1976.

Harris, N. V., Weiss, N. S., Francis, A., and Polissar, L. Breast cancer in relation to patterns of oral contraceptive use. *Am J Epidemiol* 116(4):643–651, 1983.

Hasson, H. M. Copper IUDs. *J Reprod Med* 20(3):139–154, 1978.

Hatcher, R. A., Stewart, G. K., Stewart F., et al. *Contraceptive Technology, 1984–85,* ed 12. New York, Irvington Publ., Inc., 1984.

Henshaw, S. K. and O'Reilly, K. Characteristics of abortion patients in the United States, 1979 and 1980. *Fam Plann Persp* 15(1):5–16, 1983.

Hogue, C. J., Cates, W., and Tietze, C. The effects of induced abortion on subsequent reproduction. *Epidemiol Rev* 4:66–94, 1982.

Holly, E. A., Weiss, N. S., and Liff, J. M. Cutaneous melanoma in relation to exogenous hormones and reproductive factors. *JNCI* 70(5):827–832, 1983.

Kamal, I., Ghoneim, M., Talaat, M., et al. The anchoring mechanism of retention of the copper T device *Fertil Steril* 24(3):165–169, 1973.

Kay, C. R., Progestogens and arterial disease—evidence from the Royal College of General Practitioners study. *Am J Obstet Gynecol* 142(6, pt. 2):761–765, 1982.

Keith, D., Phillips, J., and Julka, J. Gynecologic laporoscopy, in 1975. *J Reprod Med* 16(1):105–117, 1976.

Kennedy, D., quoted in Smoking risk added to birth bill warning. *Milwaukee Journal,* January 25, 1978.

Khaw, K. T. and Peart, W. S. Blood pressure and contraceptive use. *Br Med J* 285:403–409, 1982.

Koch, J. P. The experience of 372 women using the Prentif contraceptive cervical cap. Presented at the Cervical Cap Symposium, Atlanta, GA. May 7, 1983.

Koch, J. P. The Prentif contraceptive cap: Acceptability aspects and their implications for future design. *Contraception* 25(2):161–173, 1982.

Lakowski, R. and Morton, A. The effect of oral contraceptives on color vision in diabetic women. *Can J Ophthalmol* 12(2):96–97, 1977.

Lane, M. E., Arceo, R., and Sobrero, A. J. Successful use of the diaphragm and jelly by a young population: Report of a clinical study. *Fam Plann Perspect* 8(2):81–86, 1976.

Laragh, J. H. Oral contraceptive-induced hypertension—nine years later. *Am J Obstet Gynecol* 126(1):141–147, 1976.

Lee, N. C., Rubin, G. L., Ory, H., and Burkman, R. T. Type of intrauterine device and the risk of pelvic inflammatory disease. *Obstet Gynecol* 62(1):1–6, 1983.

Liu, Z. Q., Liu, G. Z., Hei, L. S., et al. Clinical trial of gossypol as a male antifertility agent, in Chang, C. F., Griffin, D., and Woolman, A. (eds.). *Recent Advances in Fertility Regulation.* Geneva, Atar SA, 1981, pp. 160–163.

Massey, G. J., Bernstein, G. S., et al. Results from the Collaborative Vasectomy Study. Presented at the 11th Annual Meeting of the American Public Health Association, Dallas, Texas. November 13–17, 1983.

Medina, J. E., Cifuentes, A., Abernathy, J. E. et al., Comparative evaluation of two methods of natural family planning in Colombia. *Am J Obstet Gynecol* 138(8):1142–1147, 1980.

Millen, A. and Bernstein, G. S. Accidental insertion of multiple IUDs. *J Obstet Gynecol* 47(3):369–373, March, 1976.

Mishell, D. Historical considerations in the development of modern IUDs: Patient and device selection and the importance of insertion techniques. *J Reprod Med* 20(3):121–132, 1978.

Mishell, D. R. and Moyer, D. L. Current status of contraceptive steroids and the intrauterine device. *Clin Obstet & Gynecol* 17(1):35–51, March, 1974.

Moghissi, K. S. Accuracy of basal body temperature for ovulation detection. *Fertil Steril* 27(12):1415–1421, 1976.

Moyer, D. L., Shaw, S. T., and Fu, J. C. Mechanism of action of intrauterine contraceptive devices, on Moghissi, K. S. and Evans, T. N. (eds). *Regulation of Human Fertility.* Detroit, Wayne State University Press, 1976.

Nofziger, M. *A Cooperative Method of Natural Birth Control,* ed 2. Summertown, Tenn, The Book Publishing Co., 1978.

Nora, A. H. and Nora, J. J. A syndrome of multiple congenital anomalies associated with teratogenic exposure. *Arch Environ Health* 30(1):17–21, 1975.

Ory, H. The noncontraceptive health benefits from oral contraceptive use. *Fam Plann Persp* 14(4):182–184, 1982.

Ory, H. W. Association between oral contraceptives and myocardial infarction. *JAMA* 237(24):2619–2622, 1977.

Paffenbarger, R. S., Fasal, E., Simmons, M. E., et al. Cancer risk as related to use of oral contraceptives during fertile years. *Cancer* 39(4 suppl): 1887–1891, 1977.

Penfield, A. J. Minilaparotomy for sterilization. *The Female Patient* 3(3):60–64, 1978.

Petitti, D. and Wingerd, J. Use of oral contraceptives, cigarette smoking, and risk of subarachnoid hemorrhage. *Lancet* 2:234–235, 1978.

Phillips, J., Hulka, J., Hulka, B., et al. American Association of Gynecological Laparoscopists, 1979 membership survey. *J Reprod Med* 26(10):529–537, 1981.

Pike, M. C., Henderson, B. E., Casagrande, J. T., et al. Oral contraceptive use and early abortion as risk factors for breast cancer in young women. *Br J Cancer* 43:72–76, 1981.

Pike, M. C., Henderson, B. E., Krailo, M. D., et al. Breast cancer in young women and use of oral contraceptives: Possible modifying effect of formulation and age at use. *Lancet* II:926–929, 1983.

Population Reports. Colpotomy—the vaginal approach. Series C, no 3, June, 1973.

Population Reports. IUDs: An appropriate choice for many women. Series B, no. 4. July, 1982A.

Population Reports. Oral contraceptives in the 1980s. Series A, no. 6. May–June,1982B.

Population Reports. Update on condoms—products, protection, promotion. Series H, no. 6. September–October, 1982C.

Population Reports. New developments in vaginal contraception. Series H, no. 7. January–February, 1984.

Ramcharan, S., Pellagrin, F. A., and Hoag, E. The occurrence and course of hypertensive disease in users and non-users of oral contraceptive drugs, in Fregley, M. J. and Fregley, M. S. (eds). *Oral Contraceptives and High Blood Pressure.* Gainesville, Fla., Dolphin Press, 1974.

Ratsula, K., Haukkamaa, M., Wichmann, K., and Luukkainen, T., Vaginal contraception with gossypol: A clinical study. *Contraception* 27(6):571–577, 1983.

Rice, F. J., Lanetot, C. A., and Garcia-Devesa, C. The effectiveness of the sympto-thermal method of natural family planning. Paper presented at the First Assembly of the International Federation for Family Life Promotion, Cali. Colombia, June 22–29, 1977.

Roe, D. Nutrition and the contraceptive pill, in Winnick, M. (ed). *Nutritional Disorders of American Women.* New York, John Wiley & Sons, Inc., 1977.

Rooks, J. B. and Cates, W. Emotional impact of D&E vs. instillation. *Fam Plann Persp* 9(6):276–277, 1977.

Royal College of General Practitioners. Further analyses of mortality in oral contraceptive users. *Lancet* I:541–546, 1981.

Rubin, G. L., Ory H. W., and Layde, P. Oral contraceptives and pelvic inflammatory disease. *Am J Obstet Gynecol* 144(6):630–635, 1982.

Sartwell, P. E. and Stolley, P. D. Oral contraceptives and vascular disease. *Epidemiol Rev* 4:95–108, 1982.

Schilling, L. H. An alternative to the use of high-dose estrogens for post-coital contraception. *J Am Coll Health Assoc* 27(5):247–249, 1979.

Shapiro, Howard. *The Birth Control Book.* New York, St. Martin's Press, 1977, p. 211.

Short, R. V. The evolution of human reproduction. *Proc R Soc Lond B* 195(1118):3–24, 1976.

Slone, D., Shapiro, S., Kaufman, D. W., et al. Risk of myocardial infarction in relation to current and discontinued use of oral contraceptives. *N Engl J Med* 305:420–424, 1981.

Study suggests possible vasectomy-impotence link. *Contraceptive Technology Update.* 4(9):101–103, 1983.

Theuer, R. and Vitale, J. J. Drug and nutrient interactions, in Schneider, H., Anderson, C. E., Coursin, D. B. (eds). *Nutritional Support of Medical Practice.* Hagerstown, Md., Harper & Row, 1977.

Tietze, C. Intrauterine devices: Clinical aspects , in Hafez E. S. E. and Evans, T. N. (eds). *Human Reproduction.* Hagerstown, Md, Harper & Row, 1973.

Trobaugh, G. E. Pelvic pain and the IUD. *J Reprod Med* 20(3):167–174, 1978.

United States Congress, Office of Technology Assessment. World population and fertility planning technologies: The next 20 years. Washington, D.C., U.S. Government Printing Office, 1982.

Valicenti, J. F., Pappas, A. A., et al. Detection and prevalence of IUD-associated *Actinomyces* colonization and related morbidity: A prospective study of 69,925 cervical smears. *JAMA* 247(8):1149–1152, 1982.

Vessey, M., Meisler, L. Flavel, R., and Yeates, D., Outcome of pregnancy in women using different methods of contraception. *Br J Obstet Gynecol* 86(7):548–556, 1979.

Vessey, M. P., Doll, R., Peto, R. et al. A long-term follow-up study of women using different methods of contraception. *J. Biosoc Sci* 8(4):373–427, 1976.

Vessey, M. P., Lawless, M., McPherson, K., and Yeates, D., Neoplasia of the cervix uteri and contraception: A possible adverse effect of the pill. *Lancet* II:930–934, 1983.

Vessey, M. P., Yeates, D., Flavel, R., and McPherson, K., Pelvic inflammatory disease and the intrauterine device: Findings in a large cohort study. *Br Med J* 282(6267):855–857, 1981.

Wade, M. E., McCarthy, P., Braunstein, G. D., et al. A randomized prospective study of the use-effectiveness of two methods of natural family planning. *Am J Obstet Gynecol* 141(4):368–376, 1981.

Weir, R. J. Effect on blood pressure of changing from high to low dose steroid preparations in women with oral contraceptive induced hypertension. *Scott Med J* 27:212–215, 1982.

Weller, R. H. and Hobbs, F. B. Unwanted and mistimed births in the United States: 1968–1973. *Family Plann Persp* 10(3):168–172, 1978.

Westrom, L. Incidence, prevalence, and trends of acute pelvic inflammatory disease and its consequences in industrialized countries. *Am J Obstet Gynecol* 17(5):509–511, 1980.

Wingrave, S. and Kay, C. R., Reduction in the incidence of rheumatoid arthritis in association with oral contraceptives. *Lancet* II:871–876, 1978.

Yuzpe, A. A. Post-coital contraception. *Int J Gyn Ob* 16:497–501, 1979.

13. *Menopause*

Humans are virtually the only species to outlive their reproductive capacities—a true biological difference of great evolutionary interest but unknown significance. In both males and females, the ability to reproduce ceases before the end of a normal lifespan, but women have a longer postreproductive period than men. When the average American woman dies at the age of 76 she has not been able to produce a baby for approximately 25 years, or one-third of her life. This female longevity is historically fairly recent; in the past, most women died before or soon after menopause.

Menopause is the final phase of woman's reproductive ability. It refers to the cessation of the menstrual periods and is considered complete after 1 year of amenorrhea. Although in some women menstruation stops abruptly, menopause is usually a gradual process, with menstrual irregularity making it very difficult to tell precisely when the periods have stopped. *Climacteric* is the name given to the gradual changeover, that period of declining ovarian function that terminates in complete menopause. It generally starts in the forties and is believed to be influenced by the same kinds of factors that affect menar-

che: race, heredity, climate, nutrition, general health, and socioeconomic status. One-half of all women experience menopause between the ages of 45 and 50, one-fourth before age 45, and one fourth after age 50. In the United States, the average age of menopause is 50, although cigarette smoking is associated with an earlier menopause (Willett, et al. 1983). About 8% of women experience premature menopause before age 40, and it is said to be delayed when menopause occurs after age 55.

CAUSE OF MENOPAUSE

The large number of ovarian follicles with which a woman is born progressively decreases throughout her lifetime. All through childhood, adolescence, and maturity, the follicles are lost by ovulation and atresia. Eventually, the total number is almost exhausted, and the ovarian hormone secretion diminishes. The decreased ovarian estrogen produced by the few follicles that are left is insufficient to inhibit the pituitary gonadotropins, and FSH and LH levels greatly increase. When the level of ovarian estrogens falls below a certain critical amount, the hypothalamic-pituitary-ovarian feedback system is altered and cycles stop. At least that is the classical explanation, although electron microscopic studies of the ovaries after menopause do not seem to support the notion that they are totally depleted of follicles. In the absence of a definitive explanation concerning the cause of the cessation of menstruation but based on present knowledge, it is reasonable to say that menopause is a normal physiological consequence of the progressive loss of follicles that started in a woman before she was born. In a sense, the climacteric starts during embryonic life and continues until death.

DECLINE OF FERTILITY

As menopause approaches, more and more of the menstrual cycles become anovulatory. Cycles in which ovulation does not occur are at first likely to be shorter, and 21-day cycles are not unusual. Any ovulatory symptoms, such as dysmenorrhea, that ordinarily accompany a progestational phase, are missing. As ovarian function further decreases, cycles then may become longer, since the endometrium is stimulated only by estrogen, and proliferation may take more time. Between the ages of 40 and 45, Sharman has estimated that only 75% of the cycles are ovulatory. After age 46, the percentage drops to 60%. Pregnancy is infrequent between ages 45 and 49, occurring in only one of 1,000 births. After age 50, the incidence drops to one in 25,000 births. There are only 15 cases of pregnancy after menopause recorded in the medical literature. It is assumed that such a conception must have resulted from the random ovulation of a surviving follicle and was not associated with a menstrual cycle.

Evidently there need be little concern about becoming pregnant after menopause has occurred, but for women in their forties whose periods have ceased, the valid questions are, "How do I know if I have reached menopause yet, and when can I stop using a contraceptive?"

The data of Wallace et al. (1979) shows that in a woman over 45, the longer the period of amenorrhea, the more likelihood there is that the menopause is permanent. Between age 45 and 49, a 6-month interval without menstruation means a 45% probability of menopause, but in a woman 53 or older, the probability is 70%. That is, after an amenorrhea of 180 days, 55% of women in their late forties, and 30% of women in their early fifties, still could expect to have one or more additional episodes of menstruation. Even after 1 year of no menstrual periods, 10.5% of women ages 45–49, 6.4% of women ages 50–52, and 4.5% of women 53 or older could have another cycle, possibly ovulatory. The probability of menopause is greater when the amenorrhea has been preceded by prior irregularity, but if avoidance of pregnancy is important to a woman, it would be prudent to use contraceptives for a *full year* after the last menstrual period.

GONADOTROPIN SECRETION IN POSTMENOPAUSE

The blood and urine levels of LH and FSH, particularly of FSH, are very high in postmenopause. In fact, the major source of human gonadotropins used to induce ovulation is the urine of postmenopausal women.

ESTROGEN SECRETION IN POSTMENOPAUSE

While gonadotropin secretion is consistently high, estrogen levels in the postmenopause are subject to very wide variation and may or may not be low. The *kind* of estrogen found in the blood and urine in postmenopause is different from that of premenopause. The major hormone produced by the ovaries during the reproductive years is *estradiol;* the estrogen found in postmenopausal women is *estrone.* Estradiol is much more biologically active than estrone. In some postmenopausal women, estrogen is maintained at a moderate level for the rest of their lives. The extent and rate of its production varies from woman to woman. The following summarizes present knowledge concerning the secretion of estrogens:

1. Estradiol and some estrone are produced by the theca cells surrounding the developing follicles during the reproductive years. The rest of the estrone is derived by the conversion of estradiol and androstenedione by peripheral tissues of the body not located in the ovaries.

2. Androstenedione is a weak androgen. It is produced by the adrenal glands and also by the stromal cells of the ovaries (connective tissue elements containing degenerating theca cells and structures from atretic follicles).

3. After the disappearance of the follicles in the ovaries, the stromal cells, under the influence of high gonadotropin secretion, produce androstenedione, which is converted to estrone by peripheral tissues, principally blood and fat tissue.

4. Even if the ovaries are removed during postmenopause, estrone levels do not decrease significantly, indicating that the adrenal glands produce estrogen, either directly or by extraovarian conversion of the precursor, androstenedione. The adrenals are the major source of this androgen in the postmenopause, contributing about 85%. If the adrenal glands are removed in a woman who has had her ovaries removed,

even then the urinary excretion of estrone does not completely disappear. The origin of this extraovarian *and* extra-adrenal estrogen is unknown.

5. The amount of estrogen in the postmenopausal woman is dependent on the percentage of androstenedione that is converted to estrone. Greater production of estrone and increased rate of conversion have been correlated with increases in age and body weight.

6. There is little evidence for postmenopausal production of estradiol by the ovaries or adrenals.

MENOPAUSAL SYMPTOMS

There are approximately 32 million women older than 50 in the United States. Their average life expectancy is 26 years beyond the menopause. To many physicians, the increasing life span in women has only meant an increase in the problems of the postmenopausal period that must be "managed," in most instances by empirically prescribing estrogen replacement therapy. But women, too, view the "change of life" with dread and anxiety because of the persistent myth that menopause in women is a very stressful and upsetting event, both physically and psychologically. This myth of menopause as illness is perpetuated by the doctors who base their assessment on their clinical experience. Physicians who have the opportunity to allay the fears and anxieties of women concerning the horrors of menopause frequently do not because they themselves believe menopause to be far more traumatic than do most women. Doctors, after all, see only the women with the most complaints.

Physicians' expectations for the menopause may also be colored by the grim descriptions that have appeared in the medical world. For example, at a 1971 government-sponsored conference on menopause and aging, a Johns Hopkins University professor of Ob/Gyn characterized menopausal women as being "a caricature of their younger selves at their emotional worst." In 1973, the past president of the American Geriatric Society declared that menopause was a "chronic and incapacitating deficiency disease that leaves women with flabby breasts, wrinkled skin, fragile bones and loss of ability to have or enjoy sex." Equally appalling accounts of the menopausal state could be readily found in medical literature of the past 20 years. Public expression of such sentiments is not as likely today, but this statement in a recent medical journal article illustrates that a physician's view of menopause is still apt to be stereotypical: "The one positive aspect of the menopause that women can look forward to is the cessation of menstruation" (Brenner, 1982). Doctors exposed to these menopausal myths during their medical training and practice may retain personal biases that could influence their evaluation of women during the climacteric.

There has been very little actual controlled and objective medical research on menopause. For example, there is a distinct lack of information concerning the numbers of women who experience a symptom-free climacteric. The one study frequently quoted was done in 1933; it indicates that of 1,000 women studied, only 15%–20% have, besides the obvious cessation of menstrual periods, no clinical symptoms at all. Evidently in these women, the hormonal and physiological changes that occur are so gradual that they are almost imperceptible.

At the other end of the spectrum are the women who suffer from almost incapacitating physical complaints and a number of psychological disturbances as well. Symptoms include hot flashes, genital atrophy, headache, backache, nervousness, poor memory, depression, insomnia, dizziness, breast problems, bloating, nausea, decreased libido, or uncontrollably increased libido. To what extent some of these symptoms are related to hormonal deficiency is not clear; they may be psychologically or sociologically induced rather than physiological in origin.

Between these two extremes are the vast majority of women. They may have some symptoms, but these are so vague or mild that they are frequently ignored.

THE ESTROGEN CONTROVERSY—IS IT HAZARDOUS OR HELPFUL?

We live in a society that places high value on youth and beauty. In our youth-oriented culture, the thought of impending menopause can be the signal of impending old age. It conjures up all the problems the aged confront in our society—illness, isolation, frequently poverty and despair, and ultimately, the mortality we all face. When the combined efforts of the pharmaceutical industry and a few physicians in the early 1960s promised that women could remain "feminine forever," healthy, young, and sexy by taking estrogen for the presumed miseries of menopause, it is not surprising that the sales of the drug quadrupled in 10 years, as millions of women were encouraged to routinely take it as a cure-all. By the mid-1970s, there were some urban areas of the country where more than half of the menopausal women were receiving estrogen (Stadel and Weiss, 1975).

Of the 22 million total annual prescriptions written for postmenopausal women, more than one-third were for Premarin, the estrogen preparation produced by Ayerst Laboratories. Premarin consists of conjugated estrogens, also called natural estrogens, although why they should be considered natural for humans is speculative, since the material is derived from the urine of pregnant mares. Premarin is supplied in four strengths: 0.3 mg (green); 0.625 mg (red); 1.25 mg (yellow); 2.5 mg (purple), which are administered cyclically; that is, 3 weeks on pills and 1 week off. In 1975, Ayerst Laboratories sold an estimated $70 million of Premarin.

Early in 1975, however, several published studies aroused suspicion that the miracle drug for the middle-aged woman was probably responsible for a marked increase in uterine cancer. In the studies, the estrogen use of women with endometrial cancer was compared with the estrogen history of matched control subjects, who were women of similar age, area of residence, and potential for development of endometrial cancer—that is, the presence of an intact uterus. There was an indication that duration of use and dosage of Premarin increased the likelihood of cancer development. In one of the studies, the researchers calculated that the risk of getting endometrial cancer was 5.6 times greater for women taking estrogens for 1–5 years than for nonusers; that it increased to 7.2 for those taking them for 5–7 years; that women who took estrogens longer than 7 years were 13.9 times as likely as nonusers to develop endometrial cancer (Ziel and Finkle, 1975).

Although the studies did not prove that estrogen therapy causes endometrial cancer, they strongly suggested a link. Ayerst Laboratories challenged the conclusions of the studies, claiming that they contained statistical flaws and that there were other factors that increased the chances of getting endometrial cancer. Many physicians were skeptical as well and indicated that the studies were not comprehensive enough and of doubtful accuracy.

At least 15 additional retrospective case-controlled studies have appeared since 1975, however, and each has substantiated that the odds of developing endometrial cancer are greater after estrogen use. The risk increases with dosage and duration of use and is greater in white women of higher socioeconomic class. Cessation of use diminishes the risk, but one report showed that the enhanced risk remained for 3 or more years after discontinuation of estrogen therapy (Shapiro et al., 1980). Endometrial cancer rates, after reaching a peak in 1975, have now declined—a decrease that closely follows the declining sales of conjugated estrogens. The diminishing incidence of endometrial cancer when women's exposure to estrogen has also been reduced strongly suggests a cause-and-effect relationship.

Other Complications of Estrogen Therapy in Postmenopause

Although the number of controlled studies on the effect of postmenopausal estrogen use on breast cancer is not as great as that for endometrial cancer, there is evidence that estrogen usage is associated with an increased relative risk of breast cancer. The magnitude of the increase is slight, but since breast cancer is a much more common disease than endometrial cancer, even a small additional risk can result in substantially more cases. One group of investigators reported that the risk was higher only for estrogen users experiencing natural menopause and was not apparent in women taking estrogen who had their ovaries surgically removed. But a history of benign breast disease apparently increased the risk for postmenopausal women, and that greater risk remained apparent in women who had had oophorectomy (Ross et al., 1980).

Henderson (1983) estimated that a menopausal woman of 50, who receives 1.25 mg of conjugated estrogens for 3 years, has increased her chances of getting breast cancer by age 75 from 6% to 12%.

Estrogen administration may cause high blood pressure, and the hypertension disappears when estrogen use is discontinued.

This effect is more often the result of taking oral contraceptives but may also occur with estrogen replacement therapy in the menopause. High blood pressure may be exacerbated by estrogen in postmenopausal women who already have hypertension (Pfeffer and Van den Noort, 1976).

There is evidence that administration of estrogen to postmenopausal women results in a twofold increased risk of gallbladder disease, and many women receiving therapy have subsequently required gallbladder surgery. A previous history of gallstones would thus be a contradiction for estrogen usage.

The whole issue of estrogen replacement in menopause is highly controversial. There are established risks, but there are also known benefits for some women. There are also many healthy postmenopausal women who are taking estrogen to provide benefits not yet clearly demonstrated. The question to be answered is whether the benefits of the drug outweigh the risks in an individual woman.

Of all the presumed symptoms of menopause, there are only two that have been established as uniquely characteristic of menopause and that are uniformly relieved by estrogen therapy. One is vasomotor instability, or "hot flashes," and the other is genital atrophy.

VASOMOTOR INSTABILITY

Hot flashes and night sweats are thought to result from disturbances in the temperature-regulating center in the hypothalamus. This special group of neurons in the preoptic anterior hypothalamus is also called the vasomotor center. The area controls body temperature via the autonomic nervous system by causing sweating and by dilation or constriction of all the small blood vessels of the skin. When body temperature increases, impulses from this area are transmitted to skin vessels and sweat glands to cause cooling by vasodilation and sweating. When body temperature is cooled, vasoconstriction occurs, and sweating is abolished.

A hot flash, or flush, sometimes preceded by a chill, is a sensation of warmth that may be perceived most intensely on the upper trunk of the body, starting in the lower chest and rising up to the head and neck. Hot flashes are a generalized phenomenon, however, and when skin temperature is measured, the greatest rise occurs over the fingers and toes, although perspiration is most apparent on the upper body. The skin of the face and neck may redden and flush, but in some women only the hands and fingers are flushed. The sensation of heat may be followed by a drenching sweat, particularly at night. Any factor that affects the temperature-regulating mechanism, like lying in bed under a blanket, sitting in a warm room, exercise, eating, or emotional stress can trigger hot flashes. They may occur only once or twice a day, or every half hour; they may last for a few weeks or continue for years. Eventually, they subside, but they can be a distressing and humiliating experience for the woman that has them severely. The stigmatization of menopausal symptoms in our culture can be a major cause of the woman's distress.

The usefulness of estrogen therapy in treating hot flashes is well established, since estrogen usually alleviates or eliminates the condition. This does not necessarily mean that lack of estrogen has caused the vasomotor disturbance. Hot flashes do not occur in women born without ovaries or with nonfunctioning ovaries, in girls who were ovariectomized before puberty, or in patients with hypofunctioning pituitary glands. If these females, however, who have never had their own source of estrogen are given estrogen treatment for more than a year and subsequently abruptly withdrawn from treatment, they will then experience hot flashes. This finding has generated a theory that vasomotor instability represents withdrawal symptoms from years of addiction to endogenous estrogen, and some researchers maintain that many of the symptoms attributed to menopause are analogous to those occurring upon withdrawal from drug dependency. It does seem far-fetched that hot flashes are the result of a woman going "cold turkey" from her own estrogen, and to other researchers the notion that women become estrogen addicts between menarche and menopause has very little credence. These investigations implicate high levels of gonadotropins as a more reasonable explanation. Because of the complexity of the hypothalamic-pituitary-ovarian axis, it is probable that there is more than lack of estrogen or abundance of gonadotropin involved in the cause of hot flashes.

Now that the quick cure-all of estrogen replacement is less often used for hot flashes, there has been greater research interest in the cause and amelioration of this symptom. The current theory is that instability of the temperature-regulating system is associated with a middle range level of estrogen as it passes from the high range of the reproductive years to the low levels occurring during menopause, and that the effect is mediated through the hypothalamus. It is now known that pituitary LH is released in bursts or pulses, and it is further known that when hot flashes occur, they are associated with a concomitant pulsatile release of LH. Since LH is stimulated by luteinizing hormone releasing hormone (LHRH), it is suspected that both LH and LHRH are implicated in triggering a hot flash. The neurons that produce LHRH are in the same area of the hypothalamus as the neurons that control body temperature. Moreover, the same neurotransmitters believed to be involved in the release of LHRH—norepinephrine and dopamine—have also been implicated in temperature regulation. It is possible that when the neurotransmitters cause the release of LHRH, they have overlap-

ping effects that alter the adjacent thermoregulatory neurons, thus causing a hot flash.

During the climacteric, the extent of estrogen deficiency differs among women and has a great deal to do with the extraovarian production of estrone. The amount of circulating estrogen is thus related to the adrenal production and peripheral tissue conversion of androgen (androstenedione), and the latter is directly correlated to body weight.

But not all heavier women are free of hot flashes, and not all thinner women get them, so it is not possible at this time to predict which women entering menopause will experience them, how severe they will be, or when they will end. An estimated 75% of the approximately 2 million women who enter menopause in a single year will have hot flashes, and only a small minority will have vasomotor symptoms severe enough to require therapy. Since the symptoms of heat and sweating are mediated through the autonomic nervous system, some women have been relieved by a prescription drug called Bellergal (Sandoz Pharmaceuticals), a combination of phenobarbital and both sympathetic and parasympathetic nervous system inhibitors. Clonidine, a drug used in treating hypertension, reduces the frequency of hot flashes but is not as effective as estrogen and may have unpleasant side effects of dizziness, nausea, and headache. When hot flashes are bad enough to interfere with daily activities, or chronically disturb sleep, they will respond well to small doses of estrogen for a short period of 6–12 months or less. Estrogen use, however, by far exceeds that which is required for the short-term management of hot flashes.

GENITAL ATROPHY

The only other symptom of the menopause experienced by some women that is definitely known to be the result of lowered estrogen levels is the gradual atrophy of the genital organs, generally appearing 10–20 years after menopause. Since the vulvar skin is more sensitive to absence of estrogen than the skin and fat of the rest of the body, the labia majora, minora, and mons pubis shrink in size, and the pubic hair may become scant. Because the strength and elasticity of the muscles and ligaments of the pelvis are also affected, cystocele, rectocele, and prolapse may occur for the first time, especially if there was previous childbirth damage. Most of the problems that women have, however, are a result of the thinning and shrinking of the vaginal and urethral epithelium and wall. The vagina shrinks in both length and width, and the epithelium atrophies. The epithelial cells lose their glycogen, the Döderlein's bacilli disappear, and vaginal secretions are no longer acid. The vagina then becomes much more prone to infections. The itching and burning that may result is called by the unfortunate term, "senile vaginitis," and pain during intercourse (dyspareunia) may occur. All of these changes can be reversed with low doses of estrogen, but local treatment with estrogen-containing suppositories or creams is also effective, and less estrogen is absorbed into the bloodstream. When atrophy is less severe, symptoms may be alleviated by the use of sterile K-Y Jelly.

It should be noted that a recent study of 52 postmenopausal women not taking estrogen replacement therapy confirmed the earlier report of Masters and Johnson concerning the beneficial effect of sexual activity on vaginal atrophy. The women, who had intercourse three or more times monthly or who masturbated had less vaginal atrophy than the women who engaged in less frequent sexual activity (here defined as intercourse less than 10 times yearly) or did not masturbate (Leiblum et al., 1983).

Despite the evidence that only hot flashes and genital atrophy are clinical symptoms unique to menopause and will respond to low doses of estrogen therapy for short periods, widespread long-term use of estrogen has continued, presumably in the hope of beneficial effects on bone loss, coronary heart disease, emotional stability, sensuality, and a youthful appearance. What evidence exists for continued use of estrogen-replacement therapy for conditions other than vasomotor symptoms and genital atrophy?

OSTEOPOROSIS

Osteoporosis is literally, "holes-in-the-bone," a condition of increased porosity, bone loss, and, therefore, increased fragility of bone. Osteoporosis is a chronic skeletal disorder associated with aging, but although peak bone mass is reached in both sexes at approximately age 35, the rate of bone loss after that is statistically greater in women than in men and seems to be associated with menopause.

In its most serious form, osteoporosis results in a predisposition to fractures. When fractures occur in the thinned and weakened vertebrae that support most of the weight in the spinal column, the result is progressive decrease in height and bending of the spine, producing swayback or humpback. Approximately one of four women older than 60 years of age has such spinal compression fractures. Hip and wrist fractures are other dangers. If a woman lives to the age of 90, her risk of hip fracture is 20%. Breaking a limb results in immobilization, which not only aggravates osteoporosis, but in the aged may result in lung collapse, pneumonia, and death.

It is uncertain at present whether osteoporosis is a natural consequence of aging or whether it is a result of estrogen deprivation. Seventy-five percent of postmenopausal women never get the severe form of the disease. It appears that some people are more susceptible than others, that those with a smaller adult bone mass have an increased vulnerability to the development of osteoporosis, although those with a greater adult bone mass seem to be protected. Blacks of both sexes, white men, tall women, and obese women have less risk of developing osteoporosis. Other factors involved in bone loss are lack of physical activity, calcium and protein deficiency, and general malnutrition.

Although estrogen administration will reverse osteoporosis in a younger woman who has had her ovaries removed, when the disease is once established in a menopausal woman, estrogens have not been shown to increase bone mass. At best, they are able to stabilize the disease and halt the progress. It also appears that estrogen prevents bone loss only for as long as it is used.

Bone growth, loss, and general metabolism are highly complex and involve the hormones calcitonin, parathormone, the steroidlike cholecalciferol (vitamin D_3) and its metabolites, the minerals calcium, phosphorus, and fluoride, and the amount of stress (exercise) to which the bone is subjected. Just how estrogen fits into the picture is not known. Estrogen replacement therapy is apparently able to inhibit bone reabsorption, but the mechanism is still obscure since investigators thus far have been unable to demonstrate the presence of estrogen-receptors in bone.

Osteoporosis is a painful, crippling, and potentially fatal disease, and its recognition, prevention, measurement, and treatment have caused some of the sharpest controversy concerning estrogen therapy. Two retrospective case-controlled studies (Hutchinson et al., 1979; Weiss et al., 1980) have demonstrated that estrogen, taken within 5 years of the menopause in the first study and taken for at least 6 years in the second study, has a protective effect against the incidence of bone fracture, reducing the likelihood of a postmenopausal woman's breaking her forearm or hip by more than 50%. Neither study had data on the other determinants of bone density, such as mineral and hormone levels. Another group of epidemiologists (Williams et al., 1982) showed that estrogen use in obese women had little or no effect on bone fracture risk but that the beneficial effect of estrogen on fractures was greatest in thin women, especially in those who smoked cigarettes. Riggs and co-workers (1982) found that fracture rates in women with diagnosed osteoporosis were reduced most effectively by a combination of calcium, fluoride, and estrogen treatment. The preliminary evidence that estrogen may prevent osteoporosis and reduce fractures is promising, and obviously the prevention is preferable to the treatment of established osteoporosis. As yet, however, there are insufficient data to warrant the use of estrogen as a prophylactic against bone loss in all women—a position advocated by many physicians. All women are estrogen deficient in the menopause, but osteoporosis is not an inevitable disease, and only one out of four white women over age 60 develops symptomatic osteoporosis. The possible benefit of estrogen treatment has to be weighed against the known risk.

In a series of studies, Heaney and co-workers (1982) found that the average intake of calcium in a group of normal women at the time of menopause was 660 mg a day, well below the recommended daily intake of 800 mg. Long-term low calcium intake, decreased physical activity with increasing age, and some of the aspects of the American diet are believed to be factors that contribute to calcium loss over the years and thereby promote the development of osteoporosis. It has been recommended by a number of investigators that an intake of at least 1 g a day before menopause and 1.5 g daily after the menopause is necessary to maintain the appropriate calcium balance for the prevention of bone loss. Getting that much from the diet alone is difficult, and supplementation with calcium carbonate, calcium gluconate, or calcium lactate pills or powder would be necessary. Some researchers have advised that such calcium supplementation should begin for women as early as age 25, and there are a number of over-the-counter pills available. The Food and Drug Administration has, however, warned against the consumption of calcium sources such as bone meal or dolomite (usually found at health food stores), because these products have been found to be contaminated with lead or other substances. There are inexpensive antacid preparations (Tums) that contain 500 mg calcium carbonate, equivalent to 200 mg elemental calcium, as the sole ingredient. Two to four tablets daily, depending on the diet, would be a palatable way to obtain the recommended total calcium requirement. There are no epidemiological data to indicate that amounts of calcium carbonate in that range have any adverse effects on stomach acid or increase the likelihood of kidney stone formation, but calcium in any form can be constipating. It may be necessary to take a stool softener along with additional calcium.

CORONARY HEART DISEASE

There is a difference in the vulnerability of men and women to coronary heart disease, but the reasons for it are unknown. In the last 30 years in the United States, the ratio of male to female deaths from heart disease at ages 45–50 was 5 to 1, but it was 2 to 1 in Italy, and 1 to 1 in Japan, indicating that the sex difference regarding heart disease is much smaller in less affluent societies. Furthermore, only white women have the sex advantage; it is seen to a much lesser extent in blacks. Even in this country, the ratio of male to female mortality rates steadily declines after age 50, principally as a result of a slower statistical acceleration with age of coronary heart disease death rates for men. Eventually, women catch up to the men, and there is a steady increment in the rate of heart disease deaths

through the menopausal years. In other words, there is no sudden increase in deaths from heart disease in women after menopause, but there are fewer men that die from heart attacks after age 50. The statistical difference that exists between white men and white women younger than 50 years of age in this country is due to a vulnerable group of *men* who get coronary heart disease and probably not to a hormonal difference between men and women.

It is known that the risk of developing coronary heart disease is greater when the blood levels of cholesterol and low-density lipoproteins (LDLs) are higher and that the risk is lessened when the high-density lipoprotein (HDL) levels in the blood are higher. There is evidence that estrogen replacement therapy increases HDLs (the "good" blood fats) and decreases cholesterol and LDLs, suggesting a possible protective effect of estrogen against heart attacks. One recent investigation did find that the death rate from all causes was one-third lower in estrogen users aged 40–69 who were followed for an average 5.6 years when compared to nonestrogen users (Bush et al., 1983). The authors had no explanation for the lowered mortality rate but speculated that it may have been due to the increased HDL levels. But trying to isolate estrogen as a factor in cardiovascular disease is virtually impossible because of all the other variables that confuse the issue. Risk factors for heart disease also include obesity, smoking, use of oral contraceptives, heredity, and general lifestyle. Moreover, there are few data on the effects of different doses and duration of use of estrogen on the risks of developing coronary disease in the menopause. It is known that taking estrogen in the larger doses found in oral contraceptives decreases HDL levels and increases the incidence of heart attack in premenopausal women, especially if they are cigarette smokers, and the increased risk of blood clot formation has clearly been shown to be associated with estrogen administration. As yet, there is no valid evidence that estrogen replacement therapy is or is not protective against cardiovascular disease, and more study is needed. The relationship is still too uncertain to warrant taking estrogen on that basis alone.

OTHER PRESUMED BENEFITS OF ESTROGEN THERAPY

Long-term use of estrogen has been advocated for emotional stability, sensuality, a youthful appearance, and a "feeling of well-being." It is for these alleged benefits that estrogen has been most overused and misused. While there are innumerable clinical observations by physicians that indicate that estrogen relieves anxiety, tension, depression, and irritability, there is no evidence from controlled clinical trials that any of these conditions are improved if hot flashes are not present. It could be that the relief of severe hot flashes and the accompanying insomnia has the secondary benefit of alleviating the psychological symptoms that the incapacitating vasomotor symptoms may have caused.

There has been no objective documentation that estrogen therapy contributes to a youthful appearance or a feeling of well-being. Estrogen cannot reverse or retard the aging process. Aging creates menopause, not the other way around. But when youthfulness, attractiveness, and the ability to reproduce are seen as important attributes for women, menopause can be a trying time for even the most stable of individuals. Menopause itself, however, does not turn an emotionally healthy, functional woman into a dysfunctional, depressed neurotic; it is generally agreed that the total behavior patterns and personality of a woman before menopause have a determining effect on her response to menopause. So-

cial and cultural factors are far more likely than hormonal deficiency to affect psychic functions.

Perhaps many women are afraid of menopause and afraid of aging. But with just cause, they are more afraid of cancer. Studies have shown that administration of estrogen increases the risk of developing endometrial cancer, breast cancer, gallbladder disease, and hypertension. Prudent and cautious use of this potentially dangerous drug is necessary.

There are two age-old maxims in medicine that appear to have been ignored in the entire estrogen therapy controversy. One is that healthy patients are best left untreated, and the other is found in the Hippocratic oath: *primum non nocere*, or "first, do no harm." Estrogens are valuable in treatment but should be reserved for women who have symptoms that are severe and incapacitating or who are identified as being at increased risk for osteoporosis. The use of estrogen in the climacteric should not be left to the physician alone; all women should be informed of the risks of treatment and given the choice to decide whether the benefits are worth the risks. After the evidence linking administration of estrogen to uterine cancer, the FDA ordered that new, stronger warnings clarifying the indications for treatment, explaining the risks, and stating the least hazardous treatment regimen be available not only to every physician prescribing the hormone but to every patient taking them. Both the FDA and the American College of Obstetricians and Gynecologists (ACOG) have recommended that the least dangerous therapeutic mode for conjugated estrogen is "cyclic administration of the lowest effective dose for the shortest possible time with appropriate monitoring for endometrial cancer." The guidelines stress that physicians should stop therapy periodically to determine whether the conjugated estrogens are still needed.

Some clinicians maintain that while there is

little question that estrogen therapy increases endometrial cancer, the risk is markedly lessened when a progestin is given along with the estrogen. It is estrogen given alone, they believe, that is unnatural and nonphysiological and produces the cellular abnormalities in the endometrium that may lead to cancer. When a progestin, such as norethindrone, norgestrel, or medroxyprogesterone acetate, is given along with the estrogen for a week to 10 days, it may protect against possible neoplastic changes caused by estrogen because it produces sloughing off of endometrial cells through withdrawal bleeding. The bleeding that occurs after withdrawal of the added progestin is the equivalent of monthly menstruation. Although some women in the postmenopause may be understandably reluctant to experience return of the menstrual periods, this is a necessary accompaniment to progestin-estrogen treatment. R. Don Gambrell, Jr., of the Medical College of Georgia, has data to show that the incidence of endometrial cancer is 434/100,000 in estrogen users, 242/100,000 in women who have never taken any hormones, and 67/100,000 in estrogen/progestin users, suggesting that adding progestin to estrogen therapy may actually be protective against uterine cancer (1983).

Besides unwanted menstruation, however, there are other side effects associated with the combination estrogen/progestin therapy. Additional or "breakthrough" bleeding in the middle of the month, bloating, irritability, and headaches may occur, and the adverse effect of progestins on blood lipids is a concern. There are still questions about the use of estrogen and progestins, and yet to be decided are which progestin to use, what dose of estrogen and progestin is of the greatest benefit, and whether other routes for administration of estrogen other than orally—vaginally, implants, through the skin—have fewer adverse effects.

It could be another 10–20 years before all the answers are in on hormone therapy and cancer. There are two manifestations of the menopause (i.e., hot flashes and genital atrophy) for which estrogen therapy has been proved effective, and the prevention of osteoporosis in older women at risk for the disease and young women who have had a surgical menopause is also an established benefit. The decision to take estrogen in the menopause should be individualized, taking into account each woman's special needs. Women must decide if what estrogen does for their symptoms outweighs what it may do to their health. Women must also be aware that they should not have estrogen therapy if they have any of the following conditions: fibroids of the uterus, endometriosis, previous history of cancer of the endometrium or breast, or any history of blood clot formation. Severe diabetes, elevated blood fat level, or high blood pressure are other contraindications. Indeed, there are a number of drawbacks to estrogen therapy in addition to those already mentioned. These include breast tenderness, weight gain, pelvic pain, vaginal bleeding, leg pain, digestive disorders, and recurrence of allergies. Any woman taking estrogen therapy should see her doctor regularly, and if she has any unscheduled vaginal bleeding, she should see her doctor immediately.

If a woman if fully aware of all the ramifications of taking estrogen menopausally and has decided that it is essential, then it is her informed choice. Using estrogen, however, is not like using saccharin. It should not be minimized as just one more potentially dangerous option.

REFERENCES

Brenner, P. F. The menopause (Medical Progress). *West J Med* 136:211–219, 1982.

Bush, T. L., Cowan, L. D., Barrett-Connor, E., et al. Estrogen use and all-cause mortality. Preliminary results from the Lipid Research Clinics Program Follow-up Study. *JAMA* 249(7):903–906, 1983.

Dixon, A. St. J. Nonhormonal treatment of osteoporosis. *Br Med J* 286(6370):999–1000, 1983.

Estrogens and endometrial cancer. FDA Drug Bulletin, February–March, 1976.

Gambrell, R. D. Menopause: Role of hormone therapy (Forum Highlights). *The Female Patient* 8(7):50(28–44), 1983.

Heaney, R. P., and Recker, R. R. Effects of nitrogen, phosphorus, and caffeine on calcium balance in women. *J Lab Clin Med* 99:46–55, 1982.

Henderson, B. E. Breast cancer, in Judd, H. L. (moderator). Estrogen replacement therapy: Indications and complications. *Ann Intern Med* 98:200, 1983.

Hirvonen, E., Malkonen, M. and Manninen, V. Effects of different progestogens on lipoproteins during post menopausal replacement therapy. *New Engl J Med* 304(10):560–563, 1981.

Hutchinson, T. A., Polansky, S. M., and Feinstein, A. R. Postmenopausal oestrogens protect against fractures of hip and distal radius. *Lancet* 2:705–709, 1979.

Laufer, L. R., Erlik, Y., Meldrum, D. R., and Judd, H. L. Effect of clonidine on hot flashes in postmenopausal women. *Obstet Gynecol* 60(5):583–586, 1982.

Leiblum, S., Bachmann, G., Kemmann, et al. Vaginal atrophy in the postmenopausal woman. *JAMA* 249(16):2195–2198, 1983.

Pfeffer, R. K., and Van den Noort, S. Estrogen use and stroke risk in postmenopausal women. *Am J Epidemiol* 103:445–446, 1976.

Riggs, B. L., Seeman, E., Hodgson, S. F., et al. Effect of the fluoride/calcium regimen on vertebral fracture occurrence in postmenopausal osteoporosis. Comparison with conventional therapy. *N Engl J Med* 306:446–450, 1982.

Ross, R. K., Paganini—Hill, A., Gerkins, V. R., et al. A case-control study of menopausal estrogen therapy and breast cancer. *JAMA* 243:1335–1339, 1980.

Shapiro, S., Kaufman, D. W., Slone, D., et al. Recent and past use of conjugated estrogens in relation to adenocarcinoma of the endometrium. *N Engl J Med* 303:485–489, 1980.

Sharman, A. Ovulation around the time of the menopause, in Bowes, K. (ed). *Modern Trends in Obstetrics and Gynecology*, 2nd series. London, Butterworths, 1955.

Stadel, B. V. and Weiss, N. Characteristics of menopausal women: A survey of King and Pierce counties in Washington, 1973–74. *Am J Epidemiol* 102:209–216, 1975.

Wallace, R. B., Sherman, B. M., Bean, J. A., et al. The probabilities of menopause with increasing duration of amenorrhea in middle-aged women. *Am J Obstet Gynecol* 135:1021–1024, 1979.

Weiss, N. S., Ure, C. L., Ballard, J. H., et al. Decreased risk of fractures of the hip and lower forearm with postmenopausal use of estrogen. *N Engl J Med* 303:1195–1198, 1980.

Willett, W., Stampter, H., Bain, C., et al. Cigarette smoking, relative weight, and menopause. *Am J Epidemiol* 117:651–658, 1983.

Williams, A. R., Weiss, N. S., Ure, C. L., et al. Effect of weight, smoking and estrogen use on the risk of hip and forearm fractures in postmenopausal women. *Obstet Gynecol* 60(6):695–699, 1982.

Ziel, H. K. *Estrogen's role in endometrial cancer. Obstet Gynecol* 60(4):509–515, 1982.

Ziel, H. K., and Finkle, W. D. Increased risk of endometrial cancer among users of conjugated estrogens. *N Engl J Med* 293:1167–1170, 1975.

14. *Health and the Working Woman*

In the last 30 years, the number of American women working outside the home has tripled. For the first time in U.S. history, and for reasons largely related to both the economic climate and the rise of the woman's movement, more than half (54%) of adult women older than 16 are employed. Women are working today because they have to; contrary to the myths, they are not bored housewives merely dabbling in the workplace. And despite depictions on TV and in the movies, neither are they all superwomen and supermom professionals, making policy in corporate boardrooms while a nanny cares for the children and a cook makes the meals. More than two-thirds of working women are divorced, widowed, single parents, the sole earner in the family, or married to men who make less than $10,000 annually.

A woman's place is clearly on the job more often than in the home; however, a woman's place is not yet on any job. Although there have been some significant increases in the number of women executives, engineers, doctors, dentists, lawyers, and college teachers, there has been little change in the number of women employed in blue-collar jobs. Pro-

gress has been made from the almost non-existent figures of a decade ago, but very few women are construction workers, electricians, plumbers, or mine workers—the better paying jobs that are still considered men's work. The majority of the nearly 47 million working women continue to do "women's work"; they are school teachers, librarians, nurses, bank tellers, or clerical workers, and they earn only around 60% of what men make.

But as women moved in significant numbers into the job market during the 1970s, it began to be evident that the issues of occupational health and safety, formerly believed to concern only men, were highly relevent to working women as well. Like men, women faced the potential hazards of job-related injury and disease, not only as industrial workers, but also in the offices, hospitals, and laboratories where most women work.

In 1970, Congress passed the Occupational Safety and Health Act that established a federal agency, the Occupational Safety and Health Administration (OSHA), and adopted improvement in the workplace environment as a national priority. Along with the Mine Safety and Health Administration (MSHA), the charge to OSHA is the promulgation and enforcement of standards to assure the greatest protection of workers from job-related injury or disease—a mandate that has embroiled the agency in controversy since its inception, since both labor and industry have frequently viewed OSHA's regulatory efforts as less than optimally beneficial. The act also created the National Institute for Occupational Safety and Health (NIOSH), the Federal preventive health agency responsible for identifying workplace hazards, developing means of preventing them, and recommending the standards and guidelines to OSHA for the protection of workers.

There are two general types of hazards associated with work: one is traumatic injury, and the other the possibility of illness or death resulting from exposure to toxic agents—chemicals, pathological microorganisms, cotton or coal dust, asbestos, heavy metals, and noise, among others. In 1982, the senior scientific staff at NIOSH developed a suggested list of 10 leading work-related diseases and injuries, ranked in order of frequency and severity (Centers for Disease Control, 1983):

1. Occupational lung diseases (silicosis, asbestosis, byssinosis from cotton dust, etc.)
2. Musculoskeletal injuries (e.g., back trouble from heavy lifting)
3. Occupational cancers (from 4% to 20% of cancer cases are estimated to be job related from exposure to carcinogens)
4. Amputations, fractures, eye loss, lacerations, and traumatic deaths
5. Cardiovascular diseases (with possible links to job-related stress)
6. Disorders of reproduction
7. Neurotoxic disorders
8. Noise-induced loss of hearing
9. Dermatologic conditions (exposure to chemical agents)
10. Psychologic disorders

During the middle 1970s, as more women moved into the work force, the sixth ranked disorder, that of adverse reproductive effects, became the focus of greater public, industry, and government interest. The nation's goal of reducing the hazards of the workplace now developed a further dimension—the need to protect the health of women of reproductive age from exposure to substances that could affect reproduction.

But the possibility that adverse effects, such as infertility or birth defects, could be caused or influenced by the working environment has resulted in a new kind of dilemma: the potential conflict between the obligation of the employer to provide for occupational health and safety and the rights of women to equal opportunity in employment. New questions of ethics, moral commitment, legality, and constitutionality have arisen. To what extent could or should another group of individuals—the potential offspring of workers exposed to mutagens or teratogens—be protected? Women had struggled for equal work and equal pay; were their reproductive capacities again to be a barrier? How should a company be able to protect itself against lawsuits brought against it if it could be proved that the next generation was harmed by something in the mother's workplace?

Some companies attempted to solve their problem by developing exclusionary policies that eliminated women of reproductive age from certain jobs. In disregard of the strong probability that any substance toxic enough to threaten a woman's reproductive health is very likely to produce a similar adverse reproductive effect (sterility) in men, they reasoned that since it was economically unfeasible to totally remove the risks, the removal of the women from the job and shifting them to a less hazardous (and lower paying) environment was the only way to protect them and their potential fetuses from hazardous substances. In one highly publicized reaction to such a job restriction policy, five women employees at the American Cyanamid Company chose in 1979 to have themselves sterilized so that they could return to their former jobs.

The fetal protectionist and exclusionary policies of some employers have been attacked by critics and challenged in the courts. The issues are complex and as yet unresolved, and the conflict and litigation can be expected to continue. Much of the difficulty in obtaining the answers to the questions raised by OSHA regulators and the representatives of labor and industry results from the paucity of valid scientific data. Little is conclusively known and even less is proven concerning the substances or conditions in the working environment that are especially hazardous to reproduction. Until recently, most of the studies by government and privately supported researchers have been devoted to other kinds of occupational safety and health hazards to men. Only lately, now that half of the nation's work force is female, has the matter of adverse reproductive effects of the workplace become, in the words of M. Donald Whorton, the researcher who first called attention to infertility in male pesticide workers, the "occupational health issue of the 1980s." In the last decade, the reports of research on work and reproductive hazards have begun to accumulate. Some studies have been well designed and documented, but some have also been based on small samples, provide incomplete information, or are in other ways methodologically flawed. It is apparent that much more research needs to be done before any definitive conclusions on cause-and-effect relationships can be drawn.

EFFECTS OF TOXIC AGENTS IN THE WORKPLACE

Some of the chemical and microbiological agents that are known or suspected to be mutagens, teratogens, or carcinogens (like DES)

that cross the placenta have already been discussed, and many of them can be encountered in the home as well as in the workplace. As previously indicated, there are specific problems associated with attempting to prove a link between exposure to a substance and an adverse reproductive effect, such as spontaneous abortion or birth defect. Congenital abnormalities occur at about a 3% rate in the general population; their origin could be genetic, environmental, or through interaction of both. Moreover, the true rate of spontaneous abortion in the general population is unknown and probably underestimated. Many miscarriages may be very early, unrecognized, and unreported; any comparisons of the rate of spontaneous abortion in working women exposed to a particular agent and the rate in nonworking or nonexposed women could be very misleading. Work is certainly not the only factor affecting the outcome of pregnancy.

But even if one keeps in mind the possible shortcomings of the studies and the difficulties in proving the link, there is evidence suggesting that there may be an increased risk of spontaneous abortion for women who work in the following kinds of occupations: research chemists or biologists and women laboratory workers who are exposed to organic solvents like benzene and toluene; hospital workers or laboratory workers who sterilize instruments with ethylene oxide and glutaraldehyde; operating room personnel (anesthesiologists and O.R. nurses and technicians) because of exposure to anesthetic agents; women working in the copper smelting industry or with soldering agents. Occupational exposure to heavy metals, pesticides, and herbicides has also been reported to be linked to increased rates of spontaneous abortion in women workers as well as in wives of exposed men.

Finnish researcher Hemminki and his colleagues were the first to suggest (1983) that the occupations of both parents may interact in contributing to an increased risk of spontaneous abortion. They found that the rate of miscarriage was greater for women working in the textile industry when their husbands worked at a large metallurgical factory than it was for women whose husbands worked elsewhere.

In general, there are more data attesting to an increased risk of miscarriage than a greater incidence of birth defects after exposure to workplace toxic agents. Exposure to organic solvents in chemical laboratories has been reported to be related to congenital malformations in the offspring of pregnant workers, and so has employment in the copper smelting industry during pregnancy. There is conflicting evidence, however, for the risks of exposure to anesthetic gases and the rate of birth defects in the children of women hospital workers.

EFFECTS OF PHYSICAL FORCES ON REPRODUCTION

A report by the Council on Scientific Affairs of the American Medical Association (1984A) considered the evidence for adverse reproductive effects as a result of exposure to such physical factors as temperature, atmospheric pressure, vibration and noise, and radiation. The found that neither low nor high atmospheric pressure have been demonstrated to produce teratogenic effects in infants, but that high altitudes were correlated with low birth weights. One report linked increased atmospheric pressure experienced by women scuba divers who dove during pregnancy with increased fetal abnormalities (Bolton, 1980), but the survey sample was small.

Increased temperature or hyperthermia has adverse effects on male fertility and is known to be teratogenic in animals when the temperature has been raised to 43°C (109°F) during embryonic development. In humans, there are data to suggest that fever higher than 104°F that lasts at least 1 day during the third to seventh week of gestation could affect the nervous system of the fetus (Pleet et al., 1981), but getting body core temperature that high through an ambient working environment is believed to be virtually impossible since no one could tolerate remaining in that hot an atmosphere for long enough.

There is no evidence that any differences in fertility or the rate of spontaneous abortion or birth defects occur as the result of exposure to occupational electronic and magnetic fields, gravity and acceleration, noise, ultrasound, vibration, or ultraviolet and infrared light or laser beams. Neither are there well-documented reports of injuries to reproduction from microwaves or radio frequency waves. Of course, ionizing radiation (X-rays) is damaging to all living tissues, including reproductive organs, and the effects are dose related. It is unlikely that today, men and women who are occupationally exposed to ionizing radiation are unaware of the potential consequences. They are more likely to take great caution to keep exposures as low as possible.

VDT RISKS TO WOMEN

Almost all secretaries and other clerical workers are women (99.2%), and most of them have had their typewriters or adding machines replaced by video display terminals. At least three-quarters of the full-time VDT users are women of reproductive age, and with 60 million VDTs in use projected by the end of the decade, women workers have raised questions about the safety and potential health risks of working at their terminals. In 1983, the National Association of Working Women, known as "9 to 5," started a national education and action campaign about VDTs to dispense information, collect reports of complaints, and encourage more research on the subject. In the 6 months after opening a toll-free hotline, 9 to 5 received more than 6,000 calls as indication of the growing concern.

Although most health complaints related to VDTs are focused on head, neck, and back pain, fatigue, eye strain, and stress, some users have worried about the reports of an increased number of spontaneous abortions occurring in women exposed to the terminals. Several locations in the United States and Canada have shown statistical oddities or "clusters"—an unusually high incidence of miscarriage—in a group of women office workers. But as yet, studies by both private researchers and the National Institute for Occupational Safety and Health all but ruled out the low-frequency radiation coming from the screens as having any adverse biologic effects. No harmful rays at any frequency have ever been found to be emitted.

The possible links between pregnancy problems and VDTs, however, continue to be studied by NIOSH, and it will take some time before anything definite is known. What has become evident, however, is that most or many of the other problems related to working with a terminal can be ameliorated by paying attention to such factors as being able to adjust the room illumination to eliminate glare, adjusting the height of the operator's chair and height of the screen, or changing the angle of the screen. Some states have passed or considered legislation that would require employers to provide ways of minimizing glare through indirect lighting or special filters, provide additional rest breaks and eye exams, and, on request, give leaves with no loss in compensation to pregnant VDT operators.

WORK DURING PREGNANCY

Another AMA Council on Scientific Affairs report on the "Effects of Pregnancy on Work Performance," released in 1984, generally reaffirmed what many women already know: that in a normal and uncomplicated pregnancy, if a woman wants to, and feels like it, there is no reason why she should not continue to work until her first contraction. In recognition of the individuality of the pregnancy experience, the report also acknowledged that there are no hard data to confirm that 6 weeks is the most appropriate postpartum period and indicated that a woman can return to work whenever she feels physically ready, whether 2, 6, or 10 weeks.

Although it was stressed that determinations that a pregnant woman can or cannot work at a particular job should be individualized on a case-by-case basis, the report did provide some guidelines to indicate the period of time during pregnancy that a healthy employee with an uncomplicated pregnancy might be expected to perform her duties without an adverse effect on her work performance or her pregnancy.

The guidelines, developed by a panel of all-male physicians, advised that women doing secretarial, clerical, professional, or managerial work can continue to work with no adverse effects on performance or pregnancy up until the onset of labor even if some standing (30 minutes at a time), stooping and bending (less than twice in an hour), stair-climbing (fewer than four times in an 8-hour shift), and frequent or repetitive lifting of 25 pounds or less were involved on the job. The recommendations indicated, however, that women whose work involves more strenuous activities than the above, such as frequent and repetitive stooping and bending, climbing of ladders or stairs, or lifting more than 25 and up to 50 pounds in weight, should discontinue working by 20–28 weeks of gestation, depending on the type and amount of activity. This option of quitting work in the fifth month is attractive but unavailable, of course, to the pregnant housewife whose similar, very strenuous activities take place at home—frequently and repetitively cleaning, washing, climbing stairs, toting heavy bags of groceries from the supermarket, and lifting a heavy toddler. Those job responsibilities were not addressed by the doctors' guidelines.

WOMEN, WORK, AND STRESS

Hans Selye, a Canadian physician and endocrinologist, first developed the concept of physiological stress and stress-related disease. Selye defined stress as the nonspecific response of the body to any demand, or stressor. The key word is nonspecific: the stressor can be positive or negative, physical or emotional. It can cause feelings of joy or pleasure; it may result in anxiety or fear—but the body sees little difference and makes no distinction among the stressors. It responds in an nonspecific and generalized way: by increased activity of the nervous system and by production of hormones from the hypothalamus, pituitary, and adrenals. These mechanisms make it possible for the body to react to the stressor, cope with it, and adjust the internal environment to make the least damaging and most appropriate response.

The stress response proceeds through three stages. The first one is the alarm reaction, with hormonal and nervous sytem effects that increase blood pressure and heartbeat, increase energy utilization, and decrease antigen–antibody reactions, thus decreasing resistance to disease. The next stage is the resistance stage. After the initial alarm, the body responses return to normal and there is no further reaction to the given stressor while the body repairs the damage. Most stressors are mild or of short enough duration to elicit only the alarm and resistance stages. But if the stressor continues and is severe enough, the body is no longer able to cope and goes into the exhaustion stage. The hormonal and

nervous system responses again reappear and can lead to tissue damage and disease. In experimental animals, even death can result.

By Selye's definition, any stimulus that produces the stress response is a stressor and can be a variety of agents—physical, chemical, or internalized as an emotion. It follows, therefore, that stressors are continually encountered and are a part of living, associated with all life experiences and activities. It is not only impossible to avoid them, but undesirable to try to avoid them. The stress response is essential to keeping us alert and motivated enough to respond to life's challenges. Ideally, according to Selye, the level of stress in our lives should be *eustress*, an amount that is positive and leads to productivity. It is the *distress*, the continual and unrelieved tensions and frustrations—the kinds of emotions that result in internal fruitless wheel-spinning—that can lead to illness. One significant aspect of stress-related disease, however, is how an individual perceives and interprets the stressing agent. Something that is devastatingly stressful for one person could be seen as unimportant to another or even pleasurable to a third. Another important determinant is the way in which one deals with the stressor. Some people will use physical releases and exercise off their stresses; other may chainsmoke or use alcohol or other drugs for release of tensions and compound the damage. Because human behavior in response to stressors is so variable, it is impossible to know for certain the extent to which stress alone is actually involved in human illness.

"Stress," however, is accepted as a known risk factor for coronary heart disease and has been implicated in the development and course of a number of other so-called male diseases, such as gastric ulcers, hypertension, and colitis. Certain occupations, too, have long been acknowledged as carrying a higher level of stress and pressure. The coronary-prone male executive was epitomized by the

Type A personality type described by cardiologists Rosenman and Friedman in 1974—the hard-driving, fast-talking, ambitious, competitive workaholic. As women began to enter the work force and take on male jobs, many in executive ranks, there were predictions that they would take on the Type A behaviors and begin to suffer the same stress-related diseases as males: more heart attacks, ulcers, and high blood pressure.

Whether the increased participation of women in male-dominated professions is going to result in their acquiring more male diseases and decrease their longevity disadvantage remains to be seen. As yet there is little evidence to warrant the assumption. For reasons that are not clearly understood, but probably are related to diet and exercise, the incidence of cardiovascular disease in both sexes is decreasing. Women still continue to live longer than men, and the gap is increasing, not narrowing. Moreover, research thus far suggests that when workers of either sex suffer apparent stress-related cardiovascular problems, it is not the executives at the top who get heart attacks, but the people in the middle and lower levels of the job hierarchy that are more subject to health problems. It appears that lack of decision-making capacity on the job can be directly correlated with the incidence of heart disease. Studies by the Metropolitan Life Insurance Company that followed successful and prominent men and women listed in *Who's Who* found that high executive men outlived other middle-class men by 29% and that executive women outlived all other women as well as working men. It is evidently not the pressures of work alone that are unhealthy; rich, successful, powerful men and women who are in control of their lives can take a lot of work-related stress and thrive. Presumably, then, the frustrations and insecurities of middle-management positions, the associate and assistant executive jobs that more women are beginning to occupy, are

more likely to result in stress-related illness. The really high-risk jobs for coronary heart disease may be at job classification levels where the psychological demands are great but the worker control is lowest—the positions where most women workers are concentrated.

But to put concerns about working women and health into appropriate perspective, it should be recognized that consistently, according to the National Center for Health Statistics, all women who work in any job classification enjoy better health than all nonworking women. They also appear to be healthier overall than working men. Although women workers tend to have more health problems of a mild nature and shorter duration, men workers have more chronic illness, more work injuries, and more life-threatening diseases (Verbrugge, 1982).

The Framingham Heart Study, in which 6,000 residents of the Massachusetts city have been observed since 1950, followed a subgroup of 387 working women and 350 nonworking housewives for the development of heart disease. The report by Haynes and Feinleib (1980) confirmed that there was no increase in the risk of heart attack for working women generally as compared with housewives who had never worked outside the home. There was, however, a higher rate of coronary heart disease in one group of women who were clerical workers—secretaries, typists, clerks, and bookkeepers—when compared with other working women and homemakers. But even in that group, analysis revealed that there was no greater incidence of heart disease in single women clerical workers; the rates were boosted by the increased incidence that occurred particularly in women who were mothers and married to blue-collar workers. The researchers discovered that the clerical workers, compared with other working women, were likely to have nonsupportive employers, little or no job mobility, and a tendency to suppress anger and bottle-up their resentments. These results lend substantiation to the theory that a routine, dead-end job with few satisfactions besides the paycheck carries a greater risk.

Lois Verbrugge's (1984) analysis of data from the National Health Interview Study, however, found that clerical women in the United States have the best overall health profile, with low rates of injury, chronic limitation, restricted activity, and average health services use. She suggested that in the Framingham clerical workers, it was the combination of an unsatisfying work situation *plus* childrearing duties *plus* financial pressures that resulted in the greater incidence of heart disease. Wolfe and Haveman's study (1983) of more than 2,300 working women further substantiated that the added burdens of childcare and housework can be detrimental to the health of working mothers, who had greater problems with illness than working women without children. The mothers reported that they generally devoted about the same amount of time to the children and the house after getting a job as before, which might provide a clue to their additional ills.

Clearly, it would appear that working, in itself, is no more stressful for women than it is for men. It is more likely that it is the "plusses," when a woman is powerless to deal with them, that could be unhealthy. Stress factors for a working women are not difficult to identify. They are built into a job when a woman is stuck in a low-paying, low-status position, when she is subjected to sexual harassment on the job, or when she trains a man for a job and then finds him promoted before she is. Work-related stress is intensified when a woman performs the same duties as a male employee or occupies a position of comparable worth and then finds out that the man makes more money than she does. Pressure can come from outside as well as on-the-job. For a working mother with preschool children, stress is when a child is sick, the sitter is sick, or the car is sick. For a woman with school-age children, it is stressful when she cannot leave work early enough to get to a parent–teacher conference before the teacher leaves; for the teacher, it occurs when she has the same problem with her own child's teacher. When a woman has to be a full-time worker and a full-time homemaker, when she has simultaneous responsibilities and conflicting role expectations, continual high levels of stress may result. Many women experience at least some of these "plusses"; a few may even be subjected to all of them. There need be less concern for the effects of job stress on the working woman's health if more ways of reducing those burdens were available: through job-sharing, flexible time scheduling, better and more accessible daycare facilities, and shared parental responsibilities, and if there were greater efforts toward eliminating the inequities, discrimination, and resultant frustration that working women experience in some jobs.

REFERENCES

Bolton M. E. Scuba diving and fetal wellbeing: A survey of 208 women. *Undersea Biomed Res* 7:183–189, 1980.

Centers for Disease Control: Leading work-related diseases and injuries—United States. *MMWR* 32:24–26, 1983.

Council on Scientific Affairs. Effects of physical forces on the reproductive cycle. *JAMA* 251(2):247–249, 1984A.

Council on Scientific Affairs. Effects of pregnancy on work performance. *JAMA* 251(15):1995–1997, 1984B.

Haynes, S. G. and Feinleib, M. Women, work and coronary heart disease: Prospective findings from the Framingham heart study. *Am J Public Health* 70(2):133–141, 1980.

Hemminki, K., Kyyronen, P., Nieme, M. L., et al. Spontaneous abortions in an industrialized community in Finland. *Am J Public Health* 73(1):32–37, 1983.

Metropolitan Life Insurance Company. Longevity of prominent women. *Stat Bull Metropol Life Ins Co* 60(1):2–9, 1979.

Pleet, H., Graham, J. M., and Smith, D. W. Central nervous system and facial defects associated with maternal hyperthermia in four to fourteen weeks gestation. *Pediatrics* 67:785–789, 1981.

Rosenman, R. H. and Friedman, M. *Type A Behavior and Your Heart.* New York, Alfred A. Knopf, 1974.

Verbrugge, L. M. Physical health of clerical workers in the U.S., Framingham, and Detroit. *Women & Health* 9(1):17–41, 1984.

Verbrugge, L. M. Sex differentials in health. *Public Health Rep* 97:417–437, 1982.

Whorton, M. D. Adverse reproductive outcomes: The occupational health issue of the 1980s. *Am J Public Health* 73(1):15–16, 1983.

Wolfe, B. and Haveman, G. E. Time allocation, market work, and changes in female health. *Am Economic Rev* 73(2):134–139, 1983.

15. *Cosmetics: The Twelve-Billion Dollar Put-On*

People probably have more awareness of their skin than of any other organ system of the body. Not that skin is the largest organ; in terms of surface area, the lungs, circulatory system, and the digestive tract are bigger. But skin is so very obvious. There it is, covering the entire body, and unlike other organs, it is completely visible. The skin functions as the only barrier between the outside environment and the body inside. It protects against thermal, chemical, and physical injury. Relatively waterproof, the skin allows the body to exist in dry air without shriveling like a raisin and permits one to soak in a tub for hours without appreciably swelling. And, because it is so abundantly supplied with nerve endings, the skin acts as one enormous sense organ, constantly receiving information from its surface for transmission to the brain.

Along with its other functions, and perhaps of greater importance to most people, skin also defines the individual. It provides humans with an identity. The "real me" inside, with all the flaws or attributes of character, remains hidden and internal; what the world sees first of me is the skin covering the contours of face and body. All people are literally sisters and brothers under the skin; without it, everyone would look very much the same.

Skin and its derivatives, hair and nails—the body's facade—are obviously the major media of sexual attraction. Enhancing their appeal has been a part of every culture for thousands of years. Eons ago, some enterprising prehistoric human probably first used oils extracted from plants to soften the skin or to smooth the hair. Perhaps, recognizing something powerful and splendid about the vivid colors of nature, primitive men and women mixed red clay or copper ore with water, daubed it on the face, and invented the first makeup. The face and body became the canvas for visible symbols, for dramatizing cultural ideas. Paint pots and implements to grind and apply eye shadow and liner have been unearthed in Egypt and dated to 10,000 years ago. Archeologists digging in Sumerian tombs have found 5,000-year-old lipsticks. Men and women throughout the ages have applied various substances to the skin to clean, perfume, or color it, camouflage or disguise it, frighten off enemies, ward off evil spirits, protect it from the weather, make it look younger, and generally to make a statement about self: look at me, this is who I want to show! There is nothing new about cosmetics; they have been in use for thousands of years.

What is new is the emergence of the cosmetics *industry*, which packages, promotes, and sells. Before this century there was very little marketing of cosmetic products. Preparations were mostly made up in the home from "recipes" that used ingredients purchased from the pharmacist. In 1849, 39 manufacturers produced a total of $355 worth of cosmetics that cost only $164 to make (Corson, 1972). Today, the cosmetics industry has annual sales estimated between $10 and $15 billion, depending on exactly which items are included in the 30,000-plus formulations on the market. The figure varies, according to whether sales of shampoos, deodorants, toothpaste, or mouthwash are contained in the totals. It is probably valid to include all such products into one mass group. All of them affect the buyers in the same way: to make ourselves better, to make ourselves more appealing. With the largest advertising budget of any commerical enterprise in America, the cosmetics industry, through the media of TV, magazines, and newspapers, offers health, success, fulfillment, new attractiveness, and a whole new life-style if only people will buy, buy, buy the products.

The primary focus of cosmetics advertising is directed at women, the greatest consumers of the products. Major cosmetic sales are in beauty aids. Women who believe they do not use cosmetics, that is, facial makeup or nail polish, still buy shampoos, hair conditioners, and toilet soaps. Even the most widely used cosmetic product, toothpaste, is sold as a beauty enhancer. (Only when it is pushed as a product for children do the mothers in TV commercials speak of toothpaste as being health related. The promotion of toothpaste may be based on sex appeal or "Crest test," depending on whether the sales pitch is to be sex or cavities. Either is geared toward

women.) And while there are indications that skin-care products designed strictly for men are becoming viewed by the industry as additional opportunity for revenue, the major profit is derived from cosmetics purchased by women.

We live in a beauty culture, a world fostered by a cosmetics industry in which stereotyped models of feminine attractiveness—the beautiful people—set the standards. Every year a new image of fashion and beauty is packaged and promoted like the products themselves, by advertising "hype," and to achieve the image one apparently must only buy the product. Looking different and somehow also looking alike, the models are tall, slim, with marvelous teeth and gorgeous hair, ever widening the gap between fantasy and reality. Real-life women are urged by the fashion magazines and the TV commercials to *get the look*, variously described as clean, natural, glowing, sun-warmed, polished, rain-wet, glossy, sheer, silky, satiny, or sensational. They are promised that if they use a certain product they can get rid of the frizzies, the greasies, split ends, and unsightly dandruff. They are persuaded to curl, color, perm, and 30-minute deep-condition their hair. They are told that they must exfoliate, clarify, moisturize, and replenish their skins, and if only they will apply this lotion or that cream, its special, unique, secret, and rare ingredients will prevent, smooth out, or eliminate the effects of aging.

Perhaps there are those who are able to perceive the whole cosmetics scene as the massive put-on it is and remain unaffected by the industry's massive put-down of women's real bodies and faces, which it literally does. But in a society where women are generally overvalued for their personal appearance and undervalued for their contributions and competence, most women are to a certain extent vulnerable to the messages. What teenage girl does not harbor a secret desire to be the subject of a "makeover," believing that the answer to looking better is out there somewhere, and the media tells us where: in a tube, a bottle, a box. Even Gloria Steinem, feminist, author, editor of *Ms* magazine, and a very attractive woman, is not immune—she streaks her hair. And feminist author Susan Brownmiller, who admittedly has an anti-makeup bias, acknowledges that she dyes her own prematurely gray hair although she considers it "a shameful concession to the wrong values." A job has been done on all of us.

All women—wealthy, poor, teenage, mature, women who work, those who stay home, the feminist woman and the self-described nonfeminist—buy cosmetic products. Cosmetic use is hardly debatable; they have been used for countless millenia, and we will undoubtedly continue to use them. Virtually all of the studies by psychologists and social scientists have reaffirmed what most people instinctively know—physical attractiveness is better than unattractiveness. Looking good makes you feel good about yourself and improves your self-esteem. Good looks can make you liked, get you viewed as a better and more intelligent person, get you hired, get you promoted, and get you elected to office. Regardless of what this implies about the superficialities of human judgment and our interactions with each other, there are valid psychological and tangible benefits to be derived from trying to enhance our appearance. Wearing makeup does not have to be a guilty concession; it can still be a choice, and as informed consumers women should at least have the knowledge to choose products that are not overpriced or unsafe to use. If a man or woman wants to use cosmetics there is every reason to do so, but there is little reason to be swayed by the often spurious claims of the manufacturer. Knowing about skin, hair, nails, and the composition of cosmetics can provide the information to understand what products can reasonably be expected to do. It should then be easier to look the way one wants to without accumulating a bathroom full of useless and expensive and possibly even dangerous mistakes.

SKIN STRUCTURE

Skin consists of two principal layers, each of which has its own subdivisions. The top cellular layer that faces the outside is the *epidermis*. Epidermis, relatively thin except on the palms of the hands and soles of the feet, also gives rise to specialized skin derivatives, the hair, glands, and nails. The surface of the skin is marked by numerous tiny ridges and furrows and many minute orifices, the openings (pores) of the sweat glands and hair follicles. The underlying and supporting *dermis*, much thicker than the epidermis, is also called the true skin, which is probably why a skin doctor is never called an "epidermatologist." The dermis contains the blood vessels to nourish the epidermis. The two layers of skin are intimately held together by tiny elevations, or papillae, of dermis that project into corresponding depressions of the epidermis. Continued strong friction or heat can cause an accumulation of fluid and the separation of the epidermis from the dermis, or a blister.

Below the dermal layer is the *subcutaneous layer* or *superficial fascia*, which anchors the skin to the underlying muscle. The subcutaneous layer is loosely constructed and contains fat that varies in amount in different parts of the body. In some areas, the fat forms a continuous layer; in hefty individuals, the layer reaches more than an inch in thickness. The fat serves as a reserve energy supply and further insulates the body and cushions it from injury. The characteristic more rounded body contours found in women result from a different distribution and, usually, a greater percentage of subcutaneous fat (Fig. 15.1).

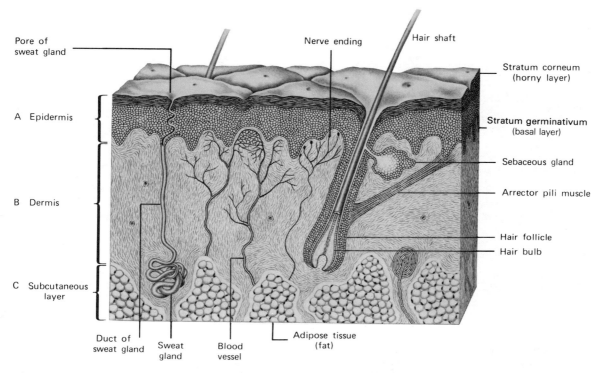

Figure 15.1. Three-dimensional diagram of the skin in cross section. (A) Epidermis; (B) dermis; and (C) subcutaneous layer.

Epidermis

The epidermis is the skin's frontier. Its function is to protect against the invasion of microorganisms, against fluid loss or gain, mechanical injury, or any of the other onslaughts of an everchanging environment. Most of this barrier is provided by 10 or 12 rows of dead cells at the surface of the epidermis called the *stratum corneum*, or horny layer. The lifeless cells of the horny layer are continually being scraped, rubbed, or worn off. As they are sloughed off, they must be replaced from below by the rapidly dividing cells in the *stratum germinitivum*, or basal layer. Actually, the entire living body is wrapped in a husk of dead cells that contain a tough insoluble

semitransparent protein called *keratin*, from a Greek word for "horn." The flat, scalelike surface cells of the stratum corneum contain soft keratin with less sulfur content than the hard keratin of the hair and the finger and toenails.

Where the epidermis is thickest, it is stratified into five layers. Figure 15.2 illustrates the epidermis and the layers within it. Rapid cell reproduction occurs in the basal layer. The new cells are pushed upward into the spiny or prickle-cell layer, so-called because the cells appear to have little spiny projections between them that represent cellular attachments. After the cells get into this layer, they stop reproducing and start producing keratin. Above the spiny layer, cells begin to accumulate granules of keratohyalin, a precursor to

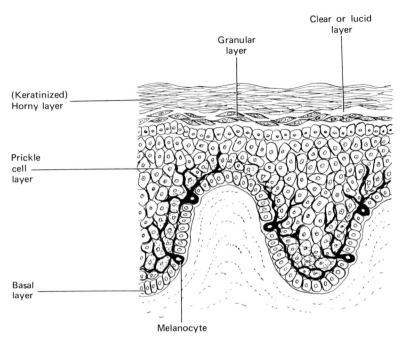

Figure 15.2. Diagram through thick skin showing all five layers. The melanocytes produce pigment to protect the cells from the damaging rays of the sun.

keratin, and are then called the granular layer. Beyond the granular layer, the cell boundaries and nuclei seemingly disappear. The cells look glassy and closely packed, forming the clear or lucid layer that contains a semifluid substance derived from keratohyalin. By the time the cells reach the top of the epidermis they are flat, scaly, and lifeless, and they are retained in the horny layer until they are shed off as little dead flakes. The rate of growth in epidermis is such that it completely replaces itself from the bottom up every 15 to 30 days, depending on the area of the body. All five layers are not evident when the epidermis is thin. In those locations, the basal layer gradually and imperceptibly passes into the horny layer.

Hair

Humans may look naked and hairless compared to other mammals, but we have about the same number of hair follicles as our primate relatives, the gorillas, chimpanzees, and orangutans. Most of them, however, give rise only to tiny colorless *vellus* hairs, or "down." Hairs are actually present almost everywhere on the body and are missing only on the palms, soles, the skin around the nails, and at the various body openings. Children have as many hairs or hair rudiments as adults, but until puberty their only coarse *terminal* hairs are on the scalp, eyebrows, and eyelashes. With maturity and the production of sex hormones, the hair follicles enlarge and become more active. Terminal hairs replace the vellus hairs in the pubic area and under the arms. In men, similar terminal hairs appear on the chest, face, shoulders, legs, and arms. Women have the same number of hairs on their bodies as men, even on their faces, but many of them are small and not as noticeable.

All hair follicles are formed before birth and arise early in the third month of fetal life. A tubular hair follicle begins as a downgrowth from the epidermis into the underlying dermis. The epidermal hair bud becomes bulb shaped at its base and develops a concavity that is invaginated by a mass of connective tissue, blood vessels, and nerves called the dermal papilla. The epidermal cells that lie directly over the papilla are known as the germinal matrix. The germinal matrix cells are analogous to the basal cells of the epidermis in that the product of their repeated cell divisions, in this case a hair, also consists of cells that become cornified and die.

Nourished by the blood vessels in the dermal papillae, the matrix cells proliferate. As they push up toward the surface of the skin they grow further and further from their source of nutrients, become progressively keratinized, and differentiate into an outer *cuticle* of scaly dead cells, a middle keratinized *cortex* with variable amounts of pigment, and in some hairs, a central medulla. The cuticle and cortex are composed of hard keratin, similar to that of nails or the feathers and scales of birds and reptiles. Medullary keratin is softer and the same as that in the horny layer of the epidermis. The medulla is poorly developed and frequently absent. It is not present in the short and fine vellus hairs, is missing from some of the hairs on the scalp, and is rarely found in blond hairs. The *shaft* of a hair is that part which extends beyond the surface of the skin. The *root*, enclosed within a tube-like follicle, expands at its base into a bulb, the only part of the hair with living, germinating cells (see Fig. 15.3).

Shed hairs that still contain some live cells can be analyzed for enzyme groups and typed, in the same manner that other body tissues can be typed for genetically determined antigens. It is also not difficult to determine microscopically if hair has been subjected to bleaching or dyeing. It is further possible to detect traces of minerals like lead, arsenic, cadmium, or mercury in hair, but such chemical analysis, although much favored in the plots of mystery novels, does not necessarily mean that the elements have been ingested in toxic amounts. Anything in contact with the hair—water, hair sprays, shampoos, dyes, permanent wave solutions—may deposit or remove minerals from the hair. Hair analysis that is supposed to reveal nutritional deficiencies or disease, or "trichanalysis" to determine which products or procedures are to be used in a hairdresser's salon are a waste of time and money.

Glands

Sweat Glands. Sudoriferous or sweat glands are epidermal derivatives located in the dermis and are widely distributed over the entire body surface except on the nail beds of the fingers and toes, the margins of the lips, and on certain parts of the external genitalia. Sweat glands empty their secretion onto the surface of the skin by way of a tiny opening. The watery sweat, or perspiration is a mixture of certain solids (mostly salts) in solution. It has a cooling effect on the body and also helps to eliminate wastes.

Sebaceous Glands. One or more sebaceous glands are always associated with a hair follicle, having differentiated during the development of the follicle by budding off from the epidermal downgrowth. There appears to be a kind of inverse relationship between the size of the sebaceous glands and the size of the hair; that is, follicles that contain the smallest, finest hairs have the largest glands. The seba-

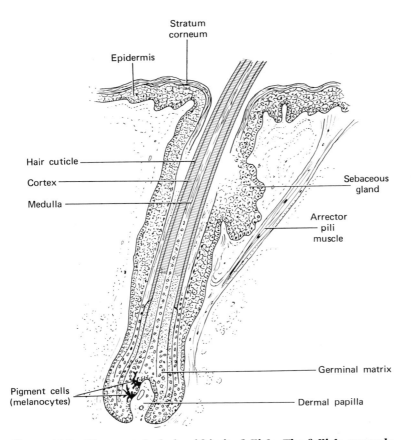

Figure 15.3. Diagram of a hair within its follicle. The follicle expands at its base to form the bulb that contains the dermal papilla. The cells of the germinal matrix lying directly over the papilla give rise to the hair. Pigment cells or melanocytes in the germinal matrix are responsible for hair color.

ceous gland drains its fatty secretion, *sebum*, into the follicle to oil the hair and pour out over the skin surface. In some locations on the body, the glands develop independently of hair follicles and open by their own duct directly onto the skin.

The hair follicle openings on the surface of the skin are commonly called "pores," although sweat gland openings are also pores. The diameters of the follicle openings are generally related to the size of the sebaceous glands they contain. People who have oily skins have larger glands and hence more conspicuous or enlarged pores. Pores are not doors, and they cannot be opened or closed by heat, by cold, or by any other method. Astringents are solutions that irritate the facial skin to cause a slight swelling and may temporarily make the pore openings appear smaller. Any slight inflammation of the skin, such as a mild sunburn, can cause a similar edema and "shrink" the pores. Of course, pores can be made less obvious by covering them with makeup.

Except in the eyelashes, hair follicles and hairs are located on a slant rather than at a right angle to the skin surface. The sebaceous glands and a small bundle of smooth muscle fibers, the *arrector pili*, are usually found on the slanting side. The arrector pili muscles are controlled through the autonomic nervous system. When the outside temperature is lowered, or some hair-raising event has occurred, the muscles contract to pull on the follicles and make the hairs erect—literally, to make them stand on end. At the same time, little mounds of skin on the surface are produced by the muscular contraction. We call them goose bumps or goose pimples.

Nails

Fingernails and toenails also develop during fetal life from downgrowths of the epidermis into the dermis underneath. The nail itself is a cornified curved plate of hard keratin that rests on a thickened surface of epidermis called the nail bed. The exposed part of the nail is the body, pink because the blood in the capillaries of the dermis shows through the nail translucence. The hidden part of the nail is the root, from which it grows. The root is covered by a curved fold of skin, the cuticle, and a half-moon shaped whitish area, the lunule, appears on the part of the body closest to the cuticle. Why the lunule is opaque and does not let the capillaries show is not clearly understood.

The half-moon is most visible in the thumbnail, becomes progressively less apparent in the other fingers, and is generally completely hidden by the nail fold on the little finger. The white spots that occasionally appear on the nails are caused by air bubbles between the keratinized cells. They are probably caused by an interruption of keratinization during growth of the nail.

Nails grow an average of 0.5 mm per week,

more rapidly in the fingernails than in the toenails, and faster in the summer than in the winter. The rate of growth is greatest in the longest finger, least in the little, and intermediate between the two in the other fingers.

Hangnails are the result of cracks or splits of the skin alongside the nails. They may stem from excessive skin dryness from frequent washing of the hands but are more usually caused by a nervous picking at the skin.

The cause of brittle and splitting nails is unknown, but it is suspected that exposure to environmental agents (detergents, solvents, nail polish remover) may be a factor. Eating large amounts of gelatin (collagen) is supposed to be helpful, but there is little real evidence that it works.

THE SKIN—COMMON SENSE AND NONSENSE

Skin Color

In fair-skinned people, much more so than in blacks, the skin color is modified by blood in the capillaries of the dermis and in the larger arteries and veins just under it. The translucent epidermis allows the blood vessels to show through resulting in a pinkish tone. On the face and neck where the epidermis is thin and the blood vessels are close to the surface, dilation of the vessels during anger or embarrassment will cause the skin to blush. Fear or anxiety may cause a constriction of the blood vessels, and the skin blanches and turns pale as it take on the color of the underlying connective tissue. Naturally rosy cheeks, or a high color in some people does not mean they are healthier, but merely that genetically they have thin skins with many blood-filled capillaries close to the surface and showing through.

The major reason for skin color differences

is owing to the activity of pigment-producing cells called *melanocytes*. Sandwiched in among the dividing keratin-producing cells in the basal layer of the epidermis, the melanocytes synthesize a yellow, brown, or black pigment called melanin. Melanin is produced by oxidation of an amino acid, tyrosine, with the aid of a copper-containing enzyme, tyrosinase, and bound to proteins inside the melanocyte to form granules called melanosomes. Melanosomes are then transferred to the other epidermal cells by the thread-like extensions or dendrites of the melanocytes.

The purpose of the melanin is to protect the skin from the harmful effects of ultraviolet rays of the sun. The pigment absorbs the damaging ultraviolet radiation that could, in excessive doses, cause sunburn and skin cancer. The more pigment produced, the darker the skin and the greater the protective power. The actual number of melanocytes is the same in everyone, no matter what the skin tone. Black, white, and Asian skins all contain similar numbers of melanocytes, about 1,000–2,000 per square millimeter, depending on the body area. Racial differences in color and individual differences in skin hue are determined genetically and are the result of greater melanin production by the melanocytes and wider dispersal of the pigment.

When fair skin receives a heavy exposure to sunlight, the melanin that is present oxidizes and darkens. This is followed by an increased production of melanin by the melanocytes and a thickening of the epidermis to reflect more of the radiation. Over a period of several days the skin becomes tanned or darker. Skins that are darker to begin with tan more readily. Well-tanned or naturally very dark skins are protected to an extent against the harmful effects of sunlight, but even black skins can become sunburned and are not immune to overexposure.

Suntans: Are They Worth It?

Every summer millions of fair-skinned people spend millions of hours trying to acquire a tan. A golden, glowing tan is highly prized and admired. It makes the teeth appear whiter, the eyes brighter, and the complexion smoother and more even. Troubled skins are likely to improve, perhaps because the sun's drying effect reduces the number of microorganisms on the skin, perhaps because of the increased blood flow to the skin. Even getting the tan can be a relaxing and pleasant experience if it takes place on a vacation beach rather than in the backyard. For some, a tanned skin is a sign of affluence, indicating leisure and the ability to afford a trip to a resort area. But while tanning may be beneficial to the psyche, in the long run it may be harmful to both appearance and health.

The notion that a suntan is healthy and beautiful is fallacious. All it means is that the skin has responded to ultraviolet radiation by stepping up its protective mechanism, a process that ultimately can not only irreparably affect the appearance of the skin but also set the stage for the development of skin cancer.

Even without trying to get an annual "terrific tan," there is a visible difference in the appearance of the skin on the face, head, neck, arms, and hands which always has been exposed to the sun when compared with the skin on the abdomen, which rarely sees the light of day. Skin that has always been protected by clothing is still "baby skin"—softer, smoother, and finer textured. If this is the normal difference between covered and uncovered skin, imagine what will happen to the skins of sun-worshippers after years of ritual tanning. The thickened epidermis and the increased production of melanin is not enough to prevent the damaging rays from penetrating down to the dermis of the skin. The cumulative effect of the sun causes irreversible damage to the resiliency of the connective tissue fibers in the dermis. Getting a gorgeous tan in one's twenties could result in a prematurely wrinkled, leathery looking skin in one's forties.

The risk of permanent effects of tanning is greatly increased when the tanning process is started with a sunburn, as it is by almost everyone. Few people have the patience to wait for the tan to come slowly, taking the sun in small doses of 10–15 minutes a day, increasing by 5 minutes daily, using an effective sunscreen all the while, and sunning only before 10 a.m. and after 2 p.m. They want a tan quickly and are willing to endure soreness, wakeful nights, and peeling to get the final bronzing. It is believed, however, that once the skin is burned initially, even moderately, the ultraviolet rays are able to penetrate into the dermis despite the tanning of the outer layers. Once the elasticity of the fibrous tissue has been damaged, the result is premature wrinkling of the skin, and nothing can reverse it. Some investigators are even convinced that any skin tanning means skin damage.

Skin Cancer

The relationship between ultraviolet radiation and skin cancer is well established. The most common forms are basal cell cancer, which rarely metastasizes, and squamous cell cancer, which occurs less frequently but is more likely to spread. A third type, melanoma, is highly dangerous but much less common. Its cause is not as clearly correlated to overexposure to sunlight, but melanomas occur more frequently in women and in people with fair complexions.

Eighty to 90% of basal and squamous cell cancers occur on the parts of the body that are exposed to the sun. In the last 25 years, the incidence of skin malignancies in this country has doubled, and it is no coincidence that this has accompanied the increase in the number of tennis players, joggers, golfers, swimmers, and sunbathers. More cases of the disease occur in southern areas of the country where there is more sun and greater exposure, and people with fair skins and blue or green eyes are particularly vulnerable. Naturally dark-skinned individuals are not immune to skin cancer, however, although the risk is less.

Obviously, there are other factors besides ultraviolet radiation that are involved in skin cancer. The disease sometimes occurs on areas of the body not exposed to the sun, and there are inveterate sunbathers who never get skin cancer. As with all cancers, there are probably genetic or environmental factors that combine with the damaging effects of the ultraviolet rays. There is no way of predicting whether frequent bouts with sunburn during youth will result in skin cancer in the later years, but there is evidence that even several short exposures to a very hot sun are capable of causing a malignancy.

Sunscreens

Sunscreen preparations are designed either to reflect the rays of the sun or to absorb them, mimicking the body's own defenses against radiation. Zinc oxide ointments or any other preparation that contains an opaque substance, such as titanium dioxide or kaolin, are very effective sun blocks; that is, they reflect the damaging rays. Most people dislike sun blocks because they minimize tanning and are not cosmetically pleasant to use, but they are necessary for those who cannot tolerate any exposure to the sun. The most effective solar radiation absorbing agents are those containing para-amino benzoic acid (PABA) or derivatives, such as octyl dimethyl PABA, but occasional allergic reactions occur when using PABA, chemically related to benzocaine. Of somewhat lesser effectiveness are preparations containing benzophenone or cinnamate compounds derived from cinnamon oil. These products are able to reduce the intensity of the ultraviolet rays by absorbing them, thus permitting a longer period of exposure, but no product, despite its claim, is able to provide "tanning without burning." If sunbathing takes place in intensive sunlight for an extended period, sunburn will occur in the same way that a burn can occur right through a deep tan.

It is important to read the label before buying any suntan oil or lotion. PreSun, Pabafilm, and Pabanol contain plain PABA. Eclipse contains a derivative, glyceryl PABA, and Block Out contains octyl dimethyl PABA, neither of which are as likely to stain clothing as the plain PABA. A product that contains vegetable fats, such as sesame seed oil, olive oil, or cocoa butter, may have some limited sunscreening properties, but will primarily only lubricate the skin. Baby oil has a mineral oil base and is virtually useless as a sunscreen.

To help consumers figure out which sunscreen to use, the Food and Drug Administration adopted a rating method called the "sun protection factor" or SPF of 2–15 that can be found on most brands of lotions or oils. The numbers in the system refer to the amount of additional exposure to the sun that can be tolerated if the product is used. For example, an SPF of 8 means that if the product is applied and not washed or perspired off, one could remain in the sun eight times as long as without the product. A person with a very fair and sensitive skin who burns after 30 minutes exposure to the sun could apply that product and theoretically stay out for 4 hours without burning but probably should not count on it. Individuals who burn easily and never tan should use a product with an SPF factor of 12 or more. For average skin that burns moderately and tans gradually, the FDA recommends using a product with an SPF of 6 for the first exposure to the sun.

Photosensitization

Some individuals will get a very severe sunburn or have a severe skin reaction characterized by itching, burning, inflammation, and skin rash even if their exposure to the sun is minimal to moderate. Such increased sensitivity to the sun is possible when certain drugs have been ingested, certain chemicals in soaps or perfumes have been applied to the skin, or if there is already the presence of a skin disorder, such as lupus erythematosus or herpes simplex, that becomes aggravated by exposure. The photoallergic response can result from a wide range of substances. Antibiotic drugs, such as the sulfonamides and their derivatives (oral diabetic drugs, thiazide diuretics) and the tetracyclines, are known to be particular offenders in some people. So are some tranquilizers and antihistamines, quinine, barbiturates, aspirin, and foods containing riboflavin. Anyone taking a drug known to be implicated in photosensitivity reactions (check the *Physician's Desk Reference* or other sources of information on drug action) must *beware of the sun*. A photosensitivity reaction can leave a permanent pigmentation of the skin (Fig. 15.4).

As in All Things, Moderation

Basking in the sunshine is great, and so are water sports, tennis, and any other outdoor athletic activities, as long as unprotected exposure does not take place during the period of the day when the maximum amount of ultraviolet radiation reaches the earth's surface. There is no reason to doubt that the admonition to get plenty of fresh air and sunshine is good advice for a healthy existence, as long as it is done wisely. One known beneficial effect of exposure to sunlight on the skin is that ultraviolet rays do stimulate production of vitamin D by the activation of 7-dehydrocholesterol. But since vitamin D can also be obtained from the diet and many foods are fortified

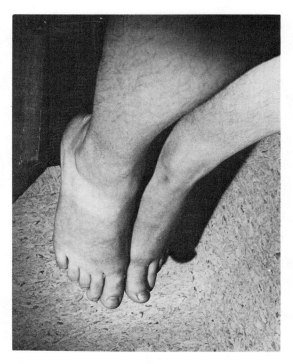

Figure 15.4. Photosensitivity reaction that occurred as a result of exposure to sunlight while taking a tetracycline antibiotic, demeclocycline hydrochloride, as a treatment for acne. Skin darkening has persisted more than a year.

with it, a normal adult can get all that is needed through proper nutrition. Sun is good, but there is no benefit to going out of one's way to get a suntan, and there is every evidence of detriment. To expose the skin unnecessarily to get it darker is inane, since the ultimate result is accelerated aging and the possibility of skin cancer. The most recent absurdity is the indoor tanning parlors. These permit the same solar risks, but inside in a booth. If today's terrific tan turns out to be tomorrow's prune, what's the point? Moreover, one can only speculate what future historians could make of this peculiarity of our culture. It is certain to appear strange that a society which frequently discriminates on the basis of dark skin color should have a large number of people who, oddly enough, seasonally and ritually, exert every effort to change their skins to brown.

Wrinkled Skin

The dermis of skin is a loose interweaving network of connective tissue proteins called collagen, elastic, and reticular fibers. The collagen and elastic fibers are responsible for the ability of the skin to stretch and contract to accommodate body movement or for changes in size or shape of the body. Generally, although there is great individual variation, skin tends to lose its resiliency with age. For example, if the loose skin of the top of the hand of a young man or woman is pinched up between the fingers, it will instantly return to normal when released. In a 85-year-old, the skin will remain wrinkled up anywhere from 5 to 15 seconds after release (see Figs. 15.5–9).

Facial skin normally wrinkles at right angles to the lines of pull of the facial muscles underneath. After years of facial expression, the lines become accentuated and exaggerated. Since there are more abundant elastic fibers in the dermis of the face and scalp than elsewhere, their natural degeneration with time, accompanied by the decrease in water content of the skin and the pull of gravity, results in the appearance of permanent wrinkles on the face. Skin aging depends to a great extent on heredity, and a good way to retain an unwrinkled skin after 40 is to be born to parents who also have smooth skins at that age. Of course, one could also try going through life without smiling, frowning, laughing, or talking—a stony-faced alternative to normal communication.

Facial lines and wrinkles are a natural consequency of aging, and they result from the

Figure 15.5. Hand of a woman of 85 (left) and hand of a woman of 35 (right). Although there is great individual variability, skin elasticity tends to be very different in these two age groups.

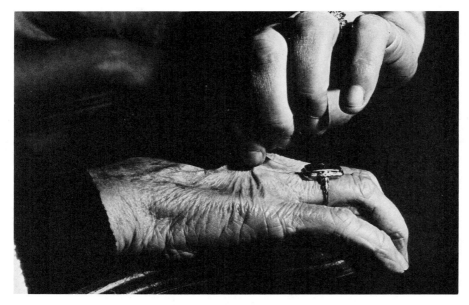

Figure 15.6. The skin on the hand of the older woman is pinched up and stretched between the fingers.

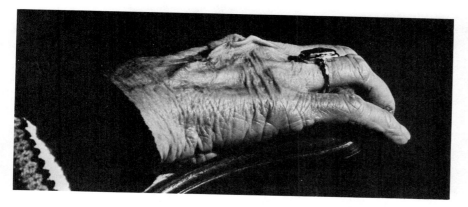

Figure 15.7. The older woman's skin remains wrinkled 10 seconds after release. The elastic and collagen fibers in the dermis have lost much of their resiliency with age.

normal deterioration of the bundles of collagen and elastic fibers in the dermis. The process can be accelerated by overly enthusiastic sunning. Heavy smoking, for some reason, will also hasten the formation of deep wrinkles around the eyes and should be further motivation, if more is needed, to stop the cigarette habit. But there is no way of preventing the eventual appearance of wrinkles—no cosmetic product that "creams" the skin rather than washes it, no matter how faithfully used, and no regimen of facial exercises massage,

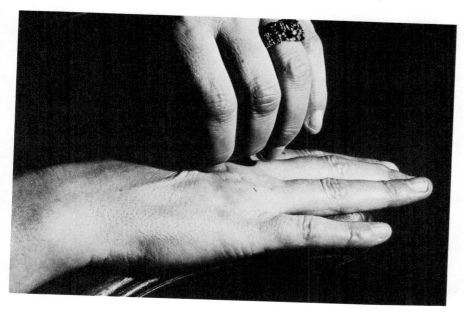

Figure 15.8. The younger woman's skin is pinched up between the fingers.

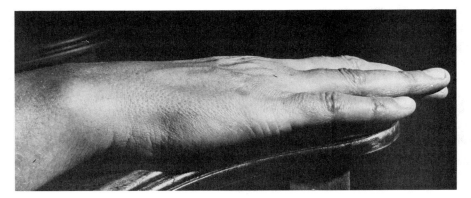

Figure 15.9. Young skin is elastic and instantly returns to normal after being released.

masks, or saunas. Facial exercises merely tone the facial muscles and have no effect on the skin above them. Vigorous facial massage could actually help to break down the fibrous connective tissue and create new lines in a hurry. There are some cosmetic aids that can form an oily film that makes tiny lines appear less apparent, but nothing except plastic surgery can cure or remove wrinkles.

Some of the most outrageous claims are implied by some of the most expensive face creams. One "extra-rich nourisher," presumably for the extra-rich woman, since it sells at $55 for 2 oz, contains a "unique kind of collagen." Its uniqueness is questionable, since it is actually the insoluble protein from the collagen fibers of animal connective tissue, which has been made soluble chemically. The collagen is present presumably to "sink into skin" and "attack the reason for fine dry lines." Note that there is no direct promise that the collagen will penetrate the dermis to rejuvenate or repair the wornout connective tissue fibers, but the name of the product and the wording of the claim tend to leave that erroneous impression. The other ingredients in this costly cream provide quite valid ways of dealing with fine dry lines because they form a barrier to water evaporation and thus to skin dryness, but the same effect can be accomplished for much less money.

Dry Skin, Oily Skin

The ideal in women's facial skin is a smooth, unblemished complexion in which the little orifices from the sweat glands and the sebaceous glands exiting through the hair follicles are not enlarged (large pores); that does not look dry, flaky, or chapped; or does not appear greasy from too much oil on the surface. Women with such skins are fortunate to have inherited the combination of genes to produce it. Many women do have dry skins, not only on the face but all over their bodies, and have either chronic or seasonal problems with itching and flaking skin. Other women have varying degrees of oiliness or possibly a combination of some areas that are dry and others that are oily. Oily skins are perceived as unsightly, but unless the condition precipitates a skin disorder, they are far less of a problem than dry skin.

Dry skin occurs when the water loss from the top layer of the epidermis occurs more rapidly than it can be replaced from the underlying tissues. The keratin in the epidermis must contain at least 10% water for the skin to be soft, pliant, and supple. If the percentage of moisture falls below this level, the cornified layer dries out, becomes brittle, and flakes off. The skin feels tight and dry, and in the extreme, the skin on the hands, for example, can crack and bleed. Water is constantly being supplied upward to the keratin from underneath. The skin retains its hydration because there is a barrier layer in the epidermis to prevent rapid movement of water; without it, moisture would evaporate off too quickly. Besides the barrier layer, two other factors prevent rapid water loss and promote water retention. One is the presence of hydrophilic (water-loving) substances in the horny layer that bind the water these cells need. These water-soluble and water-attracting substances could be removed from the skin by washing or by perspiration, but they are protected from loss by a thin film of water-insoluble sebum. Sebum, the fatty secretion of the sebaceous glands, is a complex mixture of lipids containing fatty acids (mostly palmitic and oleic acids), neutral fats or triglycerides, wax esters, cholesterol esters, and squalene, a precursor to cholesterol. This "skin fat," in combination with sweat, forms a coating on the surface and functions to help slow down the too rapid evaporation of water. Sebum also coats the projecting edges of the dead keratinized cells, smooths them down, and thus reduces the surface area for evaporation. Creams, lotions, and "moisturizers" for dry skin attempt to duplicate and enhance the functions of sebum.

Sebaceous glands are largest and most numerous on the skin of the forehead, around the nose and mouth, and over the cheekbones, attaining a density of 800 per square centimeter compared with 100 glands per square centimeter in other locations. Their number and activity are under both hereditary and hormonal control. The production of sebum is greater in teenagers, the increase resulting primarily from the secretion of androgen, but affected by estrogens and progesterone as well. The luster of youthful skin, that shiny, glossy glow so prized and promised by the cosmetic ads, is essentially caused by the same sort of sebaceous gland activity that is labeled a "too oily" skin when a woman is past adolescence. There are vast individual differences in sebum production, and these differences cause the differentiation between dry, normal, and oily complexions. Obviously, when the amount of sebum produced is neither too scanty nor too excessive, and when the rate of water loss from the surface of the skin is balanced by the rate at which water moves from the dermis into the top layers, the skin appears neither dry nor oily, but normal.

Any skin type, even naturally oily, can become dehydrated when water evaporation exceeds the capability for replacement. Environmental and hormonal factors play a role in inducing dessication of the horny layer. Water loss is greater when the temperature is elevated, the humidity is low, and the wind velocity is increased. Warm weather alone is not necessarily drying, even in lowered humidity, since sweat on the skin surface keeps it moist. A warm, dry, windy day, however, may cause a problem for those whose skins are naturally drier. Dry-skinned people can really suffer in the winter months, when the combination of cold, dry outside air and overheated, low-humidity indoor air are likely to result in the "winter itch," a dry skin condition. The daily bath or shower will also add to the difficulty in the winter, since washing the skin removes the oily sebum film. Ordinarily, this defatting of the skin surface is only temporary since the increased activity of the sebaceous glands soon renews the barrier effect of the sebum. The replacement is less rapid, however, in people who through heredity or age normally produce less sebum. Fewer, cooler, and quicker baths in the winter time would probably benefit everyone's skin. The use of emollients to replace the surface skin lipid layer and to retard evaporation will also be helpful.

Once the skin has become dry, brittle, and cracked, it will take more than the application of a lotion or cream to restore it. The most direct way to heal chapped and roughened skin, common on the hands and face during the winter months, is to rehydrate it. Moisture can only be replaced with moisture, not with fats, which have no effect on returning hard, dry keratin to softness and pliability. This observation, first made by Blank in the early 1950s, caused a complete change of direction in the treatment of dry skin. It is now known that merely attempting to replace the lost oils on the skin is not enough and that a barrier should be applied to prevent further evaporation after the skin is hydrated.

Hand lotions and creams are emulsions of water and oils; the water provides moisture to the skin and the oils are supposed to form a barrier to keep it there. But when the skin has become very dry and is painfully chapped, lotions and creams in themselves cannot add enough moisture or provide enough of a retardant to further water loss. Rather than relying on the products alone, restoring moisture can better be accomplished by soaking the hands in water, drying them, and then slathering on a thick layer of petroleum jelly (Vaseline). The skin will pick up the moisture during immersion, and the petroleum jelly barrier will retain it.

The same method of petroleum-jelly occlusion is used commerically to combat chapped lips, since various lip balms are basically petrolatum plus solidifying waxes. The straight Vaseline treatment after soaking is appropriate for the hands, since one can always go to bed with cotton gloves on. Using petroleum jelly on the face is less appealing; it messes up the sheet and is too greasy to wear under makeup. All night creams and most daytime moisturizers, however, contain mineral oil, which is liquid petroleum jelly. They also contain other ingredients in a more elegant formulation to provide similar benefits with greater esthetic qualities. The ability of any moisturizing cream or lotion to aid a flaky and chapped skin is enhanced when the skin is moistened before application. This means that putting the preparation on when the skin is still wet is going to make the barrier work more effectively.

Does everyone's skin need the protection of a moisturizer? Cosmetics manufacturers, whose business is selling products to make a profit, tend to advocate the use of moisturizers in some form as beneficial for all women. In areas where the year-round climate is hot and dry or in the northern parts of the country where the inside of most houses during the winter months is similar to a desert atmosphere, it may be that a barrier film is needed even by women with a normal skin-oil to skin-moisture ratio, but moisturizers are of no benefit to those with oily skins. They may further aggravate the skin in those who are troubled with acne.

Skin dryness or oiliness is, after all, highly subjective and variable. It may change, even in the same individual, under different conditions—with the environment, with the amount of skin cleansing, with the kind of makeup used, and to a certain extent, even with the emotional state. Unless the dryness or oiliness is very unsightly or is related to a skin disorder, it is probably overrated as a problem. Dry skin does not cause wrinkles. Heredity, aging, and exposure to ultraviolet light causes wrinkling, and no amount of creaming or moisturizing is going to prevent it. Wrinkles can be more apparent when the top layers of the skin are dehydrated and lifted up rather than lying flat to refract light evenly. Scanning electron microscopy photographs of the horny layers taken before and after the use of moisturizing lotions seem to indicate that the primary action of such products appears to be the lubricating and flattening down of the uplifted or flaking layers of cornified cells. The effect results from the oils and additives in the product and not from the moisture or water component (Garber, 1978). If the skin feels tight and looks dry, any emollient preparation, even something as mundane as Vaseline or plain mineral oil or pure solid vegetable shortening will put a fatty film on the skin surface and accomplish the same purpose.

Of course, commercial products may contain a more pleasant combination of ingredients that makes using them more agreeable than applying Spry, Crisco, or baby oil to the face. Moisturizing lotions also contain, in addition to oil and water, substances called *humectants*. Humectants, such as glycerine, sorbitol, urea, propylene glycol, polypropylene glycol, or in "organic" products, honey, are highly hygroscopic and are capable of attracting moisture. Advertisements imply that such agents have the ability to draw the water out of the air and bring it to the skin surface. It is very unlikely that humectants actually transfer water from the atmosphere to the skin. They are, in fact, nondirectional. If humectants were used alone or in high enough proportions they would be just as likely to pull the water out of the skin. The only way that a high concentration of glycerine (or glycerol, which is the same substance) can be useful for dry skin is under conditions of very high atmospheric humidity of 90% or more. Of course, if the humidity is very high, dry skin would not be a problem. The main purpose of humectants in moisturizing lotions is to bind with the water in the emulsion and therefore keep the preparation itself from drying out or shrinking through evaporation. Humectants also smooth and soften the skin surface and make the application of the lotion easier.

Skin Mythology

The public gets a lot of advice concerning facial skin care, and much of it is utter nonsense. Women are constantly being told by beauty "experts" to let the skin breathe, to let the skin rest, to nourish it, to let it thirstily soak up the oils it craves, to flush out its pores, to pat or massage it, and to coddle it with applications of all kinds of substances. During the last few years when anything "natural" or "organic" has been very popular, recipes for do-it-yourself cleansers, toners, masks, and various remedies have been appearing in magazines and newspapers. There was a time when the expression "ending up with egg on your face" meant discomfiture or embarrassment. Now it means that a mixture of egg, honey, oatmeal, and so forth has been applied to deep-clean, tone, prevent blotches, moisturize, or in other ways provide innumerable benefits.

The homemade preparations composed of vegetable oils, herbs, fruit juices, and cereals are certainly cheaper than cosmetics. They can have similar emollient or drying properties and provide the same effect as many commercial products. They can also be equally allergenic, sensitizing, irritating, or acne-producing on some skins as the chemicals in commercial cosmetics, and there is no particular benefit in buying cosmetics that contain "organic" substances. Moreover, the notion that the skin requires a little nourishing snack in the form of mayonnaise or a little drink of apple cider vinegar in preference to other emollients or astringents is patently absurd. Human epidermis does not "eat," need to be nourished, or in any way fed from the outside. Neither does it "drink," soak up oils, "breathe," or need to breathe. Skin gets all the nourishment and oxygen it needs delivered to it from the underlying blood vessels in the dermis. Little if any gaseous exchange occurs from the environment. Skin can be nourished only from the inside. Eating the eggs, oatmeal, strawberries, avocados, cucumbers, or whatever, is a better way to get a healthy and good-looking skin than by mashing the foods up in a "berry-butter-beautifier" or "protein mask."

Letting the skin breathe may refer to nonocclusion of the openings from the sweat glands and the hair follicles. Certain cosmetics are said to clog the pores. Virtually all facial makeups, powders, or moisturizing creams and lotions are occlusive to an extent, but even heavy facial makeup is unable to block perspiration, as anyone who has sat up close to the actors during a stage performance can verify. Cosmetics that contain large amounts of oil may be occluding to hair follicle openings, but thorough face washing with soap unclogs the clogged pores, and this is sufficient for most women. Some skins do have a lesser tolerance for occlusive oily makeups and respond with "acne cosmetica," a condition to be described later. But not everyone gets clogged pores or pimples from wearing makeup. At night, however, common sense indicates that makeup should always be completely removed.

Skin Care

Essential to keeping skin healthy and good looking is keeping it clean. Cleansing the skin removes the flakes or dead cells, dust or soot particles, dried perspiration, accumulated skin oils and remnants of cosmetics, and skin bacteria that act on organic materials to cause odors. Water alone does a good job of rinsing away dirt if enough of it is used for a long enough period of time in conjunction with a brush or a harsh cloth. In order to more easily remove the impurities that are embedded with skin fat in all the little skin folds and follicle openings, however, a cleansing agent should be used. The most efficient agents are those which surround or dissolve dirt parti-

cles so that water can rinse them away and not merely redistribute them.

The cosmetics industry implies that different methods of cleansing are necessary for different kinds of epidermis. Some manufacturers, with an eye toward the current interest in computer technology, have begun to use computerized skin analysis to determine the exact preparations to be purchased by every skin type. A vast array of substances for facial cleansing are on the market—morning cleansers for dry skin, evening cleansers for oily skin, creamy cleansers containing milk, scrub creams, facial baths, and various extraordinarily priced "cleansing bars."

Although many cosmetic "experts" would probably disagree, most dermatologists concur that the best method for washing any face is using soap and water. Toilet soaps are a mixture of the sodium salts of various fatty acids—mostly stearic, palmitic, and oleic—and are essentially the same fat and lye compound that has been used to remove dirt for thousands of years. Today's soaps, however, contain extra ingredients that are supposed to confer special qualities. Superfatted soaps such as Dove, for example, contain 3%–5% more fat, and transparent soaps, such as Neutrogena, contain glycerine. Many soaps claim to have deodorant effects. Since the antibacterial hexachlorophene can no longer be used in soap, most of the deodorizing effect of deodorant soaps is the same as that of any soap—the removal of dirt and body oil. Acne soaps contain sulfur and tar, which are drying agents and may be irritating to some skins. Castile soap contains coconut or olive oil, and soaps that float are inflated with air. Ivory soap smells like soap; it contains less perfume or oil than some other brands and could be less sensitizing to some skins. Some soaps are completely made of synthetic detergents and are "nonsoap" soaps, but they are neither better nor worse than any others.

All soaps get the skin clean. Choosing one

over another is a matter of personal preference. As pointed out by *Consumer Reports*, all modern soaps have been adjusted to an appropriate pH, and none is harshly alkaline. Oily-skinned individuals are able to tolerate more frequent washing whereas people with dry skins will probably want to wash no more than twice a day. Those with very dry skins may prefer a soap containing more creams or oils—the superfatted variety. Spending a lot of money on exotic beauty soaps does not guarantee that they have any particular qualities beyond cleansing power. Soap is soap.

Facial Makeup

There are many women who inherit the kind of facial skin and coloring that really needs no covering or enhancement. Through genetics, and with no particular effort on their part, they have smooth, unblemished, velvety skin that is the epitome of a beautiful complexion. Such skin is often remarkably indifferent to the way it is treated. Flawless skin is, after all, most similar in quality and texture to baby and children's skin, and no special regimen of creaming, cleansing, or any other kind of coddling with high-priced products is used on youngsters. For such lucky women, perhaps a little mascara and a lip gloss would be as much makeup as they would ever use.

Other women whose facial skin has small imperfections, whose pigmentation is not as even, whose skin texture is not as fine, may choose not to meet the world with a bare face. They may use a number of cosmetics—foundation, blusher, eye makeup, lipstick—as the most practical solution to being born with skin that is less than perfect. Although there is a prevalent belief that there is value in going without makeup, using no makeup is not necessarily better for *normal* skins. Some women, however, have unusually sensitive skins that cannot tolerate cosmetics without reacting adversely. Some women are prone to acne

eruptions long after the usual adolescent acne should have disappeared and would do better to avoid very oily makeups that will aggravate their conditions. Women who tend to have dry skin will probably choose oil-based cosmetics but may want to avoid pancake or matte finish makeups. Even though they contain more emollients, they also contain more chalk and talc that absorb oils and draw moisture out of the skin.

Probably the best way to find a number of products for personal use, given that each women has her own requirements, is to remain highly skeptical about the advertised claims and experiment around, trying to spend as little as possible. When it comes to buying cosmetics, the old axiom "you get what you pay for" is not necessarily true. In most instances, what you get, whether you spend a little or a lot, is essentially the same combination of ingredients.

"Hypoallergenic" Products

Consumers may be surprised to learn that the cosmetic labels that claim that a cosmetic or an entire line of products are "hypoallergenic," that is, less likely to cause an allergic or adverse skin reaction, are meaningless. There are no scientific studies showing that products that make such claims actually contain ingredients that have a lower potential for causing sensitivity or irritation than any other similar products not claiming hypoallergenicity. All cosmetics contain allergens for some people; to completely avoid reactions to cosmetics would require abolishing the use of all of them.

The Food and Drug Administration has been unsuccessfully trying since 1974 to issue a regulation that would require manufacturers to withdraw cosmetics labeled "hypoallergenic" from the market, unless testing on human subjects indicated that a product so labeled caused significantly fewer adverse reactions than competing products *not* making such claims. The FDA's efforts were challenged legally by Almay and Clinique, two of the largest manufacturers of alleged hypoallergenic cosmetics. When the FDA's proposal was upheld in U.S. District Court, the two firms appealed the decision. In late 1977, a Federal court ruled that the FDA's regulation was invalid because it had not demonstrated that the word "hypoallergenic" was perceived by consumers in the way described in the regulation, and that the agency's definition of the term was hence unreasonable.

Manufacturers may now continue, as they have in the past, to advertise and label their products as hypoallergenic without having to provide any substantiation. A cosmetic marketed as hypoallergenic may indeed contain fewer sensitizing ingredients, but its content may be the same as in any other similar product, a fact easily ascertained by reading the labels. A recent FDA regulation, which did not suffer the same legal fate as the hypoallergenic proposal (although this was not due to lack of effort on the part of the industry), requires that the ingredients used in cosmetics be listed on the product label in order of predominance in amount. Women who know they are allergic to certain chemicals can be afforded some protection by looking for the presence of the offending ingredient in the product. Unfortunately, "trade secrets," "fragrance," and "flavor" are categories presently exempt from the required labeling. Each of these broad classifications may in themselves contain dozens of chemicals that are capable of provoking an adverse reaction. The FDA is considering a change in the regulation to include certain flavor or fragrance ingredients.

Federal Regulation of Cosmetics

The majority of cosmetic products are probably safe for use by the majority of people. Their harmlessness is only an optimistic assumption, however, because there is actually no proof of their safety. Moreover, there is not much chance in getting that verification. Of course, cosmetics manufacturers have reputations to protect and stand to lose a great deal if the safety of their products becomes suspect. Most firms do test their products before putting them on the market, but there is no law that requires that they certify them for safety before sale, no necessity for them to report the results of their tests to the FDA, and no mandate that they inform the FDA of consumer complaints of adverse reactions to their products. Under present legislation, there is no way of knowing how many cosmetics manufacturers exist, what they are making, what the products they sell contain, or whether or not they are hazardous. An estimated 8,000 chemical ingredients are used in formulating cosmetics, and only for about 75—the color additives—is there any necessity for documenting safety and effectiveness.

This priviledged status of cosmetics, exempt from the regulations governing all other substances under the purview of the Food and Drug Administration, has its basis in the Food, Drug, and Cosmetic Act. Passed by Congress in 1938, the law defines cosmetics as articles "rubbed, poured, sprinkled, or sprayed on, introduced into, or otherwise applied to the human body for cleansing, beautifying, promoting attractiveness, or altering the appearance without affecting the body's structure or function." This means that some cosmetic products, such as deodorants or antidandruff shampoos, because they do affect body structure or function, are legally drugs and have to meet tests for safety and efficacy before marketing. But many cosmetics that by definition are excluded from regulation contain ingredients that not only can cause adverse skin reactions, but also can be absorbed, ingested, or inhaled to result in the same kind of systemic health problems as those caused by the chemicals in drugs and food additives. In the last few years, there have been reports of microorganism contamination of eye makeup and hand lotion, skin photosensitizing chemicals in bar soap, asbestos in talcum powder, formaldehyde in nail hardeners, highly skin-sensitizing feminine deodorant sprays, and carcinogens in hair dyes. No one really knows, for example, to what extent cosmetic ingredients may be absorbed through the skin, what happens to the molecules once they get into the bloodstream, or what their long-term effects may be. The safety of beauty products is a largely ignored major issue in women's (and men's) health. The apparent disinterest in cosmetic safety regulation may stem from the general assumption that there already are existing rules, codes, and ordinances controlling the sale of cosmetics. The government, after all, seemingly extends its regulatory presence into the marketing of all commodities. Under current laws, however, the only way in which an unsafe cosmetic product can be removed from the marketplace is for the FDA to prove to a court that the cosmetic is misbranded or adulterated. Unlike drugs, food, drug, and cosmetic color additives, and medical devices, cosmetics do *not* have to be pretested for safety before being sold. Millions of consumers, therefore, have become the ex post facto experimental subjects for the possible and potential hazards of cosmetics.

Consumer groups have argued that the government should have the same control over cosmetics that they have over the other items regulated by the Food and Drug Administration. Cosmetics safety reform legislation has been introduced before the Senate. When and if enacted, the bill would require premarketing certification for safety of all cosmetics. It would also make necessary the registration of all cosmetic formulations with the FDA and further require that the manufacturers report all complaints of adverse reactions that they receive.

In a move toward greater consumer protection, the FDA itself has extended its regulatory authority over cosmetics by new labeling requirements. All cosmetics manufacturers must now list ingredients on the product labels. Moreover, if a cosmetic manufacturer has not substantiated the safety of a product, the label must carry the following message: "Warning—the safety of this product has not been determined." While the regulations are of benefit to consumers, they are not as far-reaching as would be expected. Cosmetics used in beauty shops (hair dyes, permanent wave solution, and so on) need not carry warning labels. Furthermore, since the FDA has no authority under law to examine the data on which the manufacturers' claims of safety are based, the effectiveness of the regulation is obviously weakened.

Read the Labels!

While the ingredient listing on the labels in itself does not provide the kind of regulatory control over cosmetic formulation that is necessary, it is a step forward and is significant for several reasons. First, consumers have the right to know what is in the products that they buy. People are then able to avoid suspect chemicals that have been implicated as carcinogenic, have produced adverse allergic skin or systemic reactions in them, are known to be photosensitizing, or that can aggravate an existing skin condition, such as acne. Second, the listing can result in comparison shopping and promote truth in advertising. For years there have been rumors that a lipstick that costs $9.50 may contain the same 15¢ worth of ingredients as a lipstick selling for 59¢, but the expensive one has the prettier tube, the more prestigious brand name, and is sold in a department store instead of a drug or discount store. Now it is possible to actually compare the value of competing cosmetic products and to personally decide whether the costly one is worth the price difference.

Cosmetic companies maintain that the evaluation of a product's worth has little to do with the listing of ingredients. They claim that the quality and purity of the ingredients may vary among the expensive and cheaper brands, and that these criteria cannot be listed on the label. The validity of such arguments, similar to those used to justify the inordinate price differences between generic and brand-name drugs, is difficult to evaluate. It may be that some products contain commercial grade chemicals and others pure grade. It may also be true that different grades of chemicals may mean that varying amounts of degradation products or impurities are present. There is no assurance, however, that expensive products use a better grade of chemical. The only information that is available is the identity of the ingredients; whether being expensive and creatively advertised makes an individual product better or purer is controversial.

Two skin lotions, for example, are contrasted below. They both purport to perform essentially the same function. The more costly product is termed a "clarifying" lotion, which serves to remove the top layers of dead skin cells and leave skin looking "its cleanest, freshest, healthiest." The cheaper preparation is a "texture" lotion, meant to "remove dirt" as well as "stimulate and refresh skin for a smooth, glowing complexion." Since the ingredients are listed in order of predominance, the consumer can judge whether the difference between the two products warrants the fourfold difference in price.

What Do the Labels Mean?

The above example demonstrates that it is possible to compare ingredients to see how one product compares to another more expensive or cheaper brand without understanding much about the purpose or the action of the chemicals. It is going to be increasingly difficult for a manufacturer to continue to make exaggerated claims about what a product with a "miracle" or "European" formula can accomplish when all the "secrets" must, by law, be listed on the label. According to the FDA regulation, the materials in the formula must be presented in the descending order of their predominance and by established and uniform names so that the consumer is not confused or misled by the use of different terms for the same substance. The terminology was developed by the Ingredient Nomenclature Committee of the CTFA, the Cosmetic, Toiletry, and Fragrance Association, and is published in the Cosmetic Ingredient Dictionary.*

The names may be uniform and consistent, but their intelligibility to anyone without a degree in organic chemistry would appear to be doubtful. When confronted by 10 tiny lines of long and technical terms on the bottle, one's normal response might be the presumption that such a complex combination of chemicals must surely be worth every bit of the price! It is important to remember, however, that the formula listing is qualitative only, and that the concentration of the ingredients are lacking. If water is the first ingredient listed,

Lotion A ($12.50 for 12 oz)	Lotion B ($4.50 for 16 oz)
SD alcohol 40	water
water	SD alcohol 40
witch hazel	witch hazel
glycerine	propylene glycol
acetone	sodium borate
sodium borate	isopropyl alcohol
menthol	methylparaben
caramel	fragrance
D&C red #33	D&C red #33
	FD&C blue #1

* Available from the CTFA, 1133 15th Street, NW, Washington, D.C., 20005, for $45.00.

the product is mostly water, and the components near the bottom of the list are not going to be present in greater concentrations than just a few percent. Those small amounts are not likely to contribute much to the effectiveness of the product and are there mostly as stabilizers or preservatives. Moreover, after a little persistence, there is no reason to feel completely intimidated by the chemical jargon. The ingredients group themselves into certain categories. Once a person becomes accustomed to being a label-reader, the terms may begin to take on familiarity. Perhaps one may never feel comfortable with "diisopropanolamine-carbomer 041" (a preservative and emulsifying agent), but after seeing the ubiquitous "glycerine" or "paraben" on bottle after bottle, it may be no more strange than the sugar, BHA, and BHT on the breakfast cereal box.

Consider, for example, the ingredients listed in the two lotions above. The first three are water, SD alcohol 40, and witch hazel. The alcohol is one of a group of ethyl alcohols that have been denatured, that is, adulterated to prevent them from being drinkable in accordance with government regulations. Witch hazel is an extract made from the bark and leaves of a small tree. It has astringent, or puckering, action on the skin, as does the sodium borate. Astringents are mildly irritating and make the skin feel tingly. Menthol has a slight anesthetic affect and provides a cool feeling as it evaporates. Glycerine and propylene glycol are humectants; their purpose is to prevent the liquid from too rapid evaporation. Acetone and isopropyl alcohol are fat solvents, similar in function to the SD alcohol. Methylparaben is a widely used antibacterial preservative. Caramel, as most people know, is burnt sugar, and is present as a coloring agent. So are the D&C and the FD&C colors. D&C means that the color may be used in drugs and cosmetics only and not in foods; and "F" preceding the D&C means that the color has been certified for use in foods as well.

D&C red #33 belongs to a group known as azo dyes that are known to cause allergic reactions in some individuals. FD&C blue #1 is a food coloring that has been banned from use as a food additive in Europe and the United Kingdom. Both colors are members of the family of dyes known as coal-tar colors. The safety of any coal-tar color additives in foods is highly suspect. All have been implicated as carcinogenic in animals and allergenic in humans. There is controversy concerning their safety in cosmetics. The potential hazard when small quantities are absorbed through the skin or injested in lipstick is unknown. Many people do have adverse skin reactions to coal-tar colors in cosmetic products.

It might also be observed that the purpose of both of these lotions, degreasing the skin and removing or exfoliating the top layers of dead surface cells, can be accomplished as easily by washing with soap and a terry washcloth and drying with a towel. If more "clarifying" is desired, a complexion brush, a polyester fiber sponge, or a natural sponge could be used.

For someone who has experienced an allergic reaction—a skin rash, edema, itchy eyes, puffy lips—and needs to avoid other products that contain the suspected ingredient, it would be worthwhile to invest in Ruth Winter's *A Consumer's Dictionary of Cosmetic Ingredients.** This excellent handbook, also available in paperback, describes the origin, function, and safety of specific cosmetic components in nontechnical, easily understandable language. It is an indispensible and enlightening reference for anyone interested in separating out the realities of cosmetic function from the advertised claims. But since carrying a dictionary to the cosmetic counter is

* Crown Publishers, Inc., New York, 1976.

not really feasible, and since most people want to know what is in a product before they buy it, it is also possible to remember a few fundamentals about cosmetic ingredients.

Foundations. Cosmetics contain various combinations of oils, fatty acids, alcohols, preservatives, humectants, colors, fragrances, and chemicals that stabilize, emulsify, foam, reduce surface tension, and generally aid in keeping the other ingredients together or enable them to go on the skin better. In a typical foundation or makeup base, the mineral oil, lanolin oil, lecithin, and spermaceti, are all emollients; so are apricot oil, sesame oil, grape-seed oil, wheat germ oil, or any other vegetable oils, no matter how exotic they may sound. Stearic acid, palmitic acid, oleic acid, and myristic acid are all naturally occurring fatty acids. In their alcohol or salt derivatives they are used as emulsifiers or stabilizers. TEA stearate, for example, is triethanolamine stearate, an emulsifier. Talc and kaolin are chalk and clay, respectively, and they are color and covering agents. Propylene glycol, glycerine, and sorbitan or sorbitol are humectants and skin softeners. Mica, titanium oxide, iron oxides, ultramarine blue, and all the coal-tar colors provide the various shades for the foundation. Methyl, butyl, or propyl parabens are antibacterial, antifungal preservatives.

Lipsticks. All lipsticks are combinations of oil and wax with red pigments to stain the lips and a perfume that is present more to eliminate the fatty taste that may be present in the ingredients than to provide an odor. Typical organic pigments could be eosin red, (D&C red #21) orange-red (D&C orange #5) and blue-red (D&C red #27), or inorganic pigments, such as bromo-acid (tetrabromo fluorescein) and derivatives of fluorescein. Lip gloss and lip shine are composed of the same ingredients but contain more lanolin and mineral oils. Frosted lipsticks may contain guanine crystals or bismuth oxychloride for the pearlized look.

Eye makeup. Only inorganic pigments are permitted in eyeshadows, mascara, and eyeliners since coal-tar colors are prohibited from use in the eye area. The pigments are dispersed in a wax, gum, or resin along with perfume, oils, and preservatives, such as propylparaben, imidazolidinyl urea, or p-hydroxybenzoate. The major colors are iron and chromium oxide pigments, aluminum powder for silver, the ultramarines, and carmine, a crimson pigment derived from the dried bodies of female cochineal insects.

There is substantial evidence that mascara is subject to potentially dangerous bacterial contamination. Serious eye infections and even loss of vision have resulted after an accidental scratch of the cornea with the applicator from a contaminated tube of mascara. Lewis Wilson and his associates have made several reports on the incidence of ocular infections from mascara use. In testing various brands of mascara, they discovered that some supported a heavy growth of bacteria after being used only five times and suggested that many brands of mascara contain inadequate preservatives. Wilson maintains that mascara should never be used longer than 3–4 months and should be thrown away after that to avoid the danger of contamination. An old brush should never be inserted into a refill mascara container without cleaning and sterilizing it first. No eye makeup should ever be shared with another person, and one should never try on eye shadow from the tester well or pot at the cosmetic counter. If the corneal epithelium is scratched while applying mascara, Wilson suggests that the scratch be treated immediately and that both the eye and the mascara be cultured to detect the presence of specific organisms. *Staphylococcus* or *Fusarium* species have been associated with mild infections, but *Pseudomonas* has been implicated in corneal ulcers and resultant blindness (Wilson et al., 1975, 1977).

Less serious adverse effects of eye area cosmetics include sensations of stinging and burning experienced immediately or a short time after application of the makeup or clinical manifestations of response, such as puffy lids or eyes that are bloodshot and teary. Sometimes these symptoms are not the result of ingredients in the makeup but occur because of a physical irritant, such as a flake of mascara or eyeshadow or a lash extender inadvertently falling into the eye. Schorr (1981) has pointed out that allergic reactions can often be initiated by the practice of drawing eyeliner across the upper and lower inner, instead of exterior, lids. Another consequence of inner eyeliner application is the permanent deposition of black pigment with possible burning and tearing as a result. No long-term adverse effect is known, but neither is there any known treatment for the condition (Pascher, 1982).

Deodorants and antiperspirants. An average adult produces from 1 to 3 pints of perspiration a day, depending on physical activity, temperature, and humidity. Most of this water is produced by approximately 2 million *eccrine sweat glands* distributed over the body. They open directly onto the skin surface and primarily function to regulate the body temperature. Heat and nervous tension cause copious secretion of sweat, and evaporation on the body surface provides cooling. Sweat is a salt-containing dilute acid solution that is colorless and virtually odorless.

The other type of sweat glands are the *apocrine glands* that develop with puberty and are found in relatively small numbers, either singly or in clusters, under the arms, in the anogenital region, on the abdomen, and around the nipples. The mammary glands and the wax-producing glands are modified apocrine sweat glands (see Fig. 15.10).

Apocrine glands are much larger than eccrine glands and are usually associated with a hair follicle, although a few open directly onto the skin surface. The apocrine gland secretion is a milky fluid that contains more solids and does become malodorous when it undergoes decomposition by skin bacteria. The under-

arm, abdomen, and ano-genital areas of the body are, therefore, a greater potential source of unpleasant body odor. Fresh perspiration on a clean skin has little odor. Even with daily washing, however, odor can develop when apocrine sweat is no longer fresh and remains on the skin. Some form of odor control is used by an estimated 90% of women and 80% of men. Deodorants prevent odor by covering it or by destroying or inhibiting the growth of bacterial on the skin surface. Antiperspirants reduce the volume of perspiration in a localized area. Perfumes mask odors, and powders absorb moisture.

Deodorants may contain propylene glycol, salts of stearic acid, or other antibacterial agents. Not all deodorants have antiperspirant activity; if they do, they must be so labeled. No antiperspirant can stop wetness, and it would not be desirable if it did, for that would completely block the gland ducts. All antiperspirants in this country are aluminum salts. Exactly how they work is unclear. The aluminum is evidently absorbed on the top layers of keratin and the sweat gland openings. It may mechanically narrow the orifice by astringent action, increase the duct's permeability to water and thus reabsorb the sweat, or perhaps simply attract water during the crystallization of the hydrated form, alum. The most effective antiperspirant is a 10% solution of aluminum chloride, which can be purchased at the pharmacy very cheaply. Its disadvantage is its high acidity in contact with skin and its tendency to destroy clothing. Almost all commercial antiperspirants compromise with a 20% solution of aluminum chlorhydrate, sometimes complexed with zirconium salts. This is not as effective, but neither is this solution as irritating to skin or to fabric.

According to most surveys, the one category of cosmetic most likely to cause adverse reactions was the deodorant/antiperspirant group. All such products carry a warning concerning application to broken skin and indi-

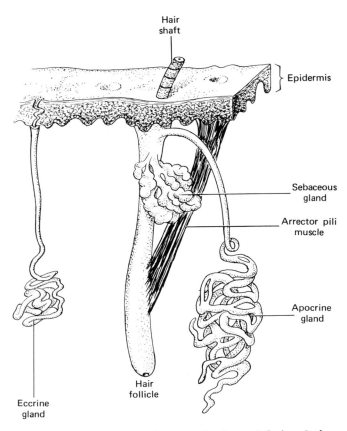

Figure 15.10. Diagram of sweat glands and their relationship to a hair follicle. The secretion of the apocrine glands, which open into hair follicles, is responsible for body odor on unwashed skin. Eccrine glands help regulate body temperature.

cate that use must be discontinued should a rash develop.

ACNE—THE MOST COMMON SKIN DISORDER

The occurrence of acne is so frequent that the disease has been called the bread and butter of dermatologists. It is said to be responsible for more office visits than any other skin problem, but even this may not accurately reflect the true incidence of the condition. Many people do not go to see a doctor at all but self-medicate with one of the hundreds of available over-the-counter preparations. At least 75% of the population suffers from acne at one time or another, predominantly during the adolescent years. Some authorities maintain that the incidence is 100% and that no one passes from puberty to adulthood without some manifestation of a blemish, bump, or blotch.

While the majority if acne cases occur during adolescence, the disorder sometimes appears to hold off until post adolescence. Some men and women get acne in their late twenties or thirties when they never had a pimple as a teenager; some are troubled with "teenage acne" all their reproductive lives, and many women are troubled with distressing eruptions of premenstrual acne.

What Does Not Cause Acne

Clearing up acne when it occurs is often not too difficult; clearing up the confusion about its onset and prevention is a different matter. Acne is not precipitated by diet or by eating chocolate, hamburgers and fries, soft drinks, or any other greasy or sweet food. Plewig and Kligman (1975) reported a study in which acne patients ate a special candy bar each day that contained 10 times the usual quantity of chocolate with no particular effect on the acne. There is also no unequivocal evidence that high fat intake results in an oilier skin. Many people are convinced, however, and know through experience that for them, eating certain foods appears to result in a flare-up. The sensible recourse is to avoid those particular items. Some individuals may have food allergies that cause or aggravate acnelike eruptions. Long-term ingestion of iodides and bromides present in some drugs or vitamins can also make existing acne worse or reinitiate it in a former sufferer, but the amount in iodide-containing foods (shellfish, peanuts, cabbage, spinach) or iodized salt is believed to be too minimal to worry about.

Despite common misconceptions, acne is nobody's fault. It has nothing to do with infrequent washing, and obsessive cleanliness is not going to help. Neither does acne result from constipation, lack of sleep, or too little or too much sexual activity. In those who are acne-prone, there may be some externally applied or internally ingested substances that are acnegenic; that is, have a greater ability to induce acne. In adults, one such factor may be "acne cosmetica" to be discussed later.

What Causes Acne?

The disease is a disorder of the sebaceous glands associated with hair follicles—the pilosebaceous unit. The gland-containing follicles that are especially prone to acne lesions are most numerous on the face and the upper trunk, those that have exceptionally large sebaceous glands and tiny rudimentary hairs. It is the production of sebum from these glands that "fuels the acne flame," as described by Plewig and Kligman.

Sebum is normally synthesized by the sebaceous glands in response to circulating hormones, and the glands are especially sensitive to androgens. Androgen cannot be said to cause acne, however. Even with the advent of improved techniques for assay in blood or urine, no consistent hormonal excess, deficiency, or imbalance has ever been determined to exist between acne sufferers and those who do not have acne. A study of women with acne found that only 12% had blood levels of testosterone above the upper limit of normal (Lucky et al., 1983). Investigators have never found any chemical difference between the sebum produced by an individual with acne and the sebum of a person who does not have acne, or even between the sebum in one part or the body of an acne victim compared with the sebum is another part. Some people get acne; others do not, and the reason is essentially unknown. One generally accepted theory is that there is a gene-determined susceptibility and an end-organ response, the end-organ being the pilosebaceous unit. At puberty when the hormone levels rise, the end-organs respond. Why, at some point in life, the end-organs stop responding is not clear. The same follicles of the face, neck, and back that may erupt in acne during adolescence cease for no apparent reason to be influenced by the identical hormonal stimuli sometime during the third decade of life. In most people, acne then spontaneously disappears.

There are other perplexing factors associated with this common skin condition. Some acnes produce scars; other acnes, equally severe, cause little or no scarring. Although it is accepted that large sebaceous glands and excessive sebum production are related to acne, some people have very oily skins and little acne, whereas others have a lot of acne but produce little oil. Still others have a lot of oil and acne. Another mystery—when women get acne later in life well beyond adolescence, the acne usually is limited to the chin.

The causes of acne are still imperfectly understood; the mechanics of its occurrence are better known.

What Happens in Acne?

Acne is always associated with the pilosebaceous unit composed of a hair follicle and sebaceous gland. In the susceptible individual, the problem is believed to arise with a *hyperkeratinization*, or abnormal piling-up of cells of the follicle lining and the wall of the gland. The dead cells accumulate in the follicle ducts and prevent the sebum from exiting onto the skin surface, and the clogged duct begins to distend as the oils continue to pour into the canal. Several kinds of bacteria and fungi normally present in the follicle and on the surface of the skin find the sebum and excellent growth medium, and the follicle canal soon bulges with sebum, dead cells, and microorganisms. The result is a blocked pore occluded with excess sebum. It is called a whitehead or comedo (plural, comedones). A closed comedo may progress to become an open comedo, or blackhead. The accumulating material gradually causes the pore open-

ing to dilate allowing the impacted mass of horny cells to protrude from the orifice. Blackheads are not caused by neglect or dirt. The black in a blackhead is the result of the dark pigment melanin which has moved up to the surface with the cells.

A closed comedo may also rupture to form an inflamed pus-filled pimple, technically called a papulopustule, which actually represents a blowout of the follicle wall. As the oils from the gland continue to clog up the follicle lining, microorganisms called *Corynebacterium acnes* that inhabit the depths of the follicle reduce the sebum through enzymatic action to form fatty acids and glycerol. These free fatty acids not only stimulate further proliferation of the keratinized cells blocking the follicle, but also act as an irritant to the follicle walls. Eventually, the walls rupture, releasing toxic oils and bacteria into the surrounding tissues (Fig. 15.11).

Treatment of Acne

Acne can be mild, moderate, or severe and is usually divided into four major grades. Grade 1 acne is the kind with just whiteheads and blackheads and never leaves scars. Grade 2 acne has visible pimples, but if further inflammation does not occur, scarring rarely results. Grade 3 includes the large inflammatory lesions, and grade 4 is the most severe form of cystic acne that has been resistant to therapy until recently. Since in acne the follicle opening is plugged with sebum and cellular and bacterial debris, much of the therapy for mild and moderate acne is based on methods to induce the ducts to remain open and to stimulate blood flow to heal the lesions. The mainstay of such treatments are the topical applications of chemicals called "exfoliants." These chemicals irritate the top layers of the skin to cause an inflammatory response that results in reddening, then thickening, and finally the scaling off or peeling of the horny layer of the epidermis. While the skin looks dry and scaly, there is actually no measurable decrease in the amount of sebum production.

Various exfoliants are used in the treatment of acne, and some of them are available without prescription. Benzoyl peroxide in a 5% or 10% concentration is the strongest of the over-the-counter agents. It can be found under such trade names as Benoxyl, Loroxide, Oxy-5, Persadox, Panoxyl, Persagel, Besnagel, Topex, Dry and Clear, and Desquam-X. Salicylic acid is also a traditional exfoliant, but according to dermatologists, its concentration must be at least 5% for effectiveness. Most commercial preparations contain less than that amount and would not be expected to be of as much value.

A very potent irritating and peeling agent is retinoic acid (tretinoin) available by prescription only. The drug produces a strong inflammatory response, and reactions can be very intense, but very effective. In a kind of double-whammy assault on severe acne, some dermatologists prescribe both benzoyl peroxide and retinoic acid, one in the morning and the other in the evening. Not all skins can tolerate the increased irritation, however.

The Food and Drug Administration in its bulletin sent to all health professionals has cautioned against the use of retinoic acid and exposure to ultraviolet light. Studies on mice treated with topical applications of the drug and simulated sunlight showed that the combination produced multiple skin tumors that continued to increase in size and number well after the experimental period was concluded. The findings suggest that retinoic acid may in some way enhance the known cancer-causing effects of prolonged exposure to the sun. Although the findings are still preliminary, the American Academy of Dermatology was concerned enough to send a letter to its physician members urging them to advise patients to minimize exposure to the sun and sunlamps while using retinoic acid.

Some of the very familiar and widely advertised topical acne remedies contain sulfur and resorcinol, two agents traditionally prescribed for topical application by dermatologists for generations. Today, doctors generally agree that sulfur and resorcinol are virtually without value in the treatment of acne. Still, there are a lot of people who are attached to Acnomel or Clearasil, despite evidence of their ineffectiveness. As *Consumer Reports* noted in a survey of acne medications, there are many over-the-counter remedies that clear up mild acne although doctors cannot explain how they work. One investigation, however, graded the ability of various preparations to *induce* acne comedones on a scale of 1 (least comedogenic) to 5 (most comedogenic). Because the active antiacne ingredient was contained in an oily acnegenic base, Contrablem, Clearasil, and Liquimat with sulfur ranked a 3. Retin-A cream and Desquam-X were rated 4, and PanOxyl ranked 5 (Fulton and Black, 1983).

Medicated cosmetics are useless either in the treatment or in the prevention of acne. They have no exfoliant activity, and any antibacterial agents in the formulation serve only to preserve the cosmetic from deteriorating. Bacteria involved in pustule formation are deep in the follicle and not on the skin surface. Reaching them requires the use of a topically applied or orally ingested antibiotic.

When acne is severe, antibiotics, primarily tetracyclines or their relatives, are prescribed in addition to the effective exfoliants. Many dermatologists are enthusiastic advocates of antibiotic therapy, but while the drugs are undeniably helpful for some acne victims, they can be both a blessing and a disaster. Vaginitis, colitis and other gastrointestinal symptoms, and photosensitivity reactions are not uncommon. There have also been reports that high doses of tetracyclines may cause a kidney-induced form of diabetes insipidus, characterized by excessive thirst and excre-

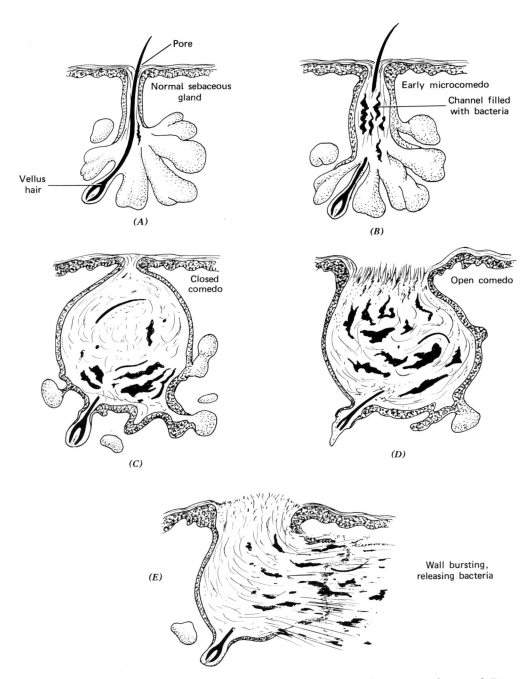

Figure 15.11. Life history of a pimple. (A) The type of pilosebaceous unit especially prone to acne; the hair follicle has a tiny vellus hair and large sebaceous glands. (B) Early development of the comedo; the follicle duct has become distended with

tion of copious quantities of dilute urine. Liver toxicity has also occasionally occurred. To avoid the systemic effects of ingested antibiotics, some doctors have tried antibiotic lotions containing tetracycline, erythromycin, clindamycin, or lincomycin. Topically applied clindamycin is reportedly better at suppressing acne bacteria in comedones than are tetracycline and erythromycin, which have the added disadvantage of producing yellow stains on the skin.

The breakthrough in treatment for the physical and psychological destruction of severe cystic acne came in September, 1982, when the Food and Drug Administration approved Accutane (Roche Laboratories), or isotretinoin, also known as 13-*cis*-retinoic acid, for the treatment of acne that is unresponsive to conventional therapy including systemic antibiotics. Accutane has been called a real miracle drug; it is not only highly effective in the treatment of severe acne, but it also appears to cure it. In most acne sufferers treated with the drug, prolonged remissions occur that continue years after discontinuance of drug therapy. Accutane has a profound effect on reducing the sebum production in the sebaceous glands, which tends to explain its initial effectiveness. But although the sebum levels tend to return to normal after the drug is stopped, the benefits continue—a result that is not clearly understood.

Almost all individuals treated with Accutane experience dryness, chapping, and itching of the skin. Forty percent get some form of conjunctivitis, and temporary contact lens

problems could result. Other symptoms during treatment, although they rapidly disappear after discontinuation, can include mild musculoskeletal complaints, skin rash or hair thinning, nonspecific urogenital or gastrointestinal symptoms, fatigue, headache, or an increased skin sensitization to sunburn. *Accutane is a known teratogen and should never be taken during pregnancy.* Moreover, it would be prudent to use a contraceptive for several months after cessation of treatment with Accutane before initiating a pregnancy. Since it is not known whether the drug is excreted in breast milk, it should not be taken by nursing mothers. Because isotretinoin is a vitamin A derivative, vitamin supplements containing additional vitamin A should be avoided during treatment.

In clinical studies of patients receiving Accutane, about 25% developed abnormal blood-lipid levels that included an elevation of triglycerides, a reduction in the high-density lipoprotein (HDL) levels, and a small increase in blood cholesterol levels. All of these changes went back to normal when the treatment was stopped, but individuals who have high blood lipid levels to begin with should be monitored during treatment. There has been one report of an increased blood calcium concentration developing as a result of Accutane therapy, but it also reversed after cessation of the drug (Valentic et al., 1983).

The effects of severe acne on the individual are so devastating that most are happy to accept the possibility of side effects and the current very high expense of treatment in order

---

keratinized cells, and a few bacteria-filled channels are evident. (C) Whitehead or a closed comedo: The pore opening is occluded with sebum; the dense keratin forms an impaction of concentric layers that dilates the follicle; there are more numerous and larger spaces containing bacteria and fungi. (D) Blackhead or open comedo: The impacted mass moves up to dilate the pore opening, and the comedo is filled with keratin, bacteria, fungi, hairs, and oils. (E) The blowout, or pus pimple: the epithelial lining of the follicle wall ruptures and releases bacteria and toxins to cause inflammation of the surrounding tissue.

to be cured. As more is learned about the mechanisms of action of isotretinoin and the optimal dosage schedules, it is probable that the drug can be used for less severe acne as well. Oral retinoids are believed to have great potential is a broad spectrum of skin diseases. They have been tested clinicaly in psoriasis and are currently being evaluated in the prevention and treatment of skin cancer.

The wide variety of treatments available for acne means that some combination of agents is bound to be effective in controlling the disorder in everyone. In most instances, however, quick improvement should not be expected. Generally, it takes about 6 weeks to 3 months before any marked effect of therapy is obvious.

Mild acne can probably be kept in check without consulting a physician, particularly if the nonprescription remedies are chosen wisely. Several studies have indicated that the daily use of a polyester web sponge called Buf-Puf (Riker Labs) for cleasing produced significant improvement in acne for well over half the individuals participating in the investigation. Most of the people in the study were also given other forms of therapy, such as exfoliants or antibiotics, but those with mild comedonal acne improved with use of the cleansing method alone. The light dermabrasion of the top layers of skin is valuable not only for acne patients but is said to improve skin texture, tone, and appearance in people whose skins are not troubled with blemishes (Durr and Orentreich, 1976; MacKenzie, 1977).

Grown-up Acne—Premenstrual Variety

There are estimates that one-third of all women experience premenstrual facial breakouts, and if very mild manifestations are included, some have placed the incidence as high as 60%. The general assumption is that in some unknown manner, progesterone is responsible for premenstrual acne, but there is no clear explanation for the increase in skin eruptions a week or so before menstruation. Some investigators claim that excessive sebum is produced in the postovulatory phase of the cycle, but Plewig and Kligman maintain that sebum production remains constant throughout. It has been observed by Williams and Gunliffe that during the luteal phase the follicle openings appeared to become smaller in diameter. This narrowing of the ducts, presumably the effect of progesterone, may aid to block the orifices and lays the groundwork for the acne flareups just before the menses. It is recognized that the amount and type of progestin in oral contraceptives are important in the effect of the pills on acne.

Opinions vary as to how or whether premenstrual blemishes can be prevented. The acne appears in women who have premenstrual weight gain and also in those who do not retain water. Some doctors prescribe salt restriction and diuretics, others believe that the elimination of water has no prophylactic effect. Hormonal therapy might help, but it seems more reasonable to treat the premenstrual acne in the same manner as any other acne—that is, by topical applications of effective exfoliants—rather than to initiate pill taking, which may not be effective for the acne anyway.

Acne Cosmetica

There is evidence that women in whom acne has persisted since adolescence, or women who have bouts with premenstrual acne may aggravate the condition by the use of certain cosmetics that contain ingredients that have the ability to induce the formation of comedones, that is, are comedogenic. The term "acne cosmetica" was coined in a 1972 paper by Kligman and Mills. The investigators described the occurrence of a mild but persistent form of acne in women that was associated with the long-term use of heavy and oily creams, lotions, and moisturizers. Acne cosmetica seems to be prevalent in women who are known to be acne-prone, but it also has afflicted women who went through their teen years without so much as a single pimple. The two workers tested samples of the leading brands of facial creams by the rabbit-ear assay method (the sensitive external ear canal of the rabbit easily responds to comedogenic chemicals after short periods of topical application) and determined that about half of the popular creams contained substances that would produce comedones in the rabbit ear canal. Fulton and his associates later (1976, 1983) expanded the tests to include a large range of foundation makeups, acne preparations, and some of the various surfactants, emulsifiers, preservatives, and miscellaneous substances used as ingredients in cosmetics.

Using a rating scale of 1 to 5 to describe substances that were least to most comedogenic, Fulton and co-workers considered that a response of 1 or 2 in the rabbit-ear assay was nonsignificant, but that any substance rated 3 or above was distinctly comedogenic. In the cosmetics tested, a rating of 3 was given to Revlon-Eterna 27, Ultima II Transparent Wrinkle Cream, Ultima II Under Make-up Moisture Lotion, Avon Ultra-Cover, Covergirl Super Sheer, Pond's Moisture Cream, Allercreme Petal Lotion and Velvet Finish, Merle Norman Tahitian Tan, Love's A Little Cover, Almay Tint-Natural, Lancome Maquimat, and Noxzema Skin Cream. Ratings of 4 were assigned to Elizabeth Arden Ardena Moisture Oil, Allercreme Matte Finish, Germaine Monteil Moisture Makeup, Max Factor Pure Magic Oilfree, Clinique Continuous Coverage, and also a product called, remarkably, Natural Wonder Aint-Acne Spot Cover. The only 5 ratings for the makeups tested were earned by Ultima II Transparent Tawny Tint, Allercreme Satin Finish, and Cover Girl Oil Control Liquid Makeup.

It was determined that cosmetic ingredients with a high potential for causing acne were fatty acids, fatty acid esters, and fatty acid alcohols, such as isopropyl myristate, isopropyl isostearate, sodium lauryl sulfate, butyl stearate, and hexadecyl alcohol. Substances found to have negligible or no comedogenic ability were candelilla wax, titanium dioxide, glycerin, caster oil, iron oxide pigments, propyl gallate, methyl paraben, propylene glycol, and mineral oil. Mineral oils, in fact, would probably cause the least problems compared with vegetable oils, especially olive oil and cocoa butter in high concentrations.

It should be emphasized that the rabbit-ear assay method is probably ultrasensitive compared with the human face. The tested substances were applied to the rabbit daily and not washed off with soap and water. Even the most powerful acnegens in the rabbit are likely to be only weakly acnegenic in the human. Women who are not acne-prone could get acne cosmetica only by providing what Fulton has called 24-hour acnegenic coverage—starting the day with a comedogenic foundation, removing it at night with a comedogenic cleansing cream, and then faithfully applying a comedogenic night moisturizer before going to bed. Following this kind of regimen for a couple of months would invite acne-form eruptions, and it makes little difference how expensive the cosmetics are. This program of cosmetic use, it might be noted, is frequently advised for dry-skin care and forms the basis of various beauty plans heavily promoted by cosmetics companies, particularly for older women.

Women who have persistent postadolescent acne, who have oily skins, or who have a tendency to premenstrual flareups, would probably respond more easily to comedogenic cosmetics. Since the more lipid or lipidlike the ingredients in the product, the more potent is its comedogenicity, such women should avoid all oily heavy cosmetics.

It would probably be better for a women with acne to avoid *all* cosmetics, but this advice, while logical, is unrealistic. A woman with a flawless complexion can more easily go without cosmetics; one with a troubled skin may have a greater need for coverup. If wearing makeup is seen as a necessity, the only answer is to try to wend one's way through the morass of "oil-free," "oil-controlled," "pore-minimizer," "oily-skin," and "water-base" foundations in order to find those that are free of offending oils and also free of comedogenic potential. Unfortunately, cosmetics are usually the exception to "what you see is what you get." A product may actually contain a lot of oil but still claim to be for oily skins or be water-based. Fulton has recommended a simple test to determine whether a makeup is actually free from oil. If a few drops of the foundation are placed on 100% cotton bond paper and left for 24 hours, an oily halo will form around any makeup containing oil. He warns, however, that the test is not foolproof; some oils may evaporate so quickly that the halo will not be apparent.

In an article on acne and cosmetics which appeared in *The Milwaukee Journal*, reporter Paula Brookmire used Fulton's test to check a number of cosmetics and determined that the following ones were both oil-free and noncomedogenic according to Fulton assay: Almay Foundation Lotion for Oily Skin, Clinique Pore-Minimizer Makeup, Clinique Gel Rouge and Bronze Gel Makeup, Corn Silk Pressed Micron Powder, Love Clean and Natural Oil-Free Gel Makeup and Gel Blusher, Maybelline Fresh and Lovely Fingertip Powder Blush, and Revlon Natural Wonder Oil-Free Pressed Powder. Note that one line of products, for example Natural Wonder, may contain cosmetics that can cause acne as well as be safe for acne. As yet there is no company that has developed an entire group of noncomedogenic cosmetics. For women with acne-prone skins, the requirements for selecting makeup are:

read the labels to check the ingredients and be certain to stay away from cosmetics containing isopropyl myristate or isopropyl isosterate if these substances are near the top of the list.

The industry has apparently not found the production of oil-free, noncomedogenic cosmetics to be that lucrative. There are fewer such cosmetics to choose from and the range of colors is much less. Women accustomed to the spreadability of other foundation creams will probably find oil-free makeup more difficult to apply because it is quick-drying. Obviously, no moisturizer, even one presumably formulated for oil skin, should be used under the oil-free foundation.

FOLLICLE FACTS AND FALLACIES

Except in certain areas, the hair on human bodies has little functional value. Around the body openings—the nose, ears, anogenital areas—the hair is protective to a certain extent, mostly acting as a filter against foreign particles, and the hair on the brows and lashes help to safeguard the eyes from sweat and specks that may obscure or interfere with vision. It is generally accepted, however, that the only reason hair on the face, the head, and under the arms has survived in the human race is probably for decorative purposes. Scalp hair in particular is a body ornament. There is certainly enough evidence from anthropology, archeology, history, folklore, and the works of poets, painters, and sculptors to attest to the fact that humans have always had a great emotional investment in the appearance of their hair.

It has been suggested that today we appear to have a greater interest than ever before in the keratin growing out of our scalp follicles. Some have even said that the preoccupation approaches that of an obsession with hair.

Whether contemporary men and women have more concern with their hair than did their ancestors is uncertain, but it is undeniable that this part of the human body does have important cosmetic significance in our culture. Perhaps our fascination with hair is inordinate, but there has never before been such an opportunity to indulge our interest. Until recently, people had not been exposed to mass media advertising that promotes the sale of an astonishing number of hair care products—more than 650 formulations of shampoos alone. When we are confronted by such a baffling array of things to do to and for our hair, it only reinforces our uneasy presentiment that something needs to be done. There are options for everyone—for those whose hair is too straight, too curly, too fine, too coarse, too oily, too dry, too dark, too light. As Clairol, the largest manufacturer of hair products, advertised: "Why not change the things you can?" (Fig. 15.12).

We certainly try. Many of us spend a lot of money and effort looking for the product or procedure that will result in the lustrous and manageable hair seen in the ads. We end up poorer and frustrated because few products appear to live up to their promises. Much expense and time can be saved by knowing about the physical and chemical structure of hair and the composition and action of hair-care products. We may find ourselves in less of a love–hate relationship with our hair when some of the myths concerning its growth, color, texture, and care are debunked. It should be remembered, however, that there are variations in hair characteristics between one person and another. There are even differences between the hair on different parts of the head in the same person. Therefore, the method or substance that "works" on one head will not necessarily produce the same effect on someone else's hair.

Figure 15.12. The market is flooded by a bewildering assortment of products for hair care.

Hair Growth

Many people are convinced that cutting hair makes it grow faster, perhaps because, in the opinion of one beautician, "it takes the weight off the ends." Hair growth, however, is cyclic. It does not grow continuously, but only during certain definite periods. During the growing phase of the cycle, the cells at the roots of the hair follicle proliferate rapidly and the hair grows longer. When a hair reaches its limit in length—characteristically short for the body hairs, longer for the scalp in both sexes and beard hairs in men—it stops growing and rests for a while, being retained in its follicle. During the resting phase, the base of the hair becomes a solid completely keratinized club-shaped mass, and it sends out little rootlets that fasten it to the remains of the follicle. After a resting period, a new bulb forms at the base of the follicle below the old clubbed hair. The new hair forms and pushes upward, loosening the old hair from its attachment and causing it to be shoved out and shed. Periods of growth and periods of rest for hair follicles are about equal, except for the hair on the face and the scalp. Each follicle, however, has its own individual rhythmic cycle. In any given area on the body, some of the hairs are actively growing, some are resting, and some are being shed. Unlike some other mammals, humans do not molt seasonally to replace the summer coat with a winter growth, but hairdressers have pointed out that people appear to lose more hair in the autumn of the year. Apparently about the time the leaves on the trees are being shed, so are the hairs on the head, but there are no quantitative published studies to substantiate the observation.

Obviously, since hair grows from the bottom up, cutting or shaving it cannot affect its rate of growth or its texture. Since all hairs on the scalp are growing at different times, however, cutting the ends periodically will even off the longer hairs and the stragglers and will probably look better. It takes about 7 weeks to grow an inch of hair on the head. Hair on the legs, which grows about 0.05 inches weekly, does not become coarser or grow faster after shaving. It does, however, become more stubbly because the soft and tapered ends of the hairs have been bluntly cut off. If those hairs were permitted to be shed and replaced, their successors would again feel soft. Since shaving, once started, is more or less continuous, hair on the legs always feels prickly.

The life span of an eyelash is about 4–5 months, and only 30 days of that time are spent in active growing. The lash is in the resting phase for the remainder of the period before being shed. Hair on the head has a much longer period of growth (anywhere from 2 to 6 years), rests for only 3 months or so, and is then shed. If uninterrupted by cutting, scalp hair can attain very long lengths in some people, a feature that is genetically determined. Of the 100,000 hair follicles on the average scalp, 90% are actively growing and 10% are resting at any given time. The normal 50–100 hairs that are shed daily to end up on the comb, the brush, or in the shower drain have, therefore, been on the scalp an average of 4 years. If a hair is pulled out during its growing period, the root will be tapering and cylindrical. If a resting hair is plucked, it will have a rounded, club-shaped root (Fig. 15.13).

A certain amount of thinning of the hair on the head in both sexes is a normal consequence of aging because there is a progressive increase in the number of follicles that are in the resting phase as one grows older. While hair evidently grows faster between the ages of 16 and 46, there is an accompanying gradual decrease in the density of the hair, most noticeably on the forehead. Actual baldness, however, is about six times more prevalent in men than in women and is a genetic trait that is sex-influenced; that is, the genetic factors become evident only when male hormone is

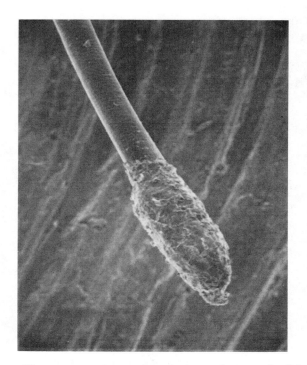 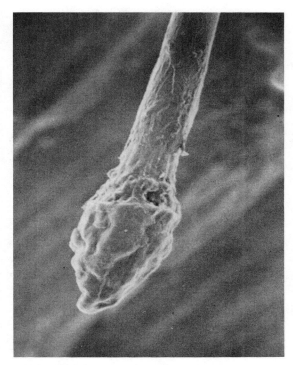

Figure 15.13. A scanning electron micrograph of the roots of two hairs. The hair on the left was pulled out during the growth phase. Note the narrow tapering end. In contrast, the clubbed root on the right is from a hair plucked during the resting stage.

present in the body. Hair loss that is mournfully excessive at an early age, beginning in some men even before their twenties, is rarely caused by any medical problem and is more likely the result of their having the gene for male-pattern baldness. Whether it happens early or late in life, men who become bald will probably find little consolation in the fact that androgens, responsible for hair growth, male secondary sex characteristics, libido, and potency are also responsible for hair loss when the genetic predisposition is present. Eunuchs (male castrates) rarely become bald. Why the same hormones that cause activity of follicles should later result in their inactivity is unknown.

The result of the lowered estrogen levels after menopause may result in unwelcome changes in hair distribution in some women—more on the face and less on the head. Facial hair may be made less obvious by bleaching or removed by tweezing or electrolysis, but little can be done for thinning of the hair on the head except to disguise the thinness with styling. Women may have increased hair loss with age, but they do not become bald.

Hair Structure

The hair shaft consists of the cuticle, the cortex, and the medulla and is almost completely composed of the protein, keratin. Between 90% and 99% of the dry weight of hair is keratin, making hairs the most durable parts of the body. Hair is tough; it survived on Egyptian mummies for thousands of years. Durable as it is, however, hair can be damaged by weathering and by various cosmetic manipulations. For 2–6 years, a hair on the head is washed, combed, brushed, exposed to ultraviolet light, extreme temperatures, and may undergo various chemical treatments from dipping in a chlorinated pool to processing with dyes and permanent wave solutions. Small wonder it may not always look good. Obviously, the gentler one treats the hair, the better it is likely to appear.

The cuticle, for example, is made up of a very thin and heavily keratinized single layer of cells. Cuticle cells look like overlapping shingles on a roof, except that their free edges point upward. Covered by a layer of lipid from the sebaceous glands, the cells move over each other to function as a flexible protective armor-plate for the hair shaft. Cuticle cells that lie flat against the cortex reflect light evenly off the surface oil film and are responsible for the quality of hair called shininess. Disrupting the cuticle to make it swell and uplift in order to tint or wave the hair, or treating the hair in a less than gentle manner—backcombing it, blowdrying it, sunning it—will result in an uneven diffraction of light. The hair no longer appears shiny and looks dull. Various conditioners can glue the cuticle scales back down and coat them with oils so that the hair again appears glossy (see Fig. 15.14).

Split ends are merely a separation of the cell layers at the ends of the hair shaft. The longer the hair, the longer it has been there to be exposed to the elements, the daily combing and brushing, and perhaps the nontender treatments used to improve its color or curl. It is inevitable that even without any chemical processing, long hair will begin to separate at the ends. The cure for split ends is to cut them off. There are conditioning products that can temporarily hold the layers together, but split ends will always reappear unless they are removed (see Fig. 15.15).

The cortex is the main bulk of the hair shaft, and it consists of long and intertwined molecules of keratin. Most studies of keratin have been made on wool fibers (because of their economic value) and feathers, and the exact structure of human hair keratin is not known. It is believed to be a keratin complex composed of a sulfur-rich nonfibrous protein matrix that lies between fibrous protein chains low in sulfur content. The fibrous parts of the keratin complex are made up of long protein molecules aligned in a characteristic alpha helix (spiral staircase) configuration. The molecular arrangement, which could be compared to lengths of rope tied and wound around each other, is generally accepted to be as follows: three alpha helices are wound around each other to form a protofibril; eleven protofibrils are twisted and coiled to make up a microfibril; and the microfibrils are aligned together into the macrofibrils within the matrix of the cortical cells. A hair may be visualized as a number of supercoils imposed on primary coils—a kind of overly twisted rope (see Fig. 15.16).

Several kinds of chemical bonds hold the keratin complex together. The most common linkage is the *hydrogen bond*, important in stabilizing and forming the coiled chain. Some of the hydrogen bonds are easily broken by wetting the hair, and this forms the basis for setting hair while it is wet. If the wet hair is stretched—over rollers, for example—the bonds reform into a new and temporary curl position when the hair subsequently dries. Hair sprays are aqueous solutions of shellac or the resin polyvinylpyrrolidone (PVP), which stiffens the hair to hold the reshape in shape. Heat will also break hydrogen bonds so that they may be realigned. Electric rollers and curling irons can set in a curl, but the result is generally not as lasting as one formed

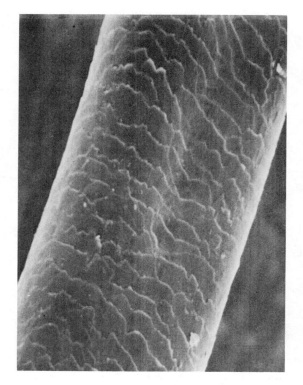 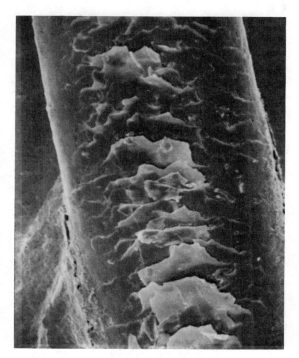

Figure 15.14. As seen with the scanning electron microscope, the cuticle cells of the unprocessed hair on the left are lying flat against the cortex. This hair will look shinier than the hair on the right, which has been both bleached and permed. Note the disrupted and uplifted cuticle cells.

through wetting and setting. Blowdrying works on the same principle. The hair style is directed into place by stretching the hair over a brush while drying it with hot air. But since wet hair in the presence of heat is highly susceptible to mechanical damage, the brush or comb that stretches the hair during a blowdrying should be used slowly and gently.

Permanent Waves

The curl formed by wetting and setting is only temporary, and lasts until it "falls," or the hair is dampened. A permanent realignment of the protein chains to form a permanent wave can be accomplished by breaking the *disulfide bond.* This sulfur to sulfur link is responsible for the strength, great stability, and cohesiveness of keratin. Water cannot disrupt the disulfide bond, and strongly alkaline thioglycolate compounds, substances with a characteristic strong and unpleasant odor, are used to rupture the bond. Thioglycolates are the essential ingredients in permanent-wave solutions.

When getting a perm, the hair is washed and set while wet on the perm rollers—the smaller the roller, the tighter the curl. Then the rolled hair is soaked with the waving lotion that swells and raises the cuticle so that the keratin of the cortex can be made softer

Figure 15.15. Split ends. This is the same overly processed hair shown on the right in Figure 15.14.

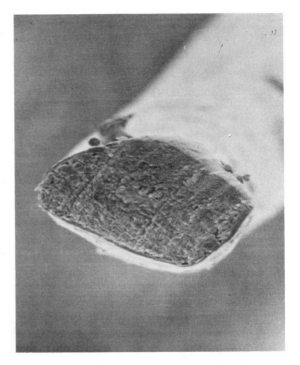

Figure 15.16. The molecular structure of the cortex is almost suggested by this scanning electron micrograph of the cut end of a hair. The keratin complex is made up of a number of macrofibrils or twisted coils set in a matrix of high sulfur proteins that links them together.

and more malleable, and its chemical bonds can be broken. After a certain period, the alkaline hydrolysis is stopped, and the bonds are reformed in the new position with the "neutralizer," usually an acidifying agent combined with an oxidizing agent, such as sodium or potassium bromate or persulfate, or most commonly, hydrogen peroxide. The neutralizer counteracts the swelling and restores the disulfide links.

Permanents applied correctly and with the right timing do not necessarily cause hair damage—at least no more damage than other cosmetic manipulations. Moreover, more expensive permanents do not guarantee better results. The solutions in the $25 or the $75 permanent may be very similar. What one

may be paying for in the more costly process is the skill and experience of the operator. Most of the problems with frizzy hair, split ends, dryness, and breakage occur as a result of an inexperienced hairdresser. Too much wave lotion left on too long or inadequately neutralized can cause a disastrous effect that is irreversible, and cutting off the perm after it grows out is the only recourse. Home permanents are usually weaker solutions than the ones used in salons, but overprocessing is also possible at home.

Failure of permanents to "take" may be caused by a hair texture that is more resistent to waving and that may require more skill in application, another reason to select an experienced hairdresser. Failure of the wave is not caused by pregnancy, menstruation, and is very unlikely to be associated with the taking of any medication. One woman said her hairdresser maintained that her perm did not take because she lacked a medulla in her blond hair, which is nonsense. The medulla, frequently absent from hair, is of little importance to the hair when it is present.

A permanent, applied correctly, will generally look best on "virgin" hair. Unless the hair has been permitted to grow out completely and the old permanent has been cut off, there will be unavoidable structural modifications to hair after the repeated chemical processing of permanents.

Hair Color

The pigment in hair is provided by the activity of melanocytes in the germinal matrix, the cells that proliferate to give rise to the hair. The pigment is deposited mainly in the cortex of the hair shaft. The color of the hair and its depth depend on the size, number, and kind of pigment granules, or melanosomes, and on the presence or absence of air bubbles in the cortex. The black-brown pigment melanin is the predominant color in deep black, brown-black, ash brown, ash blond, and platinum blond hair. Golden blond, red, or auburn hair contains greater quantities of the yellow-brown pigment, phaeomelanin. White hair is caused by a progressive reduction in the activity of the melanocytes and the eventual cessation of pigment production. White hairs mixed in with partially pigmented and fully colored hairs results in the appearance of graying. Actual gray hairs are very rare.

The eventual loss of pigment in hair is determined genetically, occurs in some people quite early in life, and is more obvious in brunettes. Graying is irreversible, and any reports of repigmentation as a result of taking various vitamins or eating a particular diet have not

been substantiated, although malnutrition, protein deprivation, or illness can cause changes in the hair that include temporary loss of pigmentation. Stress has not been proved to accelerate graying, and neither can a trauma turn hair white overnight. The only possibility that could provide a basis for all the eyewitness reports to such a happening is that a stressful experience might result in the quick loss of a number of thinner normally pigmented hairs that were in the resting stage. The thicker coarser unpigmented hairs that were present all the time would then suddenly become exposed, giving the appearance of a rapid color change. Then when the new hairs grow in, they too would be unpigmented, providing more "evidence" of the overnight phenomenon.

Coloring the Hair

Reportedly, nearly 60% of American women color their hair, but a substantial number of men are also displeased with their natural color or do not think that gray is that beautiful.

People who want to change the color of their hair can do it several different ways. A temporary rinse contains a water-soluble color that merely coats the outside of the hair and does not actually dye it or change the hair structure. The ability of the product to color the hair is, therefore, limited. It can provide highlights, lessen the appearance of graying, or intensify and darken the natural shade, but the effect is removed with the next shampoo, although the color may adhere more strongly when the hair has previously been bleached or has been otherwise processed. The coating is water-resistant, but it can be affected by rain or even perspiration, and sometimes it will rub off on the pillow or clothing. These disadvantages make the rinses hardly worth the time or money for most people.

A semipermanent tint contains dyes or stains that barely penetrate the hair shaft, but which adhere strongly enough to the cuticle to provide a color-coating that lasts through four to six shampoos. Such a product may come in many shades, but there is no possibility of lightening the hair by using it. It can only diminish graying and produce a color several tones darker than the original. Loving Care is the most widely sold product of this type. It alters hair structure somewhat, but only minimally.

Permanent dyes are coloring agents that provide a deeper and more lasting color effect. Three kinds of dyes are used: an oxidation procedure that utilizes hydrogen peroxide; vegetable hair colorings; or metallic salts. The most commonly used type that can lighten hair to a different shade and that remains until new growth appears is the oxidation dye.

The oxidation-dye procedure may be a single process in which two components, an alkaline dye and a peroxide system, are mixed, applied to the hair for 30 minutes, and then rinsed off. Usually referred to as a "tint" by the hairdresser, it remains until the "roots," actually the new growth, appear and a "touch-up" must be performed. In the touch-up, the previously dyed strands are exposed for a lesser time than in the original procedure. Obviously, the older portions of the hair shaft can become progressively altered. Single process dyeing can produce an extremely dark shade or a color several shades lighter than the natural one.

The double-process method permits any possible color desired—even going to platinum blond hair from black hair—but is hardest on the hair structure. The hair is drastically bleached or "stripped" to remove most of the color in the cortex by a strong persulfate powder and peroxide mixture. After stripping, the hair is redyed with a "toner" to the desired shade. New growth has to be stripped and toned at intervals to maintain the color. Frosting or streaking is double-processing of either random tufts of hair all over the head or of selected clusters of hair strands. The hair to be processed is pulled through a special cap with holes and wrapped with foil during the bleaching. This is a time-consuming and expensive procedure but does expose less of the total hair on the head to the chemical manipulations. It also needs to be repeated far less frequently.

Hair Dyes and Cancer

The oxidation or aniline dyes, also inelegantly known as coal-tar dyes, are the only permanent hair coloring agents that penetrate through the cuticle to the hair cortex. The molecules of the other nonpermanent dyes are too large to move through the cuticle barrier, and they remain on the hair surface. Oxidation dyes are able to enter the hair shaft only because they contain amine bases, such as paraphenylenediamine (PPD) and para-aminophenol and their derivatives. These components of the dyes are small molecules crucial to penetration of the hair cuticle. Although they are not in themselves dyes, they are dye intermediates. Once introduced into the shaft they undergo a coupling reaction with an oxidizing agent, hydrogen peroxide. This transforms the amine bases into large dye molecules too big to get washed out of the cortex, and the color is therefore locked in. At the end of the dyeing process, the excess chemicals are rinsed off when the hair is shampooed.

For a long time, paraphenylenediamine has been recognized as a strong skin sensitizer able to cause severe allergic reactions or even blindness should it accidentally get into the eyes. This substance or the other dye intermediates that are used are, however, essential to the dyeing process, and there are no substitutes. Although Congress had banned the use of coal-tar dyes in the 1938 Food, Drug, and Cosmetic Act, coal-tar *hair* dyes were exempted from the ban provided that their manufacturers include a warning statement on the bottle labels. All permanent hair-dye labels tell the consumer that adverse reactions are possible and warn that a "patch test" should be made on the skin for 24 hours before the product is used to dye the hair. Whether many people actually perform such a test at home after they buy the dye at the drugstore or supermarket is speculative. Certainly, the precaution is infrequently observed at the hairdresser's.

Allergic responses to the dyes are only part of the problem. Reports that indicated their carcinogenicity have also appeared in the medical literature since the early 1950s. It was only after the results of several animal studies were announced in 1977, however, that the Food and Drug Administration began to consider taking action against these long-used products that had been immune from federal regulation for 40 years.

In the studies, conducted under the sponsorship of the National Cancer Institute, rats and mice were fed doses of 4-methoxy-*m*-phenylenediamine, or 4 MMPD, also called 2,4 diaminoanisole or 2,4 DAA, and its sulfate, 4-methoxy-*m*-phenylenediamine sulfate, or 4MMPD-sulfate. These chemicals, coal-tar dye ingredients more likely to be found in the cool or drab tints—black, brown, or ash blond tones—had been suspected of being carcinogenic. According to the results of the rodent-feeding study, 4MMPD and 4MMPD-sulfate caused a significantly increased incidence of several kinds of cancer.

Because of these findings, the FDA, prohibited by present law from outright banning the chemicals as it did with cyclamates and attempted to do with saccharin, proposed that a warning label for the dyes containing the suspect ingredients be placed on the bottles, and also proposed that a large poster be displayed in each beauty shop that would urge customers to check the bottle labels for the ingredients. The mandated label was to state:

"Warning: Contains an ingredient that can penetrate your skin and has been determined to cause cancer in laboratory animals."

The cosmetics industry reacted to these proposals with predictable resistance, and distributed a pamphlet to all beauty salons ridiculing the animal studies. The circular, an attempt to mollify the public's concern about the safety of the hair dyes, pointed out that humans put dye on their hair and do not ingest it, and that the doses that caused cancer in rodents were the equivalent "to a person drinking more than 25 bottles of hair dye every day for a lifetime." But while people are not drinking hair dyes, evidently the chemicals are able to be absorbed through the skin of the scalp in varying amounts. It has been calculated that about 4–6 mg of dye are absorbed into the body during the coloring process (Kiese and Rauscher, 1968). While some have questioned the validity of extrapolating animal results to humans, such animal bioassays are one of the scientifically established ways of screening chemicals for carcinogenicity. Neither is the effect of the substance presumed to be related to the high dose. It is not true that virtually any substance will be carcinogenic if the exposure to it is high enough. If a substance is noncarcinogenic, it does not cause cancer at the maximum dose tolerated by the animal. Large numbers of industrial chemicals, pesticides, and food additives have been fed to animals in large doses and have not been implicated in cancer formation.

Screening studies proceed on the assumption that if the maximum tolerated dose causes cancerous tumors in a small number of test animals, a very small amount will cause some cancer in large numbers of animals or people. There is not complete agreement, however, that substances that cause cancer in animals will also cause cancer in humans, and considerable criticism has been directed at the hair-dye studies.

Also at the heart of the hair-dye controversy are the conflicting results from other cancer-risk studies. Cancers that have been reported as occurring in test animals when paraphenylenediamine was injected under the skin were not found to occur when the chemical was topically applied to the skin (Stenback et al., 1977). A report of an increased incidence of bladder tumors in male hairdressers (Anthony and Thomas, 1970) has never been confirmed in subsequent studies. As yet there is no statistically significant evidence of greater incidence of cancer among women beauticians compared with other women. It should be noted, however, that hairdressers usually wear gloves when applying hair dye, and thus are possibly less exposed to the chemicals than their customers.

One 1979 report suggested a relationship between hair dye use and breast cancer. Roy Shore and co-workers collected information concerning hair dye use as well as a number of risk factors for breast cancer from 129 breast cancer patients and 193 control women. Their data showed a relatively small association between breast cancer and the use of hair dyes or rinses among women over 50 years of age who had used the hair dyes 10 or more years before breast cancer diagnosis. The link appeared to be somewhat greater in women who were naturally in a lower risk category for breast cancer. The researchers themselves, however, pointed to methodological weaknesses in the study and indicated the necessity for further validation.

When more than 30 million women in the United States dye their hair, a possible relationship between the dye and breast cancer could have enormous health consequences, but reassuringly, there is no statistically significant evidence from any prior or subsequent investigation that confirmed the association (Kinlen et al., 1977; Hennekens et al., 1979; Starrady et al., 1979; Nasca et al., 1980; Najem et al., 1982). In a large study of 401

women with breast cancer and 625 matched controls, Wynder and Goodman (1983) found no link between hair dye use and breast cancer, or with any other factors of use such as type of product (permanent or semipermanent), duration of use, shade, or frequency of application.

The imposition of mandatory warning labels on hair-dye bottles and posters in the hairdressing salons is still pending, however, after Congress voted a moratorium on the FDA proposal in mid-1978. Moreover, despite the adverse publicity and the controversy concerning the safety of hair dyes, sales of the products did not drop after the National Cancer Institute report. People bought just as much hair dye, reported by dollar volume of sales, as they did before. Both Clairol, the top manufacturer of hair coloring products, and Loreal, the close second, prudently eliminated the suspected 4MMPD and 4MMPD-sulfate from their dye formulations in anticipation of the mandatory warning labels. Since there are more than 30 such dye intermediates and their derivatives used in oxidation hair dyes, the substitution of similar chemicals for the questionable ingredients may not have lessened the risk. It cannot be assumed, in the absence of more evidence, that the reformulated hair dyes are completely harmless. The mutagenicity of at least one of the substitutions (EMPD) has been determined (Prival et al., 1980). EMPD is used by Revlon and Jeffrey-Martin, Inc. in their hair dyes. The FDA is presently investigating other components of the products for their potential carcinogenic hazard.

Other Permanent Dyes

Metallic dyes for coloring the hair have been used for centuries. The name of one commercial preparation, Grecian Formula, implies there is a time-honored quality in its ingredients. The dyes, salts of various metals, such as lead, silver, and copper, are advertised for use on gray hair as "color restorers," as if they had some magical power that could make hair without pigment resume its natural pigmented shade. When a small amount of formula containing lead acetate, for example, is combed through the hair daily, the salt reacts with the sulfur of hair keratin to form lead sulfide, which coats the cuticle of the hair. Gradual color change from gray to yellow to brown to black are obtained with time. The result is a flat-looking unnatural color that becomes worse with time, is incompatible with oxidation dyes or with permanent waves or straightening, and is very difficult to remove. Moreover, lead colors are toxic. While they may be safe on intact skin, their use is highly questionable if any abrasions exist on the scalp.

Another coloring agent that has been known for centuries and enjoyed a new resurgence as a "natural" product is henna, an extract or powder made from the dried leaves and twigs of several varieties of a Middle Eastern bush, *Lawsonia alba*, *Lawsonia inermis*, or *Lawsonia spinosa*. The chemical substance responsible for the dyeing ability of henna is 2-hydroxyl-1, 4-naphtho-quinone, which is also prepared synthetically and is called lawsone.

Henna has been widely promoted as a completely nontoxic, nonallergenic "organic" dye that produces body, shine, highlights, and completely natural colors, as opposed to those artificial chemical dyes with all their additives. But unless the hair is black, dark brown, or no lighter than medium brown to begin with, the use of 100% henna powder can produce downright startling, highly unnatural, results. Even if one finds the shade satisfying, continued use of henna appears to alter the hair keratin, and the hair becomes stiff, dry, and brittle.

Henna is not pleasant to apply. The powder is mixed to a paste with hot or boiling water, worked into the hair, and left on for 45 minutes to $1\frac{1}{2}$ hours when it is shampooed off.

The red dye is not believed to penetrate to the hair cortex, but it does have a remarkable affinity for keratin, adhering tenaciously not only to the hair, but also to the scalp, ears, and hands. Although henna is often unpredictable, red highlights will probably be the result on naturally dark hair that has not been chemically processed. If the hair is light, more than 15%–20% gray, or has been bleached or permed, application of henna is apt to be a disaster. Henna horror stories abound—flaming red hair, tricolor hair that resulted from different tinting where the hair was more porous, and white hair that has turned orange.

In the hairdressing salons, henna is available in several colors as well as "neutral," meant only to shine and condition the hair. In one product, henna has been combined with refined proteins, pH neutralizers, conditioners," all 100% organic, of course. But whether or not neutral no-color henna conditioning actually produces the fantastic shine, body, and polish to the hair that is claimed is a matter of opinion. Henna products do provide a coating to apparently thicken the hair, but that thickening might be subjectively interpreted as coarseness rather than body, depending on the individual. There has also been some nonscientific observations that using neutral henna at home causes no shine and no discernible difference in the hair, but when the treatment takes place at the salon, the results are more satisfactory. This could be the halo effect; the results are perceived as better because it cost more to get the treatment, and it was done by a professional.

Another kind of vegetable dye is found in chamomile flowers, and a yellow brightening effect can be obtained from repeated use of chamomile packs.

Shampoos

Shampoos are meant to clean the hair, remove dirt, sebum, cosmetics, and dead scalp cells that have accumulated on the hair shafts since the last time the hair was washed. Cleaning the hair is not the only function of a shampoo, however, because there are a number of other attributes such as leaving the hair shiny, fragrant, healthy-looking, and manageable, among others, that most people want and expect from the cleansing agent. Moreover, since everyone's hair is different—curly, straight, long, short, oily, dry, dyed, bleached, or permed, some qualities produced by using a particular shampoo may be more desirable to one person than to another. My "soft and manageable" may be your "limp and lank," or what is fluffy body for my hair could be dry and wiry for yours. The only way for a consumer to differentiate among all the claims and options is to try to ignore them and to forget the promises and the gimmicks. After realistically assessing the condition of one's own hair, the best way to pick a shampoo from the dazzling number available is to start with a cheap one and keep trying until a preferred product is found.

Only a few shampoo formulas still contain soap, which performs well in softened water but leaves a dulling deposit in hard water. Ogilvie's Castile Soap is a soap shampoo, and Breck's Gold Label is part soap. Most of today's shampoos contain synthetic detergents or *surfactants*, substances that have the ability to enhance the washing ability of water by making the water "wetter," by emulsifying the dirt particles and by surrounding the dirt and oil with an electrical charge so that they may be easily rinsed off rather than being returned to the hair or the scalp. The surfactant may be either *anionic* or negatively charged, *cationic* or positively charged, *amphoteric* or both negatively and positively charged, or *nonionic*, possessing no electrical charge. Johnson's Baby Shampoo, one of the largest sellers in the country, contains an amphoteric surfactant that makes it milder—it does not sting the eyes. Its mildness makes it a less effective cleanser, however, so adults would have to use more of it more often. Most adult shampoos contain primarily anionic surfactants.

A typical liquid shampoo is likely to contain lots of water—about 60%–80%—and around 10%–25% surfactant (usually sodium lauryl sulfate), 5%–15% foam builders, and 0.1%–2% sequestering agents, substances that keep the other ingredients from precipitating out to cloud the shampoo. Other additives are varying amounts of thickening agents to give the shampoo a proper consistency so it will not run down into the eyes, and conditioning and finishing agents to control the hair and put the luster back on it. Surfactants are degreasing substances that tend to remove the luster. Small amounts of preservatives and sometimes a substantial amount of perfume (herbal, lemon, strawberry, new-mown hay, fruit salad, whatever) to add to the shampoo's appeal or to cover up the odor of the other ingredients complete the formulation. *Consumer Reports* noted that the manufacturers attempt to build product differentiation into shampoos in a number of different ways, but few of them appear to be relevant to the actual quality of the product. All shampoos are essentially the same.

Some of the antidandruff shampoos contain various antiseptics, which are hardly necessary since it has never been shown that dandruff is an infection or associated with microorganisms. Simple dandruff, the shedding of the top cornified layers of the scalp skin, is experienced by everyone to some degree and appears several days to a week after shampooing. Any shampoo will control dandruff if the hair is washed frequently enough. The medicated shampoos may delay the return of dandruff symptoms for a longer period of time, however, and for this reason may be preferred by people who do not want to wash their hair more often than once a week. Seborrheic dermatitis is a skin condition characterized by inflammation, severe itching, and flaking, and is the only kind of dandruff that could actually require the use of a special shampoo. Some skin diseases, such as psoriasis, produce dandruff-like symptoms but will not respond to antidandruff shampoos.

The shampoos formulated to control dandruff may contain potentially toxic compounds such as zinc pyridinethione (ZnPT) or selenium sulfide. Both are considered to be safe for use in shampoos. The ZnPT is not readily absorbed through the skin, whereas the selenium, which does have considerable absorption potential, presumably has only intermittent and limited contact with the scalp.

Apparently the only difference between a shampoo for dry hair and a shampoo for oily hair is in the amount of surfactant—there is less of it in the dry hair formulation. In their survey, *Consumer Reports* found that many people who characterized themselves as having dry hair preferred a shampoo for oily hair, and that an equal number of those who said they had oily hair preferred the formulations for dry hair. They concluded that the "dry" and "oily" designations were just another advertising gimmick.

"Low pH," "acid-balance," and "nonalkaline" are also qualities with little meaning or advantage other than the enhancement of product appeal. The pH is a measure of the alkalinity or acidity of a substance on a scale of 1 to 14; anything that is neither acid nor alkaline will be neutral with a pH of 7. The pH of the skin, as a result of its film of sebum and perspiration, ranges from 4.5 to 6 or 7, depending on the area of the body. Most shampoos have pH values between 5 and 8.

Unless the pH of a shampoo were as alkaline as 10 or over (Ivory soap has a pH of 9), it is certainly not going to have any lasting effect on hair or the scalp. Since the total application period of a shampoo is only a few minutes and it is rinsed off with tap water (which could be acid or alkaline, depending on how the water is treated in a particular area of the country), there is even less reason to be concerned about the shampoo's pH.

Shampoos may also contain protein, balsam, egg, beer, or other kinds of gums or mucilages to coat the hair shafts—hair may then appear thicker or shinier, have body and "bounce." An individual may or may not find that these additives make the product better; it is largely a matter of personal preference. One trouble with trying to make a shampoo perform all kinds of additional cosmetic functions is that its major features—cleaning ability and rinseability—tend to become lessened with each addition. If a conditioner is necessary, it is probably better to choose a shampoo without all the extras, and then use a separate conditioning agent after shampooing (Fig. 15.17).

Hair Conditioners

The function of hair conditioners is to reduce static electricity that causes the individual hairs to repel one another and to also restore a shiny appearance, improve the texture, and increase the manageability of processed hair. The primary active ingredients in conditioners are cationic surfactants for controlling the hair and increasing the tensile strength, polyvinylpyrrolidone, (PVP), and protein to coat the hair and add thickness for body, plus a variety of oils to provide sheen and lubrication.

The cationic surfactant is a *quaternary ammonium compound*. One of the frequently used quaternaries is benzalkonium chloride. Since any kind of hair treatment, even combing, can cause a negative electrostatic charge on the hair, the cationic agent neutralizes that charge by adhering to the hair fiber. In that way it eliminates flyaway hair and adds to the hair's tensile strength to give it more body and set-holding ability. PVP is a plastic resin and, in a similar fashion to other gums and resins in conditioners, forms a film on the hair shaft

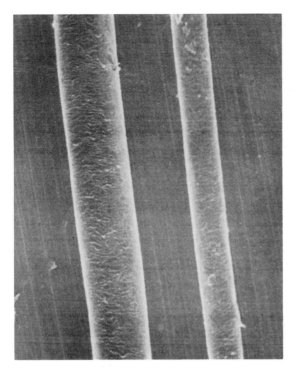

Figure 15.17. Thick and thin hairs. Hair that is thin in diameter may not necessarily be "thin" in numbers of hairs on the head. Hairs with thin shafts that are also in sparse supply would probably benefit from preparations that adhere to the hairs and "thicken" them.

to give it more substance. Both quaternary compounds and PVP stick themselves onto the hair shaft (the chemists' term is adsorption) and are not substantively incorporated into the hair itself.

The "protein" in the protein conditioners is not really protein at all. Label reading will show that the protein preparations contain "hydrolyzed animal protein," actually a number of amino acid building blocks or subunits of protein that are obtained when animal collagen or elastin—the source of the protein—is chemically treated to break it down. The amino acids supposedly penetrate the hair

shaft to incorporate themselves into the keratin protein of the hair to repair and restore damage to the protein chains—a remarkable accomplishment, if it were possible. There is no evidence that the hair is chemically changed by the addition of protein conditioners. Processed hair tends, however, to adsorb more protein onto the shaft than virgin hair.

The addition of nucleic acids to conditioners is the latest biochemical marvel. The hair preparation manufacturers are evidently hoping that by now everyone has heard of DNA and RNA but does not remember much about how they work. One label reads: "Contains proper amounts of the nucleoprotein team, DNA and RNA, which assists in the initiation of reconstruction and insures maintenance reconstruction of the hair. The nucleoproteins . . . are in a ratio of three RNA to each DNA and are actually genetically coded for the reconstruction of keratin protein." Since autolyzed yeast is the sole protein source in the list of ingredients, the conditioner appears to be using yeast DNA and RNA to synthesize human protein. This is enormously impressive. Even if the yeast nucleic acids could conceivably have any reconstructing ability, they would make yeast protein, and few people would want *that* growing out of their hair follicles. In order to make human keratin, the conditioner appears to be carrying out a successful gene transfer experiment, a procedure usually performed only in laboratories equipped to do DNA recombination. Or perhaps we are to believe that the yeast is cloning hairs? Either way, this information should certainly be shared with molecular biologists, who are obviously unaware that an astounding breakthrough in genetic engineering is occurring each time the conditioners are used.

It should not be inferred that conditioners—actually, reconditioners—are not useful. They glue the cuticle scales back down against the cortex, they tend to fill in any missing cuticle, and they serve to hold the split ends together for awhile. The public, however, should not be "scientifically" deluded into believing they have any other effect. Besides, not everyone needs a separate conditioner after shampooing , and their use will not necessarily make all hair look better. Depending on the hair style and type, it may be desirable for the hair to have a certain amount of "flyaway," and conditioning could make hair too soft or even too controlled. There is evidence to indicate that cationic surfactants are the best conditioners after hair processing and that PVP adds more material onto the hair shaft than hydrolyzed protein (Lindo and Spoor, 1974; Spoor, 1977). Most conditioners take the shotgun approach anyway and contain a little of everything.

The newer types of setting lotions that are applied after shampooing and partial drying contain cationic surfactant additives. Variously known as *sculpting lotion, contouring gel,* or *styling mousse,* the products contain as a primary ingredient a quaternary compound to coat the hair and hold it in a relatively stiff "wet look" until the hair is brushed out. The gels and mousses are reminiscent of the older Dippity-Doo type setting lotions, but they do not flake out with combing or itch.

Some have suggested that protein conditioners are a wasteful and unnecessary use of protein in a protein-starved world. While there may some justification in the argument, in this country the protein that is used is second class; it comes from animal by-products and is lacking appropriate nutrients for complete human nutrition. The use of so-called protein in hair preparations, along with all the other vitamins, minerals, and assorted biochemicals that are added to shampoos and conditioners make the products more expensive. Unless it is discovered that all those extras really do a better job than preparations without the additives, there is no point in buying them.

The aware cosmetic consumer has probably concluded by now that most formulas for hair care products are very much the same. If one finds, after wading through all the miracle claims, that a particular product does have benefit and esthetic appeal, so be it—it is a good product.

REFERENCES

Anthony, H. M. and Thomas, G. M. Tumors of the urinary bladder: An analysis of occupations of 1,030 patients in Leeds, England. *J Natl. Cancer Inst* 46:1111–1113, 1970.

Blank, I. H. Factors which influence the water content of the stratus corneum. *J Invest Dermatol* 18:433–440, 1952.

Brookmire, P. How to buy it and apply it. *The Milwaukee Journal* July 22, 1977.

Corson, R. *Fashions in Makeup.* New York, Universe Books, 1972, p. 386.

Durr, N. P. and Orentreich, N. Epidermabrasion for acne: The polyester fiber web sponge. *Cutis* 17(3):604–608, 1976.

Fulton, J. D. and Black, E. *Dr. Fulton's Step-by-Step Program for Clearing Acne.* New York, Harper & Row, 1983.

Fulton, J. E., Bradley, S., Aquandez, A., et al. Non-comedongenic cosmetics. *Cutis* 17(2):349–351, 1976.

Garber, C. A. Characterizing moisturized skin by scanning electron microscopy. *Cosmetics and Toiletries* 93(4):74–78, 1978.

Hennekens, C. H., Spelzer, F. E., Rosner, B., et al. Use of permanent hair dyes and cancer among registered nurses. *Lancet* 1:1390–1393, 1979.

Kiese, M. and Rauscher, E. The absorption of p-toluenediamine through human skin in hair dyeing. *Toxicol Appl Pharmacol* 13:325–331, 1968.

Kinlen, L. J., Harris, R., Garrod, A., and Rodriguez, K. Use of hair dye by patients with breast cancer: A case-control study. *Br Med J* 2:366–368, 1977.

Kligman, A., and Mills, O. Acne cosmetica. *Arch Dermatol* 106:843–850, 1972.

Lindo, S. and Spoor, H. J. Hair processing and conditioning, in Brown, A. G. (ed). *The First Human Hair Symposium.* New York, Medcom Press, 1974.

Lucky, A. W., McGuire, J., Rosenfield, R. L., et al. Plasma androgens in women with acne vulgaris. *J Invest Dermatol* 81(1):70–74, 1983.

MacKenzie, A. Use of Buf-Puf and mild cleansing bar in acne. *Cutis* 19(3):370–371, 1977.

Najem, G. R., Louria, D. B., Seebode, J. J., et al. Life-time occupation, smoking, caffein, saccharine, hair dyes, and bladder carcinogenesis. *Int J Epidemiol* 11(3):212–217, 1982.

Nasca, P. C., Lawrence, C. E., Greenwald, L. C., et al. Relationship of hair dye use, benign breast disease and breast cancer. *JNCI* 64:23–28, 1980.

Pascher, F. Adverse reactions to eye area cosmetics and their management. *J Soc Cosmet Chem* 33:249–258, 1982.

Plewig, G. and Kligman, A. M. *Acne Morphogenesis and Treatment.* New York, Springer-Verlag, 1975.

Prival, M. J., Mitchell, V. D., and Gomez, Y. P. Mutagenicity of a new hair dye ingredient: 4-ethoxy-m-phenylenediamine. *Science* 207:907–908, 1980.

Retinoic acid and sun-caused skin cancer. FDA Drug Bulletin, August–September, 1978.

Schoor, W. F. Eye cosmetics. *Dermatol and Allergy* 4:45–46, 1981.

Shore, R. E., Pasternack, B. S., Thiessen, E. U., et al. A case-control study of hair dye use and breast cancer. *JNCI* 62(2):277–283, 1979.

Spoor, H. F. Supplemental hair care. *Cutis* 19:299–303, 1977.

Stenback, F. G., Rowland, J. C., and Russell, L. A. Noncarcinogenicity of hair dyes: Lifetime percutaneous application in mice and rabbits. *Food Cosmet Toxicol* 15(6):601–606, 1977.

Stavrady, K. M., Clarke, E. A., and Donner, A. Case-control study of hair dye use by patients with breast cancer and endometrial cancer. *JNCI* 63:941–945, 1979.

Valentic, J. P., Elias, A. N. and Weinstein, G. D. Hypercalcemia associated with oral isotretinoin in the treatment of severe acne. *JAMA* 250(14):1899–1900, 1983.

Williams, M. and Gunliffe, W. Explanation of premenstrual acne. *Lancet* 2:1055–1057, 1973.

Wilson, L. A. and Ahearn, D. G. Pseudomonas-induced corneal ulcers associated with contaminated eye mascaras. *Am J Ophthalmol* 84(1):112–119, 1977.

Wilson, L. A., Julian, A. J., and Ahearn, G. G. The survival and growth of microorganisms in mascara during use. *Am J Ophthalmol* 79(4):596–601, 1975.

Wynder, E. and Goodman, M. Epidemiology of breast cancer and hair dyes. *JNCI* 71(3):481, 1983.

16. *Our Health in Our Hands*

Health, according to the old saying, is the one thing that money can't buy. Illness can happen to anybody; being rich has never been a guarantee against getting sick. Somewhere along the way, however, we seem to have forgotten that health is not for sale. For the last 30 years the American public has evidently been convinced that health *could* be bought for everybody, if only enough money were spent, if only enough were invested in health care—the diagnosis and treatment of disease.

Every year the costs of health care have rapidly and inexorably increased to the point where they now account for more than $1,400 per person and over 10.5% of the nation's total production of goods and services. The ever-growing share of government dollars spent on doctors, medical schools, health workers, hospitals, and technology leaves that much less for housing, education, creating new jobs, or cleaning up the air and water. The health care industry, third largest in the country, has raised our taxes and our prices as the costs of health care insurance are passed along to the consumer by the employers. As only one example, every American-made auto sold in 1983 had an added $600 to cover the expense of health insurance for the employees.

Although the bills from the expensive health care industry continue to strain the national economy, the expenditures might be justified if all Americans, regardless of their economic status, were not only receiving quality health care, but were living longer and healthier lives as a result. But increasing the spending has not produced better quality care. On the contrary, there is evidence that duplication, waste, inefficiency, and downright fraud exist within the system. Moreover, there are large segments of the population who have limited access to quality care. Even more dismaying, there has been little improvement in overall public health, despite the high costs of medical care. There has been a 10% decline in mortality from heart and blood vessel disease in the last 30 years, but deaths from cancer, chronic respiratory diseases, and cirrhosis of the liver continue to rise. In fact, in the late 1960s a health authority suggested that if the amount spent on health were either doubled or halved, it would have no significant effect on longevity.

At that time only $40 billion was spent. Now that the amount has increased nearly 10 times, that observation has sadly proved to be more accurate than expected.

Americans rely on doctors to keep them well, paying dearly for the expectation, but the most highly sophisticated medical treatment can do little for those diseases which cause most of the health problems. The major causes of death in both men and women are cardiovascular diseases (heart attack, stroke, high blood pressure, arteriosclerosis) and cancer. Their incidence, it is now being recognized, is at least partly related to certain risk factors that can be controlled by individuals themselves. These killer diseases might be partially or perhaps even totally preventable if people would take the responsibility for preventing them. Some health analysts have even called certain ailments "diseases of choice,"

implying that we can choose to be sick or we can choose to be well—it is up to us. There is mounting evidence that correlates lack of exercise and overweight to cardiovascular disease, a high fat diet with cancer, alcohol abuse with cancer and cirrhosis of the liver, and smoking with virtually all health problems. Doctors can treat disease once it has struck, but they cannot be responsible for our health. Our health may be influenced more by what we do to and for ourselves than by how much professional medical care we get.

Better health can never be bought by buying more doctors, more hospital beds, or more computerized brain and body scanners that cost more than a million dollars each. (Fig. 16.1). A better way of spending some of our health care taxes and insurance premiums might be in educating people toward the prevention of illness and the preservation of

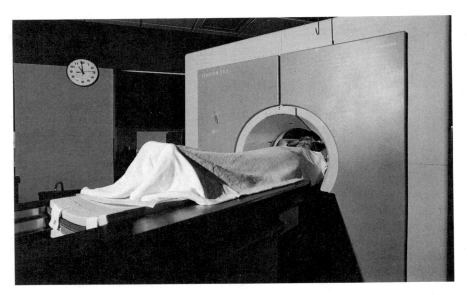

Figure 16.1. A CT (computerized tomography) scanner. It may look like a giant pencil sharpener, but this machine, used for brain and body imaging, costs between half a million and million dollars. The NMR (nuclear magnetic resonance) and PET (position emission tomography) imagers are other costly diagnostic tools that depict body structure and function.

health. It has been suggested that if people were to modify their behavior to incorporate good health habits into their lives—no smoking, drinking, and drugs in moderation or not at all, maintenance of a normal weight, getting plenty of exercise and plenty of rest—there would be a reduction in the number of premature deaths, an overall improvement in health, and a containment of the exorbitant costs of health care. John Knowles, late president of the Rockefeller Foundation and a foremost proponent of the personal responsibility for health, stated that the next major advances in the health of the American people will be determined by what individuals are willing to do for themselves and for society at large. Marvin Schneiderman, National Cancer Institute scientist, maintained that 100,000 annual cancer deaths, almost 30%–40% of the total, could be prevented by changes in people's eating, drinking, and smoking habits.

Not all Americans, however, are in a position to choose a different life-style, and the accusation that they can is sociologically naive. Perhaps the decision not to smoke or drink is a choice available to anyone, but a major part of the individual assumption of responsibility for health is obviously profoundly limited when individual living conditions include poverty, unemployment, and deprivation. The poor, needy, aged, and uneducated and disadvantaged people of this country—about 35 million of them—suffer more illness than the advantaged, but certainly they have neither chosen their lot nor their diseases. Moreover, their health-care needs are not always met. While Medicaid and Medicare are supposed to take care of those people who cannot afford to pay, these programs provide uneven and inadequate coverage in many parts of the country. It is a myth that Medicare and Medicaid offer the needy and elderly health care equivalent to that received by the middle and upper classes.

It must also be remembered that although bad health habits can contribute to the incidence of some diseases, they are not the sole determinants of all illnesses.. Our genetic constitutions are an important factor in our health—we do not choose our genes. Illness and mortality in many instances can be reflections of hazardous environmental conditions—of polluted air and water, of toxic substances or carcinogens in the food we eat, and of occupational dangers. There are many people who suffer from illness caused by factors beyond their control. They get sick and they die, but personal responsibility has played a minor role.

The benefits of taking charge of one's own health are not an illusion—they are real and measurable. Self-care can have an enormous impact on the prevention of illness, on life expectancy, on personal quality of life, on everybody's insurance premiums and taxes. The only danger lies in trying to shift the entire burden for health maintenance to the individual. We cannot do it completely by ourselves. We have to be assisted in our quest for health by improved and quality health care and by the elimination of the economic, social, and environmental conditions that contribute to disease. Taking good care of ourselves can play a great part in preserving the nation's health and in helping to control costs, but the individual's responsibility for health is an accompaniment, not an alternative, to the responsibilities of society and government for the improvement and maintenance of the public health.

GETTING HEALTHY, STAYING HEALTHY, HOLISTICALLY

Along with the rising costs of health care, there has been a rising dissatisfaction with the care delivered by the traditional medical establishment. Increasing numbers of people are searching for nontraditional methods of healing, some of which are considered non-scientific or even absurd when judged by the usual standards. It is not surprising that the public is disillusioned and disappointed by orthodox medicine. There is evidence not only of the limitations of a health-care system based on sickness rather than on health, but also of its adverse effects. Recent knowledge concerning the prevalence of iatrogenic (doctor-induced) disease, of unnecessary surgery, and of excessive and even irrational prescribing of drugs has hardly been reassuring. Faith in the system has been shaken, and people are ready for alternatives. The proponents of a new approach to health and illness believe they have the answer to the crisis in "disease-care." *Holistic* medicine, or holistic healing, stresses that health is more than the absence of disease or infirmity—it is also the feeling of "ease" rather than "dis-ease." Holistic health care searches all the dimensions of an individual's life—the physical, emotional, intellectual, spiritual, and interpersonal—to find the cause for dis-ease. Total health is seen as a balance of all factors, a wholeness of body, mind, and spirit that allows the attainment of an individual's total potential for well-being. Holism emphasizes "positive wellness"; it is more interested in prevention than treatment. The holistic approach is said to be the practice of humanistic health care. It is concerned as much with the improvement of the quality of life as with physical wellness.

The world "holistic" has unfortunately been so overused that in many instances its original meaning has all but disappeared. A recent book, for example, is entitled *Holistic Running*. Presumably, some enterprising food manufacturer will soon start marketing something called "Holistic Breakfast Bars" or "Holistic Granola" and make a fortune.

There are also hundreds of therapies that are included under the general umbrella of "holistic health" or "wellness" techniques. The practitioners of holistic medicine range from traditional physicians or other established professionals to gurus, Indian shamans, or psychic healers. Their forms of healing may have their basis in known and accepted methods of preserving health, that is, through changing or modifying unhealthy life-styles. Other modes of treatment may be mystical or experiential. Adherents of holistic health generally keep an open mind to alternative ways of relieving physical or emotional pain. Herbal healing, homeopathy, hypnotism, acupressure and acupuncture, meditation, and even laying on of hands are unscientific by traditional standards, but they sometimes are successful in producing cures when orthodox treatment has failed. Recognizing that the belief in the treatment and the practitioner of a therapy is frequently of greater importance than the content of the therapy, the sole test of the usefulness of a technique is usually pragmatic—if it works, use it.

Obviously, some of these modes of healing are of limited usefulness and downright dangerous in serious illness, injuries, or infections. Waiting to see whether nutritional therapy works when chemotherapy has a proven track record is a foolhardy course to pursue. A better test of the safety and effectiveness of a nontraditional technique is whether or not it can actively harm an individual. The decision to use an alternative method involves making the same kind of *informed* choice one uses for the usual forms of medical care. Herbal remedies, for example, may be more natural than the drugs prescribed by a physician, but they may be, nonetheless, drugs. They should be subject to the same kind of scrutiny. A nutritional regime that promises a cure for disease, instant permanent weight loss, or that "cleanses out the body" may be harmless or a waste of time; it could also be of dubious safety. Many women have learned not to be passive recipients of current medical care, but passive acceptance of the claims of the alternative and unorthodox therapies without investigation and knowledge is hardly progress. When exploring some of the unconventional pathways to good health, it would be wise to retain some of the cynicism and informed skepticism with which the standard surgery and drugs of the doctors are now being viewed.

Of all the described ways to wellness, of all the methods and techniques used to attain a healthy life, the bottom line is still: good nutrition, good exercise, good relaxation. These are the three inseparable elements of health. The rest of this chapter takes a physiological look at nutrition and physical activity with a view toward providing information that can result in better nutrition and more exercise.

NUTRIENTS

The human body is a marvel of engineering and, with just a few minor exceptions, excellently designed to carry out all of its specific functions. Similar to all machines, the body, too, needs fuel in the form of energy to propel it throughout its daily living—to keep the heart beating and the blood circulating, to inhale and exhale the air, to contract the muscle for movement, to power cell division and the synthesis of new tissue that means growth, to do all of the work of the life processes that keep the body alive and active. Humans get their energy from the carbohydrates, proteins, and fats in their diet. The energy value of the food is expressed in terms of a unit called the kilocalorie or Calorie (with a capitol "C"). A Calorie is the amount of heat that is required to raise the temperature of a kilogram of water (100 g) from 14.5° Celsius to 15.5° Celsius. The calorie (with a small "c") is really 0.001 Calories, but it is used so often in association with food that it is usually understood to mean the kilocalorie. When oxidized or burned inside

the cells of the body, 1 gram of carbohydrate provides 4 calories, 1 gram of protein also provides 4 calories, and 1 gram of fat yields 9 calories. The process by which food is combined with oxygen for the release of energy takes place in all the cells of the body and is called *cellular respiration.* The energy is not released as heat, which would be destructive and useless to the cells. Instead, it is captured to form a high-energy molecule called *adenosine triphosphate* or ATP. Then, when the cell needs energy to power its physiological work it can get it from the ATP. Cellular respiration, without which no living cell could survive, results in the production of energy as ATP and of carbon dioxide and water. The entire process looks like this:

Food molecule + oxygen $\longrightarrow$
ATP + carbon dioxide + water

Actually, the carbohydrates, proteins, and fats in the foods eaten are not in a form utilizable by cells; their molecules are too large to pass into the cells. Before they can be a source of energy, the three foodstuffs must be broken down into their smaller building block molecules. The chemical conversion of carbohydrates into simple sugar, proteins into amino acids, and fats into fatty acids and glycerol, is called digestion, and it takes place within the digestive tract.

Figure 16.2 is a diagram of the digestive system. Food taken in at the mouth is acted on within the tract by digestive enzymes produced both by the digestive tube itself and by glands located outside the tube itself and by glands located outside the tube and emptying into it. When digestion is complete, the simple sugars, amino acids, and fatty acids and glycerol are absorbed by the cells lining the small intestine. All of the sugars and amino acids and a small quantity of the end products of fat digestion get into the bloodstream and are carried to the liver first before they are

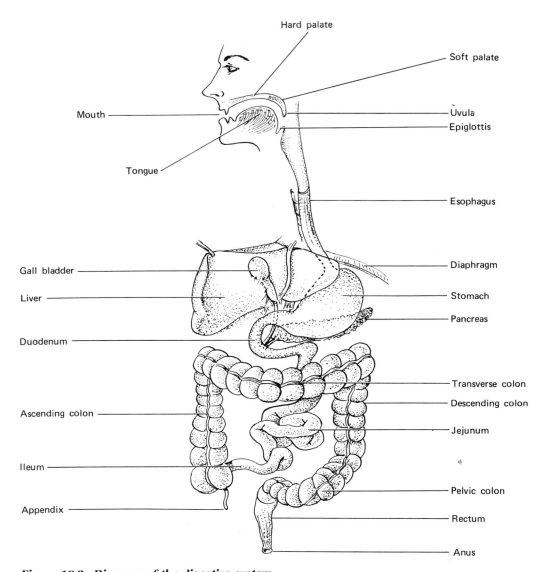

Figure 16.2. Diagram of the digestive system.

transported to the rest of the body. The majority of the fatty acids and glycerol molecules bypass the liver. They are absorbed into the lymphatic system, but also get into the bloodstream eventually via a large vein near the heart. All undigested substances left over after absorption are expelled from the body as waste or feces.

While carbohydrates, proteins, and fats provide the energy in food, other required components of the diet are minerals, vitamins, and water. These substances do not contribute energy but are no less vital for survival. All of these chemical substances on which humans and other animals rely for the continuation of life processes are called *nutrients*.

Carbohydrates

In the typical American diet, 50%–60% of the energy is provided by carbohydrates, compounds made up only of carbon, hydrogen, and water. Since carbohydrates are found abundantly in plants, they may provide as much as 90% of the energy in the diet (with adverse effects on health) in underdeveloped countries where protein is unobtainable or too expensive.

In the process called photosynthesis, green plants use the energy from the sun to combine carbon dioxide and water to form *glucose*, the basic form of carbohydrate. All the more complex carbohydrates are variations and multiples of glucose. The plant cell converts the glucose to larger carbohydrates and also to fats, proteins, and vitamins, utilizing the minerals and nitrogen obtained from the soil. The main sources of carbohydrates in the diet are tuber and root vegetables, legumes (beans and peas), cereal grains, and fruits.

There are three groups of carbohydrates, divided on the basis of the size and complexity of their molecules. Glucose is the best known and most prevalent *monosaccharide*, or simple sugar, and is the main sugar of carbohydrate metabolism and energy production in the body. "Blood sugar" is primarily glucose. Other monosaccharides are fructose or fruit sugar and galactose, a component of milk sugar.

Disaccharides are formed from two monosaccharides linked together. *Sucrose*, or table sugar, which comes mainly from the juices of sugar cane, sugar beets, or maple trees, is composed of glucose and fructose. Honey is about 50% fructose and 50% glucose plus a trace of vitamins and minerals usually consid-

ered inconsequential. *Lactose* is another disaccharide and is made up of glucose and galactose. *Maltose*, which rarely occurs in foods but is an intermediate produce of carbohydrate digestion in the body, has two glucose units.

Polysaccharides are very long chains made up of hundreds, thousands, or even millions of glucose units. The most common polysaccharides are plant starch, or *amylose* and *amylopectin*, *cellulose*, and animal starch or *glycogen*. Cellulose forms a large part of the diet, but it cannot be utilized for energy because humans lack the enzyme for breaking it down to glucose. Although cellulose cannot be digested, it does add to the mechanics of the digestive system by contributing fiber and bulk. A moderate amount of cellulose and other indigestible materials or roughage is necessary for normal bowel activity and is obtained from whole-grain cereals, and raw fruits and vegetables. An increased incidence of diverticulitis (inflammation of small pockets in the wall of the colon), cancer of the colon, and heart disease has been correlated to a low fiber intake in the diet.

Few of the food nutrients that provide energy, except when they are highly refined as in table sugar or salad oil, are ever "pure." Even flour, which is primarily starch, has a 6%–10% protein content. Dry peas and beans contain 20% protein, and soybeans are different from other legumes in that they contain even more protein and fat and less starch.

Fats and Lipids

Since fats provide more than twice the calories per gram than both carbohydrates and proteins (9 calories/g in fats versus 4 calories/g in the latter two), foods containing fats will be higher in calories and energy content. More than 40% of a typical American diet is fat, much of it coming from animal sources.

The term *fat* is usually used synonymously

with "triglycerides," "true fat," "neutral fat," or "visible fat," and it refers to body fat or adipose tissue as well as to food fats, such as butter, margarine, lard, and vegetable shortening and oils. *Lipid* is a broader term; it includes the phospholipids, the sterols such as cholesterol, and all the other substances that contain fatty acids and are insoluble in water. Triglycerides are used for cellular energy in animals and plant. The other lipids form parts of cellular membranes, are involved in the transport of triglycerides in the bloodstream, or are important in the synthesis of other important compounds. Cholesterol, for example, is the precursor for the synthesis of sex hormones, bile salts, the adrenal cortical hormones, and vitamin D.

Triglycerides

Containing only carbon, hydrogen, and oxygen atoms, a triglyceride molecule consists of three molecules of fatty acids linked to one molecule of an alcohol called glycerol, commonly known as "glycerine." The fatty acids can either be all the same kind or all different. Palmitic, oleic, and stearic are the most common fatty acids found in nature. They are found frequently in food fats and are the common fatty acids in human body fat. Palmitic and stearic are examples of *saturated* fatty acids. A saturated fatty acid contains all the hydrogen it is capable of containing, attached to the carbon atoms. In *unsaturated* fatty acids, such as linoleic, linolenic, and arachidonic, the place on a carbon atom that would ordinarily be attached to a hydrogen is unoccupied. The carbon atoms are then joined by double instead of single bonds, like this: (—C=C—). When a fat contains fatty acids with two or more double bonds between carbon atoms it is said to be *polyunsaturated*. The consistency of fat at room temperature is determined by the degree of saturation with the hydrogen atoms. Animal fats are composed largely of saturated fatty acids and are solid,

but fats from plant sources are found in a liquid state as oils because they contain unsaturated fatty acids. Margarine, Fluffo, Crisco, and Spry are vegetable oils that have been partially saturated and converted to a solid form by hydrogenation, the addition of hydrogen atoms. Cocoa butter (chocolate fat) and coconut oil are two exceptions to the unsaturated quality of vegetable fats.

Linoleic and linolenic acids are called the *essential* fatty acids because they cannot be synthesized by the body and must be obtained in the diet. The essential fatty acids are necessary precursors to the biosynthesis of prostaglandin by human tissue.

In addition to providing the source of certain necessary fatty acids, fat in the diet is also required as a carrier for the fat-soluble vitamins in the food. Without dietary fat, there would be great difficulty in absorbing vitamins A, D, E, and K from the intestine.

The fat stored in the body, found within the adipose cells of adipose tissue, is desirable and advantageous, as long as not too much is accumulated. Moderate fat deposits provide roundness and shape to the body in both males and females and is a reserve source of energy. Fat insulates the body, conserves against heat loss, and also supports and cushions the body organs. But the body's requirements for fat as a nutrient can be met by 15–25 g of food fats eaten daily. The typical diet contains 10 times that amount—110–260g—or over a quarter of a pound of fat. Much of the excess body fat in overweight individuals does not come from dietary fat, however. When the food energy ingested in the form of fat, protein, or carbohydrate exceeds the amount of energy needed for body activities, it is converted to body fat and stored in the adipose tissue.

Lipids Other Than Triglycerides

Phospholipids. Phospholipids are similar to triglycerides, but one of the fatty acids in the molecule has been replaced by a phosphate-containing substance. In the phospholipid *lecithin*, found in egg yolk and soybeans, the substance is choline, a nitrogenous base. Some of the claims for the value of high lecithin intake have bordered on the miraculous. It has been said to lower blood cholesterol levels, dissolve gallstones, prevent heart disease, and, when combined with seaweed, vitamin B_6, and cider vinegar in a low-calorie diet, emulsify and "burn up" fat to cause spot-reduction from special places. There is little or no evidence that extra ingestion of lecithin provides any of these benefits. When lecithin is digested in the body, it breaks down into glycerol, fatty acids, phosphate, and choline. The body cells are able to synthesize both choline and lecithin on their own.

Probably the most important function of phospholipids in the body is that they form part of the structure of all cells, mainly in the membrane elements.

Cholesterol. Cholesterol is a member of a third class of lipids called sterols. It is found in the fatty portions of all foods of animal origin and is, therefore, present in the diet of most people. Much of the cholesterol of the body is converted by the liver to cholic acid and combined with other substances to form bile salts, which are necessary for the digestion and absorption of fats. Cholesterol is also essential for the synthesis of hormones by the adrenal glands, the ovaries, and the testes. Along with phospholipids, cholesterol forms a part of the structural components of all body cells.

After 6 months of age, dietary consumption of cholesterol is unnecessary since the body will make what it needs from other substances, and normal adults produce about 200 mg of cholesterol daily. Most of this endogenous cholesterol is manufactured by the liver, but it is also synthesized by all the other cells of the body. When foods high in cholesterol are eaten, the production of endogenous cholesterol is somewhat but not completely cut back. Thus, the habitual ingestion of high cholesterol-containing foods in most people causes a rise in blood cholesterol levels. Common foods that contain large amounts of cholesterol are those high in saturated fats, such as butter, whole milk, and cheeses made from whole milk, and meats which are heavily marbled with fat. Egg yolks, organ meats, such as liver and kidney, and shrimp also contain appreciable amounts of cholesterol.

There is a great deal of evidence to link high blood levels of cholesterol with atherosclerosis, considered to be the leading cause of coronary heart disease and stroke. Atherosclerosis is a condition in which fatty plaques of lipids, mostly cholesterol, are deposited on and in an artery wall. The cholesterol deposits slowly build up, become fibrous and calcified, and eventually cause narrowing and hardening (arteriosclerosis) of the artery. The artery can become so narrow that it becomes closed, cutting off the supply of blood to a vital area supplied by the diseased vessel, or, part of the plaque can break off and become a thrombus or clot that blocks the vessel, which may also lead to a heart attack or a stroke.

It should be emphasized that the relationship of a high intake of saturated fats and cholesterol to high blood cholesterol levels and the incidence of heart disease and stroke is no longer controversial. It is known that a low-fat diet and exercise lowers blood cholesterol. Changing the nutritional life-style to incorporate a low-fat, low-cholesterol diet along with regular exercise is not only prudent but potentially lifesaving.

Proteins

Most of the body is composed of water, but the solid part is three-quarters protein. Without exception, every cell contains protein fundamental to cell function and structure. The 20-odd amino acids that are the structural units of proteins contain carbon, hydrogen, oxygen, and nitrogen plus small quantities of other atoms, such as sulfur and phosphorus. The proteins in food provide the amino acids necessary to build new tissue or maintain and repair the tissues already built.

Actually, the basis of all life is the protein synthesis going on within the cells. The energy-supplying role of food protein is secondary; its major importance is the support of normal growth and repair of the body's cells, tissues, and organs.

When proteins are eaten, they are digested down to amino acids, absorbed from the small intestine, and taken to all the cells of the body to be utilized in making the types of proteins needed. Protein synthesis can only take place when all of the 20 amino acids are present in an adequate supply. As long as there is enough nitrogen available (via protein food intake), the cells of the body are capable of manufacturing most of the amino acids normally present in animal proteins. Eight of the amino acids, however, cannot be synthesized in human tissues because the enzymes necessary to their formation are lacking. (Actually, our DNA lacks the genes required to direct the manufacture of those particular enzymes.) Those eight amino acids are called *essential* because they must be obtained from the diet. They include isoleucine, leucine, lysine, methionine, phenylalanine, threonine, tryptophan, and valine.* Histidine is considered a ninth essential amino acid in children. It is able to be synthesized in the body but not rapidly enough to sustain growth during childhood.

The other amino acids are just as essential in the cellular formation of protein, but they are not essential from the diet. Almost all foods eaten contain some protein, but not all of them have the same quantity and kinds of amino acids. Only animal proteins, such as

meat, poultry, fish, eggs, and dairy products, have all the amino acids, including the essential ones in the amounts and proportions appropriate for protein synthesis. For these reasons they are called *complete* or *high quality proteins.* Other foods, such as gelatin made from animal collagen and elastin, or soybeans, are rich in protein but do not contain all the essential amino acids. Most plant proteins are deficient in several of the essential amino acids, usually threonine, tryptophan, or lysine. Protein foods from vegetables and grains that lack essential amino acids are called incomplete sources of protein because they have a lower biological value than animal protein. Some plant foods have a lower percentage of protein and may be more difficult to digest since the cellulose in the plant wall resists the action of digestive enzymes. The protein present in the plant cells will probably pass right on through the tract as part of the bulk. Dried beans, for example, have to be softened by long periods of cooking to improve their digestibility.

As the prices of meat and fish and even cheese become increasingly astronomical, it is necessary to appreciate the fact that protein synthesis in the body can be supported without ingesting much animal protein or even any at all, although pure vegetarianism can result in inadequate nutrition unless foods are selected very carefully and wisely. A money-saving and nutritious way of eating mixes low-cost incomplete proteins from plant sources with complete protein from animals. Oriental cooking, for example, uses only a little meat or fish along with a lot of rice and vegetables. Another way of obtaining quality complete protein is by combining two vegetable proteins so that they supplement each other. Legumes lack some essential amino acids; cereal grains lack others. Peanut butter on whole wheat bread, however, or beans and rice together provide good nutrition. But since the body has no mechanism for storing

---

* The mnemonic device Pvt. T. M. Hill stands for the first letters of the essential amino acids if histidine is included.

amino acids in the way that it accumulates fats or carbohydrates for future use, it is important to remember to mix complementary protein foods *at the same meal*. In this way, all 20 amino acids necessary in the manufacture of protein will be present in the body cells at the same time.

VEGETARIANISM

Particularly during the last decade, large numbers of Americans, for various reasons, have decided to eliminate animal flesh from their diets. Their motives are diverse. They may believe there are harmful constituents—hormones, antibiotics, pesticide residues—in meat. They may practice a religion or special life-style that requires the exclusion of animal products. They may simply be unable to afford meat, fish, or fowl; a diet based on plant foods can cost one-quarter that of one centered on meat. They may want to make a personal contribution to the alleviation of the world protein shortage. It does, after all, take 21 pounds of grain fed to a steer to make 1 pound of beef protein.

There need not be any nutritional disadvantage to practicing vegetarianism as long as the diet is carefully selected. Depending on the foods included, a vegetarian diet may even be superior to the typical American diet, loaded with extra calories and saturated fats. The lacto-ovo-vegetarian diet, one that includes milk, eggs, and other dairy products, will provide sufficient calories, a good quality balance of essential amino acids, and adequate vitamins and minerals. It is nutritionally safe and appropriate for all age groups, including children. Some "vegetarians" eliminate only red meat and permit fish and chicken or turkey, claiming that they feel better or even smell better when beef or pork has not been eaten. Any benefits derived from the exclusion of red meat with its higher fat content would be likely to be long-term rather than immediate. It is doubtful whether the body, after digestion of animal protein to amino acids, is able to distinguish the source of those amino acids.

A pure vegetarian or "vegan" will not eat animal products in any form and chooses his/her diet from cereals, legumes, nuts, fruits, seeds and vegetables. Vegans may have a little more difficulty maintaining adequate nutrition, but with knowledge and appropriate food selection, it is still possible to stay well nourished. The vegan must make certain to get enough calories for energy needs. A pure vegetarian diet usually contains so much bulk that it is hard to eat enough to meet caloric requirements. It is also important to include protein-rich legumes in the diet and to combine them at the same meal with cereal grains. Furthermore, the diet must be supplemented with vitamin B_{12} and calcium, since B_{12} is found only in foods of animal origin, and the major source of calcium is milk. Ingestion of soybean milk fortified with vitamin B_{12} and significantly increasing the intake of the calcium-containing green leafy vegetables will provide adequate amounts of these two nutrients. Eating nuts and seeds is also important. These foods are a source of protein and vitamins and contain the fat necessary for absorption of fat-soluble vitamins from the intestine. As a safety measure it would probably be sensible for vegans to consider vitamin supplementation in tablet form.

Vegetarian extremists who eat only cereal grains or only fruits and nuts, associating this kind of restriction with some sort of purification or spiritual reawakening, are playing with fire nutritionally. The body will always put a higher priority on energy than on maintaining and repairing the body proteins. A low-calorie diet that is also low in protein, inevitable on this kind of limited regimen, is very dangerous to health. The carbohydrates and fats in the diet are insufficient to supply the energy needs so the body is forced to use proteins pulled out of the structural components of the cells for energy. Normal protein synthesis necessary for building and replacement of tissues then suffers, and the body literally wastes away.

The risks to health of a severely inadequate diet depend on the previous nutritional status and the length of time during which strict adherence to the diet takes place. An individual who has gone through infancy and childhood in a well-nourished state may not immediately suffer the consequences of nutritional deprivation. There are hundreds of millions of people in the world, however, even some in this country, who are energy-deficient and protein-deficient and never by choice. They exist in a state of semistarvation, subject to disease and early mortality because they simply cannot get enough food to lead healthy and active lives. To invite a similar condition through a self-inflicted dietary regime is irrational.

How Much Protein is Enough?

Many people are convinced of the virtues of a high-protein diet. Increased consumption of protein is said to improve health, sex life, energy, athletic performance, and to be beneficial for losing weight. Athletes ingest inordinately large amounts of steak and milk before athletic events because they believe it gives them energy. Weightlifters and body builders eat a lot of animal protein and take protein supplements. Quick weight-loss diets of the "eat-all-you-want, calories-don't-count" variety extol the advantages of high-protein, low-carbohydrate eating. "To stay in shape," claims one cookbook with better recipes than nutritional advice, "don't count calories, count protein grams . . . if you start filling yourself with prote instead of just cals, you'll look and feel better and have more *real* energy than you've ever had in your whole life".

The widespread misconception illustrated here is that "prote" is a major energy source for the body. Carbohydrates and fats are the primary providers of "real" energy, however, and calories obviously do count. As pointed out before, amino acids, unlike carbohydrates and lipids, cannot be stored in the body to any extent. Once the quantities of amino acids needed for protein synthesis have been taken up by the cells and utilized, any *additional* amino acids are converted and oxidized for energy. If the total energy intake in calories is too great, the excess amino acids will simply be converted to fat and retained in the adipose tissues.

Very high consumption of protein during dieting is based on a mistaken interpretation of the fact that it takes somewhat more energy to digest and metabolize pure protein when it is eaten than to digest and metabolize carbohydrate and fat. Protein in pure form, however, does not occur in protein foods; it is usually accompanied by fats and carbohydrates. The energy requirement to utilize protein, therefore, is not likely to be much of a factor in weight reduction. Moreover, a gram of protein contains as many calories as a gram of carbohydrate. There is no advantage to replacing carbohydrate with protein to lose weight; a balanced diet with restriction of total calories is more important.

During periods of rapid growth, such as in infancy and childhood, an increase in dietary protein is necessary for tissue building. Greater amounts of complete protein must be supplied to insure optimum growth and development. For similar reasons, more protein is also needed by a pregnant woman to synthesize new tissue and by a lactating woman for milk formation. Other circumstances under which more protein is required occur during convalescence from a debilitating illness. Recovery from surgery, for example, increases the need for protein in the diet in order to repair and heal the injured tissues. After hemorrhage, more protein would obviously be necessary to replace blood tissue. For an athlete, however, a small extra amount of protein might be utilized during athletic training as the muscles get larger, but exercise or athletic performance does not increase the requirement for protein. The additional energy needed can be met by generally increasing food intake.

Under ordinary conditions, the adult body is in a state of protein balance. The rate of protein synthesis equals the rate of normal breakdown of body protein. Healthy adult men and nonpregnant, nonlactating healthy adult women need to eat enough protein to maintain that equilibrium. The recommended daily dietary allowance for adults is somewhere between 40 g and 70 g of protein, depending on the size of the individual. The value is approximately 0.4 g protein per pound of body weight. A 125-pound woman should, therefore, eat about 50 g of protein per day.

An ounce of meat, which also contains fat and water, yields an average of 7 g of protein. An ounce of cheddar cheese also contains 7 g, and a cup of 2% milk will provide 9 g. Getting the daily required amount of protein (which already includes a liberal safety margin) is no problem, and it is obvious that most people in this and other developed countries probably already consume more protein than they need. There is every reason to insure that there is enough protein in the diet, but there is no evidence that a superabundant high-protein diet of 200 g per day confers any special advantage. Whether that much is worthless and wasted, or actually fattening depends on the total caloric intake.

VITAMINS

Life could not be sustained on chemically pure protein, carbohydrate, fat, and water only. Animals and humans need to eat *foods* because foods contain additional factors necessary for survival. These accessory food substances—the vitamins and minerals—do not contribute energy but they are nonetheless essential. They must be ingested at relatively frequent intervals throughout the life span of an individual. Because they are required in such small quantities, sometimes only in trace amounts, they are frequently referred to as micronutrients.

Vitamins are organic compounds other than amino acids, fatty acids, glycerol, or monosaccharides that are essential for normal metabolic processes. They must be components of the diet because they cannot be manufactured within body cells, or at least cannot be synthesized in the necessary amounts. The terms for these food factors was coined by a Polish biochemist who was working with a substance that prevented a disease called beriberi in chickens. Since the active compound was an amine and it was necessary to the life (vita) of the chickens, he called it a "vitamine," although presumably he could just as easily have named it "aminevite." When other such factors were subsequently discovered, it was evident that they were not chemically related and hardly any of them were amines. The final "e" was dropped, and the chemicals became called vitamins. As each one was identified, it was assigned a letter name.

Today there are 13 or 14 vitamins recognized as essential to the human diet. The higher figure includes choline, which is a vitamin for various animals although its need for humans has not been proved. Table 16.1 summarizes the 13 vitamins, their richest sources, biological role, recommended dietary allowances, and the associated deficiency diseases.

Vitamin Needs. The Recommended Dietary Allowances (RDA) are estimates of nutritional requirements. They are revised approximately every 5 years and are based on present knowledge concerning the average amounts of specific nutrients needed by the average individ-

Table 16.1 VITAMINS

Name	Rich Sources	Function	Recommended Dietary Allowances (RDA)	Deficiency	Potential Toxicity When Large Amounts Consumed
Water-soluble vitamins					
Thiamine (**B₁**)	Pork; whole grains, enriched cereal grains; legumes (lost when cereals are milled and refined)	Component of co-enzyme; energy-release	1.2–1.5 mg	Beriberi; impairment of cardiovascular, nervous, and gastrointestinal systems	None; excreted in urine
Niacin (Nicotinic acid)	Lean meats, liver; peanuts; yeast; cereal bran and germ	Component of two coenzyme systems; energy-release	15–20 mg	Pellagra; "4 Ds": dermatitis, diarrhea, depression, death	None; may cause harmless symptoms of flushing of skin, dizziness, and nausea
Riboflavin (**B₂**)	Milk; eggs; liver; kidney; heart; green leafy vegetables (lost in dehydrated vegetables)	Component of various enzymes involved in energy release	1.1–1.8 mg	Dermatitis; light sensitivity of eyes; sores at corners of mouth	None
Pantothenic acid (**B₃**)	Liver; kidney; egg yolk; wheat bran; fresh vegetables (esp. broccoli and sweet potatoes); molasses	Component of co-enzyme A; energy-release	Unknown, probably 5–10 mg	Fatigue; GI distress; personality changes; numbness and tingling of hands and feet; muscle cramps	None
Biotin	Milk; liver; kidney; egg yolk; yeast; also synthesized by bacteria in intestine	Coenzyme carrier of carbon dioxide	0.15–0.3 mg	Scaly skin; seborrheic dermatitis in infants	None

Vitamin	Sources	Functions	Recommended daily allowance	Deficiency	Toxicity
Folic acid (folacin; pteroylglutamic acid)	Green leafy vegetables; liver; kidney; lima beans; asparagus; whole grains; nuts; legumes; yeast	Blood-cell formation	0.05 mg	Anemia	Generally none; reports of folate hypersensitivity, possible neurotoxicity at 15 mg daily
Cobalamin (B_{12}; cyanocobalamin)	Only foods of animal origin; beef, liver, kidney; milk; eggs; oysters; shrimp; pork; chicken	Essential for function of all cells	0.003 mg	Anemia; nerve fiber degeneration	None
Pyridoxine (B_6)	Yeast; wheat and corn; egg yolk; liver; kidney; muscle meats (20% lost in processing grain)	Components of coenzymes for protein synthesis; central nervous system metabolism	2 mg	Anemia	None until doses of 600 mg; neurotoxicity, depression at megadoses
Ascorbic acid (C)	Citrus fruits; tomatoes; green vegetables	Collagen formation; capillary integrity; synthesis of adrenal cortical hormones; aids in iron absorption from intestine	45–80 mg	Scurvy; bleeding gums; easy bruising; swollen joints; impaired wound healing	Inconclusive; reports of kidney stones, diarrhea, nausea at megadoses over 2–5 g; interference with absorption of B_{12} and trace minerals
Fat-soluble vitamins					
A (A alcohol = retinol; A aldehyde = retinal; A acid = retinoic acid)	Whole milk; liver; kidney; cream, butter; egg yolk; yellow and green vegetables; fruits	Night vision; growth; reproduction; health of epithelial cells; cell membrane maintenance	1,000 retinol equivalents (5,000 I.U.)	Night blindness; skin lesions	Very toxic in high doses of 20–30 × requirement; effects reverse on discontinuation
D (ergocalciferol = D_2; cholecalciferol = D_3)	Fatty fish; eggs; liver; butter; fortified milk; cod-liver oil; also through exposure to ultraviolet light	Normal bone formation; promotes absorption and retention of calcium and phosphorus	300–400 I.U.	Rickets in children; osteomalacia in adults	Highly toxic in large doses; possibly fatal in children

Table 16.1 (*Continued*)

Name	Rich Sources	Function	Recommended Dietary Allowances (RDA)	Deficiency	Potential Toxicity When Large Amounts Consumed
E (alpha-to-copherol)	Wheat germ oils; other vegetable oils; beef liver; milk; eggs; butter; leafy vegetables	May function as antioxidant in tissues; possible role in prevention of cell degeneration	12–15 mg	None known in humans; sterility; muscular dystrophy; red cell fragility in animals	Reportedly nontoxic up to 1 g/day; effects of long-term megadoses unknown.
K (naphtho-quinones)	Synthesized by intestinal bacteria; also lettuce, spinach, kale, cauliflower	Essential for blood clotting	Unknown	Increased clotting time	None

ual taking into account age, sex, and reproductive status. The RDA are formulated by the Food and Nutrition Board of the National Research Council, National Academy of Sciences. The nutritional scientists who make up the board provide the recommendations that reflect the information that has been obtained through various research studies conducted on large population groups. The RDA are meant to be a helpful and practical guide to planning well-balanced diets for groups of people; they are not necessarily meant to be goals or some sort of rigid standards to be met by individuals.

The RDA for vitamins already include a very generous margin of safety to cover possible wide variations of individual needs and to provide for the possible losses of vitamins that occur in the preparation and storage of food. For this reason, getting less than the recommended amount of vitamins does not mean that the intake is too low. The allowances are already at a high level, perhaps 25%–50% higher than the determined average requirement. People who eat a *variety* of foods that includes meats, milk, eggs, and legumes, fruits and vegetables, and whole grain or enriched cereals and breads—the so-called balanced diet—can be assured that their vitamin needs are being met.

There is, however, another aspect to the suggested daily allowances to be considered. The recommendations, which take into account the individual differences among people for the amounts of vitamins or any other nutrient to be consumed each day—already exceed the requirements for most individuals. They may not, however, meet vitamin needs of certain people because real people, as distinguished from the hypothetical average person, are likely to have great *biochemical individuality* in addition to all the other differences derived as a result of their genetic makeups. This concept of biochemical individuality, (credited to Roger Williams, discov-

erer of pantothenic acid) can make the idea of recommendations based on averages inherently invalid. Some people probably have low requirements for a particular vitamin and may get along very well on a quarter of the RDA with no apparent further need. Others may have very high requirements; they could need as much as twice the daily allowance for total well-being and best performance but still show no outward symptoms of vitamin deficiency. These individuals within the total population, whose metabolism may be such that they have an increased requirement for a particular vitamin even when on a well-balanced diet, make it difficult to state unequivocally that extra vitamin supplementation is never needed with proper diet.

The question is, should people take vitamin pills as insurance just in case they happen to be the ones whose requirements are not met by a balanced diet? The only answer is—maybe. No one appears to have the definitive word. Certainly, pharmaceutical companies that manufacture vitamins would have us believe that we need supplementation to lead healthy, active lives, full of vim and vitality, and they advertise accordingly. Most nutritionists, however, maintain that taking extra vitamin pills in the presence of an adequate diet is an unnecessary expense. Furthermore, there are no scientific data to indicate the necessity for supplementation when people are eating reasonably well.

It is not too difficult to determine whether there are aspects of individual eating patterns or life-style that may increase the risk of inadequate vitamin intake. There are the meat and potatoes people—those who never learned to like anything green. They have stringent food preferences that limit the variety of foods ingested. Other individuals may be voluntarily restricting their diets in order to lose weight and are thereby unbalancing their nutrients. Some may be involuntarily restricting their normal diets because the cost of meat and fresh fruits and vegetables has become more than they can afford. Heavy smokers sometimes show reduced blood levels of vitamin D, and heavy drinkers are more likely to have poor eating habits. Other kinds of drugs, either over-the-counter (antacids, laxatives) or medically prescribed, can cause vitamin depletion by interacting with the nutrients in foods. Some of the chemotherapeutic drugs used in the treatment of cancer, certain antibiotics, and various anticoagulants can act as antivitamins or interfere with vitamin absorption, effects that could be ignored by the prescribing physicians.

Obviously, swallowing vitamin pills should never be a substitute for the optimal nutrition obtained through the wise selection of a variety of good foods. But under certain circumstances, something only individuals themselves can know, vitamin pills may be a feasible way of defending against a probable deficiency. The most sensible way to take vitamin pills, if one chooses to buy them, is to figure out which vitamins (and minerals) are potentially lacking in the diet and to supplement only the ones needed as inexpensively as possible. Since vitamins exert their effects in only small quantities, pills that contain enormous amounts in excess of the RDA are more costly and a waste of money. In some instances, an excessive intake can have adverse effects.

Reasonable quantities of the water-soluble vitamins are safe to take. If unneeded by the body, they are merely going to be excreted to provide the sewage system with vitamin supplementation. (Someone once said that the greatest source of vitamins is middle-class urine.) Excessive amounts of water-soluble vitamins have been found to be harmful, and the overconsumption of fat-soluble vitamins can be toxic. Vitamins A and D are stored in the body for long periods of time, are metabolized very slowly, and are excreted with great difficulty by way of the bile. It is virtually impossible, however, to get too much A or D

from the diet, and the only way for toxic effects to occur is by taking these vitamins in tablet or capsule form.

Carotene, found in carrots, sweet potatoes, apricots, peaches, and other yellow and orange fruits and vegetables, is the provitamin for A. It is converted to vitamin A in the body cells after ingestion. Excessive consumption of carotene will actually turn Caucasian skin yellow, especially on the palms of the hands, but it does not appear to be harmful. The yellow tinge will disappear when less carotene is consumed. Daily ingestion of more than 25,000 I.U. of A, however, will cause toxicity. Similarly, more than 50,000–75,000 I.U. of vitamin D per day will cause toxic symptoms and possibly development of kidney and bone disease.

Even consumption of far lesser amounts could increase intestinal uptake and decrease intestinal excretion of calcium. The result may be unwanted calcium deposits in soft tissues (kidney, heart, lung, arteries). Holmes and Kummerow (1983) have even suggested that the fortification of dairy products with vitamin D is an unwise practice and should be abolished since it leads most people to consume and synthesize more of the vitamin than they really require.

Natural versus Synthetic Vitamins

In almost any drugstore, there are several kinds of vitamin preparations on the shelf. If the pharmacy is part of a large chain, the "name" brand will be more expensive than the "house" brand; more costly than either are the "natural" vitamins. Vitamins labeled natural have presumably been extracted from a food rather than formulated synthetically in the laboratory. Many people, believing that natural vitamins are superior and provide special benefits, appear to be quite willing to pay the inordinately high price for the products so labeled.

All of the compounds presently considered vitamins have been isolated. Their chemical formulas are known and, with the exception of vitamin B_{12}, they can all be synthesized in the laboratory. In manufacturing synthetic vitamins the same chemicals found in nature are used. The same chemical reactions that take place in the cells are duplicated. In every way, the synthetics are identical in structure and biological activity to the vitamins occurring naturally in foods, and no form of chemical analysis or biological assay can tell them apart. Not only are there no known advantages to "natural" vitamins, but there is some question about how natural they are. In the *Journal of Nutritional Education*, pharmacologist Adolph Kamil described a visit to two manufacturers of natural vitamins in California. These companies make most of the preparations sold under the famous natural brand names in health food and drug stores across the country. He discovered that while "Rose Hips Vitamin C Tablets" contained some naturally extracted C, there was also a lot of additional synthetic ascorbic acid or C in the preparation. The reason for the addition was that natural rose hips contain only 2% vitamin C. If no synthetic were added, he was informed, "the tablet would have to be as big as a golf ball." Because the labels on the finished bottles neglected to mention the additional source of vitamin C, the consumer presumes that the product is entirely extracted from rose hips.

In a similar fashion, the natural B-vitamins were composed of large amounts of synthetic B compounds combined with small amounts that had come from yeast and other natural sources. Of all the manufactured vitamins, only vitamin E was actually completely derived from vegetable oils. It also was sold as inexpensively as the synthetic variety. As Kamil pointed out, however, the wheat germ, soy, and corn from which the E was extracted *was not* organically grown, that is, without the usual chemical fertilizers and pesticides. Moreover, the vitamin was concentrated using the usual chemical solvents and procedures, and even the gelatin capsule in which it was enclosed contained a chemical preservative to keep it from getting rancid. All in all, the result was hardly a natural or organic vitamin E.

Vitamins, whether extracted from food sources or chemically synthesized, are the same in every way except price. Anyone who wants the benefits of natural vitamins should spend the extra money to get them the most natural way—from the nutritious foods that contain them.

High-Dose Vitamin Therapy

Knowing that illness can result from the absence or insufficiency of a particular vitamin tends to promote the idea that if a little of that substance is good, a lot is going to be better. It is tempting to believe that something as easily available and relatively inexpensive as vitamin pills can have miracle powers to provide superhealth. There are many claims for the preventive and curative effects of massive doses of the various vitamins. Although there are many proponents of such megavitamin therapy, there is no really convincing evidence to confirm the advantages of excessive consumption. It should also be emphasized that when huge amounts of a vitamin are taken to benefit a nonnutritional disease, the vitamin is likely to be acting as a pharmacological agent or drug, and not in the usual manner as a vitamin.

Intake of vitamin A in excess of nutritional requirements has been claimed to improve vision, prevent infection, and protect against cancer. Greater than the RDA for vitamin D, especially when taken in codfish liver oil—the "natural" way—has been said to have all kinds of benefits. There is no scientific substantiation for the claims that large doses of A and D in the absence of deficiency are beneficial in any way. As previously indicated, the abuse of A and D is actually dangerous.

Vitamin E has been enthusiastically advocated for almost everything—leg cramps, asthma, heart disease, sterility, impotence, ulcers, burns, wound healing, high cholesterol, menopausal symptoms, and aging—to mention some of them. Again, there is at present insufficient basis for the claims, and much more research and evaluation is needed before benefits of extra E can be confirmed. Most of the belief in the effectiveness of E is derived from animal research. For example, a vitamin E deficiency leads to pregnancy failure in female rats and sterility in male rats. Somehow this has been translated to mean that the vitamin enhances the sexual potency of male humans. It is curious how people who are quite willing to accept laboratory animal studies indicating beneficial results of vitamin E will reject animal studies when they illustrate an adverse effect from substances such as artificial sweeteners.

After Nobel Prize laureate Linus Pauling wrote the book, *Vitamin C and the Common Cold*, millions of people began to take ascorbic acid in the recommended 1–5 g daily so they could prevent the most common of human ailments. An imposing 15 g daily was advised for treatment of a cold. There are not enough data to confirm or deny the benefit of high intake of C to prevent colds, although there are some indications that increased ingestion of C may reduce the severity, if not the incidence, of colds. Pauling also pursued the idea of vitamin C being effective against cancer through enhancement of natural resistance to the disease, and in collaboration with other proponents, conducted studies that reported a beneficial effect on cancer patients. One controlled clinical trial reported no benefits resulting from administration with the vita-

min (Creagan et al., 1979), and another said it actually stimulated cancer growth (Campbell and Jack, 1979), so the case for vitamin C as cancer therapy is still very controversial.

Vitamin C ingestion has been generally viewed as nontoxic, but massive consumption may not be harmless in some individuals. Although the data have been disputed, Herbert and Jacob (1974) showed that large doses of C can destroy vitamin B_{12} when the two are ingested in foods together, thus potentially interfering with the absorption and metabolism of B_{12}. A rebound scurvy effect can develop when megavitamin doses of C are withdrawn (Siegel et al., 1982), and Cochrane (1965) reported that babies born to mothers who took more than 400 mg of C during pregnancy later developed scurvy although they were fed a normal diet. The feeding of large doses of C to laboratory animals has caused diarrhea, very acid urine, and bladder and kidney stones, and gastrointestinal symptoms have occurred in some people on 1 g doses. In general, it would be wise to be cautious in taking really substantial amounts of vitamin C until more information is available. Moreover, anyone having a complete physical examination should be aware that fecal excretion of vitamin C interferes with the fecal occult blood test for colon cancer and should abstain from C ingestion for several days before the exam.

Some vitamin preparations that contain 10–15 times the RDA for the B-complex vitamins are called "stress-formula," under the assumption that such high potency in these vitamins will relieve emotional distress and depression. Since the B vitamins are necessary for the metabolic function of all cells, they are, of course, essential for the proper function of nerve cells. It is not surprising that a deficiency of any one of them—thiamine, riboflavin, niacin, pantothenic acid, folic acid, biotin, B_6, and B_{12}—would produce mental symptoms. There is no evidence to indicate that during periods of emotional stress and in the absence of deficiency, taking amounts of B vitamins in excess of the RDA is of any benefit. The idea may have originated in studies that reported that the occasional depression that occurs in women on oral contraceptives was linked to an absolute vitamin B_6 deficiency. When the vitamin was administered, the symptoms of depression were relieved (Adams et al., 1973). It is also known that a riboflavin deficiency is more frequent during periods of physiological stress (after surgery, trauma, illness) and during pregnancy and rapid growth periods in children. There are many who believe, however, that psychological stress also increases the need for the B-vitamins, but there is little corroboration for the supposition.

Orthomolecular psychiatry is a controversial method of treatment that advocates megavitamin therapy for such emotional disorders as schizophrenia, hyperactive children, childhood autism, and various kinds of neuroses and psychoses. Orthomolecular psychiatrists may still use the traditional forms of treatment but also add very high doses of water-soluble vitamins and, frequently, special diets. The usefulness of megavitamin therapy in the treatment of psychiatric disorders has been challenged by the American Psychiatric Association, and the methods and claims of its practitioners have not been well accepted by most psychiatrists.

The true believers in megavitamin dosage hailed the practice as an exciting breakthrough. The critics denounced it as faddist with the possibility of adverse effects. Megavitamin therapy may be beneficial, but, unfortunately, one can never be completely certain at present that taking doses of nutrients at 20–100 times their usual daily level does not have some potential for harm. Generally speaking, an excess or "mega" -anything taken into the body frequently turns out to ultimately have undesirable effects.

Vitamin-like Compounds

In order for a substance to be classified as a vitamin, there are certain criteria to be met. For one thing, the compound must have an established biological role, that is, it has to be a proved essential dietary requirement, the absence of which would lead to a visible deficiency. Secondly, true vitamins are not able to be synthesized in the body's cells in adequate amounts to meet physiological needs. Finally, the established vitamins are present in foods in very small quantities and exert their metabolic effects at very low levels. While these requirements may seem somewhat arbitrary, they do define, on the basis of present knowledge, the concept of what constitutes a vitamin.

Choline is a substance frequently listed with the B-vitamins although it can be synthesized in the body from the amino acid, glycine. While it is known to be a vitamin for many young animals, its need by humans has never been established. Choline in animals is necessary to prevent the accumulation of fat in their livers, but although fat infiltration of the liver is common in chronic alcoholism or in protein-calorie deficiency, feeding choline has never been successful in curing this disorder in humans. Choline, however, does have many important functions in the body. It is a component of cell membranes and of myelin, the insulating covering around the nerve fibers. It is a part of the lipoproteins involved in the transport of fat-soluble substances and is necessary for the synthesis of acetylcholine, which functions in the transmission of nerve impulses.

The sugarlike inositol and para-aminobenzoic acid (PABA), also sometimes included in the B-vitamins, are important growth factors for lower animals or microorganisms, but there is no evidence that they are vitamins for humans.

Bioflavonoids are a group of compounds that include hesperidin, rutin, flavonones, flavones, and flavonols. They are widely distributed in plant foods, found particularly in citrus and other fruits, berries, such vegetables as cabbages, brussels sprouts, and onions, and in tea, vinegar, and red wine. It is estimated that the average daily intake of flavonoids amounts to about a gram from a normal mixed diet. There is evidence that some bioflavonoids have an antibacterial and antiviral activity. There has also been research to indicate that many of the flavones appear to have a tumor-inhibiting effect. The substances also have a vitamin C-like activity, and large doses have been used therapeutically in patients with increased capillary fragility or permeability. While a bioflavonoid deficiency has been produced in animals, however, it has never been demonstrated for humans. In an extensive review of the literature concerning the bioflavonoids and their role in human nutrition, Kühnau (1976) suggested that they be considered "semi-essential" nutrients.

Popular Nonvitamin "Vitamins"

Laetrile, also known as "vitamin B_{17}," has been widely promoted as a cancer cure. Originally discovered by E. T. Krebs, the preparation is commercially prepared from apricot pits but is also found in such foods as maize, sorghum millet, lima beans, kidney beans, and the seeds or pits of apples, cherries, pears, plums, prunes, and some other fruits. Laetrile is actually a compound called amygdalin, a member of a group chemically known as cyanogens or cyanogenic glycosides. Plants containing cyanogenic glycosides are capable of producing the highly poisonous hydrocyanic acid. Cyanide poisoning is rarely caused by the foods in which amygdalin is present, however, either because the quantity of toxin is so small that it can be detoxified and metabolized or because the seeds are ordinarily not eaten. The presence of even small amounts of such cyanogens would appear to preclude Laetrile's classification as a vitamin, however, since humans have never been known to require an essential dietary poison.

Laetrile has never been shown to be effective in the treatment of cancer, and there is substantial evidence to attest to its worthlessness. In response to the claims that thousands of cancer victims had purportedly benefited from the use of Laetrile and that hundreds of thousands were being denied its use as an anticancer drug, the National Cancer Institute in 1978 recommended to the Food and Drug Administration that a controlled clinical trial of Laetrile in terminally ill cancer patients be initiated. In 1981, the results of that evaluation by four prestigious cancer research centers were reported as emphatically, conclusively, negative. The substance was administered to 178 people with advanced cancer for whom conventional therapy had ceased to be effective. No substantive benefit of Laetrile in terms of cure, improvement, pain-relief, or slowing the progress of the cancer was observed, and 105 of the study group died. Of course, the true believers were undeterred, and immediately contended that the study did not use the right kind of Laetrile.

There are relationships between diet and cancer, and considerable evidence is accumulating for the nutrition-associated basis for many types of cancer. Unfortunately, many people put two and two together and get five. Since Laetrile is commercially isolated from apricot pits that contain amygdalin, and since almonds also contain amygdalin, some pseudonutritionists have advised that eating several almonds each day will protect against cancer. Nutritionally, eating almonds in moderation is a good idea. They are protein-rich and high in levels of riboflavin and niacin, but about the only thing to be said for consuming them for cancer is that it undoubtedly beats trying to eat apricot pits.

Another highly touted nonvitamin is pangamic acid or "vitamin B_{15}." Like Laetrile, pangamic acid was also first discovered in 1951 by E. T. Krebs. It has subsequently been found to exist in various grains with a particularly high content in corn and rice.

Although many physiological functions have been attributed to pangamic acid, neither a need for the substance nor any deficiency disease resulting from its lack has ever been shown to exist in humans, and it has never been accorded status as a vitamin. Nevertheless, it is available in health food stores at 40 cents per 50 mg tablet, and customers are advised to take three to six tablets daily. Whether it is a panacea for a variety of ailments or a ripoff depends on whether one believes its fervent promoters or the Food and Drug Administration who questioned its value and safety. Dr. Robert Atkins, originator of the Atkins Revolutionary Diet and author of the *Dr. Atkins' Superenergy Diet*, believes it is a fatigue-fighter and recommends it for "superenergy." Some marathon runners and professional football players are said to depend on it. Former heavyweight champion Muhammed Ali purportedly took pangamic acid before three consecutive victories. Food-Science Laboratories, whose "Aangamik 15" is the best-selling version of pangamic acid in this country, also markets "Peppy-15," a version for dogs, and "Spur," a horse supplement.

Studies in laboratory animals have shown that pangamic acid stimulates tissue oxygen consumption, inhibits fatty liver formation, causes an increased ability to perform strenuous exercise, and decreases blood cholesterol levels. The majority of experimental work has been done by Soviet scientists, who have also widely used the substance clinically in humans. The Russians have published a number of papers attesting to the therapeutic value of pangamic acid in such diverse conditions as heart disease, hepatitis, and skin conditions. Other diseases for which the Soviets suggest treatment with "B_{15}" include aging, diabetes, gangrene, hypertension, glaucoma, alcoholism, schizophrenia, allergies, and mental retardation in children. Evidently pangamic acid is seriously regarded abroad but is seen only as a feel-good pill of dubious value in this country, pending further research. In a research review article, Stacpoole (1977) concluded that there is some evidence indicating that administration of pangamic acid may be beneficial in certain conditions in which increased tissue oxygen uptake is important, such as coronary heart disease, congestive heart failure, atherosclerosis, or during strenuous physical exercise. The definitive controlled studies to conclusively prove this benefit, however, are still lacking.

MINERALS IN HUMAN NUTRITION

Of all the 92 naturally occurring chemical elements in nature, more than 50 are found in human body tissues. Fewer than half of these are known to be essential to human body function. Those minerals known to be required in the diet at levels of 100 mg daily or more include calcium, phosphorus, sodium, potassium, chlorine, magnesium, and sulfur. Pertinent information concerning their source and function is listed in Table 16.2. The remaining mineral nutrients are called *trace elements* because they are needed in daily amounts of only a few milligrams. They are listed in Table 16.3.

There are a number of minerals found in minute quantities in the body. They include

Table 16.2   ESSENTIAL MINERALS NEEDED IN AMOUNTS OF 100 MG/DAY OR MORE

Element	RDA (adults)	Rich Sources	Functions in Body
Calcium	800 mg	Milk, cheese; leafy vegetables, legumes, nuts; whole grain cereals; bones from sardines and other canned fish	Normal bone and teeth structure, muscular contraction, blood coagulation, nerve membrane stability
Phosphorus	800 mg	Protein-rich foods	Normal bone and teeth structure, production and transfer of high energy phosphates, absorption and transportation of other nutrients, regulation of acid-base balance
Sodium	5 gm (five times more than actual physiological need)	All food, table salt	Osmotic pressure of body fluids, muscle function, permeability of all cells
Potassium	4 gms	All foods	Muscular activity, especially heart, and proper nerve function
Chlorine	Same as sodium	All foods, table salt	Osmotic pressure regulation, water balance, acid balance
Magnesium	350 mg	Most foods, especially vegetables; milk, meat, cocoa, nuts, soybeans	Enzyme activity, energy-release, nerve and muscle function
Sulfur	0.6–1.6 gms	All proteins, particularly those rich in cystine and methionine	Component of vitamins, hormones, enzyme systems, important in detoxification mechanisms

tin, nickel, silicon, and vanadium. These are elements that *may* be essential and vital in human metabolic processes, but no human requirements have been established for them as yet.

With the possible exception of iron in women who have heavy or prolonged menstrual periods, deficiencies of the trace minerals are rare. In the opinion of some researchers, however, the current popularity of high fiber diets could result in various mineral deficiencies, most notably calcium. Evidently increasing the fiber content of the diet may impair the absorption of iron, copper, and calcium from the intestine. There has also been a suggestion that a chemical reaction between phytic acid in fiber and some minerals, such as calcium, causes trace mineral inactivation.

One prevalent theory about the heartbreaking neurological disorder called Alzheimer's disease is that it is the result of aluminum toxicity because some people with the disease have increased levels of aluminum in their brains and neurons. Evidence is contradictory, however, and patients who get aluminum toxicity from long-term dialysis treatment do not get Alzheimer's disease.

Table 16.3 ESSENTIAL TRACE ELEMENTS

Element	RDA	Rich Sources	Functions in Body
Iron	10 mg (males) 18 mg (females)	Organ meats (liver, heart, kidney, spleen), egg yolk, fish, oysters, whole wheat, beans, figs, dates, molasses, green vegetables	Oxygen transport, cellular respiration
Copper	2.5 mg	Liver, kidney, shellfish, nuts, raisins, dried legumes	Enzyme component, hemoglobin formation
Cobalt	Unknown	Animal protein sources	Part of Vitamin B_{12}
Zinc	15 mg	Meat, especially liver, eggs, seafoods, milk, grain	Enzyme component, part of insulin molecule
Manganese	300–350 mg(?)	Bananas, whole grains, leafy vegetables	Normal bone structure, normal function of reproductive and nervous system
Iodine	100–140 micrograms	Fish, iodized table salt	Necessary for normal thyroid function
Molybdenum	Unknown	Beef kidney, legumes, some cereals	Enzyme component
Selenium	Unknown, probably 50–100 micrograns is adequate	Seafood, meat, grains raised in selenium-rich soil	Enzyme component, similar to vitamin E in function
Chromium	Unknown, probably 20–50 micrograms adequate	Meat, corn oil	Normal glucose metabolism
Fluorine	1–2 mg	Fluoridated water, milk	Resistance to dental caries

DIETARY GUIDES

People are born knowing how, but not what, to eat. There is no inherent instinct that makes us choose a properly balanced diet. In fact, there appears to be every evidence that left to their own devices, people will probably select diets that are inadequate to meet the body's requirements for essential nutrients. Perhaps the human body utilizes proteins, carbohydrates, fats, vitamins, and minerals, but individuals eat *food*, not nutrients. And food is not just an adequate diet. Food means Mom, home, and the family—love and emotional identity. Food is not just for hunger; it means comfort when one is upset and a re-

ward when one is being indulgent. Food is finger-lickin' good, a pepperoni pizza, a dinner party, or splurging in a restaurant. There can be any number of social and psychological considerations that affect eating and food choices, and getting good nutrition at the same time is not automatic. People have to be educated to select the foods consistent with optimal health. One practical approach to nutrition education is the use of the Basic Four Food Guide that is based on the recommended daily allowances (see Table 16.4).

The Basic Four—milk, meat, vegetable-fruit, and cereal-bread—has sometimes been criticized as being perhaps too heavily weighted toward the interests of the dairy and meat

Table 16.4 A DAILY FOOD GUIDE

Meat Group

2 or more servings of lean meat, fish, poultry, eggs, legumes

    1 serving can be: 3 oz hamburger, chicken leg, or fish
                                              2 eggs
                                              1 cup cooked dry beans or peas
                                                4 tablespoons peanut butter

Milk Group

Children: 3–4 servings
Teenagers: 4 or more servings
Adults: 2 servings
Pregnant women: 3 or more servings

    1 serving can be: 1 cup whole, skim, buttermilk, evaporated
                                                  or dry milk, reconstituted
                                              1 oz cheese such as Swiss or Cheddar
                                              4 tablespoons of cottage cheese
                                              1 cup plain yogurt

Fruit and Vegetable Group

4 or more servings, including 1 good or 2 fair sources of Vitamin C

    1 serving can be: 1 orange
                                              1/2 cup orange juice
                                              1 cup tomato juice
                                                1/2 cantalope
                                              3/4 cup strawberries
                                              1 cup cabbage
                                              1 large potato

1 good source of Vitamin A, at least every other day

    1 serving can be: 1/2 cup raw or cooked apricots, broccoli,
                                              carrots, chard, collards, cress, kale,
                                              pumpkin, spinach, sweet potatoes,
                                              turnip greens, winter squash

Bread and Cereal Group, Whole-Grain or Enriched

4 or more servings

    1 serving can be: 1 slice of bread
                                                1 ounce dry cereal

Table 16.4 (*Continued*)

	1/2 square matzo
	1/2 bagel
	1/2 cup cooked cereal
	1/2 cup cooked pasta, rice, grits, cornmeal

Other Foods as Needed to Complete Meals and to Provide Additional Food Energy and Other Food Values

1 serving can be:	1 tablespoon butter or margarine
	1 tablespoon vegetable oil

industries. As usually presented, there may be insufficient emphasis on the alternates, that legumes and nuts in appropriate combinations with cereal grains can adequately replace the protein requirement and that the milk group need can be as nutritiously met by skim milk and skim-milk cheese.

A further weakness of the Basic Four plan may be that it is no longer as useful in the selection of a well-balanced diet as it was when the educational grouping was first introduced in 1956. There have been four revisions of the recommended daily allowances since then, and a diet plan based strictly on the minimum number of servings from the four food groups has been found to be inadequate according to current allowances (King et al., 1978). But perhaps the greatest shortcoming of the Basic Four is that it is too simplistic for today's needs—that the kind of nutritional information it provides does not really match the way people are eating and the kind of education they want. In the mid-1950s it was easier to group foods into two servings from this classification and three servings from that one. Not only was there a smaller food selection in the grocery store, but more foods were staple items that could recognizably conform to the Basic Four. Today's eating habits have changed considerably. There are new categories: "health foods," "junk foods," "fast-foods," and engineered or synthetic foods. Surveys have indicated that Americans now spend one-third of their food budgets at restaurants, mostly of the fast-food chain variety. The country has become dotted from one end to the other with McDonalds, Burger Kings, Taco Bells, Arthur Treacher's, the Colonel's unbiquitous Kentucky Fried Chicken, Denny's, Sambo's—a veritable explosion of carry-out or eat-it-here quickie and relatively inexpensive places (Fig. 16.3). Mom's preparing meals from scratch takes more time, time that a working woman may not have, and convenience foods have become a major dietary inclusion for many people. The food companies, scrambling over each other to cash in on the convenience food dollar, have produced a number of products for the supermarket shelves and freezers to compete with the fast-food chains. How do these new foods fit into the Basic Four?

A typical contemporary diet may be balanced, but a new form of malnutrition—*over*, rather than undernutrition—may now be the danger. For example, a Big Mac contains all four food groups: meat, cheese, vegetable and cereal, and it also provides 31 grams of protein. Although apparently nutritiously sound, those "two-all-beef-patties-special-sauce," or its equivalent, will also provide more than 1,100 calories when eaten with large fries and

Figure 16.3. Anytown, U.S.A.

a chocolate shake. Not only are those calories excessive for many people, but many of them are in the form of saturated fats.

The concept of the Basic Four Food Groups is not without value, however. This kind of grouping of foods is an effective and easy way of learning about a nutritious diet. Since it is frequently taught at the elementary school level, it can impress children with the necessity of eating a variety of foods and getting some fresh fruits and vegetables daily. If people past grade school continue to consume dairy and protein two times per day and cereals, fruits, and vegetables three times daily, they are certain to be choosing an adequate diet. Many adult Americans, however, would

welcome and use a more complex and sophisticated approach to selecting their meals. There has been an increasing awareness that it is not so much that "you are what you eat," but that what you eat may largely determine what you are (or whether you are) 20 or 30 years from now. The payoff of today's good nutrition could be the prevention of the killer diseases that accompany aging. Consumers have a vested interest in nutritional education.

Unfortunately, accurate and definitive nutritional information is not that easy to obtain. Traditionally, nutrition has been considered the province of home economists, relegated to the food and women's pages of the newspapers. Although nutrition is now a hot item and front-page news, nutrition as a science is only slowly approaching the status of an acknowledged and respected discipline. There is a shortage of qualified nutritional scientists and educators and a lack of comprehensive and controlled research studies that would provide accurate nutritional information. At this point, many of the issues in nutrition are highly controversial. The relationship between diet and heart disease, hypertension, diabetes, and cancer is obvious to some scientists and debatable for others. About the only agreement among the experts is that many Americans probably eat too much of everything.

In the absence of unequivocal answers to the questions of how and in what manner nutrition affects health and disease, all of us become more vulnerable to the advice of the self-styled "authorities" who proclaim that they know the true way to superhealth and energy. Who, after all, does not want to be brimming with vim and vigor and glowing with vitality? A mishmash of misinformation is easily accessible through magazine articles and popular books, from promoters of various food supplements, and by word of mouth. Many of the articles are likely to contain a nugget of fact buried in a mound of hypotheses. Since there is much about nutrition that still remains to be determined, it is impossible to categorically prove or disprove all the claims. Fortunately, most of the food mythology is harmless. Because special foods or food supplements may be bought in special stores, nutrition misinformation may have its greatest effect on the pocketbook. There is always the potential of harm, however, when people take seriously such notions as "pasteurizing milk destroys its enzymes" or that "fasting cleanses the body of noxious poisons." And when all health problems are self-diagnosed as being caused by nutritional inadequacy and self-treated by nutritional or "metabolic" therapy, real danger to health exists.

Several physician "experts" have written best-selling books on weight reduction that contain unsubstantiated nutritional advice of highly dubious value, illustrating that some doctors are equally guilty of spreading nutritional nonsense. It is not usually recognized that doctors frequently know as little or less about nutritional matters than their patients. The amount of nutritional education physicians get during their professional training is very sparse, and few ever receive a formal nutrition course in medical school. Eminent nutritionist Jean Mayer, president of Tufts University, has charged that the nutrition teaching in most medical cirricula was, and still is, "miserable." Although the need to upgrade the education of physicians in nutrition has been recognized and about 100 of the 143 medical schools in the United States and Canada now offer "some sort of course," according to the Association of Medical Colleges, only 15 schools have a mandatory course incorporated into a doctor's undergraduate or graduate education.

At this time, no one has all the answers on nutritional issues. Although there is a lack of scientific data, especially in the area of clinical research, it is unlikely that anyone will discover any new pill, mineral supplement, special food, or dietary regime that is going to guarantee good health and long life. One must always be wary of claims for the exaggerated virtues of a particular food, foods, or nutrients. Any diet that does not include the necessity to *eat moderately from a variety of foods* should be viewed with skepticism.

ARE HEALTH FOODS HEALTHIER?

Depending on whom one asks, health foods could be considered all, some, or none of the following: a) minimally processed foods that contain no chemical additives; b) wheat germ, protein supplements, sunflower and pumpkin seeds, various kinds of sprouts, and other brown, bland, and uninteresting foods; c) plant foods that are organically grown in soils fertilized only with manure or compost and without the addition of pesticides; d) foods bought in a health food store and costing more money.

Since the terms "natural," "organic," and "health" are used so frequently and for so many products, it is not surprising that they have become almost impossible to define in any kind of meaningful way. Even a tobacco company proclaims that its cigarettes use only "natural flavorings." When those words are used in relation to foods, however, most people usually think of the back-to-nature foods—simpler, whole, unrefined, unfabricated. Such foods have always been available, but it is only recently that one finds that higher prices must be paid for the privilege of buying them. As many food companies have discovered, there are big bucks in the health food industry.

There is apparently little chance that food can be grown in any large-scale quantities without retaining some traces of lingering pesticides. Although no reports of illness have directly been attributed to pesticide residues on foods, the possibility of harm exists. Without insecticides and rodenticides, however, crop losses would be inevitable, and the diseases carried by insects and rodents would become health hazards. Since most of our foods contain various residues, a certain amount of pesticide contamination is impossible to avoid. In one study, 55 products purchased at a health food store contained a greater incidence of pesticide traces than similar products not called "health" or "natural" (Barrett and Knight, 1976). Obviously, the level, or toxic threshold, for a potentially hazardous substance is very important. The federal government is responsible for making certain that pesticides as well as other environmental contaminants in food do not exceed established tolerance levels for safety. The public relies on such agencies as the Agriculture Department, the Environmental Protection Agency, and the Food and Drug Administration for protection. That they could be doing a better job of checking food safety was indicated by a congressional study released in late 1978 by the House Commerce Subcommittee on Oversight and Investigations. The report, the result of a year-long investigation, concluded that the efficacy of the monitoring techniques for tolerance levels was highly questionable. For example, although random sample tests found that 8%–15% of the meats and poultry analyzed contained drugs, pesticides, and general environmental contaminants, which exceeded legal levels, the FDA recalled no meat or poultry during 1975–1977 and prosecuted no one for violating the residue standard laws. Another report released in 1982 by a Washington-based consumer group, Public Voice for Food and Health Policy, charged that the FDA inspects only 20% of the total 8 billion pounds of fish consumed annually by the public and cited further gaps in government inspection of foods for antibiotics, lead, and food dyes. This kind of informa-

tion is discouraging in terms of food safety, but the consumer might find it heartening that at least there are congressional subcommittees and consumer agencies watching the watchdogs.

Not all of the chemicals intentionally added to foods are harmful, and some are of definite benefit to the consumer. We may have been educated to know the value of whole grain products, for example, and recognize that we should eat whole wheat bread and bran muffins. An overwhelming majority of the population still prefers to eat white bread, pancakes, and English muffins. White flour is, therefore, enriched with thiamin, niacin, riboflavin, and iron to partially restore some of the nutrients lost in the milling of whole wheat. Milk is fortified with vitamin D, margarine with vitamin A, and salt with iodine. There are even some substances, such as benzoic acid, that are naturally found in certain foods at levels greater than that permitted as additives. But although there is little evidence, as some popular writers have alleged, that we are slowly being poisoned by our food supply, not all food additives have been systematically and thoroughly tested. It is possible that some are capable of posing a real threat to health.[*]

Two chemicals that are suspect are BHA and BHT. Butylated hydroxyanisole and butylated hydroxytoluene are antioxidants. They are added to fat-containing foods in order to extend their shelf life and prevent them from becoming rancid. BHA and BHT are added to cereal packaging, gum, vegetable oil, shortening, snack foods, and most oil-containing products. Improved manufacturing tech-

---

[*] Issues of food safety are highly complex, controversial, and largely unresolved. For an accurate and unbiased view of food additives, their risks and benefits, see Michael Jacobson's *Eater's Digest: The Consumers Factbook of Food Additives*. Washington, D.C., Center for Science in the Public Interest, 1982. Another reference is *Nutritional Toxicology*, volume 1, edited by John Hathcock, New York, Academic Press, 1982.

niques have largely eliminated any necessity for the addition of these antioxidants, and they provide no benefit. Studies in which animals have ingested large quantities of BHA and BHT have produced inconclusive results, but many believe that these two chemicals have not been adequately tested for carcinogenicity or other potentially adverse effects.

On the other hand, and in one more example of the inconsistencies and controversies in nutritional research, mutagenic substances for bacteria are generally assumed to be more or less carcinogenic in humans. Antioxidants such as propyl gallate and BHA have been shown to effectively inhibit mutagen formation.

Another questionable additive that may ultimately be banned is sodium nitrite. Nitrites are added to cured or smoked meats, poultry, and pickled fish to prevent botulism (a lethal form of food poisoning) and in the case of meat, to add an appealing red color and distinctive cured taste. It has been known since the late 1960s that the added nitrites combine with protein derivatives called amines to form nitrosamines, proven carcinogens (especially stomach cancer) in laboratory animals. Not only nitrosamines, but the nitrites themselves are evidently capable of causing cancer in laboratory rats. The presence of salt in these foods may be a promoting or enhancing factor for the formation of carcinogens. The government appears to be aiming toward a gradual phaseout of the use of nitrites while giving the meat industry time to develop alternate methods of preservation. Foods containing nitrites are estimated to account for about 7% of the food supply. The public could probably get along without the color or even the taste imparted by nitrites, but in the absence of other ways to prevent botulism, an outright ban on the additive at present would effectively remove such common foods as hotdogs, ham, bacon, and luncheon meats from the diet.

There are many who believe the loss would not be that great. Since these foods have a high saturated fat content, are an expensive source of protein, and contain an additive of uncertain safety, they should be avoided and eaten only occasionally.

One of the more recent additives to engender controversy is aspartame, a sugar substitute that is marketed as NutraSweet by G.D. Searle & Company. Used by an estimated 50 million Americans in one form or another, aspartame is an intensely sweet chemical combination of aspartic aid and phenylalanine that was approved by the FDA for use in dry foods and as a table sugar substitute in 1981 and expanded for use in soft drinks in 1983. By the end of that year, aspartame, mixed with less-expensive saccharin, was present in 70% of all low-calorie soft drinks. Although the FDA attested that aspartame was safe to ingest even if abused at the unlikely levels of six times ordinary soft drink consumption, some scientists still have lingering doubts about its safety.

In final analysis, it is worth looking for foods without certain additives, but these need not necessarily be purchased in a health food store or carry a premium price. The food industry has recognized that the words "All Natural, No Preservatives or Additives" are golden, and many foods now carry that designation. It is still important to read the label to make certain that unwanted substances are not in the ingredients. The supermarket, moreover, has a whole line of health foods—whole wheat flour, buckwheat groats, wheat germ and bran, dried fruits, polyunsaturated oils without antioxidants—and the cost may not be as high. The produce department is full of health foods, and so is the dairy section. If one wants specialty or more exotic items, they are likely to be found only in the specialty health food shops, but the regular stores have always had the staples that are equally as nutritious and less expensive.

JUNK FOOD

Like health food, junk food may have several definitions. To the nutrition experts, food that is excessively high in fat and sugar content relative to the amount of other nutrients, is nutritional trash, "empty calories" that will fatten instead of nourish. To the food industry, however, which spends millions of advertising dollars to entice the public to eat the high-calorie, low-nutrient, sugar-rich items, the foods are not junk at all. They are high quality snack foods—soda pop, candy, cookies, pastries, various kinds of chips and curls, and the "nutritious" sugar-coated breakfast cereals. These products are so heavily promoted that it is almost impossible for an average TV watcher to remain unsusceptible to their mouth-watering "crunchy" or "creamy-smooth" goodness. Even those who make every effort to consume nourishing food—those nutritious low-fat, plenty of fruits and vegetables, whole grain kinds of meals—cannot be blamed for the occasional lapse from virtuous eating. Almost everyone sometimes indulges in the munchies. If gastronomic heaven means Twinkies, Mallomars, or a bag of Cheetos, Neetos, Fritos, or Doritos every once in a while, it is not a very large sin. Satisfying a junk-food yen is of little significance if overall nutrition is good, and no one can expect to be totally wholesome all the time. The real problem with junk occurs when it becomes the main course, when the stuff from the vending machine or the candy stand substitutes for lunch or dinner and crowds the other nutrients out of the diet. The other obvious danger in excessive "garbaging up on junk food" is that it inevitably leads to overweight.

Real Junk and Partial Junk

Soft drinks, candy and other surgery goodies, a bagged snack that crunches or curls—these are the acknowledged high-ranking junk foods. There are some foods, however, that are high in calories and high in fat but do have redeeming nutrient value; that is, they may be viewed as just a little bit "junky." Popular fast-food items can hardly be termed nutritionally unsound, but as indicated in Table 16.5, they may contain an awesome number of calories. Many of those calories are in the form of fat, and much of it is saturated. Unless one is actively trying to gain weight, fast-food meals should be limited to the occasional lunch or dinner, and other meals that day should contain fruits, vegetables, and the whole grain foods.

Sugar—The Hidden Assassin?

Sugar, the refined white granulated sweetener that comes in 5-pound bags or 1-pound boxes from the store, is the leading empty nutrient. Sugar tastes good and provides calories, which is all that can be said for it. It is not that sugar calories are different from other calories; it is only that it is so easy to get so many of them. Being easily dissolved and concentrated in solution, large amounts of sugar can be obtained from relatively small quantities of food. Sugar-rich foods thus provide a very poor nutrient return for the amount of caloric investment. When a high percentage of ingested calories are in the form of sugar, it becomes almost impossible to get all the necessary nutrients and still take in few enough calories to maintain a reasonable weight.

It would appear to be easy to avoid excessive sugar consumption. Just take the sugar bowl off the table and stop eating and drinking soda pop, candy, cookies, sugared bakery goods, and the like. Unfortunately, people have much less control over the amount of sugar ingested than they think. According to a *Consumer Reports'* study, before 1930 only 30% of the sugar used in this country was purchased by industry while 64% went home

Table 16.5 CALORIE CONTENT OF SOME FAST FOODS

Chain Name	Menu Item	Calories
Arby's	Roast beef sandwich	430
	Super roast beef sandwich	540
Arthur Treacher's	Two pieces chicken and chips	825
	Three piece dinner (fish, chips, coleslaw)	1,100
	Fish Fillet sandwich	390
Burger King	Whopper and regular fries	850

Kentucky Fried Chicken	Three piece dinner
	   (chicken, mashed potatoes,
	   gravy, roll, coleslaw)
		825

	Three piece dinner, extra crispy	1,070
McDonald's	Egg McMuffin	327
	Filet-o-Fish	432
	Big Mac	563
	Quarter Pounder	424
	Small fries	215
	Large fries	335
	Cheeseburger	309
	Big Mac, Large fries, Chocolate shake	1,276
Wendy's	Single	470
	Single cheese	580
	Triple cheese	1,040
	Chili	230
	Triple cheese, fries, Frosty	1,410
Pizza Hut	One/half of 13-inch cheese pizza	
	(thick crust)	900
	(thin crust)	850
	One/half of 15-inch cheese pizza	
	(thick crust)	1,200
	(thin crust)	1,150
Taco Bell	Beef and bean burrito	225
	Bell burger	243
	Enchirito	391
	Frijoles	231
	Taco	240
	Tostada	206

Source: Milwaukee Nutrition Council; Wendy's International, Inc.; Burger King Corporation; McDonald's Corporation.

with the consumer in the bag or box from the corner grocery. Today, in a complete reversal of the manner of consumption, only 24% of the sugar is bought by people at the store. Sixty-five percent of the sugar is now sold to the food industry, which places an amazingly high concentration of it in processed foods. Much of the sugar consumed is unwittingly obtained in such foods as ketchup, salad dressing, canned and dehydrated soups, frozen TV dinners, canned and frozen vegetables, yogurt, or breakfast cereals. Tests performed on dry breakfast cereals have indicated that they contain an average sugar content of 25%, ranging from a low of 1%–3% in Shredded Wheat, Cheerios, and Puffed Rice up to an astounding 58%–68% in such presweetened varieties as King Vitaman, Sugar Smacks, and Super Orange Crisp (Shannon, 1974). Even a Hershey Bar only has 51% sugar content. Consider the nutrient value of an ounce of breakfast cereal that is more than two-thirds sugar; one could obtain equivalent results by pouring sugar in the bowl and sprinkling the cereal on top!

Ira Shannon and his associates analyzed more than a thousand food and drink items for sucrose (sugar) concentration and came to the depressing conclusion that nearly everything has sugar added to it. There is even sugar in salt. Sugar has become the leading food additive. Although estimates vary, it can safely be assumed that every man, woman, and child in the United States is ingesting more than 100 pounds of sugar annually, much of it inadvertently obtained in processed foods. How detrimental all this sugar consumption is to health is inconclusive, but there is a rapidly growing body of medical literature that correlates excessive sugar intake with a number of health problems.

There is little argument, for example, against the evidence that the sweet tooth is the decayed tooth. There may be other factors such as genetic susceptibility that contribute to cavities, but the causal relationship between the amount and frequency of sugar consumption and dental caries is well established. Tooth decay is not a minor problem in a country that spends $10 billion annually on dental bills. Cutting down on cavities means cutting down on all sugar and sugar-containing foods, especially on sticky or slowly dissolving sweets between meals. Brushing and flossing after every meal, or at least rinsing with warm water after eating any carbohydrate, will help. Some authorities maintain that while sugar restriction is necessary, fluoridation of the water supply is of paramount importance (Finn and Glass, 1975).

With the possible exception of the Sugar Association, almost everyone would agree that sugar should not be a major component of the daily diet. Obviously, the known high-sugar junk foods—soft drinks, candy, cookies, and rich pastries—should be indulged in only very rarely. Trying to reduce the less visible sources of sugar is more difficult, however, and requires some diligent label reading. The list of ingredients on any canned, packaged, or frozen food must reveal the contents in descending order by weight. Unfortunately, the percentage of sugar need not be given, but if sucrose, dextrose, maltose, any kind of "ose," any kind of syrup, or the words "natural sweeteners" appear high on the list or more than once, the product should be avoided.

It should also be recognized that sugar is sugar. Calling it "brown," "turbinado," or "natural" does not offer any advantage. Brown sugar is made by spraying molasses syrup onto refined sugar crystals, and the turbinado sugar so prominently displayed in health food stores is partially refined sugar. The few additional nutrients are in such minute quantities as to be negligible. Honey, like sucrose, is composed of the simple sugars, glucose and fructose, plus a minuscule amount of potassium, calcium, and phosphorus—not enough to be significant in the daily recommended allowance.

THE CASE FOR SALT REDUCTION

The average American's daily intake of salt is 2–4 teaspoons (10–20 g). This is considerably above the Recommended Dietary Allowance of 3–8 grams or 1,100–3,300 mg of sodium, and there is no reason to presume that this huge amount of intake is desirable or good for health. To the contrary, a great deal of evidence indicates that it is dangerous to health. High-salt intake is linked to hypertension or high blood pressure, an affliction affecting an estimated 60–75 million Americans and especially prevalent in blacks. Untreated, the condition can lead to stroke, heart disease, and kidney failure. A number of studies have shown that hypertension can be experimentally induced in animals fed a high-salt diet. Moreover, in parts of Japan, where the mean daily salt intake is 26 grams, 40% of the people are hypertensive and have a higher death rate from stroke. Conversely, there is a low incidence of hypertension among several groups of primitive people in which the salt intake is very low. A mainstay in the treatment of hypertension, along with the administration of antihypertensive drugs, has long been the reduction and restriction of salt. Still, the relationship between salt consumption and hypertension is highly controversial. There are a number of other factors involved in high blood pressure. But although there is no direct proof that high salt ingestion causes high blood pressure, it is reasonable to assume that cutting back on sodium and salt intake may prevent the development of hypertension in those with a genetic predisposition to the disease and, by possibly reducing blood pressure, also prevent development in nonhypertensive Americans.

The taste for a lot of salt is acquired, and we have all been conditioned to it. Up until 1977, all baby foods contained added amounts, primarily to suit the mother's taste since infants have no special like for it. It is certainly possible to become unconditioned, however, by reducing salt intake. In many people, once that takes place, most foods begin to taste too salty. Unfortunately, cutting back on salt is not that easy because salt or sodium are so ubiquitous. About one-third of the ingested salt is present in food before cooking. About 40% of daily intake comes from processed foods. The rest is added with the salt shaker. More may be ingested via drinking water in some areas, and certain medications contain considerable amounts of salt. The first way to cut down is obvious—to follow the advice of the old adage and hide the salt cellar. Cook with little or no salt, and add none at the table. Another way is to purchase foods with no added salt. Many companies have recognized the marketing potential of low-salt products, and more than 40 potato chip producers are currently promoting salt-free "natural" chips. One can also reduce the use of condiments and obviously salted processed foods, but it is undeniably more difficult to limit sodium consumption when almost everything canned, bottled, packaged, or baked contains sodium bicarbonate, monosodium glutamate, sodium benzoate, sodium nitrate, sodium citrate, sodium saccharin, sodium aluminum phosphate, and so on. Many of these sodium-containing additives are in foods for other reasons than flavor and are necessary. They are curing agents, preservatives, stabilizers, or leavening agents, and they cannot be summarily removed unless an equivalent non-sodium-containing replacement in terms of taste, safety, and quality is found. It would be easier to watch salt intake if all foods were labeled, not only with the ingredients, but with a range of sodium content as well. It would also help to have the government mandate, or at least strongly urge, that manufacturers restrict unnecessary salt addition and seek ways to reduce or replace sodium in their products. There is considerable support in Congress for such legislation, and the Food and Drug Administration in 1983 issued proposed regulations that make sodium labeling in many processed foods mandatory. These regulations, effective in mid-1985, also spell out the standards to be met by manufacturers if they market products as "low" or "reduced" sodium or "unsalted," but make no attempt to provide limits on the amount of salt in the foods.

DIETARY REFORMS

It is generally acknowledged that the nation's eating habits—its convenience foods, its snacks, and its sweets—are gnawing away at its health. We seem to have conquered malnutrition caused by deficiencies—hardly anyone gets rickets, pellagra, and beriberi anymore. We appear to have replaced these diseases with a malnutrition resulting from an oversufficiency. There is ample evidence, although not proof, that too much fat, cholesterol, sugar, and salt are implicated in the slow-developing degenerative killer diseases—heart disease, some kinds of cancers, stroke, diabetes, arteriosclerosis, and cirrhosis of the liver. But when the government attempted to provide some guidelines to sensible eating it generated the biggest nutritional controversy of the decade. The first effort came in the form of a report prepared by the Senate Select Committee on Nutrition and Human Needs. First released in 1977 and revised in early 1978, the report, called "Dietary Goals for the United States," was both praised and vehemently attacked. The Dietary Goals were an unprecedented attempt to change the present high-fat, cholesterol, sugar, and salt consumption of the American public. Essentially the goals recommended the following:

1. Reduce the overall fat consumption from the present 40%–45% in the diet to about 30% of the caloric intake.
2. Reduce the amount of saturated fat so that it amounts to only about 10% of the total energy intake; then balance that with polyunsaturated and monounsaturated fats, which should then account for about 10% of energy intake each.
3. Reduce the amount of cholesterol ingested to about 300 mg daily.
4. Reduce salt intake so that it amounts to only 5 g or 1 teaspoon per day.
5. Increase the consumption of complex carbohydrates and naturally occurring sugars from about 28% intake to about 48% of energy intake. Reduce the consumption of refined and processed sugars down to about only 10% of total intake.
6. To avoid overweight, decrease energy intake and increase energy expenditure.

The Dietary Goals were heartily endorsed by the British and were similar to 1968 Scandinavian recommendations that were subsequently incorporated into Swedish and Norwegian nutrition and food policies. They were also basically the same recommendations made for a "prudent diet" by the American Heart Association in 1973. The goals, however, were emphatically denounced by many scientists and several special interest groups. The National Livestock and Meat Board objected because of the implied recommendation to reduce consumption of meat and increase consumption of poultry and fish. The egg producers protested that reducing egg ingestion does not lower cholesterol levels in the blood.

The sugar interests said there was no scientific basis for lowering sugar intake. The National Canners Association was displeased that there was the suggestion that more fresh or frozen vegetables should be used. The American Medical Association's statement (1977) contended that there was insufficient evidence that diet was related to disease and that dietary advice should not be prescribed to the general population but is best dispensed by physicians to their individual patients.

In 1980, the United States Department of Health and Human Services and the United States Department of Agriculture, with an eye toward the vehement controversy engendered by the goals, jointly issued essentially the same message, but framed it in words carefully chosen to offend fewer lobbies. The *Seven Dietary Guidelines* said:

1. Avoid too much fat, saturated fat, and cholesterol.
2. Eat a variety of foods.
3. Maintain ideal weight.
4. Eat foods with adequate starch and fiber.
5. Avoid too much sugar.
6. Avoid too much sodium.
7. If you drink alcohol, do so in moderation.

Another report, issued in 1982 by the National Academy of Sciences' National Research Council, did little to allay the controversy and created additional disagreement. Called *Interim Dietary Guidelines on Diet, Nutrition, and Cancer*, this statement again called for reduction of fat intake by 25% to reduce breast and prostate cancer and added the avoidance of salt-cured, salt-pickled, and smoked foods to reduce the incidence of esophageal and stomach cancer. There was also the recommendation to add more fruits, vegetables, and whole grains to the diet, especially those containing vitamin C and vitamin

A, with a particular admonition to include cruciferous vegetables—cabbage, cauliflower, broccoli, and brussels sprouts, which have been identified as possible defenses against cancer. Alcohol was to be used in moderation, especially in smokers, since the combination of smoking and alcohol is associated with a greater incidence of certain cancers.

To continue the chronology of efforts to get the public to change its eating habits to reduce the incidence of disease, the American Heart Association in early 1984 reissued its low-fat, low-cholesterol recommendations and unsurprisingly, was met by instant opposition from the National Dairy Council and the Egg Producers Association. The heart association's prudent plan was bolstered, however, by the report of a federally sponsored 19-year study of 3,806 middle aged men that provided the strongest evidence to date that lowering blood cholesterol by diet and drugs reduced the risk of coronary heart disease. For every 1% drop in blood cholesterol, heart attack rates dropped 2% (Lipid Research Clinics Programs, 1984A, B). Although a cholesterol-lowering drug (cholestyramine) was used in the study along with diet, the researchers agreed that dietary changes alone would also be beneficial.

With the exception of the established association between cholesterol level and heart attack provided by the above study, it is undeniable that actual *proof* of the causal relationship between high dietary fat and breast and colon cancer, high-salt consumption and hypertension, high-sugar consumption and diabetes and heart disease is lacking. There are only epidemiological studies that suggest *correlations* between diet and those diseases. There are no guarantees that modifying the dietary habits according to the recommendations will protect against the development of specific illnesses. Heredity, age, and in some instances, sex, are acknowledged to be risk factors of higher priority. But there is no way in which we can change our genes, age, or sex—we are stuck with our characteristics. Following the guidelines is a prudent and not terribly drastic way to lower one of the risk factors—an overly rich diet. Perhaps there are only correlative data pointing to the value of the recommended ways of eating, but there have never been *any* data to suggest that reducing fat, cholesterol, sugar, and salt, and increasing fruits and vegetables and whole grains, is in any way correlated to a disease process. Eating to achieve the goals or guidelines is not an improper diet; it will result in a wholesome and nutritious diet. The recommendations constitute no threat to health and, from all indications, are likely to promote health.

It is also possible to eat one's way to better health without carrying a calculator to the supermarket or the dining table. A lot of green vegetables, fruits, whole grains, bread, brown rice, and pasta should be eaten. In moderation, lean meats, such as veal, must be chosen, and there should be an increased consumption of chicken, turkey, and fish, which contain a larger proportion of unsaturated to saturated fatty acids in their fat. Skim milk and low-fat dairy products should be used for the same reason. Only margarines and salad dressings with a high polyunsaturated-to-saturated fat ratio should be used. High-cholesterol foods—eggs, organ meats, butterfats—should be eaten only rarely, except that this goal may be eased in premenopausal women, young children, and the elderly so that they might obtain the nutritional benefit of eggs. No sugar should be added to foods, and all desserts and snacks with a high sugar content are to be avoided. Little or no salt should be added to foods, and the consumption of foods high in salt content should be reduced. The physiological requirement for salt averages less than a gram per day, but most people presently ingest anywhere from 6 to 18 grams daily.

OVERWEIGHT AND OBESITY

Another bonus of conforming to the dietary goal recommendations is that it is more difficult to become overweight. The low-fat, low-sugar, and high-fiber foods of the guide make it easier to maintain a proper weight. The standard American diet—eating on the run, calorie-rich convenience foods, and snacks—can, in contrast, make staying slender a constant struggle. We find ourselves in a double bind when it comes to food and eating. The marketing messages say *buy and eat this remarkable abundance of good-tasting high-calorie food;* the media messages on movie and TV screens and on the pages of any magazine say *thin is sexier, thin is beautiful, the best-looking body is the tall and incredibly slender body.* The media-promoted ideal, however, is psychologically destructive; it leads people to dislike their bodies and encourages them to embark on faddish, unbalanced, nonnutritious diets. We could all benefit from both marketing and media messages that say *no one has to be thin for beauty's sake, but maintaining a normal weight for size, height, and age is healthier.*

There is no evidence that being a little fat is any kind of health hazard, but being *considerably* overweight is related to illness and premature death. So is being considerably underweight. The Framingham study, a longitudinal ongoing investigation that started in 1948 of the health of about half of the population of Framingham, Massachusetts reported in 1980 that life expectancy was the best for those people weighing the slightly plump average and worst for those in both the lightest and heaviest weight groups. While there is no reason to believe that carrying around a couple of extra pounds constitutes a medical problem, the only trouble with being about even 5% overweight is that it is very likely to creep up to 10% above ideal weight. Unless that weight is taken off, the same eating patterns can keep increasing the poundage to ultimately result in obesity. True obesity is generally taken to mean a weight gain in body fat that is 20%—30% above the given standards for height and age.

A number of health problems are associated with true obesity. Excessive fatness is correlated to hypertension, coronary heart disease, thrombophlebitis, diabetes mellitus, respiratory and gastrointestinal (liver and gallbladder) disorders, pregnancy difficulties is women, and certain kinds of cancer. Obese persons are likely to have an increased incidence of osteoarthritis in weightbearing joints, have a greater number of accidents, and are at significantly greater risk when undergoing surgery. In addition to all the risks to physical health, fat people are likely to have greater emotional distress. They are stigmatized as foodaholics, gluttons who lack willpower, and they suffer not only social but even job discrimination in a society that is repulsed by fatness. Such ordinary activities as taking public transportation, food shopping, buying clothing, or eating in a restaurant without being stared at, are severely limited for them.

It is evident that most American adults are above average in weight. It has been pointed out that even the leanest Americans, as measured by the skinfold test, have more subcutaneous fat than people in other more active and healthy societies (Blackburn, 1978). According to the National Center of Health Statistics, men and women, since the 1960s, have gained an average of 1–14 pounds depending on their height. Data from the HANES (Health and Nutrition Examination Survey) study,

which examined a representative sample of persons aged 18–74 during the years 1976–1980, indicated that the average adult female of all ages is 5 feet 4 inches tall and weighs 145 pounds. The average adult male stands 5 feet 9 inches and weighs 172 pounds. Both present profiles that are a far cry from those of the "beautiful people," and their weights are not the ideal weight figures. Despite a preoccupation with dieting and an annual expenditure of billions of dollars on dietary foods, aids, "fat farms," "fat" doctors, exercise machines, and innumerable other ways and devices to lose weight, many are losers only in the war against excessive weight. Anywhere from one-quarter to one-half of the adult population is 20% overweight, obese by most standards. In a society that venerates thinness, possibly one out of every two people is fat—a statistic that should make obesity the number one health problem in the United States (Fig. 16.4).

What Causes Obesity?

The revised edition of the U.S. Dietary Goals addresses the problem when it tersely states: "To avoid overweight, consume only as much energy (calories) as expended; if overweight, decrease energy intake and increase energy expenditure." The cause and the treatment appear to be very straightforward. Overweight, which leads to obesity, results when more calories are on the energy intake side of the equation than on the energy output side. The answer to reducing overweight is simple: eat less and exercise more. It takes 3,500 calories to make a pound of fat. Even just a couple of hundred extra calories a day above the amount expended—a glass of beer and a handful of peanuts or potato chips, a doughnut and coffee, or two glasses of coke or pepsi—will put on about an ounce a day. In 6

*Figure 16.4. Fifteen million Americans are designated as obese, at least 30%
above a desirable weight.*

months there will be a gain of almost 12
pounds.

For many people, weight gain is just that
simple. They owe their overweight to good
food and a sedentary life—taking the elevator
instead of climbing stairs, driving instead of
walking, watching sports instead of participating in them. Unfortunately, while overeating may be at the root of a weight problem,
the total picture of obesity is neither clear-cut
nor simple. There is no one etiology; there is
not even one single kind of obesity common
to all fat people. Excessive fatness is a disorder that is associated with a multitude of reasons that include anatomical, neurological,
behavioral, psychological, and social factors.
Genetic factors are believed to play very little,
if any, part in human obesity, and only very
rarely are there endocrinological origins for
the problem. The bases for getting fat and
staying fat may be highly complex and different in different people, but current thinking

places the major emphases on several areas:
early overnutrition, metabolic differences,
lack of physical activity, and a combination of
social factors that include home, family, education, economic status, and all the other environmental experiences that could contribute to the development and maintenance of
overweight.

There is a good deal of research to suggest
that a lifelong problem with obesity can sometimes originate in infancy. The fat cell number–fat cell size theory postulates that early
overnutrition may stimulate the formation of
excess numbers of adipose tissue cells. Brook
(1972) studied people who became obese as
early as 1 year of age and discovered that they
showed increased numbers of fat cells as
adults. In contrast, those individuals who became obese as adults showed fat cell enlargement but little or no change in fat cell number. The idea is that the fat cell number
becomes programmed and fixed at a high

level by early overfeeding. From then on, those people with more fat cells have greater difficulty in avoiding obesity and reducing excessive weight because, it may be presumed, all those extra cells are just sitting there waiting to be filled up with fat. It is, therefore, believed that early infant nutrition and the feeding environment in which children are raised during early life are very important in weight control during childhood. Breastfeeding causes less weight gain in infants than formula feeding, and infants who are either breastfed or formula-fed gain less than babies who are started on solid foods very early. The fat baby should not be viewed as the healthiest or the best-thriving baby.

But not all fat babies become fat adults. Although the increased adipose cell size and number theory is one possible factor in obesity, there are researchers who believe it has been overemphasized. Besides, there are inherent risks in instituting nutrient restriction and weight control measures in infancy and childhood. In particular, parents and physicians should be moderate and cautious in any attempts to stringently regulate calorie intake in infants. Early nutrition plays a lifelong role in health, and, as pointed out in several reviews of the medical literature on obesity in children, the diets of babies or of children should not be overzealously restricted for fear of interfering with optimal growth (Filer, 1978; Mallick, 1983).

Some people seem to be able to maintain a stable and ideal weight no matter how much they eat; others gain weight if they even look at food, or at least it appears that way. Moreover, many of the obese do not consume any more calories than the non-obese, and some actually eat less. That observation has led to the idea that there may be some difference in metabolic efficiency that exists between the fat and nonfat, and there has been active research investigation into the possibilities. For example, the role that brown adipose tissue might play has attracted a lot of attention. White fat, the kind distributed all over the body, is mainly a storage site; brown fat is a site of heat production or thermoregulation and acts as a kind of chemical furnace in hibernating species, in other animals exposed to cold, and in newborn animals and human infants, where it is believed to be important in the maintenance of body temperature. In human babies, brown fat is found at the nape of the neck and in between the shoulder blades, but constitutes only about 1% of the total body weight. In adults, there is some evidence that brown fat still exists, but just where it is located and what function it has in controversial. The connection of brown fat with human obesity is based on studies with genetically obese rats who were shown to have a defect in their brown adipose tissue that reduced their ability to burn up calories in response to exposure to mild cold, thus contributing to their obesity. It has been speculated that a similar ineffectiveness or a lesser quantity of brown fat may be a factor in human obesity. The brown adipose tissue theory may be a lead toward solving some of the unknowns of obesity, but any practical applications of the hypothesis have a long way to go.

Another intriguing finding has been that that number and activity of the sodium-potassium pumps in the red blood cells of obese adults and adolescents is reduced (DeLuise et al., 1980, 1982). The sodium-potassium pump, a constituent of all cells, keeps the appropriate balance of those two ions inside and outside the cell membrane and takes a lot of energy in the form of ATP to maintain. If there are fewer and less active pump units in obese individuals, they would have a reduced energy requirement to fuel them and would thus not use calories as efficiently as non-obese people. The useful applications of knowing that some obesity is related to defective sodium-potassium pumps still remain to be determined.

There is yet another theory that maintains that body weight, similar to body temperature, may be self-regulated around a specific set point located in the hypothalamus despite wide fluctuations in caloric intake, and that some people may simply be set at a higher or a lower point. To overcome the set point requires constant dieting or semistarvation, but the set point can be lowered, according to the theory's proponents, by appetite-suppressant drugs (Stunkard, 1982) and exercise.

Determination of the Ideal Weight

One way of establishing the desirable weight is to compare one's self with given standards on a height-weight table. Table 16.6 is an adaptation of a frequently used chart. These data are derived from the 1979 Build and

Blood Pressure Survey conducted on large numbers of Metropolitan Life Insurance Company policy buyers. They are based on the weights for sex and height that are considered optimal for the longest life span. The people measured do not really represent a cross-section of the American population, but since body weights are statistically related to life expectancy, they do provide a way of determining an "ideal" weight. Unfortunately, there are no precise rules to tell people how to evaluate their frame size. Some individuals have a greater bone or muscle mass and will weigh more without being fat. Shoe size, hand span, large wrists, or broad shoulders may be clues, but will not always indicate heavier bones or a larger frame.

Another measure of normal weight is to use as a guideline what a man or woman weighed between the ages of 18 and 25, assuming, of

Table 16.6 DESIRABLE WEIGHTS IN POUNDS FOR ADULTS

Height (in shoes, 1-inch heels)		Males Weight (lb.) in Indoor Clothing[a]			Height (in shoes, 2-inch heels)		Females Weight (lb.) in Indoor Clothing[a]		
Ft.	In.	Small Frame	Medium Frame	Large Frame	Ft.	In.	Small Frame	Medium Frame	Large Frame
5	2	128–134	131–141	138–150	4	10	102–111	109–121	118–131
5	3	130–136	133–143	140–153	4	11	103–113	111–123	120–134
5	4	132–138	135–145	142–156	5	0	104–115	113–126	122–137
5	5	134–140	137–148	144–160	5	1	106–118	115–129	125–140
5	6	136–142	139–151	146–164	5	2	108–121	118–132	128–143
5	7	138–145	142–154	149–168	5	3	111–124	121–135	131–147
5	8	140–148	145–157	152–172	5	4	114–127	124–138	134–151
5	9	142–151	148–160	155–176	5	5	117–130	127–141	137–155
5	10	144–154	151–163	158–180	5	6	120–133	130–144	140–159
5	11	146–157	154–166	161–184	5	7	123–136	133–147	143–163
6	0	149–160	157–170	164–188	5	8	126–139	136–150	146–167
6	1	152–164	160–174	168–192	5	9	129–142	139–153	149–170
6	2	155–168	164–178	172–187	5	10	132–145	142–156	152–173
6	3	158–172	167–182	176–202	5	11	135–148	145–159	155–176
6	4	162–176	171–187	181–207	6	0	138–151	148–162	158–179

[a] For weight without clothing, deduct 5–7 lbs. (male) or 2–4 lbs. (female)

course, that the person was not fat at the time. Growth of the muscle and skeletal systems are usually completed by the age of 25; any pounds added after that are generally in fat tissue.

Some doctors use formulas to determine the ideal weight in pounds.

Woman: 100 pounds + (height in inches minus 60) times 4

Man: 120 pounds + (height in inches minus 60) times 4

Of course, most people do not need formulas or height/weight charts to determine if they are overweight. Standing nude in front of a mirror and using yourself as a standard is a better way to determine if weight reduction is necessary. If the body looks firm and not flabby, if you are happy with its appearance and feel in no way restricted in social or physical activities by body size, if the weight has remained stable for years and has not approached dangerous obesity, there is no need to conform to a statistical norm. If additional confirmation of the appropriate amount of body fat is wanted the "inch of pinch" test can be used. Used clinically, the skinfold test measures the thickness of a double fold of skin on the back of the arm pinched up between special calipers. A personal skinfold test can be made by pinching the triceps skin between the thumb and forefinger when the arm is bent, or, while standing erect, the skin on the abdomen just above the waist can be pinched up. More than a half inch to an inch of pinch between the fingers means too much subcutaneous fat. The "ruler test" can measure the deposition of abdominal fat. While lying flat on the back, a ruler is placed running down the abdomen from the bottom of the rib cage to the pubic bone. If the ruler cannot touch both areas simultaneously and tilts upward, one's shape is no longer in good shape.

Getting Rid of Excess Weight

Putting weight on is so easy; taking and keeping off unwanted pounds can be frustratingly unsuccessful. There are any number of "cures" for overweight—more than 17,000 have been published to date. It would be easy to subscribe to a diet-book-of-the-month club. Each promises a new, revolutionary, quick, painless, way of weight loss. Many of them are "doctor's" diets—Dr. Atkin's diet is low in carbohydrates and allows unlimited amounts of protein and fats; Dr. Stillman's diet is another low-carbohydrate/high-protein regime plus eight glasses of water per day; Dr. Edelstein's Woman Doctor's Diet for Women advocates that women learn to live their lives on a very low level of carbohydrates; Dr. Simeon's Human Chorionic Gonadotropin Diet Plan is based on a 500-calorie per day intake along with daily injections of HCG. (Weight will be lost on 500 daily calories with or without shots.) Dr. Tarnower's Scarsdale diet is another low-carbohydrate, high-protein plan that also specifies that particular foods— chicken, lamb chops, steak, broiled fish— must be consumed at certain meals because of the "chemical reactions" occurring between the foods. The price of lamb chops being what it is, one dieter remarked that such a regime could only be followed in Scarsdale, a wealthy New York suburb.

All of these diets work, initially. All of them do result in a fast and relatively easy weight loss, but there is a difference between weight loss and fat loss. Since these diets are nutritionally unbalanced, low in carbohydrates and relatively low in salt, the most dramatic loss for the first few days or weeks of the regimen is in body water. Although dieters usually do not stay on the diet long enough to create problems for themselves, any unbalanced diet, especially one that lacks sufficient carbohydrates for body function is potentially a hazard to health. There are risks of fatigue, nausea, and headaches, especially in the early days of the diet. More seriously, there may be calcium depletion, increased serum cholesterol, kidney failure in those predisposed to kidney disease, and gout in those predisposed to that difficulty.

Another kind of diet is high-carbohydrate, low-fat, and low-protein. The Pritikin Diet is extremely low in fats (only 10% of total calories), salt, and sugar, but very high in complex carbohydrates. It is thus as unbalanced as other weight loss diets, but perhaps in a "healthier" way. It emphasizes increased consumption of vegetables, breads, and whole grains and prohibits table fats, oils, and dairy products of any kind except skim milk, setting up the possibility of deficiencies in iron, essential fatty acids, and fat-soluble vitamins. Pritikin specifies no set calorie intake, but the bulk provided by the carbohydrate foods makes it difficult to consume too many calories. It is too strict for most people to maintain for very long.

A diet that was touted as being the one to end all diets was Dr. Linn's Last Chance liquid protein diet. It swept the country in late 1977 and was enormously popular, perhaps because its adherents did not have to face any food at all. After the Food and Drug Administration pointed out that a substantial number of deaths had occurred among people who had been on it for a number of months, it became evident that this diet might indeed be the "last," and its popularity fizzled out. Other types of over-the-counter protein-supplemented liquid diets are similar and are equally opposed by nutritionists as being not only unwise, but unsafe. The widely advertised Cambridge Diet, for example, contains 33 g of protein from milk, calcium sodium caseinate, and soy flour, 44 g of carbohydrate derived primarily from fructose, and 3 g of fat plus vitamins and minerals. It is sold as a powder in 10 flavors to be mixed with water taken three times daily in place of meals for a total of 330 calories per day, somewhat more if mixed with milk. Its safety has not been established. Any diet containing fewer than 800 calories daily will cause rapid weight loss but, if prolonged, will also result in breakdown of lean (muscle) rather than fat tissue, cause nitrogen loss from the body, and is very risky.

Speaking about the physicians who write diet books, Dr. Sami Hashim, director of the Department of Metabolism and Nutrition at Columbia University's St. Luke's Hospital has reasonably queried, "Why do these people seem to have a predilection for being experts in nutrition? Why aren't they instant experts in neurosurgery or plastic surgery?"

Aside from the potential harm in crash diets as a result of their nutritional imbalance, the other big problem with them is that the very notion of going on a diet builds in the idea of going off the diet once the weight loss has occurred. Usually, the weight is quickly regained once the individual goes back to the original eating habits. The up-and-down, gain-and-loss pattern has been called the Yo-Yo syndrome, or, as Jean Mayer has put it, the "rhythm method of girth control." Getting thin and staying thin requires a complete re-education of one's eating and exercise habits. Those calorie-laden fattening goodies that added it in the first place are always going to be available; permanent weight loss has to mean a permanent, lifetime change in behavior.

Many people, however, believe that permanent change is too difficult. They would prefer to follow a yellow brick road to weight reduction—maybe the Wizard can come up with something, perhaps a chemical drug to get rid of fat. Prescription drugs, primarily amphetamines (speed) and an amphetamine-like substance called fenfluramine, have been used for appetite suppression. They are effective initially, but their usefulness in curtailing food intake is usually short. After about 3–6 weeks, a tolerance for the appetite suppressant effect develops and more may have to be taken until one becomes hooked on the pills. Because amphetamines have such a strong potential for addiction and in view of their really unpleasant and uncomfortable side effects—insomnia, irritability, rapid heart rate, tenseness, anxiety—they are a very undesirable way to try to lose weight.

The candies to raise blood sugar before a meal, the bulk-producers to give a full feeding, the "starch-blockers" that have no affect at all on the absorption of starch calories, the antihistamines such as phenylpropanolamine (PPA), sold under the brand names like Dexatrim, Control, Prolamine, and Appedrine and advertised as the most potent nonprescription appetite suppressants, are all over-the-counter aids that have never been proven to be effective unless a restricted food intake accompanies their use. More dollars than pounds are lost when these preparations are used. Besides, phenylpropanolamine is not an innocuous drug. It has side effects similar to amphetamines, reportedly has caused high blood pressure, heart irregularities, kidney problems, and is potentially toxic to the central nervous system. Anyone with hypertension, cardiovascular disease, thyroid, or kidney problems should not use these ineffectual appetite suppressants, but there are surveys that suggest that one in three high school and college women are popping these over-the-counter diet pills, some ingesting double or triple the manufacturer's recommended dosage.*

Although experience tells us there is no such thing, the search continues for the calorie-free lunch. Proctor and Gamble has developed a synthetic fat substitute that passes right out of the intestine unabsorbed and un-

changed and allegedly looks, smells, and tastes like cooking fat. Known as sucrose polyester (SPE), the compound will need long-term clinical safety trials before it can appear on the market.

There are several drastic surgical treatments for the extremely obese adult. One is the jejunoileal bypass, commonly called "the shunt." The purpose of this operation is to reduce the absorptive area in the small intestine by connecting the jejunum, an 8-foot length of small intestine, directly to the next part, the ileum. Since the absorbing power of the small intestine is thus short-circuited, the obese person theoretically can eat more without absorbing the calories and will lose weight. In practice, the complication and mortality rate has been determined to be an appalling 10%, and serious long-term health hazards of the procedure have become evident in survivors. In many instances the life-threatening complications occurred well after the surgery. Almost all patients developed a highly distressing diarrhea, and some had problems with arthritis, gallstones, or liver and kidney damage. At great cost to their health, they did, however, lose weight.

The jejunoileal shunt is no longer considered an acceptable surgical treatment for obesity and has largely been abandoned in favor of various stomach-reduction operations. The stomach bypass and stomach stapling surgeries are used to diminish the capacity and the outlet of the stomach, thus restricting the ability to consume foods. Other and newer methods have proliferated—balloons put inside or outside the stomach, the use of constricting mesh around the stomach, or even removal of part of the pancreas. All of these procedures have been useful for weight loss in certain individuals, but again, the rate of complications is very high. The gastrointestinal tract does more than digest and absorb food. It secretes hormones and may have other, as yet undiscovered, functions. Surgical

* The Center for Science in the Public Interest wants PPA banned and has started a Hazards Clearinghouse for PPA-Diet Pills. Anyone who has had problems with the diet pills is urged to contact the Clearinghouse, C/O CSPI, 1755 S Street, NW, Washington, D.C. 20009.

intervention for obesity should be considered still experimental and should be restricted to only individuals for whom obesity is a proven health hazard, who are 100 or more pounds overweight, and for whom all other methods have failed.

One other treatment of last resort is wiring the jaws shut. The patient has to carry around a wire cutter in case of impending nausea because of the danger of vomiting and choking on the vomit. Because liquids through a straw are all that can be consumed, the method should produce weight loss. But since chocolate shakes and high-calorie liquids from a blender can also be sipped through a straw, even this combination of what appears to be a medieval torture and a very boring diet may not be successful.

The latest kind of "fat surgery" is not a treatment for obesity. It is a procedure performed by plastic surgeons to remove fat in one specific spot in people who still have good skin elasticity (under 40–45 years of age) and are no more than 10–20 pounds overweight. Called suction-assisted lipectomy, the operation uses a long, hollow tube or cannula that is inserted under the skin in a fatty area through a small, inch-long incision. The cannula is attached to a pump operating at about 1 atmosphere of pressure. As the surgeon repeatedly pokes the cannula into the fat, the pump sucks it out, leaving Swiss cheese-like patterns in the fat tissue. At the conclusion of the procedure, the cannula is withdrawn, and the skin is pushed down to collapse the holes in the fat. The operation is performed under general anesthesia and can be used in several areas in which fat accumulates, such as the inner and outer thighs, buttocks, behind the knees, under the arms, the breasts, and the abdomen. The cost, which is nonreimbursable as cosmetic surgery by insurance companies, can vary up to $5,000, depending on the surgical site and the surgeon. As reported in the *Journal of the American Medical Associ-*ation (1983), complications, beyond those associated with general anesthesia, can include burning and tearing sensations, skin pigmentation, persistent edema, and, when overzealous surgeons remove too much fat, depressions in the skin at the operative area. Nevertheless, many patients are evidently delighted with the results, and half of the abdominal suctions reportedly performed by one surgeon are on the "potbellies" in men.

The measure of success for any weight-reduction program is whether the lost pounds stay off. The unbalanced crash diets and drug therapies have not been successful in that regard, and surgery is certainly not the answer except for a few highly motivated individuals for whom all other methods have been tried. So what does work? The only method that seems to be beneficial and is, therefore, getting a great deal of attention at present, although long-term effectiveness is still unknown, is some form of behavior modification. Actually, applying principles of behavioral psychology to eating is not new. Group therapy programs such as Weight Watchers or TOPS (Take Off Pounds Sensibly) have been using reward and reinforcement techniques for more than 40 years. The difference may be the present emphasis on self-management and self-control. Basically, the idea behind behavior modification for weight loss assumes that since eating is learned behavior, weight reduction can be promoted by changing the basic eating and activity patterns that lead to overeating and underactivity. One of the main ingredients of the program involves learning to set realistic goals and appreciating that small weight losses over a long period of time are desirable. The prime component is always the emphasis on permanent changes in lifestyle. Self-help is the key to behavioral change, and the focus is on the environmental or situation control of eating. The individual is educated to new adaptive behaviors by "shaping," practicing a series of small attainable changes that ultimately bring one to the final achieved goal.

A number of books have been written as guides to the use of behavioral modification techniques in weight control. Several suggestions to aid in making the appropriate changes in habits are common to all of them. They include:

1. Keep a food diary to create awareness of what is eaten, the time it is consumed, the location where eating takes place, and the mood or feelings during the eating period.

2. Sit down at the table while eating; do not stand, watch TV, listen to the radio, or read.

3. Eat slowly. Chew each mouthful until it disappears, put down the fork in between mouthfuls if it helps, savor the food—never bolt it down. Eating too rapidly does not give the stomach a chance to sense the presence of food and signal the brain that it is becoming full.

4. Do not put the meal on a large plate so that it looks small and becomes lost; use a small plate and then put small quantities of food on the fork or spoon. Get in the habit of leaving a little food; an adult no longer has to belong to the "clean plate club." Clear the table immediately after eating, put the food away, and leave the room.

5. Never shop for food on an empty stomach.

WEIGHT LOSS THROUGH EXERCISE

We probably would all agree that a sugared doughnut is low-nutrient junk food while 8 ounces of plain yogurt is worthwhile and nutritious. Nevertheless, they both contain the same number of calories. If the calories in either a doughnut or a cup of yogurt are above an individual's energy needs, the body is going to store them as fat. It makes no difference where the extra calories come from. Overeating on yogurt, whole wheat bread, or apples and oranges instead of beer and pretzels may be a more nutritious way of gaining weight, but in the absence of an increased energy expenditure, the end result will still be an additional inch of pinch. As discussed previously, if the daily intake is a few hundred calories more than the output, the gain will amount to nearly 2 pounds a month.

Consider the average American female, for example, who, at 145 pounds and medium build, is carrying around 20 more pounds than she should or wants to have. Suppose also that her weight has remained stable for several years and that she is neither gaining nor losing weight. It can, therefore, be assumed that in a typical day, she balances the number of calories she expends by the number of calories ingested. Ms. Average female has several ways in which she could shed those 20 pounds. Knowing that there are 3,500 calories in a pound of fat, she can reduce her intake by 500 calories a day and maintain her present level of activity. In seven days she will have lost one pound ($500 \times 7 = 3,500$ calories = 1 pound). At that rate it will take 20 weeks to lose 20 pounds.

With a few simple mathematical calculations, however, she can figure out how to lose the extra weight by increasing her energy expenditure. Moving the body through 1 mile of distance will use up about 100 calories. A slow amble, a brisk walk, a jog, or an actual run expend an equal amount of energy; surprisingly, the speed is not related to the calories burned. The actual number of calories utilized varies depending on body size; the heavier the person, the more calories used. The energy expended by walking or jogging 1 mile can be calculated by multiplying the body weight by 0.73. Of course it takes much less time to jog a mile than to stroll it. But if one does nothing more than park the car or get off the bus 6 blocks away from work or school and walk the rest of the way, or somehow make a real concerted effort to get that mile in every day, a pound of fat will be lost every 35 days. In 20 months, all the excess weight will be gone without reducing any food from the diet at all.

Now the mathematics of weight loss become even more interesting. If every mile of walking uses a hundred calories, 35 miles of walking produces a one-pound weight reduction. That sounds like a tremendous amount of walking, but at a very moderate pace, an hour's walk will cover 3 miles. Three miles per day means 35 miles in a week and a half. At the end of a year, assuming there has been no change in food intake, an incredible 36 pounds will have been lost! Walking does not require special shoes or any other kind of special equipment. It is one of the best exercises known.

If our hypothetical overweight woman decides to walk 3 miles a day and also cut down 500 calories in food intake, she will lose those 20 pounds in 10 weeks.

The more vigorous the activity, the more calories that can be burned in less time. Twenty minutes of jogging is equivalent to an hour of walking. Walking will use energy as well as jogging, running, swimming, bicycling, cross-country skiing, fast tennis, or racquet ball; it just takes a lot more of the walking. *Brisk* walking, in addition to using calories, will provide physiological benefits on the cardiovascular and respiratory systems obtained by those other more active and intense exercises.

A very prevalent myth about increasing the energy expenditure is that it only increases the appetite so that one will end up gaining rather than losing weight. Any appetite stimulation as a result of exercise is likely to be a psychological rather than a physiological effect. Unfortunately, people who want to use exercise for weight reduction sometimes fall into the trap of wanting to reward themselves with extra goodies because of the virtuous exercise they are doing. Research studies have suggested that regular and moderate physical activity has no effect on increasing hunger and may actually decrease it. This is well known to many exercisers who program their daily activity into their lunch hours. They have observed that after exercising they naturally want to eat less rather than more (Fig. 16.5). There is also some evidence that the calorie-burning benefits of exercise may even extend for several hours beyond the actual exercise period.

EXERCISE AND HEALTH

Weight reduction and weight maintenance are welcome side effects of daily walking, jogging, running, swimming, cycling, tennis, rope-jumping, calisthenics, or whatever form of physical activity fits into a life-style. Perhaps not every man or woman will end up with the figure of a movie star, but regular exercisers, even if they eat more than nonexercisers, are almost always leaner.

There are physiological and psychological benefits of working with the body that go beyond just making it more beautiful. Moderate regular exercise produces *physical fitness*. Being physically fit means having the superenergy promised but not delivered by swallowing vitamin pills. It means having strength and endurance, less fatigue, fewer aches and pains, and, possibly, even fewer colds and other infections. It means having flexible joints with the stretch and elasticity in muscles, tendons, and ligaments to produce an erect posture, a good carriage, and a smooth and easy gait. For long-term health and life expectancy, perhaps the greatest benefit of physical fitness is its significant effects

Figure 16.5. Lunchtime joggers. There is no medical reason to stop exercising during pregnancy if a woman wants to continue; physical fitness benefits the mother and the fetus.

on the cardiovascular and respiratory systems.

Not all exercises have an equal effect on all components of physical fitness. Yoga does a lot for flexibility but little for the heart and lungs. Running and walking do very little for flexibility but much for strength and endurance.

The kinds of exercises that produce the greatest number of physiological benefits are those which elevate the heart and breathing rate, raise the body temperature, and increase sweat production. Such exercises are called aerobics. The term was introduced by Kenneth Cooper, a physician and major in the U.S. Air Force. His book, *Aerobics*, first published in 1968, presents the concepts underlying the benefits of physical exertion. Aerobics, performed consistently and progressively, result in a conditioning or training effect. The heart muscle becomes stronger, able to pump out more blood with each beat, and the number of beats per minute of the heart at rest decreases. There is an increase in blood volume and in the arterial blood supply, not only to the heart itself through the coronary vessels, but also to all the skeletal muscles of the body. The muscles become stronger. Their mass increases, their tone improves, and there is a more efficient exchange of oxygen and carbon dioxide. If the resting arterial blood pressure was originally elevated before the exercise program, the systolic and diastolic pressures will become decreased. Pulmonary function improves, and the oxygen uptake capacity increases. Another positive benefit of exercise of an endurance nature is that all blood lipids, such as cholesterol and triglycerides, are reduced.

Physical fitness can only occur with physical activity that is sufficiently demanding. This does not mean that the exercise must be overly strenuous, self-punishing, or boring. The whole key to choosing an exercise plan that can be maintained is to select one that

can also be enjoyed. Anything is going to be hard at the beginning, but if there is not going to be any fun in the activity after one gets into condition, the exercise plan is likely to be dropped. The body, however, cannot store the benefits of exercise anymore than it can store water-soluble vitamins. Unless the exertion takes place regularly and consistently, the physiological advantages will be lost.

Exercise classes, aerobic dancing, or working out at special exercise gyms have become very popular, especially if one can afford them, and provide a good incentive for keeping up an exercise program. The health and fitness shelves at the bookstores are crowded with celebrity-authored exercise books that tell us how to do a daily workout at home. Any of these activities are beneficial if done sensibly. Some of them, however, may not be completely without hazard. For example, belly-dancing is great fun and good exercise, but the twisting movements are very hard on the knees. Some forms of aerobic dancing are overly vigorous, and not all the instructors are that knowledgeable about the limitations of tendons, muscles, and ligaments. Exercise is supposed to make one feel good, not crippled. Some of the exercise books, records, or tapes give nutritional misinformation or suggest that pain when exercising is a good thing. It is important to remember that pain is the body's way of saying that something is wrong. A good rule is, if it *hurts*, stop doing it.

Unless an individual is over 35 years of age, has a medical problem, or has never exercised at all, a medical checkup before initiating an exercise program is not necessary as long as the activity is begun slowly. For anyone who is over 35 and not accustomed to strenuous exercise, it is sensible to have a physical examination that includes a resting electrocardiogram. Over the age of 40, according to Cooper, the exam should include a stress test, or exercising electrocardiogram. If everything checks out all right, the only barrier left is the one in the mind that says, "I'll start tomorrow," or "next week," or "it's too hot," or "too cold," or the really negative "I don't have time to fit exercise into my schedule."

One can choose to be physically active, or one can choose not to, but those who make exercise a part of their lives make the better choice for wellness. People should exercise not only because it is so good physiologically, but also because it *feels* so good. Further positive rewards of exercise are the relief of tension and strain; working out works out emotional stress as well. Those who exercise describe their feelings as being on a natural high, of having a wonderful awareness of being fully alive and in complete control of their bodies. Bodies are designed to be active; they look and they work better when they are active. It is a pity that most people miss the joy of discovering the full physical potentials of their bodies and the realization of how truly healthy they can feel.

WOMEN, EXERCISE, AND SPORTS

How many of these statements sound familiar?

Certain physical activities are simply too strenuous for women. Women can be damaged internally by exercise, although the damage may not show up for years. Besides, vigorous exercise is unfeminine. It causes unsightly bulging muscles, and women end up looking like one of those body-building addicts.

Exercise during menstruation is unsafe. The performance of women athletes is bound to be adversely affected by competing while menstruating.

Exercise causes a tilted uterus and makes it harder for a woman to conceive or deliver a child. Women should not jog, run, or compete in any athletic events while pregnant; it is too dangerous to the fetus.

Physical activity or participation in contact sports may cause injury to the breasts. A blow to the breast can cause cancer. Even jogging or running makes the breasts jiggle and bounce, and they will become stretched-out and saggy.

Women are more vulnerable to sports injuries. They are smaller, their bones are softer, and they are more knock-kneed and loose-jointed, making them especially injury-prone.

Female athletes are physically inferior to male athletes and can never compete on an equal basis. There are fundamental biological reasons why they cannot play games or excel in events as well as men.

The above are examples of the kinds of myths that have kept women sedentary. They have justified an inequality in physical education and instructional programs in school, and a lack of community recreational facilities for women when they are adults. These fallacies have discouraged women from participating in sports and from getting the kind of training and coaching needed for competitive athletics. Fortunately, most of these myths have been recognized as untrue and are slowly fading out of the picture. Unfortunately, some of them still carry enough weight in some places to influence members of the International Olympics Committee.

There is no exercise that is too strenuous or too vigorous for women to perform, and there is no such sport for which they are physically unsuited. The only barriers to participation are cultural and social.

The thought of getting muscle bulges and turning into a "jock" has deterred many young women from pursuing vigorous physical activities, but since muscle hypertrophy is dependent on the amount of androgen present, the muscles in women who strenuously exercise become firmer, smoother, and stronger and do not balloon out in size. Studies have shown that even weightlifting, which causes increases in strength and muscle mass in men, results only in significant increases in strength in women (Wilmore, 1974; Capen et al., 1961). Pam Meister, 4 feet, 11½ inches tall and weighing 105 pounds, won the Women's National Weightlifting Championship by lifting a deadweight 310 pounds after just a year of weight training, a feat that would be an impossibility for most men. Ms. Meister's muscles possess great strength, but no bulk. Unit for unit, the physiology of women's muscle—the contractile mechanism and the ability to exert force—is no different than that of men's.

All of the menstrual myths have now been disproved. Moderate exercise has no adverse effects on the menstrual cycle and may even regulate it. Exercise has a very beneficial effect on dysmenorrhea. Concerning the impairment of peak performance for women athletes during menstruation, it has been shown that women have competed, won events, and broken records during all stages of their cycles. Some have also won gold medals and championships when they were 6 months pregnant, although most stop competing after the fourth month (Zaharieva, 1972). Heavy athletic training should not be *initiated* during pregnancy, but if an athlete wants to golf or play tennis until the day she delivers, there is no medical reason why she should not. One ranking tennis player competed in tournaments through her eighth month of pregnancy, claiming it was "great exercise" and that it "rocked the baby to sleep" (Lieber, 1978). Many nonpregnant women might be deterred by such strenuous activity, but Laurie Glenn Jacobson had no difficulty in undergoing the rigors of a 21-week officer training course in the Marine Corps while pregnant. Shortly before completing the course, Laurie marched 21 miles over a rugged terrain while carrying a 7-pound rifle, a 25-pound pack, and a 5½-month fetus.

Erdelyi's study of Hungarian women athletes showed that 87% of them had shorter labors in childbirth than women in a control group. There were also fewer complications in pregnancy and delivery in athletes. There has never been any evidence that women athletes have more difficulty conceiving children, and in many instances, their performance improves after pregnancy and delivery.

As described in Chapter 4, there is one menstrual difficulty that appears in some women who undergo intense physical training. Menstrual irregularity and, in certain individuals, amenorrhea or cessation of the menstrual periods, occur in marathon runners, ballet dancers, gymnasts, skaters, and other athletes who follow very strenuous training programs. Apparently, there are no lasting consequences of the menstrual irregularities; when the training stops, the disorder disappears. Some women athletes have expressed concern that Olympic officials may believe that the condition substantiates the belief that athletic activity is harmful to women and begin to ban women from competition.

Participation in exercise or noncompetitive and competitive sports has no effect on breast sagging, stretching, or enlargement, which is primarily genetically determined. Large-breasted women may find that their breasts get in the way in contact sports, but that they do not affect their athletic skills. All that women need are the same kinds of protection for their vulnerable areas that have been devised to protect exposed and vulnerable areas of the male anatomy. Christine Laycock, associate professor of surgery at New Jersey Medical School, has designed a protective "gym-bra" to reduce the possibility of breast problems during and after sports participation. Women who run or jog may find it difficult to get enough support from today's little-nothing nylon brassieres. Bras with a hook and eye closing may rub and cause irritation.

Proper training and preparation for sports activities are necessary to reduce injuries in women, men, or children who are active participants. Female bones may be smaller than those of the male, but unless a woman is post-menopausal and has osteoporosis, her bones are neither softer nor more fragile than a man's bones. The ligaments that give the joints stability may be more lax in some women, however, making them more loose jointed. This may result in a greater susceptibility to sprains and strains. Dr. James Nicholas, team physician to the New York Jets, Knicks, Rangers, and Yankees, has written that knee sprains in female skiers, for example, can occur when the impact is less than violent. While a greater incidence of knee injuries in women may have occurred in Dr. Nicholas' private practice, Davis and colleagues' study of 147 injured skiers brought into a hospital emergency room showed that leg injuries in males outnumbered those in females by two to one. Ski injuries are likely to be more related to lack of experience, poor physical condition, and failure of the binding to release properly than to the sex of the skier. Anyone, male or female, who starts any activity and goes too far too rapidly without a slow, progressive buildup is likely to develop injuries. After a survey of 300 college athletic programs and an extensive literature search, Haycock and Gillette (1976) concluded that *well-trained* women athletes are no more prone to injury than men athletes and that the kinds of injuries sustained by women were no different than those by men.

Whether women athletes are inherently physically inferior to men athletes is still a moot point. Certainly, most of world records for events are held by men. Men, however, have been at it longer, have had more training, better coaching, better facilities, and have been expected to be better and continue to improve. As women athletes also receive these advantages and continue to gain in strength and endurance, female records may equal and even surpass male records in certain competitions. Right now the best women swimmers are swimming faster in their events than the best male swimmers were in their events 25 years ago.

Women probably will never be able to compete with men across the board. They may always be at a disadvantage in certain competitions or contact sports where sheer force and size are factors. There is no physical reason, however, why the two sexes should not compete against each other in certain sports or play together on the same team in games such as baseball where agility and timing are more important than strength. Prejudice against women as professional major league players, not their lack of ability, is likely to keep them out of pro baseball for some time. The day of the female major league player may not be that far off, however. Even the legendary Hank Aaron gave approval for women on the same team as men, indicating that he saw no reason why they should not play pro baseball.

There are some events for which women may be physiologically better suited than men. Some researchers have suggested reasons why women may ultimately be better at long distance running. In the usual marathon race, where strength and speed are most important, men are likely to be more successful. Beyond the 26-mile marathon, when stamina may be more important than strength, it has been theorized that women may excel. Not only are they lighter, but there is evidence that they may be able to burn their body reserve of fat more efficiently than men.

Marathon running may truly be a woman's sport, but the barriers against women runners have only gradually been dropped. Before 1972, women had to enter under an assumed name and sneak into the Boston Marathon when no one was looking. Today, the Boston Marathon is open to women, and the Amateur

Athletic Union has sanctioned marathon running for men and women in the same race. The AAU also sponsors a women's marathon national championship. For years there had been many women in the United States and several countries who could run a 26-mile marathon in less than 3 hours, a standard aimed for by the top male runners. But it took until the 1984 Olympics before there was any marathon event for women runners, or even a 3,000 meter (1.85 mile) competition.

Women athletes are obviously not going to be equal in all ways to men athletes in every sport, although there are bound to be similarities between some men and women athletes in strength, endurance, and even in body composition. Of course, all men athletes are not equal either, and it is evident that not everyone makes it to superstar status. Athletes of both sexes are, however, superbly fit people. They have become athletes because they have natural abilities for their particular sport, and they are endowed genetically with physical and physiological attributes that lead them to excel. That the gap between their abilities appears to be narrowing is significant, and perhaps closing it may be of benefit to women's sports. In our highly sports-conscious society, winning is what counts. When women athletes win, break records, and become champions, when they, too, are perceived as superstars, then women's sports may be closer to generating the same audiences, money, and media coverage as men's sports.

While the difference in fitness between the male athlete and the female athlete is not very great, the difference between the trained woman athlete and the average woman is much greater, and it vastly exceeds the difference between the trained male and the average man. Females, after all, have never been encouraged from the time they were children to get out there after school and play competitive games, do pushups and situps, and in general, become active and stay active. Girls are socialized into becoming the cheerleaders, not the players. But with a push from the women's movement and Title IX of the Education Amendments of 1972, which, at least until a recent U.S. Supreme Court decision, prohibited discrimination on the basis of sex in any educational institution receiving federal funds, social and cultural attitudes have changed. A girl who wants to go out for sports is finding it easier. Formerly, she might have been derided by her contemporaries as a jock. Today, she is encouraged and, to a great extent, admired. Schools that are no longer permitted to allot a tiny fraction of the athletic budget to women's sports when half the student population is female have moved toward athletic equality* (Fig. 16.6).

Physical fitness, however, is not only for athletes; it is for everyone, of any age and of either sex. Activity provides health benefits, whatever the activity and whenever it is initiated. But good exercise, like good nutrition, should be a lifetime habit to be of greatest benefit. The most logical time to establish good health habits is during childhood, in the home and in the school. Parents must demand a good exercise program in the schools for youngsters of both sexes, and they must make certain that athletic equality in training, equipment, and facilities exist. The emphasis should be on making physical activity attractive to all children, whatever their abilities. Often, programs that attempt to develop star athletes tend to discourage the less athletically gifted children. Unfortunately, too many of us have become lifetime benchsitters. We not only have to get off the sidelines ourselves,

---

* A blow to educational equity for women under Title IX was dealt by the Supreme Court decision of February 28, 1984. In the *Grove City v. Bell* case, the Court ruled that federal funds to schools that discriminate can be cut off only to the program receiving direct federal funding. Henceforth, discrimination could take place in any program not receiving federal funding.

Figure 16.6. Active women are healthier, stronger, leaner women. Exercise is not only physiologically beneficial—it feels good.

we have to see to it that all of the next generation becomes players.

REFERENCES

Adams, P. W., Rose, D. P., Folkard, J., et al. Effect of pyridoxine hydrochloride (vitamin B_6) upon depression associated with oral contraception. *Lancet* 1:897–904, 1973.

American Medical Association. Statement Submitted to Select Committee on Nutrition and Human Needs, U.S. Senate. Re: Dietary Goals for the United States, April 18, 1977.

Barrett, S. and Knight, G. Rodale presses on. *Nutr Notes* 9:6, 1976.

Blackburn, H., quoted in Obesity—an overview. *Dialogues in Nutrition* 3(1):1–10, 1978.

Brook, C. Evidence for a sensitive period in adipose cell replication in man. *Lancet* 2:624–627, 1972.

Brownmiller, S. *Feminity*. New York, Linden Press, 1984.

Campbell, A. and Jack T. Acute reactions to mega ascorbic acid therapy in malignant disease. *Scott Med J* 24:151–153, 1979.

Capen, E. K., Bright, J. A. and Line, P. A. The effects of weight training on strength, power, muscular endurance, and anthropometric measurements on a selected group of college women. *J Assoc Phys Mental Rehab* 15:169–173, 180, 1961.

Cochrane, W. A. Overnutrition in prenatal and neonatal life: a problem: *Can Med Assoc J* 93:893–896, 1965.

Creagen, E. T., Moertel, C. G., O'Fallon, J. R., et al. Failure of high-dose vitamin C to benefit patients with advanced cancer: A controlled trial. *N Engl J Med* 301:387–390, 1979.

Davis, M. W., Litman, T., Drill, F. E., et al. Ski injuries. *J Trauma* 17(10):802–808, 1977.

DeLuise, M., Blackburn, G. L. and Flier, J. S. Reduced activity of the red-cell sodium potassium pump in human obesity. *N Engl J Med* 303:1017–1022, 1980.

DeLuise, M., Rappaport, E. and Flier, J. S. Altered erythrocyte Na and K pump in adolescent obesity. *Metabolism* 31(11):1153–1158, 1982.

Erdelyi, F. J. Gynecological survey of female athletes. *J Sports Med Phys Fitness* 2(3):174–179, 1962.

Filer, L. Early nutrition: Its long-term role. *Hosp Pract* 3(2):87–94, 1978.

Finn, S. B. and Glass, R. B. Sugar and dental decay, in Bourne, G. H. (ed.). *World Review of Nutrition and Dietetics*, vol. 22. Basel, S. Karger, 1975.

Hashim, S., quoted in Burros, M. Scarsdale diet has many critics. *The Milwaukee Journal*, July 13, 1978.

Haycock, E. E. and Gillette, J. V. Susceptibility of women athletes to injury: Myths vs. reality. *JAMA* 236(2):163–165, 1976.

Herbert, V. and Jacob, E. Destruction of vitamin B_{12} by ascorbic acid. *JAMA* 230(2):241–242, 1974.

Holmes, R. P. and Kummerow, F. A. The relationship of adequate and excessive intake of vitamin D to health and disease. *J Am Coll Nutr* 2:173–199, 1983.

Kamil, A. How natural are those natural vitamins? *J Nutr Ed* 4(3):92, Summer, 1972.

King, J. C., Cohenour, S. H., Corruccini, C. G., et al. Evaluation and modification of the basic four food guide. *J Nutr Ed* 10(1):27–29, January–March, 1978.

Knowles, J. Responsibility for health. *Science* 198(4322):1103, 1977.

Kühnau, J. The flavonoids: A class of semiessential food components: Their role in human nutrition and diet, in Bourne, G. H. (ed.). *World Review of Nutrition and Dietetics*, vol. 24. Basel, S. Karger, 1976.

Lieber, J. Myths fade as women athletes gain strength. *Milwaukee Sentinal*, December 2, 1978.

Lipid Research Clinics Program. The lipid research clinics coronary primary prevention trial results. I. Reduction in incidence of coronary heart disease. *JAMA* 251(3):351–364, 1984A.

Lipid Research Clinics Program. The lipid research clinics coronary primary prevention trial results. II. The relationship of reduction in incidence of coronary heart disease to cholesterol lowering. *JAMA* 251(3):365–374, 1984B.

Mallick, M. J. Health hazards of obesity and weight control in children: A review of the literature. *AJPH* 73(1):78–82, 1983.

Mayer, J. Does your doctor know beans about food? *Fam Health* 10(4):38–39, 1978.

Nicholas, J. A. Are sports more dangerous for women? *The Female Patient* 1(1):16–20, 1976.

Nutrition gets medical students' attention. *New York Times* service. *Milwaukee Journal*, July 6, 1983.

Schneiderman, M., quoted in Bad habits tied to 100,000 deaths. *The Milwaukee Journal*, March 31, 1976.

Select Committee on Nutrition and Human Needs, U.S. Senate. *Dietary Goals for the United States*, ed. 2. Washington, D.C., U.S. Government Printing Office, 1977.

Select Committee on Nutrition and Human Needs, U.S. Senate. *Dietary Goals for the United States—Supplemental Views*. Washington, D.C., U.S. Government Printing Office, 1977.

Shannon, I. L. Sucrose and glucose in dry breakfast cereals. *J. Dent Child* 41(5):347–350, 1974.

She lugs rifle, helmet, baby. *Milwaukee Sentinel*, August 29, 1978.

Siegel, C., Barker, B. and Kunstadter, M. Conditioned oral scurvy due to megavitamin C withdrawal. *J Peridontol* 53(7):453–455, 1982.

Stacpoole, P. W. Pangamic Acid ("vitamin B_{15}"), a review, in Bourne, G. H. (ed.). *World Review of Nutrition and Dietetics*, vol. 27, Basel, S. Karger, 1977.

Stunkard, A. L. J. Minireview: Anorectic agents lower a body weight set point. *Life Sciences* 30:2043–2055, 1982.

Suction-assisted lipectomy attracting interest. *JAMA* 249(22):3004–3005, 1983.

Williams, R. J. *Physician's Handbook of Nutritional Science*. Springfield, Ill., Charles C. Thomas, 1975.

Wilmore, J. H. Alterations in anthropometric measurements consequent to a 10-week weight training program. *Med Sci Sports* 6(2):133–138, Summer, 1974.

Zaharieva, E. Olympic participation by women; effects on pregnancy and childbirth. *JAMA* 221:992–995, 1972.

Index